Fourth International Conference on Sustainable Infrastructure and Built Environment

AA003766

IOP Conference Series: Earth and Environmental Science
Volume 1065

Online
8-9 March 2022

Editors:

Irsan Soemantri Brodjonegoro
Andhika Sahadewa
Hiroshi Yamamura
Driejana
Aine Kusumawati

ISBN: 978-1-7138-6023-5
ISSN: 1755-1307

Printed from e-media with permission by:

Curran Associates, Inc.
57 Morehouse Lane
Red Hook, NY 12571

Some format issues inherent in the e-media version may also appear in this print version.

Institute of Physics
Dirac House, Temple Back
Bristol BS1 6BE UK

Phone: 44 1 17 929 7481
Fax: 44 1 17 920 0979

techtracking@iop.org

Additional copies of this publication are available from:

Curran Associates, Inc.
57 Morehouse Lane
Red Hook, NY 12571 USA
Phone: 845-758-0400
Fax: 845-758-2633
Email: curran@proceedings.com
Web: www.proceedings.com

TABLE OF CONTENTS

Introduction of Unsaturated Soil Mechanics in Indonesia to Improve the Understanding on Geotechnical Engineering-Related Infrastructure Problems 1
S. Krisnanto, E. Susila, A. Sahadewa, H. Nawir, E. Rismantojo, D. Apriadi

Application of Cellulose of Boehmeria Nivea as Natural Flocculants in Direct Flocculation 11
D. Fauzani, S. Notodarmojo, M. Handajani, Q. Helmy, T. Kardiansyah

Factors Affecting Safety Leadership of Construction Project Owners in Indonesia 19
D. V. Indrayana, K. S. Pribadi, P. F. Marzuki, H. Iridiastadi

Study on Numerical Calculation of Flow Structures in a Curved Open Channel with Advanced Depth-Integrated Models 28
F. P. Lugina, T. Uchida, Y. Kawahara

Utilization of Dredged Mud in Land Reclamation: A Comparison Study of Several Soil Improvement Methods 35
I. Wayan Sengara, Jonathan Bratanata, Sugeng Krisnanto

The Effect of Non-Linear Wave on Oil Spill Dispersion 42
M. Muin, A. B. Muslim, T. A. Puspitasari

Integrated Flood Model in 3D Non-Orthogonal Boundary Fitted Hydrodynamic Model for Ciliwung River, Jakarta 54
A. B. Muslim, M. Muin

Vessel Size and Dredging Depth Optimization for Tol Laut Program Using Genetic Algorithm: Hub Port Case Study 66
R. Sihombing, M. Muin, R. L. Tawekal

The Effect of Land Subsidence on Hydrodynamic Conditions in Pekalongan Coast 81
R. J. Martin, A. Kurniawan, E. O. Nugroho, F. S. Wijaya, I. F. Muhammad, Yola, N. Fauzi

Mechanical Characteristics of Concrete with the Addition of Integral Waterproof Using Local Aggregate East Kalimantan 92
Mardewi Jamal, M. Jazir Alkas, Budi Haryanto, Ahmad Rasyid

Earthquake Risk Study on Residential Buildings in West Jakarta Using the Event-Based Risk Analysis Method 102
G. Lambang-Goro, M. Irsyam, M. Asrurifak, I. Meilano

Microplastics in the Continuous Biofilm Reactor: Occurrence, Fate, and Removal 117
A. Nur, P. Soewondo, A. S. Setiyawan, K. Oginawati

Biological Oxygen Demand (BOD) Simulation in Sunter River Using Non Orthogonal Water Quality Model Transport 125
Rommy Martdianto, Muslim Muin, Ricky Lukman Tawekal, Arung Bahari Muslim, Heru Nur Alam

Cohesive Sediment Simulation in Kanal Banjir Barat Using MuSed3D, Non-Orthogonal Boundary Fitted Sediment Transport Model 137
Denny Yatmadi, Muslim Muin, Ricky L. Tawekal, Heru Nur Alam, Arung Bahari Muslim

The Challenges of Implementing Green Factors in Urban Greening Schemes in Indonesia 149
M. Z. Dahlan, B. Faisal, S. Chaeriyah, I. W. Hutriani, M. Amelia

System Dynamic Simulation to Determine the Effect of Water Consumption on Land Subsidence During Covid-19 Pandemic in Jakarta 156
Diki Surya Irawan, Prismita Nursetyowati, Maria Cintya Nova, Sarah Maulina

Analysis of Behavioral Factors in the Spread of COVID-19 in Coblong District, Bandung, West Java, Indonesia. 167
Katharina Oginawati, Rinaldy Jose Nathanael, Udjianna Sekteria Pasaribu, Utriweni Mukhaiyar, Afif Humam, Nisa Fadlilah Fathul Ilmi

The Use of Deflection Bowl Parameters to Evaluate the Structural Condition of the North Coast Corridor of Java Island 178
B. H. Setiadji, A. K. Indriastuti, Y. Rivani

Effects of Dammar as the Partial Substitution of Bitumen in HMA AC-WC Containing Buton Asphalt 188
E. Ngii, L. One, N. Nasrul, M. I. Udo

Study of Lerak (*Sapindus Rarak*) Biochar Application for Andosol Agricultural Soil Remediation 198
Sekar Arum, Temmy Wikaningrum

Quantitative Consideration in Selection of Breakwater Concrete Armor Unit Based on Unit's Internal Tensile Stress Response 205
Andojo Wurjanto, Adi Putra Hardaya

Numerical Modelling of Land Subsidence Effect on Coastal Sediment Transport in Pekalongan 217
F. S. Wijaya, A. Kurniawan, N. Yuanita, E. O. Nugroho, R. J. Martin, I. F. Muhammad, Yola, N. Fauzi

The Eco Enzyme Application to Reduce Nitrite in Wastewater as the Sustainability Alternative Solution in Garbage and Wastewater Problems 229
T. Wikaningrum, P. L. Anggraina

Numerical Modeling: The Effect of Land Subsidence on Hydrodynamic Conditions and Coastal Morphology in Pekalongan 240
I. F. Muhammad, N. Yuanita, E. O. Nugroho, F. S. Wijaya, R. J. Martin, Yola, N. Fauzi

Constructed Wetlands with Cyperus Alternifolius as a Sustainable Solution for Household Greywater Treatment 253
S. Qomariyah, B. Utomo, A. H. Wahyudi

Influence of Random Variable on Reliability Index of Braced Monopod Platform Based on Fatigue Analysis in Madura North Java Sea 264
Paramashanti, Rildova, V. N. B. Sucipto, A. Kurniawan

Numerical Study of Hydrodynamic and Sedimentation for Sustainable Marine Floating Cage Aquaculture in Indonesia: A Case Study in Situbondo, East Java. 275
Marsha Sabrina, Entin A. Karjadi

Economic Valuation and Potential Pollution Load Analysis of Domestic Wastewater in Greater Bandung 287
Adhitya Rakhmadi, Suharyanto, Anindrya Nastiti

The Properties of Rolled Compacted Concrete as a Preservation Material 297
A. B. Utomo, G. Primaswari

Modular Incinerator with Pre-Treatment Plant for Municipal Solid Waste Treatment in the Super-Priority Tourism Destination of Labuan Bajo Indonesia 306
Wahyu Purwanta, Pandji Prawisudha, Firman Bagja Juangsa, Ayudia Mutiara Fani, Evan Philander

Spatial Analysis of Potential Greenhouse Gas (GHG) Emissions from Household-Scale LPG Consumption in Urban and Sub-Urban Areas of Medan City During the COVID-19 Pandemic 318
I. Suryati, N. Hasibuan, M. Hanisah, L. Nadapdap, M. Ruadelia, J. Sigalingging

Role of Non-Orthogonal Hydrodynamic and Sedimentation Model for Port Facilities Analysis on Port of Kuala Tanjung 327
M. Muin, R. Sihombing, M. W. Hermawan, A. B. Muslim, H. N. Alam

Development of Seismic Map for Offshore Platform Design in Indonesia, Case Study of the North Coast of Java 340
Hendriyawan, D. D. Aditramulyadi, R. L. Tawekal, Rildova

Removal of Phthalocyanine Reactive Dye Using Plasma Corona Discharge 350
B. Handoko, S. Notodarmojo, R. Iqbal

Study of Hybrid Annuity Model on Indonesia Toll Road 358
D. T. Ramadhanti, R. H. Karsaman, S. S. Wibowo

Estimation of Extreme Rainfall Over Kalimantan Island Based on GPM IMERG Daily Data 367
A. A. Kuntoro, R. K. Hapsari, M. B. Adityawan, M. Farid, Widyaningtias, Radhika

Estimating Household Water Consumptions in the Bandung Metropolitan Area 375
Yuniati Zevi, Wika Maulany Fatimah, Yusep Ramdani, Muhammad Yusuf Habibbullah, Nurdianti Mursyida

Properties of Recycled Hot Asphalt Mixture Added with Mastic Asphalt Component 386
I. N. A. Thanaya, I. M. A. Ariawan, I. D. G. R. Yudistira, I. M. Widiyarta

Study of Clean Water and Sanitation Access and Its Relationship to Waterborne and Stunting Prevalence in Bandung Regency 397
Putri S. Kamila, Indah R. S. Salami

Numerical Study on Connections of Replaceable Shear Link with Inserted Pins 407
M. Moestopo, D. Kusumastuti, S. Wibowo

Comparative Study in Bill of Quantity Estimates on Reinforcement Works of Pile Cap, Single Pier and Double Pier of Flyover Between Conventional Methods and BIM (Building Information Modelling) 417
Atanasius Agri Permadi Putra, Natanael Indrawan Wiharno Oei, Hermawan, Budi Hasiholan

The Challenge of Assisting Stakeholders in Data-Scarce Settings: Characterizing Intermittent Water Systems Using a Citizen Science Approach 430
L. Sioné, M. R. Templeton, C. Onof

Sustainability Analysis of the Upper Citarum Watershed Based on Water Quality, Water Quantity, and Landuse Indicators 434
M. Marselina, N. M. Putri

Development of Concrete Material Technology in the Context of Supply - Demand Based on Geographical Aspects 447
R. G. K. Pradoto, B. W. Soemardi, A. Gazali, A. T. Putri, R. P. Purba, I. Mahardika

Study of Temporal and Spatial Changes in the Estuarine Area of the Mayangan River, West Java – Indonesia 458
Hendra Achiari, Rajin, Gurusu

Evaluating the Use of Google Maps as Navigation Application by Identifying Hazards and Assessing Risks Using HIRA Matrix 468
Wildi Kusumasari, Friska M. Ilmi, Estiara Ellizar, Yos Y. Rabung

Disaster and Resilient Infrastructures at Musi Riverside Settlement in Palembang 478
Bambang Wicaksono, Ari Siswanto, Widya Fransiska, Susilo Kusdiwanggo

Impact Analysis from Visual Improvement Program of Slums to Sanitation Management: Case Study Kampung Cibunut, Bandung City, Indonesia 490
I. Kusumayanti, L. P. Defi, W. M. Fatimah, P. W. Sarli, P. Soewondo

Redefining Role of Stakeholders in Restoring Hydrological Function and Conservation of Biodiversity of Nipa-Nipa Grand Forest Park, Southeast Sulawesi 499
Ira Ryski Wahyuni, Alfian Alfian, Mutmainnah Mutmainnah, La Ode Muhammad Erif, Fitria Nur Indah Djafar

Optimizing Rainwater Harvesting Systems for the Dual Purposes of Water Supply and Runoff Capture with Study Case in Bandung Area, West Java 508
Muhammad Cahyono

Analysis of Biodegradation Characteristics Based on Visual Observation and Mass Loss Percentage in Soil Burial Test of Sustainable Geo-Bag Materials as Temporary Structures in Natural Coastal Protection System 518
W. Gunawan, Rildova, N. Yuanita, A. Kurniawan, K. Kahdar, V. Herbenita

Assessment of Water Quality and Challenges for Vegetable Irrigation in Greater Bandung Area, Indonesia 525
M. Firdayati, R. Anindita, P. A. Krisnamurti, M. Handajani

Tsunami Risk Assessment in Business Continuity Planning for Palu Special Economic Zone 535
Muhammad Rizki Purnama, Mohammad Bagus Adityawan, Krishna Suryanto Pribadi, Mohammad Farid, Widyaningtias, Arno Adi Kuntoro

Groundwater Salinity and Its Effect on Elevated Concentrations of Other Contaminants in a Coastal Plain: A Perspective for Infrastructure Protection 548
A. F. Rusydi, S.-I. Onodera, S. Ioka, R. Maria, F. Firmansyah, M. Saito, W. Purwoko, D. Sukmayadi

Comparative of Material Properties Between Natural Fibers and Geo-Bag Synthetic Fibers as Sustainable Material of Temporary Structure in Natural Coastal Protection Systems 560
V. Herbenita, Rildova, N. Yuanita, A. Kurniawan, K. Kahdar, W. Gunawan

Integration of Mass Public Transport Fare in the Jakarta Area 571
S. Triana, A. Sjafruddin, R. H. Karsaman, S. Kaderi

Palu Housing Reconstruction Process: Reviewing and Learning After the 2018 Earthquake 584
R. G. K. Pradoto, A. Oktavianus, K. S. Pribadi, I. M. A. B. Rasmawan, L. D. Wulandari

Comparative Analysis of Municipal Solid Waste to RDF Pretreatment Methods in Indonesia 599
I. M. W. Widyarsana, D. Saraswati

Removal of Sulphate and Heavy Metals from Acid Mine Drainage Using Permeable Reactive Barrier Technique 608
Rissa Anungstri, Himawan Tri Bayu Murti Petrus, Agus Prasetya

The Impact of Pavement Distress to Vehicle Maneuver in Bandung City 621
Jongga Jihanny, Kardina Nawassa Setyo Ayuningtyas, Anju David, Firnandia Zahwa Salsabila, Murti Kamajaya, Difa Salmadhia

Evaluation of Pedestrian Facilities Performance 630
N. Hidayat, I. Haryanto, R. P. Isheka

Coastal Protection System Design at the Indonesian Mangrove Center in Pekalongan, Central Java – Indonesia 639
Yola, N. Yuanita, A. Kurniawan, E. O. Nugroho, R. J. Martin, I. F. Muhammad, F. S. Wijaya, N. Fauzi

Bioprocess of Oil Contaminated Water at Gathering Station Through a Consortium of Endogenous and Exogenous Bacteria 659
Sandra Madonna, Dzalika Nurperbangsari, Astrid Meilasari Sugiana, Zulkifliani

Health Risk-Based Prioritization Approaches of Pharmaceuticals in the Upper Citarum River Basin 667
R. R. Utami, I. R. S. Salami, G. W. Geerling, S. Notodarmojo, A. M. J. Ragas

Preliminary Studies of Bandung City Health System Resilience (Case Study: Covid-19 Pandemic) 678
D. Roosmini, T. F. Kanisha, A. Nastiti, S. W. Kusumah, I. R. S. Salami

Author Index

IOP Conf. Series: Earth and Environmental Science 1065 (2022) 011001 doi:10.1088/1755-1315/1065/1/011001

International Conference "Sustainable Infrastructure and Built Environment (SIBE-2022) – Challenges on Sustainable and Resilient Infrastructure and Built Environment", 8th-9th March 2022

1. Introduction

The Faculty of Civil and Environmental Engineering, Institut Teknologi Bandung, Indonesia has organized two-day International Conference on "Sustainable Infrastructure and Built Environment (SIBE-2022) – Challenges on Sustainable and Resilient Infrastructure and Built Environment", 8th-9th March 2022. The conference was held virtually owing to the COVID situation resulting into travel restrictions. The organizers were in CIBE Building, the Faculty of Civil and Environmental Engineering, Institut Teknologi Bandung, Jalan Ganeca 10, Bandung, Indonesia.

The aim of the conference was to provide a platform for exchanging ideas, information, and experiences among academics, researchers, consultants, engineers, manufactures, and postgraduate scholars. It also serves a medium to discuss and evaluate the latest research, innovative technologies, policies and new directions in infrastructure development, pollution prevention and eco-friendly technologies adapted to developing countries, and to promote cooperation and networking amongst practitioners and researchers involved in addressing sustainable and resilient infrastructure.

2. Topic and subtopic of the conferences

The main topic of the conference is "Challenges on Sustainable and Resilient Infrastructure and Built Environment". Meanwhile, the conference subtopics include the following:

a. Disaster and Resilient Infrastructures and Earthquake Engineering
b. Maritime Infrastructure and Coastal Protection
c. Sustainable Construction and Project Management
d. Water and Wastewater Engineering and Management
e. Transportation System and Engineering
f. Water Resources Engineering and Management
g. Health, Safety and Environment
h. Environmental Management

3. Committees

Steering Committee:
Ir. Edwan Kardena, Ph.D.
Dr. Nita Yuanita S.T., M.T.
Hadi Kardhana, S.T, M.T, Ph.D.
Prof. Ir. Indra Djati Sidi, M.Sc., Ph.D.
Prof. Ir. Masyhur Irsyam, MSE., Ph.D.
Ir. Cahyono, M.Sc., Ph.D.
Prof. Ir. Rudy Hermawan Karsaman, M.Sc., Ph.D.
Prof. Dr. Ir. Krishna S. Pribadi
Dr. Ir. Muslim Muin
Prof. Ir. Ricky Lukman Tawekal, MSE., Ph.D.
Prof. Ir. Mindriany Syafila, MS., Ph.D.
Prof. Ir. Puji Lestari, Ph.D.
Prof. Dr. Barti Setiani Muntalif
Prof. Dr. Ir. Rizal Z. Tamin

Scientific Committee:
Prof. Ir. Ade Sjafruddin, M.Sc., Ph.D.
Dr. Qomarudin Helmy, S.Si., M.T.
Prof. Dr. Ir. Puti Farida Marzuki
Prof. Ir. Iswandi Imran, MASc., Ph.D
Prof. Ir. Muhammad Syahril Badri Kusuma, Ph.D.
Ir. Sindur P. Mangkoesoebroto, MSEM, Ph.D
Ir. Irsan Soemantri Brodjonegoro, Ph.D
Ir. Raden Driejana, MSCE, Ph.D.
Ir. Arief Sudradjat, MIS., Ph.D.
Ir. Erza Rismantojo, MSCE, Ph.D.
Dr. Aine Kusumawati, S.T., M.T.
Dr. Benno Rahardyan, S.T., M.T.
Dr. Herto Dwi Ariesyady, S.T., M.T.
Rofiq Iqbal, S.T., M.Eng., Ph.D.
Dr.Eng. Hendra Achiari, S.T., M.T.
Mohammad Farid, S.T., M.T., Ph.D
Ahmad Mukhlis Firdaus, S.T., M.T.
Andhika Sahadewa, S.T., M.S.E., Ph.D.
Prof. Satoshi Okabe
Prof. Louis Ge
Dr. Fadjar Goembira, S.T., M.Sc.
Dr.Eng. Rudi Waluyo Prastianto, S.T., M.T.
Prof. Ir. Bambang Suhendro, M.Sc., Ph.D.
Prof. Bonaventura H. W. Hadikusumo

Organizing Committee:
Ir. Indah Rachmatiah Siti Salami, M.Sc., Ph.D.
Dr. Anindrya Nastiti, S.T., M.T., Ph.D.
Nadiyatur Rahmatikal Wasi'ah, S.T., M.T.
Juneard Stevelana Dhipantara, A.Md
Luthfika Ghassani, S.Pd.
Ir. Ima Fatima M.Eng.
Fera Astuti, S.E.
Dr. Harmein Rahman, S.T., M.T.
Ivan Sandi Darma, S.T., M.T., Ph.D.
Ir. Yuniati, M.T., M.Sc., Ph.D.
Ir. Entin Agustini Karjadi, MCE, Ph.D.
Ir. Hernawan Mahfudz, M.T.
Sugeng Krisnanto, S.T., M.T., Ph.D.
Eko Charnius Ilman, S.T., M.T., Ph.D.

Dr. Adyati Pradini Yudison, S.T., M.T.
Septian Hadi Susetyo, S.T., M.T.
Seny Damayanti, S.T., M.T.
Mayrina Firdayati, S.Si., M.T.
Izqi Yustina Ammylia Yusuf, S.T., M.T.
Dr.Eng. Febri Zukhruf, S.T., M.T.
Prasanti Widyasih Sarli, S.T., M.T., Ph.D.
Erwin Lim, S.T., M.S., Ph.D.
Dr. Muhammad Sonny Abfertiawan, S.T., M.T.
Regina Pratiwi, S. Kom
Dhendy Surya Purnama, S.Kom
Nugraha, S.T.
Dr. I Made Wahyu Widyarsana, S.T., M.T.
Widyaningtias, S.T., M.T., Ph.D.
Dr. Mohammad Bagus Adityawan., ST., M.T.
Anggita Fazrin, S.E.

Editor for IOP EES:
Dr. Irsan Soemantri Brodjonegoro
Dr. Driejana
Dr. Aine Kusumawati
Dr. Andhika Sahadewa
Dr. Hiroshi Yamamura

4. Paper Submission
The total count of paper submission was 70 for IOP EES. A peer review for selection of papers was followed and 65 (sixty-five) papers were accepted for paper in IOP EES.

5. Speakers and Sessions
Within two days, the conference was carried out in virtual mode through Zoom platform. Two plenary sessions and parallel sessions were scheduled. As many as 382 participants from different parts of world attended the different sessions of virtual conference.

The keynote and invited speakers of the conferences come from different institutions, among other:

- Coordinating Ministry for Maritime Affairs and Investment, Indonesia
- Hokkaido University, Japan
- NTU (National Taiwan University), Taiwan
- Ministry of Public Works and Public Housing, Indonesia
- Earthquake Megacity Initiative
- Ministry of National Development Planning, Indonesia
- Kyoto University, Japan
- Imperial College London, UK

IOP Conf. Series: Earth and Environmental Science 1065 (2022) 011001 doi:10.1088/1755-1315/1065/1/011001

Picture 1. Group picture of Head of Organizing Committee, Rector of ITB, and Dean of ITB

Picture 2. Group picture of of Organizing Committee in the Operating Room

Picture 3. Organizing Committee in the Operating Room

6. Participants Affiliation

SIBE 2022 was attended by participants coming from different universities/organizations as follow:

1. Institut Teknologi Bandung, Indonesia
2. Institut Teknologi Sepuluh Nopember, Indonesia
3. Universitas Bakrie, Indonesia
4. PT. LAPI GTC, Indonesia
5. Universitas Sebelas Maret, Indonesia
6. President University, Indonesia
7. Politeknik STTT Bandung, Indonesia
8. Universitas Gadjah Mada, Indonesia
9. Universitas Lampung, Indonesia
10. Hiroshima University, Japan
11. Badan Riset dan Inovasi Nasional (National Research and Innovation Agency), Indonesia
12. Politeknik Ketenagakerjaan, Indonesia
13. Universitas Kristen Maranatha, Indonesia
14. UIN Sunan Gunung Djati Bandung, Indonesia
15. Institut Teknologi Nasional, Indonesia
16. Universitas Padjadjaran, Indonesia
17. Imperial College London, United Kingdom
18. Universitas Diponegoro, Indonesia
19. Universitas Haluoleo, Indonesia
20. Politeknik Pekerjaan Umum Semarang, Indonesia
21. Universitas Udayana, Indonesia
22. Kementerian Pekerjaan Umum dan Perumahan Rakyat (Ministry of Public Works and Housing), Indonesia
23. Radboud University, The Netherlands
24. Universitas Sumatera Utara, Indonesia
25. Kementerian Perhubungan (Ministry of Transportation), Indonesia
26. Politeknik Perkeretaapian Indonesia, Indonesia
27. Politeknik Negeri Semarang, Indonesia
28. Universitas Indo Global Mandiri Palembang, Indonesia
29. Universitas Hasanuddin, Indonesia
30. PT Dinamaritama Engineering Consulting, Indonesia
31. Universitas Katolik Soegijapranata Semarang, Indonesia
32. Universitas Mulawarman, Indonesia
33. Universitas Wijayakusuma Purwokerto, Indonesia
34. Universitas Indonesia, Indonesia
35. Poltekkes Kemenkes Semarang, Indonesia
36. National Taiwan University, Taiwan

37. Komunitas Teras, Indonesia
38. Amman Mineral Nusa Tenggara, Indonesia
39. Institut Teknologi Sumatera, Indonesia

7. Funding Acknowledgement

SIBE 2022 was supported by 15 companies as follow:

1. PT DSI (Delta Systech Indonesia)
2. Penta Architecture
3. Harita Nickel
4. PT Adhi Karya (Persero) Tbk.
5. PT Brantas Abipraya
6. PT Nindya Karya (Persero)
7. PT Modern Widya Tehnical
8. PT Pitaco Mitraperkasa
9. PT Suropati Hidro Energi
10. PT Jasamarga Tollroad Maintenance
11. PT Difa Mahakarya.
12. PT Hutama Karya (Persero)
13. PT Nur Straits Engineering
14. Nippon Koei Co., ltd.
15. Adhi-Acset-MKN KSO

IOP Conf. Series: Earth and Environmental Science 1065 (2022) 011002 doi:10.1088/1755-1315/1065/1/011002

Peer Review Statement

All papers published in this volume have been reviewed through processes administered by the Editors. Reviews were conducted by expert referees to the professional and scientific standards expected of a proceedings journal published by IOP Publishing.

- Name of event:
  International Conference "Sustainable Infrastructure and Built Environment (SIBE-2022) – Challenges on Sustainable and Resilient Infrastructure and Built Environment", 8th-9th March 2022
- Type of peer review: Double anonymous
- Conference submission management system: Morressier
- Number of submissions received: 70
- Number of submissions sent for review: 67
- Number of submissions accepted: 65
- Accepted Rate (Submissions Accepted/ Submissions Received x 100): 92.9%
- Average number of reviews per paper: 2.00
- Total number of reviewers involved: 41
- Contact person for queries:
  - Name : Editor SIBE 2022
  - Affiliation : Bandung Institute of Technology – Faculty of Civil and Environmental Engineering
  - Email : aan17@itb.ac.id
    sugeng.krisnanto@itb.ac.id

Introduction of unsaturated soil mechanics in Indonesia to improve the understanding on geotechnical engineering-related infrastructure problems

S Krisnanto[1,2,*], E Susila[1], A Sahadewa[1,2] H Nawir[1], E Rismantojo[1] and D Apriadi[1,2]

[1]Geotechnical Engineering Research Group, Civil Engineering Program, Faculty of Civil and Environmental Engineering, Bandung Institute of Technology (ITB), Jalan Ganesa No. 10, Bandung 40132, Indonesia
[2]Soil Mechanics Laboratory, Faculty of Civil and Environmental Engineering, Bandung Institute of Technology (ITB), Jalan Ganesa No. 10, Bandung 40132, Indonesia
*Email: sugeng.krisnanto@itb.ac.id

Abstract. Several geotechnical engineering-related infrastructure problems cannot be quantified comprehensively using classical soil mechanics for saturated soils. This emphasizes the need of analysis using the framework of unsaturated soil mechanics (USM). The graduate course on soil mechanics for unsaturated soils is provided in Bandung Institute of Technology (ITB) to answer these challenges. The understanding on geotechnical engineering-related infrastructure problems in Indonesia is improved in the course. The understanding on the basic topics in USM is emphasized throughout the course. The balance between the eagerness for innovative design and the carefulness in the design is conveyed using the course. The laboratory testing apparatuses are being developed as a part of the course as well as a part towards the implementation of USM in Indonesia. An option of stages in introduction of USM in Indonesia is proposed.
Keywords: geotechnical engineering, graduate course, laboratory apparatus, soil mechanics, unsaturated soils.

1. Background

The task of soil mechanics is to forecast the effect upon the soil of a given system of forces [1]. Domains of analysis in classical soil mechanics are [2]: (i) Flow through porous media; (ii) Shear strength; and (iii) Volume change.

The presence of negative pore-water pressure, u_w in unsaturated soils has been understood before the mid-20th century [3,4,5,6]. The definition of matric suction has been formulated (e.g. [7,8]) and the framework of unsaturated soil mechanics (USM) has been available [7,9]. Nevertheless, to the authors' knowledge the understanding of USM has not been well established.

Several geotechnical engineering-related infrastructure problems cannot be quantified correctly using classical soil mechanics for saturated soils. These problems should be quantified using USM. Two examples of these kind of problems are:

1. Variation of slope factor of safety due to rainfall [10]. This phenomenon is related to the slope instability during rainfall as shown in figure 1. A slope failed during rainfall even without any additional load on top of the slope.

2. Volume change of the unsaturated zone was observed in Doplang subdistrict, Central Java Province, Indonesia (figure 2). Wide opening cracks occurred during dry season in August 2019 indicating a significant of volume change. This phenomenon is related to the swelling-shrinkage of soil above groundwater. Swelling-shrinkage phenomenon occurred in the zone above water level (unsaturated soil zone).

Figure 1. An example of rainfall-induced slope instability [10].

Figure 2. Wide opening crack during dry season in August 2019 in Doplang subdistrict, Central Java Province, Indonesia.

2. Current Advancement of the Unsaturated Soil Mechanics Analysis in Indonesia

Current advancement of the analysis involving USM in Indonesia includes:

1. Analysis of rainfall-induced slope instability [10]. The case is shown in figure 2 and the analysis results is shown in figure 3. In this analysis, a variation of slope during rainfall (without any additional load on top of the slope) was performed using the principles of USM.
2. Flow through unsaturated cracked soils [11] as shown in figure 4. In this analysis, saturated flow through the crack network and the seepage through the unsaturated soil matrix was analysed. The analysis of seepage through unsaturated soil matrix was performed using the principles of USM.

In the analysis of rainfall-induced slope instability (figure 3) two methods were used to calculate the factor of safety, *FS*: (i) the extended shear strength method, and (ii) the total cohesion method. In the analysis using the extended shear strength method several parameters are required: saturated-unsaturated hydraulic properties (soil-water characteristic curve (SWCC) and permeability function) and shear strength properties at saturated-unsaturated conditions. The rainfall intensity shown in figure 3(b) was used as flux boundary condition in the numerical model presented in figure 3(a). Seepage analysis was then performed to obtain the pore-water pressure, u_w in the slope. Having the pore-water pressure obtained from the seepage analysis, the shear strength was calculated using the extended Mohr-Coulomb shear strength equation [12,7]:

$$\tau_{ff} = c' + \left(\sigma_f - u_a\right)_f \tan\phi' + \left(u_a - u_w\right)_f \tan\phi^b \quad (1)$$

where $(\sigma_f - u_a)_f$ is the net normal stress state on the failure plane at failure, $(u_a - u_w)_f$ is the matric suction on the failure plane at failure, c' is the effective cohesion, ϕ' is the angle of internal friction associated with the net normal stress state variable, and ϕ^b is the angle indicating the rate of increase in shear strength relative to the matric suction. The *FS* shown in figure 3(c) was then calculated as the resisting force divided by the driving force.

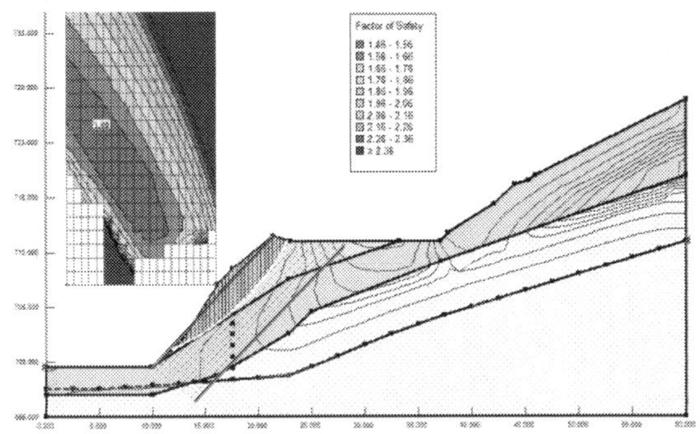

(a)

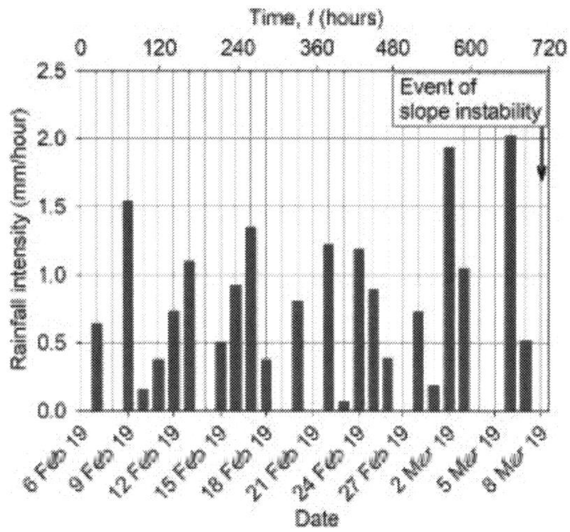

(b)

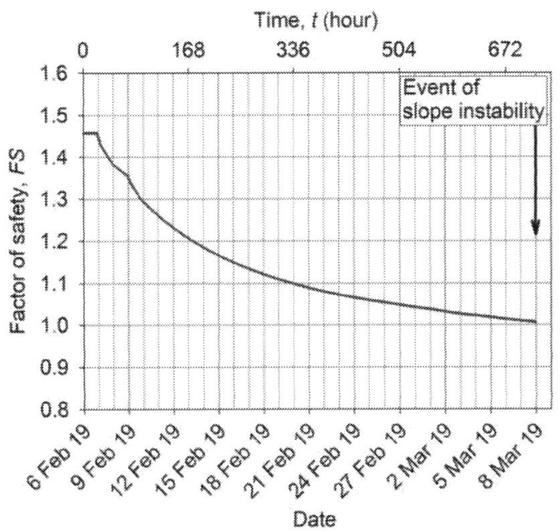

(c)

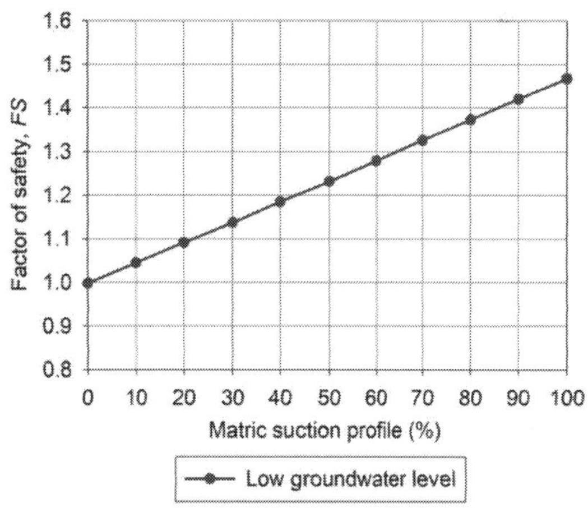

(d)

Figure 3. An example of analysis on variation of factor of safety of slope under rainfall [10]: (a) Numerical model; (b) Rainfall intensity at the slope location; (c) Factor of safety, *FS* calculated using the extended shear strength method; (d) Factor of safety, *FS* calculated using the total cohesion method.

In the total cohesion method, equation (1) is simplified as [7]:

$$\tau_{ff} = c + \left(\sigma_f - u_a\right)_f \tan\phi' \tag{2}$$

where c is the total cohesion that can be calculated as [7]:

$$c = c' + \left(u_a - u_w\right)_f \tan\phi^b \tag{3}$$

The total cohesion (equation (3)) was calculated as percentage of pore-water pressure from hydrostatic condition [7]. The *FS* (figure 3(d)) was then calculated as the resisting force divided by the driving force.

Krisnanto et al. [13] found that in the water flow through unsaturated cracked soils, two types of flow occurred: (i) Steady state flow through the crack network, and (ii) Transient flow from the crack wall towards the soil matrix. The steady state flow was modelled using flow through parallel plates [5,14,15] whereas the transient flow was modelled by representing cracks as head boundary condition (figure 4). Krisnanto et al. [16] proposed a method of averaging water content of the soil matrix part of cracked soil. Krisnanto et al. [17] developed a computer code to generate crack network numerically and modelled transient flow from the crack network to the soil matrix based on Krisnanto et al. [18]. The averaging method from [16] was performed in terms of volumetric water content, θ_w of the soil matrix part of the cracked soil. The uniqueness was found on the crack network representation on the average water content of the soil matrix part of cracked soil.

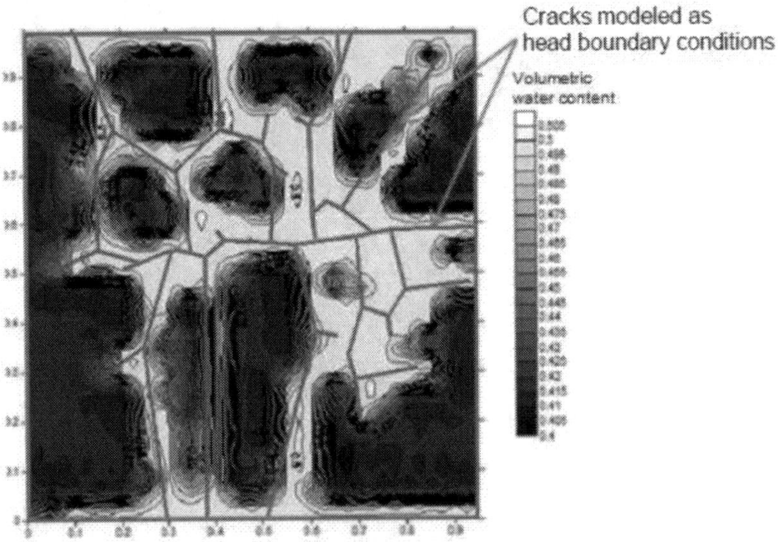

Figure 4. Analysis of flow through unsaturated cracked soil [11].

3. Important Points in Unsaturated Soil Mechanics That Need to be Understood by Engineers in Indonesia

The above explanation shows the need of a thorough understanding of USM in Indonesia. Three important points in USM that need to be understood by engineers in Indonesia involves: (i) The stress-state variable(s) for saturated and unsaturated soils; (ii) Change in paradigm in USM as compared to classical soil mechanics for saturated soils; (iii) The need for laboratory tests and field measurements for the analysis of unsaturated soils.

3.1. *The Stress-State Variable(s) for Saturated and Unsaturated Soils*

Stress state variable(s) for saturated and unsaturated soils is a fundamental knowledge in soil mechanics. Stress state variable(s) for saturated and unsaturated soils are as follows:

- Stress-state variable for saturated soils [19]: effective stress, $\sigma' = \sigma - u_w$.
- Stress-state variables for unsaturated soils [20,18]: net normal stress, $(\sigma - u_a)$ and matric suction, $(u_a - u_w)$.

3.2. *Change in Paradigm in Unsaturated Soil Mechanics*

In contrast to classical saturated soil mechanics, the following points need to be understood in USM:
1. The effect of environment, particularly the effect of weather and climate.
2. Saturated soils as a special case of unsaturated soils.
3. Function of parameters.

These requires a change in paradigm in understanding of USM.

3.3. *The Need for Laboratory Tests and Field Measurements for the Analysis of Unsaturated Soils*

As more parameters are required in USM, laboratory tests are needed. Field measurements are also needed to bring the results from laboratory test into field applications.

4. The Course of Unsaturated Soil Mechanics in ITB

The above explanation implies that a partial understanding of USM may jeopardize a design. A thorough explanation can be conveyed and disseminate using a graduate course.

4.1. *Adoption of the Course in Curriculum*

Soil mechanics for unsaturated soils (SI6227) has become a graduate course in the 2018 curriculum of Civil Engineering Program, Bandung Institute of Technology (ITB), Bandung, Indonesia [21]. The graduate course is developed to answer the challenges in geotechnical engineering-related infrastructure problems in Indonesia. The course discusses the following topics:
1. Matric suction, osmotic suction, and total suction.
2. Stress state variable(s) of saturated and unsaturated soils.
3. Suction and water content measurements.
4. Flow laws.
5. Soil-water characteristic curve (SWCC) and permeability function.
6. Measurement of permeability.
7. Steady-state water flow.
8. Pore-pressure parameters.
9. Shear strength theory.
10. Measurement of shear strength parameters.
11. Two dimensional unsteady-state flow.
12. Introduction to unsaturated volume change theory.
13. Application in engineering practice.

The above course topics show that the understanding on the basic topics in USM is emphasized throughout the course. In addition, the balance between the eagerness for innovative design and the carefulness in the design is conveyed using the course. As part of the course, the corresponding local teaching material was already developed [22].

4.2. *Laboratory Testing Facility*

Laboratory testing apparatuses are currently under development in ITB to support the graduate course of soil mechanics for unsaturated soils. Two items are currently the priority for development: (i) The measurement of SWCC using filter paper; and (ii) The measurement of shear strength using unsaturated triaxial apparatus. The high precision balance is shown in figure 5 whereas the currently developed unsaturated triaxial apparatus is shown in figure 6. In addition, the apparatuses are developed as a part towards the implementation of USM in Indonesia.

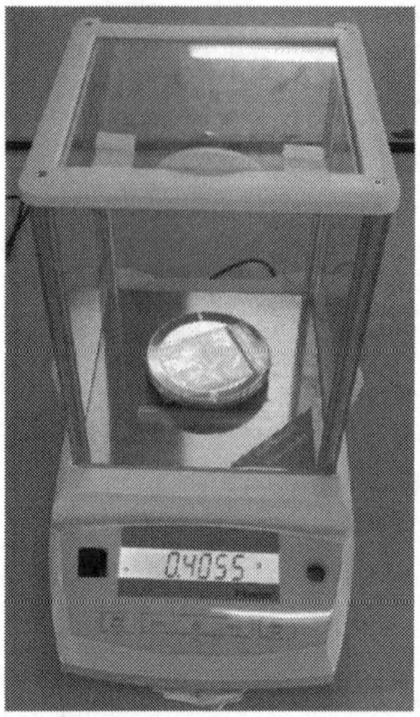

Figure 5. High precision weighing balance for suction measurement using filter paper available in Soil Mechanics Laboratory, Faculty of Civil and Environmental Engineering ITB.

Figure 6. Unsaturated triaxial apparatus available in Soil Mechanics Laboratory, Faculty of Civil and Environmental Engineering ITB.

5. Future Developments

Testing and creating database of unsaturated soil properties and instrumentation are required for the development a database for a variety of soil types in Indonesia. Local correlations are then developed from the database. This will also improve the performance of the currently assumed parameters for the analysis (e.g. [10]). An estimation used in the present analysis (e.g. [23]) can be further verified.

Instrumentations are required to assess the performance of the analysis. This step is followed by several trials of the developed local correlations on the analysis. This trial improves the developed local correlations as well as the analysis. An example of the instrumentation that shows the performance of the calculation can be found in [24].

Several "special materials" that are encountered in geotechnical engineering need to be characterized. These materials involve a material with problematic behaviour in engineering practice (e.g. [23]) or man-made materials for a specific purpose (e.g. [25]). Laboratory tests for these kinds of materials need to be performed for future development.

6. A Proposed Option of Stages in Introduction of Unsaturated Soil Mechanics in Indonesia

From the above discussion, two points can be learned:

1. The understanding of the presence of suction on the ground is not new. The presence of negative pore-water pressure, u_w in soils has been understood before the mid-20th century [3,4,5,6]. In other words, the presence of suction has been understood in the mid-20th century.
2. The application of USM involves more parameters than the application of classical soil mechanics for saturated soils.

These show the need of a systematic step from the introduction towards the application of USM particularly in Indonesia. A proposed option of stages of introduction of USM in Indonesia is shown in figure 7. The step starts with the dissemination of the knowledge through the course. This step is an initial of the series of steps towards the application of USM in engineering practice in Indonesia. This step is followed by performing laboratory tests and field monitoring. Wesley [26,27,28] and Hill [29] have initiated significant works to characterize tropical soils from Java and Sumatera, Indonesia.

The importance of SWCC in USM has been emphasized in Fredlund et al. [9] and in Krisnanto [22]. SWCC can be seen as a prospective entry point for the research and development in Indonesia (figure 8). Characterization of a particular type of soil can be started with the tests to obtain SWCC.

7. Conclusions

1. The knowledge of unsaturated soil mechanics (USM) is needed to improve the understanding on several geotechnical engineering-related infrastructure problems in Indonesia.
2. The framework of USM is available [7,9].
3. The graduate course of soil mechanics for unsaturated soils (SI6227) is available in ITB, Bandung, Indonesia.
4. Laboratory testing apparatuses are currently under development in ITB to support the graduate course of soil mechanics for unsaturated soils. The apparatuses are also developed as a part towards the implementation of USM in Indonesia.
5. Early works in the analysis incorporating USM have been performed in Indonesia as shown in figures 3 and 4.
6. An option of stages of introduction of USM in Indonesia is shown in figure 7. This also shows that more efforts are required towards the application of USM in engineering practice in Indonesia.
7. Soil-water characteristic curve (SWCC) may be considered as an entry point for the research and development of USM in Indonesia. Characterization of a particular type of soil can be started with the tests to obtain SWCC.

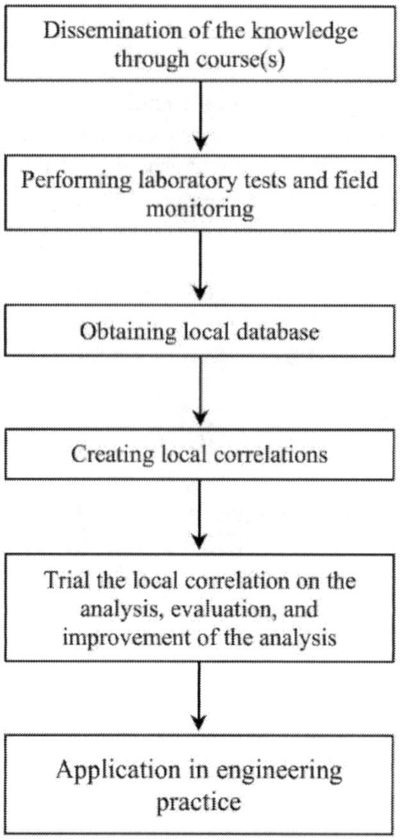

Figure 7. An option of stages of introduction of unsaturated sol mechanics in Indonesia.

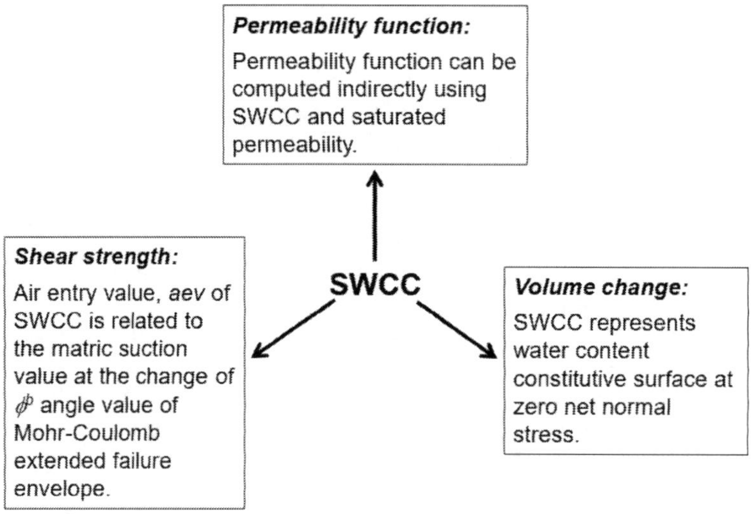

Figure 8. The relationship among soil-water characteristic curve (SWCC) to three domains of analysis in soil mechanics.

References

[1] Terzaghi K 1925 *Erdbaumechanik auf Bodenphysikalischer Grundlage* (Leipzig und Wien: Franz Deuticke)
[2] Fredlund D G 2006 *J. Geotech. Geoenv. Eng. ASCE* **132** 3 p 286-321
[3] Buisman A S K 1941 *Grondmechanica* (Bandoeng: Bandoengsche Technische Hoogeschoolfonds, Kleijene)
[4] Terzaghi K 1943 *Theoretical Soil Mechanics* (New York: John Wiley & Sons)
[5] Taylor D W 1948 *Fundamentals of Soil Mechanics* (New York: John Wiley & Sons)
[6] Terzaghi K and Peck R B 1948 *Soil Mechanics in Engineering Practice* (New York: John Wiley & Sons)
[7] Fredlund D G and Rahardjo H 1993 Soil Mechanics for Unsaturated Soils (New York: John Wiley & Sons)
[8] Terzaghi K, Peck R B, Mesri G 1996 *Soil Mechanics in Engineering Practice, Third Edition* (New York: John Wiley & Sons)
[9] Fredlund D G, Rahardjo H, Fredlund M D 2012 *Unsaturated Soil Mechanics in Engineering Practice* (New York: John Wiley & Sons)
[10] Krisnanto S, Rahardjo H, Kartiko R D, Satyanaga A, Nugroho J, Mulyanto N, Santoso P, Hendiarto A, Pamuji D B, Rachma S N 2021 *J. Eng. Tech. Sci.* **53** 5. https://doi.org/10.5614/j.eng.technol.sci.2021.53.5.4
[11] Krisnanto S and Rahardjo H 2018 *J. Eng. Tech. Sci.* **50** 4 p 566-577 https://doi.org/10.5614/j.eng.technol.sci.2018.50.4.8
[12] Fredlund D G, Morgenstern N R, Widger R A, 1978 *Can. Geotech. J.* **15** 3 p 313-321
[13] Krisnanto S, Rahardjo H, Fredlund D G, Leong E C 2014 *Eng. Geol.* **172** p 12-25 https://doi.org/10.1016/j.enggeo.2014.01.002
[14] Snow D T 1968 *J. Soil Mech. Found. Eng. Div. ASCE* **94** SM1 p 73–91
[15] Snow D T 1969 *Water Resour. Res.* **5** 6 p 1273–1289
[16] Krisnanto S, Rahardjo H, Leong E C 2011 *Unsaturated Soils: Theory and Practice 2011, Proc. 5th Asia-Pacific Conf. on Unsaturated Soils* (Bangkok: Kasetsart University) p 449-454
[17] Krisnanto S, Rahardjo H, Leong E C 2020 *Geomech. and Eng.* **21** 6, p 537-549 https://doi.org/10.12989/gae.2020.21.6.537
[18] Krisnanto S, Rahardjo H, Fredlund D G, Leong E C 2016 *Eng. Geol.* **210** p 168-179 https://doi.org/10.1016/j.enggeo.2016.06.012
[19] Terzaghi K 1936 *Proc. 1st Int. Conf. on Soil Mechs. and Found. Eng.* vol 1 (Cambridge, Massachusetts) p 54-56
[20] Fredlund D G and Morgenstern N R 1977 *J. Geotech. Eng. Div. ASCE* **103** GT5 p 447-466
[21] Surahman A, Zulkifli E, Susila E, Kusumawardhani R G, Rahman H, Harlan D 2020 *100th Anniversary of Civil Engineering Education in Indonesia* (in Indonesian) (Bandung: ITB Press)
[22] Krisnanto S 2021 *Lecture Note SI6227 Soil Mechanics for Unsaturated Soils* (in Indonesian) (Bandung: ITB Press)
[23] Krisnanto S, Muhlisah N, Ramadhani H S, Alatas I M 2020 *Proc. 24th Annual Indonesian National Conf. on Geotech. Eng.* (Jakarta: Indonesian Society for Geotechnical Engineering (ISGE/HATTI)) p 45-50 https://www.hatti.or.id/detail-pit-XXIV-2020.php
[24] Rahardjo H, Krisnanto S, Leong E C 2016 *Soils and Rocks* **39** 1 p 51-69 https://doi.org/10.28927/SR.391051
[25] Rahardjo H, Prasad A, Krisnanto S, Leong E C, Wang C L, 2019 *Geotech. Testing J. ASTM* **42** 2 p 471-484 https://doi.org/10.1520/GTJ20170404
[26] Wesley L D 1973 *Géotechnique* **23** 4 p 471-494
[27] Wesley L D 1977 *Géotechnique* **27** 2 p 125-136

[28] Wesley L D 2001 *Géotechnique* **51** 10 p 901-904
[29] Hill D 1975 *Géotechnique* **25** 2 p 417-423

Application of cellulose of *Boehmeria nivea* as natural flocculants in direct flocculation

D Fauzani[1*], S Notodarmojo[2], M Handajani[2], Q Helmy[2] and T Kardiansyah[3]

[1]Doctoral Program of Environmental Engineering, Faculty of Civil and Environmental Engineering, Institut Teknologi Bandung, Jl. Ganesha No.10, Bandung 40132, Indonesia
[2]Department of Environmental Engineering, Faculty of Civil and Environmental Engineering, Institut Teknologi Bandung, Jl. Ganesha No.10, Bandung 40132, Indonesia
[3]Center for Standardization and Services for the Cellulose Industry, Ministry of Industry of the Republic of Indonesia, Jl. Raya Dayeuhkolot No. 132, Bandung 40258, Indonesia

*corresponding author: dindafauzani@gmail.com

Abstract. The use of cellulose which is a natural polymer as a flocculant has been widely used, however, the extraction of cellulose from plants has not been widely studied, particularly extracted from *Boehmeria nivea* (ramie). In this research, the stem of ramie came from the waste of textile raw material. The isolation of α-cellulose is done through pre-hydrolysis, delignification, and bleaching processes. In this research, there are 6 types of α-cellulose that are isolated from the stem of ramie, depending on the sodium hydroxide concentration in the delignification process. The isolated α-cellulose is characterized using Fourier-transform infrared spectroscopy (FTIR) and tested its flocculation ability through jar-test apparatus with 5 gr/L kaolin suspension as synthetic water samples and turbidity as the test parameter. The coagulation-flocculation experiments were carried out without the addition of a coagulant, which is called the direct flocculation method. Data analysis was performed by One-Way ANOVA statistical analysis, with Duncan's multiple range test as the post-hoc analysis. It is concluded that cellulose is a good candidate biomaterial for natural flocculants in removing turbidity and further modification of α-cellulose with cationic grafts will be carried out for the next stage of research.

Keywords: cellulose, ramie, natural flocculant, direct flocculation

1. Introduction

The use of inorganic coagulants and flocculants and synthetic polymers has shown high-efficiency processing results, besides being easy to obtain and having economic value, so that in recent decades, these types of coagulants and flocculants are popular and widely used. However, some of these inorganic coagulants and flocculants, as well as synthetic polymers, can cause secondary pollution and cause environmental problems [1, 2, 3] The sludge produced from the coagulation-flocculation process contains BOD of hundreds of mg/l and COD of thousands of mg/l, with a ratio of both ranging from 1:15, with relatively stable characteristics, not easily decomposed biologically, containing aluminium and iron hydroxides derived from the use of conventional coagulants, inorganic and organic materials such as colloids, viruses, algae, and bacteria [1, 4]. For this reason, synthetic natural biopolymers or modifications of natural biopolymers can be alternative solutions as environmentally friendly coagulants and flocculants. Compared to traditional inorganic and synthetic polymers, modified natural biopolymers are generally non-toxic and their sludge is biodegradable so that they have minimal impact on human health [1, 2, 5]. The use of coagulants and flocculants from biopolymers is important because they are expected to improve the quality of the sludge produced in the deposition process, so that it becomes biodegradable, does not have hazardous and toxic properties such as toxicity, and does not contain heavy metal contaminants [2, 6].

Biopolymer-based coagulant and/or flocculants, due to their non-toxicity, biodegradability and ability to regenerate, can meet the growing demand for environmentally friendly materials. These biopolymers have been the focus of research around the world, for example starch-based flocculants, inulin-based flocculants, and cellulose-based flocculants [7]. Among these natural materials, cellulose, which is the most abundant polysaccharide on earth [8, 9, 10], is a prime candidate as a biopolymer for the manufacture of flocculants due to their suitable chemical structure for the bridging mechanism, high degree of polymerization, and number of potential chemical modifications [11]. Cellulose has a good water purifying effect because it has abundant free hydroxyl groups (-OH) in the chain, enabling the efficient removal of metal ions and organic matter from water for an excellent chelating effect [12, 13].

The term of direct flocculation was introduced as a water treatment process without the addition of coagulant, using cationic and/or anionic polymers offers the possibility to completely replace inorganic coagulants with water-soluble organic polymers [14]. In direct flocculation, cationic polymers with medium charge density and high molecular weight are usually used. This polymer has the dual function of neutralizing the negative charge of colloidal particles and bridging unstable particle aggregates together to form flocs [14, 15]. In the studies conducted by [16, 17, 18] the addition of polyelectrolytes as flocculants was carried out at the rapid mixing stage followed by slow mixing.

To reduce the use of conventional coagulants and flocculants that produce secondary pollution and sludge that are not environmentally friendly, it is important to synthesize biomaterial composites or natural polymers made from plants. Meanwhile, Ramie (*Boehmeria nivea*) is a large biomass potential in Indonesia and can be used as an alternative product other than textile raw materials to replace cotton. Ramie stem waste also contains cellulose which can be isolated and modified into cationic and anionic biomaterial composites for use as a flocculant, up to 70% compared to other sources of cellulose from plants. The aims of this research are to characterize Ramie cellulose which will be used as flocculants, and to know the processing efficiency of Ramie cellulose as flocculants in direct flocculation.

2. Materials and methods

α-cellulose of *Boehmeria nivea* was obtained with an isolation method adapted from previous studies [19, 20, 21, 22, 23] and was done as described in Fauzani et al. (2021) [24]. In general, the processes are pre-hydrolysis with acetic acid, delignification with sodium hydroxide, and bleaching process with sodium chlorite. In the delignification process, the alkaline solution used was sodium hydroxide (NaOH) with various concentrations of 5%, 10%, 15%, 17.5%, 20%, 25% (w/v) with distilled water as the solvent. Delignification was also carried out by heating over a water bath at 105 °C for 1 hour. The sample was then filtered through filter paper and rinsed using distilled water until the pH was neutral. The filtrate contained in the solution is beta-cellulose and gamma-cellulose, while the residue left on the filter paper is α-cellulose which will be used in this study.

2.1. Coagulation-flocculation experiments

The 6 types of α-cellulose of *Boehmeria nivea* (see Table 1) were used as a flocculant in jar testing using kaolin suspension (5 gr/L). Stirring was carried out using a magnetic stirrer at 200 rpm (rapid mixing) for 3 minutes, followed by stirring at 40 rpm (slow mixing) for 7 minutes. Without the addition of coagulation, in return, the 6 types of α-cellulose were added with the concentration of 10 ppm in the rapid mixing process, expected to act as both coagulant and flocculant. The parameter analysed is turbidity and sampling of the supernatant was taken at 5, 10, 15, 30, 60, and 90 minutes of clarification time. The removal efficiency is calculated using the following equation (1). T_0 is the initial turbidity concentration and T_i is the supernatant turbidity concentration at a given clarification time.

$$Removal\ efficiency = \frac{T_0 - T_i}{T_0} \times 100\% \quad (1)$$

Table 1. Coagulation-flocculation experiments

Sample	NaOH concentration	Dose of PAC as coagulant	Dose of α-cellulose as flocculant
C0	-	20 ppm	-
C1	05.0%	-	10 mg
C2	10.0%	-	10 mg
C3	15.0%	-	10 mg
C4	17.5%	-	10 mg
C5	20.0%	-	10 mg
C6	25.0%	-	10 mg

2.2. Analysis and characterization of α-cellulose

Fourier-transform infrared spectroscopy (FTIR) is an instrumentation used to characterize organic compounds by looking at their constituent functional groups. The results of FTIR absorption show various absorptions of functional groups belonging to α-cellulose. FTIR analysis was done using Bruker Alpha II Platinum ATR. The jar-test analysis was done in triplicates and statistically analysed using One-way ANOVA with Duncan's multiple range test as the post-hoc analysis. The statistical analysis was done with Microsoft SPSS.

3. Results and discussion

3.1. Coagulation-flocculation experiments

The direct flocculation experiment was carried out, which is the of the six α-cellulose samples as flocculants without the addition of coagulant in the rapid mixing process. As a control, synthetic water samples were used with the addition of PAC as a coagulant with the optimum dose in the previous studies of 20 ppm, and without the addition of α-cellulose, which was indicated by C0+PAC.

Test of Homogeneity of Variances

		Levene Statistic	df1	df2	Sig.
Removal_4	Based on Mean	7.374	41	84	.000
	Based on Median	1.224	41	84	.216
	Based on Median and with adjusted df	1.224	41	8.519	.402
	Based on trimmed mean	6.546	41	84	.000

ANOVA

Removal_4

	Sum of Squares	df	Mean Square	F	Sig.
Between Groups	707.361	41	17.253	258.437	.000
Within Groups	5.608	84	.067		
Total	712.969	125			

Figure 1. The results of the test of homogeneity of variances and the test of One-way ANOVA.

Table 2. Duncan multiple range test results (% removal).

Duncan[a]														
		Subset for alpha = 0.05												
Cellulose	N	1	2	3	4	5	6	7	8	9	10	11	12	13
C5+10min	3	91.2567												
C2+5min	3		92.2567											
C1+5min	3			92.9767										
C5+5min	3			93.1067										
C4+5min	3			93.3567	93.3567									
C3+5min	3				93.6933	93.6933								
C2+10min	3					94.0133	94.0133							
C4+10min	3					94.0400	94.0400							
C4+15min	3						94.1900							
C6+10min	3						94.1900							
C6+5min	3						94.2333							
C1+10min	3						94.2667							
C2+15min	3						94.2900							
C1+15min	3						94.3700							
C5+15min	3						94.3967							
C3+10min	3						94.4033							
C6+15min	3						94.4300							
C3+15min	3							95.9300						
C0+PAC+5min	3								96.5033					
C0+PAC+10min	3									97.1167				
C4+30min	3										97.8200			

Table 2. (continued)

C0+PAC+15min	3										97.8500			
C2+30min	3										98.2267	98.2267		
C0+PAC+30min	3											98.3133	98.3133	
C1+30min	3											98.3233	98.3233	
C3+90min	3											98.3333	98.3333	
C5+30min	3											98.3400	98.3400	
C6+30min	3											98.3733	98.3733	
C2+90min	3											98.4900	98.4900	98.4900
C4+60min	3											98.5400	98.5400	98.5400
C1+90min	3											98.5500	98.5500	98.5500
C4+90min	3											98.5867	98.5867	98.5867
C1+60min	3											98.6333	98.6333	98.6333
C2+60min	3											98.6400	98.6400	98.6400
C0+PAC+60min	3											98.6500	98.6500	98.6500
C5+60min	3											98.7133	98.7133	98.7133
C6+60min	3											98.7333	98.7333	98.7333
C3+30min	3												98.7900	98.7900
C5+90min	3													98.8933
C0+PAC+90min	3													98.9033
C3+60min	3													98.9100
C6+90min	3													98.9867
Sig.		1.000	1.000	.092	.114	.124	.104	1.000	1.000	1.000	.071	.051	.067	.055

Means for groups in homogeneous subsets are displayed.

a. Uses Harmonic Mean Sample Size = 3.000.

Figure 1 shows the results of the statistical analysis using SPSS software. The test of homogeneity of variances shows that based on the median, the p-value given is 0.214 and 0.402 (>0.05), which means the data could be analysed with One-way ANOVA. The data from Figure 1 Table ANOVA shows a significance value (Sig.) or p-value less than 0.05, which can be stated that statistically, there is a significant difference between the data groups.

Table 2 shows the results of Duncan's Post-Hoc test by displaying the average value of the percentage of turbidity removal. Based on Duncan's Post-Hoc test, statistically, there was no significant difference between the six α-cellulose samples and the control. This can be seen in Table 1 in column 13. The shortest precipitation time was given by the α-cellulose sample C3 (15% NaOH), which was 30 minutes. Based on Duncan's Post-Hoc test, statistically, it can be concluded that the six α-cellulose samples have the potential for direct flocculation applications to replace conventional coagulants, namely PAC.

3.2. Characterization of α-cellulose

The results of FTIR absorption showed various absorptions of the functional groups of ramie plant α-cellulose as shown in Figure 2. From the FTIR spectrum (Table 3), it can be seen that the α-cellulose functional group has peaked in as where CMCC is Commercial Microcrystalline [25, 26].

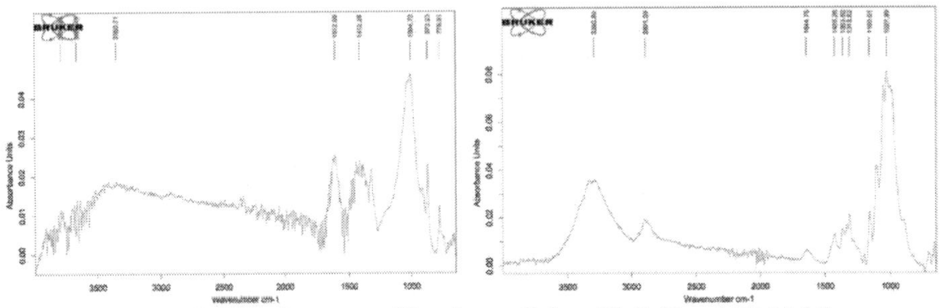

Figure 2. FTIR Spectrum of Ramie α-cellulose C3 (left) and C4 (right)

Table 3. FTIR spectrum and functional groups in α-cellulose

Functional group (Vibration)	Wavelength (cm^{-1})		
	C3	C4	CMCC
–OH	3360	3290	3344
C–H (*stretch*)	-	2891	2895
H_2O	1612	1644	1633
$-CH_2$	1412	1426	1428
C–O–C (*glycosidic*)	1004	1160	1162
C–H (*rock*)	873	898	897

The peak intensity band at 3300 cm−1 is associated with O–H stretching vibrations. The bands at 2800 and 1400 cm^{-1} are characteristic of C–H strain and –CH2 buckling, respectively. The peaks at 1600 and 900 cm^{-1} are associated with the H–O–H stretching vibrations of the water absorbed in carbohydrates and the C–H deformation vibrations of the cellulose [19, 26].

Liew et al. (2015) stated that the peak intensity at 1500 cm^{-1} of the spectrum was associated with C=C stretching vibrations in the aromatic ring of lignin. However, samples of α-cellulose C3 and C4 did not show C=C stretching in this area. This indicates that the lignin is removed properly through chemical processes, namely the prehydrolysis process using acetic acid and the delignification process using NaOH. Spectra related to the functional groups containing lipids, proteins, and nucleic acids (1800-1050 cm^{-1}) were also not seen after acid and base treatment. The peak at 1700 cm^{-1} representing the ester bond of the carboxylic groups of ferulic and p-coumaric acids in hemicellulose was not seen and could be attributed to the removal of hemicellulose through the α-cellulose isolation process. This fact also shows that hemicellulose and lignin are removed during the chemical process and the original molecular structure of cellulose can be preserved [19, 26, 27].

Figure 3 shows the comparison between α-cellulose C4 and cellulose swab. The peak at 900 cm^{-1} in α-cellulose C4 is associated with –C–H glycosidic deformation, ring vibration, and –O–H bending. This character implies glycosidic bonds between the anhydroglucose units in cellulose. The increase in peak intensity at 1160 cm^{-1} confirmed that the cellulose content increased due to alkaline treatment. Comparison between cellulose swab and α-cellulose C4 showed that both celluloses had identical peaks. The peak at 700 cm^{-1} for -cellulose C4 is related to the aromatic hydrogen C–H [19, 28]. Based on the compared spectra, it can be confirmed that the α-cellulose C4 obtained is identical to commercial cellulose swabs.

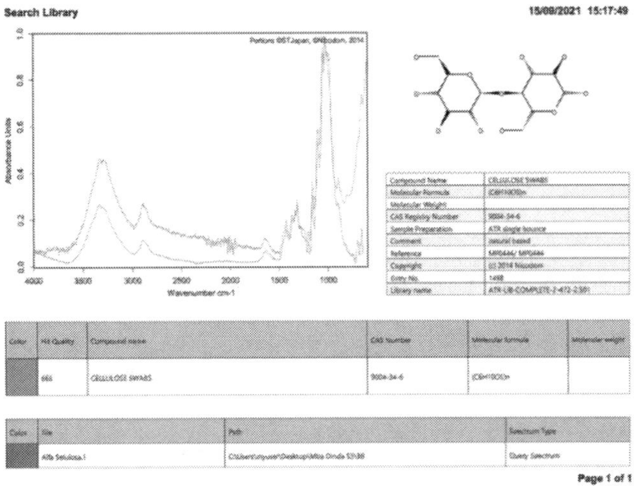

Figure 3. This figure shows the comparison between α-cellulose C4 and cellulose swab. Comparison between cellulose swab and α-cellulose C4 showed that both celluloses had identical peaks.

4. Conclusion

This research isolated the α-cellulose from agro-waste of *Boehmeria nivea*, specifically the stalks. There are 6 types of ramie α-cellulose, depending on the NaOH concentration on the delignification process. The 6 types of ramie α-cellulose were then used as a flocculant in jar testing using kaolin suspension (5 gr/L) as the synthetic wastewater, with turbidity as the research parameter. The study compares the addition of α-cellulose and the addition of PAC as a conventional coagulant. The addition of α-cellulose is done in rapid mixing, diminishing the use of coagulation, which is a method called direct flocculation. Data analysis was performed by One-Way ANOVA statistical analysis, with Duncan's multiple range test as the post-hoc analysis. Characterization of ramie α-cellulose is done with FTIR that shows various absorptions of functional groups belonging to α-cellulose. Based on Duncan's Post-Hoc test, statistically, there was no significant difference between the six α-cellulose samples and PAC. Statistically, it can be concluded that the 6 types of α-cellulose samples have the potential for direct flocculation applications to replace conventional coagulants, namely PAC. It shows that cellulose is a good candidate of biomaterial for natural flocculants for turbidity removal. Further modification of α-cellulose with cationic grafts will be carried out for the next stage of research.

5. References

[1] Kurniawan SB, Abdullah SR, Imron MF, Said NS, Ismail NI, Hasan HA, Othman AR and Purwanti IF 2020 *Int. J. Environ. Res. Public Health* **17** 9312
[2] Wang JP, Yuan SJ, Wang Y and Yu HQ 2013 *Water Res.* **47** 2643-8.

[3] Xing W, Ngo HH, Guo W, Wu Z, Nguyen TT, Cullum P, Listowski A and Yang N 2010 *Sep. Purif. Technol.* **72** 140–6
[4] Bratby J 2016 *Coagulation and flocculation in water and wastewater treatment* IWA publishing
[5] Adnan O, Abidin ZZ, Idris A, Kamarudin S and Al-Qubaisi MS 2017 *Environ. Sci. Pollut. Res.* **24** 20104-12
[6] Kaith BS, Jindal R, Jana AK and Maiti M 2010 *Biores. Technol.* **101** 6843-51
[7] Chen K, Lv W, Chen W, Wang Y, Zhang Y, Zhang X and Yao J 2016 *BioResources* **11** 5918–30
[8] Poletto M, Pistor V and Zattera AJ 2013 *Cellulose-Fundamental Aspects* **2** 45–68
[9] Thakur VK, Thakur MK and Gupta RK 2013 *Carbohydr. Polym.* **98** 820–828
[10] Zainal SH, Mohd NH, Suhaili N, Anuar FH, Lazim AM and Othaman R 2021 J. Mater. Res. Technol. **10** 935–952
[11] Kono H 2017 *Resource-Efficient Technol.* **3** 55–63
[12] Hokkanen S, Bhatnagar A and Sillanpää M 2016 *Water Res.* **91** 156–173
[13] Pellizzer L 2016 *Synthesis of cellulose-based flocculants and performance tests* (Portugal: University of Coimbra)
[14] Chong M 2012 *Advances in Water Treatment and Pollution Prevention* (Netherlands: Springer) pp 201–30
[15] Lee C S, Robinson J and Chong M F 2014 *Process. Saf. Environ. Prot.* **92** 489–508
[16] Ahmad AL, Chong MF and Bhatia S 2008 *Chem. Eng. J.* **140** 86–100
[17] Ginos A, Manios T and Mantzavinos D 2006 *J. Hazard. Mater.* **133** 135–142
[18] Sarika R, Kalogerakis N and Mantzavinos D 2005 *Environ. Int.* **31** 297–304
[19] Liew F K, Hamdan S, Rahman M R, Rusop M, Lai J C H, Hossen M F and Rahman M M 2015, *J. Chem.* **2015** 1–6
[20] Nawangsari D 2019 *J. Farmasi Indonesia* **16** 67–72
[21] Nawangsari D, Chaerunisaa A Y, Abdassah M, Sriwidodo S, Rusdiana T and Apriyanti L 2018 *Indones. J. Pharm. Technol.* **5** 55–61
[22] Sumada K, Tamara P E and Alqani F 2011 *J. Teknik Kimia* **5** 434–8
[23] Vandamme D, Eyley S, Van d M G, Muylaert K and Thielemans W 2015 *Biores. Technol.* **194** 270–5
[24] Fauzani D, Notodarmojo S, Handajani M, Helmy Q and Kardiansyah T 2021 *IOP Conf. Ser. Earth Environ. Sci.* **896** p 12056
[25] Chaouf S, El Barkany S, Jilal I, El Ouardi Y, Abou-salama M, Loutou M, El-Houssaine A, El-Ouarghi H, El Idrissi A and Amhamdi H 2019 *J. Water Process Eng.* **31** p 100807
[26] Suciyati SW, Manurung P, Sembiring S and Situmeang R 2021 *J. Phys. Conf. Ser.* **1751** p 012075
[27] Agwuncha SC, Owonubi S, Fapojuwo DP, Abdulkarim A, Okonkwo TP and Makhatha EM 2021 *Mater. Today: Proc.* **38** 958-63
[28] Nguyen HD, Mai TT, Nguyen NB, Dang TD, Le ML and Dang TT 2013 *Adv. Nat. Sci.: Nanosci. Nanotechnol.* **4** p 015016

Factors affecting safety leadership of construction project owners in Indonesia

D V Indrayana[1], K S Pribadi[2], P F Marzuki[3] and H Iridiastadi[4]

[123] Faculty of Civil and Environmental Engineering, Institut Teknologi Bandung, Ganesha St., Bandung, 40132, West Java, Indonesia
[4] Faculty of Industrial Engineering, Institut Teknologi Bandung, Ganesha St., Bandung, 40132, West Java, Indonesia

Abstract. Construction is an industry with a high possibility of risk. The construction industry accounts for the highest number of work accidents globally, with more than 60,000 fatal accidents per year. A construction safety culture through mature safety leadership can be the fundamental factor to improve safety performance. Safety leadership is the process of interaction between leaders and followers to achieve organizational safety goals. In many cases of work accidents in construction projects in Indonesia, the leadership of the project owner is rarely discussed. It is more often determined that the responsible stakeholder is the contractor. Leadership by the project owner is the main factor for generating new and effective safety management measures. This study aims to determine the factors that influence the Safety Leadership of construction project owners in Indonesia. This study uses the Delphi round technique to analyze experts' judgments to avoid biases. This study results in obtaining several factors and indicators that are considered to have an influence on the safety leadership of construction project owners in Indonesia.

1. Introduction

The construction industry has a high possibility of fatality risk (Oh et al.. 2021). Globally, construction becomes one of the most harmful industries, with over 60.000 fatality incidents per year (Albert et al. 2020). The construction workplace comprises various heavy equipment and hazardous materials, whereas each component can produce risks to the workers. The complexity of stakeholder engagement makes the inherent construction project risk is higher. Contractors, consultants, designers, and workers are working together in a high-risk workplace. Therefore, the safety culture of stakeholders is considered as the main driver to determine safety performance (Esterhuizen & Martins, 2016). Diversity level of safety culture from different stakeholders causes the lower possibility of proper safety performance (Filho & Waterson, 2019). For instance, safety implementation enforcement by the contractors will not be as effective as expected if the workers do not have the awareness to do so. Many causative factors of safety culture have been discussed in the past few years. Esterhuizen & Martins (2016) stated that safety leadership is the most influential factor. Another correlation between safety leadership and safety culture was reviewed by Oswald & Lingard (2019). Safety leadership is considered a predecessor factor of safety culture, while safety performance is the output of the safety culture fulfillment. Therefore, the first objective to meet by the stakeholders is safety leadership.

Safety leadership is one of the organizational leadership sub-systems that focuses on safety (Daniel, 2015). Specifically, in Indonesia, a decent leadership will decrease 30,9% of safety risk in a construction project (Machfudiyanto, et al. 2019). There are many other arguments and findings discussing the importance of leadership in a project. Nevertheless, one big issue to be answered: which stakeholders who have the biggest responsibility to own a decent leadership. Many researchers have focused on the contractor to assess the leadership (Oswald & Lingard, 2019). Contractors become the main object since they directly interact with the workplace and workers. If there is an incident, contractors always are the stakeholder to be inspected. There is one stakeholder with higher power from the contractors, i.e., the project owner. They are allowed to give an order to be obeyed by the stakeholders in a workplace. However, research with focus on project owner is relatively fewer than contractors.

Construction project owners are responsible for direct and indirect causative factors of safety performance. The owner's commitment and motivation directly affect the safety performance (Olanipekun, et al. 2017). The project owner has an essential duty to ensure every clause in the contract related to safety (Nabi, et al. 2020). Contractor selection is also the responsibility of the project owner. They need to assess the contractor before signing the contract, whether they have a good safety plan or not (Karakhan, et al. 2018). Therefore, the project owner also plays an essential part in the non-performing contractor from a safety perspective.

This paper aims to identify the factor affecting the safety leadership of construction project owners. It is expected to give a different perspective on obtaining well-performed safety in construction projects. It is also hoped that the high-level safety culture of stakeholders occurs through safety leadership to ensure safety performance, particularly in Indonesia.

2. Literature review

2.1. Safety leadership

There are many previous research discussing safety leadership. Zhang, et al. (2018) define safety leadership as an interactive process between leader and workers to achieve organizational safety. It led to processes where the leaders needed to mentor the workers, rather than only giving instruction (Fernández-Muñiz, et al. 2017). Safety leadership is a specific behavior from a leader to lead the workers toward safety objectives (Griffin & Hu, 2013).

Simmons, et al. (2020) explore the factor affecting safety leadership globally. There are 19 leadership competencies to be met by a leader, i.e., communication ability, responsible, professionalism, critical thinking, big thinking of the mind, ambition, self-awareness, modesty, teamwork, people focus, time management, adaptability, management ability, quality control, computer ability, risk management, assertiveness, general knowledge, and economic principles understanding. These factors are globally acceptable by the construction industry. Yet, adjustments are needed to refine the most influential factors in the Indonesian construction industry. There are many specific conditions in the Indonesian construction industry, e.g., different cultures of workers' hometowns in a workplace yield to other mentoring methods. Hence, particularly in this case, the communication ability of a leader becomes very crucial of a leader in Indonesia's construction industry.

Conscientiousness positively impacts workers' safety compliance (Xia, et al. 2021). Meanwhile, agreeableness has a negative impact. Indonesia's characteristics of high-diversity culture may result in reverse results. The agreeableness may have a positive effect. The diversity led to a sensitive issue if the workers come from different city or island from all over Indonesia as a comer. Therefore, contextualize safety leadership in Indonesia is crucial to provide a unique perspective to enhance safety performance.

2.2. Construction project owner contribution to safety performance

Safety performance models are directly affected by at least four factors (Huang & Hinze, 2006), i.e.: 1) a proper contractor selection; 2) a clear statement of safety requirement in a contract; 3) implement safety objectives in project; 4) project owner participation in safety management. The project owner contributes a big portion on the 1st and 2nd factors, while the 4th clearly states that project owner participation is important. The 3rd factor is the full responsibility of the contractor if the 2nd factor is fulfilled.

Project owners' leadership in a construction project is the foremost requisite to effective safety management (Wu, et al. 2015). A decent leadership of project owners will lead to a change of safety perception, motivation, and behavior of stakeholders into a better form. Olanipekun, et al. (2017) also mentioned that project owners' commitment enhances project performance, including safety. Therefore, good leadership of project owner generates several conditions to ensure the safety performance.

3. Method

Factors affecting safety leadership have been discussed broadly by previous research. Nevertheless, contextualization of the factors into Indonesia's construction industry is critical. Safety leadership causative factors are considered non-quantitative aspects, so it needs to be determined by an appropriate method. Experts' judgement from the stakeholders of Indonesia's construction industry is required. To analyze the judgment, delphi round technique is often used to solicit opinions through different processes and controls (Bhandari & Hallowell, 2021). This method, through its analysis techniques, e.g.: anonymity, multiple rounds, and controlled feedback, can effectively decrease social biases that threaten the validity and reliability of the results.

Invited Indonesia's construction industry stakeholders for this paper consist of various backgrounds. There are a total of 7 experts, including 1) construction project owner of a railway; 2) government (2 experts); 3) academics (3 experts); 4) senior vice president of a big contractor company. The experts' composition was selected by considering that every point of view had to be represented by at least an expert. The Delphi technique is beneficial when the participants must be preserved to assure the validity of the results (Hallowell, 2009). To ensure validity, some steps are carried out, i.e.:

- The researcher gives a brief explanation of the safety leadership affecting factors and indicators in Indonesia
- The researcher invites the experts to give a score on a scale of 1 (lowest) – 10 (highest) on each indicator
- The researcher guides the discussion to reach a score consensus on each factor, whereas the minimal score to proceed as the safety leadership causative factors is 7
- If there is no consensus on a factor, the researcher invites the experts to the next round of scoring and discussion
- If the consensus is below the number of 7, the factors or indicators, therefore, are excluded from the analysis
- Repeat the above steps to all factors and indicators

The Delphi technique was performed through a discussion forum led by a moderator. Due to the Covid-19 pandemic, the discussion was held through video conference and limited to 2 hours to ensure the effectiveness. All factors had been discussed effectively within the limited time and were expected to represent the construction owners' safety leadership.

4. Safety leadership factors

A proper leadership of construction project owner is justified by 5 factors, i.e.: 1) psychology; 2) sociology; 3) participative; 4) dialogue intensity to stakeholders; 5) technical knowledge. Each factor consists of several indicators to represent it. The detail of factors and indicators showed in Table 1.

Table 1. Indicative affecting factors of safety leadership

Factor	Notation	Indicators of project owner's safety leadership	References
Factor 1: psychology	**1.1**	Commitment in safety implementation	Orlando, et al. (2019) & Levovnik (2020)
	1.2	Persistence, perseverance, and consistency	Ma, et al. (2018)
	1.3	Reliable and honest	Simmons, et al. (2020)
	1.4	Communication ability	Zhang, et al. (2020) & Rudnicka (2016)
	1.5	Ability to receive induvial respect from workers	Wu, et al. (2017)
	1.6	Ability to influence	Rudnicka, A. (2016)

Factor	Notation	Indicators of project owner's safety leadership	References
Factor 1: psychology	1.1	Commitment in safety implementation	Orlando, et al. (2019) & Levovnik (2020)
	1.2	Persistence, perseverance, and consistency	Ma, et al. (2018)
	1.3	Reliable and honest	Simmons, et al. (2020)
	1.4	Communication ability	Zhang, et al. (2020) & Rudnicka (2016)
	1.5	Ability to receive induvial respect from workers	Wu, et al. (2017)
	1.6	Ability to influence	Rudnicka, A. (2016)
Factor 2: sociology	2.1	Known by the stakeholders to provide safety motivation	Wu, et al. (2017) & Oswald & Lingard (2019)
	2.2	Participate in problem solving	
	2.3	Assure the stakeholders to have adequate resources to implement safety	Ojuola, et al. (2020)
	2.4	Ability to care	
	2.5	Assure the safety policy is obeyed by the stakeholders	Orlando, et al. (2019)
Factor 3: participative	3.1	Develop reward and punishment with objective parameters	Ghuzdewan & Damanik (2019)
	3.2	Implement the reward and punishment	Ghuzdewan & Damanik (2019)
	3.3	Routinely attend physically in project workplace	Orlando, et al. (2019)
	3.4	Routinely review the stakeholders' safety performance	
	3.5	Effectively delegate the task of project owner to internal team if the representative is unable to attend	Wu, et al. (2017)
Factor 4: dialogue intensity to stakeholders	4.1	Assure the public informed by the project	Ghuzdewan & Damanik (2019)
	4.2	Known by the public around the project workplace	
	4.3	Assure the resources to mitigate impact to surrounding public	Ali, et al. (2020)
	4.4	Participate in problem solving of project impact, including environmental impact	Ali, et al. (2020) & Wu, et al. (2017)
	4.5	Actively conduct discussion with stakeholders related to social and environment	Wu, et al.. (2017) & Oswald & Lingard (2019)
	4.6	Routinely review the environmental impact	
Factor 5: technical knowledge	5.1	Ability to develop safety performance objectives	Ali, et al. (2020) & Wang, et al. (2017)
	5.2	Routinely review the construction method from safety perspective	Wu, et al. (2017)
	5.3	Ability to perform creative and innovative way in safety implementation	Ali, et al. (2020)
	5.4	Ability to be consulted by the stakeholders regarding safety issues	Ali, et al. (2020)
	5.5	Ability to control the project workplace environment	Wang, et al. (2017)

5. Discussion

5.1. Delphi technique results

The Delphi technique effectively provided a solid answer since there are no rejected factors by the experts yet there was one factor with editorial adjustment. However, there was one excluded indicator from the analysis. The others result in an editorial adjustment.

Psychology factor consist of 6 indicators. There were no change and adjustment on factor and indicators except one. The fourth indicator has editorial adjustment. The communication ability was changed to 'ability to be a role model'. This factor describes how the construction project owner has a personal attributes to become a good leader.

Commitment has the highest score of 10 meanwhile the others have similar score of 9. This factor becomes the fundamental aspect of being a good safety leader in construction projects. Therefore, the project owner should fulfil this factor before the other factors.

There were two adjusted indicators on the sociology factor. The second and third indicators were changed respectively became 'ability to mentor the stakeholders in problem-solving' and 'ability to control and supervise'. This factor is the expansion from the previous as this factor describes how a leader interact with others. As presented by the indicators, this factor requires a leader to care, mentor, and supervise others, especially workers. If the construction project owners have fulfilled these two first factors, they may proceed to the next factor to have decent safety leadership.

Indicators in the third factor has major adjustment. All the indicators were changed based on experts' consensus. Reward and punishment were the most important factors to be implemented in Indonesia according to experts' judgement. Meanwhile, the construction project owner is also expected to physically attend to the construction project workplace in a regular basis. Therefore, project owner's participation is hoped to boost up the safety performance by the stakeholders.

The fourth factor has the editorial adjustment to 'communicative'. Experts were arguing that a construction project owner should own the ability to promote good things of the projects to both internal and external stakeholders. Unlike the indicator in the first factor, which also has a term of communication, this factor focuses on external communication to the surrounding society as well as environment-related stakeholders. There is also an indicator that further in the next research will be excluded. Experts agreed not to include the sixth indicator, i.e.: 'Routinely review the environmental impact'. A construction project owner is expected to ensure the safety of impacted people, especially in a radius of 300-500 meters from the site. This factor was considered more challenging to be owned if the first and second factors have yet to be fulfilled.

The fifth factor has the highest score on every indicator. All of them were scored with 10. Technical knowledge is the most required aspect to be owned by the contractor. It is acceptable if the project owner doesn't have high technical knowledge. However, they need to select a good contractor with decent technical knowledge. All experts agree if the construction project owners have good technical knowledge, it would be very beneficial for the project. Therefore, this factor is considered as the most important and advanced factor to be owned by the construction project owner.

Table 2. Factors affecting safety leadership of construction project owners in Indonesia

Factor	Notation	Indicators of project owner's safety leadership	Delphi technique score (out of 10)
Factor 1: psychology	**1.1**	Commitment in safety implementation	10
	1.2	Persistence, perseverance, and consistency	9
	1.3	Reliable and honest	9
	1.4	Ability to be a role model	9
	1.5	Ability to receive induvial respect from workers	9
	1.6	Ability to influence	9

Factor	Notation	Indicators of project owner's safety leadership	Delphi technique score (out of 10)
Factor 2: sociology	**2.1**	Known by the stakeholders to provide safety motivation	9
	2.2	Ability to mentor the stakeholders in problem solving	9
	2.3	Ability to control and supervise	9
	2.4	Ability to care	10
	2.5	Assure the safety policy is obeyed by the stakeholders	9
Factor 3: participative	**3.1**	Ability to develop reward & punishment policy in project	10
	3.2	Ability to develop standard operating procedure as implementation method of the reward & punishment policy	9
	3.3	Perform the walkthrough & safety patrol routinely	9
	3.4	Perform the performance evaluation of safety	8
	3.5	Persuasion ability	8
Factor 4: communicative	**4.1**	Actively inform the surrounding society about the project	10
	4.2	Known by the surrounding society and provide a contact person	9
	4.3	Report the achievement of environmental risk mitigation to the authorized stakeholders	10
	4.4	Ability to mediate the external project issue	9
	4.5	Persuasion ability to surrounding society	9
Factor 5: technical knowledge	**5.1**	Ability to develop safety performance objectives	10
	5.2	Routinely review the construction method from safety perspective	10
	5.3	Ability to perform creative and innovative way in safety implementation	10
	5.4	Routinely provide advice to the stakeholders regarding safety issues	10
	5.5	Ability to conduct a safety audit and investigation	10

5.2. How the project owner implements the safety leadership

Every project was highly unique and dynamic, including the construction sector (Enshassi, A. et al.., 2009). This was because the dynamic and changeability of a project were directly correlated with the level of risk, specifically safety (Orlando et al.., 2019). Contractor often became the most blamed stakeholder if there was any incident since they were directly related to the construction site. However, the project owner also plays an important role. According to Huang & Hinze (2006), an active project owner's participation, specifically in the site, was one of the four factors used to determine safety quality in safety management. The other three factors in this process include (1) selecting a competent contractor, (2) the existence of safety requirements in the contract, and (3) implementing safety objectives in the project.

Safety culture is a fundamental factor in improving safety performance (Esterhuizen & Martin, 2016; Oswald & Lingard, 2019). Meanwhile, safety culture is strongly influenced by safety leadership (Li et al.., 2020). Safety leadership is the primary requirement for effective safety management, particularly for the construction project owners (Wu et al.., 2015). Therefore, the determinants of safety performance will further be discussed by exploring a term, namely safety leadership. It is a sub-system of leadership where the leadership behavior focuses on safety (Daniel, 2015).

Several studies have shown that leadership directly affects the implementation of effective safety management (Levovnik et al.., 2020). From the above discussions, it can be concluded that safety performance is directly affected by the safety culture. Further, it is formed by decent leadership of the related stakeholders.

6. Conclusion

Safety becomes a more important factor to be concerned by the stakeholders nowadays. The economic growth has affected to the escalation of construction projects globally, particularly in Indonesia, where infrastructure development is the government's priority. The high possibility of risks was causing incidents that were reported annually by the authorized stakeholders. Safety culture in Indonesia's construction workplace is a crucial factor. To assure the safety culture, stakeholders need a decent safety leadership – not only the contractors as the most observed stakeholder in the construction industry, but also the project owner.

Many factors affect, direct and indirectly the safety leadership. Researchers have also been discussed on the affecting factor of safety leadership. Nevertheless, Indonesia's construction industry contextualization is needed since several specific conditions are considered. At least, project owners need five factors to apply a decent safety leadership, i.e.: 1) psychology; 2) sociology; 3) participative; 4) communicative; 5) technical knowledge. Experts' judgement from the delphi technique method results in affecting factor of safety leadership. The highest scored factor is technical knowledge. Although it is the most advanced factor for the project owner, experts argued it is crucial for safety performance. Project owners also need good communication skills, both to the external and internal project stakeholders.

Further works are required. This paper only provides the safety leadership causative factors and a general description of how the project owner uses safety leadership to improve safety performance. The factors can be converted into an applicative model, e.g., the safety leadership maturity model. It would be a decent method to assess the construction project owner in Indonesia. A government construction project may become the best option to implement those factors. A well-structured project team in the government body provides an enormous advantage to assess the project owner objectively.

7. Acknowledgments

Authors wishing to acknowledge assistance or encouragement from colleagues and special work by technical staff from the organizations.

8. References

[1] Albert A, Pandit B, Patil Y, and Louis J 2020 Does the potential safety risk affect whether particular construction hazards are recognized or not? *Journal of Safety Research* **75** 241-250

[2] Ali A, Wang H, Soomro M A & Islam T 2020 Shared Leadership and Team Creativity: Construction Industry Perspective *Journal of Construction Engineering and Management* **146** 04020122

[3] Bhandari S and Hallowell M R 2021 Identifying and controlling biases in expert-opinion research: Guidelines for variations of Delphi, nominal group technique, and focus groups *Journal of Management in Engineering* **37** 04021015

[4] Daniel L A 2015 *Defining safety leadership and associated behaviours within the Australian construction industry* (New South Wales: Southern Cross University)

[5] Enshassi A Mohamed S & Abushaban S 2009 Factors affecting the performance of Construction projects in the Gaza Strip *Journal of Civil Engineering and Management* **15** 269–280

[6] Esterhuizen W & Martins N 2016 The factor structure of a safety leadership assessment tool for the mining industry *Journal of Contemporary Management* **13** 1–26

[7] Fernández-Muñiz B, Montes-Peón J M & Vázquez-Ordás C J 2017 The role of safety leadership and working conditions in safety performance in process industries *Journal of Loss Prevention in the Process Industries* **50** 403-15

[8] Filho A P and Waterson P 2018 Maturity models and safety culture: A critical review *Safety Science* **105** 192–211

[9] Ghuzdewan T & Damanik T 2019 Analysis of accident in Indonesian construction projects *MATEC Web of Conference 2019*

[10] Griffin M A & Hu X 2013 How leaders differentially motivate safety compliance and safety participation: The role of monitoring inspiring and learning *Safety Science* **60** 196-202

[11] Hallowell M R 2009 Techniques to minimize bias when using the delphi method to quantify construction safety and health risks *Construction Research Congress 2009*

[12] Huang X & Hinze J 2006 Owner's Role in Construction Safety *Journal of Construction Engineering and Management* **13** 164-73

[13] Karakhan A A, Rajendran S, Gambatese J & Nnaji C 2018 Measuring and Evaluating Safety Maturity of Construction Contractors: Multicriteria Decision-Making Approach *Journal of Construction Engineering and Management* **14** 04018054

[14] Levovnik D & Gerbec M 2020 Role of Leadership Types in Managers' Commitment to Safety *European Safety and Reliability Conference and Safety Assessment and Management Conference 2020*

[15] Li M Zhai H Zhang J & Meng X 2020 Research on the relationship between safety leadership, safety attitude and safety citizenship behavior of railway employees *International Journal of Environmental Research and Public Health*, **17** 1-17

[16] Ma Y, Wu C, Fang D & Wang C 2018 Safety Leadership Effectiveness Assessment of Project Managers in the Construction Industry: A Case Study of China *Proceeding of Construction Research Congress 2018*

[17] Machfudiyanto R A, Latief Y & Robert 2019 Critical Success Factors to Improve SafetyCulture on Construction Project in Indonesia IOP *Conference Series: Earth and Environmental Science 2019*

[18] Nabi M A, El-Adaway I H & Dagli C 2020 A system dynamics model for construction safety behavior Procedia *Computer Science* **168** 249-56

[19] Oh T K, Kwon Y J, Oh B H, Gwon Y I, Yoon H K 2020 Suggestions for Safety Coordinator's Roles at Each Construction Stage (Client, Designer, Supervisor, and Contractor) to Improve Safety and Health Activities in South Korea *Safety Science* **133** 1-11

[20] Ojuola L, Mostafa S & Mohamed S 2020 Investigating the role of leadership in safety outcomes within oil and gas organsiations safety and reliability *Safety and Reliability* **392** 121-33

[21] Olanipekun A O, Xia B P, Hon C & Darko A 2018 Effect of Motivation and Owner Commitment on the Delivery Performance of Green Building Projects *Journal of Management in Engineering* **341** 1-14

[22] Orlando A G S, Lima G B A & Abreu C G S 2019 Assessment of Maturity Level: a Study of Qhse Culture *Revista Produção e Desenvolvimento* **5** 1-17

[23] Oswald D & Lingard H 2019 Development of a frontline H&S leadership maturity model in the construction industry *Safety Science* **118** 674–86

[24] Rudnicka A 2016 How to Managset Sustainable Supply Chain? The Issue of Maturity Scientific *Journal of Logistics* **124** 203-11

[25] Simmons D R, McCall C & Clegorne NA 2020 Leadership Competencies for Construction Professionals as Identified by Construction Industry Executives *Journal of Construction Engineering and Management* **14** 04020109

[26] Wang L, Xue X & Wu H 2017 A Framework for Developing an Engineering Management Theory Maturity Model *ICCREM 2016*

[27] Wu C, Li N & Fang D 2017 Leadership improvement and its impact on workplace safety in construction projects: A conceptual model and action research International *Journal of Project Management* **358** 1495–1511

[28] Wu X, Liu Q, Zhang L, Skibniewski M J & Wang Y 2015 Prospective safety performance evaluation on construction sites *Accident Analysis and Prevention* **78** 58-72
[29] Xia N, Tang Y, Li D & Pan A 2021 Safety Behavior among Construction Workers: Influences of Personality and Leadership *Journal of Construction Engineering and Management* **14** 04021019
[30] Zhang L, Chen H, Li H, Wu X & Skibniewski M J 2018 Perceiving interactions and dynamics of safety leadership in construction projects *Safety Science* **106** 66-78
[31] Zhang R P, Lingard H & Oswald D 2020 Impact of Supervisory Safety Communication on Safety Climate and Behavior in Construction Workgroups *Journal of Construction Engineering and Management* **14** 04020089

Study on numerical calculation of flow structures in a curved open channel with advanced depth-integrated models

F P Lugina[1], T Uchida[1] and Kawahara Y[2]

[1]Graduate School of Advanced Science and Engineering, Hiroshima University, JP.
[2]Resilience Research Center, Hiroshima University, JP.

*Corresponding author: d206823@hiroshima-u.ac.jp

Abstract. A better understanding of flow structures distribution in rivers is crucial to determine the safety degree of rivers. A practical and reliable model is required to overcome the issue of the long computational time of a three-dimensional calculation model and the lack of computation detail of a two-dimensional calculation model for flow structures distribution simulation in rivers. This paper presented an advanced depth-integrated numerical calculation method called the bottom velocity calculation (BVC) to reproduce the flow structures in a curved open channel. BVC method is an integrated multiscale simulation of flows in rivers that can evaluate vertical distributions of velocities and bottom velocity distributions by introducing depth-averaged horizontal vorticity and horizontal momentum equations on a water surface. It has several models, such as simplified bottom velocity calculation (SBVC) with shallow water assumption and general bottom velocity calculation (GBVC) method without the assumptions. The advantages of the BVC method, including SBVC and GBVC models, are validated in this paper using experimental datasets of a curved open channel and compared to two-dimensional and three-dimensional models. The results show that the BVC method has good reproducibility to simulate the flow structures distribution in the channel.

Keywords: curved open channel, depth-integrated model, secondary flwo, flow structures

1. Introduction

A better understanding of flow structure distribution in rivers is crucial to determine the safety degree of the river since it can assess velocity distribution and sediment transport patterns for bank protection. One of the flow structure characteristics in curved open channels is secondary flow. A secondary flow modifies streamwise velocity's transverse and vertical distribution [1–3]. The transverse water surface gradient and centrifugal force imbalance induce secondary flows in curved open channels. As a result, water flows outward at the surface but inward near the riverbed, leading to erosion along the outer bank and deposition at the inner bank [4,5].

Numerical models have been widely used in modeling river hydrodynamics because of their cost-effectiveness over experiments and field measurements after being validated with experimental datasets. A practical and reliable model is required to overcome the long computational time of the three-dimensional calculation model (3DC) and the lack of computational detail of the two-dimensional calculation model (2DC).

Several depth-integrated models have been proposed to solve those problems, i.e., long computational time and lack of computational detail. One is the bottom velocity calculation (BVC)

method [6–8]. BVC method is an integrated multiscale simulation of flows in rivers that can evaluate vertical distributions of velocities and bottom velocity distributions by introducing depth-averaged horizontal vorticity and horizontal momentum equations on a water surface. This study aims to present the advantages of the BVC method in reproducing flow structures distribution in a curved open channel and compare it to 2DC and 3DC models.

2. Methods

2.1. Experimental conditions

The configuration of the physical model of the experiment conducted by de Vriend [5] is shown in figure 1. Measured water surface elevation and velocity along the channel at the cross-sections are available. The experiment conditions are presented in table 1. The k_s value was obtained from a linear regression of the vertical velocity distribution at the upstream end using the logarithmic law of the wall [9], $k_s = 0.0005$ m, which is equivalent to 0.011 Manning's coefficient.

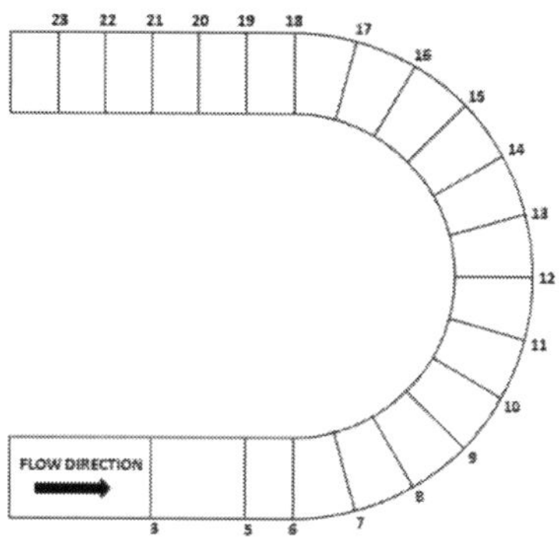

Figure 1. Experimental channel.

Table 1. Experiment condition for flow measurement.

Bed slope	Discharge (m^3/s)	Downstream water depth (m)	Width (m)	Channel length (m)	Inner radius r (m)	Outer radius R (m)
0.0	0.18	0.189	1.7	23.35	3.4	5,1

2.2. Numerical models

For the numerical models, a BVC method was applied together with 2DC and 3DC models to analyze the advantages of depth-integrated models in reproducing flow structures in a curved open channel. As a non-hydrostatic 3DC model, NaysCUBE [10,11] was used. The solver is available in the public domain within the iRIC river software package (www.i-ric.org). The detail of the BVC models and the 2DC model are summarized in table 2. The calculation domain includes the entire experimental channel with a computational mesh $dx = 0.1$ m and $dy = 0.1$ m (*dx*, *dy*: along the channel and transverse direction, respectively), and $dz = 0.02$ for NaysCUBE. The experimental discharge and water depth are used as upstream and downstream boundary conditions.

The BVC methods were developed based on equation (1), in which the bottom velocity is evaluated using the water surface velocity, depth-averaged horizontal vorticity, and depth-averaged vertical velocity.

$$\delta u_i = u_{si} - u_{bi} = \varepsilon_{ij}\Omega_j h + \left(\frac{\partial Wh}{\partial x_i} - w_s \frac{\partial z_s}{\partial x_i} + w_b \frac{\partial z_b}{\partial x_i} \right) \tag{1}$$

where i, j: $1(x)$, $2(y)$, $x_3 = z$: vertical axis; ε_{ij} : Levi-Civita symbol; u_{bi} : bottom velocity; u_{si} : water surface velocity; Ω_j : depth-averaged vorticity in the j direction; h : water depth; W: depth-averaged of vertical velocity w, w_s and w_b : vertical velocity on the water surface and bottom, respectively. The vertical distributions of the horizontal velocities are expressed by a cubic function [Eq. (2)] using the depth-averaged velocity U_i, velocity differences $\delta u_i\left(\delta u_i = u_{si} - u_{bi}\right)$, $\Delta u_{ij}\left(\Delta u_{ij} = U_{si} - U_i\right)$, and dimensionless depth η. A detailed explanation of the BVC method is explained in the Appendix.

$$u_i = \Delta u_i\left(12\eta^3 - 12\eta^2 + 1\right) + \delta u_i\left(-4\eta^3 + 3\eta^2\right) + U_i \quad (2)$$

The 2DC model consists of depth-integrated continuity and horizontal momentum equations. The SBVC (simplified bottom velocity calculation) methods with shallow water assumption employ hydrostatic pressure distribution and neglect the variation in vertical velocity. SBVC2 assumes the equilibrium condition of water surface flow, while SBVC3 calculates the momentum equations for water surface flows. GBVC3 (general bottom velocity calculation) is not restricted by the shallow water assumption and calculates non-hydrostatic pressure distribution and variation in vertical velocity.

Table 2. Unknown variables and equations.

Calculation method	Unknown variable	Governing equation	Equation number
GBVC3, SBVC3, SBVC2, 2DC	h	Depth-integrated continuity equation	(A7)
GBVC3, SBVC3, SBVC2, 2DC	U_i	Depth-integrated horizontal momentum equation	(A8)
GBVC3, SBVC3, SBVC2	u_{bi}	Depth-integrated definition equation of horizontal vorticity	(1)
GBVC3, SBVC3, SBVC2	Ω_i	Depth-integrated vorticity equation	(A1)
GBVC3, SBVC3	u_{si}	Horizontal momentum equation on water surface	(A5)
GBVC3	dp_b	Depth-integrated vertical momentum equation	(A9)
GBVC3	W	Double integrated continuity equations	(A10)

3. Results

3.1. Water surface profile

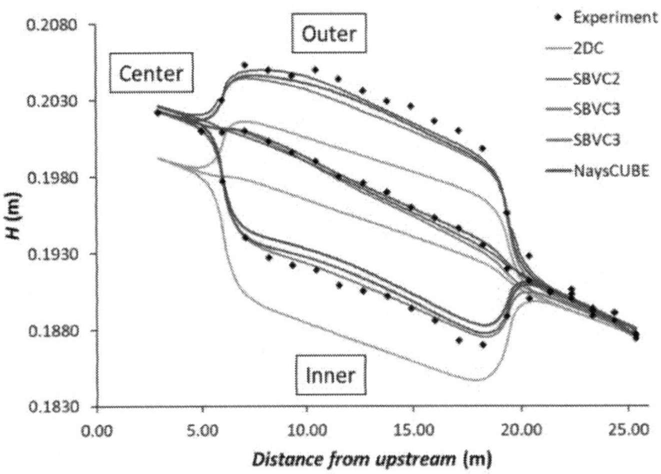

Figure 2. Water surface profile along the channel

Figure 2 shows the comparison of water surface profile along the outer, center, and inner bank between experimental datasets with numerical models consisting of 2DC, SBVC2, SBVC3, GBVC3, and NaysCUBE. The water surface profile is uniform in the transverse direction in the section upstream of the curved part. Once the flow enters the curved part, the water surface profile increases at the outer wall and decreases at the inner wall because of the centrifugal force effect. After leaving the bend, the water surface profile once again becomes uniform.

All the models except the 2DC model can reproduce the water surface profile of experimental datasets. The 2DC model shows an underestimation since it does not consider variations in the vertical velocity distribution, so it cannot reproduce the flow resistance increased due to secondary flow. In our prior investigation [12], the 2DC model needs a coarser equivalent k_s value to fit experimental datasets. The advantage of models of the BVC method considering the effects of secondary flow due to channel bend can be confirmed compared to experimental datasets and the 3DC model, i.e., NaysCUBE. SBVC2 has slightly underestimated the water surface profile compared with SBVC3 and GBVC3 because SBVC2 applied quadratic polynomial velocity distribution while SBVC3 and GBVC3 applied cubic polynomial velocity distribution in the vertical direction. Consequently, SBVC3 and GBVC3 produce a similar pattern of water surface profile.

3.2. Depth-averaged velocity distribution

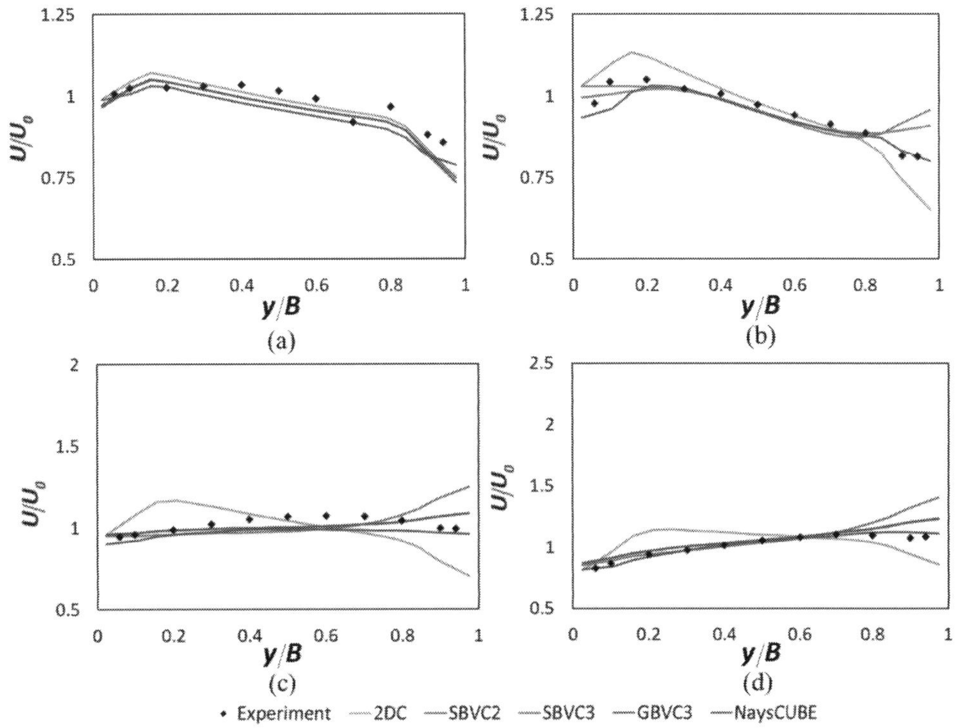

Figure 3. Lateral distribution of the depth-averaged velocity. (a) CS-6, (b) CS-9, (c) CS-15, (d) CS-18

Figure 3 shows the comparisons of the depth-averaged velocity along the streamwise direction at several cross-sections (CS). The vertical axis represents the depth-averaged velocity U normalized by U_0, which is the calculated averaged velocity for each cross-section. The horizontal axis represents the distance from the left bank proportionate to the channel width (i.e., 0 is the inner bank and 1 is the outer bank).

The comparison was made at the entrance section of the channel bend (CS-6), at the early bend (CS-9), at the late bend (CS-15), and at the bend exit (CS-18). At the entrance, the velocity near the inner bank increases [figure 3(a)]. At the beginning of the bend [figure 3(b)], the maximum velocity is located near the inner bank. This increasing velocity near the inner wall is induced by the longitudinal pressure gradient related to the water surface gradient shown in figure 2. In the late bend [figure 3(c)], the velocity near the inner bank gradually decreases. Finally, when the flow is in the bend exit [figure 3(d)], the maximum velocity is located at the outer bank owing to the transverse convection of the momentum transfer by the secondary flow.

The 2DC cannot consider the secondary flow effect, and there is no shifting of the high velocity from the inner to the outer walls. The BVC method can reproduce the acceleration of the depth-averaged velocity near the inner bank before entering the bend and the shifting of the maximum velocity toward the outer bank after leaving the bend. SBVC2 shows high velocity along the outer bank compared to SBVC3 and GBVC3, and this is attributed to the lack of consideration of the horizontal momentum equation on the water surface. SBVC3 and GBVC3 show a similar profile. However, GBVC3 considers the distribution effect of the vertical velocity component, and the difference is not pronounced since the effect of secondary flow in this channel is not strong enough. NaysCUBE becomes the best model to reproduce a similar profile as experimental datasets. NaysCUBE is a fully three-dimensional model that considers the non-hydrostatic pressure distribution and the non-linear turbulent model.

4. Conclusion

In this study, the advantages of the BVC method in reproducing flow structures in the curved-open channel can be confirmed. The method has better results than 2DC that cannot consider the effect of secondary flow. It has also had a good agreement compared to experimental datasets and NaysCUBE, a fully three-dimensional model.

Appendix

The governing equations are solved in the general coordinate system. The bottom velocity [equation (1)] was evaluated by the depth-integrated horizontal vorticity equations [equation [A.1)] and the water surface velocity equations [equation (A.5)].

$$\frac{\partial \Omega_i h}{\partial t} = ER_{\sigma i} + P_{\omega i} + \frac{\partial h D_{\omega ij}}{\partial x_j} \tag{A.1}$$

where $ER_{\sigma i}$: expansion-rotation term of the vertical vorticity $\left(ER_{\sigma i} = u_{si}\omega_{\sigma s} - u_{bi}\omega_{\sigma b}\right)$, $\omega_{\sigma s}$ and $\omega_{\sigma b}$: rotation of u_{si} and u_{bi}, $P_{\omega i}$: production term of vorticity on the bottom vortex layer, and $D_{\omega ij}$: horizontal vorticity flux due to advection, expansion-rotation, dispersion, and turbulence diffusion expressed by equation (A.2).

$$D_{\omega ij} = \Omega_j U_i - \Omega_i U_j + \overline{\omega_j' u_i'} - \overline{\omega_i' u_j'} + \frac{v_t}{\sigma_\omega}\frac{\partial \Omega_i}{\partial x_j} \tag{A.2}$$

where $\omega_i' = \omega_i - \Omega_j$, $\sigma_\omega = 1$. The vorticity distribution is given by the differential of velocity equation with respect to z. The production terms $P_{\omega i}$ are given by equation (A.3).

$$P_{\omega i} = C_{p\omega} v_{tb}\left(\omega_{bei} - \omega_{bi}\right)/h \tag{A.3}$$

where $C_{p\omega} = \kappa/\alpha$, $\alpha = \kappa/6$, $\kappa = 0.41$, $v_{tb}: v_t$ at the bottom converted into depth-averaged scale by assuming $v_{tb} = \max\left(a^2 h^2 \omega_b, a^2 h^2 \omega_{be}\left(= \alpha c_b u_b h\right), v_t\right)$, ω_{bi}: horizontal vorticity at the bottom; ω_{bei}: equilibrium ω_{bi} for u_b given by equation (A.4).

$$\omega_{bej} = 2\frac{\varepsilon_{ij3} c_b u_{bi}}{\kappa h}\ln\left(1 + h/\delta z_b\right) \tag{A.4}$$

The water surface velocity is solved by equation (A.5) for SBVC3 and GBVC3.

$$\frac{\partial u_{si}}{\partial t}+u_{sj}\frac{\partial u_{si}}{\partial x_j}=-g\frac{\partial z_s}{\partial x_i}+P_{si} \tag{A.5}$$

where P_{si}: production term due to the shear stress acting on the thin water surface layer δz_s, represented by equation (A.6).

$$P_{si}=\frac{2\nu_t}{h^2}\left\{12C_{ps}\left(u_{sei}-u_{si}\right)-\left(3\delta u_i-6\Delta u_i\right)\right\} \tag{A.6}$$

where $C_{ps}=3$.

The water depth and depth-averaged velocity are calculated with depth-integrated continuity [equation (A.7)] and momentum equations [equation (A.8)], considering the vertical velocity distribution.

$$\frac{\partial h}{\partial t}+\frac{\partial U_i h}{\partial x_i}=0 \tag{A.7}$$

$$\frac{\partial U_i h}{\partial t}+\frac{\partial U_i U_j h}{\partial x_j}=-gh\frac{\partial z_s}{\partial x_i}-\frac{\tau_{bi}}{\rho}+\frac{\partial hT_{ij}}{\rho\partial x_j}-\left(\frac{\partial hdp_0}{\rho\partial x_i}+\frac{dp_b}{\rho}\frac{\partial z_b}{\partial x_i}\right) \tag{A.8}$$

where z_s: water level, z_b: bed level, U_i: the depth-averaged horizontal velocity in the i direction, g: gravity, τ_{bi}: the bed shear stress, and T_{ij} the horizontal shear stress due to turbulence and vertical velocity distribution. The pressure p is expressed using pressure deviation from hydrostatic pressure distribution as $p=\rho g(z_s-z)+dp$, $dp_0=dp_b/2$, dp_b: bottom pressure deviation, dp_0: depth-averaged pressure deviation. The 2DC method does not account the pressure distribution and assumes a velocity distribution for uniform flows. While the SBVC2 method assumes quadratic velocity distribution, SBVC3 and GBVC3 employ the cubic velocity distribution (2) as the vertical distributions of the horizontal velocities, using the depth-averaged velocity, water surface velocity, and bottom velocity. GBVC3 was expanded to more general cases without the shallow water assumption for 2DC and SBVC methods. The bottom pressure deviation from the hydrostatic pressure distribution is given by depth-integrating the vertical momentum equation.

$$\frac{dp_b}{\rho}=\frac{\partial hWU_j}{\partial x_j}+\tau_{bj}\frac{\partial z_b}{\partial x_j}-\frac{\partial h\tau_{jz}}{\partial x_j} \tag{A.9}$$

where τ_{bj}, τ_{jz}: shear stress acting on the bottom and x_j-plane, respectively. To evaluate equation (A.9) and the second term in the right side of equation (1), the vertical velocity equation double-integrated the continuity equation over water depth is given by equation (A.10):

$$Wh=h\left(\frac{\partial z_m}{\partial t}+U_j\frac{\partial z_m}{\partial x_j}\right)+\frac{\partial}{\partial x_j}\left\{h^2\left(k_2\Delta u_j+k_1\delta u_j\right)\right\} \tag{A.10}$$

where $z_m=(z_s+z_b)/2$, $k_1=1/10$, $k_2=1/20$. k_1 and k_2 depend on the velocity distribution function, which is derived by taking the double integral of the continuity equation along the vertical direction [6-7].

References

[1] Falcon M 1984 Secondary flow in curved open channels *Ann. Rev. Fluid Mech.* **16** 179–93
[2] Hooke R L B 1975 Distribution of sediment transport and shear stress in a meander bend *J. Geol.* **83** 543–65
[3] Nikora V and Roy A G 2012 Secondary Flows in Rivers: Theoretical Framework, Recent Advances, and Current Challenges *Gravel-Bed Rivers Process. Tools, Environ.* 1–22
[4] Bathurst J C, Thorne C R and Hey R D 1977 Direct measurements of secondary currents in river bends *Nature* **269** 504–6
[5] de Vriend H J 1979 *Flow measurements in a curved rectangular channel* (Netherlands)
[6] Uchida T, Fukuoka S, Papanicolaou A (Thanos) N and Tsakiris A G 2016 Nonhydrostatic quasi-

3D model coupled with the dynamic rough wall law for simulating flow over a rough bed with submerged boulders *J. Hydraul. Eng.* **142** 04016054

[7] Uchida T and Fukuoka S 2014 Numerical calculation for bed variation in compound-meandering channel using depth integrated model without assumption of shallow water flow *Adv. Water Resour.* **72** 45–6

[8] Uchida T and Fukuoka S 2019 Quasi-3D two-phase model for dam-break flow over movable bed based on a non-hydrostatic depth-integrated model with a dynamic rough wall law *Adv. Water Resour.* **129** 311–27

[9] Nezu I and Nakagawa H 1993 *Turbulence in open-channel flows* (Netherlands: CRC Press)

[10] Kimura I, Uijttewaal W S J, Hosoda T and Ali M S 2009 URANS computations of shallow grid turbulence *J. Hydraul. Eng.* **135** 118–31

[11] Shimizu Y, Nelson J, Arnez Ferrel K, Asahi K, Giri S, Inoue T, Iwasaki T, Jang C L, Kang T, Kimura I, Kyuka T, Mishra J, Nabi M, Patsinghasanee S and Yamaguchi S 2020 Advances in computational morphodynamics using the International River Interface Cooperative (iRIC) software *Earth Surf. Process. Landforms* **45** 11–37

[12] Lugina F P, Uchida T and Hatono M 2021 Effect of channel meander on flow resistance *J. Japan Soc. Civ. Eng. Ser. B1 (Hydraulic Eng.* **77** I_865-I_870

Utilization of Dredged Mud in Land Reclamation: A Comparison Study of Several Soil Improvement Methods

I Wayan Sengara[1], Jonathan Bratanata[1], and Sugeng Krisnanto[1*]

[1]Civil Engineering Program, Faculty of Civil and Environmental Engineering, Bandung Institute of Technology (ITB), Jalan Ganesa No. 10, Bandung 40132, Indonesia

*Email: sugeng.krisnanto@itb.ac.id

Abstract. Utilization of dredged mud in land reclamation will become more applicable in the future due to the increase of the need for space and the rising cost of granular fill materials. Dredged mud mainly consists of soft clay and has very low bearing capacity and high compressibility. Therefore, soil improvement is urgently required. In this paper, the effectiveness of using dredged mud as the fill material in land reclamation was analyzed for a case study of land reclamation. The performance of several soil improvement methods in using such reclamation material was calculated and compared. The soil improvement methods considered in this study were prefabricated vertical drain (PVD) with vacuum preloading, cement mixed dredged mud, and rigid inclusion. The load transfer mechanism in each method was analyzed using analytical and numerical methods. In addition, an elementary cost analysis was performed to compare the cost of the methods based on Indonesian prices. It was found that for the considered case study, cement mixing achieved the highest bearing capacity with the shortest time but with the highest cost. For the same increase in bearing capacity, PVD with vacuum preloading provided the lowest cost. This study provides a preliminary step towards the implementation of effective soil improvement for land reclamation in Indonesia.

Keywords: dredged mud, prefabricated vertical drain (PVD), rigid inclusion, soil-cement mixed, soil improvement, vacuum preloading.

1. Introduction

Land reclamation has become more challenging due the depleting availability and rising cost of granular fill materials. A more economical and environmentally friendly option that should be increasingly considered is the use of soft dredged soils. Soft soil deposits are available in abundance on the seabed or riverbed and are usually disregarded in land reclamation projects. These deposits of soft clay were also reported to be available in large quantities in the coastal regions of many expanding coastal cities that need large amount of fill materials to satisfy their reclamation works [1]. Some examples from past projects include the land reclamation works carried out at the Central Japan International airport [2] and the Brisbane, Australia port expansion [3].

This dredged mud slurry usually consists of fine soils that have low shear strength, high compressibility, and extremely high water contents. Since the top surface is too soft and has a low bearing capacity. Another challenge is forming a working platform and selection of machinery to conduct soil improvement works [4].

This paper presents a comparison study of several soil improvement methods utilization of dredged mud in land reclamation. The soil improvement methods considered in this study were prefabricated vertical drain (PVD) with vacuum preloading, cement mixed dredged mud, and rigid inclusion. The engineering performance and cost analysis are performed in the study.

2. Methodology

Three methods of soil improvement for soft fill material are considered in this study. The first method is prefabricated vertical drains (PVD) with vacuum preloading. An important development of the preloading technology is the idea of vacuum preloading, which was first proposed by Kjellman [5]. Vacuum consolidation combined with PVDs is one of the most popular methods for soil improvement [4-8]. Different from fill preloading, vacuum preloading utilizes atmospheric pressure on an air-tight system installed in the field by applying vacuum inside the system [9]. One distinct advantage of preloading using vacuum instead of utilizing preloading by fill is that the increase in undrained shear strength happens without a greatly reduced possibility of undrained bearing capacity failure during the preloading operation [10]. When the vacuum pressure is not sufficient, fill may be added on top of the system to increase the preloading pressure [11].

The second method is ex-situ cement mixing, in which soil and a chemical binder are mixed with the help of the turbulent flow generated within the pipeline during transportation by compressed air before being placed [12].

The third method is rigid inclusion. For soil that is reinforced with rigid inclusions, the columns will carry a higher stress than the soil under an equal strain condition due the difference in modulus between the columns and the soil that surrounds it. Therefore, soil improved with rigid inclusions has a condition between equal strain and equal stress [11]. Through a load transfer platform (LTP), the load from the structure in this system is distributed to a network of columns which distributes the load to the bearing layer. A LTP made of well compacted granular material is usually required above the rigid inclusions to distribute the loads through arching between the rigid inclusions and the surrounding soils.

The soil data that was used in this analysis originated from the Port of Brisbane, Australia project [3]. While the methods of construction and prices for each component were based on the resources available in the Indonesian market or could feasibly be shipped to Indonesia for use in the reclamation and soil improvement works.

3. Results of Engineering Analysis

3.1. PVD with vacuum preloading

A limitation to the use of vacuum preloading is that the vacuum pressure is limited to only 70-80 kPa. Depending on the target strength, combining vacuum preloading with additional fill surcharge might be necessary. The ratio of the undrained shear strength to the effective vertical stress (the s_u/σ_v' ratio) [e.g. 13-15] can be used to evaluate the value of undrained shear strength.

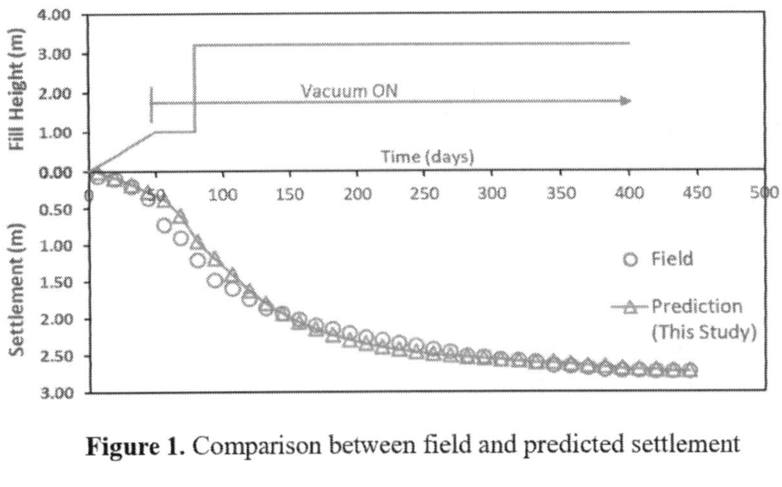

Figure 1. Comparison between field and predicted settlement

Figure 1 shows a comparison between field and predicted settlement monitoring data. During the initial period of vacuum preloading, the prediction model tends to underestimate the total settlement that occurs but were still found to agree with the field monitoring data. The process that occurs during vacuum preloading is assumed to be one dimensional based on the single-well model [16]. But due to the characteristics of dredged mud with very high water contents, some studies based in experimental models or numerical simulations have shown the clogging effects on the soil surrounding the PVD because of the lateral migration of soil particles [17-19]. Future research on the effects of clogging and the formation of soil piles around the PVD should be conducted to better assess the behavior of PVD with vacuum preloading systems on Indonesian dredged mud soils.

3.2. Cement stabilization

Since there is cement that is mixed into the soil, the phase diagram of the soil changes and its properties can be adapted. Though these characteristics are different from ordinary clay, stabilized soil is assumed to be the same as ordinary cohesive soil having ductile characteristics in the current design standard [7]. In the design methods for bearing capacity and earth pressure, the calculation is similar to the calculation with ordinary soils.

Due to the physical and mechanical properties of stabilized soil being affected by numerous factors, including soil properties, binder type, mixing and placement conditions, and curing conditions, it is difficult to directly estimate the soil strength by laboratory mix test only. The field mixing and curing conditions are quite different from the standard laboratory test conditions. Field stabilized soil also has a relatively larger strength variability [12]. A large part of field stabilized soil strength should be larger than the design strength, while the percentage of field stabilized soil strength larger than the design strength can be defined as a probability function.

In this study, calculations of the unconfined compression strength (q_{uL}) were performed based on the pneumatic flow mixing method [12]. Calculations using this method uses Ordinary Portland Cement (OPC) as the binder. The calculations were performed by varying water content, w and liquid limit, LL values. The results of calculation in this study are shown in Figures 2 and 3. As shown in Figures 2 and 3, it was found that the lower the water to cement ratio of the cement slurry (W/C) and subsequently the higher the cement content (c^*), the q_{uL} of the cement stabilized soil will increase exponentially. For a variety of natural water content (w_n) with the same W/C ratio or c^*, it was also found that the lower the natural water content, the unconfined compression strength was relatively higher. Though the natural water content of dredged mud is usually quite high, it could still be smaller than the liquid limit. If that were to happen, then some additional water would need to be added to the dredged mud in order to obtain a sufficient fluidity when being pumped through the pipes.

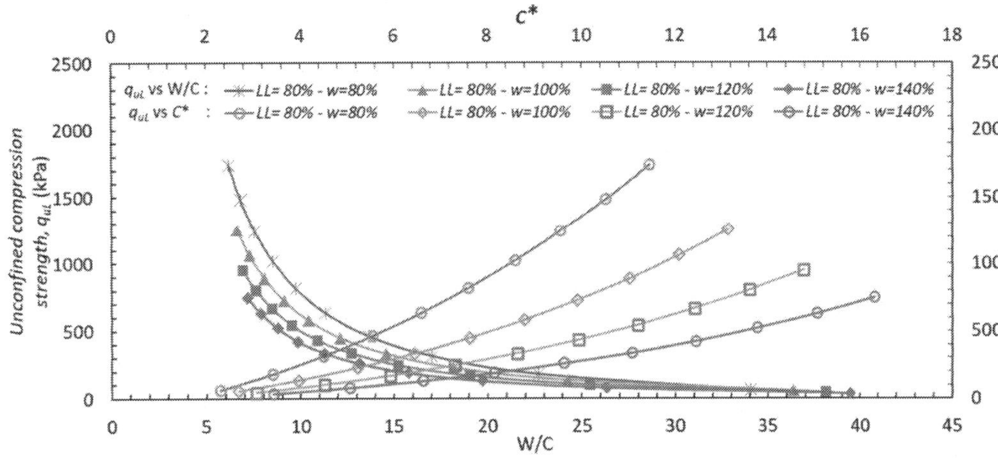

Figure 2. Comparison of q_{uL} with different water content values

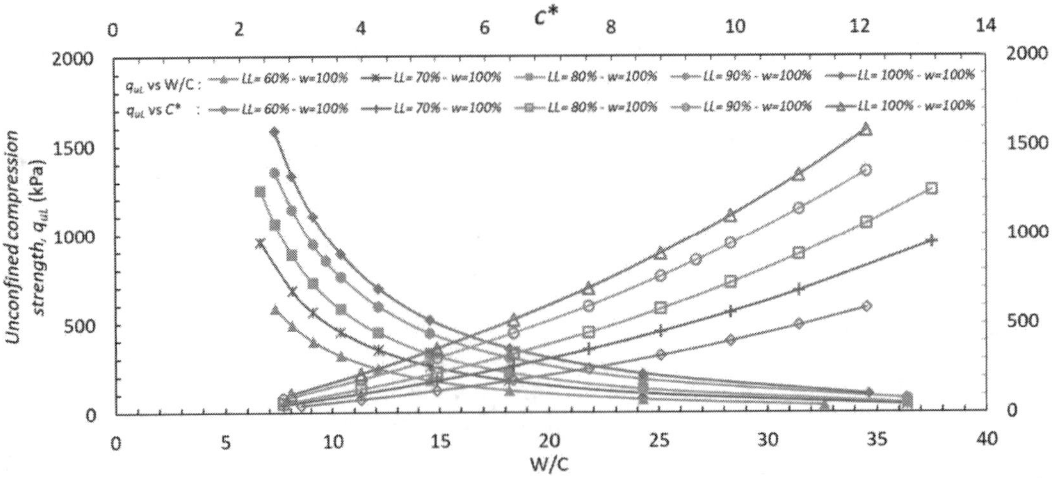

Figure 3. Comparison of q_{uL} with different liquid limit values

Cement mixing has the potential of achieving the highest bearing capacity while also necessitating the shortest time due to the integration of the soil improvement with soil placement, but there should be a strict quality control for the field mix design to achieve the desired soil strength.

3.3. *Rigid inclusion*

As prevously mentioned, an important aspect to consider when designing using rigid inclusoin is the thickness of the LTP can have a great impact on the pattern of surface deformation that occurs. An unfavorable affect that occurs when the LTP height is too small is that the load from the structure above will not form a full arch and leads to differential settlement between the area where load is transferred to the rigid inclusion piles and the surrounding soil.

Several recommendations for the critical embankment height have been proposed and incorporated into various design codes, such as the Hewlett and Randolph model [21] were adopted by British Standard (BS8006)[22], the wedge-shape soil arching model developed by Carlson [23] was adopted in the Nordic Guidelines (NGG) [24], and the concentric arches model proposed by Van Eekelen et al. [25] was adopted in the Dutch Design Guideline (CUR 226) [26]. The assumptions and formulas used for the determination of critical LTP height adopted by the different design codes are very different from each other [27].

From the analysis that has been conducted, the answer is not as simple as choosing a recommendation used in a design code. Due to the different soil conditions, each individual case should be evaluated to get the critical LTP height. As shown in Figure 4, a couple factors do stand out: (1) The efficiency (percentage of load above the soft soil that is transferred to the rigid inclusion piles) never reaches a value of 100%, the load that is transferred to the soft soil between the rigid inclusion piles increases with the LTP or embankment height, (2) Critical LTP height depends greatly on the spacing and pile diameter, though it was found that reducing the spacing between piles has a larger impact on the overall efficiency compared to increasing the pile diameter.

As shown in Figure 5, due to the diference in settlement that occurs between the rigid inclusion piles and the surrounding soils, a negative skin friction will also occur until a neutral point where the settlement of the two elements is equal. This results in the bearing capacity provided by the rigid inclusion piles on dredged mud being highly dependent on the end bearing capacity. For this reason, the depth at which there is a good soil layer beneath the soft soil and dredged mud should be considered.

Rigid inclusions as a method of increasing the bearing capacity of a reclamation site using dredged mud as the fill material was found to be effective. Though further research should be conducted regarding the increase of the soil's lateral capacity due to the rigid inclusions.

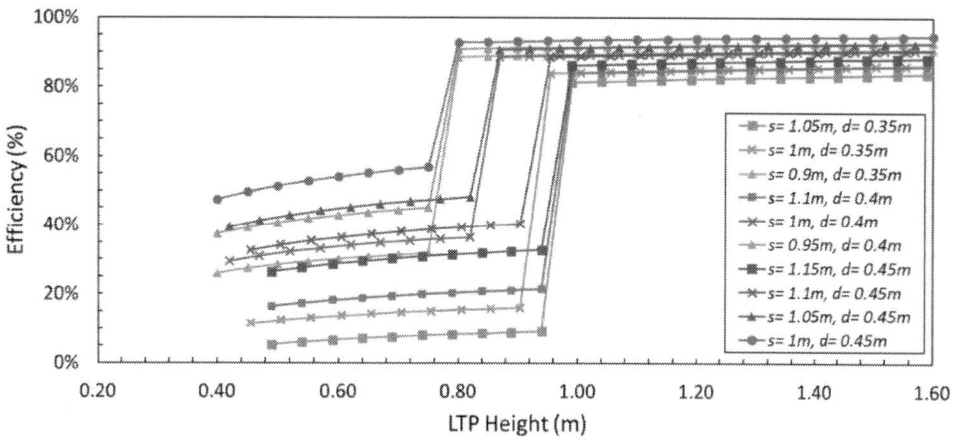

Figure 4. Variation of efficiency with LTP height

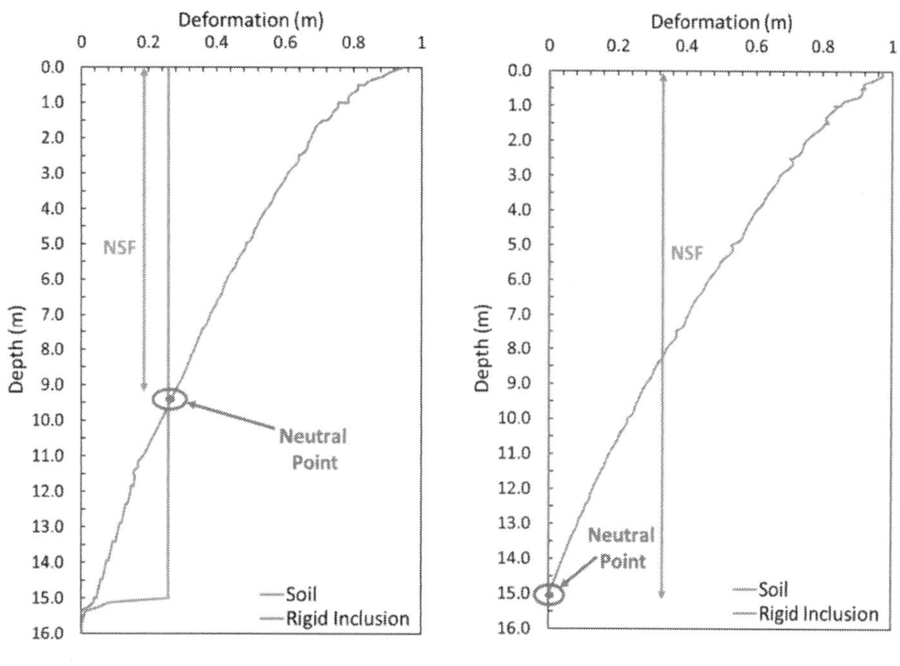

Figure 5. Surface deformation with depth for (a) sand layer, or (b) bedrock beneath pile

4. Results of Cost Analysis

To achieve and promote the use of dredged mud for the utilization in reclamation projects, the soil improvement method conducted should also be economical and cost-effective. Elementary cost analysis was performed to compare the cost of the methods for each method of soil improvement for a variety of dredged mud layer thicknesses and compared to a conventional scenario which utilizes granular fill material. In every scenario, the final soil strength was made to be similar. The prices for each component were based on the resources available in the Indonesian market or could feasibly be shipped to Indonesia for use in the reclamation and soil improvement works.

A summary of the results can be seen in Figure 6. Regardless of the thickness of dredged mud layer combined with the underlying soft clay layer, the PVD combined with vacuum preloading method is the most economical method overall. The cost gap from IDR 29.5 to 89.4 billion gives the dredged mud fill material along with PVD and vacuum preloading method a significant advantage over the use of conventional granular fill. In the future, this gap will rise due to the decreasing availability and rising cost of granular material suitable for reclamation works. The cost gap compared to the other soil improvement methods is IDR 24.8-108.9 billion. Though the main weakness of this method is the long duration compared to the other methods discussed which may drive up the overall cost, it is made up for by the more cost-effective equipment needed to conduct the soil improvement works, making it very much feasible to be conducted in Indonesia. Although the cement mixing and rigid inclusion methods have the potential to be used depending on the scenario with similar costs with the conventional method.

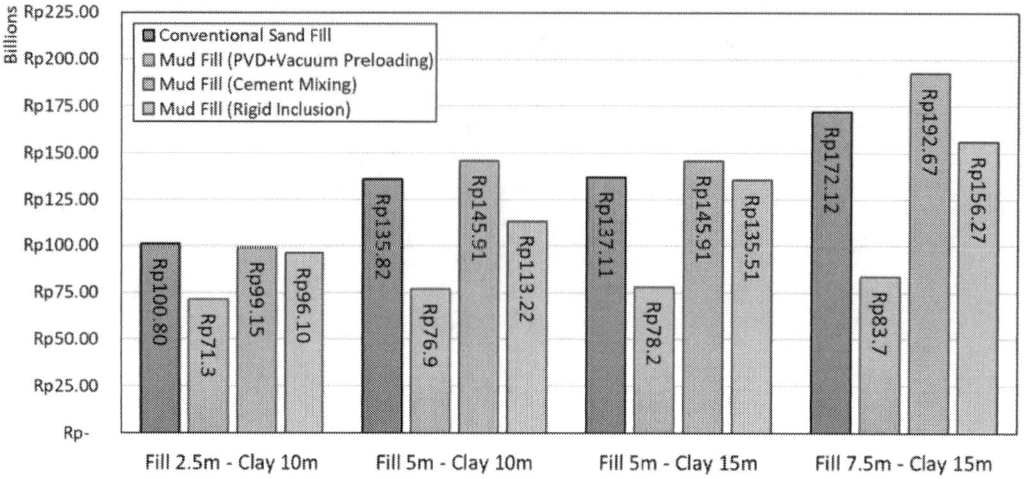

Figure 6. Total cost of each soil improvement method and scenario

5. Summary

The solutions for land reclamation using dredged mud as the fill material along with three methods of soil improvement are discussed. While effective as a method of soil improvement, the use of cement stabilization in a large scale land reclamation project is expensive. Rigid inclusions were also found to be effective but requires a hard and stable soil layer at a relatively shallow depth. Among the 3 methods studied, cement mixing has the potential of achieving the highest bearing capacity while also needing the shortest amount of time to complete but also the highest cost. For the same increase in bearing capacity, the most economical method per unit of area is PVD with vacuum preloading or combined vacuum preloading and fill surcharge.

6. References

[1] He J J, Chu J, Tan S K, Vu T T and Lam K P 2017 Sedimentation behavior of flocculant-treated soil slurry *Marine Georesources & Geotechnology* **35** p 593 - 602

[2] Satoh T and Kitazume M 2002 Sea reclamation with dredged soft soil for Central Japan International Airport *Proc. Nakase Memorial Symp.* (Yokosuka) p 311-318

[3] Ameratunga J, De Bok C, Boyle P and Berthier D 2010 Ground improvement in Port of Brisbane (POB) clay *CaseHistory ISSMGE Bulletin* **4** p 28-54

[4] Ganesalingam D 2011 Geotechnical properties of reconstituted dredged mud *Proc. Pan-Am CGS Geotech. Conf.* (Toronto)

[5] Kjellman 1952 *Consolidation of clayey soils by atmospheric pressure* (Boston: Massachusetts Institute of Technology)

[6] Chu J, Yan S and Yang H 2000 Soil improvement by the vacuum preloading method for an oil storage station *Geotechnique* **50** p 623-632

[7] Chen J 2016. Evaluation of effective depth of PVD improvement in soft clay deposit: a field case study *Mar. Georesour. Geotechnol* **24** p 420-430
[8] Indraratna B, Rujukiatkamjorn C, Baral P and Ameratunga J 2019 Performance of marine clay stabilised with vacuum pressure: based on Queensland experience *J. Rock. Mech. Geotech* **11** p 598-611
[9] Indraratna B, Rujukiatkamjorn C and Balasubramaniam A 2005 Analytical and numerical modeling of soft soil stabilized by prefabricated vertical drains incorporating vacuum preloading *Int. J. Geomech* **5** p 114-124
[10] Mesri G and Khan A 2011 Increase in shear strength due to vacuum preloading *Proc. Pan-Am CGS Geotech. Conf.* (Toronto)
[11] Han J 2015 Principles and Practices of Ground Improvement (New Jersey: John Wiley and Sons)
[12] Kitazume M and Terashi M 2017 The Penaumatic Flow Mixing Method (Florida: CRC Press Boca Rato)
[13] Skempton A W 1957 The planning and design of the new Hong Kong Airport *Proc. Institution of Civil Engineer* **7** p 305–307
[14] Mesri G 1975 New engineering procedures for stability of soft clays *J. Geotech Eng Div ASCE* **101** p 409-412
[15] Krisnanto S 2021 Theoretical equations for the ratio of undrained shear strength to vertical effective stress for normally consolidated saturated cohesive soils *J. Teknik Sipil* (Bandung: Institut Teknologi Bandung) **28** p 241-252 https://doiorg/105614/jts20212831
[16] Indraratna B, Chong R, Fox J and Rujukiatkamjorn C 2017 Large-strain vacuum-assisted consolidation with non-darcian radial flow incorporating varying permeability and compressibility *J. Geotech. Geoenviron.* **143**
[17] Deng Y 2019 Colloid effect on clogging mechanism of hydraulic reclamation mud improved by vacuum preloading *Can. Geotech. J.* **56** p 611-620
[18] Zhang F 2020 Laboratory and field experimental study on the vacuum preloading method with pumping and discharging *Advances in Civil Engineering*
[19] Wu Y, Qin P and Yang J 2013 Consolidation test and numerical simulation of particles movementto ultra-soft soil bearing vacuum load *J. Eng. Geol.* **483** p 116-122
[20] Kang G, Tsuchida T and Arthapaththu A 2016 Engineering behavior of cement-treated marine dredged clay during early and later stages of curing *J. Eng. Geol.* **209** p 163-174
[21] Hewlett W and Randolph M 1988 Analysis of piled embankment *J. Ground Eng.* **21** p 8-12
[22] British Standard BS8006 2010 Code of practice for strengthened/reinforced soils and other fills (London: British Standards Institution)
[23] Carlson B 1987 Armerad jord beräjningsprinciper för banker på pålar (Linköping: SGI Linköping)
[24] Nordic Handbook 2002 Nordic handbook reinforced soils and fills (Stockholm: Nordic Geotechnical Society)
[25] Van Eekelen S J M, Bezuijen A and van Tol A F 2013 An analytical model for arching in piled embankments *J. Geotext. Geomembr.* **39** p 78-102
[26] Dutch Design Guideline CUR 226 2016 Dutch design guideline piled embankments (London: Taylor and Francis)
[27] Lai H, Zheng J, Zhang R and Cui M 2018 Classification and characteristics of soil arching structures in pile-supported embankments *J. Computers and Geotech.* **98**

The Effect of Non-Linear Wave on Oil Spill Dispersion

M Muin [1*], A B Muslim[1] and T A Puspitasari[2]

[1]Ocean Engineering Research Group, Faculty of Civil and Environmental Engineering, Institut Teknologi Bandung, Jalan Ganesha No 10, Indonesia
[2]Department of Ocean Engineering, Faculty of Civil and Environmental Engineering, Institut Teknologi Bandung, Jalan Ganesha No 10, Indonesia
*Corresponding author: muslimmuin@gmail.com

Abstract. On 31 March 2018, an oil spill accident polluted Balikpapan Bay. A failure occurred in a pipeline running from Penajam to Balikpapan Oil Refinery, East Kalimantan, Indonesia. Leaking pipeline caused the total estimated volume of leaked oil to approximate 44,000 Barrels of crude oil. MoTuM was used to simulate the dispersion of oil in the area. The modeling results indicate that a strong dynamic tidal current in the Bay controls the oil's movement. The MoTuM software is suitable for spill combating, contingency plans, and backtracking. MoTuM is developed in Windows System, which integrates 3D Non-Orthogonal Boundary Fitted Ocean Hydrodynamics Model, Trajectory, Fates, Stochastic, Backtracking in Geographic Information System. This paper presents a further study of the effect of non-linear waves on the dispersion of oil in the Bay. The simulation results were validated by comparing the model with a satellite image. The agreement between the result of simulation and satellite image is excellent and shows that the non-linear wave is an essential factor for oil spill dispersion.
Keywords: *non-linear wave, oil spill modelling, mangrove forcing, balikpapan bay, 3D hydrodynamics model*

1. Introduction

One of the environmental disasters in the oil & gas industry is the oil spill. ITOPF and UNCTAD reported that oil spills have decreased since 1970, but the main issue is that the volume of oil spilled has increased, as shown in Figure1 [1]. As a result, if there is an oil spill incident, an oil spill response is required as a set of guidelines. One of the parts of the oil spill response is oil spill dispersion modeling. MoTuM is the most appropriate software to simulate oil dispersion in very complex geometry regions. The software is suitable for spill combating, contingency plans, and backtracking from an oil spill incident. MoTuM is developed in Windows system based on 3D Non- Orthogonal Boundary Fitted Ocean Hydrodynamic Model [2] and integrated with Geographic Integration System (GIS).

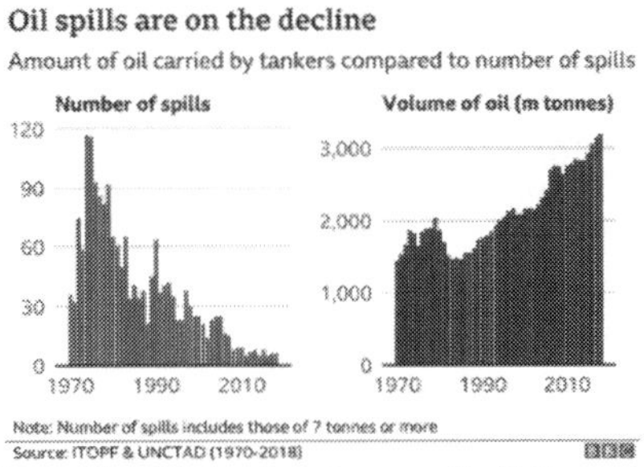

Figure 1. Number of spills and volume of oil from 1970 to 2010 [1]

A non-linear wave can create a mean current, or the term is residual current. Dean and Dalrymple [3] stated that non-linear waves could also create a net current that moves back and forth, which can affect the dispersion of pollutants, or in this case is oil.

There is no research about the non-linear wave effect to oil spill dispersion on Balikpapan Bay. In this study, we found that the non-linear wave significantly impacts the oil spill movement. The paper presents the comparison result between oil spill dispersion with and without non-linear wave forcing. The simulation with non-linear waves has a better result based on the satellite image.

2. MoTuM Description

MoTuM or Model Tumpahan Minyak is developed in Windows System, which is suitable for oil spill mitigation, contingency plans, oil spill risk assessment, and backtracking of oil sources. The software has been registered as intellectual property in Indonesia (HAKI) since 2005. The models were integrated with Geographic Information System (GIS) and the 3-dimensional hydrodynamic model developed by Muin or the 3D Non-Orthogonal Boundary Fitted Ocean Hydrodynamic Model [2]. MoTuM comes with a model setup (grid, open boundary, and other setups.), so it is user-friendly software. The model successfully tracks the spill source for the Balikpapan Spill Case in 2004 and the Montara Blowout Case in 2009. The simulation results in Montara match 26 field observations (satellite images and flight observation). Since MoTuM is also integrated with GIS, the user can predict where the oil spill disperses on the GIS Database. The state-of-the-art MoTum is shown in Figure 1, where the software can be used as Decision Support System (DSS).

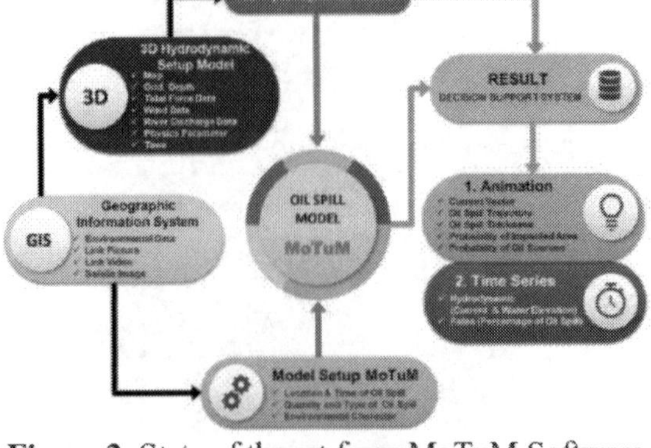

Figure 2. State of the art from MoTuM Software

2.1. Hydrodynamics Model

The MuHydro3D model is a boundary-fitted ocean hydrodynamic model using a non-orthogonal curvilinear coordinate technique in a spherical coordinate that can compute tidal elevations, velocities, salinity, and temperature distributions in the computational domain at the same time. The non-orthogonal grid system matches the shoreline and bathymetric feature boundaries for accurately representing areas with complex coastal geometry. The grid system can be adjusted in a high resolution for the area of interest and lower resolution at other locations for computational efficiency. MuHydro3D can also be run in either two or three dimensions depending on the complexity of the area of study. A detailed description of the model with associated test cases can be found in Muin [2], Muin, and Spaulding [3] [4]. The model has undergone extensive testing against analytical solutions. Anticipating the importance of mangroves as a driving force, in 1999, Muin included the water exchange between wet/dry area (mangrove) and main channel in the computation [5].

The mathematical basis for the model begins with continuity and momentum equations. Both equations were developed with the underlying assumption that the flow is incompressible, variation of water mass density negligible except when multiplied with the earth's gravity (Boussinesq approximation), and vertical acceleration negligible compared to gravitational acceleration (hydrostatic approximation). This ocean circulation model developed by Muin [2] was used by an American consultant under the name BFHydro widely applied worldwide, such as Providence River [6], Cone Bay [7], Bay of Fundy [8], New York Harbor [9], and many more.

2.2. Non-Linear Wave Model

$$\bar{u}(x,\eta) = \frac{1}{T}\int_0^T u(x,\eta)\,dt = \frac{a^2 k\sigma}{2} = \frac{(ka)^2 C}{2} \quad (1)$$

Dean & Dalrymple presented non-linear properties derivable from small-amplitude waves. The velocity asymmetry indicates that more fluid moves in the wave direction under the wave crest than in the trough region; there is a mean water transport. We employ this approach using the Eulerian Frame, a fixed point to measure the mean flux of the mass in the model [10]. The mean transport of water is shown in Equation (1), where k is wave number, a is wave amplitude and C is wave celerity. The non-linear wave creates an additional current to the hydrodynamic models.

2.3. Oil Spill Model

MoTuM has four oil spill models: Trajectory, Fates, Stochastic, and Backtracking [11]. The trajectory model calculates the movement of the oil spillet due to ocean currents, wind, random diffusion. Ocean currents were obtained from a three-dimensional marine hydrodynamic model (MuHydro3D); the forcings are tidal, winds, density differences, and rivers. The marine hydrodynamic model is calibrated by comparing the simulation results and observations. Wind data can be obtained from direct measurements in the field or from the nearest meteorological station. The position and age of each spillet are traced in the windows system with the embedded Geographic Information System (GIS) to facilitate the analysis of simulation results.

The Fates model calculates the amount of oil that has evaporated, trapped along the coast, and is still in the offshore area. The evaporation of the oil depends on the type of oil and the temperature differences. The empirical formula for evaporation of several oils was obtained from the literature and laboratory experiments for condensate oil. From the simulation results of the Fates Model, the user can estimate the amount of oil that must be cleaned in the waters and beaches. The Fates model also calculates the oil thickness for each spillet.

The Stochastic model calculates the area likely to be polluted by an oil spill. Areas of possible oil pollution can be presented in terms of time and percentage. If the Probability Areas are presented in time, the simulation results will be very useful in combating oil spill clean-up strategies. The results of the Stochastic Model in percentage will help the user conduct an oil spill risk study for several spill scenarios. Risk is the product of the Probability of Oily Area multiplied by the Consequences. Consequence's value is obtained from the study of the Environmental Sensitivity Index.

The Backtracking Oil Spill Sources model is used in determining the possible source of an oil spill if oil contamination is found at a location in the waters or the coast. The possible spill source areas in the time scale are presented in the tanker's position. The user can determine the ship responsible for the oil spill if the ship's position matches the time scale of the model results.

3. Area of Study Description

Balikpapan Bay is in East Kalimantan Province, Indonesia. It has 3,656.97 kilometers of shoreline [12]. Figure 3 shows the map of Balikpapan Bay. The mangroves forests in Balikpapan Bay are presented in Figure 4. Balikpapan Bay industry is located along the shoreline in the Bay, including a lumber industry, oil drilling and refining, capture fisheries, aquaculture, and harbor support services

Balikpapan Bay has a dominant, strong tidal current where the contaminants can be advected upstream. Additionally, it has a significant net outflow during the wet season and mesotidal characteristics, with a tidal range of 3 meters (mix to the semi-diurnal regime) [13]. Dynamics water exchange between the mangrove and the main channel plays an essential role in the current.

Figure 3. Map of Balikpapan Bay [14]

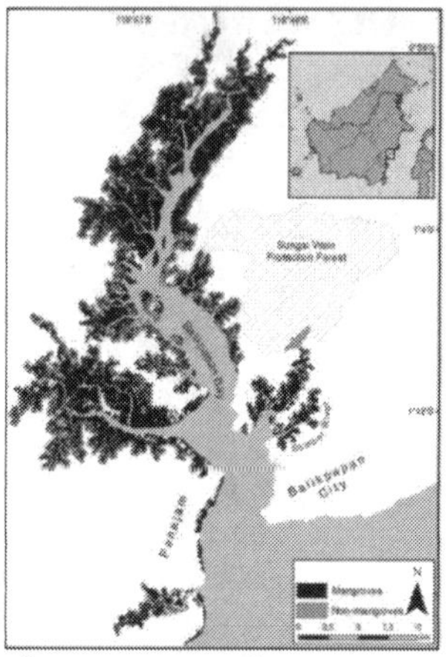

Figure 4. Distribution map of mangrove and non mangroves in Balikpapan Bay [15]

4. Balikpapan Oil Spill Case

The oil spill tragedy in Balikpapan happened on 31 March 2018, caused by the tug of a ship's anchor carrying 74,000 tons of coal that were passing by. The anchor ship caused the seabed pipeline rupture, which carried crude oil from the Single Point Mooring (SPM) Lawe-lawe Terminal to CDU IV Pertamina RU V Balikpapan [16]. The total oil spilled in the incident was estimated at 40,000 barrels with a polluted ocean area of about 7,000 hectares from the coast on the side of Balikpapan and North Penajam Paser (PPU) of about 60 kilometers [17]. This Oil Spill also claimed the death toll from a devastating fire. Figure 5 shows the oil along the beach of Balikpapan City.

IOP Conf. Series: Earth and Environmental Science 1065 (2022) 012006 doi:10.1088/1755-1315/1065/1/012006

Figure 5. Oil spill in Balikpapan Bay [18]

5. Hydrodynamics Model Setup

5.1. Computational Domain

The computational grid on this paper is shown in Figure 6. The non-orthogonal boundary fitted grid consists of 4036 water cells, 151 boundary cells, and 26 river/mangrove cells. The size on Balikpapan Bay is much finer than on the offshore or Makassar Strait. The grid system in Balikpapan covers all tidal affected waters

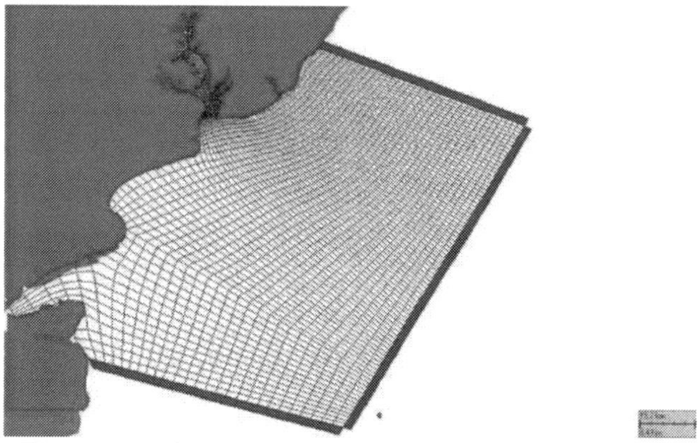

Figure 6. Grid computation of Balikpapan oil spill case

5.2. Bathymetry

The water depth ranges from 2 to 59 meters, as shown in Figure 7. The deepest depth is about 32 meters across from North to Southeast. The first gap of the Bay is located at the mouth of the Bay right in the water on the west side of Balikpapan City. The second slit crosses to the north from the Eastern part of Penajam Harbor upstream Semoi River – Sepaku on the Northside of the Bay Balikpapan [19].

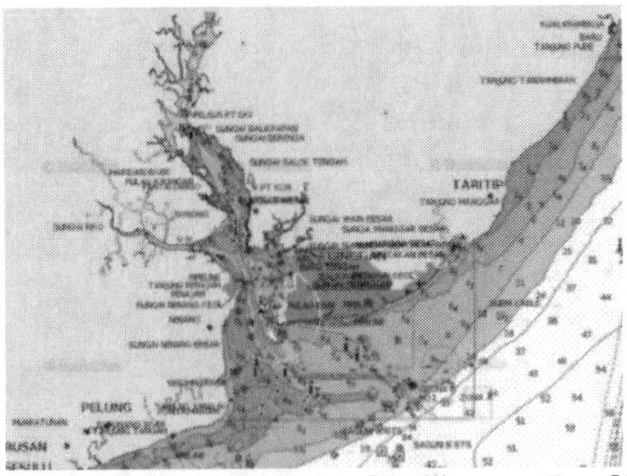

Figure 7. Bathymetric Map of Balikpapan Bay [20]

5.3. Open Boundary

For minimizing the effect of open boundary on the model, the best strategy is to apply the boundary model farthest from the spill oil location. It is also chosen to extend the suggested model domain over a larger region to eliminate possible boundary conditions. The tidal component along the open boundary is obtained from ORI World Tidal Model.

5.4. River Flow

Balikpapan Bay has a characteristic of seasonally freshwater from the rivers that flow into the Bay [21], where the total average river flow from four tributaries is 143 m^3/s. However, the river flow in this model is 120 m^3/s because the spill occurs in a transition season. The flows discharge value remains constant along the simulation time.

5.5. Wind and Wave

The wind data used for the model was obtained from Sepinggan Airport Meteorology Station, which is at the mouth of Balikpapan Bay. The wind speed and direction are obtained every 6 hours, and the directions are dominantly from the Southwest. The wave parameters for the model are hindcasted from wind data. The hindcasted results show that the wave height is about 2 meters and 6 seconds wave periods [22].

6. Oil Spill Model Setup

The location of the oil spill is at Refinery Processing V PT. Pertamina, close to the beach in North Penajam Paser Regency, East Kalimantan. The exact position of the oil spill is a Longitude of 116° 47' 15.6" East and Latitude of 1° 14' 40.7" South, shown in Figure 8. The oil spill occurred on 31 March 2018 at 01.20 AM Central Indonesia Time (zone). The type of oil spilled in Balikpapan Bay is crude oil with a viscosity of 75 csT with the amount of 44000 barrels. The duration of oil spill occurrence is about 9 hours.

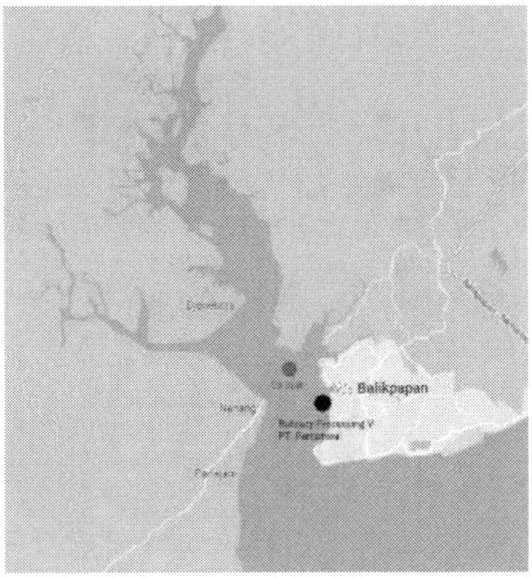

Figure 8. Oil spill location on Balikpapan Bay 2018 [14]

7. Calibration of Hydrodynamic Model

The success of a model application depends on the credibility of model predictions. The credibility of a model depends on the careful comparison of model predictions with the observation data. The observation of water elevation and current data is obtained from a primary survey on a port design project in Karingau for Petrossea by PT LAPI ITB [22]. Figure 9 shows the comparison between the predicted and observed water elevations where the Root Mean Square Error (RMSE) is 4 cm. The comparison between the model and data is excellent.

We also compared the water current velocity from the model and observation data in Kariangau, as shown in Figure 10. The comparison between predicted and observed ocean currents shows a good agreement with RMSE of 0.54 m/s. The tidal current in Kariangau is very strong, which is about 1.5 m/s. The current is driven by water exchange between the mangrove and the main channel. Neglecting the dynamic of mangrove and grid resolution will cause an underestimate of ocean current. Therefore, the current simulation results by Nur [21] are too small compared to the observation results because the grid system does not reach upstream and ignores the water exchange of mangroves.

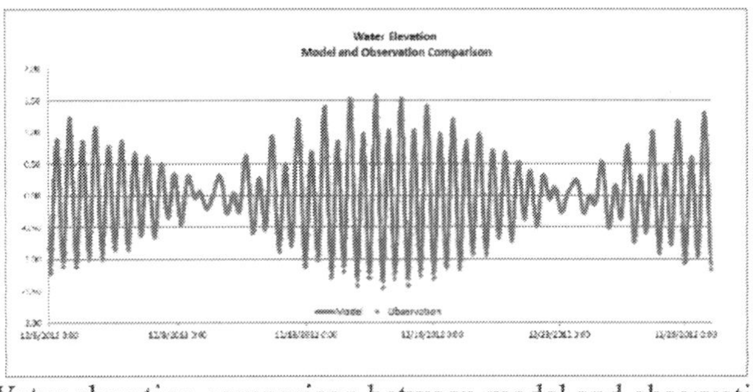

Figure 9. Water elevation comparison between model and observations [22]

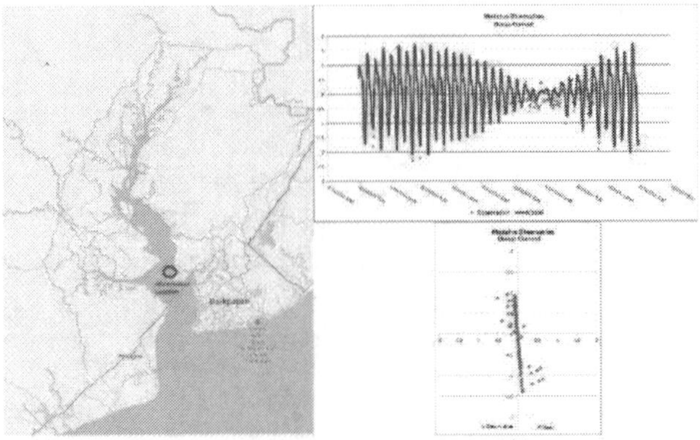

(a) (b)

Figure 10. (a) observation location and (b) Water current comparison between model and observations in time series and scatter plot [14] [22]

8. Oil Spill Model and Satellite Image Comparison

The comparison between oil spill model prediction and satellite images are presented in Figure 11 to 13. The satellite images from 1 April to 3 April 2018 are provided by the National Institute of Aeronautics and Space Indonesia (LAPAN). Both models and satellite images indicate that the spill can spread to the upstream region of Balikpapan Bay, indicating strong tidal currents in the area. It also shows that the spill will not spread further offshore to Makassar Strait. Most of the oil remains in Balikpapan Bay and spreads along the East Coast of Kalimantan. The simulation results clearly show that the model with a non-linear wave is in better agreement with the satellite image because the non-linear wave creates a mean current to the coast

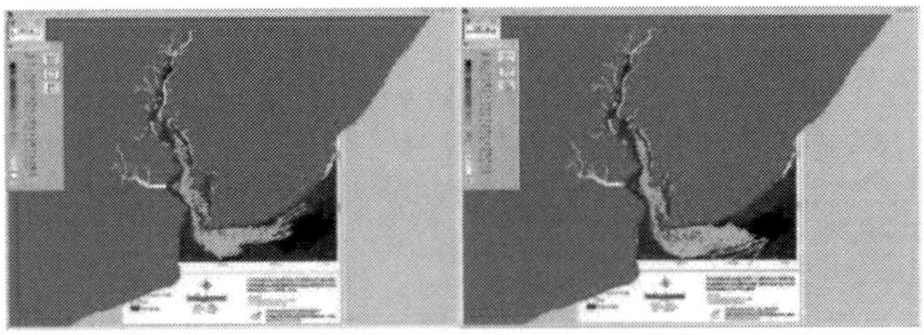

(a) (b)

Figure 11. Comparison model result (a) without non-linear wave forcing and (b) with non-linear wave forcing on 1 April 2018

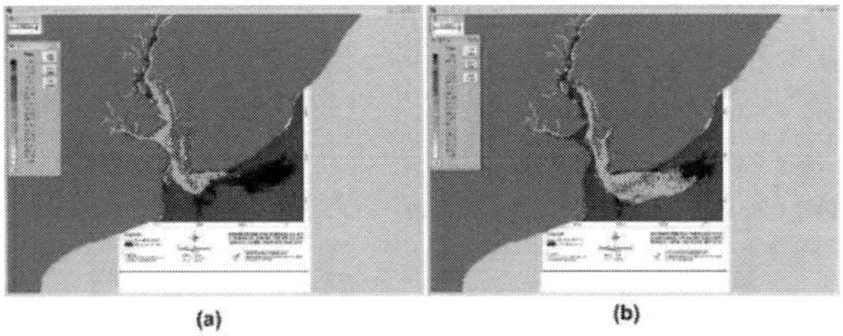

Figure 12. Comparison model result (a) without non-linear wave forcing and (b) with non-linear wave forcing on 2 April 2018

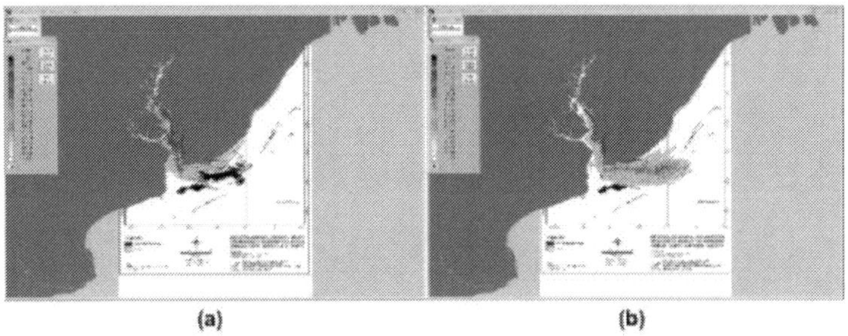

Figure 13. Comparison model result (a) without non-linear wave forcing and (b) with non-linear wave forcing on 3 April 2018

9. Discussion

The geometry and bathymetry of Balikpapan Bay are very complex and challenging, so a flexible grid system must be used in this area. The Non-Orthogonal Curvilinear Coordinate Technique is very appropriate to be applied. The orthogonal Model will not produce an accurate simulation because it is impossible to generate an orthogonal grid system in complex geometry. The results of this study clearly show that water exchange to mangroves and non-linear waves play a significant role in the spread of the oil spill in Balikpapan. This aspect cannot be ignored in the simulation.

It is essential to see the risk area in an oil spill; therefore, the Stochastic Model is used to obtain the probability of oily area. The simulation results are presented in Figure 14. It shows that Balikpapan Bay and the East Coast of Kalimantan are the most at-risk areas with the same sensitivity value. It is unlikely that the oil spill will spread to deep waters in the Makassar Strait.

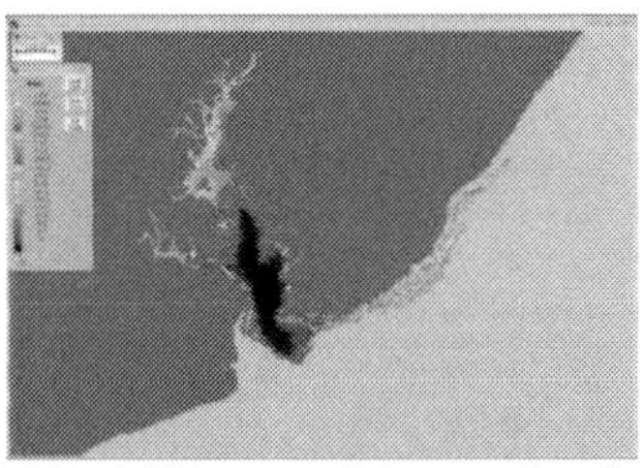

Figure 14. Probability of Impacted Area after five days of the spill

Figure 15 shows the oil spill trajectory during the high tide, while Figure 16 shows the oil spill trajectory during ebb tide. From the figure, we can conclude that a non-linear wave creates more oil spill dispersion than without forcing a non-linear wave. The figure also shows that most of the oil spill remains on the Bay. The probability simulation shows that it contradicts Hopley's opinion [13] that pollutants will be pushed out of the Bay, which will only be true during heavy rain.

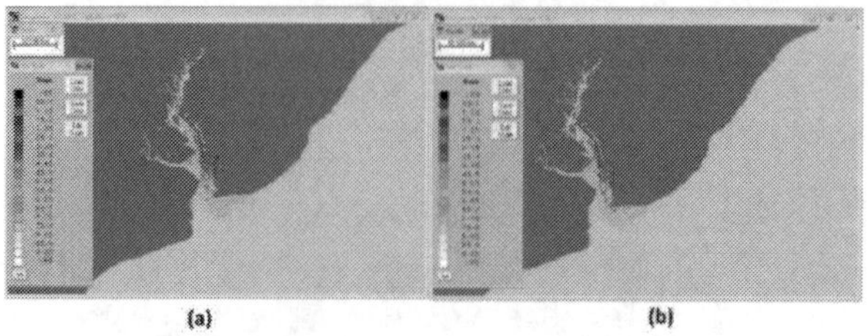

Figure 15. Trajectory model during high tide (a) without non-linear wave forcing and (b) with non-linear wave forcing

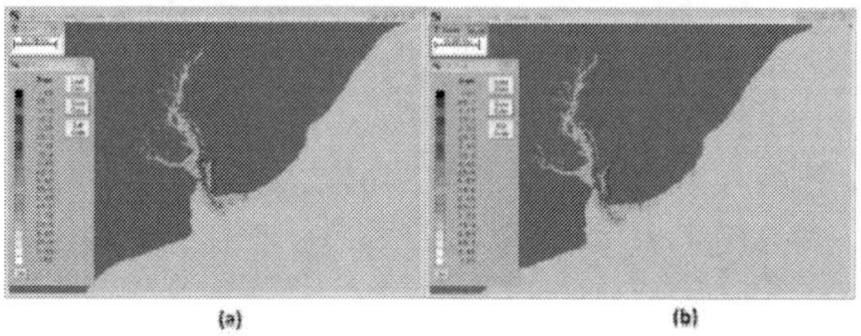

Figure 16. Trajectory model during ebb tide (a) without non-linear wave forcing and (b) with non-linear wave forcing

10. Conclusion

The key to the success of the oil spill simulation in Balikpapan is the accuracy of the hydrodynamic model that includes mangroves as a computational area and non-linear waves. Results of simulation are consistent with satellite images. A strong tidal current caused the spill to spread to Balikpapan Bay's upstream region. Non-linear waves cannot be ignored in oil spill simulation because the mean current will transport the spill to the coast.

Stochastic Model simulation results show that most oil spills were in the Bay 5 days after the incident and spread along the east coast to the north. Alternating tidal currents in the Bay cause this. The problem of the oil spill in Balikpapan was quickly resolved because the oil did not spread to the offshore deepwater of the Makassar Strait

11. References

[1] Mwai P 2020 Mauritius oil spill: Are major incidents less frequent? - BBC News *BBC*
[2] Muin M 1993 *A Three-Dimensional Boundary-Fitted Circulation Model in Spherical Coordinates* (University of Rhode Island)
[3] Muin M and Spaulding M 1996 Two-Dimensional Boundary-Fitted Circulation Model in Spherical Coordinates *J. Hydraul. Eng.* **122** 512–21
[4] Muin M and Spaulding M 1997 Three-Dimensional Boundary-Fitted Circulation Model *J. Hydraul. Eng.* **123** 2–12
[5] PT. LAPI-ITB 1999 *Application of 3-D Boundary Fitted Hydrodynamics and Sediment Transport Model in ADA Estuary*
[6] Muin M and Spaulding M 1997 Application of Three-Dimensional Boundary-Fitted Circulation Model to Providence River *J. Hydraul. Eng.* **123** 13–20
[7] Zigic S, Makarynskyy O, Langtry S and Westbrook G 2009 A Numerical Modelling Study for the Proposed Increase in Barramundi Production, Cone Bay, Western Australia *Estuar. Coast. Model.* 428–46
[8] Sankaranarayanan S and McCay D F 2003 Three-Dimensional Modeling of Tidal Circulation in Bay of Fundy *J. Waterw. Port, Coastal, Ocean Eng.* **129** 114–23
[9] Sankaranarayanan S 2005 A 3D boundary-fitted barotropic hydrodynamic model for the New York Harbor region *Cont. Shelf Res.* **25** 2233–60
[10] Dean R G and Dalrymple R A 2000 *Water Wave Mechanics for Engineers and Scientist* ed P F Liu (Singapore: WORLD SCIENTIFIC)
[11] Muin M 2016 *Manual Model Tumpahan Minyak (MoTuM)* (Bandung)
[12] Kementrian Kelautan dan Perikanan (KKP) 2019 *DATABASE WILAYAH KERJA BPSPL PONTIANAK* (Pontianak)
[13] Hopley D 1999 *Geological & Geomorphological Input Into Coastal Management with special reference to Balikpapan Bay, East Kalimantan* vol 7209596
[14] OpenStreetMap 2022 Open Street Map
[15] Toulec T, Lhota S, Soumarová H, Putera A K S and Kustiawan W 2020 Shrimp farms, fire or palm oil? Changing causes of proboscis monkey habitat loss *Glob. Ecol. Conserv.* **21**
[16] Fatmawaty D 2020 Environmental Pollution Liability Analysis Due to Oil Spill (Case Study: Oil Pipe Leak in Balikpapan Bay) *Bumi Lestari J. Environ.* **20** 14
[17] Kehutanan K L H dan 2018 Report of the Team for Handling Oil Spills in the Waters of Balikpapan Bay, Balikpapan City and North Penajam Pasir Regency, East Kalimantan Province 1–13
[18] Kahfi K 2018 Balikpapan oil spill: What we know and don't know - National *Jakarta Post*
[19] Soeyanto E and Arifiyana 2018 Dynamics of Sedimentation Process at Riko River Estuary, Balikpapan Bay *Oseanologi dan Limnol. di Indones.* **3** 63–72
[20] NAVIONICS Navionics ChartViewer
[21] Nur A A, Radjawane I M, Suprijo T and Mandang I 2020 Numerical Modeling of Currents Circulation in Balikpapan Bay during Oil Spill Event on March 31, 2018 *IOP Conf. Ser. Earth Environ. Sci.* **618**
[22] PT. LAPI ITB 2014 *Marine Facility Design POSB Extension Project*

Integrated Flood Model in 3D Non-Orthogonal Boundary Fitted Hydrodynamic Model for Ciliwung River, Jakarta

A B Muslim[1*] and **M Muin**[1]

[1]Ocean Engineering Research Group, Faculty of Civil and Environmental Engineering, Institut Teknologi Bandung, Jalan Ganesha No 10, Indonesia
[*]Corresponding author: arungbahari15@gmail.com

Abstract. The flood model (MuFlood) had been developed and applied as a flood early warning system in Jakarta, Indonesia. The low land Jakarta faces flood threat because of a significant land-use change in the upstream region, especially in the Katulampa catchment area. Katulampa, Bogor, is located 350 meters above Mean Sea Level (MSL). The distance from Katulampa to Manggarai Gate in Ciliwung River is approximately 91 km. The flood arrival time is about 12 hours. The output of MuFlood is as input for the ocean hydrodynamics model (MuHydro3D) using the 3D Non-Orthogonal Boundary Fitted. The paper presents the integrated model results to simulate flood events from Katulampa and the local catchment area of Jakarta. The comparison between observation and model prediction from the MuFlood hydrodynamic model is excellent for the 21-22 September 2020 case. MuHydro3D predicted that from 21-22 September 2020 case cause the water level increases up to 19 cm on West Flood Canal (KBB), and the closer the bay, the water level will gradually smaller and as same as the tide.
Keywords: *unsteady, flood modelling, 3D hydrodynamics model, non-orthogonal boundary fitted, ciliwung river*

1. Introduction
The flow in the Ciliwung River is unsteady. The assumption that inflow equals outflow could not be applied in the Ciliwung River. The differences between inflow and outflow cause control volume to change during flood events. The Ciliwung River (Figure 1) flows from the Katulampa at an elevation of 350 meters above sea level to the Manggarai Gate with an elevation of 10 meters above sea level. At the Manggarai Gate, most of the Ciliwung river water is diverted to the West Flood Canal (KBB). A small portion is diverted to Ciliwung Lama for flushing. During the

flood event, Manggarai Gate is opened to accelerate water level decline in flooded areas such as Rajawati and Kampung Pulo.

The flood area in the Ciliwung River has long been a national polemic. Whether the Ciliwung River must be widened, or catchment area on upstream must be improved to reduce the peak flow of the flood discharge. In this paper, we will not discuss this polemic. This paper only presents the simulation results from MuFlood and MuHydro3D in the Ciliwung River and the West Flood Canal (KBB), when river discharge from Katulampa is extreme. Finally, this model package is expected to be part of the decision support system in flood management and an early warning system to overcome and minimize casualty floods in Jakarta.

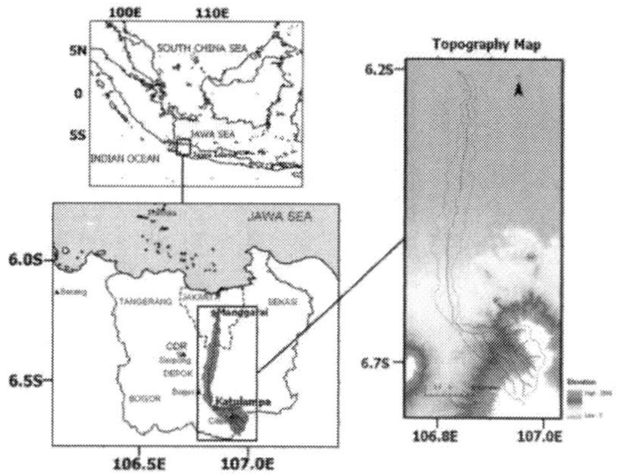

Figure 1. Map of Ciliwung River [1]

Good flood early warning systems must be integrated with a hydrodynamic model, river, and drainage system GIS Database of Jakarta Province. Also, accurate rainfall data is one of the critical successes of hydrodynamic simulation results. The problem is that rain is a stochastic process, not a deterministic process. It cannot be predicted accurately.

Rainfall can be obtained from the results of field measurements and radar images. The results of rain measurements in the upstream area can certainly estimate flooding in the downstream area. The main problem is flooding due to the local rain. The measurement results are too late to be used as an Early Warning System. Therefore, the prediction of the meteorological model becomes very important. BMKG (Meteorology, Climatology, and Geophysical Agency), a meteorological institution from the Indonesian government, has two rain prediction models, which are given on their website [2]. The data from meteorological models by BMKG will become one of the inputs of MuFlood and MuHydro3D.

To this day, developing a hydrodynamic model in the Ciliwung River is very challenging because the river's flow is unsteady. One of the software to simulate hydrodynamic simulation in the river is HEC-RAS developed by the USA [3]. The demerit from this software is that the unsteady model is not stable [3] [4]. MuFlood

(Developed by Muslim Muin [5][6][7][8] and application was funded by Jakarta Provincial Government) is one-dimensional unsteady flow model software to simulate the hydrodynamic of the Ciliwung River. MuFlood has successfully predicted the flood area of the Ciliwung river, and the simulation results show an excellent comparison with the observation data.

Hydraulic models have been widely constructed to simulate the flood in the river. According to Sampson [9], these models vary from a coupled model of 1D with flood plain storage models by de Paiva [10], 1-D channel with simplified 2-D flood models by Biancamaria and da Paz [11][12]. Another model is in 2D models with sub-grid representation along the minor channels by Neal [13]. Unfortunately, all these model approaches neglected the tidal dynamics of the bay. There is no model that can simulate the time series of the flood water height to match the observations data. Since the arrival time of the flood is important; therefore new approach of the model is proposed in this study as early warning system in Jakarta. In addition, the geometry of river system in downstream of Jakarta is very complex, so it requires a flexible and non-orthogonal grid system, where MuHydro3D is suitable in this region.

This paper will discuss the results of the modeling approach for one of the cases when the Katulampa Dam was in a high alert level-1 position on September 21, 2020, using MuFlood. The simulation results were compared with observations of water elevation at the Manggarai Gate. The hydrodynamic results of MuFlood become hydrodynamic parameter input of MuHydro3D to simulate water level along the downstream area or West Flood Canal (KBB).

2. Model Description

In this study, a hybrid model approach is proposed to solve the large gap between the river flood model and the ocean hydrodynamics model. A one-dimensional model approach was employed for areas not affected by tides (Ciliwung River) and three-dimensional models for downstream areas (KBB). The output of the one-dimensional model is used as input for the three-dimensional model. It does not mean that the three-dimensional model cannot be used for upstream areas; this approach minimizes the computational time in an early warning system.

2.1. MuFlood

MuFlood is a one-dimensional model. The input for each cell is riverbed elevation, cross-section, manning coefficient, wet/dry area, time series of Inflow/Outflow. The wet and dry areas in computation to calculate the water height in the flooding area were included in the model. The dynamic of water exchange between the main channel and wet/dry area is the critical factor in simulation. The Mannings coefficient for each cell is tuned in to match observation.

The rainfall inflow/outflow time series for each cell is obtained from rainfall prediction data from the BMKG or from observations data which are BMKG radar and Catchment Area Data. The contour of radar intensity is converted to rainfall intensity (mm/hour) for each catchment area. The discharge from each catchment area is obtained using the Rational Method, and the user can set the appropriate Runoff Coefficient if the user wants to use their calculations. In this case, the user inputs the rainfall Inflow/Outflow time series for each cell.

For floodplain areas, the user enters data on the area, elevation, discharge coefficient from and to the flooded area. Water from the river will enter the flooded area when the water level in the river exceeds the elevation of the flooded area.

2.2. MuHydro3D

The 3D boundary fitted hydrodynamic model, namely MuHydro3D, was developed by Muslim Muin, Ph. D, based on his dissertation [14][15][16]. The 3D Non-Orthogonal Boundary Fitted Hydrodynamic Model solves continuity and momentum equations in a spherical coordinate on the non-orthogonal boundary fitted grid to simulate a time-varying field of surface elevations and 3D ocean currents. The basic equations are written in spherical coordinates to represent large model areas accurately. The sets of governing equations, with dependent and independent variables transformed from spherical to curvilinear (boundary fitted) coordinates. The vertically averaged (external mode) is solved by the semi-implicit technique.

In contrast, the vertical structure (interior mode) is solved using the implicit technique in vertical diffusion and the explicit technique for other equations. The matrix is solved using the sparse matrix technique to predict the surface elevation. This technique is robust when the time step is no longer limited by gravity wave speed. Momentum and conservation of mass equations are conducted using the Non-Orthogonal Curvilinear Coordinate Technique to solve three-dimensional and semi-implicit spaces of time. This technique accommodates a huge time step that implies a brief simulation time and accurate results even in very complicated geometries. River currents could be obtained at any time desired so that the flood simulation can be estimated for a long simulation period. The non-orthogonal boundary fitted ocean hydrodynamic model developed by Muin [3] was used by an American consultant under the name BFHydro, which had been applied worldwide, in New York Harbor [17], Bay of Fundy Canada [18], and Cone Bay, Western Australia [19].

3. Model Setup

3.1. Model Setup MuHydro3D

MuHydro3D was applied to simulate water elevation from Jakarta Bay to the affected tidal river. The grid system is finer in the river area, where the bay area has a coarse system since our interest is the river area. Figure 2 shows the grid system of MuHydro3D, which has 13250 water cells, 200 boundary cells, and 46 river cells. The bathymetry that was used for this simulation is shown in Figure 3. The water depth in the computational domain was acquired from the Navionics chart viewer. The time step of the model is 10 minutes.

Figure 2. The grid system in Jakarta Bay and rivers

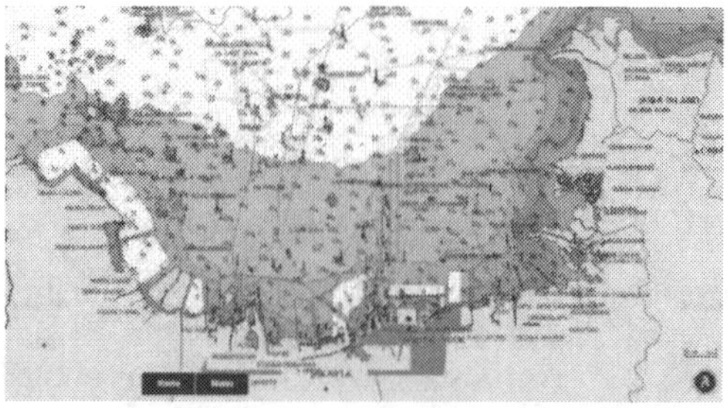

Figure 3. Bathymetry that was used on MuHydro3D [20]

3.2. Model Setup MuFlood

The cross-sectional data of the Ciliwung River is very limited. Therefore, in this study, the elevation and width of the river are obtained from Google Earth. The approach in this study is not perfect but due to budget limitations and world conditions in the covid-19 pandemic, using this approach is a compulsion. Figure 4 shows the riverbed elevation of the Ciliwung River from upstream (350 m) to downstream (2 m). Figure 5 shows the river width profile of the Ciliwung River, which ranges from 10 meters to 45 meters. The mild slope section and narrowing width are the flooding region (Rajawati, Kampung Melayu).

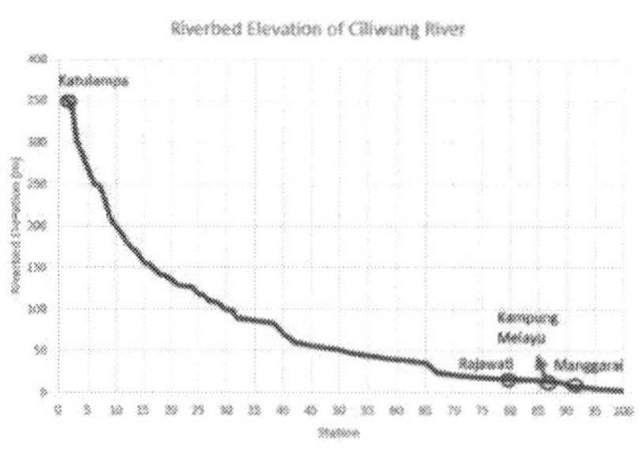

Figure 4. Riverbed elevation of Ciliwung River

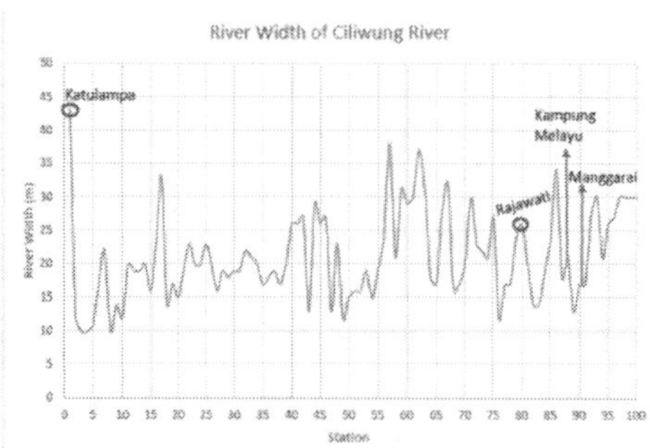

Figure 5. River Width of Ciliwung River

4. Model Calibration

4.1. Model Calibration MuFlood

The MuFlood software is calibrated by changing the Manning Coefficient for each cell depending on the physical condition of the river and observational data. In general, and according to the literature, the Manning Coefficient ranges from 0.25 to 0.35. MuFlood is very sensitive to the Manning coefficient. In addition to comparing the water depth with field data, the arrival time of floods from Katulampa was also compared, ranging from 11 to 13 hours.

For model calibration, we used the case of flooding from Katulampa on September 2, 2020. The flow from Katulampa is derived from the formula above the weir. The peak discharge at Katulampa at that time was around 261 m^3/s. There was no heavy rain in the Catchment Area between Katulampa and Manggarai. So, this simulation is an example of a case for extreme flood from upstream without local rain

The comparison between the predictions of the model and the observations is presented in Figure 6. The simulation results show that the model successfully predicts the water level of Katulampa at 250 cm, which is classified as high alert level-1 or in Bahasa Indonesia is Siaga 1. It can be seen that the arrival time of the Katulampa flood is almost the same as the observation, which ranges from 11 to 13 hours. The predicted water level on Manggarai Floodgate is 780 cm, while the observation is 760 cm, where this water level is classified as high alert level-3 or in Bahasa Indonesia is Siaga 3. This difference is quite acceptable due to the complexity of the physical processes in the Ciliwung River.

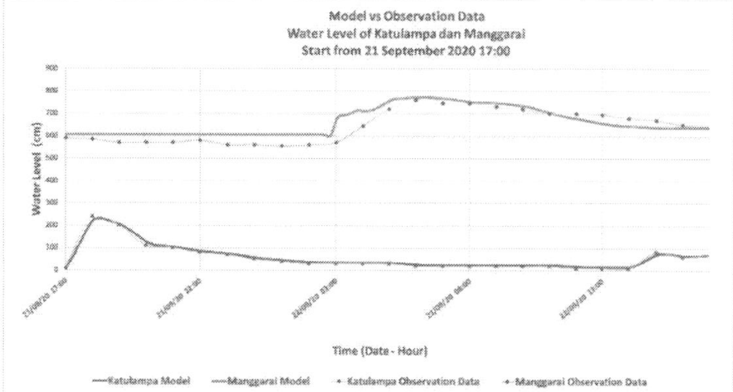

Figure 6. Comparison between model prediction and observation in Manggarai and Katulampa from 21 – 22 September 2020

4.2. Model Calibration MuHydro3D

The calibration is done by comparing the simulation results and observation data in Pasar Ikan, as shown in Figure 7. The figure shows a good comparison with the observation data. We can conclude that the 3D hydrodynamic model is well developed and can be applied to predict water elevation.

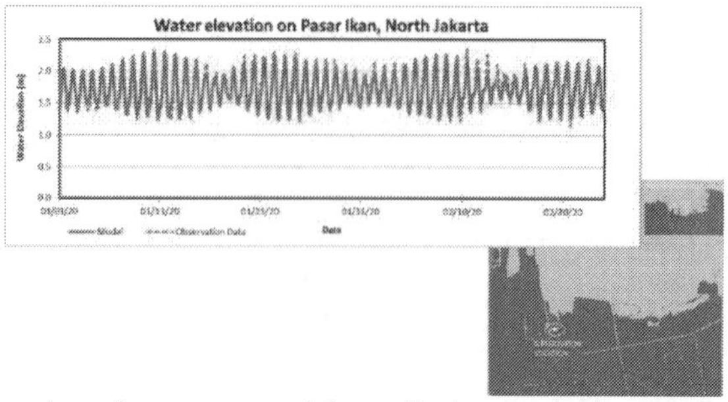

Figure 7. Comparison between model prediction and observation data of water elevation on Pasar Ikan, North Jakarta

5. Results and Discussion

5.1. Hydrodynamic Results from MuFlood

The flood area is presented in Figure 8, where it shows that Cikoko, Pengadengan, and Cawang experienced a flood of 0.1 m. In contrast, the other section of the Ciliwung River area did not experience any flood. The model indicates that on Rajawati and Kampung Melayu, which these areas frequently experienced a flood, but in this scenario, the two areas did not experience any flood.

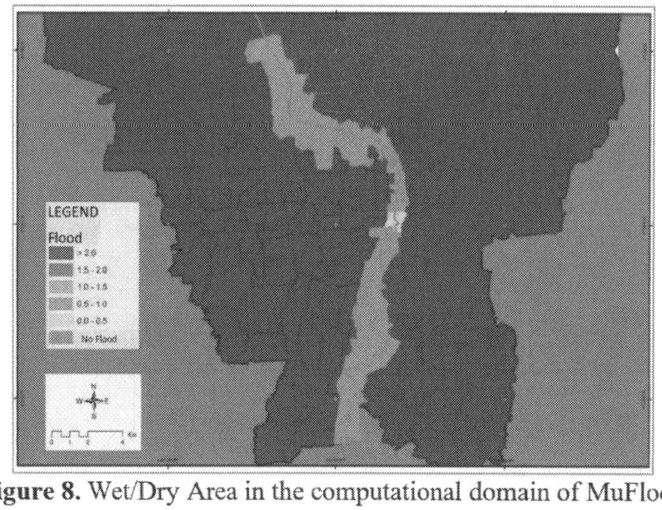

Figure 8. Wet/Dry Area in the computational domain of MuFlood

Figure 9 shows the water depth maximum along the Ciliwung river on the 21-22 September 2020 scenario. The figure shows that the water depth maximum varies from 1.25 m to 4 m. The flood occurs when the water depth exceeds the embankment height. Then, the water exchange between the river cell and wet/dry is calculated in the model.

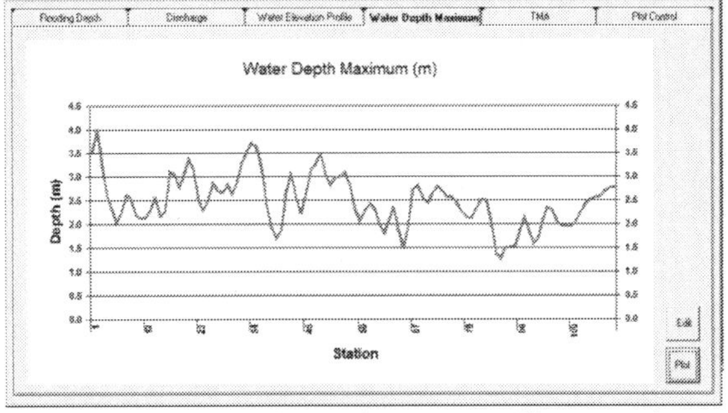

Figure 9. Water depth maximum on Ciliwung River

Figure 10 shows the water elevation profile from stations 70 to 85 for 2 hours simulation time. The rainfall at Katulampa starts at time = 60 Hr. The flood arrives at Station 70 in time = 64.4 Hr, or it takes 4.4 hours for Katulampa Flood to reach Station 70. The flood also arrives at the station at station 85 in time = 66.4, or it takes 6.4 hours for Katulampa Flood to reach station 85.

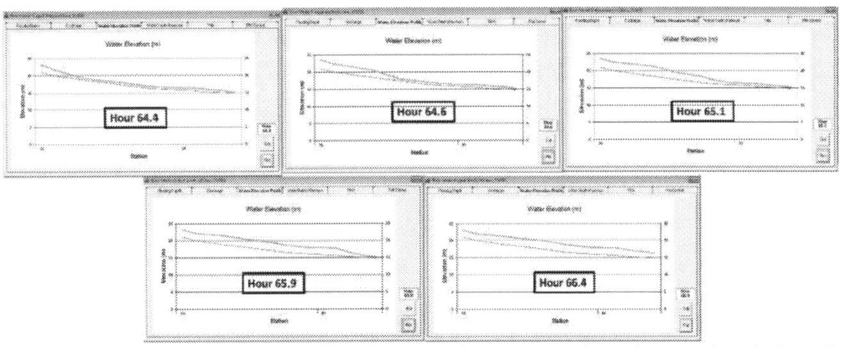

Figure 10. Water elevation profile along stations 70-85 for 2 hours simulation time

Figure 11 shows the water discharge from Katulampa and Manggarai. The simulation shows interesting results, and it is necessary to point out that the maximum flood discharge in Manggarai is only 85 m^3/s, while the maximum discharge in Katulampa is 261 m^3/s. This is because the flow of the Ciliwung River is being choked in the narrowing area of the river. Therefore, we can conclude that the steady flow assumption cannot be applied in the Ciliwung River.

The simulation result also shows that flood duration in Manggarai is much longer than in Katulampa. Again, this is clearly due to the narrowed river width on several locations along the flooded area of the Ciliwung River. Suppose the suffocation of this flow is eliminated by widening the river. In that case, the result is that the peak of flood discharge downstream will increase dramatically and result in severe flooding in the downstream area.

IOP Conf. Series: Earth and Environmental Science 1065 (2022) 012007 doi:10.1088/1755-1315/1065/1/012007

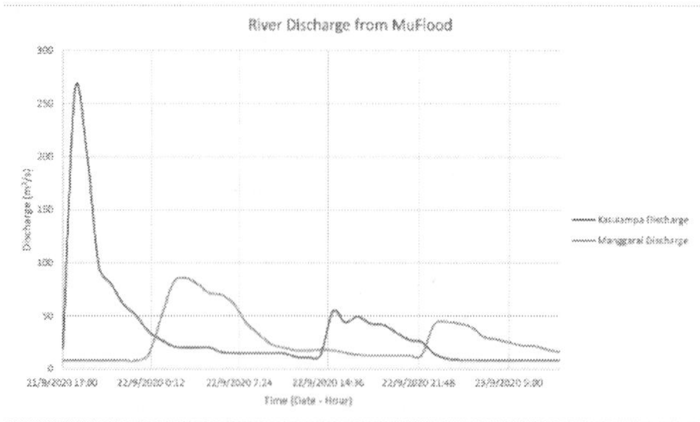

Figure 11. Water discharge of Katulampa and Manggarai from MuFlood

5.2. Integrated MuFlood and MuHydro3D Hydrodynamic Results

After the validation and MuFlood simulation, we integrated MuFlood hydrodynamic results into the MuHydro3D. The location of the river input is shown in Figure 12. The scenario in the integrated model is the same scenario from the validation model, which is a flood that occurred from 21-22 September 2020. In this section, we also compare the water elevation on KBB with river discharge from MuFlood and without the river discharge, and the comparison will show whether KBB occurred a flood or no flood at all.

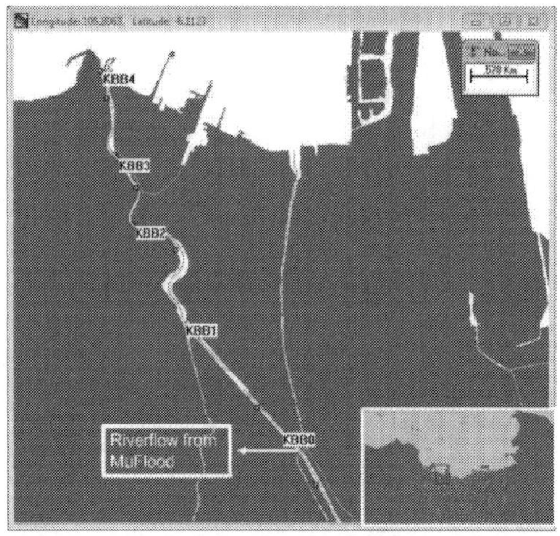

Figure 12. Observation station on MuHydro3D

Figure 13 shows the river charge of Manggarai from the software MuFlood. The river discharge from Katulampa arrived at Manggarai Floodgate in 6 hours. The value discharge is 85 m^3/s. After 6 hours, the river discharge started to decrease and increase again after 30 hours, the Katulampa rain occurred. This discharge becomes the river input on MuHydro3D and simulates the water elevation along the KBB channel.

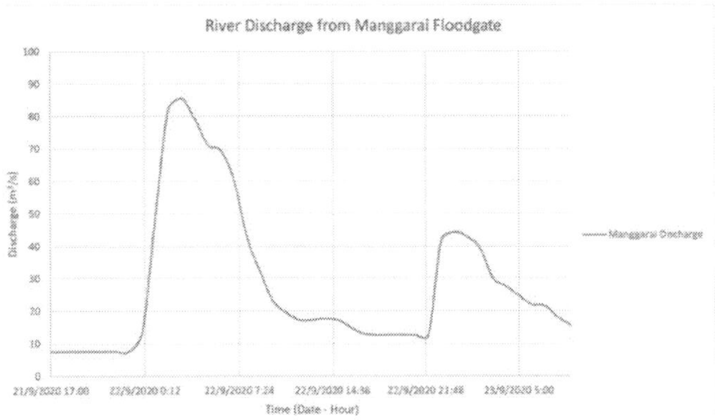

Figure 13. Discharge of Manggarai Floodgate on 21 – 22 September 2020

The water elevation from KBB0 to KBB4 is shown in Figure 14. The figure shows that the flood river discharge increases water elevation by about 0.194 meters or 19 cm increment in KBB0's water level. However, in KBB1's, the water level only increases by 0.103 m or 10 cm. We did not show the water level at KBB2 until KBB4 because the model prediction shows that the water level at each station will gradually decrease and become as same as the water level using tide forcing only.

From the MuHydro3D simulation results, we can conclude that the flood from 21-22 September 2020 only increases the water level by 19 cm in the tidal affected area at KBB0 and 10 cm at KBB1. The water level will gradually decrease, and the water level will be as same as the tidal.

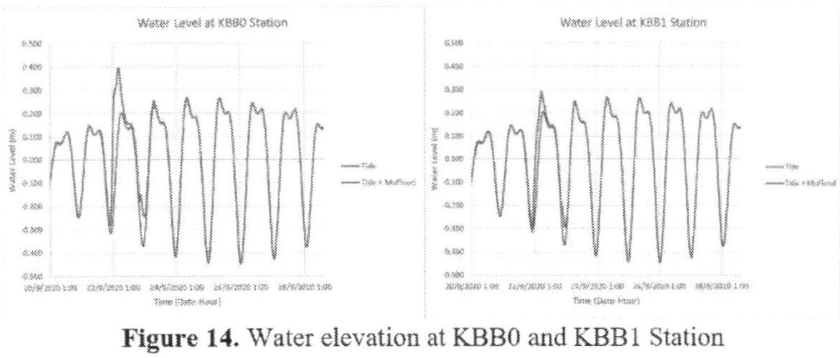

Figure 14. Water elevation at KBB0 and KBB1 Station

6. Conclusion

Flood analysis in the Ciliwung River cannot use the steady flow assumption. The flow peak in each cell will not be the same as in the steady flow analysis. For this reason, we introduce an unsteady flow model, which we named MuFlood. This model has been used for several cases of flooding from upstream in Jakarta [5][6][7][8]. The output from MuFlood will be the input for the ocean hydrodynamics model (MuHydro3D) in the West Flood Canal (KBB). The results of the MuFlood model simulation had been compared with observations in Manggarai Station with satisfactory results. The flood case from upstream on 21

September 2020 was chosen as the model calibration because this case is classified as extreme, where the water level at Katulampa is 250 cm, which is classified as high alert level-1 or in Bahasa Indonesia is Siaga-1. The hybrid model (MuFlood/MuHydro3D) simulation results are used as a flood an early warning system in Jakarta, Indonesia.

Flood Early Warning Jakarta is accurate for cases of flooding sent from upstream because the flow from Katulampa is very measurable, and the arrival time is around 11 hours. The problem is the Flood Early Warning due to local rain; for that, MuFlood must use the rainfall predictions from meteorological models. Until now, there is not a single city in the world that can predict rainfall accurately. Besides that, the inflow from each catchment area along Ciliwung needs to be investigated to obtain the proper runoff coefficient.

From the application of the model in Cliwung, the model result shows that the narrowing river that occurs along the Ciliwung is a savior for Jakarta from the threat of more severe flooding because KBB's capacity to flow water into the sea is very limited. For further research, more model setups will be used to simulate other than the Ciliwung river, such as Cipinang, Krukut, Sunter, etc.

7. References

[1] Sulistyowati R, Hapsari R I, Syamsudin F and Mori S 2014 Rainfall-Driven Diurnal Cycle of Ciliwung River : Overview and Future Prospect **4**

[2] BMKG (Meteorology, Climatology and G A 2022 *BMKG Signature (System for Multi Generation Weather Model Analysis and Impact Forecast)* (Jakarta)

[3] USACE 2016 HEC-RAS River Analysis System User ' s Manual 171

[4] Tigor Oktaga A, Suripin S and Darsono S 2016 Comparison of Unsteady Flow and Steady Flow Modeling Results in City Floods *Media Komun. Tek. Sipil* **21** 35

[5] Muin M 2021 *Monthly Report TGUPP (February 2021)* (Jakarta)

[6] Muin M 2021 *Monthly Report TGUPP (April 2021)* (Jakarta)

[7] Muin M 2021 *Monthly Report TGUPP (September 2021)* (Jakarta)

[8] Muin M 2021 *Monthly Report TGUPP (October 2021)* (Jakarta)

[9] Sampson C C, Smith A M, Bates P D, Neal J C, Alfieri L and Freer J E 2015 A high-resolution global flood hazard model *Water Resour. Res.* **51** 7358–81

[10] De Paiva R C D, Buarque D C, Collischonn W, Bonnet M P, Frappart F, Calmant S and Bulhões Mendes C A 2013 Large-scale hydrologic and hydrodynamic modeling of the Amazon River basin *Water Resour. Res.* **49** 1226–43

[11] Biancamaria S, Bates P D, Boone A and Mognard N M 2009 Large-scale coupled hydrologic and hydraulic modelling of the Ob river in Siberia *J. Hydrol.* **379** 136–50

[12] Paz A R da, Collischonn W, Tucci C E M and Padovani C R 2011 Large-scale modelling of channel flow and floodplain inundation dynamics and its application to the Pantanal (Brazil) *Hydrol. Process.* **25** 1498–516

[13] Neal J, Schumann G and Bates P 2012 A subgrid channel model for simulating river hydraulics and floodplain inundation over large and data sparse areas *Water Resour. Res.* **48** 1–16

[14] Muin M 1993 *A Three-Dimensional Boundary-Fitted Circulation Model in Spherical Coordinates* (University of Rhode Island)

[15] Muin M and Spaulding M 1996 Two-Dimensional Boundary-Fitted Circulation Model in Spherical Coordinates *J. Hydraul. Eng.* **122** 512–21

[16] Muin M and Spaulding M 1997 Three-Dimensional Boundary-Fitted Circulation Model *J. Hydraul. Eng.* **123** 2–12
[17] Sankaranarayanan S 2005 A 3D boundary-fitted barotropic hydrodynamic model for the New York Harbor region *Cont. Shelf Res.* **25** 2233–60
[18] Sankaranarayanan S and McCay D F 2003 Three-Dimensional Modeling of Tidal Circulation in Bay of Fundy *J. Waterw. Port, Coastal, Ocean Eng.* **129** 114–23
[19] Zigic S, Makarynskyy O, Langtry S and Westbrook G 2009 A Numerical Modelling Study for the Proposed Increase in Barramundi Production, Cone Bay, Western Australia *Estuar. Coast. Model.* 428–46
[20] NAVIONICS Navionics ChartViewer

Vessel Size and Dredging Depth Optimization for *Tol Laut* Program using Genetic Algorithm: Hub Port Case Study

R Sihombing[1*], M Muin[1], R L Tawekal[1]

[1]*Ocean Engineering Research Group, Faculty of Civil and Enviromental Engineering, Institut Teknologi Bandung, Jalan Ganesha No 10, Indonesia*

*Corresponding author: rajin@itb.ac.id

Abstract. *Tol Laut* is an Indonesian government containerized sea freight program which is aimed at reducing the price disparity, as well as extending logistic distribution from developing regions. The program can be enhanced by determining the optimum dredging depth of port navigation channel and the optimum size of vessel. The required channel depth is affected by the size of vessel, which is represented by its draft, on the respected hub port. The significance of vessel size and dredging depth in accommodating the container demand will be evaluated by considering the physical environmental conditions such as surface water elevation which affected by tidal conditions. Genetic Algorithm is utilized in this study since it showed robust results on optimizing port performances problems that have many dependent variables and solutions. As a hub port within *Tol Laut* Program, Port of Belawan is selected as a case study due to its strategic location. Several demand scenarios were run using the algorithm, resulting several channel depth-vessel size pairs as the solution. In conclusion, the feasibility of the dredging activity relies significantly on the growth of the demand.
Keywords: genetic algorithm, optimization, tol laut, vessel size, dredging depth.

1. Introduction

Logistics costs are significantly influenced by the operational costs of ship shipping. The average berthing time in Indonesia is 40 hours, compared to Singapore with an average 21 hours average berthing time [1]. This phenomenon is strengthened by *Bappenas* [2], which stating that the amount of waiting time has an impact on uncertainty and delay of ship schedules, leading to a low loading factor. Schedules delays can be caused by uncertainties in berthing, loading, and unloading time on port [3]. Environmental conditions, port congestion, or labour strike could also affect vessel schedules [4]. Hence, the port infrastructure and operation are found to be depending on environmental conditions [5]. Environmental conditions such as tides, waves, wind, and sedimentation are important factors in infrastructure design and port operational [6,7]. Tidal conditions could affect port operational patterns such as the allocation of berths and scheduling of ship arrivals [8].

Port of Belawan is strategically located on the northeast coast of Sumatera. Based on Port of Belawan Masterplan, container demand at this port reached 895,281 TEUs by 2015 and predicted to increase in the upcoming year. In accordance with the masterplan, the container demand was forecasted to grow to 1,657,659 TEUs by 2022 [9]. This high demand could be accommodated by increasing vessel size in line with increasing the vessel draft. In the case of 1,000-10,000 DWT container vessels, the draft is in the range 3.1-8.1 meters.

According to Port Authority of Belawan, the channel depth estimation of Port of Belawan was 6 mLWS (Lowest Spring Tide) [9], so that the vessels with more than 6 meters draft could not directly enter the port through the navigation channel. This type of vessel must be put on hold, waiting for the tidal condition to allow the entering process. The waiting time for the tidal conditions will be burdened by the vessel and considered as vessel cost. The other way to overcome this issue is by dredging activities to reach a minimum depth of 8.1 mLWS, so that large vessels do not rely on tidal condition to enter the port. The dredging cost estimation varies from 45,000-150,000 IDR[1] per meter-cubic depending on dredging vessels and methods.

Genetic Algorithm is well-proven to solve port optimization problems which has numerous possible solutions [10–13]. The objective is to determine the minimum expected cost resulting from vessel waiting time cost and dredging cost. Genetic Algorithm model is generated to optimize the vessel size and dredging depth with tidal consideration.

The study covers research data on Section 2, including vessel data, calculation of vessel operational costs, calculation of dredging costs, and the overview of Port of Belawan. Part 3 covers the methodology and problem formulation used on this study, Part 4 covers the model results, and Part 5 includes conclusions and discussions.

2. Research Data

2.1. Operational Cost

The cost calculation on public service activities of vessels is regulated on *Peraturan Menteri Perhubungan Laut No. 55 Tahun 2019.* Vessel operating costs consist of fixed costs and variable costs. Fixed costs consist of vessel crew salaries, accommodation, freshwater, vessel insurance, vessel fumigation, vessel depreciation, daily maintenance of vessel, and annual docking cost. Variable costs consist of fuel cost, lubricant cost, mineral water for passengers, crew insurance, cargo insurance, marketing cost, and overhead cost [14]:

a. Vessel Crew Salary

According to the International Transport Workers Federation, crew salaries are calculated as follows:

$$\text{crew salary} = \text{number of days in contract} \times \text{number of crew} \times \text{daily crew fee} \quad (1)$$

b. Accomodation

The cost of accommodation is calculated as follows:

$$\text{accommodation cost} = \text{number of days} \times \text{number of crew} \times \text{daily accommodation} \quad (2)$$

Daily accommodation cost for vessel crew is IDR50,000.

c. Fresh Water for Crew.

The amount of fresh water for vessel crew is set at 200 liters per day. Cost of freshwater is calculated by the following formula:

$$\begin{aligned}\text{freshwater for crew cost} &= \text{number of days in contract} \times \text{number of crew} \times \text{fresh water} \\ &\quad \text{requirement per person per day} \times \text{fresh water price per liter}\end{aligned} \quad (3)$$

d. Vessel Insurance

The cost of vessel insurance calculated by 1%-2.5% vessel price. This cost covers for hull and machinery insurance, vessel frame, oil pollution, and other government insurance.

e. Vessel Fumigation

The annual cost of vessel fumigation is IDR10,000,000-IDR25,000,000, depends on size of vessel and fumigation material.

[1] Notes: IDR = Indonesian Rupiah, 1 USD is equal to 15,000 IDR

f. Vessel Depreciation
Vessel depreciation is considered for non-state owned vessels and calculated as follows:

$$\text{vessel depreciation} = \frac{5\% \times \text{vessel price}}{20 \text{ years}} \quad (4)$$

g. Daily Vessel Maintenance
Vessel maintenance cost incurred for repair and daily maintenance. This variable is considered if the vessel operates for a period of one calendar year. The cost of vessel maintenance is calculated as follows:

$$\text{daily maintenance cost} = \text{GT} \times \text{repair and maintenance cost} \quad (5)$$

Vessel maintenance cost is IDR1,000,000 per GT per year.

h. Annual Docking Cost
The annual vessel maintenance cost incurred for maintenance activities on the vessel such as carrying out the vessel to the vessel yard for cleaning the keel of the vessel. The annual maintenance budget plan is calculated as follows:

$$\text{annual docking cost} = \text{GT} \times \text{average cost per GT} \times \text{vessel age factor} \times \text{GT factor} \quad (6)$$

i. Fuel Cost
The cost of vessel fuel is calculated as follows:

$$\text{fuel cost} = \text{voyage frequency} \times \text{duration of engine operates per voyage} \times \text{daily fuel consumption} \times \text{fuel price per ton} \quad (7)$$

Duration of operating engine per voyage is subtraction between duration of one round voyage and duration of vessel docked at the port.

j. Lubricant Cost
The cost of lubricant is calculated as follows:

$$\text{lubricant cost} = \text{voyage frequency} \times \text{duration of engine operates per voyage} \times \text{daily lubricant consumption} \times \text{lubricant price per liter} \quad (8)$$

Duration of engine operates per voyage is subtraction between duration of one round voyage and duration of vessel docked at the port. Oil consumption is calculated as follows:

$$\text{lubricant consumption} = \frac{1}{0.89} \times 2 \text{ gr/HP/hr} \times 24 \times \text{MCR} \times 10^{-3} \times 1.05 \quad (9)$$

k. Fresh Water for Passengers
Fresh water consumption for passengers is 150 liters per day. The cost of freshwater for passengers is calculated as follows:

$$\text{fresh water for passenger} = \text{voyage frequency} \times \text{number of passenger} \times \text{requirement per person per day} \times \text{fresh water price per liter} \quad (10)$$

l. Crew Insurance
Life insurance premium for crew is given at 2% of passenger and cargo revenue. Passenger revenue is calculated as follows:

$$\text{frevenue} = (\text{number of passenger} \times \text{passenger tariff}) + (\text{cargo weight} \times \text{cargo tariff}) \quad (11)$$

m. Cargo Insurance
The cargo insurance cost is provided at 2% of the income of passengers and cargo.

n. Marketing Cost
Marketing cost is set at 2% of cargo and passenger revenue. Marketing costs are used for banners, billboards, advertisements, video clips, and other promotions.

o. Overhead Cost
Overhead cost is set at 5% of fixed cost.

The vessels used in operational cost calculation are cargo and container vessels which specifically operated by the *Tol Laut* program. Dimension of *Tol Laut* vessels according to Biro Klasifikasi Indonesia, Marine Traffic Website and *Layanan Pengadaan Secara Elektronik* are listed in Table 1:

Table 1. Dimension and capacity of vessel [15–17]

Vessel Name	LoA (m)	L_{BP} (m)	B (m)	H (m)	D (m)	GT	TEUs	DWT	Power (kw)	V avg (knots)	Vessel Price (IDR)
Logistik Nusantara 1	126.1	113.8	20.0	10.4	6.0	7738	800	10822	4064	7.7	546,698,000,000
Logistik Nusantara 2	80.1	77.3	15.0	7.2	4.8	3050	380	3893	2653	7.7	216,852,100,000
Logistik Nusantara 3	80.1	77.3	15.0	7.2	5.8	3040	380	3901	2690	7.4	217,233,000,000
Logistik Nusantara 4	90.0	77.3	15.0	7.2	4.8	3013	380	3901	2653	8.4	217,233,000,000
Logistik Nusantara 5	129.6	121.3	19.0	11.7	6.6	8407	870	11184	8158	10.0	563,930,500,000

2.2. Dredging Cost
Dredging cost analysis is conducted from productivity calculation analysis and unit price analysis. More detailed explanation for both analyses will be discussed in the following sections.

2.3. Case Study: Port of Belawan
Port of Belawan located in Medan, North Sumatera, Indonesia, is the third largest port in Indonesia after Port of Tanjung Priok in Jakarta and Port of Tanjung Perak in Surabaya. This port serves the non-oil commodities, covering exports and imports activities using containers. The number of vessels served in this port is more than 4,000 vessels a year [9].

Port of Belawan has a channel of 14.5 NM / 26.8 km length, 100 m width, -6 mLWS depth, and 2.6 m tidal range. Based on these channel profile, the vessel with more than 6 meters draft is needed to put on hold for the tidal enables the channel. According to Master Plan of Belawan Port, the container projection is shown in the Figure 1:

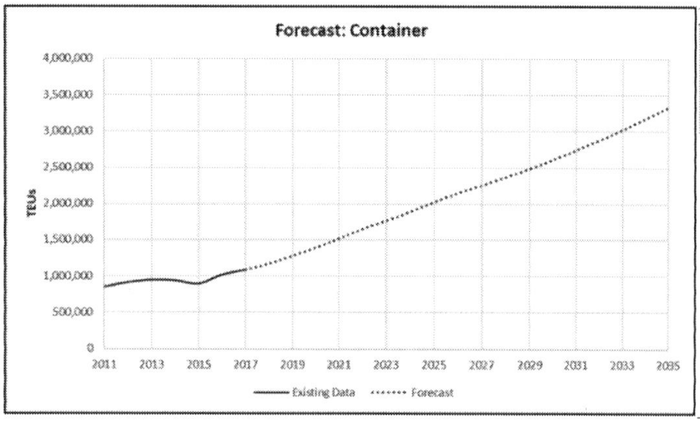

Year	TEUs
2011	847,748
2012	912,418
2013	946,377
2014	937,700
2015	895,281
2016	1,016,206
2017	1,087,340
2018	1,174,328
2019	1,280,071
2020	1,395,219
2021	1,520,788
2022	1,657,659
2023	1,773,695
2024	1,897,854
2025	2,030,704
2026	2,152,546
2027	2,260,173
2028	2,373,182
2029	2,491,941
2030	2,616,493
2031	2,747,255
2032	2,884,618
2033	3,028,849
2034	3,180,291
2035	3,339,306

Figure 1. Container Projection [9].

The Belawan Masteplan demand data will be used as an optimization scenario. Three scenarios using 2011, 2016, and 2021 demand are conducted to determine the relationship between demand, vessel size, and dredging depth.

3. Methodology

This study is conducted by quantitative approach on Port of Belawan. Belawan Port data is used to see the operation and technical aspect of the port. These data were obtained by field survey and desk study.

3.1. Genetic Algorithm

Genetic Algorithm is a metaheuristic method that applies the principles of Natural Selection and Natural Genetic. Generally, the optimization problems are solved by this method. Genetic Algorithm consists of three operators, namely: reproduction and selection; crossovers; and mutations. The advantage of using genetic algorithm are as follows:

a. Can represent port performance problems that have many dependent variables and solutions.
b. It is common to solve stochastic optimization model [11,12].
c. "Naturally" will eliminate individuals with low Fitness Value (FV).
d. Because of the mutation operator, the solutions obtained will vary.
e. Can produce a fairly good solution in a short time

More detailed Genetic Algorithm approach on port optimization problems may be found on "Berth scheduling model at multiple marine container terminals with tidal considerations"[8], "Container fleet sizing and empty repositioning in liner shipping systems"[11], and "Multi-objective optimization for planning liner shipping service with uncertain port times"[12].

Mathematical model equation for vessel size and dredging depth optimization with tidal conditions consideration can be arranged as follows:

$$Z_{cost} = V_{cost} + I_{cost} \tag{12}$$

$$V_{cost} = (B_t + W_t) \times V_{op} \tag{13}$$

$$I_{cost} = D_{dredge} \times Dr_{cost} \tag{14}$$

$$D_i = D + \textstyle\sum_{k=1}^{k} R_k \cos(\omega_k t - \phi_k) \tag{15}$$

$$W_t = \Delta T\,(D_i < D_v) \tag{16}$$

$$D_{dredge} \leq (\alpha \times D_v) - D \tag{17}$$

$$Z_{cost}, V_{cost}, I_{cost}, V_{op}, DR_{cost} \geq 0 \tag{18}$$

$$S_t, B_t, W_t \geq 0 \tag{19}$$

$$\Delta T \geq 0 \tag{20}$$

$$\alpha \geq 0 \tag{21}$$

where

Z_{cost} : total cost
V_{cost} : total vessel cost
V_{op} : total vessel operating cost
I_{cost} : total dredging cost

V_{pp}	: vessel operational cost
B_t	: berthing time (hour)
W_t	: waiting time (hour)
D_{dredge}	: dredging depth (m)
Dr_{cost}	: dredging cost per meter
D_i	: dredging depth at time i (m)
D	: water depth (m)
D_v	: vessel draft (m)
R_k	: amplitude of the k-th constituent
ω_k	: angular frequency of the k-th constituent
t	: time
ϕ_k	: phase angle of k-th constituent
ΔT	: range of waiting time (hour)
α	: dredge depth coefficient

Optimization flowchart is shown in Figure 3-4. The calculation starts from data sets initiations which consists of determining the combination of ships, total cargo, environmental conditions, and calculating the arrival waiting time as shown in Figure 2. The next step is calculating the waiting time for departure, total ship operating cost, and total dredging cost as shown in Figure 3. Finally, the Genetic Algorithm begins from selection operator, crossover, and mutation as shown in Figure 4. The termination criteria of Genetic Algorithm is when the iteration reaches 100 times.

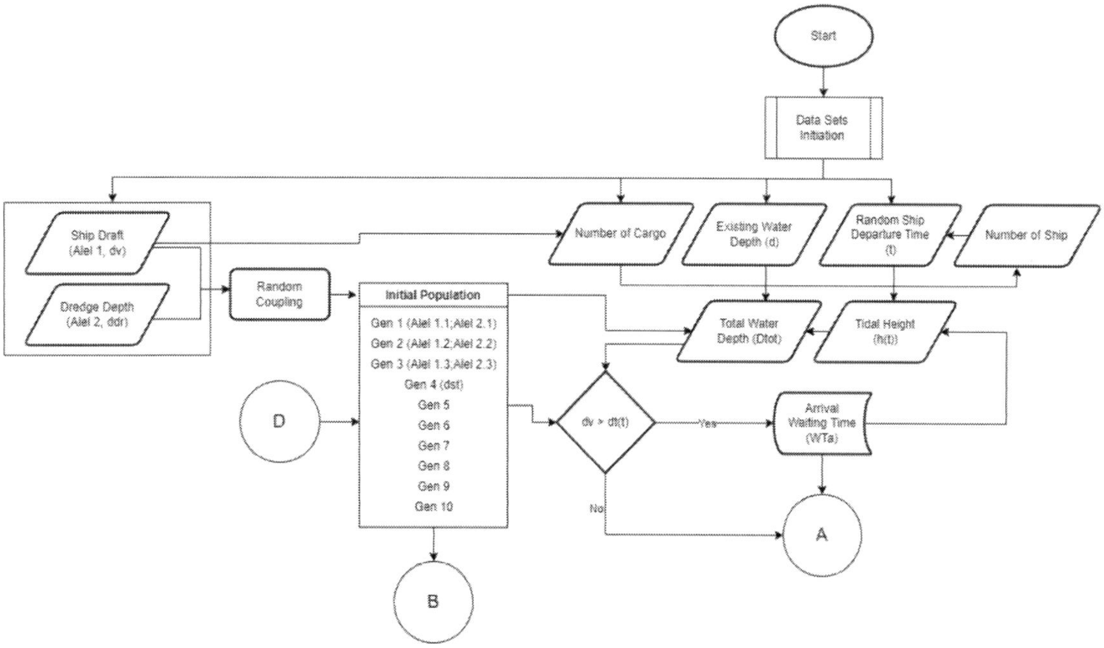

Figure 2 Model flowchart (1).

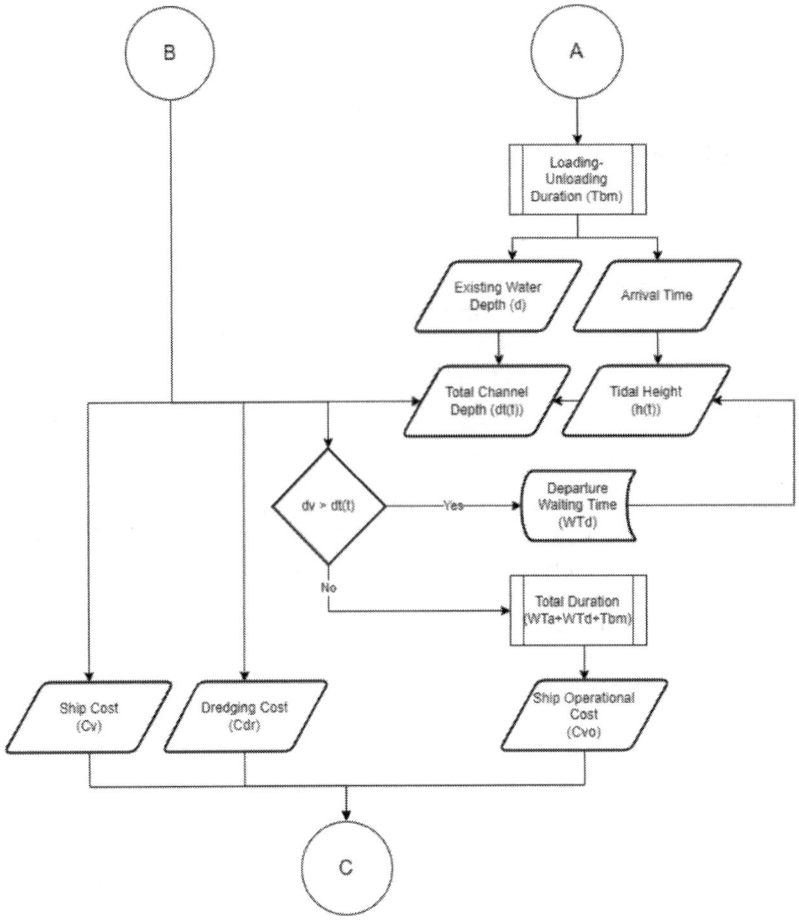

Figure 3. Model extended flowchart (2).

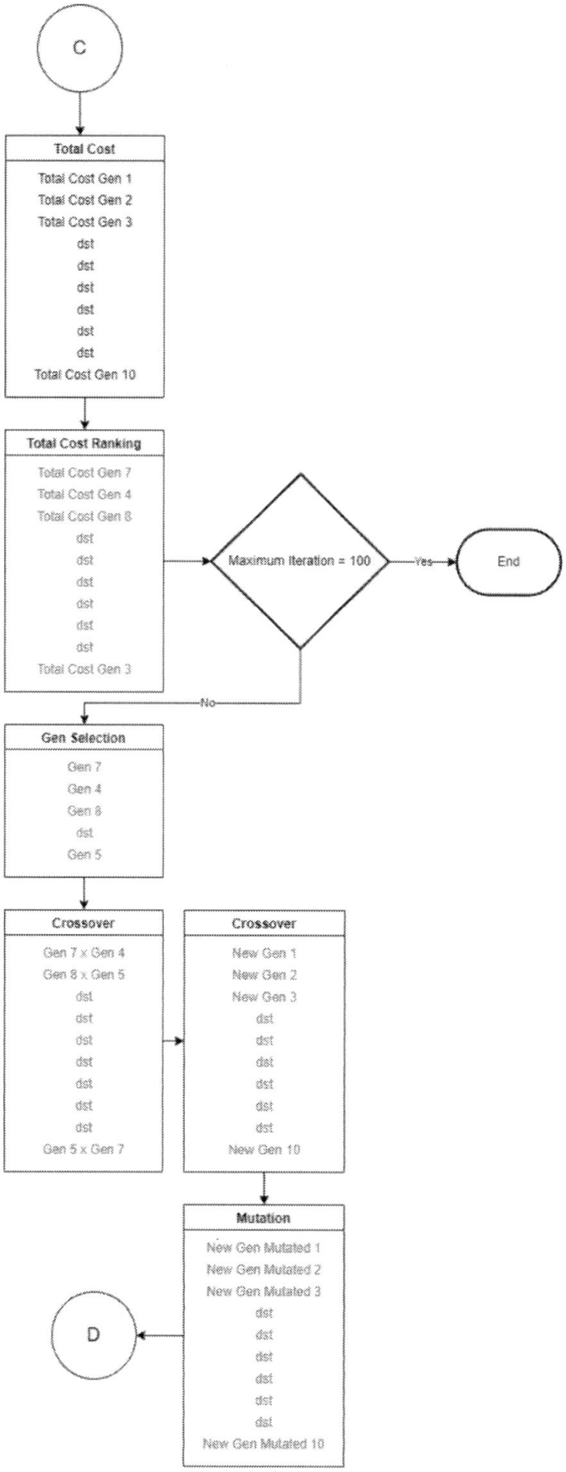

Figure 4 Model extended flowchart (3).

Tidal data is obtained from the least-square equation as follows.

$$y(t)=\sum_{k=1}^{k} R_k \cos(\omega_k t-\phi_k) \quad (22)$$

where

$y(t)$	: water level i-th
R_k	: amplitude of the k-th constituent
ω_k	: angular frequency of the k-th constituent
t	: time
ϕ_k	: phase angle of k-th constituent

By including the amplitude and period of the tidal constituents (M_2, S_2, N_2, K_2, K_1, O_1, P_1, M_4, and MS_4), the equation is as follows.

$$y(t)= R_{M_2} \cos(\omega_{M_2} t-\phi_{M2}) + R_{S_2} \cos(\omega_{S_2} t-\phi_{S2}) + R_{N_2} \cos(\omega_{N_2} t-\phi_{N2})$$

$$R_{K_2} \cos(\omega_{K_2} t-\phi_{K2}) + R_{K_1} \cos(\omega_{K_1} t-\phi_{K1}) + R_{O_1} \cos(\omega_{O_1} t-\phi_{O1}) + R_{P_1} \cos(\omega_{P_1} t-\phi_{P1}) +$$

$$R_{M_4} \cos(\omega_{M_4} t-\phi_{M4}) + R_{MS_4} \cos(\omega_{MS_4} t-\phi_{MS4}) \quad (23)$$

3.2. Operational cost

The operational costs for *Tol Laut* vessels are shown in Figure 5.

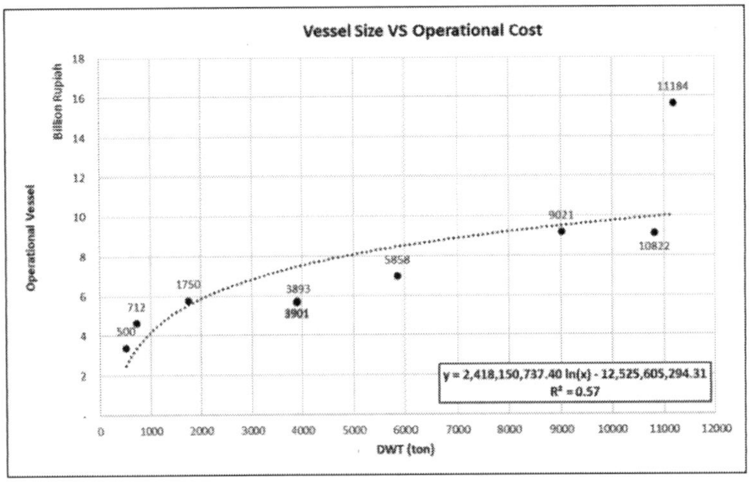

Figure 5. Relationship between vessel size and operating cost.

The operational cost calculation for each *Tol Laut* vessel is plotted into a graph to obtain operating costs data for 1,000-10,000 DWT vessels. The relationship between vessel size and operating cost is approximated using the equation of logarithmic trendline with the R^2 value of 0.57. This value is considered adequate due to the data limitation on vessel price [18]. Thus, the operational cost per hour of 1,000-10,000 DWT container ship in general are shown in Table 2:

Table 2. Operational cost per hour for each container ship with DWT 1,000-10,000.

DWT (ton)	TEUs	Draft (m)	Operational Vessel Cost per hour (IDR)
1,000	70	3.6	5,900,000
2,000	130	4.1	8,200,000
3,000	200	4.6	9,500,000
4,000	260	5.1	10,500,000
5,000	320	5.6	11,300,000
6,000	390	6.1	11,900,000
7,000	450	6.6	12,400,000
8,000	510	7.1	12,800,000
9,000	580	7.6	13,200,000
10,000	640	8.1	13,600,000

Table 2 shows the operating cost of 10,000 DWT vessels are lower than the operating costs of 10 units of 1,000 DWT vessels. The operational cost per hour for each container vessel will be one of the data inputs for the Genetic Algorithm.

3.3. Dredging Cost

Few assumptions which used on dredging cost unit price analysis are as follows:

a. Backhoe vessels is used as dredger equipment.
b. Distance between dredging location and dumping area is 12 Nmil (22.2 km)
c. Dredging volume 500,000 m^3
d. 1 USD = 14,000 IDR

The dredging cost per meter cubic can be obtained by dividing the total dredging cost with total dredging volume. The dredging unit price analysis is shown in Table 3:

Table 3. Dredging cost with 25-ton backhoe as equipment.

Unit Price Analysis	Unit	Unit Price (IDR)	Volume/Day	Total Price (IDR)
Non-hopper fuel and lubricant for operational consumption	liter/day	62,426,000	31	1,935,206,000
Daily fuel operational of barge per day	m^3/day	31,213,000	93	2,902,809,000
Operational freshwater usage	m^3/day	393,600	31	12,201,600
Barge freshwater usage	m^3/day	196,800	93	18,302,400
Dredging incentive of non-hopper vessel	person/day	48,390,000	50	2,419,500,000
Dredging incentive of barge	oh	20,000,000	150	3,000,000,000
Dredger crew salary	oh	87,612,000	50	4,380,600,000
Barge crew salary	oh	4,380,600	150	657,090,000

Unit Price Analysis	Unit	Unit Price (IDR)	Volume/Day	Total Price (IDR)
Dredger accommodation	day	10,200,000	50	510,000,000
Barge accommodation	day	2,100,000	150	315,000,000
Non-hopper fuel and lubricant for mobilization and demobilization	liter/day	62,426,000	12	749,112,000
Barge fuel and lubricant for mobilization and demobilization	liter/day	31,213,000	36	1,123,668,000
Non-hopper mobilization and demobilization freshwater usage	m^3/day	769,200	12	9,230,400
Barge mobilization and demobilization freshwater usage	m^3/day	384,600	36	13,845,600
			Total (IDR)	18,046,565,000
			Tax 10%	1,804,656,500
			Total Dredging Cost (IDR)	**19,851,221,500**
			Total Dredging Cost ($/m³)	**2.8**

The dredging area in Port of Belawan is determined on channel and turning basin with less than 6 meters depth. Total dredging area of Belawan Port is 308,000 m^2. Dredging cost for various depth will be one of the data inputs for the Genetic Algorithm. The result of dredging cost for Port of Belawan is shown in Table 4:

Table 4. Dredging cost for each water depth.

Dredging Depth (m)	Dredging Cost (IDR)
0.1	1,198,736,000
0.2	2,397,472,000
...	...
4.9	58,738,064,000
5.0	59,936,800,000

4. Result and Discussion

The optimization analysis for the projected container data is carried out to observe the sensitivity of container demand amount against the ship selection and dredging depth result. In the Genetic Algorithm, the vessel size and the dredging depth are gene representations. The two genes will be randomly generated to create the initial population. Individuals with adequate gene combinations will be retained, while others will be modified to provide greater reproductive opportunities for better individuals. Reproduction process is conducted using two operators which are cross-over and mutation. In this study, the maximum number of iterations is 100 with a maximum population of 200 individuals. The elite ratio used to form new individuals is set at 5% of the maximum population and the mutation rate is 6%. One cycle of evolution or iteration in Genetic Algorithms forms a generation. The regeneration process will be repeated until the desired number of generations is reached. Individuals formed from the process of mutation and cross-over are considered as the population.

Based on Figure 1, the demand of Belawan Port containers in 2011, 2016, and 2021 respectively are 847,748 TEUs, 1,016,206 TEUs, and 1,520,788 TEUs. The average cost is an average cost from every solution on each generation. While the the best cost is minimum cost among every solution on each generation. The best generation of Genetic Algorithm optimization results is listed in Table 5, Table 6, Table 7, and Figure 6.

Table 5. Average cost and total cost of Genetic Algorithm optimization results in 2011.

Generation	Vessel Size (TEUs)	Dredging Depth (m)	Average Cost (IDR)	Best Cost (IDR)
		2011		
1	580	1.4	304,207,077,000	211,638,400,000
2	580	1.4	268,667,115,000	211,480,000,000
3	580	1.3	241,569,915,000	211,117,200,000
...	...	...	...	...
99	640	1.8	211,314,906,000	204,248,000,000
100	640	1.8	208,928,793,000	204,248,000,000

Table 6. Average cost and total cost of Genetic Algorithm optimization results in 2016.

Generation	Vessel Size (TEUs)	Dredging Depth (m)	Average Cost (IDR)	Best Cost (IDR)
		2016		
1	580	1.5	322,485,226,000	250,848,800,000
2	580	1.4	285,496,853,000	250,222,000,000
3	580	1.4	277,514,691,000	249,971,200,000
...	...	...	...	...
99	640	1.8	272,708,748,000	240,288,000,000
100	640	1.8	258,409,146,000	240,288,000,000

Table 7. Average cost and total cost of Genetic Algorithm optimization results in 2021.

Generation	Vessel Size (TEUs)	Dredging Depth (m)	Average Cost (IDR)	Best Cost (IDR)
		2021		
1	580	1.6	494,034,033,000	366,936,000,000
2	640	2.1	425,390,030,000	350,472,000,000
3	640	1.9	401,251,202,000	348,728,800,000
...	...	...	...	...
99	640	1.9	354,816,474,000	348,212,000,000
100	640	1.9	375,017,004,000	348,212,000,000

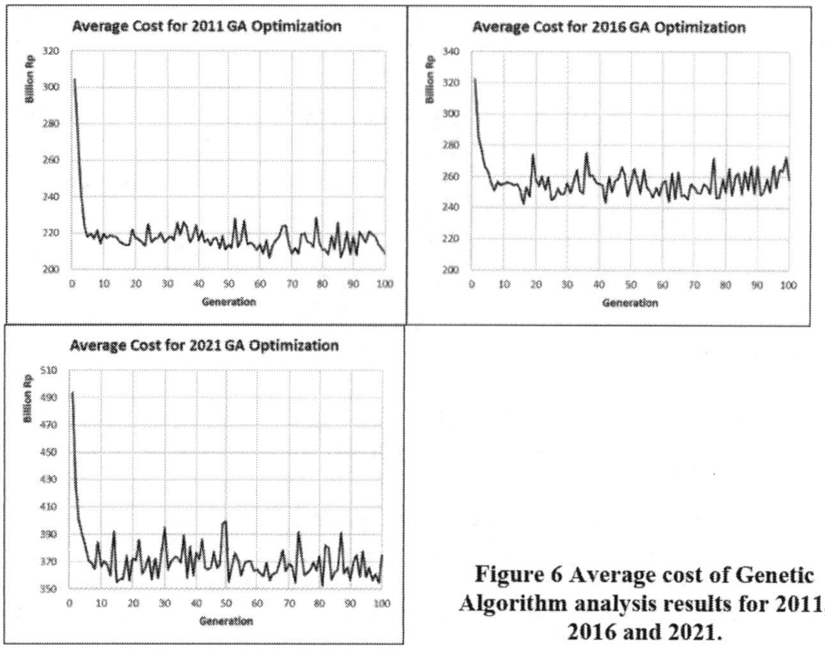

Figure 6 Average cost of Genetic Algorithm analysis results for 2011, 2016 and 2021.

Based on Figure 6, the average cost shows that regeneration process is relatively stable after the 10th generation. The average costs on 2011 scenario are in the range of IDR210,000,000-230,000,000, while the average costs on 2016 scenario are in the range of IDR240,000,000-270,000,000, and the average costs on 2021 scenario are in the range of IDR350,000,000-390,000,000.

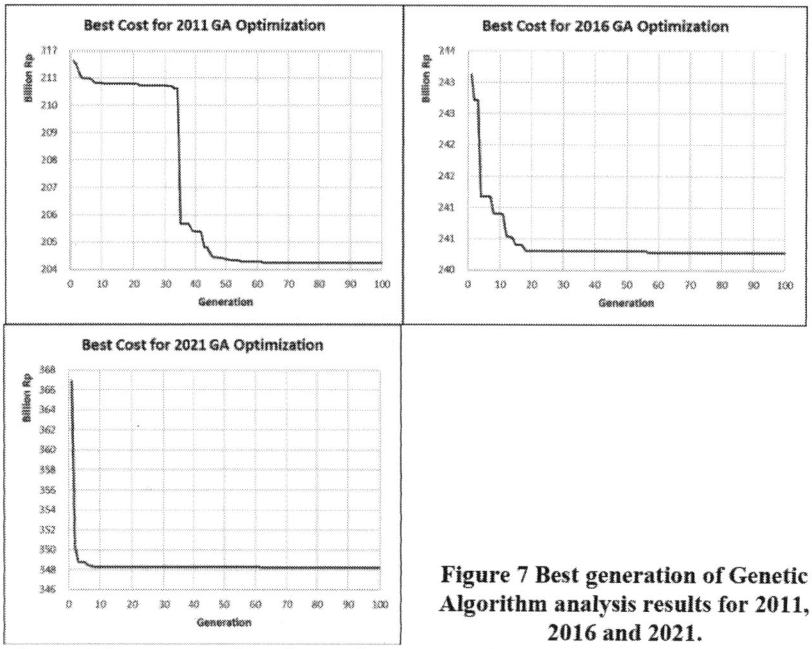

Figure 7 Best generation of Genetic Algorithm analysis results for 2011, 2016 and 2021.

Based on Figure 7 best generation shows the most optimum combination between vessel size and dredging depth with minimum cost. The solution for 2011 and 2016 is a vessel of 640 TEUs (10,000 DWT, draft 8.1 m) and dredging depth of 1.8 meters. The solution for 2021 for the demand is a vessel of 640 TEUs (10,000 DWT, draft 8.1 m) and dredging depth of 1.9 meters.

5. Conclusion

This paper determines the minimum expected cost resulting from vessel waiting time cost and dredging cost for 3 cases scenarios on Belawan Port:

- In 2011, for 847,748 TEUs container demand, the optimum vessel and dredging depth are 640 TEUs and 1.8 meters with the cost of IDR 204,248,000,000.
- In 2016, for 1,016,206 TEUs container demand, the optimum vessel and dredging depth are 640 TEUs and 1.8 meters with the cost of IDR 240,288,000,000.
- In 2021, for 1,520,788 TEUs container demand, the optimum vessel and dredging depth are 640 TEUs and 1.9 meters with the cost of IDR 348,212,000,000.

Based on the above results, the dredging depth does not meet the draft of the selected ship. In details, the 2011 and 2016 scenarios with dredging depth 1.8 meters produces maximum depth of 7.8 mLWS, and in the 2021 scenario with the dredging depth 1.9 meters generates a maximum water depth of 7.9 mLWS. These conditions could require the selected ship with a draft of 8.1 meters to wait for tidal assistant to safely enter the port area.

Demand plays an important role in determining the combination of vessel size and dredging depth. The 2021 scenario result shows that the dredging depth will increase by 0.1 meters from the 2011 and 2016 scenarios. These results imply that in the future Belawan port may require dredging above 1.9 meters if the demand continues to grow. To sum up, the feasibility of the dredging activity relies significantly on the growth of the demand.

6. References

[1] Asariotis R, Benamara H and Moho-Naray V 2017 *UNCTAD Research Paper No. 18: Port Industry Survey on Climate Change Impacts and Adaptation* vol 18

[2] Kementerian Perencanaan Pembangunan Nasional Republik Indonesia/Badan Perencanaan Pembangunan Nasional Republik Indonesia 2019 *Kajian Policy Paper Potensi Kerjasama Pemerintah dengan Badan Usaha Terkait Operasional Kapal Pada Jalur Tol Laut*

[3] Notteboom T E 2006 The Time factor in liner shipping services *Maritime Economics and Logistics* **8** 19–39

[4] Vernimmen B, Dullaert W and Engelen S 2007 Schedule unreliability in liner shipping: Origins and consequences for the hinterland supply chain *Maritime Economics and Logistics* **9** 193–213

[5] Ng A K Y, Andrews J, Babb D, Lin Y and Becker A 2018 Implications of climate change for shipping: Opening the Arctic seas *Wiley Interdisciplinary Reviews: Climate Change* **9**

[6] Koetse M J and Rietveld P 2009 The impact of climate change and weather on transport: An overview of empirical findings *Transportation Research Part D: Transport and Environment* **14** 205–21

[7] Yang Z, Ng A K Y, Lee P T W, Wang T, Qu Z, Sanchez Rodrigues V, Pettit S, Harris I, Zhang D and Lau Y yip 2018 Risk and cost evaluation of port adaptation measures to climate change impacts *Transportation Research Part D: Transport and Environment* **61** 444–58

[8] Dadashi A, Dulebenets M A, Golias M M and Sheikholeslami A 2017 A novel continuous berth scheduling model at multiple marine container terminals with tidal considerations *Maritime Business Review* **2** 142–57

[9] Menteri Perhubungan Republik Indonesia 2012 *Peraturan Menteri Perhubungan Republik Indonesia Nomor PM 21 Tahun 2012: Rencana Induk Pelabuhan Belawan* (Jakarta)

[10] Chen P, Fu Z, Lim A and Rodrigues B 2004 Port yard storage optimization *IEEE Transactions on Automation Science and Engineering* **1** 26–37

[11] Dong J X and Song D P 2009 Container fleet sizing and empty repositioning in liner shipping systems *Transportation Research Part E: Logistics and Transportation Review* **45** 860–77

[12] Song D P, Li D and Drake P 2015 Multi-objective optimization for planning liner shipping service with uncertain port times *Transportation Research Part E: Logistics and Transportation Review* **84** 1–22

[13] Ilie A, Oprea C, Olteanu S, Dinu O and Ruscă F 2019 An Heuristic Model for Port Optimization *Procedia Manufacturing* vol 32 (Elsevier B.V.) pp 975–82

[14] Kementerian Perhubungan Republik Indonesia 2019 *Peraturan Menteri Perhubungan Laut Nomor 55 Tahun 2019 tentang Komponen Biaya dan Pendapatan yang Diperhitungkan dalam Kegiatan Pelayanan Publik Kapal Perintis*

[15] MarineTraffic 2021 Live Map Voyage Information

[16] Kementerian Perhubungan Republik Indonesia 2010 *Peraturan Menteri Perhubungan Laut Nomor 70 Tahun 2010 tentang Standar Biaya Tahun 2011 di Lingkungan Kementerian Perhubungan*

[17] Kementerian Perhubungan Republik Indonesia 2021 Layanan Pengadaan Secara Elektronik

[18] Falk R F 2014 *A Primer for Soft Modeling Partial Least Squares Path Modeling View project Effects of Intellectual Ability and Personality on Self-Concepts of Gifted Children using PLS View project*

The effect of land subsidence on hydrodynamic conditions in Pekalongan coast

R J Martin[1*], A Kurniawan[2], E O Nugroho[3], F S Wijaya[1], I F Muhammad[1], Yola[1], and N Fauzi[1]

[1] Ocean Engineering Master Program, Faculty of Civil and Environmental Engineering, Bandung Institute of Technology, Bandung 40132, Republic of Indonesia
[2] Coastal Engineering Research Group, Faculty of Civil and Environmental Engineering, Bandung Institute of Technology, Bandung 40132, Republic of Indonesia
[3] Water Resources Engineering Research Group, Faculty of Civil and Environmental Engineering, Bandung Institute of Technology, Bandung 40132, Republic of Indonesia

*reinard.martin01@gmail.com

Abstract. Due to land subsidence at Pekalongan, some areas at Pekalongan coast submerged at a rate up to 10 cm/year. This study was conducted to examine the effect of land subsidence on the hydrodynamic conditions in Pekalongan through a numerical model simulation using Delft3D. There are 2 scenarios for the numerical model, scenario 1 in 2016 and scenario 2 in 2021 after the occurrence of land subsidence at a rate of 10 cm/year. Model calibration was carried out on water level elevation and current velocity by comparing the simulation data and field data based on the variation of the Chezy number. The calibrated Chezy number parameter was 60 $m^{0.5}/s$. The sensitivity test is carried out by changing the input parameters such as water level elevation, river discharge, and wave height. The sensitivity test results shows that the most sensitive parameter is wave height. The comparison of the simulation results of the two scenarios shows that the current velocity decreased by about 0.05 m/s to 0.1 m/s and the wave height increased by about 0.05 meters to 0.15 meters after land subsidence occurred. Land subsidence causes the waters in the study area to become deeper, and the results of the study show that current velocity tends to decrease and wave height tends to increase because of the land subsidence.
Keywords: *numerical model, hydrodynamic, land subsidence, Delft3D, Pekalongan*

1. Introduction

Tidal flooding is a natural phenomenon and has been a concern for coastal countries and archipelagic countries [1]. In Indonesia, tidal flooding has occurred in several cities on the northern coast of Java, such as Jakarta, Pekalongan, Semarang, and Demak [2]. Pekalongan area has experienced tidal flooding since 2008 thus disrupting the lives of people living in the coastal areas of Pekalongan City [3]. The most widely inundated lands were residential lands with an area of 55.50% of the total land area, and irrigated rice fields with an area of 32.81% of the total land area [4]. In response to this problem, Pekalongan City government built a sea dike with length of 7.2 kilometer and height of 3 meter to prevent tidal flooding on Pekalongan City [5]. Then in 2009, Pekalongan City government allocated funds for the construction of embankments in areas prone to tidal flooding such as Pekalongan Regency.

Pekalongan Regency is a coastal area that is also experiencing severe tidal flooding problems. These tidal flooding problems have damaged the area where residents live and caused it to be flooded. One of the hypotheses regarding the cause of this tidal flood is land subsidence [6]. It is known that in Pekalongan, wells are used to fulfil water needs. The use of these wells can accelerate the rate of groundwater subsidence, thus causing land subsidence [7]. This land subsidence is thought to be one of the factors causing tidal flooding in Pekalongan Regency. Therefore, it is necessary to conduct a study that examines the effect of land subsidence on the hydrodynamic conditions in Pekalongan coast.

The study on the effect of land subsidence on the hydrodynamic conditions in Pekalongan coast was carried out using a numerical modelling method. Numerical modelling is done using Delft3D software with Flow module for hydrodynamic simulation [8], and Wave module for wave transformation simulation [9]. The data used in the modelling are shoreline data, bathymetry data, tidal data, wind data, and wave data.

2. Methodology

The study was conducted in the Pekalongan coastal area, which is around the Mangrove Park Pekalongan as shown in Figure 1.

Figure 1. Study location.

The methodology is structured in systematic stages starting from the beginning of the study to the end of the study which ultimately results in answering the research objectives. The study workflow is given in Figure 2.

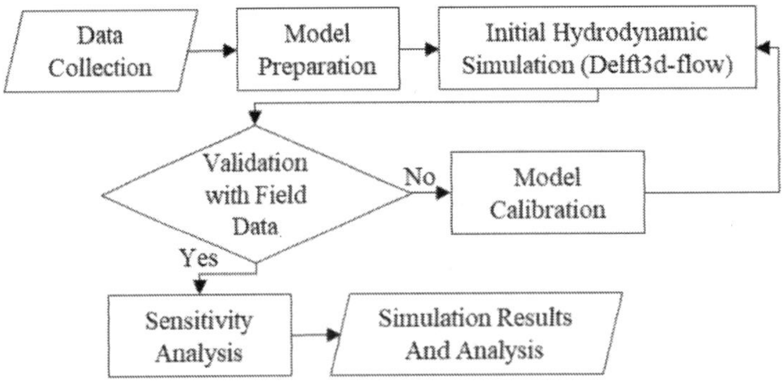

Figure 2. Study workflow.

2.1. Data Collection

The study started with environmental data collection. Environmental data collection consists of primary data and secondary data. Primary data was obtained from a field survey in the coastal area of Pekalongan by the FCEE ITB (Faculty of Civil and Environmental Engineering Institute Technology of Bandung) team in 2021 on several area points as shown in Figure 3. Field survey was carried out at several points in the river area for water level observation, bathymetry survey, and current observation; and beach area for water and sediment sample. The survey plan in Figure 3 can be distinguished by 3 different color points, namely red point for rivers, green point for beaches, and black point for current and wave measurement using ADCP. Primary data collected includes bathymetry, tidal elevation, wind speed, wave, current velocity, and sediment (TSS). Secondary data is obtained through satellite imagery and open source data. Secondary data collected includes shoreline from Google Earth, bathymetry from GEBCO, tidal elevation from TPXO 7.2, wind speed from Era5, wave from Era5, and current velocity from TIPS NOVELTIS.

Figure 3. Location and observation points of field survey.

2.1.1. Bathymetry Map. The bathymetric survey was carried out on April 6 - 9, 2021. The survey results showed that the depth of Pekalongan waters reached 9 meters with LWS datum. Bathymetric data from the survey was combined with bathymetric data from GEBCO for a depth of 10 meters to complete the data. The slope of the Pekalongan coastal profile is 0.005 - 0.016.

2.1.2. Tidal Data. Observations of water level are carried out every hour on April 6-10, 2021, at observation points MA02, MA07, MA08, MA10, and MA12. In Figure 4 it can be seen that the water level at each observation point has a similar pattern, but there is a gap in water level of 0.2 - 0.4 meters at different observation points. This indicates the occurrence of land subsidence.

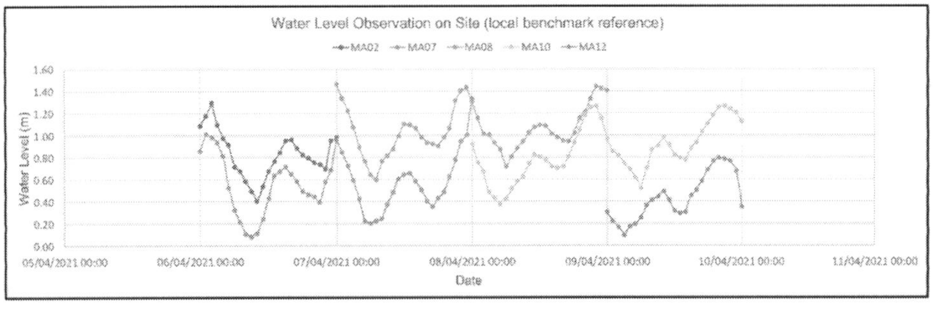

Figure 4. Water level observation.

2.1.3. Current Observation. Observations of sea currents were carried out at 10 minute intervals on 6 - 9 April 2021 at the ADCP observation point as shown in Figure 3. Observations were carried out using the ADCP Argonaut-XR. From the observations, it is known that the speed of ocean currents ranges from 0 - 0.6 m/s.

2.1.4. Wave Observation. Observations of ocean waves were carried out at 10 minute intervals on April 6, 2021, 15:00 WIB to April 09, 2021, 17:10 WIB at the ADCP observation point as shown in Figure 3. Observations were carried out using the ADCP Argonaut-XR. From the observations, it is known that the average sea wave height is 0.7 meters with a maximum value of 1.1 meters and a minimum value of 0.5 meters. The wave period ranges from 4 - 7 seconds.

In addition to surveys, wave data is also obtained from Era5 data. When Era5 wave data in April 2011 with survey wave data in April 2021 were compared, there is a difference of 0.5 meters in the average wave height value. This indicates the occurrence of climate change [10].

2.1.5. Sediment type and TSS. From the results of sediment observations and grain size analysis it is known that the coastal sediments of Pekalongan consist of sand, stone fragments, and gravel with the average d50 value of 0.187 mm. While the average d50 value of the river is 0.27 mm with some silty clay, mostly sand, stone fragments, and gravel.

2.2. Model Preparation
To perform modelling in Delft3D software, a grid or guideline is required. The grid was created at Delft3D-RGFGRID, using shoreline data as the land boundary. Then the depth data is inputted using the Delft3D-QUICKIN module, where triangular interpolation is also performed on the module to interpolate the depth data.

2.3. Simulation
In the early stages of modelling, tides are used as boundary conditions with the Delft3D – Flow module. The result of this stage is a simulation of the pattern of the direction and speed of the current as well as changes in tidal elevation in Pekalongan with variations of bottom roughness. This bottom roughness variation is identified by 5 different Chezy Number values, namely 55 $m^{0.5}/s$, 60 $m^{0.5}/s$, 65 $m^{0.5}/s$, 70 $m^{0.5}/s$, and 75 $m^{0.5}/s$. The modelling results will be calibrated with field data in Pekalongan. The data will be calibrated using the Root Mean Square (RMSE) method, which is the calculation of the smallest error value (difference between the two data). The data used for calibration is tidal data and current velocity, and by considering the variation of the Chezy Number, the smallest RMSE value will be determined. Then sensitivity test is carried out by making changes to the input parameters such as water level elevation and wave height then seeing the results of the current velocity due to these changes. After that, numerical modelling will be carried out for both scenarios in 2016 and 2021.

3. Numerical Model
In this study, Delft3D software will be used for numerical modelling. Delft3D is a software that can be used for modelling hydrodynamics, sediment transport, coastal morphology, and water quality for aquatic environments [8, 9].

3.1. Model Configuration
The domain covered the whole nearshore area of Mangrove Park Pekalongan, with a grid resolution of 21 – 460 meter using the rectangular grid. Bathymetric data, specifically at the nearshore area at depth 9 meter, was taken from the hydrography survey in April 2021, while the offshore site at depth of 10 meter and deeper was taken from GEBCO. The computational grid, bathymetry of the model area, and the bathymetry survey plan are shown in Figure 5.

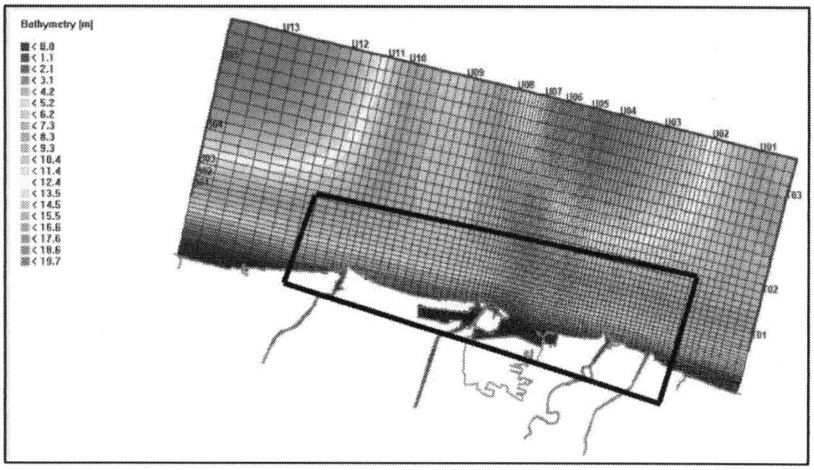

Figure 5. Computational grid, bathymetry of the model area, and bathymetry survey plan (on the black lines).

There are 4 open boundaries i.e., Eastern, Northern, Western and Southern as can be seen in Figure 5. The water level was prescribed along the open boundary for the model for flow computation. The Eastern, Northern and Western boundaries were generated using global tide model from Delft3D Dashboard. For sediment concentration at the open boundary, it was assumed to be 0.0001 kg m^{-3} (sea water). The southern boundary consists of rivers. The river discharge was input based on the observed data from the Department of Public Work of Indonesia. The concentration was generated uniformly from survey data in 2021 of 1.3 kg m^{-3}.

3.2. Model Calibration

Model calibration is done by comparing the modelling results with the field and calculating the error value using the RMSE method with variations of 5 chezy numbers, namely 55 $m^{0.5}$/s, 60 $m^{0.5}$/s, 65 $m^{0.5}$/s, 70 $m^{0.5}$/s, and 75 $m^{0.5}$/s. From the results of comparison between water level of the field data with water level of modelling results as shown in Figure 6 and Table 1, it was found that the chezy number 60 $m^{0.5}$/s had the smallest RMSE value, amounting to 8.72%.

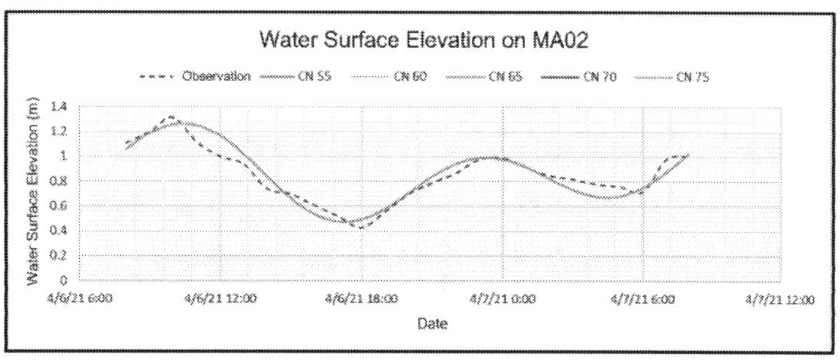

a)

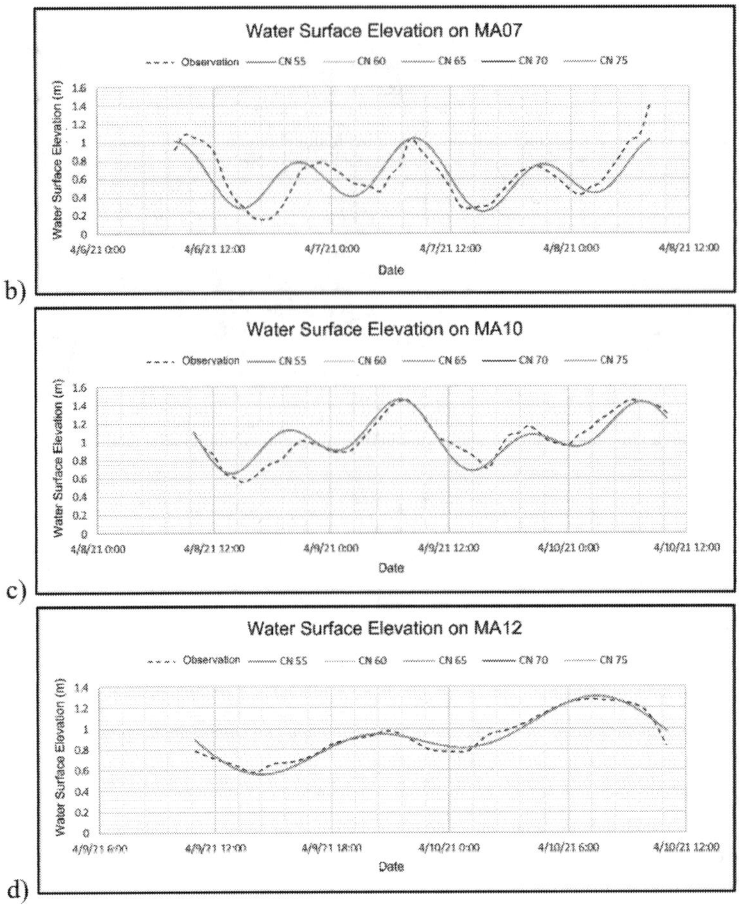

Figure 6. Time series of Water Surface Elevation Comparison between the Data and the Model for a) MA02, b) MA07, c) MA10, d) MA12

Table 1. Comparison of RMSE water level value.

Chezy Number ($m^{0.5}/s$)	RMSE WATER LEVEL (%)					
	ADCP	MA02	MA07	MA10	MA12	Average
55	3,659	7,173	16,792	10,818	5,198	8,728
60	3,665	7,137	16,819	10,794	5,182	**8,719**
65	3,663	7,269	16,847	10,825	5,177	8,756
70	3,674	7,243	16,850	10,772	5,127	8,733
75	3,674	7,088	16,860	10,843	5,202	8,733

From the comparison of the field data flow velocity with the modelling results as shown in Table 2, it was found that the chezy number 65 $m^{0.5}/s$ had the smallest RMSE value, of 11.402%. So, for further modelling, chezy number 60 $m^{0.5}/s$ will be used from the results of the calibration of the water level.

Table 2. Comparison of RMSE current velocity value.

Chezy Number ($m^{0.5}/s$)	RMSE Current Velocity (%)
55	11,664
60	11,603
65	**11,402**
70	11,541
75	11,538

3.3. Sensitivity Analysis

The results of the hydrodynamic numerical modelling will then be verified by performing the sensitivity analysis of a specific, particularly on their effects on coastal flow pattern. The model output suggests that the wave parameter is sensitive. It can be seen in Figure 7 where there is comparison on the result of current velocity modelling with variation of wave height.

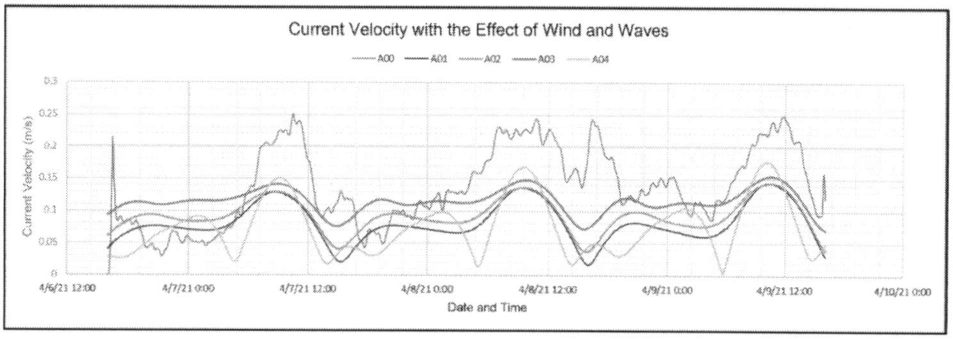

Figure 7. Result of current velocity modelling with variation of wave height.

Information

- A00: Current velocity measurement result
- A01: Current velocity with significant wave height 1.216 m and period 6 s
- A02: Current velocity with significant wave height 1.5 m and period 6 s
- A03: Current velocity with significant wave height 2 m and period 6 s
- A04: Current velocity with wave time varying original

Therefore, it can be deemed consistent with previous findings that the climate changes affect the hydrodynamic condition in the area of interest.

3.4. Scenarios

In this study, simulations were carried out for 2 scenarios, namely before the occurrence of land subsidence and after the occurrence of land subsidence.

The scenario after the land subsidence occurs is taken in the existing condition, namely in 2021. The scenario before the land subsidence occurred was taken in 2016, where the bathymetry of the existing condition is added by 50 cm around the Mangrove Park Pekalongan as shown in Figure 8.

Figure 8. Area of 0.5 meter increment in bathymetry for scenario before land subsidence occurs.

4. Result and Analysis

To investigate the effect of land subsidence, the spatial distribution of current velocity in the area of interest is analysed. It is noted that the analysis is carried out at the maximum current velocity during falling and rising tide on the Spring Cycle as can be seen in Figure 9 and Figure 10.

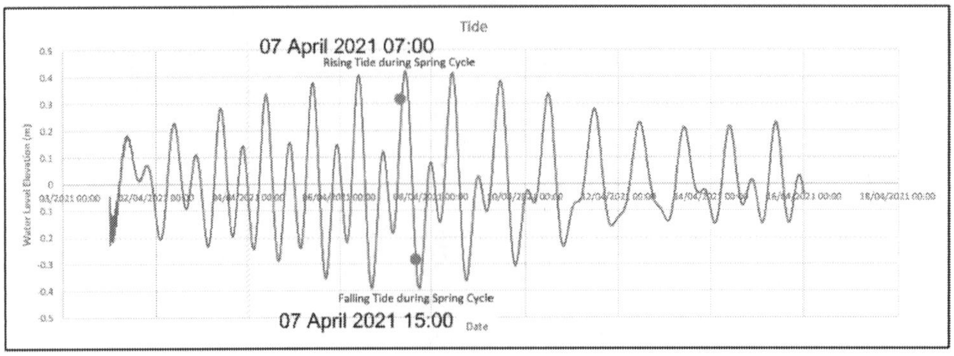

Figure 9. Rising and falling tide during spring cycle.

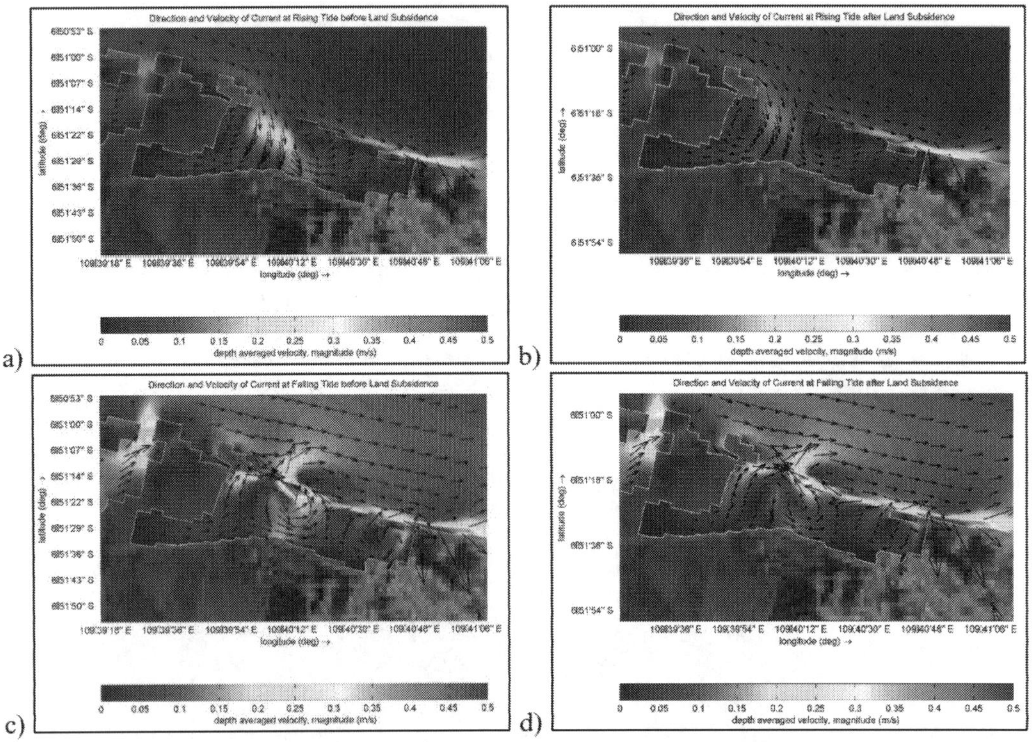

Figure 10. Spatial distribution of maximum current speed showing at **a)** rising tide during Spring Cycle before land subsidence; **b)** rising tide during Spring Cycle after land subsidence; **c)** falling tide during Spring Cycle before land subsidence; **d)** falling tide during Spring Cycle after land subsidence.

Based on Figure 10, there is a trend where the current velocity is getting lower after land subsidence. This can be explained as follows; Land subsidence causes the waters in the study area to become deeper, and it is known that the deeper the water, the lower the current velocity. This can be seen by visualizing cross sections as shown in Figure 12 and Figure 13, based on transections in Figure 11.

IOP Conf. Series: Earth and Environmental Science 1065 (2022) 012009 doi:10.1088/1755-1315/1065/1/012009

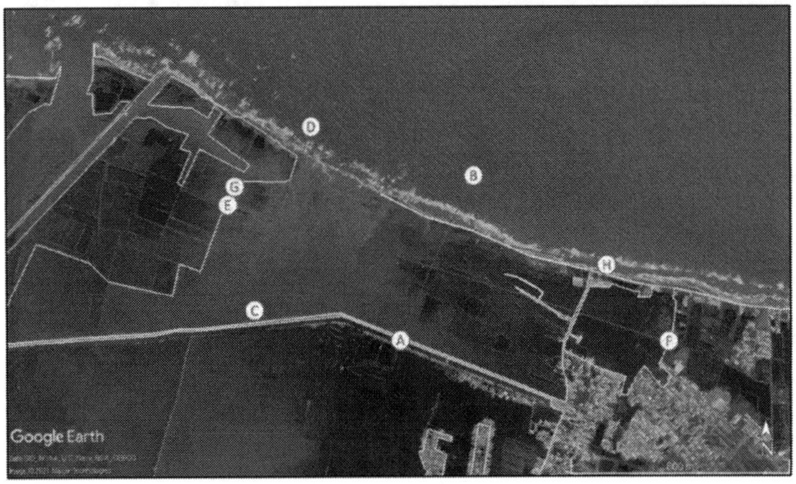

Figure 11. Transect for cross-section outputs.

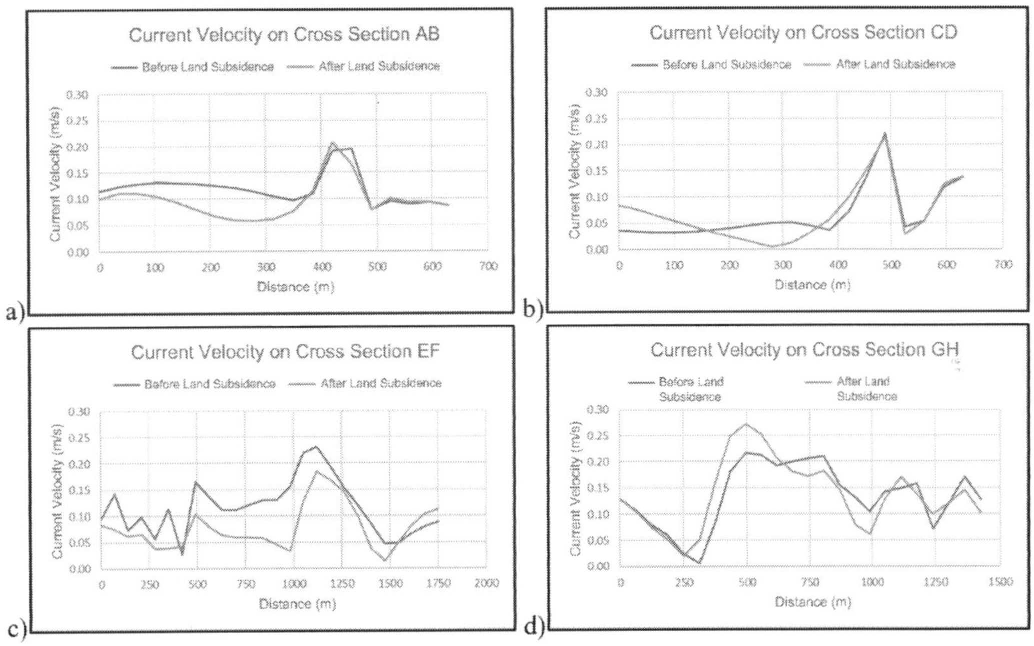

Figure 12. Current speed cross section profile for after and before land subsidence showing at **a)** AB; **b)** CD; **c)** EF; **d)** GH; based on transections in Figure 10.

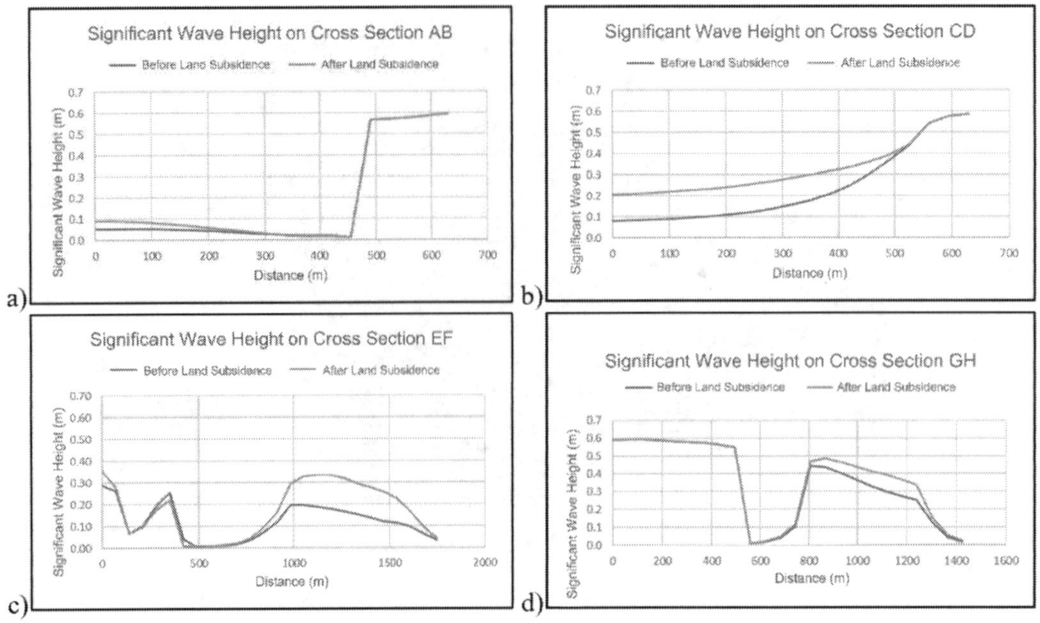

Figure 13. Significant Wave Height cross section profile for after and before land subsidence showing at **a)** AB; **b)** CD; **c)** EF; **d)** GH; based on transections in Figure 10

From Figure 12 and Figure 13, the average current velocity is lower, and the average significant wave height is greater after land subsidence, compared to before land subsidence. This happens because Pekalongan waters are getting deeper because of land subsidence. The current velocity decreased by around 0.05 m/s to 0.1 m/s and the wave height increased by around 0.05 meters to 0.15 meters after land subsidence occurred.

5. Conclusion

From the visualization of the spatial domain and the visualization of cross-sections and longitudinal sections for current velocity, a trend is seen where after land subsidence, the value of current velocity gets smaller. This is in accordance with the theory that the shallower the water, the greater the speed of the current. Vice versa, where the deeper the water, the smaller the current speed. The difference in current velocity before and after land subsidence is about 0.05 m/s to 0.1 m/s.

From the cross-sectional and longitudinal visualization for wave height, it can be seen that the average wave height after land subsidence is greater than before land subsidence. This is because the waves can propagate further in areas with deeper water depths, so the wave height seen in the section is larger after land subsidence than before land subsidence. The significant difference in wave height before and after land subsidence is about 0.05 meters to 0.15 meters.

The water level elevation does not show a significant effect due to land subsidence. For each cross section, the water level elevation after land subsidence is lower than before land subsidence, because the lower the Pekalongan land is, the higher the water level elevation is determined. The difference in water level before and after land subsidence is about 0.005 meters to 0.01 meters.

Climate change impact at Pekalongan is indicated by increased significant wave height between secondary data (ECMWF) and wave observation survey. This is relevant to what Young and Ribal (2019) explained where climate change can cause small increases in significant wave height with larger increases in extreme conditions that affected to the magnitude of coastal erosion, overtopping and coastal flooding.

Acknowledgements
This work was funded by the Indonesia Collaboration Research Program grant from Bandung Institute of Technology that was initiated by Ministry of Education and Culture Republic of Indonesia.

References

[1] D. Sutrisno, J. Pariwono, J. Rais, T. Kusumastanto, *Dampak Kenaikan Muka Laut pada Pengelolaan Delta: Studi Kasus Penggunaan Lahan Tambak di Pulau Muaraulu Delta Mahakam*, Jurnal Ilmiah Geomatika, **11(1)** (2005)

[2] M.A. Marfai, *Impact of sea level rise to coastal ecology: A case study on the northern part of Java island, Indonesia*, Quaestiones Geographicae, **33(1)** 107 – 114 (2014)

[3] Pusat Krisis Kesehatan Kementerian Kesehatan RI, 19th January 2009, *Perkembangan banjir di 6 kabupaten/kota di Prov. Jawa Tengah*, Pusat Krisis Kesehatan Kementerian Kesehatan RI, https://pusatkrisis.kemkes.go.id/perkembangan-banjir-di-6-kabupaten-kota-di-prov--jawa-tengah [in Bahasa]

[4] M.A. Marfai, D. Mardiatno, A. Cahyadi, F. Nucifera, H. Prihatno, *Pemodelan spasial bahaya banjir rob berdasarkan skenario perubahaan iklim dan dampaknya di pesisir Pekalongan*, Jurnal Bumi Lestari, **13(2)** 244 – 256 (2013)

[5] A.R. Maharlika, S.P. Hadi, Kismartini, A.L. Hoya, *Tidal flooding and coastal adaptation responses in Pekalongan city*, in Proceedings of the 6th International Conference on Energy, Environment, Epidemiology, and Information System (2020)

[6] M.B. Pratama, *Tidal Flood in Pekalongan: Utilizing and Operating Open Resources for Modelling*, in Proceedings of the 2nd EPI International Conference on Science and Engineering (2019)

[7] E. Chaussard, F. Amelung, H. Abidin, S. Hong, *Sinking cities in Indonesia: ALOS PALSAR detects rapid subsidence due to groundwater and gas extraction*, Remote Sensing of Environment, **128** 150 – 161 (2013)

[8] Deltares, *Delft3D-FLOW: Simulation of multi-dimensional hydrodynamic flows and transport phenomena, including sediments*, Netherlands (2009)

[9] Deltares, *Delft3D-WAVE: Simulation of short-crested waves with SWAN*, Netherlands (2007)

[10] A. Ribal, I.R. Young, *33 years of globally calibrated wave height and wind speed data based on altimeter observations*, Scientific Data, **6(77)** (2019)

[11] A.S. Budiman, A.F. Koropitan, I.W. Nurjaya, *Pemodelan Hidrodinamika Arus Pasang Surut Teluk Mayalibit Kabupaten Raja Ampat Provinsi Papua Barat*, **3(2)** 146 – 156 (2014)

[12] M. Fahmi, T.M. Hafli, *Simulasi Numerik Perubahan Morfologi Pantai akibat Konstruksi Jetty pada Muara Lambada Lhok Aceh Besar menggunakan Software Delft3D*, Jurnal Teknik Sipil Universitas Syiah Kuala, **8(2)** (2019)

[13] O.R. Sianturi, S. Widada, I.B. Prasetyawan, F. Novico, *Pemodelan Hidrodinamika Sederhana berdasarkan Data Hidro-Oseanografi Lapangan di Teluk Lampung*, Jurnal Oseanografi, **2(3)** 299 – 309 (2013)

[14] H. Susiati, B.A. Subki, H. Ajiwibowo, *Hidrodinamika dan Transport Sedimen di Tapak PLTN Semenanjung Muria*, in Proceedings of the Seminar Keselamatan Nuklir (2011)

[15] A. Syarifudin, M.S. Imanuddin, B. Simanjuntak, *Model Hidrodinamika dan Sediment Transport pada Muara Sungai Musi Palembang, Sumatera Selatan*, in Proceedings of the International Commision on Irrigation and Drainage (2016)

[16] I. Syahputra, H. Yuliana, *Pemodelan Kecepatan Arus dan Tinggi Gelombang pada Rencana Breakwater Kolam Pelabuhan Meulaboh dengan menggunakan Program Delft3D*, Jurnal Teknik Sipil Unaya, **5(2)** 59 – 67 (2019)

[17] K.B. Widhi, E. Indrayanti, I.B. Prasetyawan, *Kajian Pola Arus di Perairan Teluk Lampung menggunakan pendekatan Model Hidrodinamika 2-Dimensi Delft3D*, Journal of Oceanography Universitas Diponegoro, **1(2)** 169 – 177 (2012)

Mechanical Characteristics Of Concrete With The Addition Of Integral Waterproof Using Local Aggregate East Kalimantan

Mardewi Jamal[1*], M Jazir Alkas[1], Budi Haryanto[1] and Ahmad Rasyid[1]

[1)] Civil Engineering Departent, Engineering of Faculty, University of Mulawarman, Jl.Sambaliung No.9 Kampus Gunung Kelua, Samarinda

*wie_djamal@yahoo.com

Abstract. Nowadays, there are various ways to deal with leaks in concrete. One way that is often used to deal with leaks is to provide a waterproofing or waterproofing layer on the surface of the floor or wall. Along with technological developments, research on the quality of concrete waterproofing continues to be improved. One of them is the use of Concrete Integral waterproofing, which is a mixture of concrete that is added with integral waterproofing to minimize water penetration into the concrete, thereby increasing the concrete's impermeability and protecting the concrete against leakage. The purpose of this study was to determine the effect of adding Integral waterproofing to the compressive and tensile strength of concrete and to determine the optimum level of addition of Integral waterproofing. The aggregates used are Palu coarse aggregate and Kalimantan local fine aggregate to utilize the natural resources in East Kalimantan. Addition of Integral Waterproofing into the concrete mixture with a percentage of 0%, 0.5%, 1% and 2% of the cement weight. Concrete testing was carried out at the age of 28 days by testing its mechanical properties, namely compressive strength and tensile strength. Based on research results, for concrete that uses Mahakam fine aggregate, the highest compressive strength value at 2% addition is 34.73 MPa, and the highest tensile strength value at 2% addition is 2.38 MPa. And for concrete that uses Sambera fine aggregate, the highest compressive strength value at 2% addition is 30.91 MPa, and the highest tensile strength value at 1.5% addition is 2.34 Mpa.

Keyword : Integral Waterproof, mechanical characteristic, concrete

1. Introduction

Currently, there are various ways to deal with leaks. Along with the modernization of the spatial layout of an area, the function, compressive strength, value and economic life of a construction can change quickly, maximum maintenance with the aim of getting a high compressive strength of concrete. The method that is often used to overcome this leak is to provide a waterproofing or waterproofing layer on the surface of the floor or wall. Waterproofing that is often used in the field is waterproofing coating and waterproofing membrane. The two waterproofings are applied to the surface of the structure which serves to close the pores of the concrete surface so that the concrete becomes impermeable. However, problems that often occur due to the influence of weather and time, the waterproofing layer will be damaged so that water can still pass through the cracks in the damaged part and leaks occur again. Therefore, repairs to concrete should not only be carried out on the outside, but also must be carried out from within the concrete itself.

Along with technological developments, research on the quality of concrete waterproofing continues to be improved. One of them is the use of Concrete Integral waterproofing, which is a mixture of concrete that is added with integral waterproofing to minimize water penetration into the concrete, thereby increasing the concrete's impermeability and protecting the concrete against leakage. Integral waterproofing was created on the basis of the nature of concrete which under normal conditions is still able to be passed by water so that in the manufacture of concrete an additional material is needed into the concrete mix which can make the concrete more watertight.

The purpose of this study was to determine the effect of adding Integral waterproofing to the compressive and tensile strength of concrete and to determine the optimum level of addition of Integral waterproofing. The fine aggregate used is local East Kalimantan aggregate, namely Mahakam fine aggregate and Sambera fine aggregate. The use of integral waterproofing is 0%, 0.5%, 1%, 1.5 and 2% of the cement weight.

2. Concrete

Concrete is a construction material that is commonly used for buildings, bridges, roads, and others. Concrete is a homogeneous unit. This concrete is obtained by mixing fine aggregate (sand), coarse aggregate (gravel), or other types of aggregate and water, with Portland cement or with other hydraulic cements, sometimes with the addition of additives that are chemical or physical in a certain ratio, until it becomes a homogeneous whole. The mixture will harden like rock. Hardening occurs due to a chemical reaction between cement and water.

Making concrete is actually simple, just mixing the basic ingredients to form a plastic mixture as we often see in the manufacture of simple buildings. But if you want to make good concrete, in the sense of meeting more stringent requirements due to higher demands, it must be carefully considered. ways to obtain fresh concrete mix and obtain good hard concrete. Good fresh concrete is fresh concrete that can be mixed, can be transported, can be poured, can be compacted, there is no tendency for the separation of gravel from the mortar or separation of water and cement from the mixture. mortar. Good hard concrete is concrete that is strong, durable, water-resistant, wear-resistant, and has little shrinkage [12].

2.1 Mechanical Characteristic of Concrete

2.1.1. Compressive Strength of Concrete

According to SNI 03-1974-1990, the compressive strength of concrete can be a criterion or parameter for the resistance of concrete. The compressive strength of concrete is the ability of concrete to withstand or receive a compressive axial force per unit area which can be written in the formula:

$$Fc = \frac{P}{A} \tag{1}$$

Where : Fc = compressive strength of concrete (Megapascal), P = Maximum load (kg), and A = Cross-sectional area of the test object (cm^2)

2.1.2 Tensile Strength Of Concrete

According to SNI 2491-2014, to evaluate the shear resistance of concrete and to determine the distribution length of the reinforcement. In addition, the split tensile strength is generally greater than the direct tensile strength and lower than the flexural strength (modulus of repture). The magnitude of the split tensile strength of concrete is calculated using the following equation:

$$Fct = \frac{2P}{\pi LD} \tag{2}$$

Where : Fct = Tensile Strength Of Concrete (Mpa), P = Maximum load (kg), L = Cylindrical test object length (mm) and Diameter of cylindrical specimen (mm)

2.2. Waterproofing

Waterproofing is a chemical compound used to reduce water absorption and permeability in concrete (Nugraha, Paul, et al. 2007). According to Neville (1996) water-resistant additives have several actions but the most prominent effect is to make the concrete hydrophobic. Furthermore, according to Hewlet (1996) stated that these chemical compounds can reduce the water absorption of normal concrete and make the concrete more impermeable to water. The composition of waterproofing admixtures usually consists of silicate, blast furnace slag, pozzolan, bettonite, talc. Water-reppelent can reduce the effect of capillary forces and therefore reduce the degree of saturation of the concrete. Water-reppelent agents are calcium stearate, butyl stearate, other soaps, oil emulsions, and so on [8].

2.2.1 Types of Waterproofing

Waterproofing is commonly known as a layer that has resistance to water seepage on the roof and not concrete, Steward Treatment Plant (STP), Ground Water Tank (GWT), basement walls and others. Types of waterproofing are as follows :

a. *Waterproofing membrane,* _is a type of waterproofing that uses sheets made of chemical monomers, ethylene, propylene mixed with rubber. Usually used for dak structures or concrete roofs. The membrane sheets are arranged according to the requirements of the entire surface of the protected structure.
b. *Waterproofing Coating,* is a type of waterproofing that is used to protect the surface of walls, tubs, tanks and can also be used for wooden surfaces.
c. *Waterproofing integral,* is a modification in the field of waterproofer. Integral waterproofing was developed because of the nature of concrete which cannot withstand water infiltration or pressure so that additional useful materials are needed to make concrete more watertight or impermeable to water. The types of integral fluids are as follows :

1. An integral liquid that can increase the slump of concrete and contains a plasticiser that has a hydrophobic system or substances that are insoluble in water so that the concrete becomes denser and watertight.
2. Integral Crystallization is a hydrophilic system that uses a high-quality added mixture and has easy application of concrete and has the function of waterproofing and protecting concrete from corrosion [6].

3. Methodology

3.1. Time and Location of Research

The time of this study began in September 2021 until December 2021. The place of compressive strength research and testing was carried out at the Civil Engineering Laboratory, Faculty of Engineering, University of Mulawarman Samarinda.

3.2. Research Stages

The first stage is testing the materials of coarse aggregate, fine aggregate, and cement. After the material testing data is obtained, it is continued with the manufacture of Concrete Mix Design which refers to SNI 03-2834-2000, material preparation, concrete material mixer, slump inspection, printing of test objects and let stand for a day. soaked in the pool for 28 days. The next stage is testing the compressive strength and split tensile strength of the concrete then analysis and discussion. The materials used are:

1. The cement used is Portland cement type 1 which functions as a binder for other filler materials.
2. There are 2 fine aggregates used in this study, namely Sambera and Mahakam . fine aggregates
3. The coarse aggregate used is Palu aggregate
4. Damdex is the admixture material used in this study.

3.3. Specimen

The test object used was cylindrical with a diameter of 15 cm, and a height of 30 cm as many as 60 specimens divided into 2 types of specimens, 30 specimens using sambera fine aggregate and 30 specimens using Mahakam fine aggregate. The integral waterproofing used is the "Damdex" brand. The proportions of the integral waterproofing mixture are 0%, 0.5%, 1%, 1.5%, and 2% by weight of cement. The tests carried out in this research are compressive strength testing based on SNI 03-1974-1990 and split tensile strength testing of concrete based on SNI 03-2491-2002. The compressive and tensile strength tests were carried out after 28 days.

4 Results And Discussion

4.1. Concrete Compressive Strength

4.1.1. Compressive Strength Of Concrete Using Fine Aggregate of Mahakam

The results of the compressive strength test of concrete using Mahakam sand fine aggregate can be seen in table 1 and figure 1 below.

Tabel 1. Compressive Strength Of Concrete Using Fine Aggregate of Mahakam

Speciment	Compressive Strength (MPa)	Average (MPa)
Normal Concrete	35,43	38,64
	39,46	
	41,02	
Concrete WP 0,5 %	35,37	34,57
	36,80	
	31,55	
Concrete WP 1 %	32,98	33,50
	31,07	
	36,46	
Concrete WP 1,5 %	33,25	34,39
	31,48	
	38,43	
Concrete WP 2 %	36,59	34,73
	30,80	
	36,80	

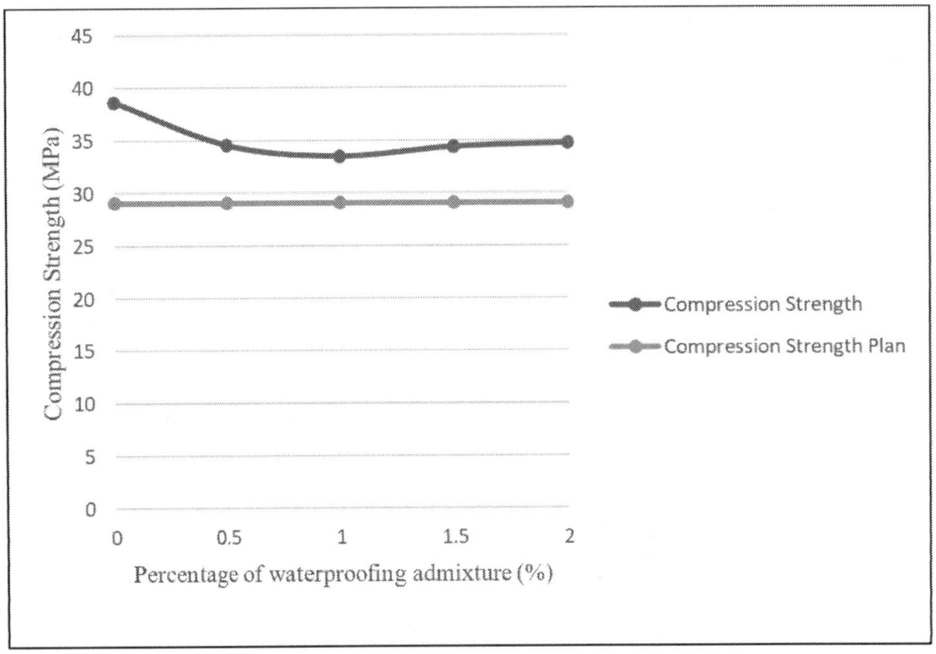

Figure 1. Value of Concrete Compressive Strength for Several Variations of Waterproofing Using Fine Aggregate of Mahakam

Based on the table and graph above, the compressive strength values for all percentages of integral water proofing exceed the design strength, and the highest compressive strength value for adding waterproofing using Mahakam fine aggregate is in the addition of 2%, which is 34.73 Mpa.

4.1.2. Compressive Strength Of Concrete Using Fine Aggregate of Sambera

The results of the compressive strength test of concrete using Sambera sand fine aggregate can be seen in table 2 and graph 2 below.

Tabel 2. Compressive Strength Of Concrete Using Fine Aggregate of Sambera

Speciment	Compressive Strength (MPa)	Average (MPa)
Normal Concrete	30,87	31,30
	29,78	
	33,25	
Concrete WP 0,5 %	30,26	29,55
	29,03	
	29,37	
Concrete WP 1 %	21,67	27,44
	31,69	
	28,96	
Concrete WP 1,5 %	33,73	30,82
	29,92	
	28,82	
Concrete WP 2 %	32,78	30,91
	31,14	
	28,82	

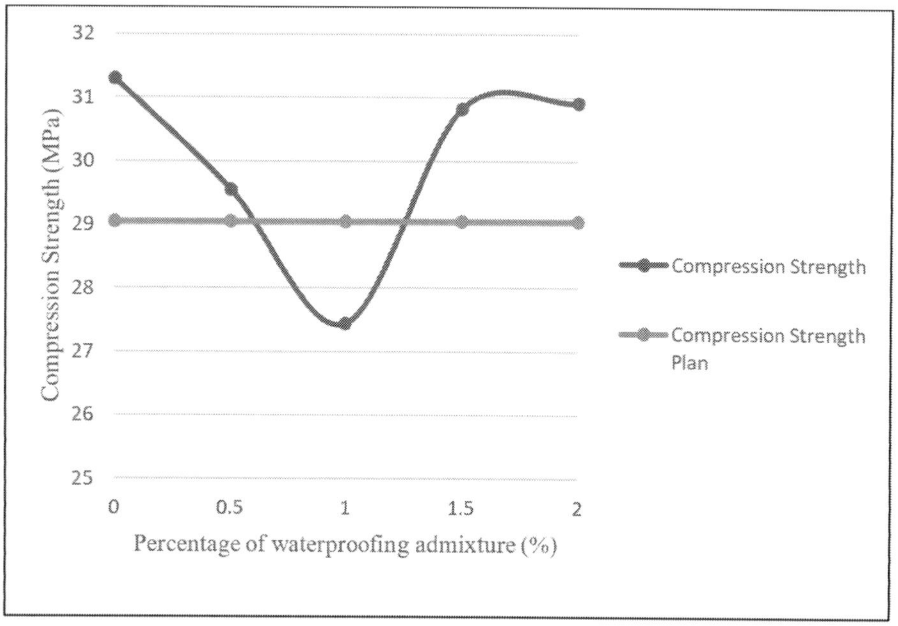

Figure 2. Value of Concrete Compressive Strength for Several Variations of Waterproofing Using Fine Aggregate of Sambera

Based on the table and graph above, the compressive strength value of the concrete added with integral water proofing exceeds the design strength except for the percentage of 1%. The decrease in

IOP Conf. Series: Earth and Environmental Science 1065 (2022) 012010 doi:10.1088/1755-1315/1065/1/012010

the average compressive strength value at a level of 1% is caused because one specimen has a very low value compared to another, it can occur due to less than optimal mixing or poor compaction. this affects the average value at the level of 1%.

The highest compressive strength value for the addition of waterproofing using Mahakam fine aggregate is the addition of 2%, which is 30.91 MPa. The compressive strength value for Sambera fine aggregate is lower than for Mahakam fine aggregate.

From the results of the compressive strength of concrete for fine aggregates using Mahakam sand or Sambera sand, the optimum level of addition of Integral Waterproofing is 2%. And it can be stated that the concrete using local aggregates from East Kalimantan is suitable for use in the manufacture of waterproofing admixture concrete with a design compressive strength of f'c 29.05.

4.2. Split Tensile Strength

4.2.1. Tensile Strength Of Concrete Using Fine Aggregate of Mahakam

Table 3. Tensile Strength Of Concrete Using Fine Aggregate of Mahakam

Speciment	Tensile Strength (MPa)	Average (MPa)
Normal Concrete	2.28	2.44
	2.35	
	2.70	
Concrete WP 0,5 %	2.18	2.17
	2.18	
	2.15	
Concrete WP 1 %	2.87	2.29
	1.90	
	2.11	
Concrete WP 1,5 %	2.43	2.24
	2.21	
	2.36	
Concrete WP 2 %	2.15	2.38
	2.28	
	2.72	

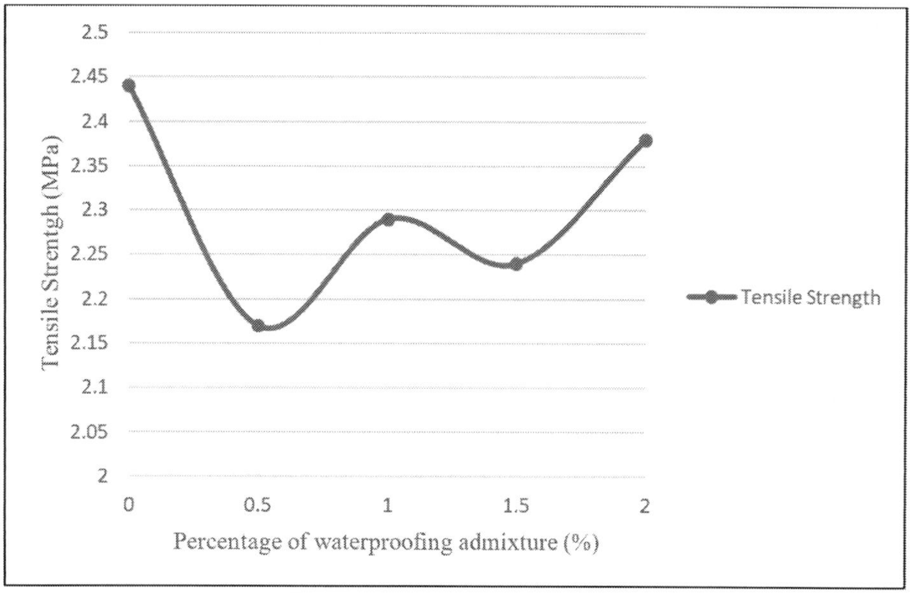

Figure 3. Value of Concrete Tensile Strength for Several Variations of Waterproofing Using Fine Aggregate of Mahakam

From table and graph above, the highest tensile strength value of concrete with the addition of integral water proofing using Mahakam fine aggregate is the addition of 2% with a value of 2.38 MPa. The results of the tensile strength of concrete do not show an increasing graph as in the compressive strength value. this can happen because the test does not use a tensile strength machine but by using a modified compressive strength machine. However, the maximum compressive strength value is the same as the compressive strength value, which is 2%.

4.2.2. Tensile Strength Of Concrete Using Fine Aggregate of Ssambera

Table 4. Tensile Strength Of Concrete Using Fine Aggregate of Sambera

Speciment	Tensile Strength (MPa)	Average (MPa)
Normal Concrete	2.75	2.34
	2.14	
	2.12	
Concrete WP 0,5 %	1.97	1.96
	2.01	
	1.94	
Concrete WP 1 %	1.92	1.99
	2.04	
	2.01	
Concrete WP 1,5 %	2.42	2.34
	2.38	
	2.21	
Concrete WP 2 %	1.83	1.84
	1.90	
	1.81	

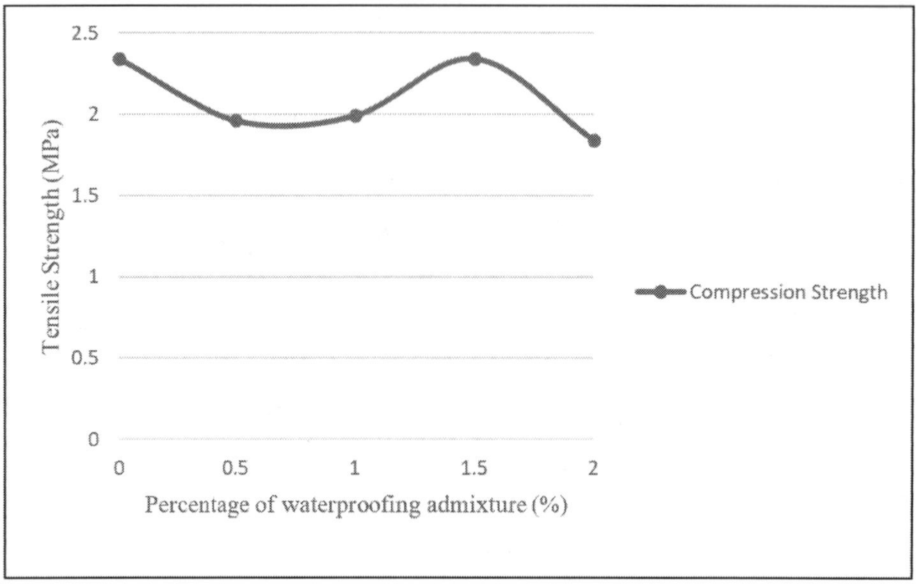

Figure 4. Value of Concrete Tensile Strength for Several Variations of Waterproofing Using Fine Aggregate of Sambera

From the table and graph above, the highest tensile strength value of concrete with the addition of integral water proofing using Sambera fine aggregate is the addition of 1.5% with a value of 2.34 MPa. In contrast to mahakam sand, the graph shown on the sambera sand shows the optimum value at a level of 1.5%, also has a maximum value that is different from the compressive strength value which has a maximum value in the addition of 2%.

The same as the compressive strength value, the tensile strength value using Mahakam fine aggregate has a better value than the sambera fine aggregate.

5. Conclusion

Based on research results, the addition of integral waterproofing to concrete affects the value of compressive strength and tensile strength for concrete using Mahakam and Sambera fine aggregates. For concrete that uses Mahakam fine aggregate, the highest compressive strength value at 2% addition is 34.73 MPa, and the highest tensile strength value at 2% addition is 2.38 MPa. And for concrete that uses Sambera fine aggregate, the highest compressive strength value at 2% addition is 30.91 MPa, and the highest tensile strength value at 1.5% addition is 2.34 MPa. From the results, the optimum level of addition of Integral Waterproofing is 2%, and it can be stated that the concrete using local aggregates from East Kalimantan is suitable for use in the manufacture of waterproofing admixture concrete.

6. References

[1] ACI 363R-92. State of The Art Report on High Strength Concrete.

[2] Asroni A. 2010. *Reinforced Concrete Beams and Slabs* Yogyakarta: Graha Ilmu.

[3] BSN. 2002. *Standard Procedures for Calculation of Concrete Structures for Building*(SNI 03-2847-2002). Jakarta, Indonesia.

[4] BSN SNI 03-2491-2002, 2001. *Test method for split tensile strength of concrete*. Badan Standarisasi Nasional, Bandung.

[5] Dipohusodo, 1994. *Reinforced Concrete Structure*. Yogyakarta : Gramedia Pustaka Utama.

[6] I Made Jaya dkk, November 2017. *" Comparison of Compressive Strength and Split Tensile Strength Between Normal Concrete And Integral Waterproofing Concrete"*. Jurnal Logic. Vol. 17 NO. 3.

[7] Neville, AM (1996). *Properties of Concrete,* Longman. USA.

[8] Nugraha, Paul. & Antoni. (2007). *Concrete Technology from Material, Manufacture, to High Performance Concrete*. Yogyakarta : ANDI.

[9] Sinergia. 2018. Know the Benefits of Waterproofing in Building Construction. https://sinergia.id/mengenal-waterproofing

[10] SNI T-15-1990-03. *Procedures for Making Normal Concrete Mixtures*, Departemen Pekerjaan Umum, Yayasan Lembaga Penyelidikan Masalah Bangunan, Bandung.

[11] SNI 03-1974-1990. Test Method for Compressive Strength of Concrete. Badan Standarisasi Nasional (BSN).

[12] Tjokrodimuljo, K., 1996, *Concrete Technology*. Jurusan Teknik Sipil, Fakultas Teknik Universitas Gadjah Mada, Yogyakarta.

Earthquake Risk Study on Residential Buildings in West Jakarta using the Event-Based Risk Analysis Method

G Lambang-Goro[1]*, M Irsyam[2], M Asrurifak[3], I Meilano[4]

[1] Civil Engineering Department., State Polytechnic of Semarang, Semarang, Indonesia, 50275
[2] Faculty of Civil and Environmental Engineering, Bandung Institute of Technology, Bandung, Indonesia, 40132
[3] Research Centre for Disaster Mitigation, Bandung Institute of Technology, Bandung, Indonesia, 40132
[4] Faculty of Earth Sciences and Technology, Bandung Institute of Technology, Bandung, Indonesia, 40132

*Corresponding email: garupla@gmail.com

Abstract. UNISDR (United Nations Office for Disaster Reduction) states that Indonesia is in the third-highest rank after Japan and the United States for the average economic loss due to earthquakes. Previous earthquake risk studies conducted by the National Agency for Disaster Management (BNPB) in 2011 renewed in 2016 both in the form of risk indexes have yet taken into account the vulnerability function of buildings, so that economic loss has not been presented quantitatively. This research aims to analyse earthquake risk in urban areas (West Jakarta Municipality) using the Event-Based Risk Analysis method by applying several factors including: earthquake source updates, GMPE, site-specific, exposure and building vulnerability functions. Hazard and risk analyses have been conducted for the probability of exceedance (PoEs) 10% and 2% in 50 years. Moreover, risk analysis was conducted for residential buildings exposure. The results predict that the lowest total building loss due to the earthquake per district is around 8.1 million USD and the highest loss is 15.5 million USD at a PoEs 10% in 50 years. Whereas for the PoEs 2% in 50 years, the lowest total loss was 14.86 million USD whilst the highest total loss was 25 million USD.

Keywords: Event-Based Method, Loss Analysis, Residential Buildings, West Jakarta

1. Introduction

Various countries data compiled by UNISDR (United Nation Office for Disaster Reduction) stated that the highest number of fatalities was the result of geophysical disasters (earthquakes, tsunamis and volcanoes) although those disasters occurred in low frequency. Meanwhile, Indonesia was in the third highest position after Japan and the United States for the average economic loss due to the earthquake [1].

The significance of earthquake hazard studies in Indonesia has been started in 1983 and continues to develop until now with various improvements, both in the aspect of updating earthquake source data, selecting the latest GMPE (Ground Motion Prediction Equation) and its analysis methods. The latest earthquake hazard and source maps were published in 2017. Backward before the year 2017, studies on the development of Indonesia's earthquake hazard map for the period of 2500 years have been carried out. That study was a revision of SNI-03-1726-2002 which uses a total probability method with three-dimensional (3-D) earthquake sources [2], [3]. The latest study was an earthquake hazard analysis study with a combination of probabilistic and deterministic analysis applying the latest earthquake sources and GMPE data in the context of developing a 2017 Indonesia earthquake map. Besides, the use of relevant information from active fault studies using trench excavation and carbon dating methods, earthquake centre relocation, strain analysis (GPS), base map data including SRTM-30, IFSAR, LiDAr and other new available data [4], [5], [6], [7].

Meanwhile, first studies related to earthquake risk were carried out by the National Agency for Disaster Management (BNPB) in 2011. Further development in 2016 taking into account the site-specific soil condition [8], [9]. The resulted risk map is presented in the form of a risk index (low, medium and high risk) but does not taking the vulnerability of the building, so it does not present the actual loss value. Later study on seismic risk of Jakarta was conducted for two types of residential buildings (Unreinforced Masonry and Confined Masonry) accommodating site specific and building fragility functions.[10].

Jakarta as the capital city of Indonesia has a very dense population, ranged from 12,000 to 20,000 km^{-2}. As a consequence, it has a very dense buildings particularly residential building. This study was conducted in one of the municipalities of Jakarta, i.e., West Jakarta. West Jakarta was chosen due to its highest concentration of small-scale industries. It has eight districts namely Cengkareng, Grogol Petamburan, Kali Deres, Kebon Jeruk, Kembangan, Palmerah, Taman Sari and Tambora, also including Jakarta's Chinatown and Dutch colonial landmarks. Categories of residential buildings in the area of study were unknown (UNK) type of residential building in slum area to the more than 5 storeys flat buildings with reinforced concrete (CR) material type.

Jakarta has also high risk and prone to disaster especially earthquake. Although not very devastating, within a decade the earthquake catalogue of BMKG (Indonesia Meteorology and Geophysics Agency), reported some earthquakes event sourced in Banten Province and West Java was felt in Jakarta with MMI IV to V, for example: Lebak Banten Eq. 2018, 6.4 Mw, Tasikmalaya Eq. 2009, 7.3 Mw, West Java Eq. 2007, 6.9 Mw, South of Java Eq. 2006, 6.8 Mw, West Java Eq. 1999, 6.0 Mw, West Java Eq. 1997, 6.0 Mw [11].

Considering those occurrences and the lack of any recent study in seismic risk, it is an urge to conduct a more adequate earthquake risk in an effort to mitigate and estimate the loss of building damage. Here, risk analysis study carried out using the updating of earthquake sources, GMPE's, site-specific soil conditions, exposure, and vulnerability of buildings.

2. Methods

This research objectives were analysing seismic risk in term of structural loss of the residential building in West Jakarta Municipality. In general, four main stages for conducting seismic loss were apply in this study: 1) seismic hazard analysis, 2) exposure data, 3) vulnerability function and 4) seismic risk analysis. The calculation conducted using open-source software OpenQuake Engine.

Seismic Hazard Analysis

This study using the earthquake source catalogue of PUSGEN (the National Earthquake Study Centre) which are used for the preparation of the 2017 earthquake map [12]. The catalogue data were sorted to choose certain earthquakes sources that have influenced the site of interest within a maximum radius of 300 km. Seismic sources included in the calculation were subduction earthquake sources, shallow faults, and background sources. Influenced subduction sources for Jakarta area were the Sunda Strait megathrust and the southern part of Java megathrust. Furthermore, about 12 active faults throughout western to central parts of Java, Sumatera-Sunda Graben Fault and Sumatera-Ujung Kulon Fault were included as well.

Aside of source model, hazard analysis also requires ground motion model. The ground motion model or GSIM (Ground Shaking Intensity Model) is a model for calculating ground motion at a certain location according to its specific rupture property. In a simple case, such as when a single earthquake source applied, the ground motion model corresponds to GMPE. Whereas when predicting a more complex case such as PSHA (Probability Seismic Hazard Analysis) input model, the ground motion model contains a set of GMPE/Logic Tree; in which each GMPE is considered for each tectonic region [13], [14].

In this study, GMPE Logic Tree was used following updated GMPE's in the 2017 Indonesian Earthquake Hazard Map presented in Figure 1 [12]. These updates were for active shallow crustal and shallow background modelling GMPEs by Boore et al, (2014), Campbell & Bozorgnia (2014) and Chiou-Young (2014). Other updates for the source of the subduction interface (megathrust) were combined by Atkinson & Boore (2003) SInter, Zhao et al., (2006) SInter and Abrahamson et al., (2015) SInter. Meanwhile, the source of subduction inslab was updated from Atkinson & Boore (2008) SSlab, Atkinson & Boore (2008) SSlab Cascadia and Geomatrix 1993 SSlab (2008) [15], [16], [17], [18].

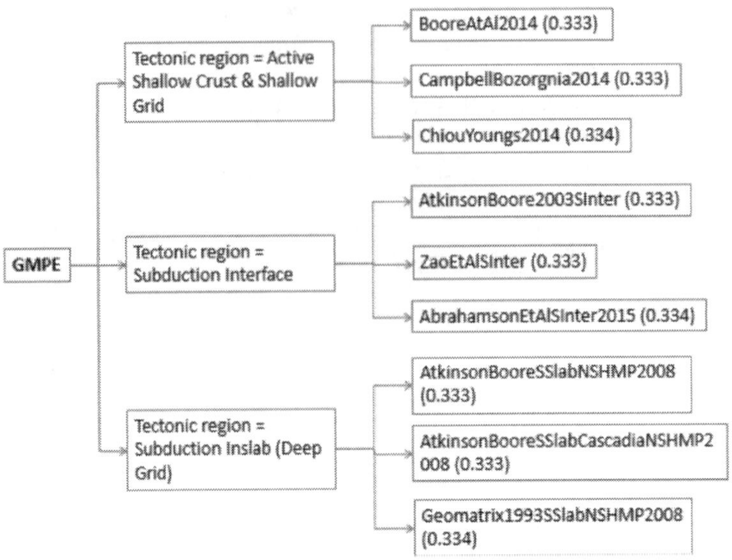

Figure 1. GMPE Logic Tree using in this study

Event-Based PSHA methods is used to compute seismic hazard and risk analysis. Besides source and ground motion models, Event-Based requires input of site-specific parameters from each site of interest as follows: typical shear-wave velocity at a depth of 30m from the surface (Vs30) and geological considerations/basin effects that determined as parameters of rock depth with a shear-wave velocity of 1,000 m.sec-1, ($Z_{1.0}$) and rock depth with a shear wave velocity of 2,500 m sec^{-1} ($Z_{2.5}$). Vs30 for the study area was obtained from the US Geological Survey [19]. The data had a grid resolution of 30 arc-seconds (0.00833 degrees), calculated based on topography and verified by field measurements in various countries worldwide. Both $Z_{1.0}$ and $Z_{2.5}$ were calculated using the formula of Chiou & Youngs (2013) [20]. Vs30 data from USGS reported that soil class in West Jakarta is soft soil (SD) with Vs30 between 180 – 255.48 m sec^{-1}. In general, from the southward to the northward the soil type is getting softer.

Event-Based PSHA basically performs the Monte Carlo integration where the calculation was conducted through random sampling from a magnitude pdf (probability density function) to get a single magnitude. Furthermore, from the magnitude, the earthquake probability is calculated using the earthquake recurrent Guttenberg Richter (GR) model and/or Characteristic (CH) model. The next step was to calculate the length and width of the rupture based on the magnitude and mechanism of the earthquake using the scaling law. Then rupture centroid is determined randomly which must be inside the rupture boundary. Moreover, the probability of each event and rupture can be estimated as a hazard by considering site amplification according to its specific site [21].

In the calculation of hazard analysis, the rate of exceeding of a given ground motion value $\lambda(GM > gm)$ in a region with *Ns* seismic sources is as follows:

$$\lambda(GM > gm) = \sum_{i=i}^{N_s} v_i \int M \int R\, P[GM > gm|m,r] f_M(m)\, f_R(r|m) dm\, dr$$

where v_i stands for the average rate of magnitude exceedance threshold for source i, $P[GM > gm|m,r]$ can be derived from the ground motion prediction model, $f_M(m)$ represents the probability density function for magnitude, and $f_R(r|m)$ stands for the probability density function for the source-to-site distances conditional on a magnitude m. The output of hazard analysis are hazard curves and hazard maps.

Exposure

This seismic risk analysis was conducted for residential buildings in West Jakarta urban area. Exposure data of residential buildings in West Jakarta collected from the website of the Jakarta Satu Peta, https://jakartasatu.jakarta.go.id. Categories and number of the residential buildings were sorted and presented in Table 1. Meanwhile site location of every building was collected by searching in Jakarta Satu Peta website, https://jakartasatu.jakarta.go.id/portal/apps/sites/#/public

Table 1. Residential Building Category by Jakarta Satu Peta used in West Jakarta Municipality

Nu.	Categories	Code	Numbers
1	Very small house	HU1	128
2	Nursing Homes and Orphanages	HU10	16
3	Guest House	HU11	5
4	Pavilion	HU12	2
5	Official residence	HU13	47
6	Small house	HU2	3782
7	Medium House	HU3	4084
8	Big House	HU4	1482
9	Flat House	HU5	106
10	Flats	HU6	56
11	Public Flats	HU7	16
12	Dormitory	HU8	27
13	Boarding House	HU9	547
	Total		10298

Building taxonomy is one attribute needed in exposure data. In this study the building taxonomy developed by GEM (Global Earthquake Model) was used in the exposure model and presented in Table 2.

Table 2. Building Taxonomy of Residential Buildings Developed by GEM used in West Jakarta Municipality

Nu.	Building Taxonomies	Description
1	CR/LDUAL+DUH/HBET: 6-12	Reinforced Concrete, Dual Frame Lateral Load Resistant System, High Ductility, 6 to 12 storeys
2	CR/LFINF+DUH/H:3	Reinforced Concrete, In-Fill Frame, High Ductility, 3 storeys
3	CR/LFINF+DUH/H:4	Reinforced Concrete, In-Fill Frame, High Ductility, 4 storeys
4	CR/LFINF+DUH/H:5	Reinforced Concrete, In-Fill Frame, High Ductility, 5 storeys

5	CR/LFINF+DUM/H:3	Reinforced Concrete, In-Fill Frame, Medium Ductility, 3 storeys
6	CR/LFINF+DUM/H:4	Reinforced Concrete, In-Fill Frame, Medium Ductility, 4 storeys
7	CR/LFINF+DUL/H:5	Reinforced Concrete, In-Fill Frame, Low Ductility, 5 storeys
8	CR/LFINF+DUL/H:3	Reinforced Concrete, In-Fill Frame, Low Ductility, 3 storeys
9	CR/LFINF+DUL/H:4	Reinforced Concrete, In-Fill Frame, Low Ductility, 4 storeys
10	MCF/LWAL+DUL/H:1	Confined Masonry with lateral load resistant system = wall, Low Ductility, 1 storey
11	MCF/LWAL+DUL/H:2	Confined Masonry with lateral load resistant system = wall, Low Ductility, 2 storeys
12	MCF/LWAL+DUL/H:3	Confined Masonry with lateral load resistant system = wall, Low Ductility, 3 storeys
13	MCF/LWAL+DUM/H:1	Confined Masonry with lateral load resistant system = wall, Medium Ductility, 1 storey
14	MCF/LWAL+DUM/H:2	Confined Masonry with lateral load resistant system = wall, Medium Ductility, 2 storeys
15	MCF/LWAL+DUM/H:3	Confined Masonry with lateral load resistant system = wall, Medium Ductility, 3 storeys
16	MUR/LWAL+DNO/H:1	Unreinforced Masonry with lateral load resistant system = wall, No ductility, 1 storey
17	MUR/LWAL+DNO/H:2	Unreinforced Masonry with lateral load resistant system = wall, No ductility, 2 storeys
18	UNK/H:1	Unknown, 1 storey
19	W+WBB/LPB+DNO/H:1	Wood and Bamboo, lateral load resistant system = pole and bar, No Ductile, 1 storey
20	W+WLI/LWAL+DUL/H:1	Wood and Light wood, lateral load resistant system = wall, Low-Ductile, 1 storey
21	W+WO/LN/H:1	Wood and Others, no lateral load resistant system, 1 storey
22	W+WO/LN/H:1+HFAPP:1	Wood and Others, no lateral load resistant system, 1 storey and base floor height from ground = 1m
23	ME+MEO/LN/H:1	Metal and other metal, no lateral load resistant system, 1 storey

Google Map and Street View features were used to match the categories of Jakarta Satu Peta building taxonomy (Table 1) to GEM building taxonomy (Table 2) for every building. The replacement cost of the buildings was averaged from structural building cost in the area of study, assuming that the replacement cost covering the price of a new building in accordance to the new code of seismic resistant building. Replacement cost data was adopted from previous study by GEM in 2019.

Table 3. Replacement Cost (Structural) of Residential Buildings (GEM, 2019)

Building Taxonomy	Replacement Cost (structural) in USD	Building Taxonomy	Replacement Cost (structural) in USD
CR/LDUAL+DUH/HBET:6-12	4,725,000	MCF/LWAL+DUM/H:1	71,837
CR/LFINF+DUH/H:3	393,750	MCF/LWAL+DUM/H:2	107,756
CR/LFINF+DUH/H:4	1,050,000	MCF/LWAL+DUM/H:3	215,511
CR/LFINF+DUH/H:5	1,640,625	ME+MEO/LN/H:1	19,923
CR/LFINF+DUL/H:3	120,100	MUR/LWAL+DNO/H:1	19,923

CR/LFINF+DUL/H:4	574,697	MUR/LWAL+DNO/H:2	29,885
CR/LFINF+DUL/H:5	897,964	UNK/H:1	19,923
CR/LFINF+DUM/H:3	393,750	W+WBB/LPB+DNO/H:1	19,923
CR/LFINF+DUM/H:4	1,050,000	W+WLI/LWAL+DUL/H:1	40,033
MCF/LWAL+DUL/H:1	40,033	W+WO/LN/H:1	19,923
MCF/LWAL+DUL/H:2	60,050	W+WO/LN/H:1+HFAPP:1	19,923
MCF/LWAL+DUL/H:3	120,100		

The surveys revealed that the percentages of reinforced concrete (RC) buildings with higher than 6 storeys is 0.63%, RC buildings with 3 to 5 storeys = 16.84%, confined masonry (MCF) buildings with 1 to 3 storeys = 49.10%, unreinforced masonry (MUR) buildings with 1 to 2 storeys = 26.54%, buildings with wood and bamboo construction (W and WBB) = 4.3%, the rest which were non-permanent buildings = 2.60%. Examples of those building taxonomy mentioned were illustrated in Figure 2.

(a)

(b)

(c)

(d)

(e) (f)

Figure 2. Some residential buildings in West Jakarta (a) CR/LDUAL+DUH/HBET:6-12, (b) CR/LFINF+DUH/H:5, (c) MCF/LWAL+DUM/H:2, (d) MCF/LWAL+DUM/H:3 (e) MUR/LWAL+DNO/H:1 (f) W+WBB/LPB+DNO/H:1

Vulnerability

Vulnerability factor of the building is an important part of the earthquake loss assessment. The building vulnerability function is the relationship between the ratio of losses to the intensity measure level of earthquake. The vulnerability model can be derived from empirical methods, where losses from past earthquake events at a particular location were related to the intensity level of ground motion at the related location; or obtained through combining the fragility function with the consequence function.

The fragility function describes the probability of exceeding the performance limit of the structure at a certain level of damage to the level of intensity measure. It can be derived from expert judgement, empirically using observed data, or numerically by modelling the typological behaviour of a given asset when experiencing an increase ground motion.

Meanwhile, the consequence functions is a function that describes the distribution of the probability of loss at a given level of performance. This function is generally derived empirically. In this study, the vulnerability function uses secondary data from previous studies adjusted for the taxonomy of buildings (Table 2) in the exposure data [22], [23]. The vulnerability model used in the calculations taken from https://platform.openquake.org/vulnerability/

Seismic Risk Analysis

The advantage of using probabilistic Event-Based approach, is the possibility of using spatial correlation of the ground motion or the correlation in the vulnerability of assets of the same building class, then the losses from all the assets can be summed, thus allowing the calculation of ALs (Annual Looses) whilst compiling loss exceedance curves representing the entire collection of the assets. For this reason, Event-Based approach is preferred for the estimation of losses for portfolio of buildings, which is a common metric used in the development of risk mitigation measures. In spite of these advantages, this approach may require the generation of a large number of SESs (Stochastic Event Set) to obtain convergence in the risk results. This makes it a less attractive methodology from a computational perspective.

In each gmf (ground motion field), the level of intensity measurement at a given location is combined with a vulnerability function, wherein the loss ratio is sampled randomly for each asset contained in the exposure model. The loss ratios sampled for assets are given a taxonomy classification in different locations. This classification may consider to be independent or fully correlated [21].

The distribution of loss occurrences for a particular asset is calculated using all ground motion fields, which leads to a loss ratio histogram which is then converted to a cumulative histogram by calculating the cumulative number of events for each loss ratio interval. The exceeded rate of each loss ratio is calculated by dividing the cumulative number of events by the number of stochastic event sets multiplied by the length of each event set. Assuming a Poisson distribution of the event model, the probability of exceeding each loss ratio is then calculated [21].

The output of earthquake risk analysis were included the Aggregated Loss Table and the Aggregated Loss Curve. The Aggregated Loss Table is a table of building structural losses based on the type of material analysed within time of risk investigation. Whereas the Aggregated Loss Curve is the loss value of the building structure due to earthquake in a certain return period. In this study, the loss curve is presented based on the type of structural material, per sub-district. Furthermore, the total loss map for 500y and 2500y return period was produced based on total loss which is derived from loss curves.

3. Result and Discussion

Seismic Hazard

For a reliable statistic result, seismic hazard was analysed using Event-Based PSHA method with SES (stochastic event set) 200,000 years [24]. Example of some resulted hazard curves for the acceleration spectrum in period of 0 second (PGA), 0.3 second SA(0.3) and 0.6 second SA(0.6) of some locations in Cengkareng District were shown in Figure 3.

The peak acceleration at ground surface for PGA was about 0.35g and 0.59g for probability of exceedance (PoEs) 10% and 2% in 50 years (PoEs 0.2% and 0.04% in 1 year), respectively (Figure 3.a). Seismic intensity at ground surface for SA(0.3) were around 0.7g and 1.1g for PoEs 10% and 2% in 50 years, respectively (Figure 3.b). Whilst the peak acceleration at ground surface for SA(0.6) were about 0.55g and 0.95g for PoEs 10% and 2% in 50 years, respectively (Figure 3.c).

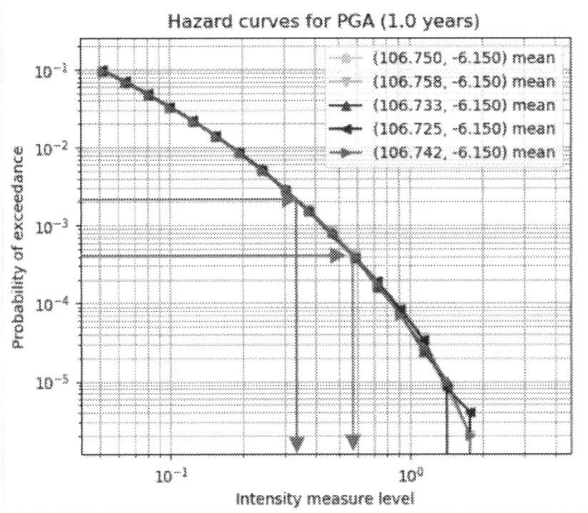

(a)

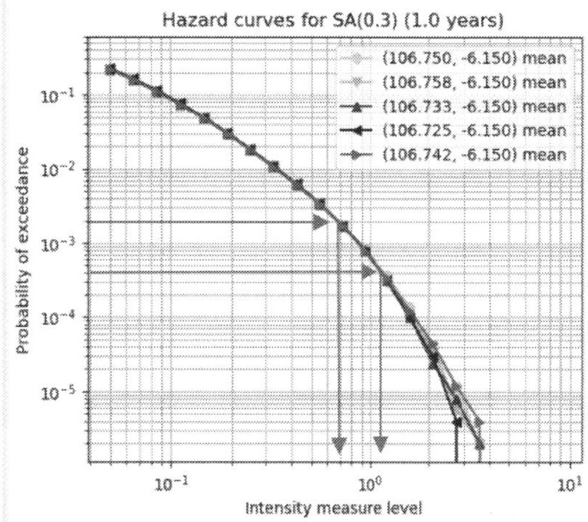

(b)

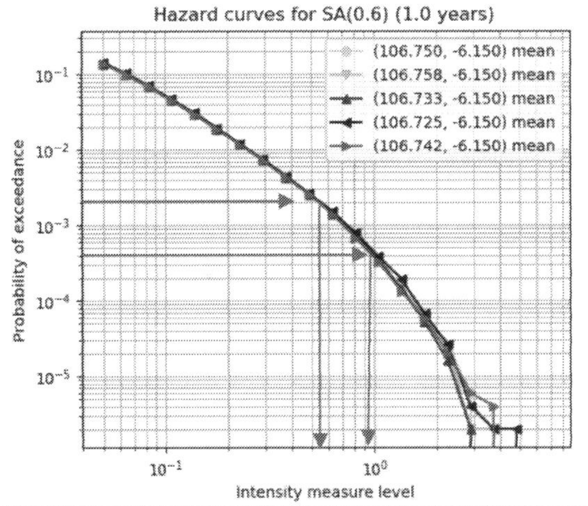

(c)

Figure 3. Hazard curve in some locations in Cengkareng District, West Jakarta

Aside of hazard curves, this study presents hazard map for peak acceleration at ground surface for period of 0 second (PGA) depicted in Figure 4; shown that for 500years return period, the ground acceleration was 0.32g in the northern part to 0.36g in the southern part of West Jakarta Municipality (Figure 4.a). Whilst for return period of 2500years, the peak acceleration at ground surface was around 0.54g in the northern part to 0.60g in the southern part (Figure 4.b). Those two maps showed that the acceleration increasing from the north to the south gradually due to the seismic sources are mostly come from southern part of the city. The result of present study has been in concordance to hazard analysis of Jakarta in previous study, that the distribution of peak ground acceleration (PGA) at bedrock ranges from 0.18g – 0.22g and from 0.33g – 0.39g for 10% and 2% probability of exceedance in 50 years [10]; considering that soil classes of Jakarta are mostly SD with amplification factor about 1.5 for PGA = 0.1g to 0.2g and 1.25 for PGA = 0.3g to 0.4g [25].

Hazard map for peak acceleration at ground surface for period of 3 second SA(0.3) is presented in Figure 5. It is shown that for 500years return period, the ground acceleration was 0.63g in the northern part to 0.7g in the southern part of West Jakarta Municipality (Figure 5.a). Whilst for return period of 2500years, the peak acceleration at ground surface was around 1.10g in the northern part to 1.26g in the southern part (Figure 5.b).

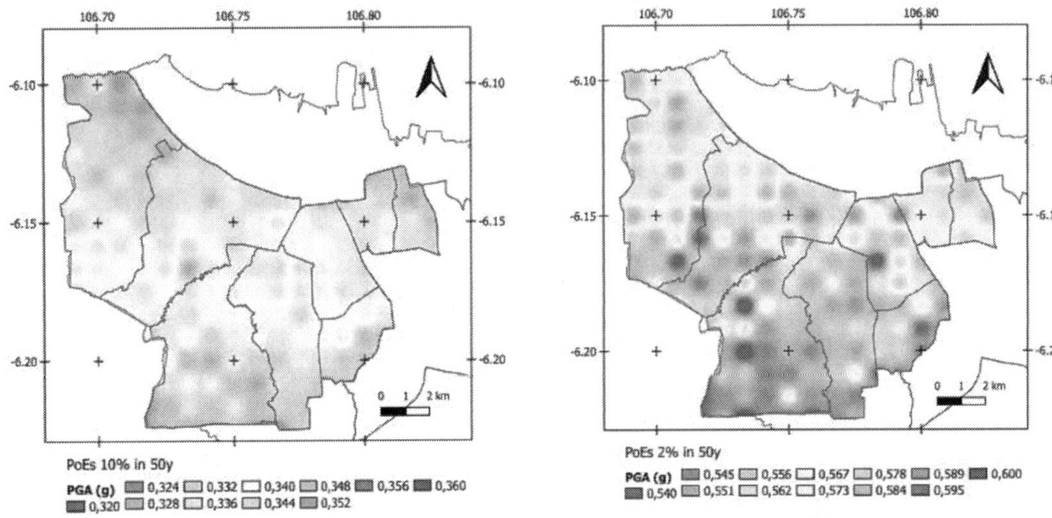

(a) (b)

Figure 4. The peak acceleration at ground surface for period 0 second (PGA) for probability of exceedance (a) 10% and (b) 2% in 50 years

Hazard map for peak acceleration at ground surface for period of 3 second SA(0.3) is presented in Figure 5. It is shown that for 500years return period, the ground acceleration was 0.63g in the northern part to 0.7g in the southern part of West Jakarta Municipality (Figure 5.a). Whilst for return period of 2500years, the peak acceleration at ground surface was around 1.10g in the northern part to 1.26g in the southern part (Figure 5.b).

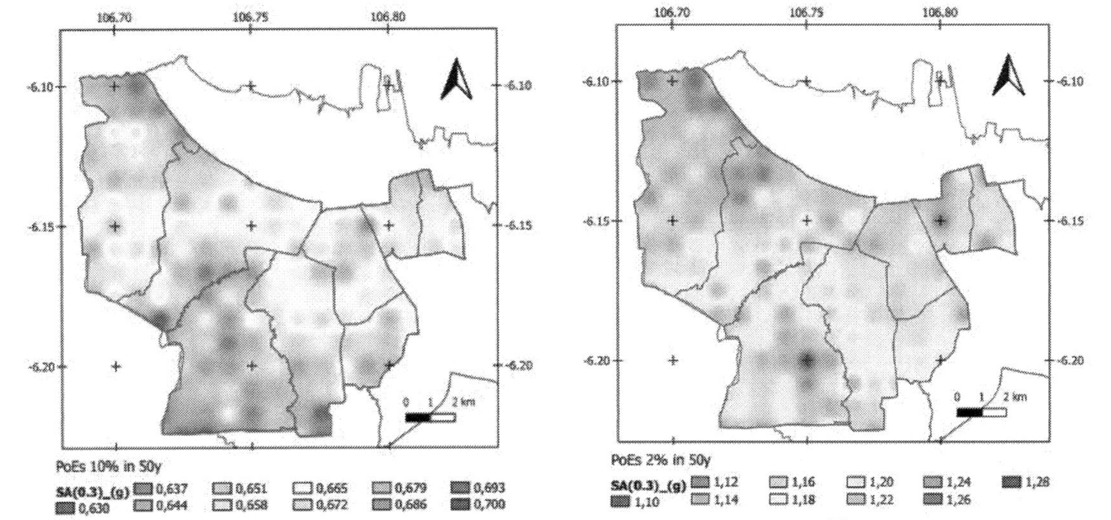

(a) (b)

Figure 5. Peak acceleration at ground surface for period 3 second, SA(0.3) for probability of exceedance (a) 10% and (b) 2% in 50 years

Hazard map for the peak acceleration at ground surface for period of 6 second SA(0.6) presented in Figure 6. The map showed that for 500years return period, peak acceleration was 0.5g in the northern part to 0.55g in the southern part of West Jakarta Municipality (Figure 6.a). Whilst for return period of 2500years, the peak acceleration was around 0.94g in the northern part to 1.10g in the southern part (Figure 6.b) respectively.

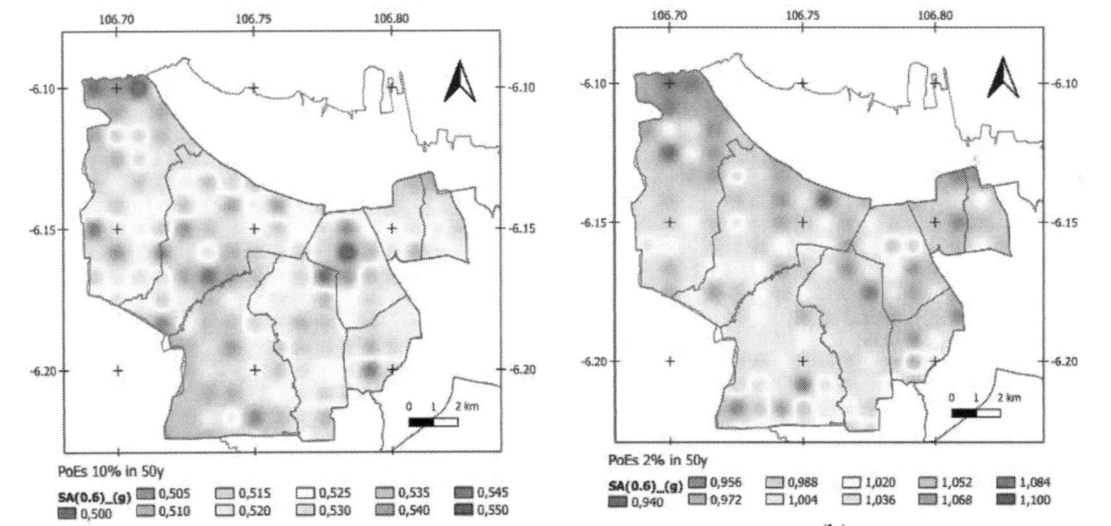

(a) (b)

Figure 6. Peak acceleration at ground surface for period 6 second, SA (0.6) for probability of exceedance (a) 10% and (b) 2% in 50 years

Seismic Risk

The last study of seismic risk for Jakarta conducted previously [10] has yet incorporated structural loss. Therefore, this study may be considered as the first to calculate risk in terms of structural loss for this particular area, since no other study may be compared to validate.

The result of loss analysis was the aggregated loss-mean table which described loss-mean value of a number of buildings in exposure data aggregated by material type and district. Table 4 is the resulted example of aggregated loss of buildings in Cengkareng District for 1 year (annual) risk investigation time. The table shows that house with wood/ light wood (W+WLI), wood and bamboo (W+WBB) and unknown material (UNK) have the highest loss ratio (0.13% - 0.14%). Buildings with structure material of reinforced concrete (CR) and confined masonry (MCF) had low loss ratio (0.02% - 0.05%), means that both CR and MCF are less vulnerable than W+WLI, W+WBB and UNK buildings.

Table 4. Annual Loss Ratio Aggregated by Material Type and District for Cengkareng District)

Loss type	Name of district	Material type	Loss value (USD)	Exposed value (USD)	Loss ratio
structural	CENGKARENG	ME+MEO	961.97	737,160.00	0.13%
structural	CENGKARENG	MUR	5,089.57	7,401,480.00	0.07%
structural	CENGKARENG	MCF	15,514.40	77,540,500.00	0.02%
structural	CENGKARENG	UNK	103.21	79,693.00	0.13%
structural	CENGKARENG	W+WLI	1,652.28	1,201,000.00	0.14%
structural	CENGKARENG	W+WBB	1,710.68	1,314,930.00	0.13%
structural	CENGKARENG	CR	75,560.40	155,213,000.00	0.05%
structural	CENGKARENG	W+WO	599.63	458,235.00	0.13%

Another advantages of Event-Based Risk Analysis output is that the loss-mean curve can be set individually or aggregated, such as on the base of material type and district as well. Aggregated loss curves can be produced by setting various combination of aggregated parts in the configuration file. The aggregated loss curve illustrated the relationship of return period to economic loss. Figure 7 shown as result example of aggregated loss curve for CR (reinforced concrete), MCF (confined masonry), MUR (unreinforced masonry) and W+WLI (wood and light wood) buildings in Cengkareng District.

The aggregated loss-mean curve for CR buildings shows that structural loss was around 6.6 million USD for 500 years return period and about 14 million USD for a return period of 2500 years (Figure 7.a). For MCF buildings, the economic loss was nearly 2.0 million USD and 5.0 million USD due to earthquakes for a return period of 500 years and 2500 years respectively (Figure 7.b). Loss estimation of MCF houses is less than CR buildings because the total number and the replacement cost of the CR buildings are higher than the MCF. The loss of the MUR building is about 600,000 USD for the earthquake with a return period of 500 years and around 1.2 million USD for a return period of 2500 years (Figure 7.c). Whereas for W+WLI houses, economic loss was 180,000 USD for the return period and 350,000 USD for a return period of 2500 years (Figure 7.d).

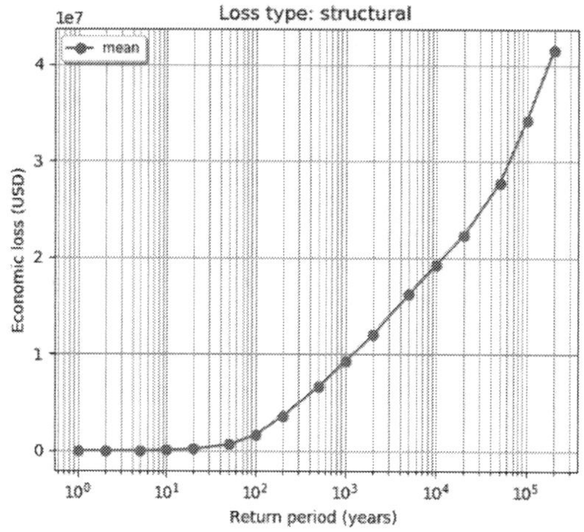

(a)

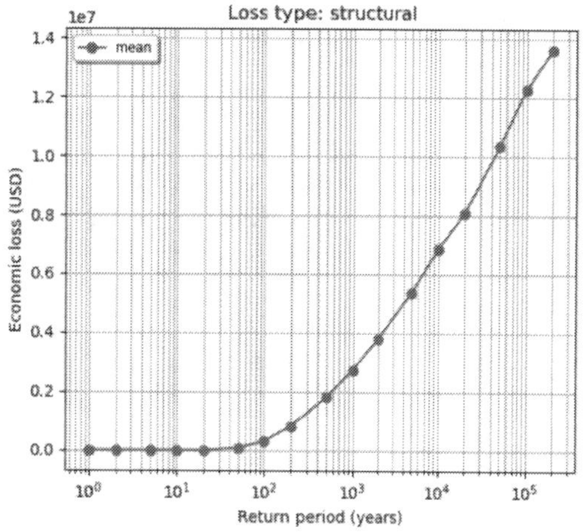

(b)

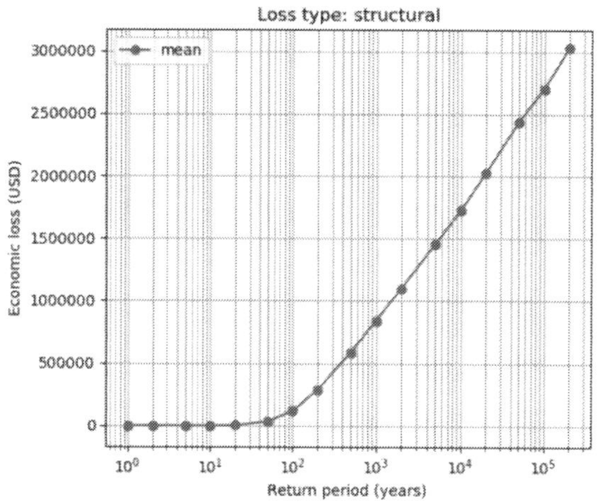

(c)

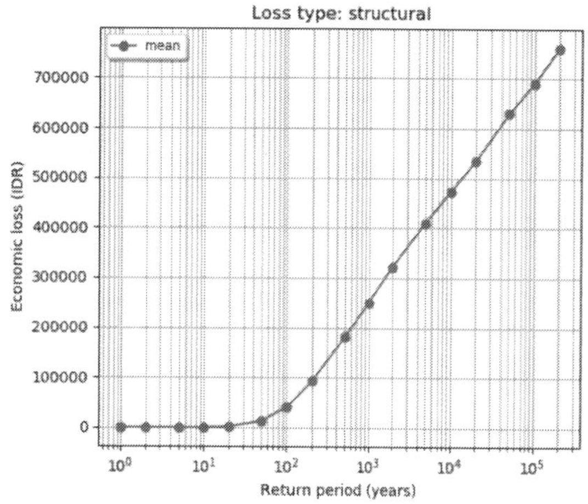

(d)

Figure 7. Aggregated Loss curve mean for (a) CR buildings (b) MCF buildings (c) MUR buildings (d) W+WLI buildings in Cengkareng District

All those loss values were the sum of the loss of structural buildings due to seismic intensity in the area of the assets, wherein that seismic intensity illustrated in the hazard curves of the buildings in the same return period (see Figure 3).

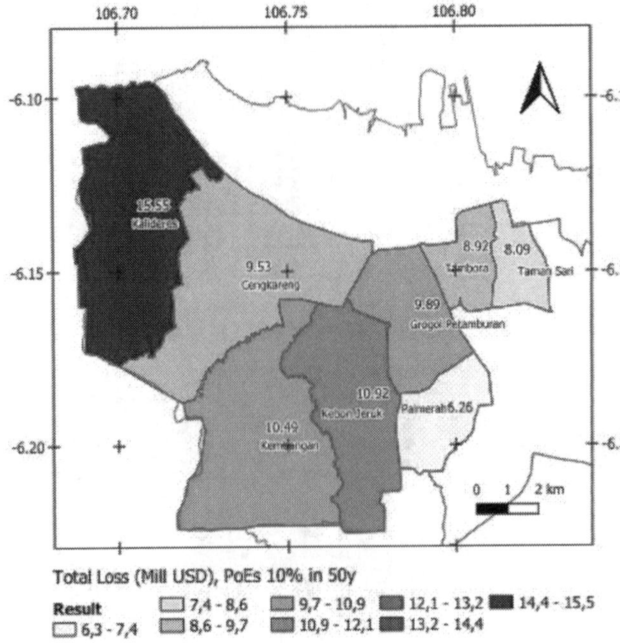

Figure 8. Total loss map of residential buildings in West Jakarta Municipality due to earthquake in return period of 500years

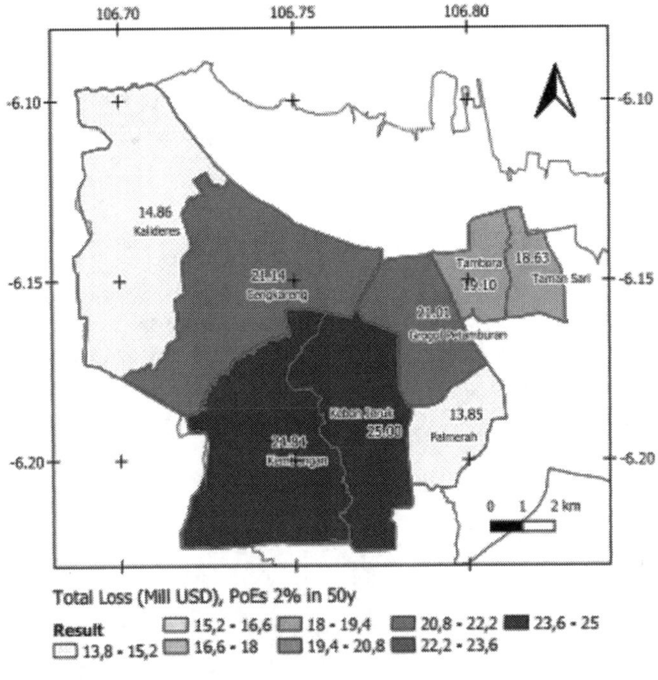

Figure 9. Total loss map of residential buildings in West Jakarta Municipality due to earthquake in return period of 2500years

The aggregated loss curve in Figure 7 can also be used to determine aggregated loss-mean value at every return period. On that account, a total loss map for a certain period can be determined. Risk maps of West Jakarta for return period of 500 years and 2500 years earthquakes were presented in Figure 8 and Figure 9 respectively. The map showed that for return period of 500 years, the total loss of residential building in West Jakarta were predicted within the lowest range of around 6.26 million USD at Palmerah District to the highest cost of about 15.5 million USD at Kali Deres District (Figure 8).

These values have been deducted from total loss-mean of >10.000 individual asset calculated from vulnerability curve of each building taxonomy with three-sigma truncation level (99.7%). Earthquake with return period of 2500 years in West Jakarta, resulting in the total loss of residential building ranged from the lowest around 13.85 million USD at Palmerah District to the highest of about 25.0 million USD at Kebon Jeruk District (Figure 9).

4. Conclusion

Having been using indirect survey for exposure data and adopted vulnerability functions to the assets being considered, risk analysis in the present study been calculated and developed. Hazard and risk analysis using Event-Based approach has the advantage to produce an annual loss ratio and an aggregated loss curve as well, so the programme may be used to compile a loss map according to the earthquake return period. In turn, the resulted annual loss ratio is also useful for building insurance purposes, while the loss map can be used for mitigation efforts particularly in the preparation of reconstruction funds by the ministry or government offices responsible to disaster management.

5. References

[1] UNDRR. 2019 *Global Assessment Report on Disaster Risk Reduction.* https://gar.undrr.org/report-2019

[2] Asrurifak M, Irsyam M, Budiono B, Triyoso W and Hendriyawan H 2010 *Development of spectral hazard map for Indonesia with a return period of 2500 years using probabilistic method* Civil Engineering Dimension, 377 12(1), 52–62

[3] Irsyam M, Asrurifak M, Ridwan M, Aldiamar F, Sengara I W, Widiyantoro S, Triyoso W, Hilman D, Kertapati E, Meilano I, Suhardjono, Hendriyawan, Simatupang P, Muhammad I, Murjanto D and Hasan M 2013 *Development of Map of Maximum Considered Earthquake Geometric Mean (MCEG) PGA for Earthquake Resistance Building Design in Indonesia*. Proceedings of the 18th International Conference on Soil Mechanics and Geotechnical Engineering

[4] Daryono M R, Natawidjaja D H, Sapiie B, & Cummins P 2019 *Earthquake geology of the lembang fault, West Java, Indonesia*. Tectonophysics, 751, 180–191.

[5] Irsyam M, Cummins P R, Asrurifak, M, Faizal L, Natawidjaja D H, Widiyantoro S, Meilano I, Triyoso W, Rudiyanto A, Hidayati S, Ridwan M, Hanifa N R, & Syahbana A J 2020 *Development of the 2017 national seismic hazard maps of Indonesia*. Earthquake Spectra, I (25), https://doi.org/10.1177/8755293020951206

[6] Marliyani G, Arrowsmith J & Whipple K 2016 *Characterization of slow slip rate faults in humid areas: Cimandiri fault zone, Indonesia.* Journal of Geophysical Research: Earth Surface, 121(12), 2287–2308.

[7] Natawidjaja D H 2018 *Major Bifurcations, Slip Rates, and A Creeping Segment of Sumatran Fault Zone in Tarutung-Sarulla-Sipirok-Padangsidempuan, Central Sumatra, Indonesia.* Indonesian Journal on Geoscience, 5(2), 137–160.

[8] Badan Nasional Penanggulangan Bencana 2013 *Indeks Risiko Bencana Indonesia (IRBI) 2013*, Direktorat Pengurangan Risiko Bencana BNPB, Jakarta

[9] Badan Nasional Penanggulangan Bencana 2016 *Risiko Bencana Indonesia (RBI)*, Direktorat Pengurangan Risiko Bencana BNPB, Jakarta

[10] Irsyam M, Hutabarat D, Asrurifak M, Imran I, Widiyantoro S, Hendriyawan, Sadisun I, Hutapea B, Afriansyah T, Pindratno H, Firmanti A, Ridwan M, Haridjono S W dan Pandhu R 2015 *Development of seismic risk microzonation maps of Jakarta city*, Geotechnics for Catastrophic Flooding Events – Iai (Ed)© 2015 Taylor & Francis Group, London, ISBN 978-1-138-02709-1

[11] Pusat Gempa Bumi dan Tsunami BMKG, 2019 *Katalog Gempa Bumi Signifikan dan Merusak 1821 – 2018*, Kedeputian Bidang Geofisika BMKG, Jakarta

[12] PusGen. 2017 *Peta Sumber dan Bahaya Gempa Indonesia Tahun 2017 (1st ed.)*. Puslitbang Perumahan dan Pemukiman

[13] Crowley H & Silva V 2013. *OpenQuake Engine Book: Risk v1.0.0.*, Pavia, Italy: GEM Foundation

[14] GEM 2019 *The OpenQuake-engine User Manual.Global Earthquake Model (GEM) OpenQuake Manual for Engine version 3.6.0.* doi: 10.13117/GEM.OPENQUAKE.MAN.ENGINE.3.6.0, 187 pages

[15] Boore D M & Atkinson G M 2007 *Boore-Atkinson NGA Ground Motion Relations for the Geometric Mean Horizontal Component of Peak and Spectral Ground Motion Parameters*. Report PEER 2007/01, May 2007,

[16] Campbell K W, & Bozorgnia Y 2007 *Campbell-Bozorgnia NGA Ground Motion Relations for the Geometric Mean Horizontal Component of Peak and Spectral Ground Motion Parameters*. Report PEER 2007/02, May2007

[17] Chiou B S J, & Youngs R R 2008 *An NGA Model for the Average Horizontal Component of Peak Ground Motion and Response Spectra*. Earthquake Spectra, 24(February 2008)

[18] Abrahamson N, Gregor N, & Addo K 2016 *BC Hydro ground motion prediction equations for subduction earthquakes*. Earthquake Spectra, 32(1), 23–44.

[19] Wald D J, & Allen T I 2007 *Topographic slope as a proxy for seismic site conditions and amplification*. Bulletin of the Seismological Society of America, 97(5), 1379–1395.

[20] Chiou B S J and Youngs R R 2013. *Update of the Chiou and Youngs NGA Ground Motion Model for Average Horizontal Component of Peak Ground Motion and Response Spectra*, PEER Report 2013/07

[21] GEM 2021 *The OpenQuake-engine User Manual. Global Earthquake Model OpenQuake Manual for Engine version 3.11.2.,* doi: 10.13117/ GEM.OPENQUAKE.MAN.ENGINE. 3.11.2, 208 pages

[22] Martins L, Silva V 2018 *A Global Database Of Vulnerability Models For Seismic Risk Assessment*, 16EECE

[23] Martins L, Silva V 2020 *Development of a fragility and vulnerability model for global seismic risk analyses*, Bull Earthquake Eng (2020). https://doi.org/10.1007/s10518-020-00885-1

[24] Silva V 2017 *Critical Issues on Probabilistic Earthquake Loss Assessment*, Journal of Earthquake Engineering, https://doi.org/10.1080/13632469.2017.1297264

[25] Tim Penyusun SNI Badan Standardisasi Nasional (BSN) 2019 *SNI 1726-2019: Tata Cara Perencanaan Ketahanan Gempa untuk Struktur Bangunan Gedung dan Non Gedung*, Badan Standardisasi Nasional (BSN), Jakarta

Acknowledgments
Thanks to National Earthquake Study Centre (PuSGeN) and Global Earthquake Model (GEM) for supporting data and software in this research.

Microplastics in the continuous biofilm reactor: Occurrence, fate, and removal

A Nur[1,3*], P Soewondo[2], AS Setiyawan[2], K Oginawati[2]

[1]Doctoral Program of Environmental Engineering, Faculty of Civil and Environmental Engineering, Institut Teknologi Bandung, Jl. Ganesha No.10, Bandung 40132, Indonesia

[2]Department of Environmental Engineering, Faculty of Civil and Environmental Engineering, Institut Teknologi Bandung, Jl. Ganesha No.10, Bandung 40132, Indonesia

[3]Department of Environmental Engineering, Faculty of Engineering, Universitas Andalas, Kampus Limau Manis, Padang 25164, Indonesia

*ansihanurst@gmail.com

Abstract. The occurrence of microplastics in a continuous biofilm reactor during the process over a period of 53 days poses a concern for the quality of the treated water. The aim of this study seeks to expand the knowledge of the dynamic of microplastics inside biofilm reactors by investigating their occurrence, fate, and removal. A 91.5 L continuous laboratory-scale reactor consists of four compartments, anoxic-1, anoxic-2, aerobic (A^2O), and sedimentation. It was fed with raw domestic wastewater from a residential scale of the wastewater treatment plant in Bandung, Indonesia. In order to investigate the occurrence of microplastics inside a biofilm reactor, it is crucial to consider into account all treatment stages. The results showed that the abundance of microplastics in the influent and the final effluent were 810 MP/L and 10 MP/L respectively. The A^2O process in this study demonstrated a high potential for microplastics removal from residential wastewater, in which it could be removed 99.18% of microplastics. The highest abundance was retained in the sludge. The remaining was found in the biofilm and in the final effluent. Microplastics less than 500 μm in size with the abundant microplastics types of fiber, and fragments were detected most frequently in the samples. Moreover, polyester (PES) and cotton were the dominant polymer types in the biofilm reactor detected by attenuated total reflection Fourier transform infrared spectroscopy (ATR-FTIR).

1. Introduction

Microplastics are microscopic particles that are less than 5 mm in size and are an emerging environmental problem [1]. Microplastics, both primary and secondary, are found in the environment. Microbeads (facial scrub cleansers, toothpaste, shower gel, cleanser) were the first primary microplastics that can be directly discharged into wastewater through human activities [2]. Secondary microplastics generate from environmental mechanical, photo-oxidative, physicochemical, and biotic processes cause bigger plastics to break down, deteriorate, and fragment into sizes smaller than 5 mm, [3]. Publications on microplastics have involved the ability concern in the food chain, as they may be consumed by people and cause health problems [4, 5]. The toxicity of microplastics in water has been shown, and their harmful impact on the aquatic environment has been widely reported.

A growing number of studies focus on microplastics in influent, effluent, and sludge in wastewater treatment plants (WWTPs) as the potential release of microplastics into the aquatic environment [6]. Although removal effectiveness can exceed 99 % in some cases, a large number of MPs are still released into the environment via WWTPs, contributing to aquatic ecosystem pollution. Despite the fact that these WWTPs were designed to remove organics and nutrients instead of microplastics, numerous investigations have demonstrated that they may remove microplastics [7, 8, 9]. Therefore, it is urgent to study the occurrence of microplastics through different treatment technologies and understand the mechanism of removing microplastics before they enter water bodies and accumulate in the environment over time [7].

Biological wastewater treatment has been the most extensively employed technology to remove organic contaminants and nutrients in recent years, because of its high treatment efficiency and cost-effectiveness [10, 11]. Attached growth systems, such as fixed-bed biofilm reactors (FBBRs) or moving bed biofilm reactors (MBBRs), offer a viable alternative for removing organic and nutrient contaminants from wastewater, such as residential wastewater [12]. Although the previous research on the occurrence and removal microplastics at full-scale WWTPs has been extensively investigated, their occurrence and removal by laboratory-scale biofilm reactors remain almost unknown. The aim of this study was to examine the occurrence, fate, and removal efficiency of each different treatment unit on a lab-scale biofilm reactor.

2. Materials and methods

2.1. Experimental setup

A lab-scale biofilm reactor, anoxic – anoxic – aerobic (A^2O), was set up and a 4-stage treatment process (shown in figure 1), was consisted of anoxic-1 (Anx-1) working volume 30.4 L, anoxic-2 (Anx-2) 16.75 L, aerobic (Ae) 31 L and sedimentation (Se) 13 L.

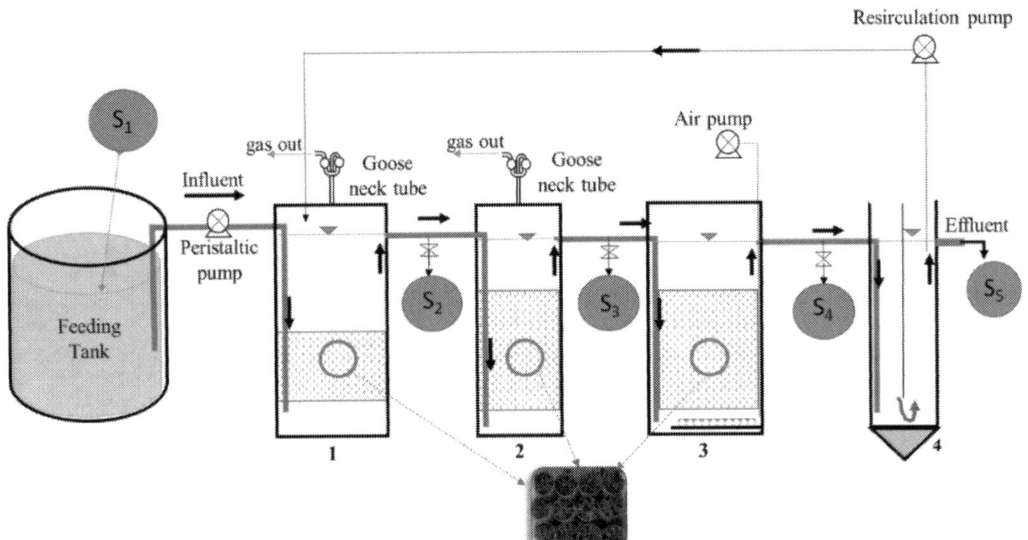

Figure 1. Schematic diagram of biofilm reactor A^2O process (1) anoxic-1, (2) anoxic-2, (3) aerobic, (4) sedimentation.

Each reactor (A^2O) was filled with bio-carrier module made from polyethylene terephthalate (PET) plastic bottle waste. The filling ratio of each A^2O were 40%, 80%, and 80% respectively. PET as attached growth media has some advantages such as lightweight, inertness, large surface area, and can attach bacteria well [13, 14, 15, 16, 17]. PET has SSA 555.78 m^2/m^3 [18]. The A^2O reactor was made of plexiglass with a total working volume of 91,5 L. Air pump was installed at the bottom of the aerobic reactor. The influent was pumped continuously 24 h from the feeding tank into anoxic zone by a peristaltic pump (Watson Marlow 323 Germany). The wastewater then flowed to the next reactor by gravity. The reactor was equipped with a recirculation system, the liquid from the sedimentation has flowed into the anoxic-1. To obtain anoxic conditions, dissolved oxygen (DO) less than 0.5 mg O_2/L, N_2 gas was injected into the reactor to remove DO in wastewater for 15-20 minutes [18]. The N_2 gas injection pressure was set at ± 4.5 lbf/in^2. Anoxic reactors had a glass goose-neck tube at the top of the reactor to release gases generated during the denitrification process. During the experiment period, the operating temperature was kept in a constant temperature room and pH in 6-8. Sampling points were S1, S2, S3, S4, and S5, collected when the reactor reached steady-state. The reactor run continuously for 24 hours at a 36-hour HRT with a discharge of 2.53 L/hour with a 200% recirculation ratio.

2.2. *Raw wastewater*
The raw wastewater source was collected from the inlet tank at the residential WWTP in Neglasari district RT 9 RW 2 (6°53'31.8"S 107°38'27.2"E in Bandung), West Java Province, Indonesia. The WWTP handles 300 inhabitants and the effluent discharge to Cidurian River. The main characteristics of influent wastewater COD, TN, and TP were 621.28 ± 25.73 mg/L, 56.20 ± 2.08 mg/L, and 6.36 ± 0.30 mg/L respectively.

2.3. *Sample processing*
The processing of samples followed the method described in Nur et al. [18]. To avoid contamination, all equipment used in the sampling protocol was thoroughly rinsed with tap water before use. Various sample processing for identification and quantification methods for microplastics were employed. There were 3 types of samples processing for microplastics; wastewater, sludge, and biofilm. To remove natural organic from samples, 30% H_2O_2 solution was added to the samples and digested using a magnetic stirrer under heat 60^0 C until H_2O_2 fully evaporated [19]. Artificial wastewater was filtered using a glass microfiber filter Whatman GF/C and dried in an oven heat 60 ^{0}C for 30 minutes [20]. Different processing from wastewater, sludge and biofilm samples were dried in an oven heat 60 ^{0}C over 24 h. When it dried, mixed with distilled water and digested with 30% H_2O_2 [21]. The next step was using density separation by using $ZnCl_2$ solution (density 1,5 g/cm^3) [22] for 24 hours. The supernatant of both was filtered under vacuum onto a glass microfiber filter paper (Whatman GF/C).

2.4. *Microplastic identification and quantification*
The identification of microplastics followed previously described in Alam et al. [20], Hidalgo-Ruz et al. [23], and Nur et al. [18]. Filter papers were observed by visual observation using an Olympus CX-21 light binocular microscope with 100x magnification equipped with Optilab Viewer 3.0 software. Observations on the surface of the filter paper in a zigzag manner. The measurement of particles by using Image Raster 3.0 software. The chemical composition of the suspected particles above 500 μm were analyzed with imaging at Attenuated Total Reflection-Fourier Transform Infrared (ATR-FTIR) (Bruker Alpha II Platinum ATR).

3. Results and discussion
3.1. *Microplastic removal efficiency*
A total of 810 ± 10 MP/L were identified as micropastics in the feeding tank/inlet of the A^2O reactor (untreated domestic wastewater from residential WWTP). Figure 2 shows the evolution microplastics removal mechanism in different step treatment processes in A^2O reactor over 53 days after the steady-state reached.

It could be seen clearly that a great proportion of microplastics removal in the effluent of anoxic-1 was 72.43%, from 810 to 223,33 MP/L. The removal was considerably bigger in anoxic 2, decreasing from 223,33 to 66,67 MP/L, or 91.77%. The removal in aerobic was likewise greater by 98.35 %, in addition, there was a little increasing removal in sedimentation effluent or biofilm reactor effluent. As a result, there were 0.82% microplastics in the biofilm reactor effluent. Finally, the effluent of wastewater samples gave a removal efficiency of 99.18%. Despite the fact that this reactor was designed to eliminate organics and nutrients from domestic wastewater, this A^2O configuration can remove up to 98% of microplastics, which is in line with the previous studies [7, 8, 24].

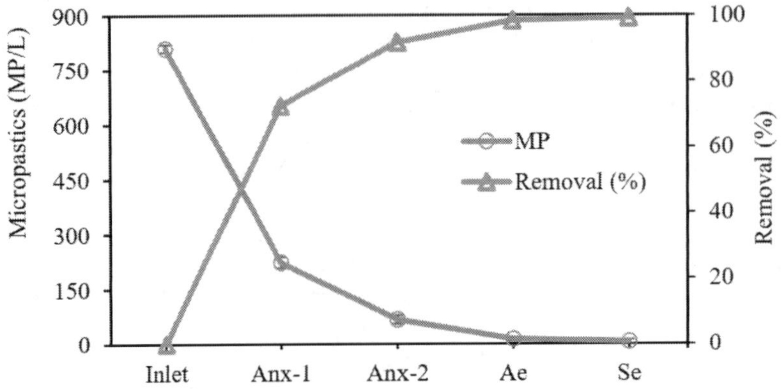

Figure 2. The occurrence and removal of microplastics in the effluent of each treatment A^2O reactor.

3.2. The occurrence and fate of microplastics in the reactor

Figure 3 illustrates the occurrence and fate of microplastics in each stage of the biofilm reactor (wastewater, sludge, and biofilm). Samples were taken on the last day of reactor operation. The concentration of total microplastics in the influent was 10 MP/L and the final effluent concentration of total microplastics was 10 MP/L. Microplastics detected in wastewater, particularly in anoxic-1, anoxic-2, aerobic, and sedimentation effluent ranged from 27.57 to 0.82%. Microplastics more typically occurred in anoxic reactor biofilms than in aerobic reactor biofilms. Microplastics flowed from bottom to top through the PET module (up-flow), smaller particle sizes and low density were trapped and stuck in the PET module (figure 4) and some of them were adsorbed [24]. The adsorbed microplastics in the biofilm were 7.96-9.24% at the start-up process in the anoxic batch reactor [18] and accumulated on the biofilm in this investigation. The highest microplastics concentration was observed in the sludge phase 491.67 12.58 MP/kg (w/w), 358.33 16.07 MP/kg (w/w), and 155.00 21.79 MP/kg (w/w) at Anx-1, Anx-2, and Ae. This was due to that most microplastics particles were trapped with suspended particles and settled as sludge on the bottom of the reactor. In the biofilm rector, microplastics concentration decreased from Anx-1 to Anx-2, and Ae and Se (effluent) continuously flowed via the PET module.

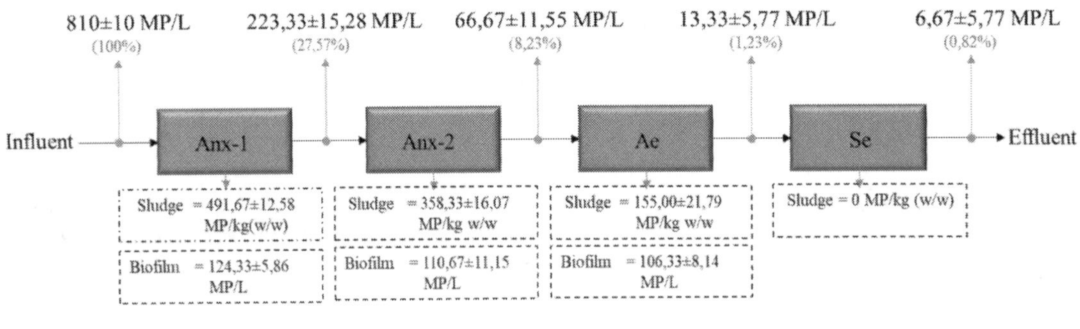

Figure 3. The occurrence of microplastics in the biofilm reactor

Figure 4. Microplastic trapped and stuck inside the PET module (adapted from Wang et al. [29])

3.3. *Influence of the microplastics shape and size on the removal of microplastics*

Figure 5 shows the size distribution of microplastics in wastewater (figure 5a), biofilm (figure 5b), and sludge (figure 5c). The figure shows that the presence of microplastics with diameters of 500 μm was higher in all samples. Fiber and fragment forms dominate over other forms, accounting for about 66.82% and 31.09%, respectively. Fibers and fragments were the most common type of MPs detected in the influent and effluent of WWTPs, which was similar to previous research [6, 7, 9]. Transparent was the most dominant color observed under the microscope. The significance of color in this investigation is the simplicity with which suspected microplastics particles may be identified.

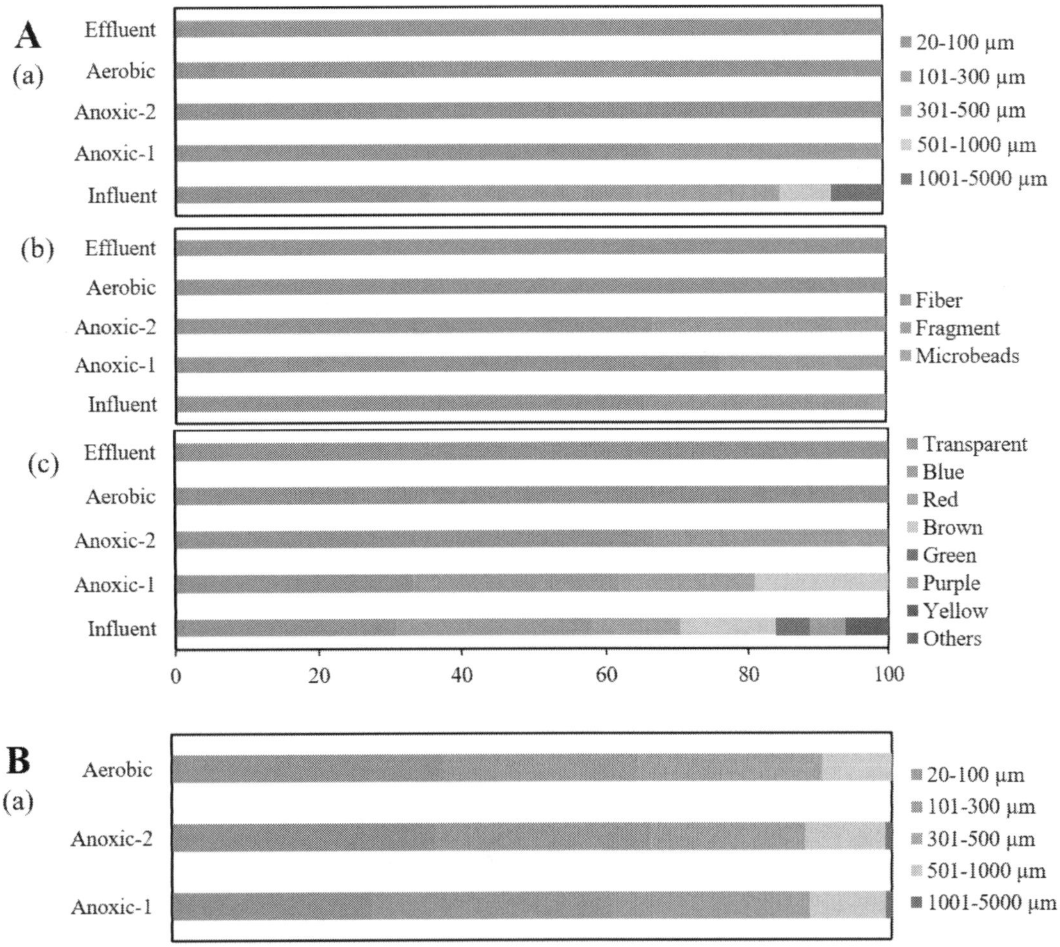

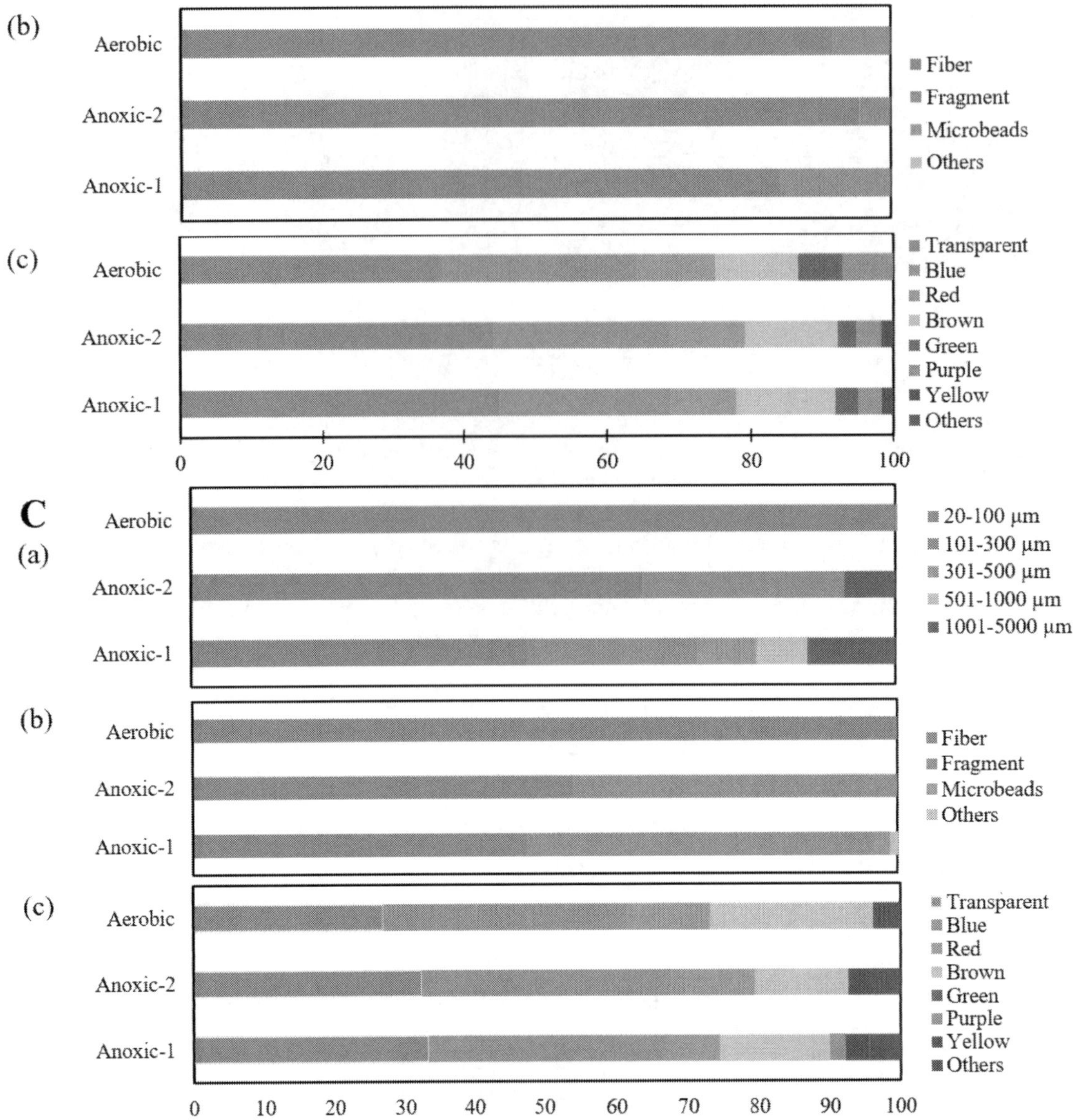

Figure 5. Microplastics distribution in A. Wastewater, B. Biofilm, C. Sludge (a) shape, (b) size, (c) color

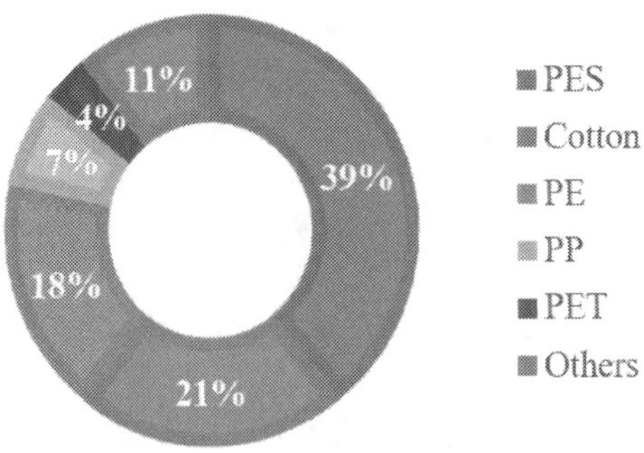

Figure 6. Pie chart of microplastics polymer distribution in the biofilm reactor

Fibers were easily trapped during treatment due to settling [29]. Fragments, on the other hand, showed great removal effectiveness throughout the secondary treatment phase. Fragments with a lamellar structure gradually agglomerated and were ingested by the activated sludge [9]. The microplastics that entered the biofilm treatment (biofilter) unit were less in particle size and density. Biofilm or biofilter technology continues to have the best microplastic removal performance [6, 24]. Biofilter technology combined physical and biological purification processes, with biofilm filtration and adsorption serving as the primary mechanisms for microplastics removal.

Polymers identified by FTIR were polyester (PES) as the most abundant (39%), followed by cotton (21%), polyethylene (PE, 18%), polypropylene (PP, 7%), and polyethylene terephthalate (PET, 4%). In addition, 11% of total microplastics were identified as other polymer materials, like polyamide (PA), low-density polyethylene (LDPE), and wool (figure 6). Domestic washing machine discharges with a large number of fibers released and related to wastewater from the manufacturing of synthetic clothes and natural fibers from cotton and linen were explained as the source of fibers found in WWTPs [25]. Meanwhile, fragments are produced by the degradation of plastic bags, packaging products, and tires erosion [7]. PES, PET, and cottons are synthetic textiles and cloth washing activities that are the main sources of household microplastics [23]. PE and PP are widely used in the manufacture of plastic products, including plastic food bags, and plastic bottles likely originated from kitchen products [9]. In this study, the PET polymer spectrum from bottle products was not observed under FTIR, indicating a positive sign for the use of PET as a biofilm-carrier media. Despite the fact that PET scrap/cutting can be fragmented at a very slow rate in an aqueous environment [30]. It means that the use of PET as a biofilm carrier is safe when used in the full-scale residential WWTPs, it will not contribute to microplastics pollution to the aquatic environment.

4. Conclusion

The total amount of microplastic in residential wastewater entering the reactor was 810 MP/L. The biofilm reactor (A^2O set-up) is effective in removing microplastics (99.18%). The highest removal efficiencies could reach 98.35% in aerob reactor, followed by 72.43 – 91.77% in anoxic reactor. The highest of the microplastics were trapped in the sludge phase, while the rest were adsorbed in the biofilm, trapped, and stuck inside the PET module. Microplastics less than 500 μm was dominant size in all samples. The treatments removed all sizes and shapes of microplastics. Transparent microplastics were the majority of the microplastics in the biofilm reactor. PES, cotton, PE, PP, PET, PA, and wool were the microplastics polymer detected by FTIR, of which PES and cotton were a large portion.

5. Acknowledgment

This research was supported by Riset Unggulan ITB 2021.

6. References

[1] Eerkes-medrano D, Thompson RC and Aldridge DC 2015 *Water Res.* **75** 63–82
[2] Napper IE, Bakir A, Rowland SJ and Thompson RC 2015 *Mar. Pollut. Bull.* **99** 178–185
[3] Penalver R, Arroyo-Manzanares N, Lopez-García I and Hernandez-Cordoba M 2020 *Chemosphere* **242** p 125170
[4] Cox KD, Covernton GA, Davies HL, Dower JF, Juanes F and Dudas SE 2019 *Environ. Sci. Technol.* **53** 7068–7074
[5] Revel M, Châtel A and Mouneyrac C 2018 *Environ. Sci. Health* **1** 17–23
[6] Liu F, Nord NB, Bester K and Vollertsen J 2020 *Water* **12** p 1085
[7] Carr S A, Liu J and Tesoro AG 2016 *Water Res.* **91** 174–182
[8] Talvitie J, Mikola A, Koistinen A and Setala O 2017 *Water Res.* **123** *401-407*
[9] Lares M, Ncibi MC, Sillanpaa M 2018 *Water Res.* **133** 236-246
[10] Zhang Y, Zhao X. Zhang X, Peng 2015 *Water Supply* **15** 442-455
[11] Xu S, Jiang Q 2018 *J. Clean. Prod.* **194** 416-424
[12] Vandith V, Setiyawan AS and Soewondo P 2018 *MATEC Web of Conferences* **147** p 04004

[13] Wąsik E and Chmielowski K 2017 *Ecological Engineering* **106** 378-384
[14] Ahn KH, Hwang JH, Song KG, Jung HY, Cho ES, Lim BR, Dezotti and Kim KS 2004 *Metals and Materials International* **10** 167-170
[15] Espinoza K, Fernandez C, Perez J, Benalcazar D, Romero D and Lapo B 2019 *Rev. Tec. Ing. Uviversitas Zulia* **42** 67-75
[16] di Biase A, Maciej SK, Tanner RD, Jan AO 2019 *J. Environ. Manage.* **247** 849–866
[17] Dorji U, Tenzin U, Dorji P, Pathak N, Johir MAH, Volpin F, Dorji C, Chernicharo, CAL, Tijing L, Shon H and Phuntsho S 2021 *Process. Saf. Environ. Prot.* **148** 370-381
[18] Nur A, Fauzi M, Soewondo P, Setiyawan AS and Oginawati K 2022 *GEOMATE Journal* **22** 63-70
[19] Nur A, Soewondo P, Oginawati K and Setiyawan AS 2020 *Online International Symposium Sanitation of Value Chain (SVC)*
[20] Alam F, Sembiring E, Muntalif BS and Suendo V 2019 *Chemosphere* **224** 637 – 645
[21] Copper DA and Corcoran PI 2010 *Mar. Pollut. Bull.* **60** 650-654
[22] Coppock RL, Cole M, Lindeque PK, Queiros AM and Galloway TS 2017 *Environ. Pollut,* **230** 829-837
[23] Hidalgo-Ruz V, Gutow L, Thompson, RC and Thiel M 2012 *Env. Sci. Technol.* **46** 3060-3075
[24] Liu W, Zhang J, Liu H, Guo X, Zhang X, Yao X, Cao Z and Zhang T 2019 *Environ. Int.* **146** p 106277
[25] Sun J, Dai X, Wang Q, Van Loosdrecht MCM and Ni BJ 2019 *Water Res.* **152** 21-37
[26] Kang P, Ji B, Zhao Y and Wei T 2020 *Sci. Total Environ.* **745** p 140943
[27] Hamidian AH, Ozumchelouei EJ, Feizi F, Wu C, Zhang Y and Yang M 2021 *J. Clean. Prod.* **295** p 126480
[28] Okoffo ED, O'Brien S, O'Brien JW, Tscharke BJ and Thomas KV 2019 *Environ. Sci.: Water Res. Technol.* **11** 1908-1931
[29] Wang Z, Sedighi M and Lea-Langton AR 2020 *Water Res.* **184** p 116165
[30] Nur A, Setiyawan AS, Oginawati K and Soewondo P 2022 *GEOMATE Journal* **22** 10-16.

Biological Oxygen Demand (BOD) Simulation in Sunter River Using Non Orthogonal Water Quality Model Transport

Rommy Martdianto[1], Muslim Muin[2], Ricky Lukman Tawekal[2], Arung Bahari Muslim[3], Heru Nur Alam[4]

[1]*Doctoral Program of Ocean Engineering, Institut Teknologi Bandung, Indonesia*
[2]*Department of Ocean Engineering, Institut Teknologi Bandung, Indonesia*
[3]*Ocean Engineering Research Group, Institut Teknologi Bandung, Indonesia*
[4]*Lembaga Afiliasi Pengembangan dan Industri (LAPI-ITB), Bandung, Indonesia*

Corresponding author: ***rommy.m193@gmail.com***

Abstract. Water pollution problem in Jakarta has been going on for a long time. The existence of some rivers that are not directly connected to upstream in the tidal area can speed up the occurrence of water pollution and slow the circulation of polluted waste that affects water quality in river-estuary. Since 2010 after Kanal Banjir Timur (KBT) operated, Sunter River streamflow was divided by KBT, changes flow rate and water quality pattern. The study area is the main channel of the Sunter River which has 17 km lengths includes Duren Sawit, Pulo Gadung, Kelapa Gading, Plumpang, Rawa Badak and Tanjung Priok Area where the average water load is 624 mg/L. The simulation is divided into 2 scenarios with a flowrate configuration of 5 m^3/s and 100 m^3/s (with a raining condition of 1 hour). MuQuald3D, a three-dimensional water quality model, has been employed in this research. The merit of this model is that it can accurately predict water pollution load within the surface layer and the bottom layer and can be shown in contour profile which is effective and efficient guidance to solve the water quality problems along Sunter river-estuary. The model results show the effect of salinity will gradually reduce BOD concentration, and the existence of permanent construction around estuary (breakwater, port) changes pollutant distribution to the West and East area and increases the time of decomposition in waters, especially in low flowrate conditions.

Keywords: BOD, Non-Orthogonal Boundary Fitted, Water Quality Modeling, Jakarta, Sunter River

1. Introduction

Estuaries are part of the productive environment on earth, including coastal areas with many components of the aquatic environment compared to the components of the freshwater environment. There is an interaction of land-sea influences in the estuary, representing the transition area between land and sea with unique characteristics, including the transition of fresh water and seawater. Jakarta Bay as a form of the estuary correlates with the urban environment in Jakarta. Sunter River channels in the study area not connected directly to upstream cause flowrate reduction, and lowing circulation of waste pollution will affect water quality in the northern area of Jakarta. This research will simulate water quality dynamics in rivers into Jakarta Bay to update maximum BOD load limit based on hydrodynamic simulations regarding water quality.

Jakarta as one of high city population in the world has complexity environmental problems which gradually cause environmental degradation. This actual condition cause research topic to be important to speed up efforts of water pollution control in Jakarta. Influence of urban drainage system modification in Jakarta should be follow by review for terms of water pollution control policy, which is still oriented to land use, watersheds, and have not yet considered river-estuaries at northern side of Jakarta.

Several previous studies have been conducted about water pollution problems and the hydrodynamic application such as Traiano Canal where salinity pattern indicates intrusion affected all branches of the canal [1]. Another research conducted to Southwest Coast of India [2] shows that some waste is discharge into the estuary where oxygen levels can be maintained by high and periodic photosynthetic activity as the estuary flushing process. So, in this we conduct a research of BOD simulation in Sunter River using application called MuQual3D. The application is developed by Muin [3], [4] where the model based on non-orthogonal boundary fitted technique. The advantages of this model are the model grid arrangement can be applied on complex geometric and fast computational time.

2. Description of Study Area

Study area characteristic for water quality problem in Jakarta shown in **Figure 1**. The study area is located in the Jakarta area, bordered by the West - East canal, not directly connected to the upstream, influenced by tides, which causes water quality problems. Sunter river section after KBT channel crosses Cipinang, Jatinegara, Pulo Gadung, Kelapa Gading, Koja, and Tanjung Priok areas, with the dominant land use around river as residential, market, industrial, and strategic infrastructure in the downstream segment.

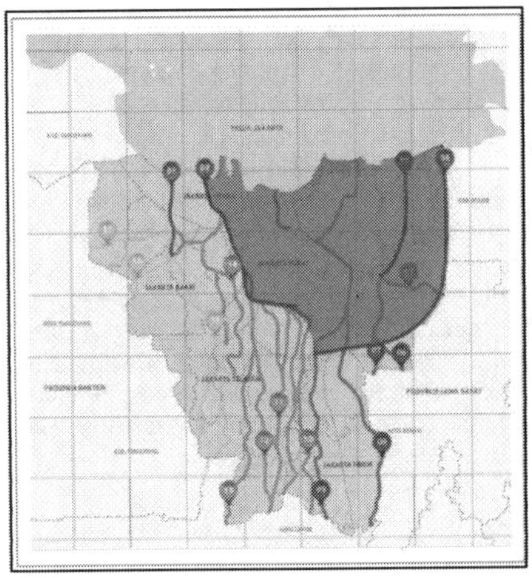

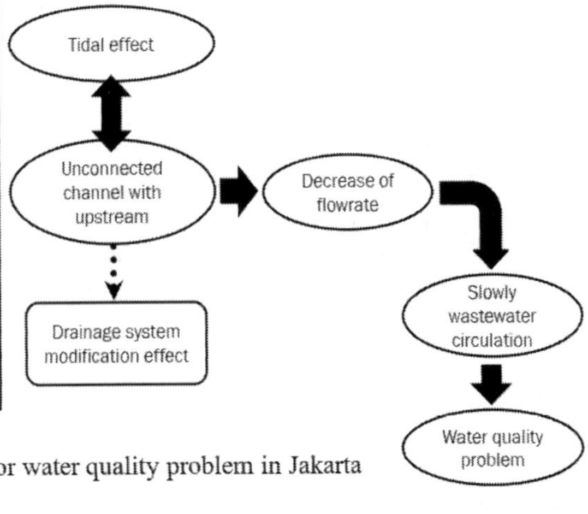

Figure 1. The characteristic of study area for water quality problem in Jakarta

The high density of buildings and residents along the Sunter River channel based on the characteristics of the study area causes a high percentage of domestic waste as the main component of the distribution of BOD in the study area. Along the tidal zone of the Sunter River, there are 4 inlets which are channels that originated from many residence criteria from luxury (examples : Kelapa Gading, and Pulomas), until slum residence area (examples : Koja, Rawa Badak, Tugu, Sunter) as high dense population along Sunter river, as shown in **Figure 2**.

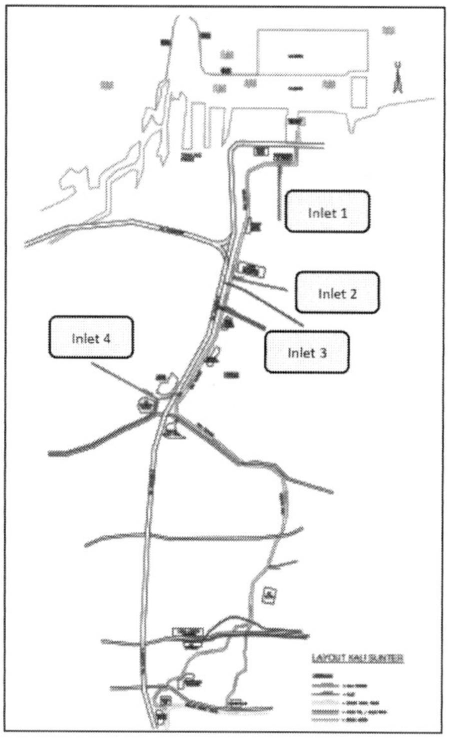

Figure 2. Sunter river segment after KBT branching channel as the study area

Based on **Figure 2**, continue to estimate BOD from Sunter River segment by considering the allocation of pollution loads and river flows. The tabulation for BOD load shown in **Table 1** as follows:

Table 1. Waste load allocation calculation Sunter River

City	Districts	Villages	Area (km^2)	Popullation (Org)	FED BOD (gr/org/hr)	mg/day	Q (L/day)	C (mg/L)
East Jakarta	Pulogadung	Pulo Gadung	1,92	40335	160	6.453.600.000	7.776.000	830
		Jati	2,16	34576	160	5.532.160.000	7.776.000	711
		East Pisangan	1,79	43184	160	6.909.440.000	7.776.000	889
		Cipinang	1,53	37426	160	5.988.160.000	7.776.000	770
		Kayu Putih	4,37	48289	160	7.726.240.000	7.776.000	994
North Jakarta	Kelapa Gading	West Kelapa Gading	4,53	42838	160	6.854.080.000	12.096.000	567
	Koja	Koja	3,27	38424	160	6.147.840.000	12.096.000	508
		North Rawa Badak	1,33	37614	160	6.018.240.000	12.096.000	498
		South Rawa Badak	1,02	49132	160	7.861.120.000	12.096.000	650
		Lagoa	1,57	63492	160	10.158.720.000	12.096.000	840
		North Tugu	3,32	73833	160	11.813.280.000	12.096.000	977
		South Tugu	2,68	54334	160	8.693.440.000	12.096.000	719
	Tanjung Priok	Tanjung Priok	5,54	43846	160	7.015.360.000	38.880.000	180
		Kebon Bawang	1,73	64759	160	10.361.440.000	29.030.400	357
		Sungai Bambu	2,36	37811	160	6.049.760.000	27.648.000	219
		Sunter Jaya	4,58	78028	160	12.484.480.000	22.118.400	564

According to the waste load calculation in **Table 1**, the average waste load was around 624 mg/L. Meanwhile, based on DKI Jakarta Governor Regulation 582/1995, the BOD limit for river flow is 20 mg/L. Ministry of Environment stated that based on regulation 112/2003, BOD limit for domestic waste load is 100 mg/L. So the problem of BOD content in Sunter River needs to be addressed to prevent widespread water pollution due to increased BOD concentration.

3. Measurement Parameters

Measurement along dry season for temperature and salinity was done in 21 September 2021 by water quality measurement tools Horiba, along the segment of Sunter River. Continue with DO-BOD laboratory tests have also been done in 21 February 2022. The scheme of the river line for measurement and laboratory test are shown in **Figure 3**.

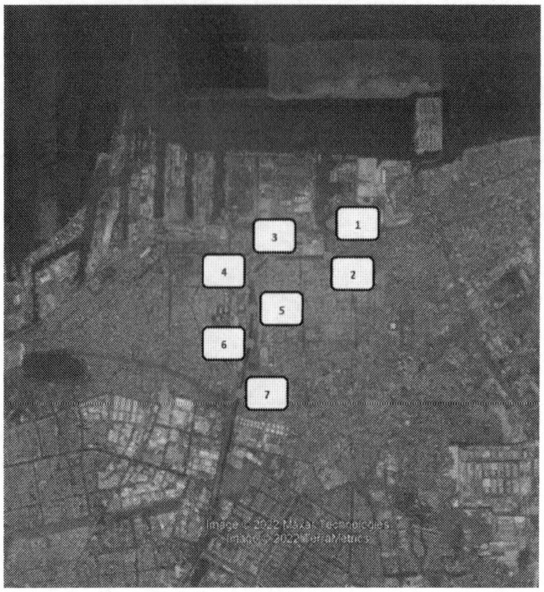

Figure 3. Measurement point along Sunter river

Monitoring points to identify water quality parameters value in the Sunter River are shown in **Table 2**.

Table 2. Measurement of water quality sample Sunter River

Observation Date : 11 September 2021			Observation Date : 21 February 2022	
Point	**Temperature (C)**	**Salinity (ppt)**	**DO (mg/L)**	**BOD (mg/L)**
1	-	-	3	11
2	31.3	2.8	-	-
3	30.5	3.2	-	-
4	30.5	2.8	-	-
5	29.9	2.6	-	-
6	30	2	-	-
7	29.5	0.4	-	-

According to the laboratory result in February 2022, the Sunter River sample shows that the DO values 3 mg/L and for BOD is 11 mg/L. The measurement method to measure DO is based on SNI 06-6989.14-2004 and for BOD based on SNI 6989.72-2009.

4. Model Setup

4.1 Water Quality Model Description

The mathematical model used in this research is a model developed by Muin [3]. The governing equation of this model is derived in spherical coordinate to accommodate the rugged shape of the earth. Momentum and conservation of mass equations are conducted using the Non-Orthogonal Curvilinear coordinate technique to solve three-dimensional (3D) problems. This technique accommodates a very large time step that implies a brief simulation time and accurate results even in areas with very complicated geometries.

MuQual3D predicts the dynamic of conservative and non-conservative pollutant transport and dispersion of the discharged material. The equation for conservation of substance and the equation of state are also required to take into account salinity, and temperature. These equations are as follows:

Conservation of Substance

$$\frac{\partial q}{\partial t}+\frac{u^c}{r}\frac{\partial q}{\partial \xi}+\frac{v^c}{r}\frac{\partial q}{\partial \eta}+\omega\frac{\partial q}{\partial \sigma}=\frac{4}{D^2}\frac{\partial}{\partial \sigma}\left(D_v\frac{\partial q}{\partial \sigma}\right)$$
$$+\frac{D_h}{r^2J^2}\left[\left(\frac{\theta_\eta\theta_\eta}{\cos^2\theta}+\phi_\eta\phi_\eta\right)\frac{\partial^2 q}{\partial \xi^2}-2\left(\frac{\theta_\xi\theta_\eta}{\cos^2\theta}+\phi_\xi\phi_\eta\right)\frac{\partial^2 q}{\partial \xi\partial \eta}+\left(\frac{\theta_\xi\theta_\xi}{\cos^2\theta}+\phi_\xi\phi_\xi\right)\frac{\partial^2 q}{\partial \eta^2}\right]-Kq \quad (1)$$

Equation of State

$$\rho = function\ of\ (S,\Theta) \quad (2)$$

Where:

ϕ,θ = longitude, latitude
t = time
g = gravitational acceleration
ρ' = vertical density difference ($\rho=\bar{\rho}+\rho'$)
$\bar{\rho}$ = vertically-averaged mass density of water
ρ_o = basin-averaged mass density of water
ρ_s = water density at the surface
ω = vertical velocity in σ-coordinate
σ = sigma stretching coordinate
ζ = water surface deviation from mean level
h = water depth
D = $h+\zeta$
D_v = vertical eddy diffusivity
D_h = horizontal eddy diffusivity
r = radius
R = earth's radius
S = salinity (ppt)
Θ = temperature in centigrade
q = concentration of substance
Ω^n = concentration of n class of suspended sediment material
u^c, v^c = contravariant velocities in the (ξ,η) directions, respectively
J = Jacobian = $\phi_\xi\theta_\eta - \phi_\eta\theta_\xi$
K = Decay Rate

4.2 Model Scenarios

Scenarios for BOD simulation along tides area of Sunter River described in **Table 3** as follows:

Table 3. Model scenarios

Model Scenario	Configuration	Description
Scenario 1	Sunter River flow in normal condition 5 m^3/dt	Only existing flowrate in along the main river.
Scenario 2	Sunter River flow has additional flow up to 100 m^3/dt	An additional flowrate added because rain occurs around the study for 1-hour duration and happened on the 3rd day of the simulation

5. Model Simulation

The model will simulate BOD transport along the tidal area of Sunter River and use 5 points monitoring, representing the main channel of Sunter River from BKT branching channel to estuary. Monitoring points position as shown in **Figure 4**.

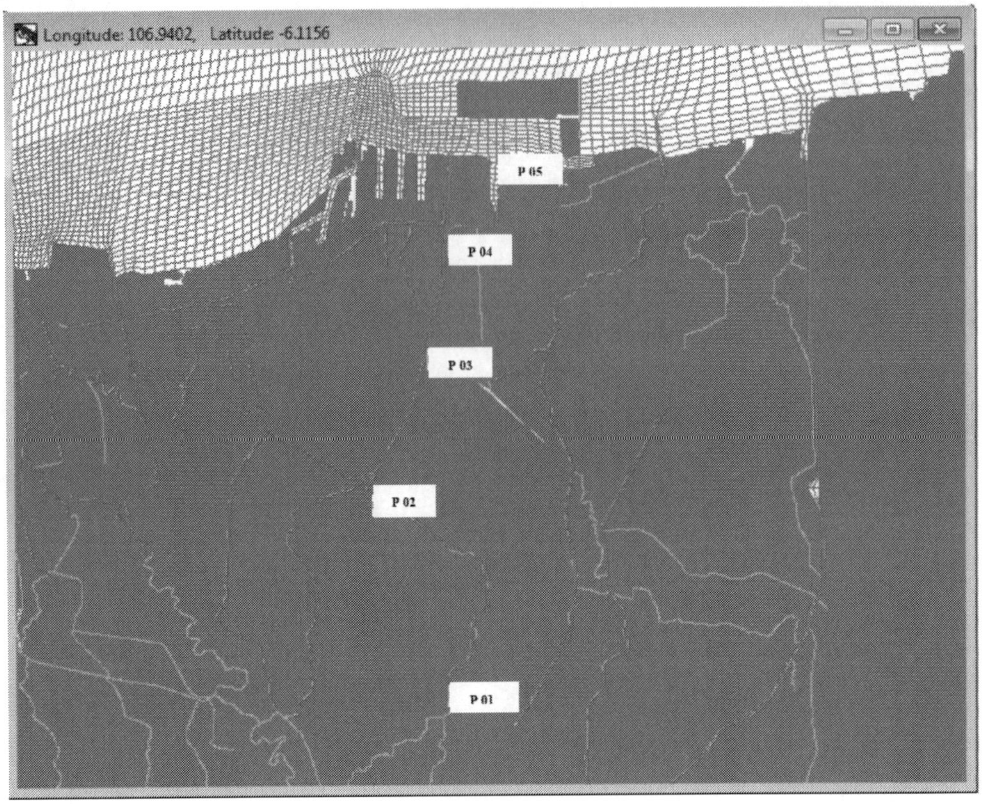

Figure 4. Monitoring points in model simulation

For comparison of BOD distribution along the study area, the model result for all monitoring nodes and all scenarios will combine profiles at the surface and bottom. BOD profile for **Scenario 1** as shown in **Figure 5** – **Figure 9**, contour profile BOD distribution **Scenario 1** as shown in **Figure 10** – **Figure 11** continue BOD profile **Scenario 2** shown in **Figure 12** – **Figure 16**, contour profile BOD distribution **Scenario 2** as shown in **Figure 17** – **Figure 18**.

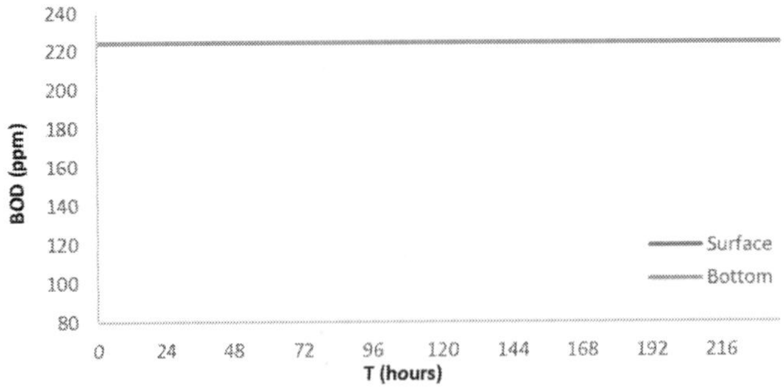

Figure 5. BOD profile P1 Scenario 1

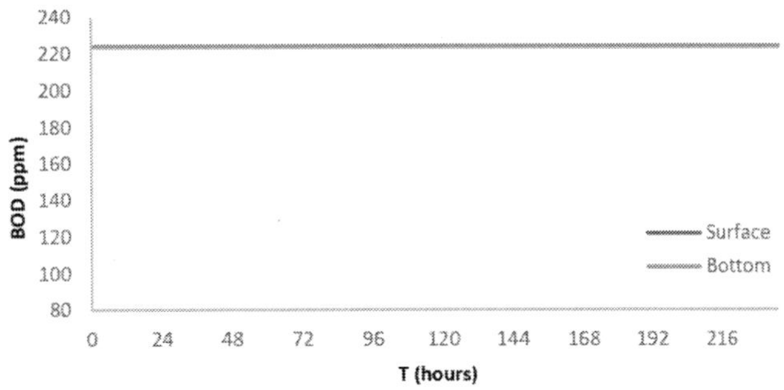

Figure 6. BOD profile P2 Scenario 1

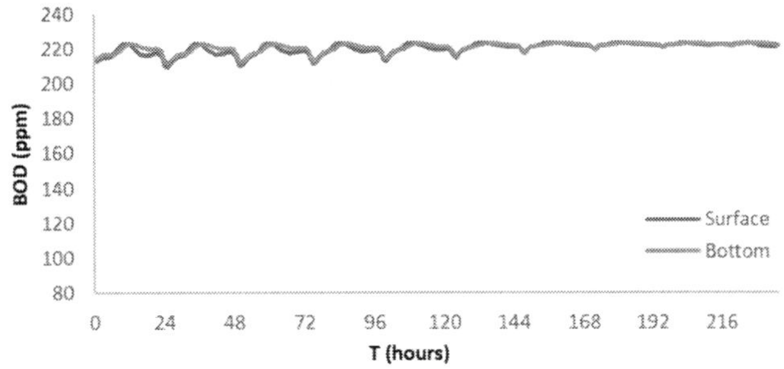

Figure 7. BOD profile P3 Scenario 1

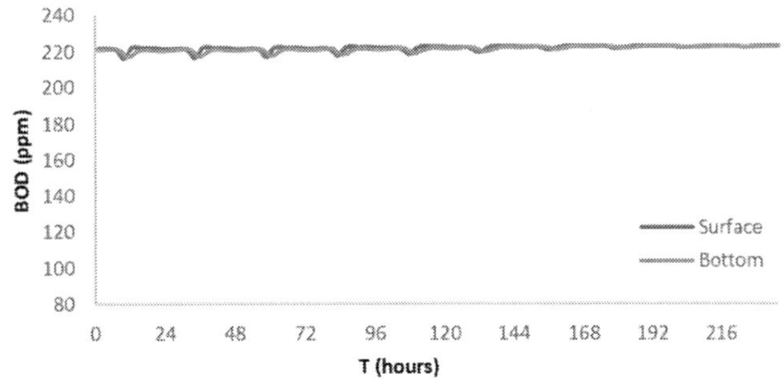

Figure 8. BOD profile P4 Scenario 1

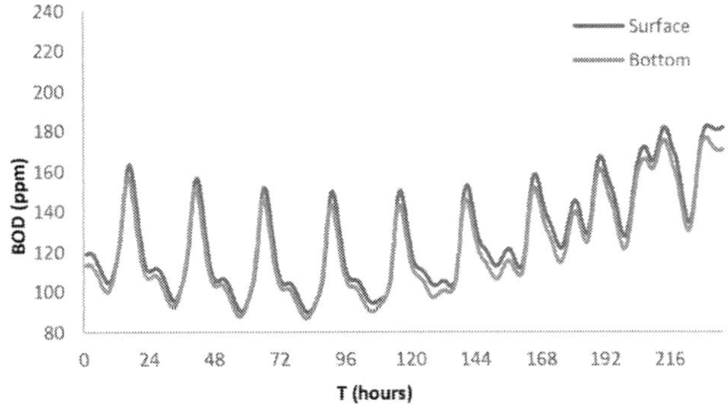

Figure 9. BOD profile P5 Scenario 1

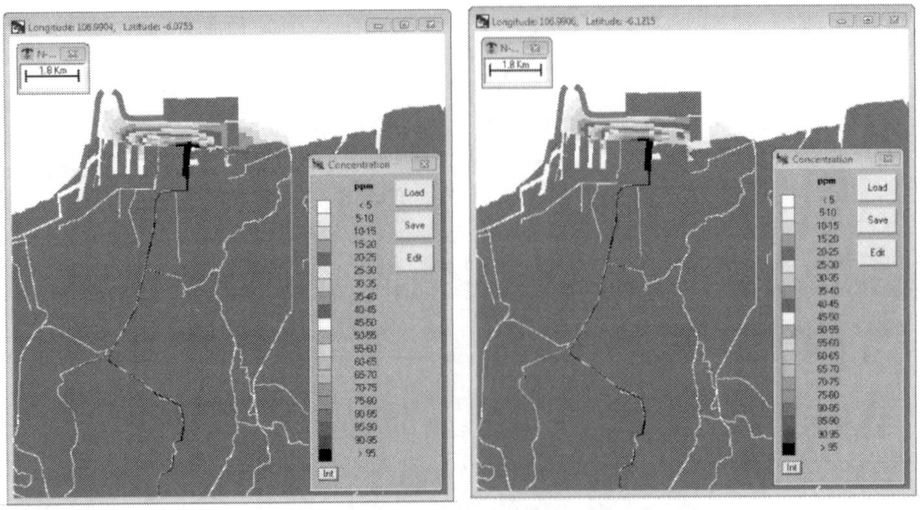

Figure 10. Contour profile surface layer for tidal condition Scenario 1

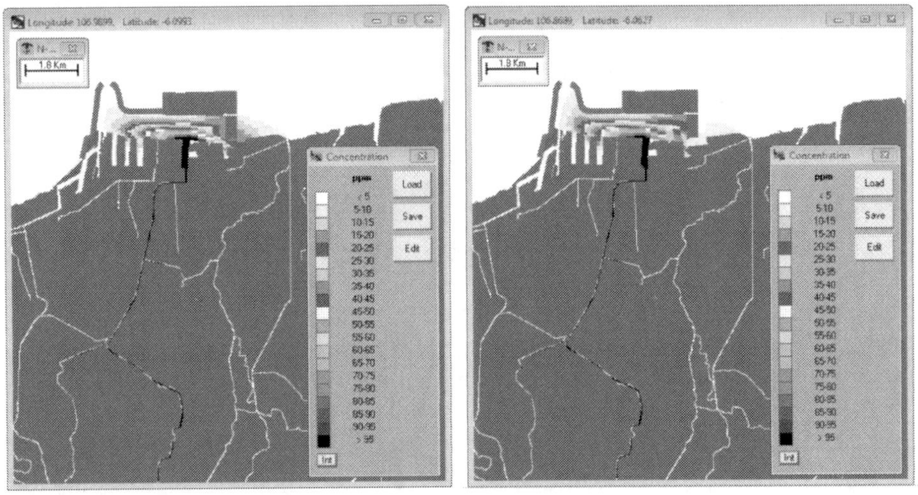

Figure 11. Contour profile bottom layer for tidal condition Scenario 1

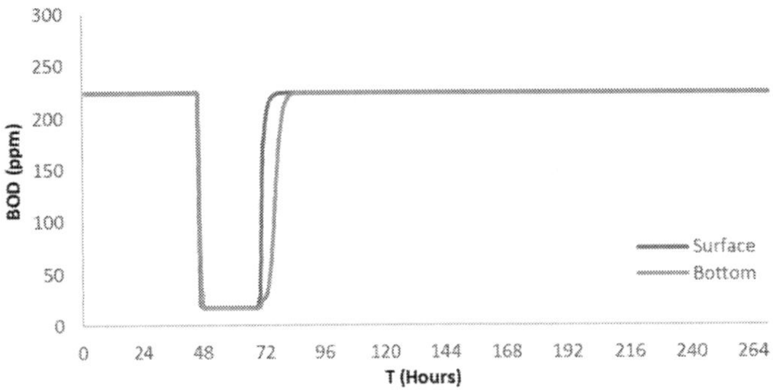

Figure 12. BOD profile P1 Scenario 2

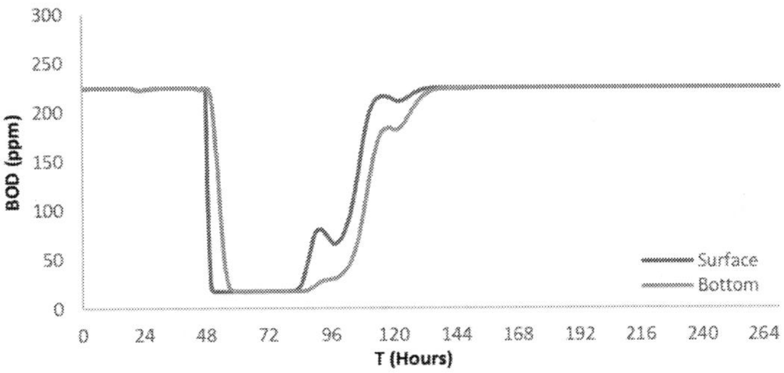

Figure 13. BOD profile P2 Scenario 2

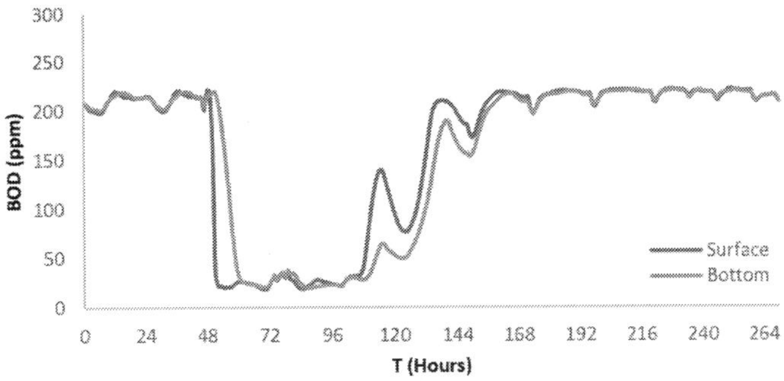

Figure 14. BOD profile P3 Scenario 2

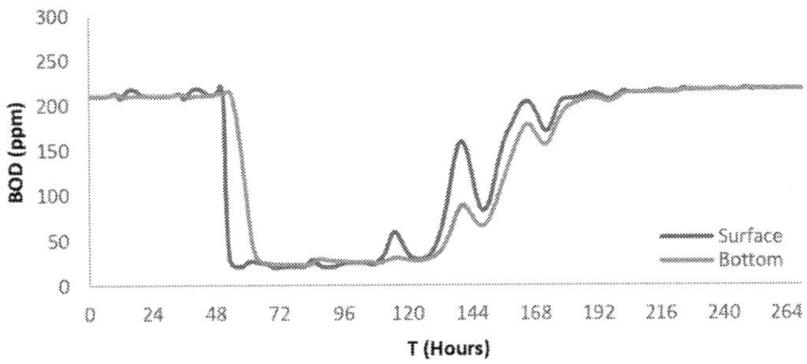

Figure 15. BOD profile P4 Scenario 2

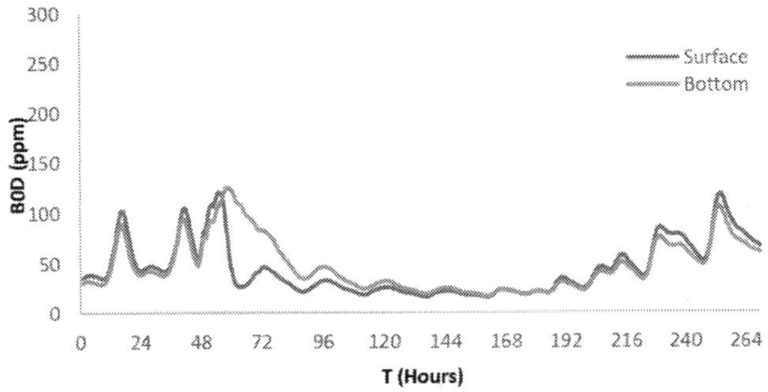

Figure 16. BOD profile P5 Scenario 2

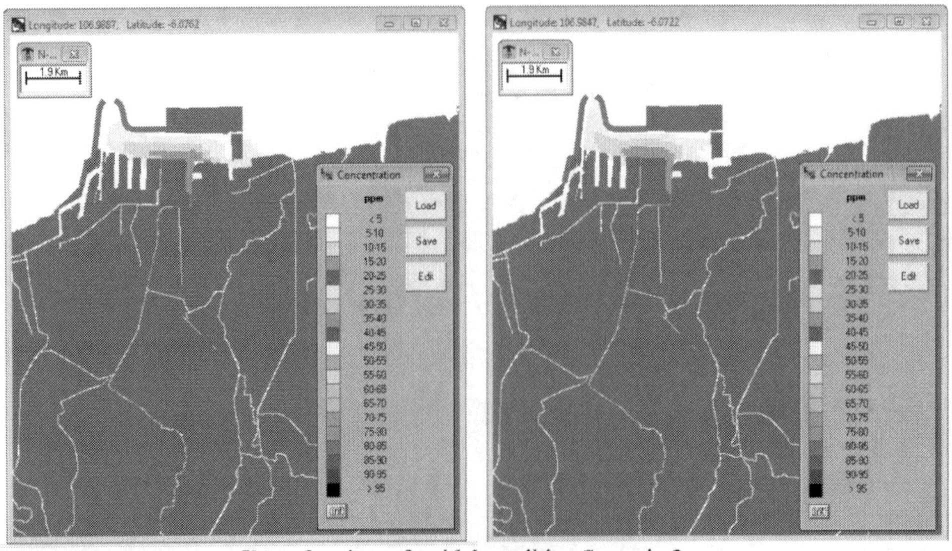

Figure 17. Contour profile surface layer for tidal condition Scenario 2

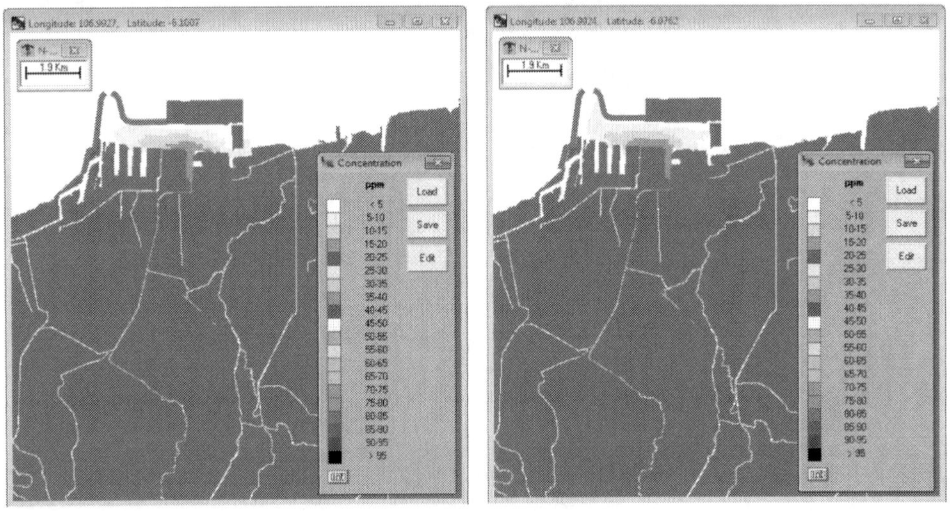

Figure 18. Contour profile bottom layer for tidal condition Scenario 2

6. Model Validation

Model validation by comparing salinity and BOD value concentration according to measurement-laboratory test and model at downstream Sunter River, as shown in **Table 4** and **Table 5**.

Table 4. Comparison salinity measurement and model result in Scenario 1

Location	Salinity Scenario 1 (ppt)	
	Measurement	Model
Downstream	2-3.2	2.2

The comparison Salinity value in **Table 4** found that the range of the model result is in the range of measurement data.

Table 5. Comparison BOD laboratory test and model result in Scenario 2

Location	BOD Scenario 2 (mg/L)	
	Laboratory Test	Model
Downstream	11	15,1

The comparison BOD value in **Table 5** found that the model results is less than laboratory test result. This condition is caused by the unstable rainy season and quantity of rainfall intensity around the study area (estuary), which needs a more specific model scenario to consider dynamic rainfall intensity.

7. Model Result

In **Scenario 1**, modelling that has been composed simulate pollutant concentration of 624 mg/L in river flow of 5 m^3/sec in the study area causes a BOD content of 224 ppm on the surface and bottom of point P1. Conditions at point P2 are relatively the same as P1, and the concentration of BOD gradually fluctuated at points P3 and P4 by 214 – 222 ppm due to the influence of salinity. The end of the Sunter River flow is point P5, with a fluctuating BOD concentration interval of 100 – 165.8 ppm on the surface. The effect of pollutant distribution in the study area is visualized in the contour profiles shown for tidal conditions on the surface and bottom, as shown in **Figure 9** and **Figure 10**, which show a fairly dense BOD concentration along the tidal area in the Sunter River. The concentration gradually decreases in the estuary and spreads to another side of the estuary due to permanent constructions. With scenario 1 simulation, it can be seen that the effect of salinity will gradually reduce the BOD concentration, and permanent construction in the estuary will replace radius of distribution of pollutant concentrations and increase time of decomposition of pollutant concentrations in estuary waters. If Sunter River flowrate in low conditions such as in Scenario 1, cause concentration of BOD in the estuary is quite high (> 95 ppm according to the model result interval) thus result in water quality problem, especially around the Tanjung Priok port breakwater.

For **Scenario 2** modelling that has been composed simulate pollutant concentration 624 mg/L in river flow which increases discharge up to 100 m^3/sec due to rain for 1 hour at the 3rd to 4th hours cause decrease in BOD concentration. At point P1 there was decrease in concentration of BOD on the surface and bottom at the 3rd hour from a concentration of 224 ppm to 17 ppm which lasted for 1 hour, and BOD concentration increased again. Condition of point P2, the effect of additional flow increasing duration for the BOD concentration is at a value of 17 ppm for 10 hours, which gradually increases BOD concentration again. At points P3 and P4 there was already an effect of salinity, thus increasing the transition time for increasing BOD concentrations both on the surface and bottom of about 1 hour. The end Sunter River flow is point P 05 with fluctuating BOD concentrations at 28.8 – 100.7 ppm on the surface. The effect of pollutant distribution in the study area is visualized in the contour profile shown for tidal and tidal conditions on the surface and bottom as shown in **Figure 17** and **Figure 18**, which shows BOD concentrations ranging from 10-25 ppm along tidal area in Sunter River. The concentrations gradually decrease in the estuary and spread to the East and West sides of the estuary due to permanent constructions. Scenario 2 shows that the influence of additional flow and salinity will gradually reduce BOD concentration, and permanent construction around the estuary will replace the radius range of distribution of pollutant concentrations and increase the decomposition time of pollutant concentrations.

8. Discussion

Around the estuary of the Sunter River, there are several strategic infrastructures that important in national and international activities. However, from another perspective, degradation of the aquatic environment is also essential to resolve carefully and precisely.

Important infrastructures such as breakwaters and jetties with massive construction create difficulties in breaking down high pollutant concentrations, especially in low flowrate conditions, caused by river discharge or low tide conditions. From **Figure 19** it can be seen that the distribution of pollutant concentrations from Sunter River has slowed to decompose due to massive infrastructure such as breakwaters and jetties located perpendicular to the end of river. The modelling results also show that pollutant concentrations tend to spread to the East and West sides of the estuary with concentrations grading from > 95 ppm to 10 ppm. The sources of pollution from upstream of Sunter River must also be managed appropriately according to the area's characteristics to reduce the accumulate water quality problems in the estuary, as shown in **Figure 19**.

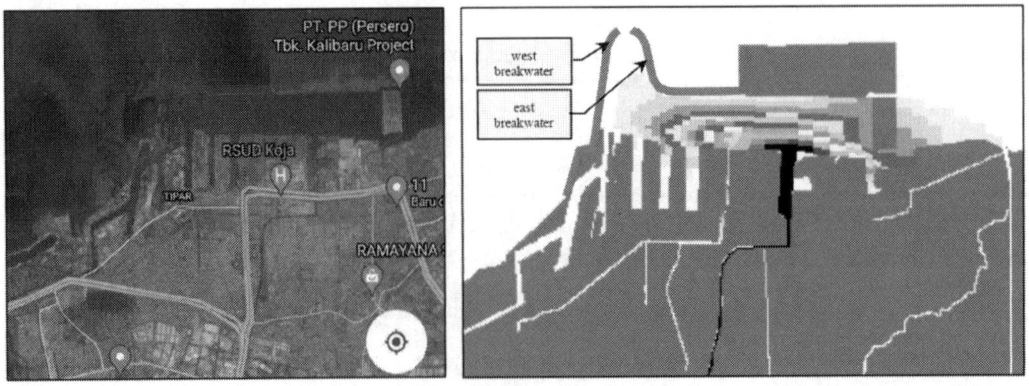

Figure 19. Comparison of existing condition via satellite image and contour profile from the model

9. Conclusion

Waste pollution in tidal area of Sunter River must be controlled as part of efforts to water pollution control in Jakarta. Application of MuQual3D model with Non Orthogonal Boundary Fitted principle used in research can represent the physical condition of Sunter River-estuaries with complex geometries and several urban channel branches. Results of 3D modelling are displayed in the form water quality profile parameters on surface and bottom as a representation of the Z-direction profile and contour profile as a representation of the X-Y plane profile. The complete result from X-Y-Z profile perspective shows good accuracy as a technical guidence in water pollution control efforts in Sunter River-estuary, considering complexity problems along river-estuaries.

The modelling scenarios conclude that the salinity levels and fluctuation discharge inflow can affect the distribution of BOD concentrations along river-estuary. The permanent constructions such as the port of Tanjung Priok breakwater, and the expansion port of Kali Baru, which is located perpendicular to the estuary, cause the pollutant from Sunter River to be blocked and change the pollutant distribution to the West-East direction. This matter can create a potential for new pollutant concentrations along the eastern coastline of Jakarta Bay.

10. Acknowledgements

Authors wish thank you to Lecturers of Ocean Engineering, Ocean Engineering Research Group Faculty of Civil and Environmental Engineering, Bandung Institute of Technology (ITB), which support working research topic, since data development, field observation, hydrodynamic modeling and intensive assistance during research process. The authors also wish thank you to Muslim Muin, PhD who allowed us to use MuQual 3D and comprehensive discussion with Research Team Ocean Engineering ITB.

References

[1] Fabio Manca - Giuseppe Capelli - Francesco La Vigna - Roberto Mazza - Annalisa Pascarella, Wind-induced salt-wedge intrusion in the Tiber river mouth (Rome–Central Italy)
[2] K.S. Anila Kumary, P.K. Abdul Azis and P.Natarajan, 2007, Water quality of the Adimalathura Estuary, southwest coast of India
[3] Muslim Muin, Malcolm Spaulding, 1997, Three Dimentional Boundary Fitted Circulation Model
[4] Muslim Muin, Malcolm Spaulding, 1997, Application of 3D Boundary Fitted Circulation Model to Providence River
[5] Province DKI Jakarta (Map of main rivers and channels in Jakarta)
[6] Robert Thomann, John Mueller, 1987, Principles Water Quality Modelling and Control, Harper & Row Publishers, New York USA
[7] SNI 06-6989.14-2004 tentang Air dan Air Limbah-Bagian 14: Cara Uji Oksigen Terlarut cara Yodometri
[8] SNI 6989.72-2009 tentang Air dan Air Limbah-Bagian 72: Cara Uji Kebutuhan Oksigen Biokimia *(Biological Oxygen Demand/BOD)*

COHESIVE SEDIMENT SIMULATION IN KANAL BANJIR BARAT USING MUSED3D, NON-ORTHOGONAL BOUNDARY FITTED SEDIMENT TRANSPORT MODEL

Denny Yatmadi[1], Muslim Muin[1], Ricky L Tawekal[1], Heru Nur Alam[1],Arung Bahari Muslim[1]
[1]Ocean Engineering Research Group Faculty of Civil and Environment Engineering,Institute Technology Bandung, Indonesia

denny.yatmadi@gmail.com ; muslimmuin@gmail.com ; ricky.tawekal@gmail.com ; herusdc@yahoo.com ; arungbahari15@gmail.com

Abstract. Fine-grained cohesive sediments dominate sedimentation in the lowland area. In the Density Induced Current process, where seawater intrusion occurs, the fine-grained cohesive sediment will be easily flocculated by saltwater and settles rapidly. There are complex problems related to sedimentation in the rivers of Jakarta in the downstream area, which is influenced by tides and the dominance of cohesive sediment. Due to the complex process in the estuary, salinity intrusion will affect the settling velocity. And then, the flocculation process, the river's geometry, will also affect sediment deposition. A proper model is needed to simulate the sedimentation in this area. This study aimed to investigate the distribution of cohesive sedimentation using 3D hydrodynamic and sediment transport model called MuSed 3D. This model will be applied to the Kanal Banjir Barat (KBB) river, Jakarta. The model result shows salinity values in the range of field observations for TSS and salinity. Salinity model present from 1 till 10 ppt and TSS present from 9,8 until 14,2 ppm. This study concludes that the dispersion of sediment cohesive on river and estuary affected by Density Induced Current also depends on salinity and TSS value.

Keyword: Cohesive sediment, salinity, TSS, Density Induced Current, MuSed 3D

1. Introduction

Sedimentation in the downstream river and the bay is dominated by fine cohesive sediments carried by currents from the middle to the downstream. The Density Induced Current process at the estuarine where seawater intrusion occurs, the fine-grained cohesive sediment will be easily flocculated by saltwater and settles in several parts of the river. Determination of settling velocity is an important and challenging factor in studying the sedimentation characteristics of rivers and estuaries. Unlike sand, the settling velocity for cohesive soils is strongly influenced by salinity.

There are several problems related to sedimentation in the rivers of Jakarta in the downstream area. These problems are tides, a dominance of cohesive fine grains, density induced current process in the estuary, salinity which affect the settling velocity of the grains also the flocculation process, the complex geometry river that affect the flow of the estuaries, and land subsidence which will impact the sediment deposition. This study investigates the distribution of cohesive sedimentation at Kanal Banjir Barat (KBB) river. With a minimal salinity content, the cohesive grains are in a separate state and will and will experience flocculation along with the increase in salinity levels [1]. River Estuary worldwide are very dynamic complexes where ocean and land events are important drivers; any modifications in either of these will impact the coastal water body [2]. Mubarak stated that [2] the silting speed is highly dependent on the competition between the forces of the sea (tides), wave, wind, and river discharge.

The tidal flow also affects the small-scale ripples of the suspended load sediment transport, both indirectly and directly[3]. Otherwise, sediments that occur in river areas are influenced by complex natural processes. Due to erosion, sediment entry into rivers is greatly enhanced by human activities such as deforestation, land-use change, agricultural activities, and inappropriate river management [4]. Field measurements of sediment transport, erosion, and deposition in three-dimensional frameworks are rare in the literature when available are only for short periods and within limited estuarine areas [5].

Furthermore, we need a 3D hydrodynamic model so the movement pattern of the cohesive sediment of the river can be adequately predicted. The 3D Ocean Hydrodynamics model used in this study is the Non-Orthogonal Boundary Fitted technique developed by Muin [6]. This model has been widely used both in Indonesia and abroad, such as Providence River [6], San Fransico Bay [7], Bay of Fundy [8]. The Non-Orthogonal Boundary Fitted (NBOF) technique, also known as the Non-Orthogonal

Curvilinear Coordinate Technique used in this model, is derived from the basic Spherical Coordinate equation. Muin also used this model to simulate Drilling cutting at Bituni Bay [9] to assess the impact of drilling activity on shallow water bodies.

2. Salinity, Settling Velocity and Total Suspended Sediment (TSS)

Turbulence and suspended sediment concentration, salinity, dissolved organic matter, flocculation ability, and floc growth rate impact settling velocity [10]. When examining the deposition of estuarine sediments, salinity is a variable that must be considered [11]. According to Muin, settling velocity is affected by salinity and sediment concentration. The relationship between settling velocity between salinity and sediment concentration is shown in Figure.1 [9].

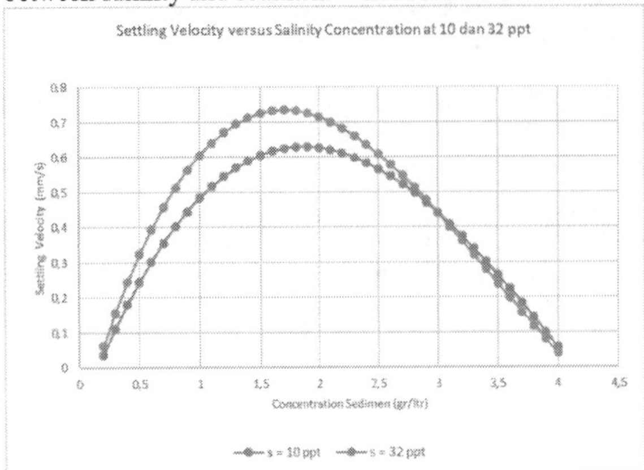

Figure 1. Relationship between Salinity, Settling velocity and Sediment Concentration [9]

Turbidity is the cloudiness of water caused by suspended particles like clay and silts, chemical precipitates like manganese and iron, and organic particles like plant debris and organisms [12]. Suspended sediments are initiated into coastal and inland waters by various sources, including river discharge, dredging activities, resuspension events, and tidal currents [13]. Sediment loads are critical for maintaining sediment accretion rates and preserving the integrity of sediment-maintained geomorphic features such as river deltas, marsh and mangrove wetland platforms, etc. [13].

3. Density Induced Current

Seawater intrusion is defined as the mass transport of saline waters into zones previously occupied by fresher waters [14]. Although the river flow rate was relatively high, the saltwater intrusion was occurred [15]. According to water column stratification or vertical salinity structure, estuaries can be classified as salt wedge, strongly stratified, weakly stratified, or vertically mixed [16].

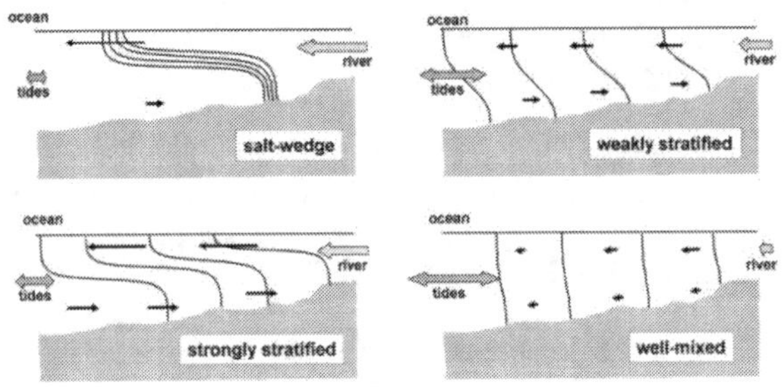

Figure 2. Classification of estuaries based on the vertical structure of salinity [16].

This classification considers between buoyancy forcing from river discharge and mixing from tidal forcing [15]. In the study area, the KBB river flows to Jakarta Bay, which is considered the shallow waters influenced by tides, river discharge, and salinity. A dominant parameter that sufficiently affects the existing flow pattern is a current controlled by density (Density Induced Current/DIC) in these types of waters.

4. Description of Area Study

The area of this study is the Kanal Banjir Barat (KBB) River which is part of the downstream area in the Ciliwung River. The upstream of KBB is started from Manggarai water gate then continuous along Sultan Agung Street, Tanah Abang, and then goes to Jakarta Bay as the downstream with a total length of about 16,5 km (google maps). The upstream was located at -6.208109, 106.848180 and the downstream area ended at -6.103388, 106.767268 was constructed at 1913 to divert the water flowing from the upstream Ciliwung river flows into downstream of Jakarta bay.

Figure 3. BanjirKanal Barat
Source :Google maps [17]

From a hydrology point of view, KBB is part of the Ciliwung watershed and has discharge based on Jakarta Integrated Flood System around 330 m^3/s

5. Research Data

The research data contains two types of data. The first type of information is primary data used for this research collected from the field observations and laboratory tests. Field observations were conducted to assess the salinity and temperature by collecting a water sample and sending it directly to the laboratory. The sample will then be tested in a laboratory to determine the Total Suspended Sediment value and Turbidity. Secondary data originated from supporting data from Google and Google Earth photos.

5.1 Field Observation of Salinity and Temperature

Observations and field tests for salinity were carried out on 30 October 2021. The starting point of observation (zero point) starts from the estuary area goes to inland waters. The parameters observed were temperature and salinity. Figure 4 shows the observation data point and salinity along the KBB River.

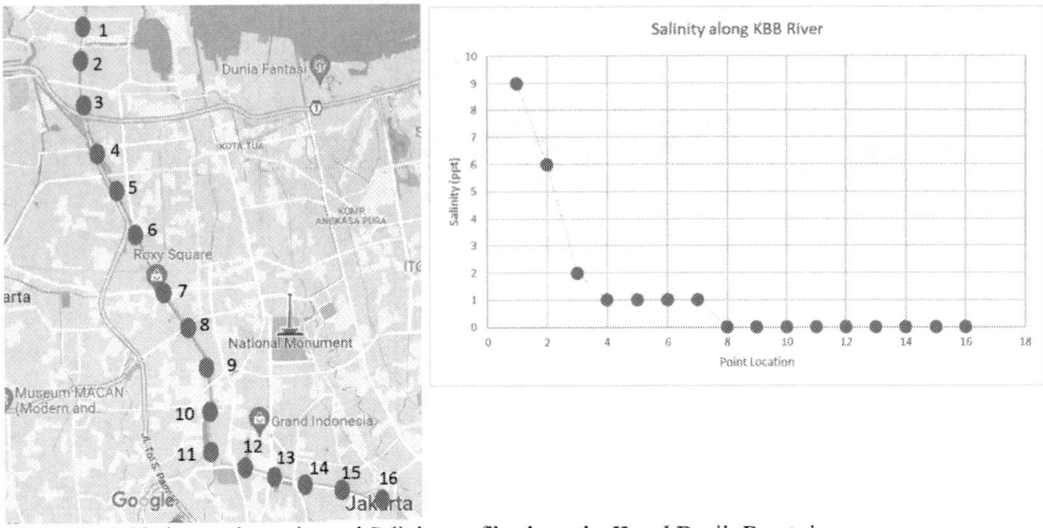

Figure 4. Field observation point and Salinity profile along the Kanal Banjir Barat river
Source : Field observation

From the in situ observations it can be seen that close to the upstream, the salinity value decreases because the influence of the discharge from the river is still strong enough to push the flow towards the estuary. The density-induced current effect in the KBB river occurred from the estuary and then became low value at point 8, around 8 kilometers.

5.2 Laboratory Result

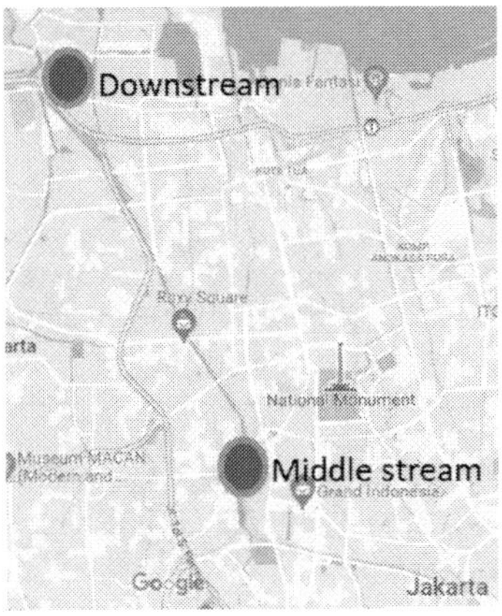

Figure 5. Location TSS Sampling
Source : Field observation

For laboratory testing, the parameter being tested is Total suspended sediment (TSS). Figure 5 shows the location of the sample for Total suspended sediment (TSS). The sample was taken from the site on 30 Oct 2022. Table 1 shows TSS results in the downstream and middle of the Kanal Banjir Barat river.

Table 1. TSS result from Laboratory

Location	Parameter	Unit	Result	Method
Downstream	Total Suspended Sediment	mg/l	3	UP.IK.21.01.07 (Spektofotometre)
Middle stream	Total Suspended Sediment	mg/l	15	UP.IK.21.01.07 (Spektofotometre)

Source : Laboratory test (sample from KBB river)

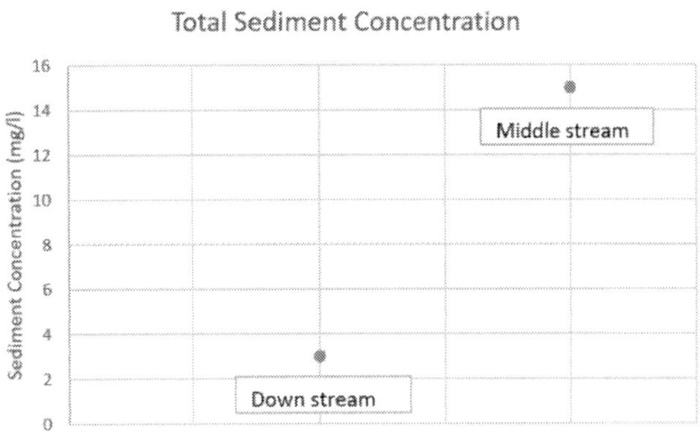

Figure 6. TSS result from KBB river
Source: laboratory result

Figure 6 shows that the river hydraulic parameters (current, cross-section, slope) affect the sediment concentration in the upstream and middle areas. The upstream area is higher than in the middle and downstream areas due to changes in the river's hydraulics.

6. Methodology

6.1 Hydrodynamic Model

In order to simulate the sediment transport processes occurring in estuarine and coastal waters the equations to be solved include fluid continuity, momentum equations in three dimensions, sediment transport equations and turbulence closure equations for the determination of the eddy viscosity and diffusivity [18]. The 3D Ocean Hydrodynamics model used in this study uses the Non-Orthogonal Boundary Fitted technique developed by Muin [6]. This model has been widely used both in Indonesia and abroad, such as Providence River [6], San Fransico Bay [7] , Bay of Fundy[8]. The Non-Orthogonal Boundary Fitted (NBOF) technique, also known as the Non-Orthogonal Curvilinear Coordinate Technique used in this model, is derived from the basic Spherical Coordinate equation. The development stage of the non-orthogonal curved coordinate marine hydrodynamic model in spherical coordinate was further developed to simulate tsunami dispersion and sediment transport. The velocity of the gravitational wave does not limit the time step of the model. The computing domain is divided into three regimes. This model can simulate both cohesive and non-cohesive sediments. Erosion, deposition, and dispersal of suspended sediments are simulated both offshore, in shore.

6.2 Sedimentation Model

The sediment transport model used in this research is MuSed 3D with a Non-Orthogonal Boundary Fitted 3D model. This model has also been widely applied to complicated cases and has good results.

Suspended sediment equation

$$\frac{\partial \Omega^n}{\partial t} + \frac{u^c}{r}\frac{\partial \Omega^n}{\partial \xi} + \frac{v^c}{r}\frac{\partial \Omega^n}{\partial \eta} + (\omega - \omega^n{}_s)\frac{\partial \Omega^n}{\partial \sigma} = \frac{4}{D^2}\frac{\partial}{\partial \sigma}\left(D_v \frac{\partial \Omega^n}{\partial \sigma}\right) + \text{Erosion}^n \quad (1)$$

$$+\frac{D_h}{r^2 J^2}\left[\left(\frac{\theta_\eta \theta_\eta}{\cos^2\theta} + \phi_\eta \phi_\eta\right)\frac{\partial^2 \Omega^n}{\partial \xi^2} - 2\left(\frac{\theta_\xi \theta_\eta}{\cos^2\theta} + \phi_\xi \phi_\eta\right)\frac{\partial^2 \Omega^n}{\partial \xi \partial \eta} + \left(\frac{\theta_\xi \theta_\xi}{\cos^2\theta} + \phi_\xi \phi_\xi\right)\frac{\partial^2 \Omega^n}{\partial \eta^2}\right]$$

Deposit sediment equation

$$\frac{dm_d}{dt} = \omega_s C\left(1 - \frac{\sigma_b}{\sigma_d}\right) \text{ for } \frac{\sigma_b}{\sigma_d} < 1 \quad (2)$$

$$\frac{dm_d}{dt} = 0 \text{ for } \frac{\sigma_b}{\sigma_d} > 1 \quad (3)$$

Where:

$\frac{dm_d}{dt}$ = deposit sediment

C = Concentration sediment

σ_b = Bottom shear stress

σ_d = Crtitical shear stress for deposit sedimen (m/s)

ω_s = Settling velocity (m/s)

IOP Conf. Series: Earth and Environmental Science 1065 (2022) 012014 doi:10.1088/1755-1315/1065/1/012014

6.3 Erosion of Sediment

Understanding the incipient motion of sediment particles in sediment transport is vital to understanding the incipient threshold of different properties of sediment under various flow conditions [19].

Sediment Erosion equation

$$Erosi^n = E_0\left(\left(\frac{\sigma_b}{\sigma_e}\right)^n - 1\right)\frac{\rho_s}{h_{cell}} \text{ for } \frac{\sigma_b}{\sigma_e} > 1 \quad (4)$$

$$Erosi^n = 0 \text{ for } \frac{\sigma_b}{\sigma_e} < or = 1 \quad (5)$$

Where :

Erosi^n	=	erosion rate
σ_e	=	Critical shear stress of sediment
E_0	=	erosion rate coefficient
D	=	h+ ζ (m)
ω	=	vertical velocity inσ - coordinate(m/s)
ω_s	=	settling velocity inσ - coordinate (m/s)
τ_s	=	surface shear stress (Pa)
τ_b	=	bottom shear stress (Pa)
S	=	salinity (ppt)
Θ	=	temperature in centigrade
n	=	the class of suspended sediment
q	=	concentration of conservative substance such as S orΘ
Ω^n	=	concentration of n-th class of suspended sediment material

6.4 Model Set Up

Figure 7 shows the description flow of the model set up on the model. Figure 8 shows the grid model for the KBB river to simulate the sedimentation on the Kanal Banjir Barat (KBB) River.

(a) Grid System

For modeling purposes, the area of study is divided into computational grids. The open boundary on the model is placed as far from the study site as possible, which can reduce the effect of the open boundary on the simulation model results. The grid used for the presented area is shown in **Error! Reference source not found.**.

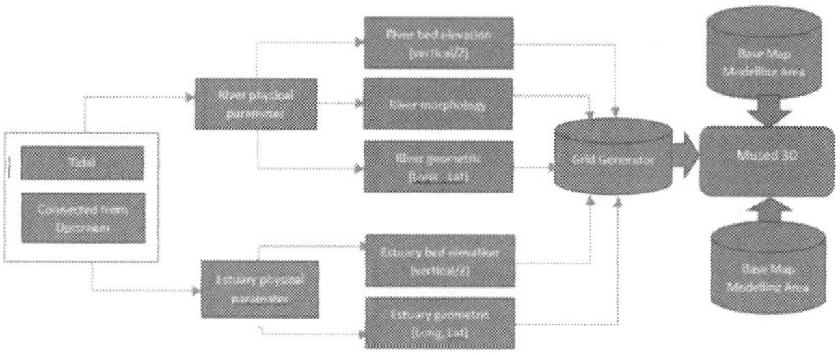

Figure 7. Model Set Up
Source : Mused 3D Hand book

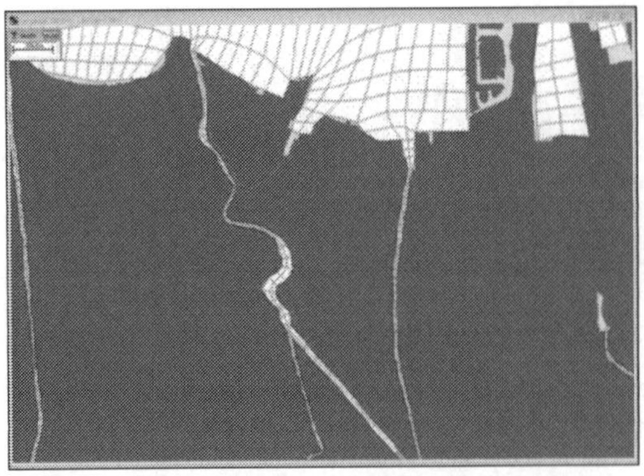

Figure 8. Grid model for KBB river
Source : MuSed 3D Model

(b) Open Boundary

Seven main tidal components (M2, S2, N2, O1, K1, P1, and Q1) are used to calculate the tidal elevation along the open boundary. The tidal components are obtained from the larger computational domain in Jakarta Bay.

(c) River Boundary

At upstream will be used water discharge and TSS concentration as input in the computing area.

Table *2* shows modelling scenarios used in this study for normal conditions.

Table 2. Scenario of model

	Q (m $^3/_s$)	TSS (ppm)
Normal	5	15

6.5 Model Calibration

The success of the model application depends on the credibility of the model results. This credibility is obtained by comparing the model results with measurements. The purpose of calibration is to select the model coefficients to match the observed data. Table 3 shows the comparison of salinity between the measurement and the model at the locations of points 1 and 2. Table 4 shows the TSS concentration comparison between model and observation at location 2. Table 3 and Table 4 show salinity and TSS result between model and observation. The model shows good agreement because the observation value is in the same range as the model.

Table 3. Comparison of Salinity between Model and Observation

Location	**Salinity (PPT)**	
	Model	**Observation**
1	1 - 17	6

2	1 – 26,6	9

Table 4. Comparison of TSS between Model and Observation

Location	TSS Concentration (mg/l)	
	Model	Observation
2 (Downstream)	1 – 9,8	3

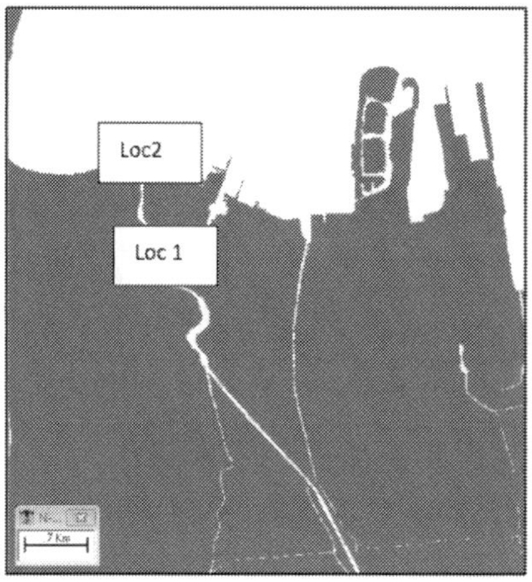

Figure 9. Location of Model and Measurement Comparison
Source : MuSed 3D Model

7. Model Simulation Result

The model using MuSed3D Non Orthogonal Boundary Fitted was implemented on the KBB river, as previously stated by employing one scenario, normal discharge with 5 m^3/s.

7.1 Salinity result

MuSed results for salinity are visualized in the figure below:

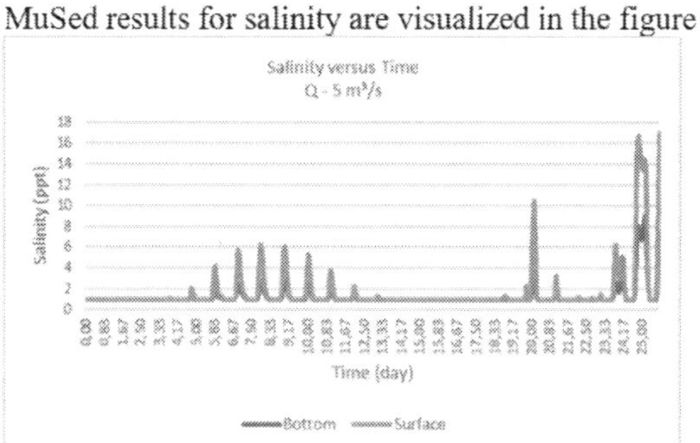

Figure 10. Salinity on location 1
Source : MuSed 3D Model

IOP Conf. Series: Earth and Environmental Science 1065 (2022) 012014 doi:10.1088/1755-1315/1065/1/012014

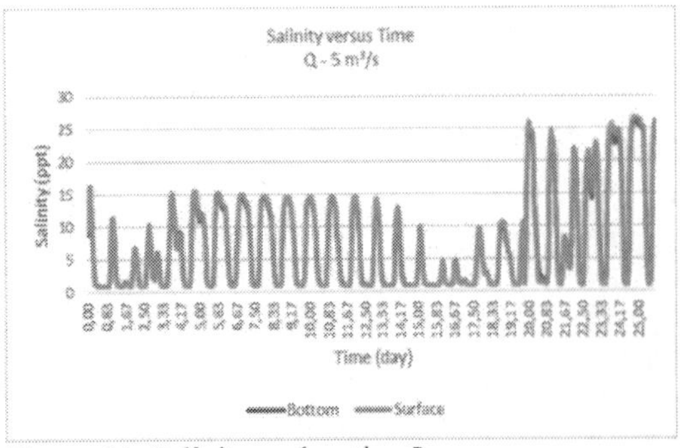

Figure 11. Salinity on location 2
Source : MuSed 3D Model

From Figure *10* until Figure *11* , it is clear that the salinity value varied with time, with the maximum being 26 ppt and the minimum is 1 ppt. It is reasonable to state that the salinity value is affected by density-induced current.

7.2 Total Suspended Solid (TSS) Result

MuSed modeling for TSS are visualized in *Figure 12* and *Figure 13* below.

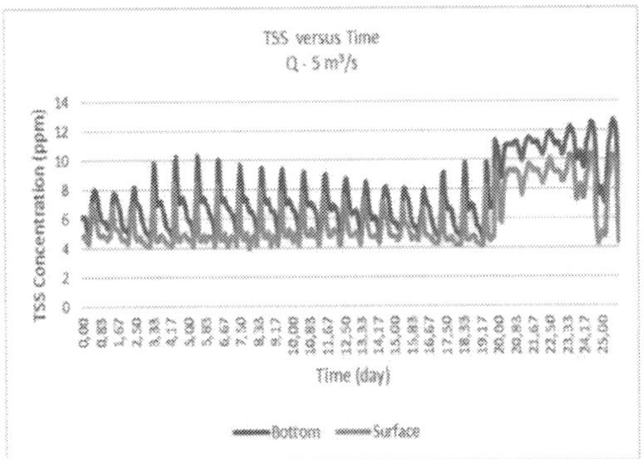

Figure 12 TSS on location 1
Source : MuSed 3D Model

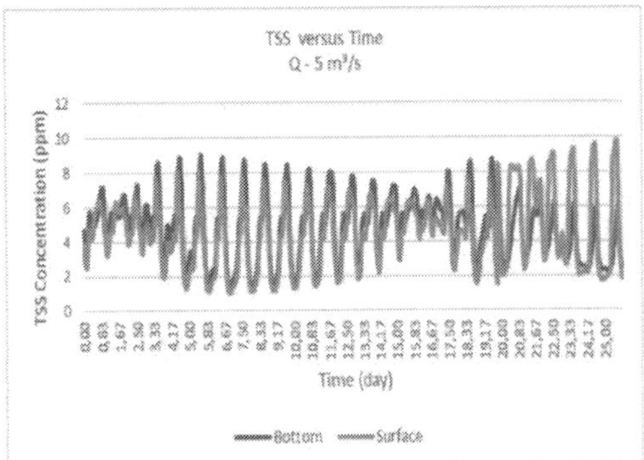

Figure 13. TSS on location 2
Source : MuSed 3D Model

The density induced current might also affect the TSS value. From Figure 12 until Figure 13, the TSS profile value varied with time at a different location. This TSS's value describes cohesive sediment dispersion in the estuary. Dispersion of cohesive sediment for normal discharge with 5 m^3/s can be presented at Figure 14 below:

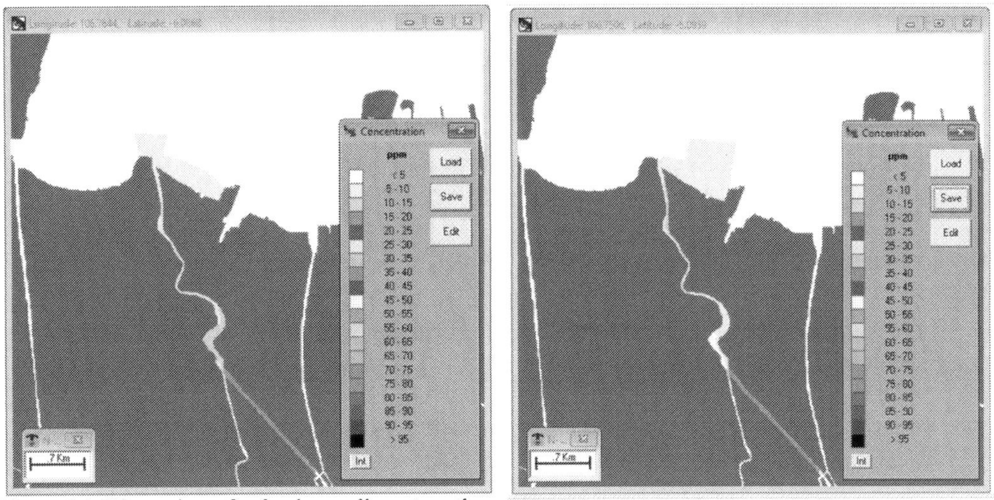

Figure 14. Dispersion of cohesive sediment on the estuary
Source : MuSed 3D Model

8. Conclusion

The value of salinity, TSS, and the river's cohesive sediment distribution are affected by density-induced current. Figure 12 and Figure 13, shows TSS result between model and observation was in good agreement because the observation value is inside the range of the model results. Further investigation is required, especially for Total Suspended Solid (TSS) because the TSS profile will affect the cohesive sediment dispersion. Also for salinity result as shown from Figure 10 until Figure 11, modelling result in a good agreement as well. Laboratory experiments are required to determine other parameters than salinity and TSS, such as critical shear stress and erosion rate, which also affect sediment dispersion.

Reference

[1] L. van Rijn *et al.*, "EFFECT OF CHANNEL DEEPENING ON TIDAL FLOW AND SEDIMENT TRANSPORT," *Photosynthetica*, vol. 2, no. 1, pp. 1–13, 2018, [Online].

[2] M. Mubarak, R. Rifardi, A. Nurhuda, and A. Sulaiman, "The properties of Sediment transport mechanism in River Mouth of Rupat Strait," no. December, pp. 6–12, 2020, doi: 10.20944/preprints202012.0531.v1.

[3] L. Brakenhoff, R. Schrijvershof, J. van der Werf, B. Grasmeijer, G. Ruessink, and M. van der Vegt, "From ripples to large-scale sand transport: The effects of bedform-related roughness on hydrodynamics and sediment transport patterns in delft3d," *J. Mar. Sci. Eng.*, vol. 8, no. 11, pp. 1–25, 2020, doi: 10.3390/jmse8110892.

[4] B. Choubin, H. Darabi, O. Rahmati, F. Sajedi-Hosseini, and B. Kløve, "River suspended sediment modelling using the CART model: A comparative study of machine learning techniques," *Sci. Total Environ.*, vol. 615, pp. 272–281, 2018, doi: 10.1016/j.scitotenv.2017.09.293.

[5] W. Liu, M. Hsu, and A. Y. Kuo, "Modelling of hydrodynamics and cohesive sediment transport in Tanshui River estuarine system , Taiwan," vol. 44, pp. 1076–1088, 2002.

[6] B. M. Muin and M. Spaulding, "(A," pp. 2–12, 1997.

[7] S. Sankaranarayanan and D. F. McCay, "Three-Dimensional Modeling of Tidal Circulation in Bay of Fundy," *J. Waterw. Port, Coastal, Ocean Eng.*, vol. 129, no. 3, pp. 114–123, 2003, doi: 10.1061/(asce)0733-950x(2003)129:3(114).

[8] S. Sankaranarayanan and D. F. McCay, "Application of a two-dimensional depth-averaged hydrodynamic tidal model," *Ocean Eng.*, vol. 30, no. 14, pp. 1807–1832, 2003, doi: 10.1016/S0029-8018(03)00014-3.

[9] M. Muin, "Integration Lagrangian Sediment Transport Into Non-Orthogonal Ocean Hydrodynamics Model to Simulate Drilling Cutting and Mud Dispersion in Indonesia Sea Water," 2013, vol. All Days.

[10] Z. Xue, R. He, J. P. Liu, and J. C. Warner, "Modeling transport and deposition of the Mekong River sediment," *Cont. Shelf Res.*, vol. 37, pp. 66–78, 2012, doi: 10.1016/j.csr.2012.02.010.

[11] L. I. Portela, S. Ramos, and A. T. Teixeira, "Effect of salinity on the settling velocity of fine sediments of a harbour basin," *J. Coast. Res.*, vol. 165, pp. 1188–1193, 2013, doi: 10.2112/si65-201.1.

[12] WHO/FWC/WSH/17.01, "WATER QUALITY AND HEALTH - REVIEW OF TURBIDITY : Information for regulators and water suppliers," pp. 1–10.

[13] S. V. Balasubramanian *et al.*, "Robust algorithm for estimating total suspended solids (TSS) in inland and nearshore coastal waters," *Remote Sens. Environ.*, vol. 246, no. July 2019, p. 111768, 2020, doi: 10.1016/j.rse.2020.111768.

[14] J. J. Bear and H.-D. A. Cheng, "Seawater Intrusion," *Model. Groundw. Flow Contam. Transp.*, pp. 593–636, 2010, doi: 10.1007/978-1-4020-6682-5_9.

[15] F. Manca, G. Capelli, F. La Vigna, R. Mazza, and A. Pascarella, "Wind-induced salt-wedge intrusion in the Tiber river mouth (Rome-Central Italy)," *Environ. Earth Sci.*, vol. 72, no. 4, pp. 1083–1095, 2014, doi: 10.1007/s12665-013-3024-5.

[16] C. Issues, E. P. Edited, and V. E. More, "De fi nition and classi fi cation of estuaries," pp. 1–11, 1963.

[17] "https://www.google.co.id/maps/search/banjir+kanal+barat+jakarta/@-6.1599851,106.7991998,12.63z?hl=en&authuser=0." .

[18] P. Taylor, B. Lin, and R. a Falconer, "Numerical modelling of three- dimensional suspended sediment for estuarine and coastal waters Numerical modelling of three-dimensional suspended sediment for estuarine and coastal waters Modélisation numérique tridimensionnelle de sediments en suspension ," no. March 2013, pp. 37–41, 2010.

[19] C. Shu, G. Tan, P. Chen, J. Wang, and P. Lv, "Experimental study on incipient shear stress of consolidated cohesive sediment," *Arab. J. Geosci.*, vol. 13, no. 18, 2020, doi: 10.1007/s12517-020-05978-4.

The challenges of implementing green factors in urban greening schemes in Indonesia

M Z Dahlan[1,2], B Faisal[1,2], S Chaeriyah[2], I W Hutriani[2] and M Amelia[2]

[1] Master Program of Landscape Architecture, Institut Teknologi Bandung, Indonesia
[2] Research Group of Urban Landscape Hub, Bandung, Indonesia.

E-mail: mzd19@itb.ac.id

Abstract. Recently, new planning tools have emerged to support urban landscape planners in achieving sustainability through optimizing the urban greening scheme. Some cities have approved the green factor tool to increase the quantity and quality of urban green spaces. This greening scheme is considered suitable for a relatively dense city where challenging to provide adequate urban green areas. This short communication challenges the usefulness of the green factor tool and how it relates to the existing urban green space planning procedures through a qualitative case study in Jakarta, Indonesia. The result shows that while the green factor tool performs well, improvements could be made primarily in convincing stakeholders regarding the new paradigm in calculating green areas in relatively dense urban areas. Furthermore, trials in actual cases need to be carried out to examine the formula's effectiveness. Regarding the usefulness of this tool, we conclude that while increasing the number of green areas, improving its quality is more necessary to build resilient infrastructures.

Keywords: Green space, Green factor, Urban landscape, Greening scheme, Urban planning

1. Introduction

Recently, new planning tools have emerged to support urban landscape planners in achieving sustainability through optimizing the urban greening scheme. The Green Factor tool, with varying names applied, has appeared in many cities around the world to increase both the quantity and quality of urban green spaces. The Green Factor tool offers a numeric value for the ratio between the built area and green areas within a selected area. A green factor aims to improve the ecological sustainability of the built environment by increasing the total green area [1]. The City of Berlin has started this Green Factor tool by implementing the Biotope Area Factor in 1994 to reforest the city more effectively and efficiently [2]. Furthermore, the City of Malmo in Sweden adopted the scheme in 2005, the City of Helsinki developed the Helsinki Green Factor in 2012 [3], and the City of London began implementing the London Urban Green Factor in 2018 [5,23].

Jakarta is a megapolitan city in the tropics with a high population density, limited availability of green space, and an expensive land price, which results in the difficulty of providing green open spaces [3,4]. To increase the quantity and quality of green open spaces, Jakarta developed the green factor tool named Jakarta Green Factor (JGF). The initial development of JGF was undertaken in 2018 [5]; the development process involved a comprehensive, transdisciplinary collaboration between policymakers, landscape practitioners, and researchers were adopted from Biotope Area Factor Berlin, can quantify ground-level garden space, permeable undeveloped space, and the contribution of building-integrated greenery (green roofs, walls, facades, etc.). JGF is beneficial in increasing the ecological value in areas with limited green open space, incredibly dense urban areas, increasing the environmental value of the private property, and considering the potential of building-integrated greenery [2,4,6].

Many green factor tools have been developed and implemented through research projects, and their analysis has primarily focused on presenting the tool rather than evaluating its use and effectiveness [1,2]. The usefulness and effectiveness of these green factor tools need to be assessed by testing the green factor tools on specific sites and actual cases. JGF also needs to be assessed to evaluate its use and effectiveness. This short communication presents a descriptive-qualitative case study in Jakarta to ask how the JGF is being used and how this tool is implemented through the specific sites (contextual) and actual cases.

2. Method

A descriptive-qualitative study [7,8] was conducted from May to November 2021 to discover the challenges of implementing green factors in Indonesia through case study methods within some selected green open spaces in Jakarta. Case study methods [9] facilitate the exploration of a phenomenon within its context using various data sources. This is a valuable method to develop theory, evaluate programs, and develop interventions. A case study is an appropriate method to gain concrete, contextual, in-depth knowledge about a specific real-world subject.

This method consists of the following four steps to ensure the best possible outcome: (1) Choose the cases and state how data is to be gathered and which techniques for analysis will be used, multiple sources and data analysis is recommended; (2) Build a theoretical framework, case study usually have some connection with theory in the field, this way the case study is not just an isolated description but is integrated into existing knowledge about the topic; (3) Identify and collect data, multiple collections will strengthen the study, the aim is to gain as thorough an understanding as possible of the case and its context; and (4) Describe and analyze the case, to bring together all the relevant aspects to give as complete as possible of the subject.

In this study, we used the Jakarta Green Factor to evaluate its function in increasing the quality and quantity of green open space in Jakarta. The Pulir Park and a section of pedestrian path on Sudirman Street were selected as evaluation objects due to data availability. The primary data regarding a detailed site plan were obtained from the Forestry Department of Jakarta Province, which will calculate and spatially analyze the size of the area and the elements of the landscape. Then, the result was discussed based on the spatial planning and design approach to show the importance of the space in the context of the surrounding environment. This approach functions in providing added value to the quality of the area. Finally, the contextual details about the case study connect it back to the literature and theory and discuss how it fits into green factors in Urban Greening Schemes (UGF) in Indonesia.

3. Results and Discussion

3.1. Jakarta Green Factor

The development of the Jakarta Green Factor (JGF) mainly aims to improve the quality of green open space. JGF are components in developing green open space, both in the form of elements and typologies. JGF becomes a constant value to assess the ecological function of each component of green open space. Assessing the ecological function for each green space element can be the basis for evaluating the green quality for each green open space typology in Jakarta. The use of JGF is beneficial for: (1) improvement of the ecological value in areas that lack green open space and public open space, especially in dense urban areas; (2) increase the ecological value of the private property; (3) develop potential green spaces for vertical areas and roofs in built areas. The development of JGF aims to become a binding and administratively registered city regulation. As the first step for Jakarta in implementing the UGF scheme, JGF adapted the Berlin Area Factor [2] method to weight the land surface type based on 6 (six) aspects: (1) rainwater infiltration and storage capability; (2) soil conservation capability; (3) evapotranspiration capability; (4) ability to adsorb or absorb dust and pollutants; (5) capacity to increase biodiversity; and (6) ability to create comfortable spaces.

The JGF weighs each surface's green open space quality on the land, vertical, and building surfaces with different values based on the related scientific references. Furthermore, the value of green factors is determined using the expert judgment method if not found in related references. The JGF consists of a value scale of 0 to 3 on 15 land surface forms in the context of green open space (Figure 1). the JGF was developed through the following considerations: (1) JGF is determined based on the ecological value of land surface with two assessments, namely conventional green open space and new approach of green open space where more focused on increasing its quality; (2) in case of the conventional assessment, the land surface is only valued by 0 for all pavement types and by 1 for all vegetated surface forms; and (3) in case of a new approach, all the pavement type are valued in more detail based on the level of porosity and permeability of the land surface. The vegetation area is further assessed based on the parameters of evapotranspiration, adsorption/absorption of dust and pollutants, and the capability to increase the sense of comfort [2].

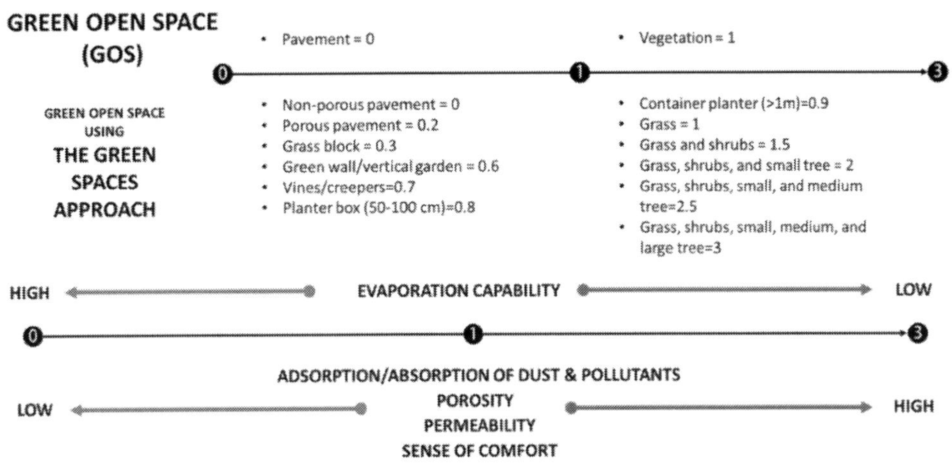

Figure 1. The consideration on the weighting of Jakarta Green Factor

3.2. Implementation of Jakarta Green Factor

Two case studies were selected regarding the implementation of JGF, namely, in the case of a park and a pedestrian path. Each green open space element in the chosen park and pedestrian path was evaluated how design meets the JGF calculation to enhance the quality and quantity of green open space.

3.2.1. Case Study on Park. The implementation of JGF in Pulir Park as one of the parks in West Jakarta (Figure 2), with an area of 5,672 m^2, shows an increase in the value of green open space quality. Calculation of the green open spaces that function as a plant growing area and a pavement surface with an ecological contribution as a water catchment area shows increasing quality. The calculation of the pavement surface using the green factor values increases the green values. As the conventional calculation of green open space shows that only the plant growing area, which is only 3,635 m^2 or 60% of the total area, was calculated. This calculation does not use factor/multiplier values, so the assessment only focused on the existing surface area.

The calculation method using JGF shows an increase in green open space quality based on the vegetated area where grass, shrubs, small to big trees grow optimally. It also offers the increasing quality of the hardscape in porous pavement surfaces such as permeable asphalt, porous concrete, paving blocks, and others. This method also calculated mixed elements of hardscape and vegetation such as grass blocks, green walls/vertical gardens, vines/creepers, planter boxes, and container planters.

Calculation of green open space quality uses different values according to the element types that show their ecological contribution. The way to calculate it is by multiplying the area of each green open space element by the JGF value. The study results show that the JGF calculation of hardscape elements on paving blocks that can absorb water with an area of 531.13 m^2 and a factor value of 0.3 contributes to improving the ecological quality of green open space equivalent to 159.3 m^2. The calculation of mixed elements of hardscape and vegetation on a grass block with a surface area of 215.7 m^2 and a factor value of 0.5 increases the quality of green open space equivalent to 107.9 m^2. The calculation of mixed elements of hardscape and vegetation in a vines area with a surface area of 11.7 m^2 with a factor value of 0.7 increases the quality of green open space equivalent to 8.2 m^2. The calculation of the vegetation element in the form of a grass area of 2,873 m^2 with a factor value of 1 increases the quality of green open space equivalent to 2,873 m^2. The calculation of vegetation elements in the form of grass, shrubs, and small trees covering an area of 625 m^2 with a factor value of 2 increases the quality of green open space equivalent to 1,250 m^2. Calculation of vegetation elements in the form of grass, shrubs, small trees, medium trees, and large trees covering an area of 1,342.4 s m^2 with a factor value of 3 increases the quality of green open space equivalent to 1,342.4 m^2. Based on all JGF calculations, there is an increased overall green open space quality value, equal to 5,740.8 m^2 with a percentage of 101.2% of the total park area or higher than the parking area itself (Table 1).

Figure 2. Site plan and 13 main facilities in Pulir Park: (1) main entrance; (2) parking area; (3) bike parking; (4) security post, mosque, and toilet; (5) labyrinth; (6) plaza; (7) playground; (8) multifunctional sports field; (9) central shelter/stage; (10) small shelter; (11) outdoor fitness; (12) reflexology path; and (13) pedestrian path

Table 1. Calculation of green open space using JGF on Pulir Park

No.	Green Open Space's Elements	JGF Code	Area (m^2)	Conventional Green Open Space Coefficient	Green Open Space Area (m^2)	Green Open Space Coefficient (JGF)	Green Open Space Area - JGF (m^2)
A	Hardscape						
1	Non-porous surface	JGF -1	912	0	0	0	0
2	Porous asphalt	JGF-2	0	0	0	0,2	0
3	Porous concrete	JGF-3	0	0	0	0,2	0
4	Paving Block	JGF-4	531.13	0	0	0,3	159.3
5	Exposed soil or/and rocks	JGF-5	0	0	0	0,4	0
B	Hardscape – Vegetation (Mixed Elements)						
1	Grass Block	JGF-6	215.7	0	0	0,5	107.9
2	Green Wall/Vertical Garden	JGF-7	0	0	0	0,6	0
3	Vines/Creepers	JGF-8	11.7	0	0	0,7	8.2
4	Planter Box (Dimension: 50-100cm)	JGF-9	0	0	0	0,8	0
5	Container Planter (Dimension >100 cm)	JGF-10	0	0	0	0,9	0
C	Vegetation (Softscape)						
1	Grasses	JGF-11	2,873	1	2,873	1	2,873
2	Grasses and Shrubs	JGF-12	0	1	0	1.5	0
3	Grasses, Shrubs, and Small Trees	JGF-13	625	1	625	2	1,250
4	Grasses, Shrubs, Small and Medium Trees	JGF-14	0	1	0	2.5	0
5	Grasses, Shrubs, Small, Medium, and Big Trees	JGF-15	447.47	1	447.47	3	1,342.4
D	Other						
1	Security post, mosque, and toilet		56	0	0	0	0
	Total Area		5,672				
	Total Green Area				3,635		5,740.8
	Jakarta Green Index (JGI)				0.6		1.0

3.2.2. Case Study on Pedestrian Path. The implementation of JGF in one of the pedestrian lanes on Sudirman Street, covering an area of 720 m^2, shows an increase in quality (Figure 3). Implementing JGF in the selected pedestrian lane calculates the open space area, including a green surface that functions as a planting area and a pavement surface that serves as a pedestrian circulation path. The calculation of the pavement surface and the calculation using the green factor values increase the green values.

Conventional green open space calculation shows that only plant growing area is only 115 m^2, with a percentage of 16% of the total area Sudirman Street pedestrian lane calculated (Table 2). This conventional calculation does not use factor/multiplier values, so the assessment is only based on the existing surface area. The green open space calculation method with JGF shows an increase in green open space quality based on the calculation of the green open area for plant growth and the analysis of the open area in the form of pavement that absorbs water. The green open spaces as a place to grow trees, shrubs, and groundcover area of 115 m^2 multiplied by a JGF value of 2.5 so that the green quality is equivalent to 287.5 m^2. The porous paving block pavements area of 505 m^2 multiplied by a factor value of 0.3 so that the green quality is comparable to 151.5 m^2. JGF calculation of the green open space shows that the overall green quality value is equivalent to 439 m^2 with a percentage of 60% of the total area of the pedestrian path segment on Sudirman Street.

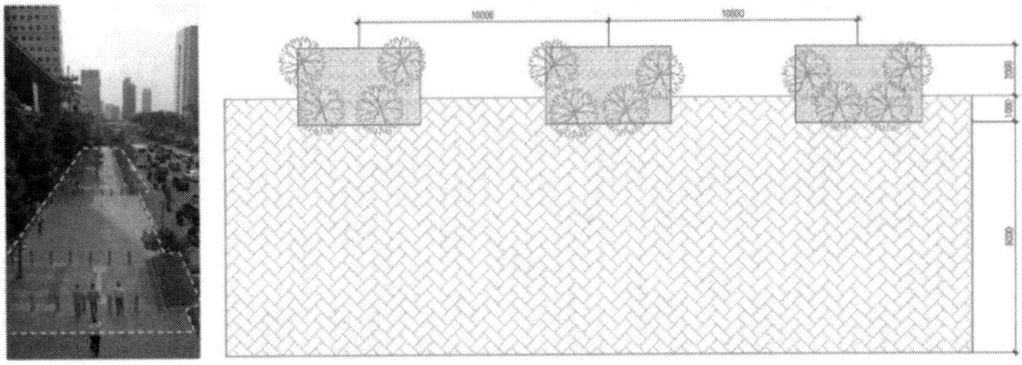

Figure 3. A segment pedestrian path of Sudirman Street

Table 2. Calculation of green open space using JGF on the pedestrian path

Green Open Space's Element	Area (m^2)	JGF Code	Green Open Space Coefficient (JGF)	Conventional Green Open Space Calculation (m^2)	JGF's Green Open Space Calculation (m^2)
Built area	0	Other	0	0	0
Open space area	-		-	-	-
• Green open space (planted with vegetation)	115	JGF-14	2.5	115	287.5
• Non-green open space (hardscape using porous materials: paving block)	505	JGF-4	0.3	0	151.5
Total area (m^2)	720			115	439
Percentage (%)				16	60
Jakarta Green Index (JGI)				0,16	0,6

This study shows that applying the green factor can increase the quantity of green open space, such as through calculation of the mixed elements (grass block or vertical greenery with vines/creepers). However, this approach is focused not only on increasing the quantity but also on improving the quality of the existing green open space [1,2,3], especially according to green open space's ability to conserve water, soil, air, and biodiversity [10,11,12]. Further, giving a weight of green factor for each land surface type and element is expected to provide a fair measurement. This can be proven based on the Jakarta Green Index (JGI) as a standard target for developing green open space in Jakarta [5]. The different results on JGI, where the green index for calculation using green factor higher than the conventional method, indicates various quality on green open space.

4. Conclusion

This short communication challenges the usefulness of the green factor tool and how it relates to the existing urban green space planning procedures through a qualitative case study in Jakarta, Indonesia. The result shows that while the green factor tool performs well, as a pilot project, improvements could be made primarily in convincing stakeholders regarding the new paradigm in calculating green areas in relatively dense urban areas. Furthermore, trials in actual cases need to be carried out to examine the formula's effectiveness. Regarding the usefulness of this tool, we conclude that while increasing the number of green areas, improving its quality is more necessary to build resilient infrastructures.

References

[1] Juhola Sirkku 2018 Planning for a green city: The Green Factor tool. Urban Forestry & Urban Greening p 254-258
[2] City of Berlin (no date) BAF – Biotope Area Factor (Retrieved from https://www.berlin.de/sen/uvk/en/nature-and-green/landscape-planning/baf-biotope-area-factor/ in April 2021)
[3] Elina Järvelä 2014 Developing A Green Factor Tool For The City Of Helsinki. the *Climate-Proof City – Tools for Planning* (ILKKA)
[4] City of Helsinki 2017 *Developing a Green factor tool for City of Helsinki* (Retrieved from https://ilmastotyokalut.fi/files/2014/11/Developing_Helsinki_Green_Factor_Summary_13032014.pdf in April 2021)
[5] Jakarta Provincial Park and Forest Service 2018 Green Open space Masterplan Report
[6] Kruuse A 2011 *GRaBS Expert 6: The Green Space Factor and the Green Points System* (London: TCPA) (Retrieved from https://www.redfrogforum.org/wp-content/uploads/2020/04/125-GRaBS-Expert-Paper-6-the-green-space-factor-and-the-green-points-system.pdf in April 2021)
[7] Creswell JW 2014 *Research Design: Qualitative, Quantitative, and Mixed Method Approaches* (4th ed.) (CA: Sage Publication)
[8] Yin RK 2013 *Case Study Research: Design and Methods* (London: Sage Publication).
[9] Baxter Pamela and Susan Jack 2008 Qualitative Case Study Methodology: Study Design and Implementation for Novice Researchers. The Qualitative Report Volume 13 Number 4 December p 544-559
[10] Freitas SR, Momm S, Piper JM and Denaldi R 2013 Integration of urban and environmental policies in the metropolitan area of Sao Paulo and in Greater London: the value of establishing and protecting green open spaces *Int. J. Urban Sustain. Dev.* **5 (1)** 10.1080 pp 89-104
[11] Tzoulas K, Korpela K, Venn S and Yli-Pelkonen V 2007 Promoting ecosystem and human health in urban areas using green infrastructure: a literature review *Landsc. Urban Plan* **81 (3)** 10.1016 pp 167-178
[12] Jim CY and Chen WY 2008 Assessing the ecosystem services of air pollutant removal by urban trees in Guangzhou (China) *J. Environ. Manage* **88 (4)** 10.1016 pp. 665-676

System Dynamic Simulation to Determine The Effect Of Water Consumption on Land Subsidence During Covid-19 Pandemic in Jakarta

Diki Surya Irawan, Prismita Nursetyowati, Maria Cintya Nova, Sarah Maulina

Environmental Engineering, Universitas Bakrie, Jakarta - Indonesia

diki.surya@bakrie.ac.id, prismita.nursetyowati@bakrie.ac.id, mariacintyanova@ymail.com, sarahmaulinaaix7@gmail.com

Abstract. Excessive groundwater extraction is believed to be one of the main factors for land subsidence which may be caused by tidal flooding due to the position of the surface which is lower than sea level. Covid-19 pandemic that has occurred in Indonesia since March 2020 has caused changes in water consumption patterns which derives from piped water and groundwater. There are many offices and industries that implement work from home (WFH) makes many buildings have a declining occupancy rate. With the decrease in the occupancy rate of the WFH policy, there will be a possibility that groundwater consumption from high-rise buildings that draw groundwater from deep aquifers can be reduced. This research is in the form of modelling and simulation that is used to build a level of understanding on a whole system as well as the interrelationships and interactions between its constituent variables. The purpose of this research was to determine the effect of groundwater consumption during Covid-19 pandemic on land subsidence in Jakarta using the dynamic system simulation method. The results showed that the work from home policy reduces groundwater consumption by 64.7%. In addition, the reduction in groundwater consumption during the Covid-19 pandemic caused land subsidence in Jakarta slows down and the rate of land subsidence in Jakarta decreased from 3.7 cm/year to 1 cm/year.

Keyword: covid-19, dynamic systems, groundwater, land subsidence, work from home

1. Introduction

Water is a basic need for life because every living thing needs water. Urban areas that have a high population growth rate make the need for clean water increases. Sources of raw water are obtained from surface water and ground water. Water that is above the ground surface such as rivers, lakes, seas, swamps, etc. is an example of surface water. Groundwater is a source of raw water that is in the soil layer. Groundwater is usually used to meet human needs such as for drinking, cooking, bathing, washing, etc. which can be taken through wells or sucked using a pump.

Covid-19 pandemic that has occurred in Indonesia since March 2020 has caused changes in the pattern of consumption of clean water from piped water and ground water. Work from home (WFH) policy implemented by most offices and industries in Jakarta have prevented many buildings from operating. In addition, new habits that increase hand washing and bathing activities in order to carry out health protocols also change the pattern of clean water consumption. According to research conducted by the Indonesia Water Institute in 2020, the pattern of community use of clean water has increased by 3 times compared to conditions before the pandemic.[1]

High-rise buildings are one of the causes of the increase in groundwater consumption because many high-rise buildings in Jakarta still use groundwater pumps that take groundwater from deep aquifers (>100m depth). The government took several steps to prevent the illegal extraction of groundwater by high-rise buildings. One of them is by calling on the residents of Jakarta to stop using groundwater and switch to using piped water, which PAM Jaya will expand its network of. In addition, with the decline in building occupancy rates due to the WFH policy, there will be a possibility that groundwater consumption from high-rise buildings that draw groundwater from deep aquifers can be reduced.

2. Method

This research was conducted using system dynamic simulation. The approach method in this research consists of three steps. The first step uses systems analysis techniques to develop a general structural model for groundwater and subsidence systems in Jakarta. This general model is used to establish the causal relationships between the various variables systems needed to create a functioning system dynamics model. Subsequently, the general model was adapted to simulate land subsidence. Relevant system variables for each system are identified based on the observed correlations. The data used was secondary data that has been collected from several government agencies in Jakarta. The flow chart of the research is shown in Figure 1.

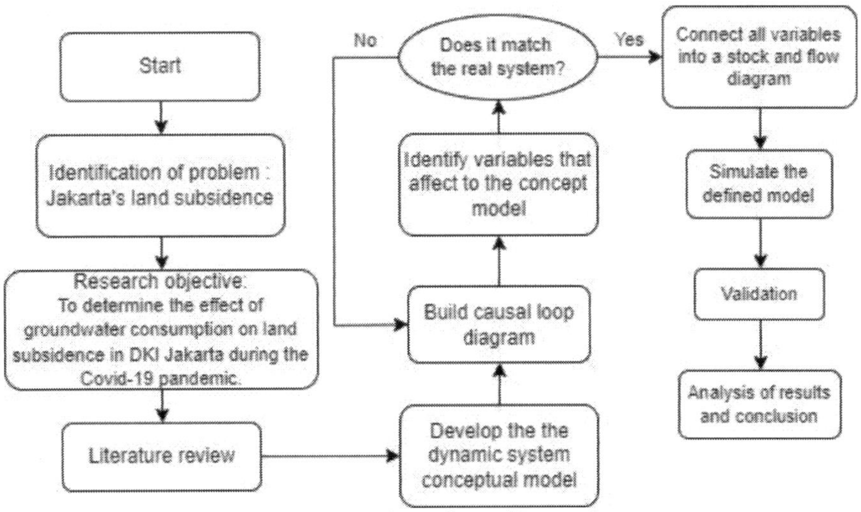

Figure 1. Flow Chart of Research

3. System Dynamic

System dynamics is a continuous simulation developed by an MIT (Massachusetts Institute of Technology) Professor named Jay W. Forrester (MIT) in 1956. System dynamics is a method used to describe, model, and simulate a dynamic system (from changing from time to time). The dynamic system method is a methodology for understanding complex problems. The dynamic system method studies problems from a system point of view, where the system elements interact with each other in a feedback relationship so as to produce a certain behavior. The interactions in this structure are translated into mathematical models which are then simulated with the help of computers to obtain their historical behaviour [2].

The software used in this research is Vensim PLE 8.2.1, which is software that can be used for the development, analysis and optimization of a sustainable model. Vensim is a modelling tool that allows users to conceptualize, document, simulate, analyze, and optimize dynamic system models. Vensim provides an easy and flexible way to build simulation models from causal loop diagrams and stock flow diagrams. By linking words (variables) with arrows, the relationships between the variables in the system are entered and recorded as cause-and-effect relationships. This information will be used in the equation editor to help create simulation models, users can perform model analysis during the model building process by looking at the causes and uses of variables and loops involving variables. After the model is completed, Vensim can explore the behaviour of the model [3].

4. General Model Development

The general model is a conceptual model that describes the variables of land subsidence system that are related to each other. In making the general concept of a Causal Loop Diagram (CLD), the most important thing is to identify and relate each variable that affects the land subsidence of Jakarta. Then this model will be developed into a stock and flow model. Table 1 describes the variables of the general model.

Table 1. General model system variables and descriptions.

Variable	Description
Groundwater Stock	The amount of groundwater in the system
Rainfall Intensity	Water entering system from rainwater
Green Open Space	Percentage of green open space per area of Jakarta
Water Company Production	Total production of fresh water produced by Jakarta Water company
Total Water Demand	Total need for fresh water for all residents of Jakarta
Population	Population of Jakarta in 2019 and 2020
Number of Buildings	Amount of buildings with minimum height of 150 meters according to the emporis.com website
Soil Load	The total weight received by the ground surface throughout Jakarta
Ground Level	Jakarta ground level in meters

Three main components in general model are groundwater sub-system, population sub-system, and soil load sub-system. Groundwater sub-system is a reservoir of water the ground surface. The population sub-system describes population growth that affects the amount of water demand and building needs in Jakarta. The soil load sub-system affects the subsidence of the land surface due to the load above the ground surface.

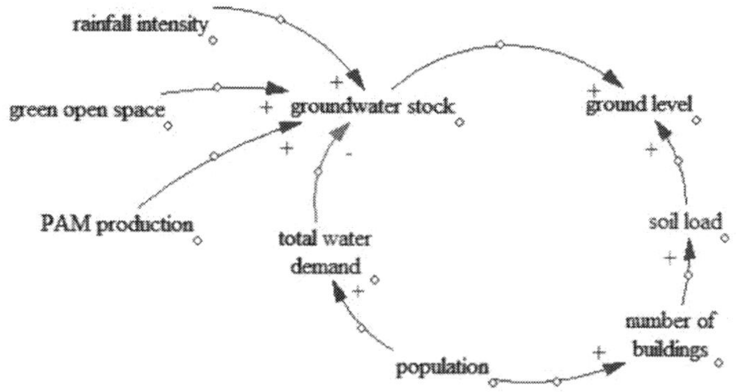

→ : indicates a directly proportional relationship between variables
→ : indicates an inverse relationship between variables

Figure 2. Causal Loop Diagram of Land Subsidence

Given the above, the CLD for the general model of land subsidence. The arrows in Figure 2 show how the relationship between variables is. The symbol "+" indicates a directly proportional relationship between variables. The sign "–" indicates a change between variables whose relationship is inversely related. These variables and connections form the structure of the system.

5. Stock and Flow Diagram

Stock and flow diagram for the Jakarta land subsidence was developed based in the general model structure. This diagram uses two types of variables named stock (level) and flow (rate). The level represents the state of the system at any time. In system-level engineering, it is better known as a state variable system. Level is an accumulation in the system. The equation of a rate variable is a policy structure that explains why and how a decision is made based on the information available in the system. Rate is the only variable in the model that can affect the level. This stock and flow diagram is a more detailed description of the previous system shown by the causal loop diagram because this diagram focusing to the effect of time on the relationship between variables, therefore each variable is able to show the accumulation results for level variables, and variables which the rate of system activity per period of time is called the rate [4]. Figure 3 shows a graphical representation of the stock and flow diagram.

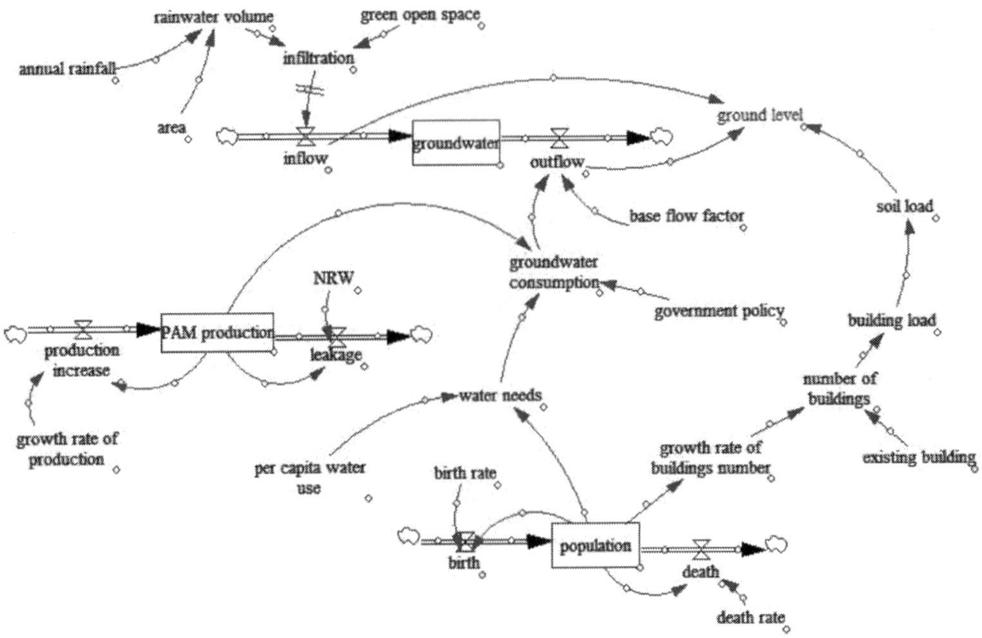

Figure 3. Stock and Flow Diagram of Land Subsidence

From the diagram above, it can be seen that the ground level is affected by groundwater inflow, groundwater outflow, and soil load. The soil load comes from the accumulated building load. Soil load can affect the level of the ground because if the load on the ground surface increases, the pressure on the ground surface will also increase. If it is assumed that initially the forces acting on the ground surface are balanced, or it is said that the sum of the forces is zero. The existence of this building load provides additional gravity on a part of the earth that can change the balance [5].

The increasing population will increase the need for water and buildings. The high demand for water can be met by producing water through PAM Jaya. However, PAM Jaya has not been able to fulfil all the needs of clean water so that some people still have to rely on groundwater as water source. If groundwater extraction is not in balance with the volume of incoming groundwater, it can cause land subsidence due to consolidation of the soil layer.

After develop land subsidence system using stock and flow diagrams, then each variable in the system will be formulated. The purpose of model formulation is to describe the model in a mathematical form so that it can represent a real system. The formulation in this study was obtained from previous studies that have been carried out previously. The model formulation is carried out in 2 versions, namely the 2019 version which is a land subsidence model before the Covid-19 pandemic occurred. The second version is a model formulation that uses 2020 data to describe the current state of the Covid-19 pandemic. The model formulation was made with the help of Vensim PLE 8.2.1 software. The formula model can be seen in Table 2

Table 2. Model Formulation

Variable	Unit	Model Formulation Year 2019 (Before Pandemic)	Model Formulation Year 2020 (Without WFH Scenario)	Source
Annual rainfall	meter	0.127504	0.22714	[6], [7]
Area	m^2	662330000	662330000	[6]
Rainwater volume	m^3	annual rainfall*area	annual rainfall*area	[6]
Green open space	%	0.024	0.024	[8]
Infiltration	m^3	0.6*(rainwater volume*green open space)	0.6*(rainwater volume*green open space)	
Inflow	m^3	DELAY1(infiltration, 0.041)	DELAY1(infiltration, 0.041)	[9]
Groundwater	m^3	inflow-outflow	inflow-outflow	
Outflow	m^3	groundwater consumption*base flow factor	groundwater consumption*base flow factor	[9]
Base flow factor	-	0.257	0.257	
Growth rate of production	%	0.0276	0.0276	[10]
Production increase	m^3	PAM production*growth rate of production	PAM production*growth rate of production	[10]
NRW	%	0.424	0.424	[10]
Leakage	m^3	PAM production*NRW	PAM production*NRW	[10]
PAM production	m^3	Product increase-leakage	Product increase-leakage	[10]
Groundwater consumption	m^3	(water needs- PAM production)*government policy	(water needs- PAM production)*government policy	[11]
Water needs	m^3	population*water use per capita	population*water use per capita	[12]
Water use per capita	m^3/people/year	43.8	43.8	[13]
Government policy	-	1	1	
Birth rate	-	0.000474	-0.000196	[6], [7]
Birth	People	population*birth rate	population*birth rate	[6], [7]
Death rate	-	0.011195	0.00219	[6], [7]
Death	People	population*death rate	population*death rate	
Population	-	10557810	10557810	[6], [7]
Existing building	Unit	124	124	[14]

Variable	Unit	Model Formulation Year 2019 (Before Pandemic)	Model Formulation Year 2020 (Without WFH Scenario)	Source
Growth rate of buildings	-	(population/85144)/100	(population /85144)/100	[14]
Number of buildings	Unit	Existing building+(existing building*growth rate of building)	Existing building+(existing building*growth rate of building)	[14]
Building load	Ton	number of buildings*52000	number of buildings*52000	[5]
Soil load	Ton	building load*0.8	building load*0.8	[15]
Ground level	meter	IF THEN ELSE(outflow>inflow, soil load/1.45e+06 , 0)	IF THEN ELSE(outflow>inflow, soil load/1.45e+06 , 0)	

6. Result

After creating a dynamic system model, the next step is to perform a simulation. The model that has been formulated will be simulated with the help of Vensim PLE 8.2.1 software. Simulations run for 50 years. After the simulation is run, the resulting output is in the form of a graph of the value of land subsidence from period to period. The simulation results of land subsidence values can be seen in Figure 4

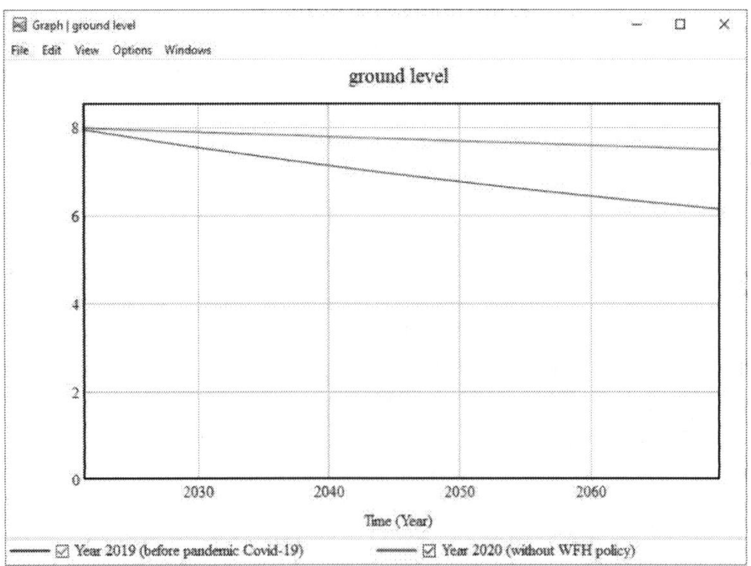

Figure 4. Simulation Results of Variable Land Subsidence Value

In Figure 4. can be seen that the graph of the land surface from the data obtained has a downward trend even though the elevation of the decline is not significant. The simulation results in 2020 have an average ground level of 7.745 meters, this value is higher when compared to the average simulation results of ground level in 2019 which is 6.978 meters. The average value of the simulated land subsidence rate for the 2019 model year is 3.7 cm/year. This value is higher than the average rate of subsidence in the simulation model year 2020, which is 1 cm/year. From this comparison, it can be seen that during the Covid-19 pandemic, there was a slowdown in the rate of land subsidence in Jakarta compared to the previous year.

After the model is simulated, the next step is to develop a scenario. Scenario development is done by changing the parameters that have an influence on the model. The parameter scenario development used in this final project is a work from home policy scenario. As we know that the implementation of Large-Scale Social Restrictions (PSBB) in 2020 has caused 39% of workers to WFH in full since it was implemented [16]. This reduces building occupancy and automatically reduces clean water consumption. From this scenario, a model is developed by changing the value of the government policy variable.

The scenario value of the government policy variable is obtained from the percentage of 100% building occupancy minus the percentage of workers who do WFH which is obtained from survey data from the BPS with the following calculations:

Model Scenario = Existing Data – BPS Data
= 100% - 39%
= 61%

Initial Value Scenario Model = 61/100 = 0.61

Changes in the model formulation in the work from home policy scenario can be seen in Table 3.

Table 3. Changes in the Formulation of the WFH Policy Scenario

Scenario	Variable	Formulation
Existing	Government policy	1
Work From Home (WFH) Scenario	Government policy	0.61

The results of the work from home policy scenario simulation are displayed in graphical form. The simulation results of groundwater discharge for government policy variables can be seen in Figure 5.

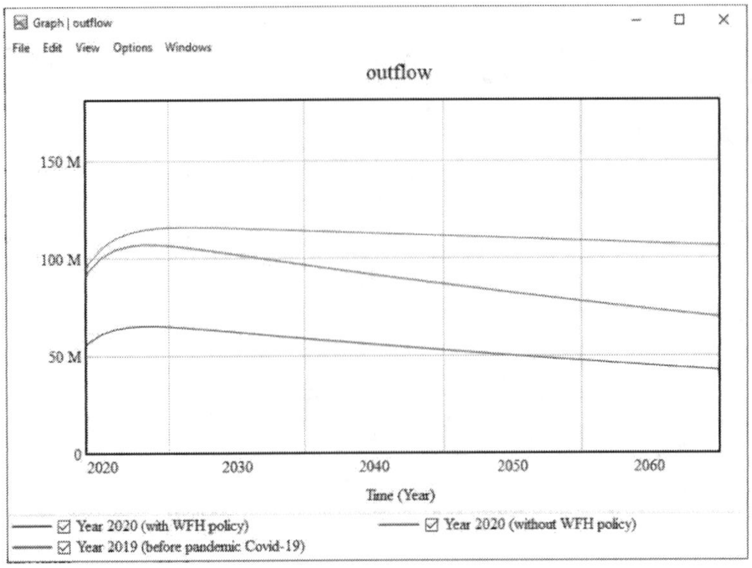

Figure 5. Groundwater Outflow Simulation Results for WFH Policy Scenario

From Figure 5, it can be seen that the volume of groundwater out in the work from home policy scenario has the lowest value compared to 2019 and 2020. This is because in the work from home policy scenario, groundwater consumption is reduced due to decreased building occupancy compared to the previous condition before the covid-19 pandemic.

After seeing the simulation results from the groundwater variable out, then the work from home policy scenario is simulated on the land surface variable. The results of the simulation of the land surface for government policy variables can be seen in Figure 6.

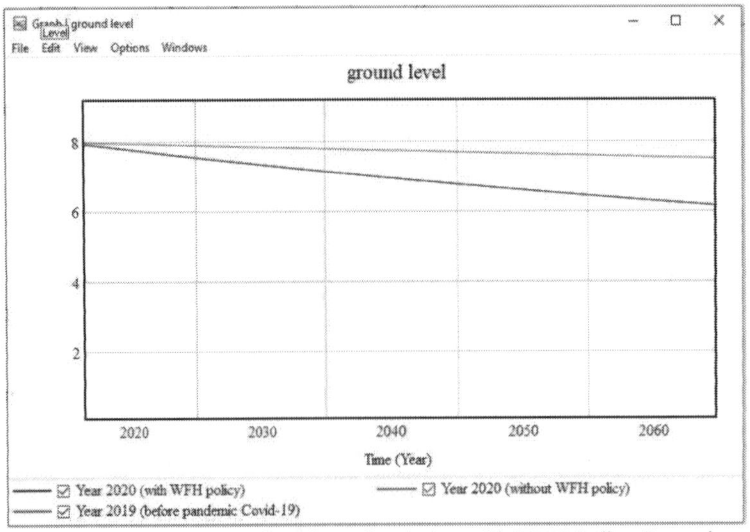

Figure 6. Result of Land Surface Simulation with WFH Policy Scenario

IOP Conf. Series: Earth and Environmental Science 1065 (2022) 012016 doi:10.1088/1755-1315/1065/1/012016

From the graph above, it can be seen that the land surface in the work from home policy scenario has the same value as in 2020. This is because the government policy variables do not directly affect the land surface. However, with the work from home policy, it is able to affect the decrease in groundwater consumption so that the rate of land subsidence becomes slower.

7. Discussion

According to an interview conducted by researchers with the Groundwater Conservation Agency (BKAT) conducted on January 19, 2020, it is known that groundwater has only 17% effect on land subsidence. The main thing that affects land subsidence is the heavy load of buildings on the ground. The heavy load of this building affects as much as 80% of the land subsidence. While the nature and type of soil naturally affect 3% of land subsidence [15].

The statement from BKAT can also be an answer to why the results of the simulation of the land surface in 2020 (without WFH) and in 2020 (WFH scenario) have the same land surface value. Although the WFH scenario affects groundwater consumption, the WFH scenario does not directly affect land subsidence because groundwater only affects 17% of land subsidence. So that it can be seen that in order to increase the government's efforts to reduce the rate of land subsidence, it is not only necessary to limit the consumption of ground water, but also to make policies regarding permits to construct high-rise buildings which have a large impact on land subsidence in Jakarta.

Building loads that affect land subsidence as much as 80% are influenced by the large number of high-rise buildings in Jakarta in the form of office buildings, residential/apartments, and hotels. The increase in the number of high-rise buildings is directly proportional to population growth. So that it can be seen that if the population growth of Jakarta is greater, the number of additional buildings will be even greater. However, according to BPS data, there has been a decline in the population of Jakarta in 2020. This has slowed population growth and the number of additional buildings has also decreased. The number of additions to the reduced number of buildings causes a reduction in building loads and soil loads. Because the soil load has a major effect on land subsidence, so with a reduction in soil load, the rate of land subsidence will also decrease.

Although the major factor affecting land subsidence is the load of the buildings above the land and groundwater consumption only affects 17% of land subsidence, the groundwater consumption factor is still interesting to study because the use of groundwater is easier to control by the Jakarta government, considering that the government cannot do many things related to the consolidation factor due to the loading of existing buildings. In the other hand WFH policy have caused buildings that use groundwater to be decreased the occupancy rate. If the WFH policy can continue to be applied, it is expected to reduce groundwater consumption from buildings that use groundwater.

8. Conclusion

The decline in groundwater consumption during the Covid-19 pandemic has slowed down land subsidence in Jakarta compared to the previous year. According to the system model simulation, the rate of land subsidence in Jakarta during the Covid-19 pandemic, either without or with the WFH policy, is 1 cm/year, this land subsidence rate has decreased compared to the period before the Covid-19 pandemic which had a rate of 3.7 cm/year. The subsidence model can be detailed by adding geological variables such as the rate of soil consolidation to make the model more valid. The proposed model development can be expanded by adding other policies such as restricted groundwater use or improved PAM services. Adding insight into system dynamics and technical use of software in order to obtain a more accurate model with valid structures and equations so that it can be a solution related to a better system in the future.

9. Acknowledgments
This research was supported by Research and Development Center Universitas Bakrie, grant number 261/SPK/LPP-UB/IX/2020. We thank several agencies that allowed us to use the data in this research including Groundwater Conservation Agency (BKAT), Jakarta Water Company, Jakarta Water Resources Institution, Central Bureau of Statistics and others who prefer to be anonymous.

References
[1] Indonesia Water Institute, "Pola Konsumsi Air Bersih Masyarakat Selama Pandemi Covid-19," 2021.
[2] M. Aminudin, "Simulasi Model Sistem Dinamis Rantai Pasok Kentang Dalam Upaya Ketahanan Pangan Nasional," *Univ. Islam Negeri Syarif Hidayatullah*, 2014.
[3] Z. Akbar, "Penerapan Metode Simulasi Sistem Dinamik untuk Mengoptimalkan Profit dalam Jaringan Supply Chain Produk Sayur," 2017.
[4] A. Firmansyah and E. Suryani, "Model Sistem Dinamik Untuk Pengembangan Smart Economy (Studi Kasus: Kota Surabaya)," *J. Tek. ITS*, vol. 6, no. 2, 2017.
[5] J. M. Manik and M. D. Marasabessy, "Tenggelamnya Jakarta Dalam Hubungannya Dengan Konstruksi Bangunan Beban Megacity," *MAKARA Sci. Ser.*, vol. 14, no. 1, pp. 69–74, 2011.
[6] Badan Pusat Statistik Provinsi Jakarta, "Provinsi Jakarta Dalam Angka 2020."
[7] Badan Pusat Statistik Provinsi Jakarta, "Provinsi Jakarta Dalam Angka 2021," 2021.
[8] Dinas Pertamanan dan Hutan Kota, "Data Rekap Luas Ruang Terbuka Hijau Per Kotamadya Di Jakarta," Jakarta, 2019.
[9] L. E. Hutabarat, "Studi Penurunan Muka Tanah (Land Subsidence) Akibat Pengambilan Air Tanah Berlebihan Di Jakarta," *Kumpulan Karya Ilmiah Dosen Universitas Kristen Indonesia Delapan Windu*. pp. 360–374, 2017.
[10] Badan Pusat Statistik Provinsi Jakarta, "Number of Customers Production and Water Volume Sold Service Company in Jakarta Province 2019." [Online]. Available: https://jakarta.bps.go.id/statictable/2021/09/20/298/jumlah-pelanggan-produksi-dan-kubikasi-air-terjual-perusahaan-air-minum-pam-di-provinsi--jakarta-2019.html.
[11] D. Primadian, "Pengembangan Model Sistem Dinamik Terhadap Ketersedian Air Bersih Di Kabupaten Kutai Timur Provinsi Kalimantan Timur," *J. Ind. Eng. Manag.*, vol. 1, no. 2, p. 16, 2017.
[12] G. Bates, M. Beruvides, and C. B. Fedler, "System Dynamics Approach to Groundwater Storage Modeling for Basin-Scale Planning," *Water (Switzerland)*, vol. 11, no. 9, pp. 1–18, 2019.
[13] Badan Standardisasi Nasional, "Sumber Daya Air Spasial," *SNI 19-6728.1 Penyusunan Neraca Sumber Daya*, pp. 1–32, 2002.
[14] Emporis International, "Emporis Buildings Information." [Online]. Available: https://www.emporis.com/search/buildingsOver150m/jakarta.
[15] R. I. H. Sulistyawan, "Permasalahan Penurunan Tanah (Land Subsidence), Air Bersih dan Air Tanah," 2021.
[16] Kementrian Perencanaan Pembangunan Nasional/Bappenas, "Tanggapan atas Kajian Pola Konsumsi Air Bersih Masyarakat Selama Pandemi Covid-19," *Databoks Data Stories*, p. 2020, 2020.

Analysis of behavioral factors in the spread of COVID-19 in Coblong District, Bandung, West Java, Indonesia.

Katharina Oginawati[1] , *Rinaldy Jose Nathanael[1], Udjianna Sekteria Pasaribu[2] , Utriweni Mukhaiyar[2] , Afif Humam[2] , Nisa Fadlilah Fathul Ilmi[2]

[1]Environmental Management Technology Research Division, Faculty of Civil and Environmental Engineering, Institut Teknologi Bandung

[2]Statistics Research Division, Faculty of Mathematics and Natural Sciences, Institut Teknologi Bandung, Indonesia

*Corresponding author e-mail: rinaldyjose@students.itb.ac.id

Abstract. This study aims to assess community habits regarding the spread of COVID-19 in Coblong District. The research method used is by conducting a survey to the respondents. Respondents who were included in the study were 50 people from 38 houses. Respondents consisted of 30 COVID-19 survivors and 20 respondents as a control group who had never been infected with COVID-19. The preventive behavior studied consisted of 6 aspects, namely wearing masks, washing hands with soap, maintaining distance, avoiding crowds, reducing mobility, and room disinfection. The results obtained are that all preventive behavior factors have a negative correlation with COVID-19 cases except for the level of mobility which has a positive or unidirectional relationship between a person's level of mobility and COVID-19 cases with a value of 0.306. Spearman correlation coefficient r were investigated for each elements of preventive behavior on cases and four behaviors found to be significantly correlated with cases ($p<0.062$). Logistic regression was used to further investigate and predict the infection risk with significant factors as predictors. The logistic regression model showed that 33.2% of cases in Coblong are caused by its relationship with preventive behaviors.

Keywords: covid-19, behavior, preventive, regression, risk

1. Introduction

The World Health Organization (WHO) has issued various recommendations for preventive measures against COVID-19 for the general public [1]. Some of them are keeping a distance from other people, wearing masks when in public places, applying personal hygiene and maintaining environmental health such as disinfecting frequently touched surfaces [2]. Efforts to prevent the disease are related to the transmission routes of COVID-19. However, in reality, not all of these recommendations are carried out by the community for various reasons, some of which are lack of awareness of the dangers of COVID-19, not getting used to new habits, and lack of confidence that these recommendations help prevent the transmission of COVID-19, especially use of masks in public places [3, 4].

One of the efforts to control the COVID-19 pandemic outbreak in Indonesia is the Large-Scale Social Restrictions (PSBB), which is defined as the restriction of certain activities of residents in an area suspected of being infected with COVID-19 in such a way as to prevent the possible spread of COVID-19 (PP RI No. 21 year 2020) [9]. The regulation states that PSBB includes restrictions on activities in social facilities such as schools, workplaces, places of worship, and other social facilities. Efforts to control the COVID-19 pandemic outbreak in the form of implementing PSBB are closely related to social network engineering with isolation which makes the opportunity for virus spread between network members through ties/contacts to be lower [5, 6].

The city of Bandung is the capital city of West Java, the province in the top three with the highest number of COVID-19 cases in Indonesia [7]. As of January 2021, the positive cases of COVID-19 in the city of Bandung have reached more than 11000 cases and have killed more than 220 people [8]. The city of Bandung has 30 sub-districts with different levels of population density, ranging from 3.75 to 40 thousand people per km^2. The number of positive cases in one sub-district in Bandung City as of February 2021 ranged from 100 to 700 cases. Therefore, this study aims to assess people's habits regarding the spread of COVID-19.

2. Methods

2.1 Study location

This field study regarding the study of residential environmental factors and preventive behavior against the incidence of COVID-19 was carried out from March to April 2021 in Coblong District, Bandung. The reason for choosing Coblong Sub-district as the research location is based on data obtained from the Bandung City COVID-19 Information Center in March 2021, Coblong District is the sub-district with the second highest total positive cases in Bandung City and the sub-district with the highest total recovered in Bandung City. In addition, Coblong District is also one of the sub-districts in Bandung with a fairly high population density with a density of 15,617.37 people/km2 (Bandung in Figures, 2020).

2.2 Data collection

Data collection is done by conducting interviews with respondents. Respondents who were included in the study were 50 people from 38 houses. Respondents are COVID-19 survivors. In addition, 20 respondents as a control group had never been infected with COVID-19.

Questionnaires and interviews were conducted for each respondent. Questionnaires and interviews were conducted to obtain general information regarding personal data and history of COVID-19 residents of Coblong District. From the questionnaires and interviews, it can be seen about the respondent's personal data such as age, weight, height, and occupation. In addition, data were also obtained regarding comorbidities, transportation facilities, suspected sources of transmission, duration of healing from COVID-19, 5M behavior before being positive for COVID-19, area of the house, number of occupants in the house and the total number of occupants in the house who were positive for COVID-19.

The preventive behavior factors studied consisted of 6 aspects, namely wearing masks, washing hands with soap, maintaining distance, avoiding crowds, reducing mobility, and room disinfection. Data on preventive behavior habits were obtained by means of direct interviews with respondents, both survivors and control respondents. To obtain the effect of preventive behavior regardless of environmental factors, data on preventive behavior were taken only from the first survivors from each house. Each of these preventive behaviors was associated with COVID-19 cases by univariate, bivariate, and multivariate analysis. The preventive behavior habits of the respondents were divided into 4 levels, ranging from very rarely to always (1-4). COVID-19 cases were obtained from the respondent's status, whether they were survivors or controls.

2.3 Data analysis

Data analysis was carried out using statistical analysis. A Chi-square test of independence was performed to examine the relationship between two categorical/nominal variables. The main focus of this test is to investigate how significant the cases in Coblong is related to preventive behavior level carried out by public. Spearman Rank correlation test was carried out to determine the value of the correlation coefficient. This correlation coefficient value describes how close the relationship between preventive behavioral factors and COVID-19 cases. Logistic Regression analysis was carried out to show the relationship between the dependent and independent variables. In this study, the independent variable is each of behavioral factors. Cases in Coblong is the dependent variable. By logistic regression, a predictive equation of COVID-19 infection risk can be formed with dependent (predictors) variables.

3. Results and discussion

3.1 Wearing a Mask

Wearing a mask is one of the recommended health protocols during the COVID-19 pandemic. Table 1. shows the level of mask wearing habits of control respondents and survivors before being exposed to COVID-19.

Table 1. Habits of Using Masks

Category	Using mask	
	Cases	control
1 (rarely)	0	0
2 (sometime)	3 (10%)	0
3 (often)	5 (16,67%)	2 (10%)
4 (always)	22 (73,33%)	18 (90%)
Total	30	20

In the table above, it can be seen that most of the respondents, both from the case and control groups, have a habit of always wearing masks outside the home (the survivors group is 73.33% and the control group is 90%). Then, in the survivor group there were 5 people (16.67%) who had the habit of frequently wearing masks and for the control group only 2 people (10%) had the habit of frequently wearing masks. However, in the group of survivors there are still 3 people (10%) whose habit of wearing masks is poor outside the house before being exposed to COVID-19.

In the results of the chi-square test which shows the significance of the habit of wearing masks for COVID-19 cases, a P-value of 0.247 is obtained. This significance number is more than 0.05, meaning that the statistical calculation results between the habit of wearing masks and COVID-19 cases in Coblong District are not significant. Thus, it can be stated that there is no significant relationship between the habit of wearing masks and cases of COVID-19 in Coblong District.

The Odd Ratio value is obtained from the results of cross tabulation using Microsoft Excel. The results of the cross tabulation, obtained OR for wearing a mask of 2,045, which means that someone who only 'often' wears a mask has a 2,045 times risk of getting from COVID-19 compared to someone who 'always' wears a mask when outside the home. This risk value can be higher for the 'rare' and 'very rare' categories.

3.2 Washing Hands with Soap

Washing hands with soap is one of the health protocols recommended during the COVID-19 pandemic. Table 2. shows the level of habit of washing hands with soap from control respondents and survivors before being exposed to COVID-19.

Table 2 Habits of Washing Hands with Soap

Category	Hands with Soap	
	Cases	Control
1 (rarely)	0	0
2 (sometime)	7 (23,33%)	0
3 (often)	10 (33,33%)	5 (25%)
4 (always)	13 (43,34%)	15 (75%)
Total	30	20

In the table above, it can be seen that most of the respondents, both case and control groups, had a habit of always washing their hands with soap (survivors group was 43.34% and 75% for control group). Then, in the survivor group there were 10 people (33.33%) who had the habit of frequently washing their hands with soap and for the control group there were 5 people (25%) who had the habit of frequently washing their hands with soap. However, in the survivor group there were still 7 people (23.33%)

whose habit of washing their hands with soap was still rare before being exposed to COVID-19.

In the results of the chi-square test which shows the significance of the habit of washing hands with soap for COVID-19 cases, a P-value of 0.029 is obtained. This significance figure is less than 0.05, meaning that the statistical calculation results between the habit of washing hands with soap and cases of COVID-19 in Coblong District are significant. Thus, it can be stated that there is a significant relationship between the habit of washing hands with soap and cases of COVID-19 in Coblong District.

The Odd Ratio value is obtained from the results of cross tabulation using Microsoft Excel. The results of the cross tabulation showed that the OR for washing hands with soap was 2,307, which means that someone who only 'frequently' washes their hands with soap has 2,307 times risk of getting from COVID-19 compared to people who 'always' wash their hands with soap. This risk value can be higher for the 'rare' and 'very rare' categories.

3.3 Keeping physical distance

Keeping a distance is one of the health protocols recommended during the COVID-19 pandemic. Table 3. shows the level of habit of keeping a distance from control respondents and survivors before being exposed to COVID-19.

Table 3. Habits of Keeping physical Distance

Category	Personal distance	
	Cases	Control
1 (rarely)	1 (3,33%)	0
2 (sometime)	7 (23,33%)	1 (5%)
3 (often)	10 (33,33%)	8 (40%)
4 (always)	12 (40%)	11 (55%)
Total	30	20

In the table above, it can be seen that most of the respondents, both case and control groups, have a habit of always keeping their distance from other people (40% for survivors and 55%). Then, in the survivor group there were 10 people (33.33%) who had the habit of often keeping their distance and for the control group there were 8 people (40%) who had the habit of often keeping their distance. However, in the survivor group there were still 7 people (23.33%) whose habit of keeping their distance was still rare before being exposed to COVID-19 and in the control group there was still 1 person (5%) who still rarely kept their distance. In the group of survivors, there is still 1 person (3.33%) whose habit of keeping their distance is very rare.

In the results of the chi-square test which shows the significance of the habit of keeping a distance from COVID-19 cases, a P-value of 0.27 is obtained. This significance number is more than 0.05, meaning that the statistical calculation results between social distancing habits and COVID-19 cases in Coblong District are not significant. Thus, it can be stated that there is no significant relationship between social distancing habits and COVID-19 cases in Coblong District.

The Odd Ratio value is obtained from the results of cross tabulation using Microsoft Excel. The results of the cross tabulation, obtained OR for keeping a distance of 6,416 which means that someone who 'rarely' keeps a distance has 6,416 times the risk of getting from COVID-19 compared to someone who 'always' keeps a distance. The value of this risk can be greater for the 'very rare' category.

3.4 Avoiding the Crowds

Avoiding crowds is one of the health protocols recommended during the COVID-19 pandemic. Table 4. shows the level of crowd-avoiding habits of control respondents and survivors before being exposed to COVID-19.

Table 4. Habits of Avoiding Crowds

Category	Avoiding Crowds	
	Cases	Control
1 (rarely)	0	0
2 (sometime)	7 (23,33%)	2 (10%)
3 (often)	14 (46,67%)	6 (30%)
4 (always)	9 (30%)	12 (60%)
Total	30	20

In the table above, it can be seen that most of the respondents from the control group (12 people) have a habit of always avoiding crowds when outside the house (60%). However, in the group of survivors there were only 9 people (30%) who had the habit of always avoiding crowds when they were outside the house before being exposed to COVID-19. Then, in the survivor group, there were 14 people (46.67%) who had the habit of frequently avoiding crowds and for the control group as many as 6 people (30%) had the habit of frequently avoiding crowds. However, in the two groups of survivors and control, there were still 7 people (23.33%) and 2 people (10%) whose habit of avoiding crowds was still rare.

In the results of the chi-square test which shows the significance of the habit of avoiding crowds for COVID-19 cases, a P-value of 0.101 is obtained. This significance figure is more than 0.05, meaning that the statistical calculation results between the habit of avoiding crowds and COVID-19 cases in Coblong District are not significant. Thus, it can be stated that there is no significant relationship between the habit of avoiding crowds and COVID-19 cases in Coblong District.

The Odd Ratio value is obtained from the results of cross tabulation using Microsoft Excel. The results of the cross tabulation, obtained OR for avoiding crowds of 4,666 which means that someone who 'rarely' avoids crowds has 4,666 times risk of getting from COVID-19 compared to people who 'always' avoids crowds. The value of this risk can be greater for the 'very rare' category.

3.5 Mobility

Reducing mobility is one of the recommended health protocols during the COVID-19 pandemic. Table 5. shows the level of mobility of control respondents and survivors before being exposed to COVID-19.

Table 5. Mobility Level

Category	Mobility Level	
	Cases	Control
1 (rarely)	9 (30%)	9 (45%)
2 (sometime)	5 (16,67%)	8 (40%)
3 (high)	4 (13,33%)	1 (5%)
4 (highest)	12 (40%)	2 (10%)
Total	30	20

In the table above, it can be seen that in the survivor group, most of the respondents (12 people, 40%) still have a very high level of mobility, while in the control group, there are only 2 people (10%) who have a very high level of mobility. Then, as many as 4 people (13.33%) from the survivor group had high mobility and in the same category, there was 1 person (5%) from the control group. For respondents with low and very low mobility levels, in the case group there were 5 people (16.67%) and 9 people (30%), respectively, while in the control group there were 8 people (40%) and 9 people, respectively (45%). In the control group, most of the respondents had very low mobility levels during the pandemic.

In the results of the chi-square test which shows the significance of the level of mobility for COVID-19 cases, a P-value of 0.047 is obtained. This significance figure is less than 0.05, meaning that the statistical calculation results between mobility levels and COVID-19 cases in Coblong District are significant. Thus, it can be stated that there is a significant relationship between a person's level of mobility and COVID-19 cases in Coblong District.

The Odd Ratio value is obtained from the results of cross tabulation using Microsoft Excel. The results of the cross tabulation, obtained the OR for the mobility level of 0.166, which means that someone with a 'very low' level of mobility is 0.166 times at risk of getting from COVID-19 compared to a person with a 'very high' level of mobility. In other words, people with "very high" mobility are six times more likely than people with "very low" mobility to develop COVID-19.

3.6 Room Disinfection

Room disinfection is one of the health recommendations recommended by WHO during the COVID-19 pandemic. Table 6. shows the room disinfection habits of control respondents and survivors before being exposed to COVID-19.

Table 6. Room Disinfection Habits

Category	Disinfection Habits	
	Cases	Control
1 (rarely)	8 (26,67%)	2 (10%)
2 (sometime)	6 (20%)	3 (15%)
3 (often)	9 (30%)	6 (30%)
4 (always)	7 (23,33%)	9 (45%)
Total	30	20

In the table above, it can be seen that in the control group respondents, most (9 people, 45%) had the habit of always disinfecting the room. In the survivor group, there were only 7 people (23.33%) who had the habit of always disinfecting the room. In the survivor group, most (9 people, 30%) had the habit of frequently disinfecting the room, while in the control group there were 6 people (30%) who had the habit of frequently disinfecting the room. However, in the group of survivors, there were still 6 people (20%) and 8 people (26.67%) who rarely and very rarely did room disinfection before being exposed to COVID-19, while in the control group, there were still 3 people (15%) and 2 people (10%) rarely and very rarely do room disinfection.

In the results of the chi-square test which shows the significance of room disinfection habits for COVID-19 cases, a P-value of 0.309 is obtained. This significance number is more than 0.05, meaning that the statistical calculation results between room disinfection habits and COVID-19 cases in Coblong District are not significant. Thus, it can be stated that there is no significant relationship between room disinfection habits and COVID-19 cases in Coblong District.

The Odd Ratio value is obtained from the results of cross tabulation using Microsoft Excel. The results of the cross tabulation showed that the OR for room disinfection was 5.143, which means that someone who 'very rarely' disinfects a room has 5.143 times the risk of getting from COVID-19 compared to someone who 'always' disinfects a room.

3.7 Correlation and Regression Analysis

The correlation coefficient shows the linear relationship between two variables as estimated by the sample correlation coefficient r. The sample determination coefficient R2 states the proportion of the overall variation in the dependent variable, which can be explained by a linear relationship with the value of the independent variable.

Table 7. Spearman Correlation Result

Parameter	Correlation coef. (r)	Determination coef. (R^2)	Sig.
Washing Hands	-0.357	0.127	**0.011**
Reducing Mobility	-0.306	0.094	**0.03**
Avoiding Crowds	-0.297	0.088	**0.037**
Room Disinfecting	-0.266	0.071	**0.062**
Keeping Distance	-0.222	0.049	0.121
Wearing Mask	-0.215	0.046	0.133

Table 8. Logistic Regression Result

Predictors	B value	Sig.
(Constant)	6.186	0.049
Washing Hands	-1.090	0.058
Avoiding Crowds	-0.536	0.319
Reducing Mobility	-0.369	0.272
Room Disinfecting	-0.354	0.254

Based on **Table 7**, There are especially four behaviors with exceptionally small correlation significance, showing that washing hands with soap, reducing mobility, avoiding crowds, and disinfecting were significantly correlated with COVID-19 case in Coblong District (α = 0.062). Moreover, the results showed that all preventive behaviors studied had a negative correlation value. These results are in accordance with the initial hypothesis that the protective or preventive factors carried out by public will protect from/reduce the occurrence of COVID-19 cases. This result promotes the importance of preventive behavior on a community during COVID-19 pandemic. In epidemiology triad, the behavior is one of the intervention/control applied to the host element.
Based on the results of logistic regression in **Table 8**, an equation predicting the risk of COVID-19 infection in Coblong District based on behavioral factor can be formed. The equation are as follow.

$$P = \frac{1}{1 + e^{-(\beta_0 + \beta_1 x_1 + \beta_2 x_2 + \cdots + \beta_4 x_4)}}$$

B used in equation above are the predictors' B value in **Table 9** while the X used to predict is the compliance rate value ranging from 1-4 for each behavioral factors. P is in percent. Step-wise regression is used, based on the significance level on **Table 8**. Predictors excluded in model are wearing mask and keeping distance. For example, if all of preventive measures are well done with 'always' category, 4 is substituted to X and the P value result is 0.0387 or 3.87%. This value predicts the risk of getting COVID-19 in Coblong District with maximum effort of preventive measures. The Nagelkerke pseudo-R square value is found to be 0.332. It can be interpreted that 33.2% of the total variation (overall) of COVID-19 cases in Coblong District can be explained by preventive behavior factor.

4. Conclusion

It was found that preventive behavior that had a significant relationship with COVID-19 cases in Coblong was mainly washing hands and reducing mobility. However, this does not mean that only washing hands and reducing mobility are two preventive behaviors needed to control the pandemic situation. On the contrary, the results emphasize the importance of society to adapt multiple new habits. Preventive behavioral factors that have a relationship or depend on cases of COVID-19 are hand washing and mobility. All preventive behavior factors have a negative correlation with COVID-19 cases except for the level of mobility which has a positive or unidirectional relationship between a person's level of mobility and COVID-19 cases.

Acknowledgement

The authors would like to thank all team members, especially the research group of the Environmental Management Technology, Faculty Civil and Environmental Engineering and Statistics Research Division, Faculty of Mathematics and Natural Sciences, Institut Teknologi Bandung. The authors wish to thank Penelitian Unggulan ITB 2021 that funded and supported this study.

References

[1] World Health Organization. (2020). *Advice on the use of masks in the context of COVID-19: interim guidance, 5 June 2020* (No. WHO/2019-nCoV/IPC_Masks/2020.4). World Health Organization.

[2] Vardoulakis, S., Sheel, M., Lal, A., & Gray, D. (2020). COVID-19 environmental transmission and preventive public health measures. *Australian and New Zealand Journal of Public Health*.

[3] Feng, S., Shen, C., Xia, N., Song, W., Fan, M., & Cowling, B. J. (2020). Rational use of face masks in the COVID-19 pandemic. The Lancet Respiratory Medicine, 8(5), 434–436. https://doi.org/10.1016/s2213-2600(20)30134-x.

[4] Machida, M., Nakamura, I., Saito, R., Nakaya, T., Hanibuchi, T., Takamiya, T., ... & Inoue, S. (2020). Incorrect use of face masks during the current COVID-19 pandemic among the general public in Japan. *International journal of environmental research and public health*, *17*(18), 6484.

[5] Mona, N. (2020). Konsep Isolasi dalam Jaringan Sosial untuk Meminimalisasi Efek Contagious (Kasus Penyebaran Virus Corona di Indonesia). Jurnal Sosial Humaniora Terapan UI. https://journal.vokasi.ui.ac.id/index.php/jsht/article/view/86/0

[6] Oginawati, K., Nathanael, R. J., Pasaribu, U. S., Mukhaiyar, U., Humam, A., Ilmi, N. F. F., & Susetyo, S. H. Analysis of the Effect and Role of Indoor Environmental Quality in the COVID-19 Transmission. Aerosol and Air Quality Research, 22, 210339.

[7] Syuhada, K., Wibisono, A., Hakim, A., & Addini, F. (2021). Covid-19 risk data during lockdown-like policy in Indonesia. *Data in Brief*, *35*, 106801.

[8] Marhamah, E., & Jaya, I. G. N. M. (2020). Modeling positive COVID-19 cases in Bandung City by means geographically weighted regression. *Commun. Math. Biol. Neurosci.*, *2020*, Article-ID.

[9] Muhammad, R. (2021). *ANALISIS TERHADAP PERATURAN PEMERINTAH NOMOR 21 TAHUN 2020 TENTANG PEMBATASAN SOSIAL BERSKALA BESAR DALAM RANGKA PERCEPATAN PENANGANAN CORONA VIRUS DISEASE 2019 (COVID 19)* (Doctoral dissertation, Ilmu Hukum).

The use of deflection bowl parameters to evaluate the structural condition of the North Coast Corridor of Java Island

B H Setiadji[1,*], A K Indriastuti[1], and Y Rivani[1]

[1] Department of Civil Engineering, Faculty of Engineering, Universitas Diponegoro
Jl. Prof. Soedarto, S.H., Tembalang, Semarang, Indonesia 50275

* Corresponding author e-mail: bhsetiadji@ft.undip.ac.id

Abstract. Monitoring the structural condition of the pavement is a very important process to be carried out periodically because it can estimate the damage that occurs as early as possible. However, this process is still constrained due to the limited non-destructive pavement response measuring instrument available and the lack of expertise in analyzing the structural condition of the road from the results of field measurements. Currently, the deflection bowl parameter method is one of the methods developed in research in South Africa to provide a more user-friendly method in determining the indication of the bearing capacity of the pavement layers. The use of this method in Indonesia is still limited, so this study aimed to evaluate the suitability of its use on roads in Indonesia. For this purpose, deflection data was collected from several arterial roads on the North Coast of Java Island. The study resulted that the deflection bowl parameter method generally can be applied to roads in Indonesia. However, the method has to be used with caution because of the complexity of the pavement structure layers on the North Coast Road of Java Island, especially for the layers that have undergone major repairs. In addition, the benchmarking process for the deflection bowl parameter criteria also only applied to roads with medium traffic load repetitions (up to 50 million ESAL), so upgrading the limit of the traffic load of the parameters was required so that the parameters could be used for roads with very high traffic load repetition as the one at the North Coast corridor.

Keywords: Deflection bowl parameters, Pavement layers, Structural condition, Traffic load

1. Introduction

In the Indonesian Government's 2021 Work Plan, one of the ten national development priorities is infrastructure, connectivity, and maritime affairs. Among infrastructures, one of the most widely built at this time is road infrastructure. In particular, road infrastructure has an iterative life cycle, starting from the planning, construction, operational, maintenance/rehabilitation, reconstruction, and then returns to the operational stage and so on. Of all these processes, the maintenance/rehabilitation stage is the stage that has the most prolonged continuity period compared to the other stages. Thus, the evaluation of structural conditions of the road pavement carried out to date is to determine whether the road still has an adequate level of performance to withstand traffic loads.

To evaluate the structural condition of the pavement, people usually measure the response, in the form of deflection values, of the loads acting on the pavement. The deflections and the layer thicknesses are two essential data for assessing pavement structure health, conducted using a backcalculation procedure [1]. The thickness of the pavement layer becomes problematic in determining the layer moduli because, without adequate layer thickness data, the backcalculation results of the layer moduli will be inaccurate. In addition, the backcalculation procedure generally requires a specialist in this field [2].

As an alternative to assessing the structural condition of the roads, a procedure was developed by Horak and Emery [3] by introducing the deflection bowl parameters. This procedure can be used to evaluate the structural condition of pavement without worrying about the two constraints above. The parameters are determined based on the relative comparison between 2 offset values or different distances from the center of the load. The parameters only require deflection data to determine the structural condition of the road pavement.

This procedure is very attractive to be used; therefore, this study aims to evaluate the procedure by implementing the procedure at several road segments along the North Coast corridor of Java Island to see the possibility to use the procedure in assessing the structural condition of the corridor.

2. Deflection Bowl Parameters

The parameters are the main part of the procedure to present the structural conditions of the road pavement proposed by Horak and Emery [3] which is based on the relative comparison between the deflection values within a certain distance from the center of the load with predetermined condition criteria. The parameters are calculated from the measured FWD deflection (D_r) produced by a certain level of load, which is dropped onto the road surface where geophones are placed at various offsets, r (in mm), from the center of the load. In the study, the FWD geophone measurements were placed at D_0 (below the center of the FWD loading plate, which has a diameter of 300 mm), 200 mm (D_{200}), 300 mm (D_{300}), 600 mm (D_{600}), 900 mm (D_{900}), 1200 mm (D_{1200}), 1500 mm (D_{1500}), and 1800 mm (D_{1800}). Several parameters then were developed based on a single or a pair of deflections to represent the structural condition of the pavement structure. There are several deflection curve parameters [2,3], four of which have a close relationship with the four-layer pavement structure: maximum deflection (D_0), base layer index (BLI), middle layer index (MLI), and low layer index (LLI). The formula of the last three indices is as follows.

$$BLI = D_0 - D_{300} \tag{1}$$

$$MLI = D_{300} - D_{600} \tag{2}$$

$$LLI = D_{600} - D_{900} \tag{3}$$

where: D_0, D_{300}, D_{600}, D_{900} are the deflections at sensor offsets of 0, 300, 600, and 900 mm, from the load center, respectively.

The calculation results of the deflection bowl parameters will be meaningless if there are no criteria to express the meaning of the resulting values. Therefore, Horak and Emery [3] introduced three levels of pavement structural conditions that can be used as an indication to predict the pavement structural condition at any time under certain conditions. An example of the criteria for pavement structural conditions on the road with a traffic load of 3 million equivalent standard axle load (ESAL) is shown in Table 1.

Table 1. Deflection Curve Parameter Criteria [3]

Structural conditional rating	Deflection bowl parameters			
	Do (μm)	BLI (μm)	MLI (μm)	LLI (μm)
Sound	< 400	< 200	< 100	< 55
Warning	400 -750	200 – 500	100 – 200	55 – 100
Severe	> 750	> 500	> 200	< 100

3. Research Methodology

There are three main steps in this study:

a. Collecting data by conducting a field survey. This study selected ten road segments in the North Coast corridor of Java Island as the case. They are (see Fig.1): MT Haryono road (Pemalang) (006.11K), Letjen Suprapto road (Pemalang) (006.12K), road between the boundary of Pemalang city and the boundary of Pekalongan city (006), Batang main road (Pekalongan) (007.18K), Sudirman road (Batang) (007.1AK), Plelen road (North) (010), road between the boundary of Batang municipality and the boundary of Weleri (009), Weleri ring road (012.11K), West main road (Kendal) (013.12K), and Walisongo road (Semarang) (014.12K). Besides deflection data, the primary survey also recorded road deterioration across the road evaluated. Traffic load data was also collected for supporting information.
b. Calculating the deflection bowl parameters (D_0, *BLI*, *MLI*, and *LLI*) from the deflection data. Evaluate the traffic loads passed through the road segments and group the road segments based on the category of traffic loads.
c. Benchmarking the values of the parameters against the working chart [3] and determining the condition of the road segments based on the condition criteria level on the chart. Evaluate the benchmarking results to determine the suitability of the parameters in assessing the road condition in the North Coast corridor.

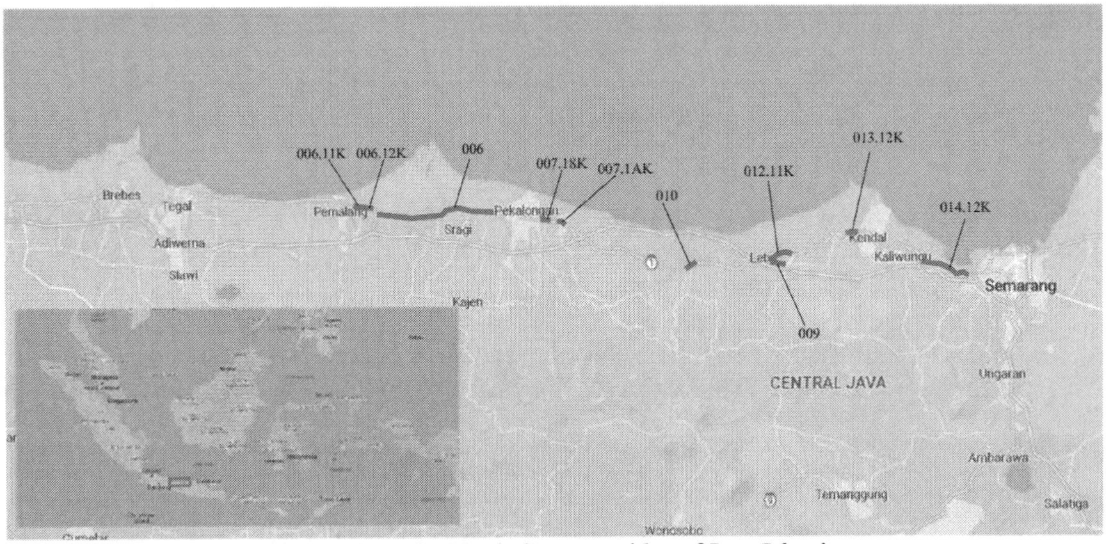

Fig.1 Road segments evaluated along the North Coast corridor of Java Island

4. Results and Analysis

The initial analysis that has been carried out illustrated that the use of raw deflection data might provide an initial indication of the bearing capacity of the layer to withstand the loads. Large deflections are generally used to indicate that the bearing capacity of the layer is low and vice versa. However, in a road pavement system, one layer is bound to another layer by a continuous interface, so looking at a glance at the deflection produced by one sensor cannot provide a comprehensive picture of the bearing capacity of the layer at that point. Similarly, the pattern of the magnitude of the deflection in a deflection curve can also indicate how the pavement system responds to the load that occurs.

Before benchmarking the deflection curve parameters for each road segment based on traffic load repetitions, it is necessary to understand the behavioral condition categories defined for each deflection bowl parameter, as presented in Table 2.

Table 2. Category of behavior conditions of deflection bowl parameters [4]

Behavior state	Range of deflection bowl parameter (μm)			
	D0	*BLI*	*MLI*	*LLI*
Very stiff	< 200	< 10	< 10	< 10
Stiff	200 – 400	10 – 200	10 – 100	10 – 50
Flexible	400 – 600	200 – 400	100 – 150	50 – 80
Very flexible	> 600	> 400	> 150	> 80

The definitions of behavioral conditions are [4]:

a. Very stiff: pavement behavior is primarily controlled by a layer having a high modulus (> 2,000 MPa) acting as a concrete slab
b. Rigid (stiff): pavement behavior is controlled by a layer having a sufficiently high modulus (> 500 MPa). Several layers may experience cracks in the form of blocks with a diameter greater than 2 m
c. Flexibility: pavement behavior is controlled by the material of the granular layer (foundation layer). The surface layer modulus values are in the range of 200 to 500 MPa. If there is a layer stabilized with cement (i.e., cemented treated base/CTB), it can crack in the form of smaller blocks.
d. Very flexible: the behavior of the pavement is controlled by a granular, flexible layer; the modulus value of the layer is usually a low modulus (<200 MPa). Pavement tends to be susceptible to water ingress (with cracks on the pavement surface)

Considering Table 2 and assuming that the foundation layer is included in the granular foundation layer, the level of the structural condition of the deflection curve parameters of all evaluated roads can be presented in Table 3.

Table 3. The level of structural condition behavior of the deflection curve parameter of the evaluated road segments

No.	Road segment (Road ID)	Behavior condition	Deflection bowl parameter(mm)			
			D0	BLI	MLI	LLI
1	MT Haryono road	Sound	< 0,190	< 0,080	< 0,035	< 0,019
	(Pemalang)	Warning	0,190 – 0,750	0,080 – 0,400	0,035 – 0,200	0,019 – 0,100
	(006.11K)	Severe	> 0,750	> 0,400	> 0,200	> 0,100
2	Letjen Suprapto road	Sound	< 0,190	< 0,080	< 0,035	< 0,019
	(Pemalang)	Warning	0,190 – 0,750	0,080 – 0,400	0,035 – 0,200	0,019 – 0,100
	(006.12K)	Severe	> 0,750	> 0,400	> 0,200	> 0,100
3	Road between the boundary of Pemalang city	Sound	< 0,180	< 0,077	<0,033	< 0,018
	and the boundary of	Warning	0,180 – 0,750	0,077 – 0,400	0,033 – 0,200	0,018 – 0,100
	Pekalongan city (006)	Severe	> 0,750	> 0,400	> 0,200	> 0,100
4	Batang main road	Sound	< 0,190	< 0,080	< 0,035	< 0,019
	(Pekalongan)	Warning	0,190 – 0,750	0,080 – 0,400	0,035 – 0,200	0,019 – 0,100
	(007.18K)	Severe	> 0,750	> 0,400	> 0,200	> 0,100
5	Sudirman road	Sound	< 0,190	< 0,080	< 0,035	< 0,019
	(Batang)	Warning	0,190 – 0,750	0,080 – 0,400	0,035 – 0,200	0,019 – 0,100
	(007.1AK)	Severe	> 0,750	> 0,400	> 0,200	> 0,100
6	Plelen road (North)	Sound	< 0,190	< 0,080	< 0,035	< 0,019
	(010)	Warning	0,190 – 0,750	0,080 – 0,400	0,035 – 0,200	0,019 – 0,100
		Severe	> 0,750	> 0,400	> 0,200	> 0,100

7	Road between the boundary of Batang municipality and the boundary of Weleri (009)	Sound	< 0,210	< 0,090	< 0,040	< 0,022
		Warning	0,210 – 0,750	0,090 – 0,400	0,040 – 0,200	0,022 – 0,100
		Severe	> 0,750	> 0,400	> 0,200	> 0,100
8	Weleri ring road (012.11K)	Sound	< 0,175	< 0,076	< 0,032	< 0,017
		Warning	0,175 – 0,750	0,076 – 0,400	0,032 – 0,200	0,017 – 0,100
		Severe	> 0,750	> 0,400	> 0,200	> 0,100
9	West main road (Kendal) (013.12K)	Sound	< 0,210	< 0,090	< 0,040	< 0,022
		Warning	0,210 – 0,750	0,090 – 0,400	0,040 – 0,200	0,022 – 0,100
		Severe	> 0,750	> 0,400	> 0,200	> 0,100
10	Walisongo road (Semarang) (014.12K)	Sound	< 0,175	< 0,076	< 0,032	< 0,017
		Warning	0,175 – 0,750	0,076 – 0,400	0,032 – 0,200	0,017 – 0,100
		Severe	> 0,750	> 0,400	> 0,200	> 0,100

In the analysis, the ten road segments are grouped based on the number of repetitions of the axle load and the number of test points, namely:

a. Group 1, roads with a total load repetition of about 25 million ESAL
b. Group 2, roads with a total load repetition of about 32 - 35 million ESAL
c. Group 3a, roads with a total load repetition of about 38 - 41 million ESAL, all roads have test points less than 40
d. Group 3b, roads with a total load repetition of about 38 - 41 million ESAL, all roads have test points more than 40

The plots of four deflection bowl parameters and their respective criterion threshold for all groups are depicted in Figs. 2, 3, 5, and 6.

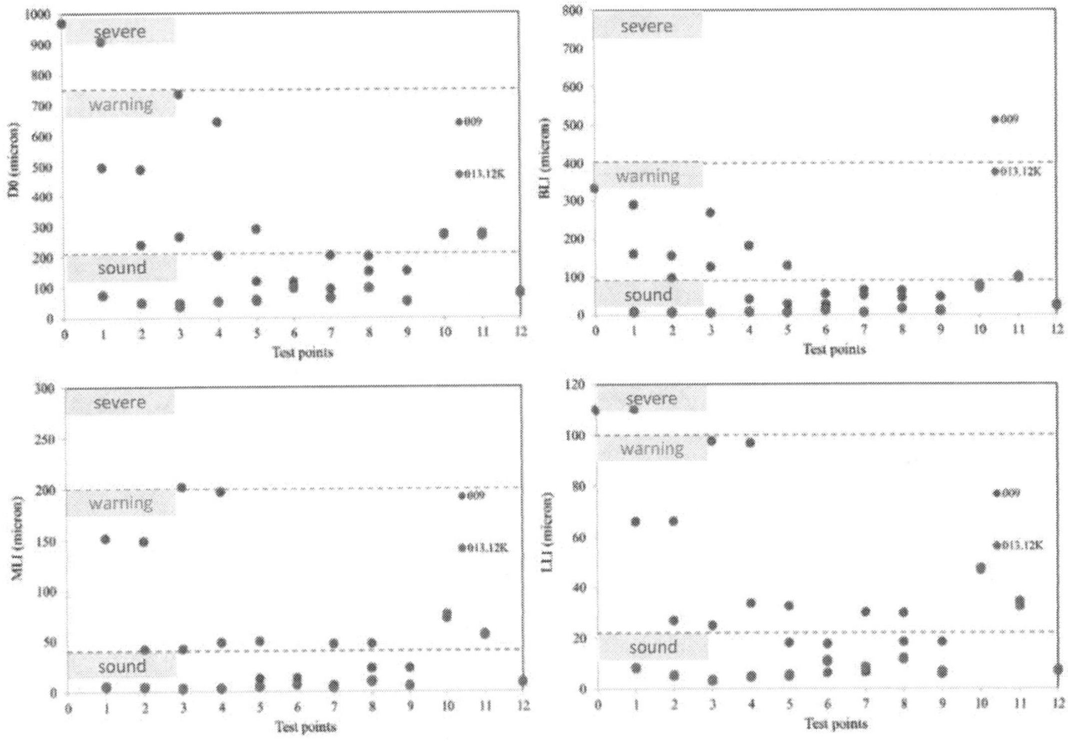

Fig.2 The plots of the deflection curve parameters on roads of group 1

In Fig. 2, two roads, i.e., the road between the boundary of Batang municipality and the boundary of Weleri (009) and West main road (Kendal) (013.12K), have a different trend. Road segment 13.12K is in good condition (sound condition) for all layers, and according to Table 3, all layers of this road section are very stiff. However, on the contrary, road segment 009 has a different trend along this segment. Some test points of this segment (test points 1 to 3) show indications of problems in the subgrade to the surface, while at other test points, the average is in good condition (sound criterion) or fair condition (warning condition).

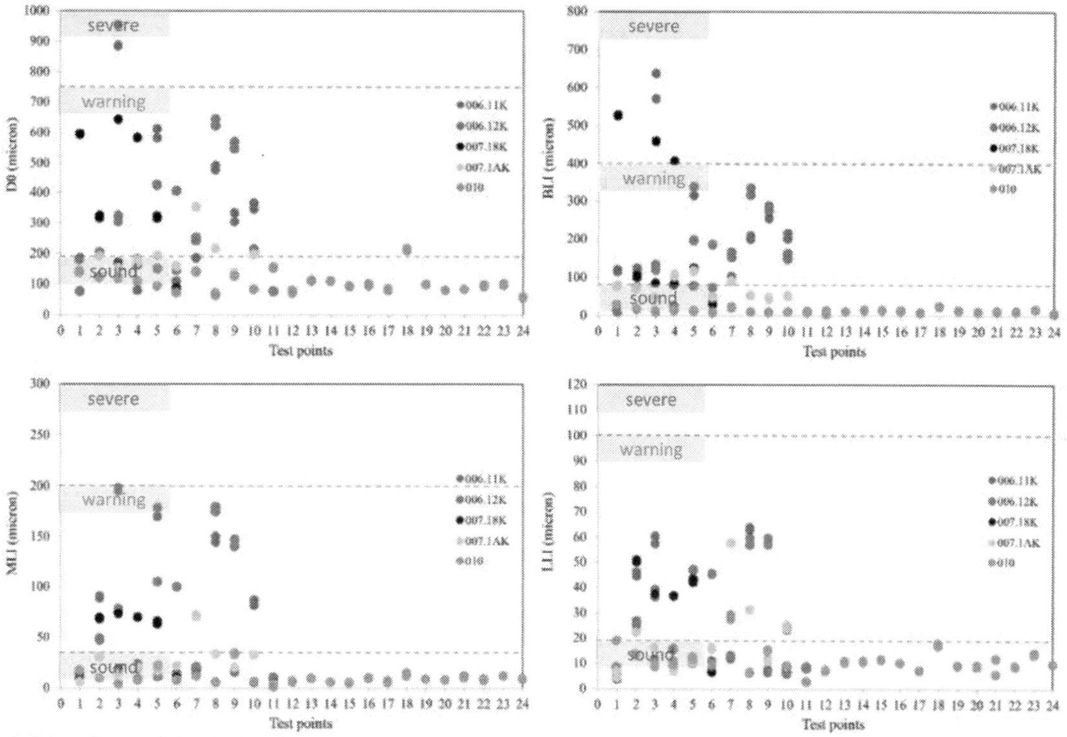

Fig.3 The plots of the deflection curve parameters on roads of group 2

Fig. 3 shows the trend of deflection curve parameters for road sections with load repetitions between 32 – 35 million ESAL. This group consists of 5 roads: MT Haryono road (Pemalang) (006.11K), Letjen Suprapto road (Pemalang) (006.12K), Batang main road (Pekalongan) (007.18K), Sudirman road (Batang) (007.1AK), and Plelen road (North) (010). In this group, road segment 010 shows very good condition (or sound condition). In contrast, road segment 006.11K has one point, i.e., test point 3, that consistently shows a large deflection (i.e., in the severe area, especially parameters D_0 and *BLI*), as seen visually in the field in the form of a fatigue crack (Fig. 4(a)), which allows water to enter through the cracks so that it weakened the bearing capacity. Fig 4(b) shows that a pothole has already developed with preliminary cracking at test point 5, which could lead to alligator cracking. For the other three roads, the average is in a warning condition. This condition needs attention because it can worsen if it is not maintained properly, such as through road preservation work.

Figure 5 shows the trend of the deflection curve parameter of two roads that have a traffic load of 38 – 41 million ESAL: Weleri ring road (012.11K) and Walisongo road (Semarang) (014.12K). Road segment 014.12K has a relatively high level of variability. There are some deflections at the test points that are quite stiff for all deflection bowl parameters, and some others in a warning condition. Interestingly, there are 2 test points (test points 10 and 28) that need to be monitored because they are located consistently in severe/warning conditions and show weak conditions from the surface layer to the subgrade. Another road segment, road segment 012.11K, does not have a clear pattern, where the plots of the parameter values are randomly distributed under sound and warning conditions.

Fig.4. Road conditions as the results of the field survey: (a) and (b) are road segment 006.11K at test points 3 and 5; (c) and (d) are road segment 006 at test points 5 and 28.

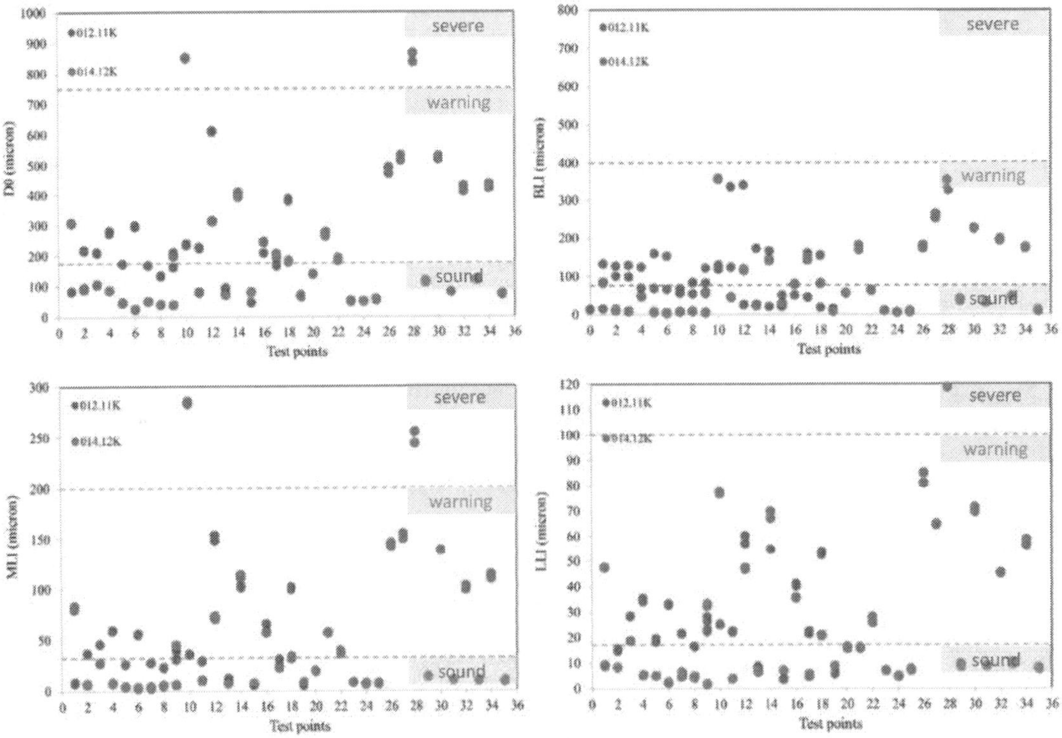

Fig.5 The plots of the deflection curve parameters on roads of group 3a

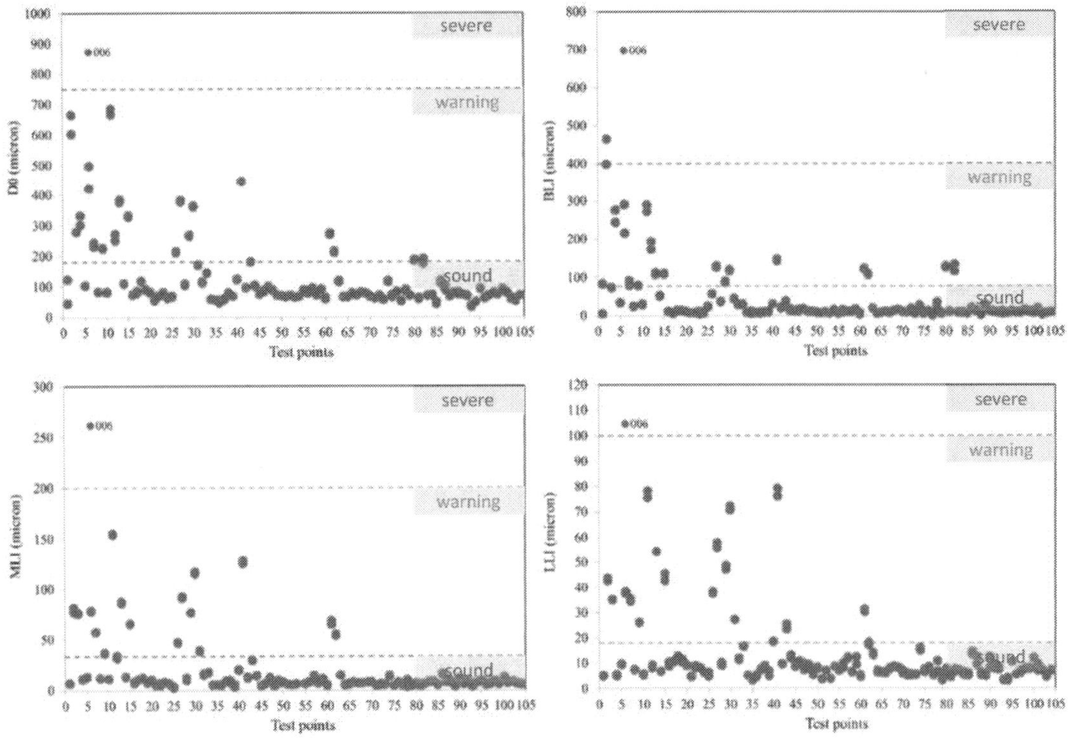

Fig.6 The plots of the deflection curve parameters on roads of group 3b

Fig. 6 presents the trend of the road segment having the same traffic load as the traffic load in Fig. 5, but it has a much higher number of test points, i.e., 104 locations, compared to the road section in Fig. 5 only have less than 40 test points. This figure is a plot of the parameter values of 1 road segment, the road between the boundary of Pemalang city and the boundary of Pekalongan city (006). The test points on this road segment are divided into two parts: test points 1 – 45 and test points 46 – 104. Some of the deflections that occur in the range of the first part (test points 1 – 45) indicate that the bearing capacity has varied conditions, as shown by their location in the warning and sound condition area. The different illustrations of road conditions in the first part of this segment can be seen in Fig. 4(c) and 4(d), representing conditions at test points 5 and 28. Meanwhile, almost all deflections in the second part (test points 46 – 104) show good conditions for all parameters, although not all show very stiff behavior conditions.

It can be summarized from the analysis of the figures above that deflection bowl parameters are very useful to detect the presence of structural damage in the pavement structure, which is not visible using only a road condition survey. To enable to predict the implementation time of a cost-effective maintenance/rehabilitation work, the measurement of the structural response of the pavement along the monitored road segment has to be carried out regularly to determine the rate of damage that occurs. This structural response can be used to evaluate how fast the behaviour conditions change (from sound to warning, and from warning to severe). The implementation time of road handling can be estimated by determining a criterion, in the terms of the percentage of the road segment in warning/severe condition, as a trigger to carry out certain treatment. For example, the application of road preservation when most of the roads with parameter D_0 in a warning condition, or the application of a certain depth of layer scarifying when it is found that many locations on that road with parameters D_0 and *BLI* in warning or severe conditions. Unfortunately, this criterion cannot be presented in this paper due to the limitations of the deflection data series that can be collected and the lack of information on the implementation of maintenance/rehabilitation works that has been carried out on the evaluated road sections.

Further analysis can be carried out to improve the parameters' effectiveness by reviewing the procedure as follows.

a. The layer thickness of the pavement structure is still required in deflection bowl parameters, although it only plays a small role in the procedure because the thickness is needed to determine which deflection data will be used for the BLI, MLI, and LLI parameters. It is very important for cases where several overlay works are done for different years, such as on the North Coast corridor. Ones need to be careful in determining the layers used in the analysis whether the structure consists of the following: (i) surface layer; (ii) asphalt treated base (ATB) layer (assuming the old layer is used as the foundation layer); (iii) granular base layer; and (iv) subgrade, or other layers. It requires a review by comparing the *BLI* values for ATB or the granular base layer. Setiadji [5] stated that thick pavement layers required outer geophones to be used for the parameters.
b. If one selects an inappropriate foundation layer (e.g., bound base layer instead of unbound base layer), the criterion for layer conditions becomes stricter. As stated by Horak [5], the sound condition for D_0 of a pavement with a granular or unbound base layer is smaller than 500 microns, but it can be less than 400 microns (for the ATB layer) or smaller than 200 microns (for CTB layer).
c. Horak and Emery [3] recommended a chart firstly developed by Maree and Bellekeens [7] that related the parameters to traffic levels and base types. This chart was used to determine the condition criteria of all parameters of the ten evaluated road segments (see Table 3). However, the chart limits the number of repetitions to a maximum of 50 million ESAL, whereas for the North Coast corridor, the load repetition is very likely to reach more than 200 million ESAL (at the design life of 20 years). Therefore, the chart needs to be used with caution because of the relatively short design life (approximately 5-year design life). So, after five years, the surface layer needs to be scrapped and replaced by a new asphalt layer, and the layers below can be assumed as a bound base layer as far as they have not deteriorated. And then, the prediction of pavement condition may be performed for the next 5-year life cycle.
d. The lower bound for the criterion of severe conditions needs to be investigated further because the determination of severe conditions is based on very flexible behavior conditions. The results of this study are useful for time effectiveness and cost-efficiency if it can be seen that severe conditions occur faster than when conditions are very flexible.
e. The results of deflection bowl parameters are required to be validated using field survey results to ensure that the parameters produce accurate predictions.

5. Conclusion

Two conclusions can be made: firstly, the deflection bowl parameters so far are still relevant as an indication of the structural condition of a pavement. It is recommended that the parameters are required to be updated, especially for a thick road pavement of more than 900 mm, because the last deflection on the deflection bowl used is the deflection at a distance of 900 mm. Further development of the chart that relates the parameters to the traffic loads is also required to enable the use of the chart for the road pavement with load repetition of more than 50 million ESAL. Secondly, validation is also important to be performed on the results of the parameters by using field survey results to ensure that the deflection bowl parameters could produce an accurate prediction.

Acknowledgment

This research was financially supported by The Faculty of Engineering, Diponegoro University, Indonesia, through Strategic Research Grant 2021.

References

[1] Yang X, Guan J, Ding L, You Z, Lee VCS, Mohd Hasan MR, et al. Research and applications of artificial neural network in pavement engineering: A state-of-the-art review. Journal of Traffic and Transportation Engineering (English Edition). 2021 Dec 1;8(6):1000–21.

[2] Horak E. Benchmarking the structural condition of flexible pavements with deflection bowl parameters. Journal of the South African Institution of Civil Engineering. 2008;50(2):2–9.
[3] Horak E, Emery S. Falling weight deflectometer bowl parameters as analysis tool for pavement structural evaluations. In: 22nd Australian Road Research Board (ARRB) International Conference [Internet]. Brisbane, Australia; 2006. Available from: https://www.researchgate.net/publication/242495926
[4] Committee of State Road Authorities (CSRA). Guidelines for rehabilitation design of flexible pavements. Technical Recommendations for Highways 12 (TRH 12). Pretoria; 1997.
[5] Setiadji B. H. Application of deflection bowl parameters for assessing different structures of road pavement. In: MATEC Web of Conferences. EDP Sciences; 2018.
[6] Horak E. Aspects of deflection basin parameters used in a mechanistic rehabilitation design procedure for flexible pavements in South Africa. [Pretoria, South Africa]; 1988.
[7] Maree JH, Bellekens RJL. The effect of asphalt overlays on the resilient deflection bowl response of typical pavement structures [Internet]. Pretoria: Dept. of Transport Chief Directorate National Roads; 1991. (Research report (South Africa. Department of Transport)).

Effects of Dammar as the partial substitution of bitumen in HMA AC-WC containing Buton Asphalt

E Ngii[1], L One[2], N Nasrul[2] and M I Udo[3]

[1]Engineering Faculty, Haluoleo University, Kendari-Indonesia
[2]Civil Engineering Department, Haluoleo University, Kendari-Indonesia
[3]Students of Civil Engineering Department, Haluoleo University, Kendari - Indonesia

Corresponding authors: laone@uho.ac.id

Abstract. Petroleum asphalt is the residue of fossil fuel distillation. The flexible pavement material generally uses petroleum asphalt. The world's need for bitumen increases from year to year to pavement materials. Meanwhile, the availability of bitumen as a non-renewable resource is limited. To meet the world's need for bitumen in the future, the utilization of natural polymers, such as a dammar, can be the best solution. This study aims to analyze the influence of dammar on the Marshall stability of HMA AC-WC containing Buton Granular Asphalt (BGA). This study was carried by using Marshall testing on the Asphalt Concrete Wearing Courses (AC-WC) containing Buton asphalt of 10%. Bitumen content is in all specimens amount 6,3%, bitumen content of BGA 50/30 is 22,48%, and the variation of dammar content is as the substitution of bitumen of 2.5-10%. Before mixing all material, it was first conducted the mixing of dammar and penetration at the temperature of 90°C to get a modified asphalt. The use of 10% dammar in asphalt mixtures meets the requirements of Bina Marga specifications. Utilization of Dammar content of 10% with BGA content of 10% can substitute petroleum asphalt by 45.68%.

Key word: Buton Asphalt, Dammar, HMA AC-WC

1. Introduction

Hot mix asphalt (HMA), an asphalt mixture type for flexible pavement, is commonly used in Indonesia. One type of HMA is an Asphalt Concrete-Wearing Course (AC-WC). It is a non-structural layer located on the surface of the road pavement. It is directly related to the vehicle wheels. The binder of HMA generally still uses petroleum asphalt. It is the residue from the petroleum distillation process. The need for petroleum asphalt in flexible pavement construction increases within good economic conditions and regional development in Indonesia. To reduce dependence on oil asphalt, the Government of the Republic of Indonesia has encouraged the use of Buton asphalt organized in the Regulation of the Minister of Public Works No. 18/PRT/M/2018.

Several studies have shown that the advantages of using it can be increased stability, modulus of elasticity, increased free compressive strength, good rutting resistance (increased dynamic stability), reduced bleeding problems, and sufficient flexibility. Meanwhile, the weakness of using Buton Asphalt is the release of higher granules and tends to be more fragile, but it is still in requirement (Affandi, 2010)[1]. In addition, the higher the use of Buton Asphalt can cause the asphalt mixture to be stiffer and the void in the mix to be greater (One L et al, 2020)[2].

Petroleum asphalt and Buton asphalt are non-renewable natural resources and are limited in number. The increasing demand for world asphalt will impact a scarcity of asphalt availability in the future. The utilization of natural polymers as renewable natural resources is one of the innovations to overcome the need for road pavement binders. There are several types of natural polymers in Indonesia. One of them having enormous potential is dammar. The research results related to the usage of dammar as a partial substitute for petroleum asphalt are still relatively few. However, the results of this study indicate that the use of dammar can improve the performance of asphalt mixtures. The use of dammar can increase the stability of the asphalt mixture (Sefrus T., 2013)[3].

This study aims to know the influence of using dammar as a partial substitute for petroleum asphalt in the AC-WC containing Buton Asphalt. The performance of asphalt mixture using dammar should be better than the HMA performance without using dammar.

2. Literature Review

2.1. *Characteristics of Buton asphalt*

Indonesia has natural asphalt forms rocks located on Buton Island, Southeast Sulawesi province. This natural asphalt is called Buton asphalt. It contains two main elements, bitumen and minerals. The bitumen of Buton asphalt has high asphaltene content and low maltene content. Here is cause oxidation process due to aging Buton asphalt under natural conditions. While mineral of Buton asphalt is more contain alkali particles (Jinjin S, et al. 2016)[4].

One of the common types of Buton asphalt products on the market is type B-50/30. This type of Buton asphalt forms granules containing 20-30% bitumen and penetration of about 50 (0.1 mm). Referred to Bina Marga specification-2018, the physical characteristics of Buton asphalt Type B 50/30 as pavement material shown in table 1 follows.

Table 1. The requirement of physical characteristics for Buton asphalt type B-50/30

Properties	Requirement
Passing sieve size of 3/8" (9,5 mm)	100%
Bitumen content	≥ 20%
Water content	4% ≥
Bitumen Characteristic, extraction of Buton Asphalt	
Solubility in TCE	*≥ 99 %*
penetration at 25°C 100 g, 5 seconds	*40-70 (0.1 mm)*
Softening Point	*≥ 50 °C*
Ductility at 25°C	*≥ 100 cm*
Specific gravity	*≥ 1.0*
Weight loss (with TFOT)	*2% ≥*
Penetration after Loss of Heating (LoH) at 25°C, 100 g, 5 seconds	*≥ 54 (0.1 mm)*

2.2. *Characteristics of dammar*

Mulyono N. et al. (2011)[5] showed that dammar has 67 organic compounds categorized into four groups. Firstly, carbon tetracyclic (30 compounds, 49.57%). Secondly, pentacyclic (3 compounds, 2.56%). Thirdly, C15 compounds (11 compounds, 17.09 %). And the last, other groups (23 compounds, 18.26%). Organic compounds in cat-eye dammar having more than 5% are Brasikasterol ($C_{28}H_{46}O$) of 20.23%, Epoxyacetylgedunine ($C_{26}H_{32}O_7$) of 9.56%, and Veridiflorol ($C_{15}H_{26}$) of 5.34%. While Pregnana-3,11,20-trion ($C_{21}H_{30}O_3$), Andros-5-en-3,17-diol,4,4-dimethyl-,diacetate (3β,17β) ($C_{25}H_{38}O_4$), and Pregna-4,6-Dien-3,20-dione,6,17-dimethyl ($C_{23}H_{32}O_2$) having organic compound content of 3-4%.

The physical characteristics of cat-eye dammar are melting point 71°C, a weight loss at 105°C for 18 hours 0.84%, and density of 0.68 g/cm3 (Kasim A. et al., 2020)[6]. The softening point of cat-eye dammar is 68.33-85.25°C, and the viscosity is 4.03-26.25 cP (Mulyono N. et al., 2009)[7]. The difference in the dammar softening point causes the difference in the content of essential oils in the dammar. Dammar that has just harvested has a higher volatile oil. The dammar stored for a long time causes the essential oil to evaporate, oxidize, and the dammar will harden. Here causes the softening point of the dammar to increase. The longer the dammar is stored, the higher the oxidation reaction rate occurs. So the softening point of dammar is higher (Sari RK., 2002)[8].

2.3. *Performance of asphalt mixture*

Marshall Characteristic and volumetric are two factors that influence the performance of asphalt mixture. Marshall Characteristic consists of Marshal Stability (MS), flow, and Marshall quotient (MQ). Volumetric consists of Void Mineral Aggregate (VMA), Void Filled Bitumen (VFB), and Void in Mixture (VIM). Calculating of MS, MQ, VIM, VMA, and VFA use equation (1)-(5).

$$MS = \frac{1000\ x\ P\ x\ k}{g} \tag{1}$$

$$MQ = \frac{MS}{F} \tag{2}$$

$$VIM = \frac{(Gmm - Gmb)x\ 100}{Gmm} \tag{3}$$

$$VMA = \frac{(100(Gsb - Gmb) + GmbxPb)}{Gsb} \tag{4}$$

$$VFB = \frac{((VMA - VIM)x100}{VMA} \tag{5}$$

Where: MS is Marshall Stability (kg), MQ is marshall quotient (kg/mm), VIM is Void in Mixture (%), VMA is Void Mineral Aggregate (%), VFB is Void Filled Bitumen (%), VMA is Void Mineral Aggregate (%), P is the compressive strength (kN), k is the correction of specimen thickness, g is the coefficient of gravity (9.81 m/s2). F is Flow (mm), Gmm is the maximum specific gravity of the mixture after compacting, Gmb is the bulk specific gravity of the mixture after compaction, Gsb is the bulk specific gravity of all aggregate, Pb is aggregate content to mixture weight (%).

Referring to the Bina Marga specifications–2018, the requirement of physical characteristics for Modified AC-WC are shown in table 2 follow.

Table 2. The requirement of physical characteristics for modified AC-WC

Property of Mixture	Requirement
Effective bitumen content	≥ 5.9 %
Total of blows each face	75 blows
Void in Mixture	3-5%
Void Mineral Aggregate	≥ 15%
Void Filled Bitumen	≥ 65%
Marshall Stability	≥ 1000 kg
Flow	2-4 mm
Marshall Quotient	≥ 250 kg/mm
Remain Marshall Stability after immersion for 24 hours at a temperature of 60°C	≥ 90%
Dynamical stability	≥ 2500 pass

3. Materials and Methods

3.1. *Materials*

3.1.1. Petroleum Asphalt - Grade 60/70. The asphalt used in this study has met the requirements of The Indonesia National Standard especially for penetration of asphalt at 25°C, the specific gravity of asphalt, and ductility as shown in table 3 following.

Table 3. The physical properties of Petroleum Asphalt - Grade 60/70

Properties	Testing Method	Specification	Result
Penetration of asphalt at 25°C	SNI 03-3639-2002	60-70 (0.1 mm)	65 (0.1 mm)
Specific gravity of asphalt	SNI 2441:2011	> 1,0	*1.03*
Ductility	SNI 2432:2011	> 100 cm	*133.5 cm*

3.1.2. Buton Asphalt – Grade B 50/30. Buton asphalt content used in this study was 10% of the total weight. Here refers to the Bina Marga specification recommendation that the use of Buton asphalt - grade B 30/50 is 7-10%. The results of testing the properties of Buton Asphalt include Bitumen content of 22.48%, water content of 0.64, and sieve analysis of Buton Asphalt mineral shown in table 4 follows.

Table 4. The sieve analysis of Buton asphalt mineral

Sieve Size	Retaining (%)	Passing (%)
3/8"	0.00	100.00
No.4	9.05	90.95
No.8	25.47	74.53
No.16	49.52	50.48
No.30	60.51	39.49
No.50	71.01	28.99
No.100	86.98	13.02
No.200	92.26	7.74
P A N	100.00	0.00

3.1.3. Aggregate. Coarse aggregate and fine aggregate used are from stone crushers produced by Moramo quarry in Kendari, Southeast Sulawesi. The physical properties of coarse and fine aggregate used in AC-WC specimens have met the Indonesia National Standard specifications. The physical properties of aggregate shown in table 5 follow.

Table 5. The physical properties of Aggregate

Properties	Testing Method	Specification	Result
Courses Aggregate			
Bulk specific gravity of 2.65	*SNI 03-1969-2008*	> 2.5	2.65
Water Absorption	*SNI 03-1969-2008*	< 3 %	1.27
Abrasion	*SNI 03-2417-2008*	< 40 %	31.1
Fine Aggregate			
a bulk specific gravity	*SNI 03-1969-2008*	*> 2.5*	*2.69*
Water Absorption	*SNI 03-1969-2008*	*< 3 %*	*1.24*
Sludge Content	*SNI 03-4142-1996*	*< 5 %*	*4.23*

3.1.4. Dammar. The type of dammar used is a cat's eye dammar (*Shoerra Javanica*). The use of dammar as a partial substitute for petroleum asphalt grade 60/70. The physical properties of the dammar are water content of 0.60%, and the bulk specific gravity of 1.08;

3.1.5. Modified petroleum asphalt. The Optimum Asphalt Content of AC-WC specimens containing 10% Buton Asphalt was 6.3%. The amount of petroleum asphalt needed in the optimum asphalt content is 46.7 grams. The use of dammar varies 0-10% of optimum asphalt content. Before Petroleum Asphalt mixed with aggregate is first modified with dammar according to the required variety of dammar. Petroleum asphalt modified with dammar is by mixing petroleum asphalt and dammar at the temperature of 90-95°C. One L. et al. 2021[9] showed that the physical properties of petroleum asphalt modified with dammar include penetration and ductility shown in table 6.

Table 6. Penetration and ductility of modified petroleum asphalt

Dammar content (%)	Penetration (0,1 mm)	Ductility (cm)
0	65	>100
2,5	64	>100
5,0	63	>100
7,5	52	>100
10	42	>100

3.2. Target of aggregate gradation.

The combined aggregate gradation uses the ideal gradation target for AC-WC. The combined aggregate gradation includes the gradation of Buton asphalt minerals. The target of gradation of AC-WC shown in Table 7 follows.

Table 7. The gradation target of AC-WC

Sieve Size	Specification (% passing)	Ideal Gradation (% passing)	Retaining (%)
3/4"	100	100	0
1/2"	90-100	95	5
3/8"	77-90	83.5	11.5
No. 4	53-69	61	22,5
No. 8	33-53	43	18
No. 16	21-40	30.5	12,5
No. 30	14-30	22	7,5
No. 50	9-22	15.5	6,5
No. 100	6-15	10.5	15
No. 200	4-9	6.5	4.0
PAN	0	0	6,5

3.3. Specimens.

The dimension target of Marshall cylindrical specimens is a height of 63.5 mm or a specimen weight of 1200 grams. They had the optimum asphalt content of 6.3%. The material composition of the specimen without using dammar consisted of 10% or 120 grams of Buton Asphalt, 46.7 grams of petroleum Asphalt grade 60/70, and 1124 grams of aggregate. The use of dammar as a substitute for petroleum asphalt is 2.5-10% interval 2,5% of the optimum asphalt content. The materials used were mixed at a temperature of ±160°C and compacted with 2x50 blows at a temperature of ±150°C. The materials composition of the AC-WC is shown in table 8 follows.

Table 8. The materials composition of the AC-WC

Sieve Size of Aggregate/ Material	Dammar Content				
	0 %	2.5 %	5 %	7.5 %	10 %
Aggregate	1,031.3	1,031.3	1,031.3	1,031.3	1,031.3
1/2"	56.0	56.0	56.0	56.0	56.0
3/8"	129.3	129.3	129.3	129.3	129.3
No. 4	244.6	244.6	244.6	244.6	244.6
no.8	187.0	187.0	187.0	187.0	187.0
No. 16	118.3	118.3	118.3	118.3	118.3
no.30	85.3	85.3	85.3	85.3	85.3
no.50	63.3	63.3	63.3	63.3	63.3
No. 100	41.4	41.4	41.4	41.4	41.4
no.200	40.0	40.0	40.0	40.0	40.0
Fillller	66.0	66.0	66.0	66.0	66.0
Dammar	0.0	1.9	3.8	5.7	7.6
Petroleum Asphalt	48.7	46.8	44.9	43.0	41.1
Buton Asphalt	120.0	120.0	120.0	120.0	120.0
Total (gr)	1200	1200	1200	1200	1200

3.4. *Sample Testing.*
The Marshall test refers to the RSNI M-01-2003 regarding the hot asphalt mixture testing method using the Marshall apparatus. The principle of Marshall stability testing is the resistance of the specimen to plastic flow. The specimens tested were previously immersed in a water bath under the temperature of 60°C for 30 minutes.

4. Results and Discussion

4.1. *Volumetrics Parameter of HMA AC-WC containing dammar and Buton Asphalt*

4.1.1. Void in Mixture. Void in Mixture (VIM) in the specimens of HMA AC-WC met the requirement of the Bina Marga - 2018 specification, which is 2- 4%. The relationship between VIM and dammar content is shown in Figure 1 follows.

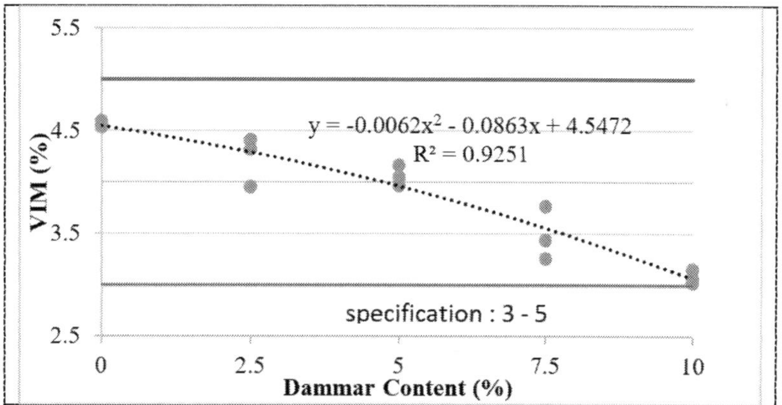

Figure 1. Relationship between VIM and Dammar content

4.1.2. Void Mineral Agregat. The increasing of dammar content in HMA AC-WC containing Buton asphalt is due to a decrease in Void Mineral Aggregate (VMA). HMA AC-WC using dammar met the VIM requirements of the Bina Marga - 2018 specification at least 15%. The relationship between VMA and dammar content is shown in Figure 2 follows.

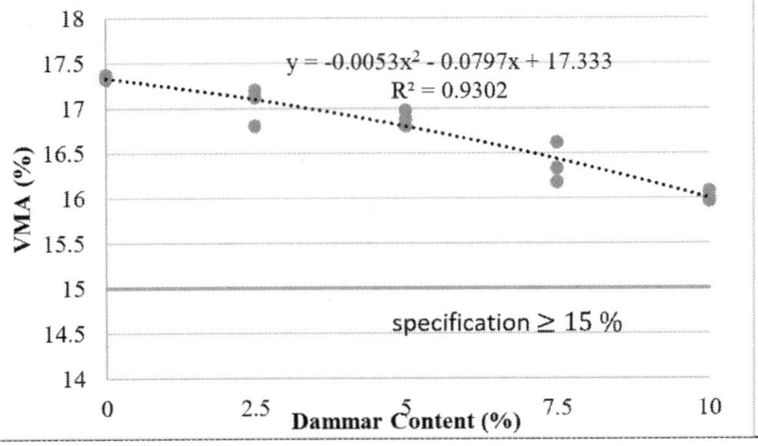

Figure 2. Relationship between VMA and Dammar content

4.1.3. Void Filled Bitumen. The addition of dammar content can increase the voids in the Void Filled Bitumen (VFB). HMA AC-WC using dammar met the VFB requirements of the 2018 Bina Marga specification, which is a minimum of 65%. The relationship between VFB and dammar content is shown in Figure 3 follows.

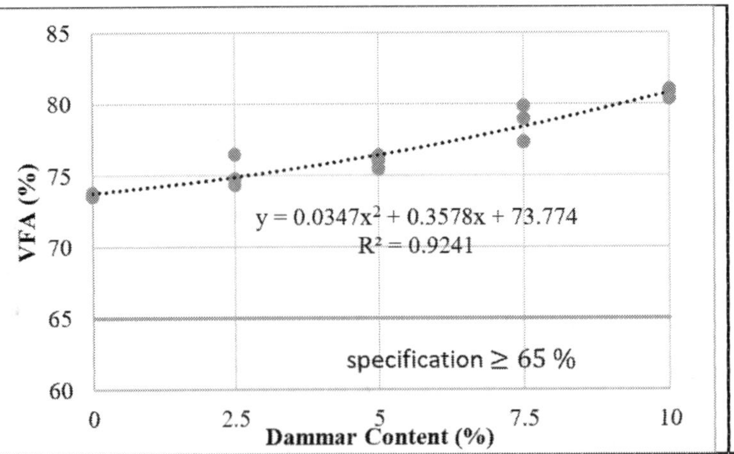

Figure 3. Relationship between VFB and Dammar content

At the mixing temperature and compaction temperature, the viscosity of the dammar visually is lower than that of petroleum asphalt. Here is due to increasing the absorption of asphalt in the aggregate and filling the void in the mixture. Thus, the addition of dammar content in HMA AC-WC containing Buton Asphalt causes decreasing of VIM, decreasing of VMA, and increasing VFB.

4.2. *Marshall Stability, Flow, and Marshall Quotient*

4.2.1. Marshall Stability. The use of Damar in AC-WC containing Buton Asphalt up to 10% increases the Marshall stability. The Marshall stability met the Bina Marga -2018 specifications, which is a minimum of 1000 kg for heavy traffic. The relationship between Marshall stability and dammar content in HMA AC-WC containing Buton asphalt of 10% shown in Figure 4 follows.

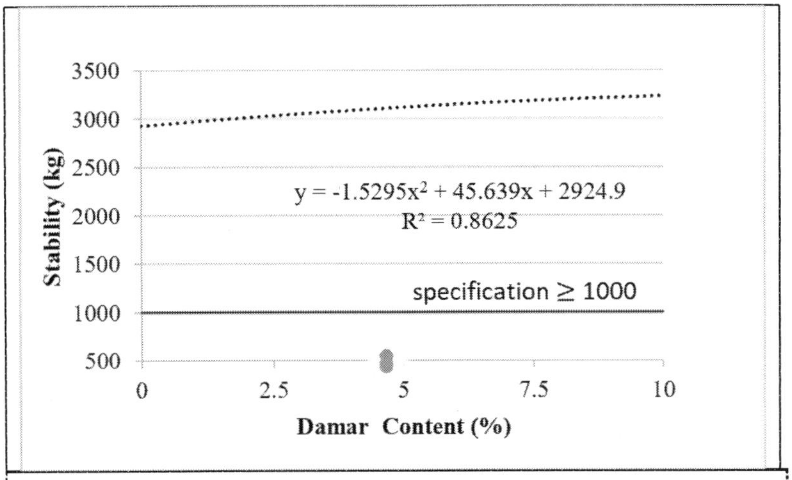

Figure 4. Relationship between stability and dammar content

4.2.2. Flow. Increasing the use of dammar in HMA AC-WC could decrease flow. It shows that increasing the use of dammar is due to the mixture more brittle. However, the use of dammar as a partial substitute for petroleum asphalt up to 10% still meets the flow requirement of the Bina Marga - 2018, a minimum of 2 - 4 mm. The relationship between flow and dammar content is shown in Figure 5 follows.

.

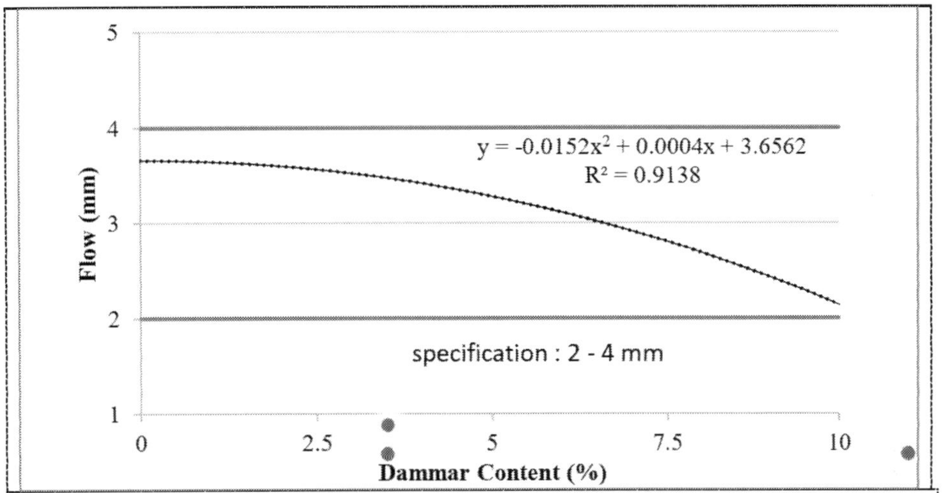

Figure 5. Relationship between flow dan dammar content

4.2.3. Marshall Quotient. The addition of dammar as a partial substitute for petroleum asphalt causes increasing the Marshall quotient (MQ) or the mixture stiffness. The Marshall quotient of all specimens has met the requirements of the Bina Marga – 2018 specification for HMA AC-WC modified, which is at least 250 kg/mm. The relationship between Marshall Quotient and dammar content of HMA AC-WC containing Buton Asphalt is shown in Figure 6 follows.

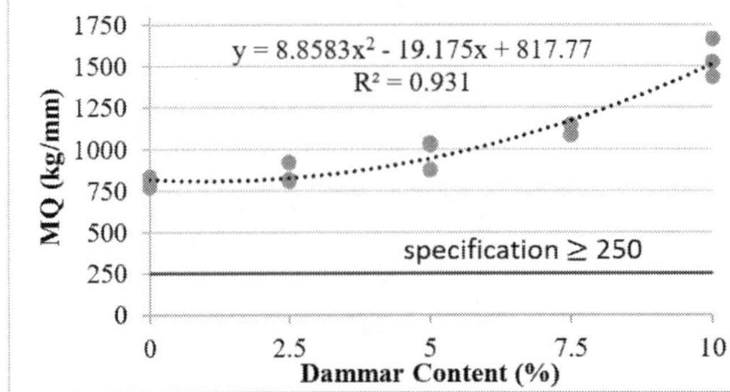

Figure 6. Relationship between MQ and dammar content

At the mixing temperature and compaction temperature, the viscosity of the dammar visually is lower than that of petroleum asphalt. Here is due to increasing the absorption of asphalt in the aggregate and filling the void in the mixture. Thus, the addition of dammar content in HMA AC-WC containing Buton Asphalt causes decreasing of VIM, decreasing of VMA, and increasing VFB

4.3. Material composition in asphalt mixture

The use of dammar up to 10% still meets the requirements of the Bina Marga-2018 specification for AC-WC containing 10% Buton asphalt. The composition of materials used in the AC-WC at an optimum asphalt content of 6.3% is 10% dammar content of the optimum asphalt content and the use of 10% Buton asphalt of the specimen weight. The dry weight composition of the material is 1 kg Dammar: 5 kgs petroleum asphalt: 16 kgs Buton asphalt: 82 kgs Coarse aggregate: 46 kgs Fine aggregate, and 9 kgs Filler. Meanwhile, the material composition based on the volume is 1 m3 of Dammar: 6 m^3 petroleum asphalt: 8 m^3 Buton asphalt: 33 m^3 Coarse aggregate: 18 m^3 Fine aggregate, and 4 m^3 Filler..

5. Conclusion

- Increasing the use of dammar in AC-WC can improve the Marshall characteristics. They include stability, Flow, Marshall Quotient, Void in Mixture, Void Mineral Aggregate, Void in Mixture requirement and Void Filled Bitumen. All Marshall Characteristics meet the Bina Marga-2018 specifications for heavy traffic.
- The increasing of dammar content in the modified asphalt causes the VIM, the VMA decreases, and the VFB increase. Here is due to the addition of dammar at the mixture temperature and compaction temperature causes the viscosity of modified Petroleum asphalt to decrease.
- The addition of dammar content as a substitute for petroleum asphalt tends to increase stability, to increase Marshall Quotient, and to decrease flow. Here is due to the increasing dammar at testing temperature causes the modified petroleum asphalt penetration to become lower. Here shows that the asphalt mixture becomes stiffer. But it still meets the requirements of the Bina Marga - 2018 specification.
- HMA AC-WC containing 10% dammar and 10% Buton Asphalt can substitute petroleum asphalt by 45.68%.

6. Future Work

It is necessary to carry out further research, including:

- Durability analysis of AC-WC containing asphalt Buton and dammar
- Performance of the AC-WC containing asphalt and dammar under cyclic load.

7. Acknowledgments

The acknowledgments address the Chancellor of The Haluoleo University that has provided financial support to pay the registration fee for the author's participation in the 4th Sustainable Infrastructure and Built Environment (SIBE) 2022 - Bandung.

References

[1] Affandi F 2010 *Pengaruh Asbuton Semi Ekstrasi pada Campuran Stone Mastic Asphalt* Puslitbang Jalan dan Jembatan

[2] Sefrus T 2013 *Analisis Penambahan Getah Damar Batu pada campuran Asphalt Concrete – Binder Course (AC-BC)* Majalah Teknik Simes, Vol. 7 No. 1

[3] One L, Tjaronge M W, Irmawaty R and Hustim M 2020. *Effect of Buton Granular Asphalt Gradation and Cement as Filler on Performance of Cold Mix Asphalt using Limestone Aggregate* Journal of Engineering Science and Technology Vol. 15, No. 1 (2020) 419 – 433

[4] Jinjin S, Yingbuao W, Jinyan L and Zhoa W 2016 *A modification mechanism of Buton natural rock asphalt in a matrix asphalt and asphalt mixture* Proceedings of the Fourth Geo-China *International* Conference 155-162

[5] Mulyono N, Wijaya C H, Fardiaz D and Rahayu W S 2012 *Identifikasi Komponen Kimia Damar Mata Kucing (Shirea Javanica) dengan metode Pirolisis-GC/MS* Jurnal Natur Indonesia Vol. 14 No. 2 Page: 155-159

[6] Kasim A, Permata D A and Malrianti Y 2020 *Karakterisasi Damar Pesisir Selatan dan Aplikasinya untuk Pembuatan Pernis* Jurnal Teknologi Pertanian Andalas Vol. 24(2)

[7] Mulyono N 2009. *Ekstrak damar untuk bahan pengeruh dan fosforilasi damar untuk bahan pemberat* Disertasi Institut Pertanian Bogor

[8] Sari R K 2002 *Isolasi dan identifikasi komponen bioaktif dari damar mata kucing (Shorea javanica K. Et V* Tesis Institut Pertanian Bogor

[9] One L, Haruzin A, Udo M A and Parizia M T 1989 *Investigasi Parameter Penetrasi Aspal Pen 60/70 yang dimodifikasi dengan Damar* Stabilitas Jurnal Ilmiah Teknik Sipil Vol. 9 No. 2. Page: 63-71

Study of Lerak *(Sapindus Rarak)* Biochar Application for Andosol Agricultural Soil Remediation

Sekar Arum[1], Temmy Wikaningrum[2]

Environmental Engineering, President University
Jl. Ki Hajar Dewantara, Kota Jababeka, Cikarang Baru, Bekasi, Jawa Barat 17750

sekar.arum@student.president.ac.id, temmy@president.ac.id

Abstract. Andosol is a type of soil that has a high nutrient content and is suitable for agricultural land. However, the use of pesticides in farming to increase and protect production from physical defects can cause a decrease in andosol soil quality. This study aims to determine whether the application of lerak biochar can improve the quality of andosol soil on the parameters of pH, organic carbon and total nitrogen based on the Soil Quality Standard of the Indonesian Agricultural Environmental Research Institute in 2009. The experiment was carried out by applying 0.805 g of biochar to 2 kg of andosol soil that already add by urea, KCl, insecticides and fungicides in standard dosages. The results showed that the quality of the andosol soil increased from the acidic category to slightly acidic in the pH category (6.01). The medium category became high for organic carbon (4.85%) and the medium category became high for total nitrogen (0.58%). This statement is evidenced by the results of the t-test which shows that there is no difference between the average value of each parameter and the respective standards. This means that the value meets the existing standard category.

Keywords: Andosol Soil, Lerak Biochar, pH, Total Nitrogen, Organic Carbon

1. Introduction

Based on data released by the Ministry of Environment and Forestry (MoEF) in 2018 in Indonesia there were 14,000,000 hectares of critical land. [1] One of the causes of soil damage is the agricultural sector which is not environmentally friendly. The agricultural sector is one of the backbones of the Indonesian economy. Agriculture and vegetable plantations are the main food commodities with production demands that continue to increase in line with the increase in population and industrial development. This encourages farmers to increase the use of inorganic fertilizers at each harvest season to get abundant yields with good products. Fertilizer applications tend to be uncontrolled and cause problems with the nutrient content in the soil. [2]

Agricultural areas in Indonesia are dominated by andosol soil types. Andosol soil is a blackish brown soil type, has a variety of characteristics, has low to high soil productivity and is suitable for agriculture. [3] Although andosol soils are suitable for agriculture, intensive application of inorganic fertilizers for a long period of time can cause residue accumulation in the andosol soil because plants generally cannot absorb 100% of chemical fertilizers. The remnants of chemical fertilizers left in the soil, when exposed to water will bind to the soil, so the soil will stick together and hard. Besides being hard, the soil also becomes acidic or has a low pH. This condition makes nutrient-forming organisms in the andosol soil die or their population decreases.[4] Therefore, it is necessary to make efforts to fix the problem of nutrient availability in andosol soil sample, one of which is the use of organic matter.

Currently, there is a technology called biochar. Biochar is a solid organic material such as carbon that is conserved from solid organic through incomplete combustion or limited oxygen supply. Biochar provides a habitat for soil microbes and is resistant to their attack. Generally, biochar persists in the soil for a relatively long time maintaining a balance of carbon-nitrogen. Biochar can retain the water and nutrients more available to plants. [5] Biochar, which is now known, is generally made from solid organic materials that are difficult to decompose, such as corn cobs, coconut shells, felt,wood and others.[6]

In this study, the author tries to find an alternative raw material using lerak fruit. Lerak *(Sapindus rarak)* is a plant known for its fruits which can be used as traditional detergents, but the saponins in lerak also have the potential as insecticides and pesticides (biopesticide). [7] So that motion has the potential tobe used as a good alternative raw material for biochar. In its application as an agricultural soil remediation agent, lerak biochar is applied togetherwith farmers' organic and inorganic fertilizers with appropriate doses to increase productivity and the availability of nutrients for plants.[8] The authors wanted to find out whether the application of lerak biochar can improved andosol agricultural soil quality on several biochemical properties (pH, total nitrogen, organic carbon). The study was conducted on real agricultural soils of the andosol order with acid pH, low organic carbon and low total nitrogen content due to intensive fertilization.

2. Research Method

2.1. Time and Place

The experiment period was conducted on June - July 2021 at Banjarnegara, Central Java. The result of this experiment was analyze at laboratory scale in Laboratory of Soil and Land Resources, Jendral Soedirman University, Purwokerto, Central Java.

2.2. Research design

This research is a field research with 1 control, 1 treatment compositions and 1 repetitions. So that there are 3 samples are obtained which are presented in the following table:

Table 1. Research design

Sample	Code	Spacing
Control	C	Urea 4.025 g KCl 1.543 g /pot Fungicide 0.123 g/pot Insecticide 3.36 g/pot
Fertilizer mix with Lerak Biochar	B1	Urea 4.025 g KCl 1.543 g /pot Fungicide 0.123 g/pot Insecticide 3.36 g/pot Lerak Biochar 0.805 g/pot
Fertilizer mix with Lerak Biochar	B2	

Th research uses 4 type of fertilizer, this kind of fertilizer is accordingly to fertilizer that commonly uses by Dieng farmers. [9] The treatment dose was based on recommendations from the Agricultural Environmental Research Institute, 2018. [10]

2.3. Soil sample preparations

The soil sample used was andosol soil originating from Dieng agricultural area, taken a two different points, namely at coordinates 7.202584185 109.85363211E and 7.188630195 109.88309496E on Pekasiran and Kepakisan, Batur, Dieng, Banjarnegara, Central Java. [11] Soil was taken at a depth of 0-20 cm from the soil surface. The soil used in laboratory research to check soil characteristics is 0.5 kg air dry weight, while for planting celery plants as much as 6 kg air dry weight. Then the soil was applied to the treatment by mixing the soil with lerak biochar according to the treatment dose.

2.4. Initial soil analysis

The soil used in this research is soil of the order andosol. Initial soil analysis was carried out to determine the nutrient content (pH, total nitrogen, organic carbon) of the soil before the treatment was applied. All soil criteria in this study were adjusted to the criteria based on the Soil Research Institute (2009).[12]

Table 2. Characterization of Andosol Soil Sample

Parameter	Values	Criteria
Organic Carbon (%)	2.143	Low
Total Nitrogen (%)	0.236	Low
pH	5.24	Acid

The results of the data analysis above come from Laboratory of Soil and Land Resources, Jendral Soedirman University, Purwokerto, Central Java. The results of the basic soil analysis (Table 1) showed that several criteria (pH, total nitrogen, organic carbon) were different from the general andosol soil conditions. Andsol soil in general has slightly acidic to neutral soil pH and high content of organic carbon and total nitrogen.[13]

2.5. Preparation of lerak (Sapindus Rarak) biochar

Lerak biochar is produced through a pyrolysis process using the Soil Pit method overnight at a temperature up to 200 degrees celsius. Soil pit is a traditional way of making biochar by using a 4-sided hole as a kiln. The traditional method is relatively simple and easy to practice. [6]

2.6. Lerak (Sapindus rarak) biochar analysis

The basic analysis was carried out to determine the characteristics of the lerak biochar used as a treatment in the study. The quality of lerak biochar is adjusted to the biochar standardization quality set by International Biochar Initiative (IBI). The laboratory result shows that lerak biochar have organic carbon and total nitrogen values 29.485% and 0.336% respectively. But for the pH it is have a value which were 6.22.

2.7. Preparation of celery seeds, planting and maintenance

The plant used is celery. Planting was carried out in pots containing the initial andosol soil sample. Each planting hole contains 5 celery seeds to anticipate the seeds that fail to grow. At first week after planting fertilizer and lerak biochar were applied according to the treatment dose. The maintenance of celery is carried out by watering and weeding the surrounding weeds. The experiment time was set for 30 days, following the conditions applied in the minimum biochar implementation which is 30-60 days.[14]

2.8. Observation parameters (Variables)

The parameters chosen for this experiment are pH, organic carbon and total nitrogen. Soil observations included analysis of pH H_2O 1:5 with Electrometry method, organic carbon with Kjeldahl method, and total nitrogen with Spectrophotometry method which was carried out on initial andosol soil samples and soil that had been treated with lerak biochar for 30 days.[15]

2.9. Data analysis

Data analysis was performed using Microsoft Excel. This study uses a two-sample variance test at 5% level to prove that each of the main parameters experienced an increase in soil quality after the addition of lerak biochar according to the Soil Quality Standard of the Agricultural Environmental Research Institute Indonesia in 2009. For the t-test, the null hypothesis is that the mean of all populations is the same (in this case it meets the expected standard ($P > 0.05$)

3. Result and Discussion

3.1 Soil Quality Result

3.1.1 pH

The application of lerak biochar has a significant effect on increasing andosol soil pH close to normal. The initial pH was 5.24 which is categorize in acid condition. After 30 days of experiments, the pH of soil sample with lerak biochar tratment has increase in value. The quality ofpH average value after treatments is 6.01 and categorized in slightly acid condition. The quality meets the quality standard of pH soil for celery which is 5.8-6.7.[16] The increase in pH occurred because biochar was able to increase the concentration of OH^- ions in the soil. The increase in pH occurs through several mechanisms, including (1) oxidation of organic acids in the decomposition process to produce OH^- ions; (2) decarboxylation of organic acids using H^+ ions; (3) Inorganic ammonification which produces OH^- ions; and (4) Al^{+3} binding by organic acids.[17]

3.1.2 Organic Carbon (%)

The initial organic carbon value was 2.143% which is categorize in low condition. After 30 days of experiments, the organic carbon value of soil sample with lerak biochar treatment has increase in value. The quality of organic carbon average value after treatments is 4.851% and categorized in high condition. This condition indicates that the andosol soil has improved in terms of organic carbon quality. The increasing of organic carbon value, it is based on the fact that biochar contains a high carbon content. Carbon is a food source for soil microorganisms so that the presence of organic carbon in the soil will stimulate activities as well as reactions that require the help of microorganisms. This is because biochar is dominated by C-stable compounds which can reduce C loss due to decomposition by soil microorganisms. [18]

3.1.3 Total Nitrogen (%)

The initial total nitrogen value was 0.236% which is categorize in low condition. But after treatment, the total nitrogen value of andosol soil sample with lerak biochar treatment has increase in value. The quality of total nitrogen average value after treatments is 0.5755% and categorized in high condition. The ability of biochar to increase soil moisture and pH stimulates the process of N mineralization and nitrification which causes plant uptake to increase. [19] The surface oxide on biochar is also effective in absorbing NH^{4+} and NO^{3-}. Biochar can increase CEC and good absorption capacity so that it can affect the function of the soil to retain nutrients. [20]

Table 3. Soil Quality (Experimental Result)

Parameter	Control	Criteria[c]	B1[a]	B2[b]	Average (B1:B2)	Criteria[c]
pH	5.25	Acid	6.16	5.89	6.01	Slightly Acid
Organic Carbon (%)	2.989	Medium	5.100	4.602	4.851	High
Total Nitrogen (%)	0.341	Medium	0.6	0.551	0.575	High

[a] Fertilizer mix with lerak biochar treatment 1.
[b] Fertilizer mix with lerak biochar repetition.
[c] Soil Quality Standard of the Agricultural Environmental Research Institute in 2009.

3.2 *T-test result*

3.2.1 pH
The t-test on the pH parameters compared with slightly acid standard showed that there was no significant difference between variable 1 and variable 2 ($P > 0.05$), meaning it was true that the addition of lerak biochar in andosol soil samples increased the quality of pH in andosol soil to slightly acid.

Table 4. t-test pH Result

Means			P one tail *(Alpha= 0.05, df=2*
Variable 1 (Lerak biochar treatment)	Variable 2 (Soil Quality Standard) Value	Variable 2 (Soil Quality Standard) Standard Criteria	
6.01	6	Slightly Acid	0.476[a]
6.01	5	Acid	0.011[b]

[a] $p > 0.05$ (no significant difference) between treatment and standard.
[b] $p < 0.05$ (significant difference) between treatment and standard.

3.2.2 Organic Carbon (%)

Both data of medium and low standard shows that the value of ($P < 0.05$) means that there is a significant difference between the organic carbon value with medium standard, and the organic carbon value with low standard. This indicates that the organic carbon value does not meet the medium and low standard but meets the high standard. This t-test result shows that the addition of lerak biochar can improve the andosol soil quality in the parameter organic carbon from low (initial sample) to high.

Table 5. t-test Organic Carbon Result

Means (%)			P one tail *(Alpha= 0.05, df=2*
Variable 1 (Lerak biochar treatment)	Variable 2 (Soil Quality Standard) Value	Variable 2 (Soil Quality Standard) Standard Criteria	
4.851	4.2	High	0.06[a]
4.851	2.7	Medium	0.0065[b]
4.851	1.5	Low	0.0027[b]

[a] $p > 0.05$ (no significant difference) between treatment and standard.
[b] $p < 0.05$ (significant difference) between treatment and standard.

3.2.3 Total Nitrogen (%)

The data of medium and low standard shows that the value of ($P < 0.05$) means that there is a significant difference between the total nitrogen value with medium standard and total nitrogen with low standard. This indicates that the total nitrogen value does not meet the medium and low standard but meets the high standard. This t-test result shows that the addition of lerak biochar can improve the andosol soil quality, in parameter total nitrogen from low (initial sample) to high.

Table 6. t-test Total Nitrogen Result

Means (%)			P one tail *(Alpha= 0.05, df=2*
Variable 1 (Lerak biochar treatment)	Variable 2 (Soil Quality Standard)		
	Value	Standard Criteria	
0.5755	0.63	High	0.07[a]
0.5755	0.35	Medium	0.0057[b]
0.5755	0.15	Low	0.001[b]

[a] $p > 0.05$ (no significant difference) between treatment and standard.
[b] $p < 0.05$ (significant difference) between treatment and standard.

4. Conclusion

In conclusion, the lerak biochar application can improve the andosol soil quality in 30 days become slightly acid category for pH (6.01), high category for organic carbon (4.85%) and high category for total nitrogen (0.58%) based on Soil Quality Standard of the Agricultural Environmental Research Institute Indonesia in 2009 and proven by the t-test result.

References

[1] Kementerian Lingkungan Hidup 2018 *Area and Distribution of Indonesia's Critical Lands* KLHK statistic
[2] Tim Kerja Pemulihan Dieng 2014 *Kawasan Lahan Kritis Dieng* **1** pp 1-10
[3] Kristiyanto, Kristiana Riajeng and Netty Demak H Sitanggang 2018 *Konservasi Lahan Pertanian Berbasis Ekologi Di Kawasan Dataran Tinggi Dieng Wonosobo* ISBN : 978-602-5614-35-4
[4] Palupi Annisa Auliani 2014 *Salah Olah dan Kelola, Lahan Dieng Kritis* **2** pp 1-5
[5] Lehmann, J., Rillig, M.C., Thies, J., Marsiello, C.A., Hockaday, W.C. and Crowley, D 2011 *Biochar effects on soil biota e A review Biochem* p 1812-1836
[6] Neneng L. Nurida, Achmad Rachman and S. Sutono 2015 *Biochar PembenahTanah yang Potensial* (Balai Penelitian Lingkungan Pertanian) p 67-69
[7] Sumartini 2016 B*iopestisida untuk Pengendalian Hama dan Penyakit Tanaman Aneka Kacang dan Umbi* **11** pp 1-8
[8] Temmy Wikaningrum Kajian Berkelanjutan Pengelolaan Lingkungan Kawasan Industri Studi Kasus di Kawasan Industri Jababeka Bekasi **1** pp 1-8
[9] Balai Penelitian Lingkungan Pertanian 2018 *The Dosage of Urea and BiocharApplication* (Dapertemen Pertanian) p 1-16
[10] Young Anthony 2014 *Andosol (Andisols). Tropical Soil and Soil, Cambridge University Press. Environmental* **6** pp 109-113
[11] Umi Haryati 2014 *Soil Physical Characteristics of Highland Vegetable Farming Area, It's Relationship with Land Management Strategy* ISSN 1907-0799
[12] Balai Penelitian Lingkungan Pertanian. 2009 *Standar Baku Mutu Tanah Pertanian". Badan Penelitian dan Pengembangan Pertanian. Departemen Pertanian* p 200-207
[13] Ghefira Rahimah, M.Iqbal, Mia Nur Aida, Nibras Hanif and Trixie Almira 2014 *Tanah Andosol*
[14] Widowati, Asnah and Sutoy 2012 *Pengaruh Penggunaan Biochar Dan Pupuk Kalium Terhadap Pencucian Dan Serapan Kalium Pada Tanaman Jagung* .Buana Sains **12** pp 83-90.
[15] Balai Besar Litbang dan Sumber Daya Pertanian 2017 *Petunjuk Teknis Edisi 2 (AnalisisKimia Tanah, Tanaman, Air, Dan Pupuk* ISBN 978-602-8039-21-5
[16] Tim Prima Tani 2011 *Petunjuk Teknis Budidaya Seledri*. Balai Penelitian Tanaman Sayuran, Pusat Penelitian dan Pengembangan Holtikultura, Badan Penelitian dan Pengembangan Pertanian p 1-2

[17] Widyantika, S.D and Prijono, S 2019 *Pengaruh biochar sekam padi dosis tinggi terhadap sifat fisik tanah dan pertumbuhan tanaman jagung pada Typic Kanhapludult.* Jurnal Tanah dan Sumberdaya Lahan **6** pp (1): 1157-1163
[18] Kanthle, A. K., Narendra,K.,L., Sangeeta,L. and K.Tedia 2015 *Biochar Impact On Nitrate Leaching As Influenced By Native Soil Organic Carbon in An Inceptisol Of Central India* Elsevier. India p 1-10
[19] Diva Ariella Herhandini, Retno Suntari and Ania Citraresmini 2021 *Effect of Rice Husk Biochar and Compost Application on Soil Chemical Properties, Maize Growth, and Phosporus Uptake on an Ultisol* **8** pp 385-394
[20] Syarifudin, Y., Pata'dungan, S. dan Isrun 2020 *Serapan fosfor tanaman jagung manis (Zea mays saccharata Sturt) akibat pemberian pupuk kandang ayam dan pupuk SP-36 pada Entisols Sidera* **1** pp: 77-88

Acknowledgments

Firstly I would to thank to Allah Subhanahu Wata'ala, because of his blessed, helped, and guided, we can running all the step for finished this journal. Foremost, I would like to express my sincere gratitude to President University for the continuous support of our research and writing of this journal.

Quantitative consideration in selection of breakwater concrete armor unit based on unit's internal tensile stress response

Andojo Wurjanto[1] and Adi Putra Hardaya[2,*]

[1]Institut Teknologi Bandung, Bandung, Indonesia 40132
[2]Dinamaritama Engineering Consulting, Bandung, Indonesia 40123

*Corresponding author e-mail: adiputrahardaya@gmail.com

Abstract. Concrete armor units are essential for the protection of rubble mound breakwaters, yet their selection is generally not based on structural integrity considerations. In this study, the authors used static finite element simulations, as implemented by Ansys software, to compare the structural integrity of five armor units: Accropode, A-Jack, BPPT-Lock, tetrapod, and dolos. All armor units have a mass of around 11,500 kg and have the same quality of concrete at f'_c of 30 MPa. The study involves two different loading cases: Armor units were (1) arranged on a breakwater slope subjected to their own weight and (2) placed on a horizontal ground subjected to idealized wave attacks. In Case #1, the armor units were modelled fixed to the ground and bonded to each other. In Case #2, the wave attack was modelled as a unidirectional and uniform pressure. From the simulation results, the authors compare the highest tensile stresses which could be found on the corner regions of the armor units—their most vulnerable region. In both cases, armor units with stocky shapes experience lower tensile stresses compared to those with slender shapes. These results confirm that Accropode, BPPT-Lock, and tetrapod are structurally safer than dolos and A-Jack.

Keywords: breakwater, concrete armor units, structural integrity, Accropode, BPPT-Lock

1. Introduction

Concrete armor units protect the outer layers of rubble mound breakwaters by dissipating the energy of wave attacks. In coastal protection practice, an ample number of choices for concrete armor units are available. In Indonesia, cubes and tetrapods are popular as their usage do not carry financial royalty. Dolos and the original A-Jack were once popular before they were empirically found to be structurally vulnerable—the modified A-Jack was then devised to overcome this particular weakness. The most popular armor unit used in breakwaters throughout the world is the Accropode [1]. Recent practice has also introduced new armor units, such as the BPPT-Lock, which was developed in Indonesia. Selection of concrete armor units are mostly focused on their hydrodynamic stability and less on structural integrity. However, considering that concrete is a material which can withstand huge compressive stress but is vulnerable to tensile stress, it is also important to consider the structural integrity of armor units.

Some recent studies have been pointed in this direction. An advanced numerical model—the combined finite-discrete element method (FEMDEM)—has allowed researchers to analyze an armor unit's inner structure and the multi-body interactions between armor units. FEMDEM was used to simulate prototype drop tests of cube and cubipod, to simulate drop tests and pendulum tests of dolos, and to simulate collisions between Core-Loc under gravity [2, 3]. Through these simulations, fracture behaviors of these armors were analyzed. Furthermore, FEMDEM was used to simulate the static stresses of 242 Core-Loc armor units stacked on a breakwater slope in a dry condition. The stability of these armor units when subjected to vibration loads under different configurations was also investigated [4].

There have been experimental studies concerning the effects of waves on armor units. In one study, a Core-Loc armor was placed on a breakwater slope and were outfitted with measuring instruments to quantify the forces and pressures exerted on the armor when subjected to wave attacks [5]. (OpenFOAM, a computational fluid dynamics software, was used to validate the experiments numerically.) Other studies on waves did not involve concrete armor units but instead aimed to develop equations to estimate wave pressure. In 1984, based on measurements of waves striking a vertical wall in full-scale experiments, Equation (1) was developed to estimate the maximum wave impact pressure p_1 [6].

$$p_1 = \lambda \rho C_b^2 T \tag{1}$$

In Equation (1), λ is the wave impact pressure coefficient, ρ is the density of seawater, C_b is wave celerity, and T is wave period. Equation (1) and other wave pressure equations were studied in a 1/20 scale experiment of waves striking a vertical wall. It was found that Equation (1) overestimated the wave forces acting on the vertical wall [7]. Equation (1), therefore, can be used as a *conservative* estimate of wave pressure.

The present study is not new in initiative; instead, it aims to contribute to coastal protection practice by comparing the structural integrity of widely constructed and promoted concrete armor units in Indonesia. Five kinds of concrete armor unit are compared: Accropode, original A-Jack, BPPT-Lock, tetrapod, and dolos. As an armor unit's individual mass is the most important parameter to measure the degree of protection of breakwaters, the mass of all armors is identically chosen at 11,500 kg; except for dolos, whose mass is 12,000 kg. The concrete quality is also identical at f'_c of 30 MPa. Structural integrity is measured by the magnitude of tensile stress that happen on the corner regions of these five armors—arguably their most vulnerable regions.

This study involves *two loading cases* on the armor units—both of which are *static* loadings. In Case #1, armor units were stacked on a 1:1.5 slope with a vertical height of 8.95 m and were subjected only to their own weight. In Case #2, armor units were subjected to their own weight, to hydrostatic pressure, and to wave attack whose pressure is defined by Eq. 1. Both cases were simulated through finite element analyses as implemented by Ansys 2021 R2 software suite. Ansys Mechanical APDL solver—utilized under the Static Structural module—was used. From each case, the armor unit with lower tensile stress response is deemed as structurally safer. Hence, the results of this analysis provide coastal protection professionals with added quantitative structural integrity consideration in selecting suitable kind of concrete armor units for their applications.

This article starts with a description of the general finite element setup applied to the two armor-unit loading cases (Section 2). This is followed by a description of the setup and the results of Case #1 (Section 3) and Case #2 (Section 4). The results of both cases are then discussed (Section 5), followed by the authors' conclusion (Section 6).

2. General finite element simulation setup

SolidWorks was used to make three-dimensional models of all armor units. The concrete properties of all armor units are presented in Table 1. Standard earth gravity (g = 9.80665 m s^{-2}) was applied in all loading cases in the direction of the negative y-axis. Due to concrete's vulnerability to tensile stress, the stress shown in all simulation results is the *maximum principal stress* (σ_1). A positive value of σ_1 signifies tensile stress; a negative value, compressive stress.

In Ansys Mechanical, armor units were modelled using unstructured tetrahedral mesh with a resolution of 4. To ensure the mathematical validity of simulation results, mesh convergence tests were performed by reducing the mesh element size, thereby increasing the number of mesh elements. Four stress probes were placed on an armor unit; the stress results of these probes were then plotted against the number of mesh elements (see Section 3.2 and 4.2). For all other mesh parameters, Ansys default settings were used. The minimum value of mesh quality in all simulations was maintained above 0.1. Figure 1 shows one mesh from each loading case.

Table 1. Concrete properties of armor units.

Concrete Property	Value
Density	2,200 kg m^{-3}
Young's modulus	30 GPa
Poisson's ratio	0.2
Ultimate compressive strength	30 MPa
Ultimate tensile strength	3.0 MPa

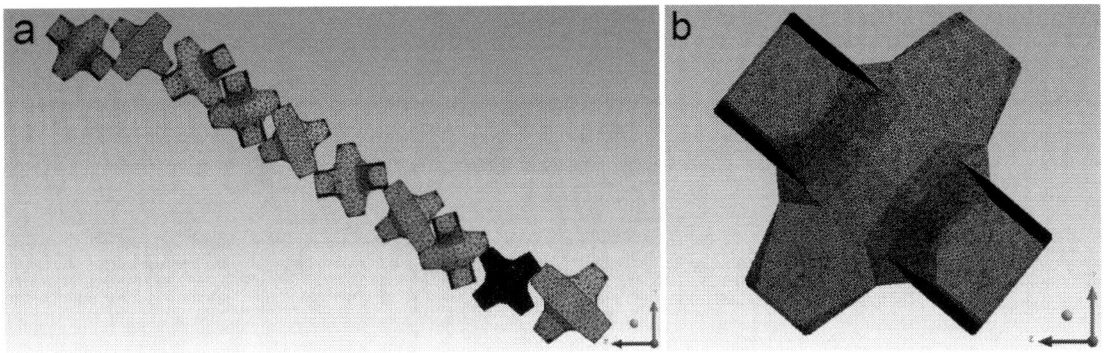

Figure 1. Mesh of (a) Accropode Conf. 2 from Case #1 and (b) Accropode Pos. 2 from Case #2.

3. Case #1: armor self-weight

3.1. Finite element simulation setup

In this loading case, all armor unit contacts with the ground were modelled as *fixed support*, and all contacts between armor units were modelled as *bonded contact*. There were six configurations in total: two for Accropode and one for the remaining armor units. These configurations are shown in Figure 2.

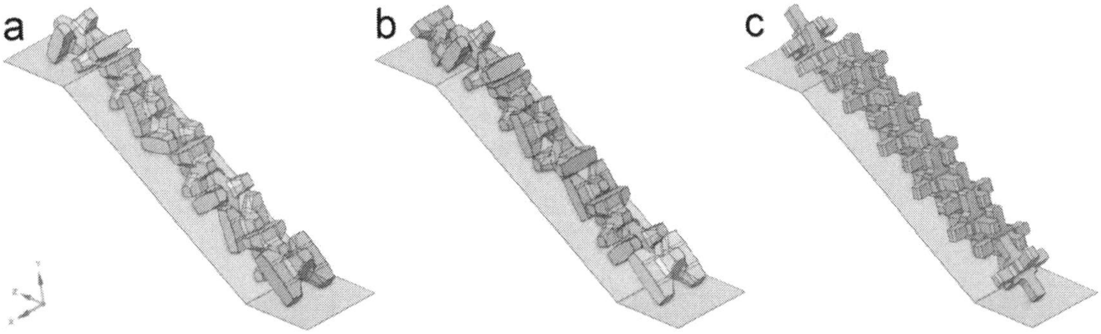

Figure 2. Armor unit configurations for Case #1. (a) Accropode Conf. 1, (b) Accropode Conf. 2, (c) A-Jack, (d) BPPT-Lock, (e) Tetrapod, (f) Dolos.

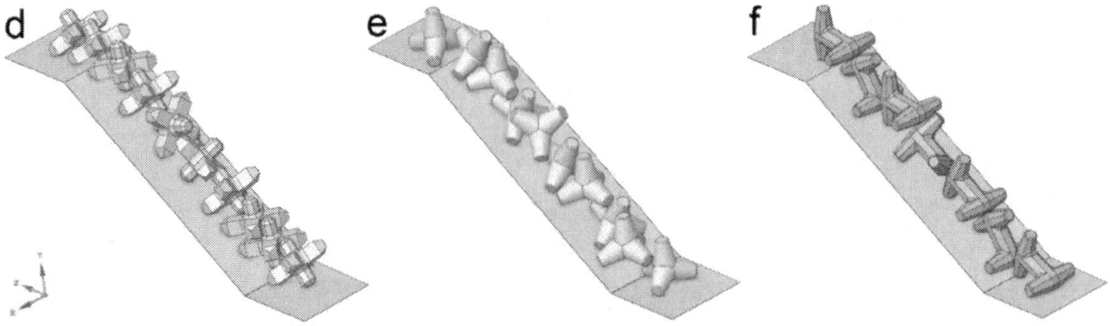

Figure 2. (Continued)

3.2. *Mesh convergence*

Mesh convergence test was performed on Accropode Configuration 2. As shown in figure 1(a), mesh refinement was done only for the lowermost armor unit on the breakwater slope, and on this armor was placed the four stress probes. Figure 3 shows the stress probes results as mesh elements were increased. The results did not differ significantly for mesh elements exceeding 3 million—this shows that mesh convergence has been reached. Thus, 3.8 million mesh elements (with an element size of 0.023 m for the lowermost armor unit) were used for all simulations in this loading case.

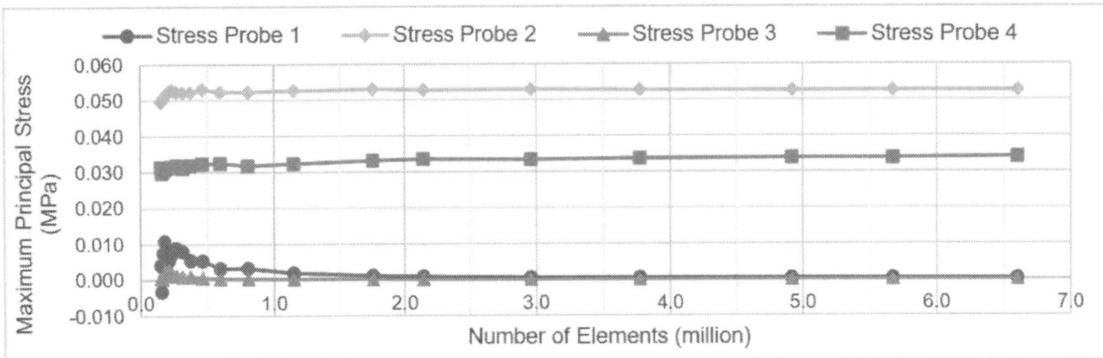

Figure 3. Mesh convergence test result for Case #1, Accropode Conf. 2.

3.3. *Results*

For each configuration, stress results are shown only for the lowermost armor units on the breakwater slope. (These armors are circled in Figure 4.) To compare tensile stresses at the *corner regions* of each armor unit, longitudinal sections were taken at places where the highest corner tensile stress could be found. The stress distributions of these sections are shown in Figure 4, along with the locations of the maximum corner tensile stress.

4. Case #2: wave attack

4.1. *Finite element simulation setup*

A total of seven armor unit positions were simulated in this loading case: two positions each for Accropode and BPPT-Lock, and one for all the others. Ground support on armor units was treated as idealized fixed support: the support acted on armor-unit surfaces as shown by the black hatched lines in Figure 5. Hydrostatic pressure acted on all other surfaces that were not modelled as fixed support, with the free surface of water defined at the highest point of the armor on the y-axis. The magnitude of hydrostatic pressure p_h on armor-unit surfaces was calculated using Equation (2), where ρ is the density of seawater, g is gravitational acceleration, and h_s is the depth of a point located on an armor-unit surface.

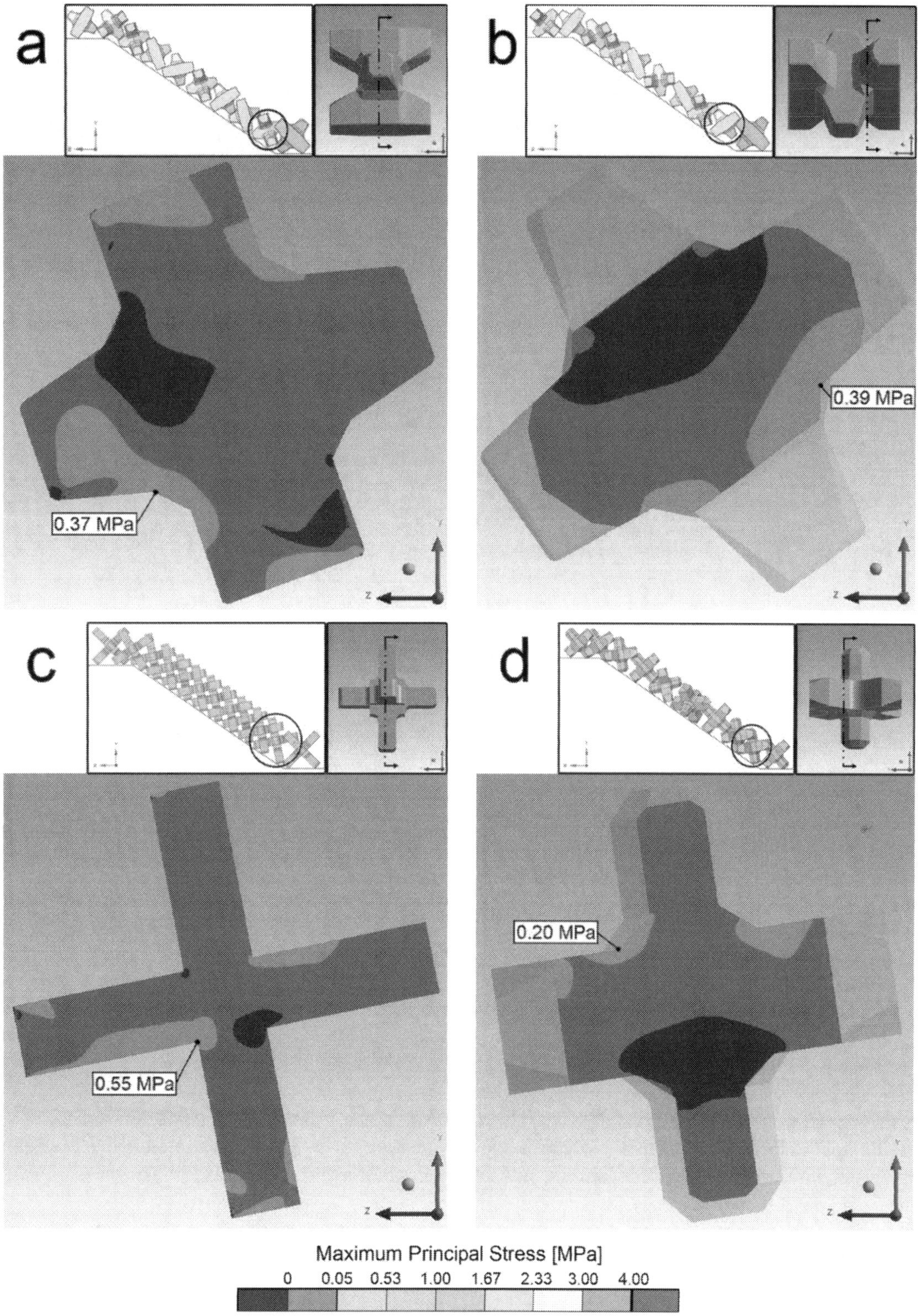

Figure 4. Case #1: Stress distributions of circled armor units, shown in longitudinal sections. For each section, the value and location of the maximum corner tensile stress are shown. (a) Accropode Conf. 1, (b) Accropode Conf. 2, (c) A-Jack, (d) BPPT-Lock, (e) Tetrapod, (f) Dolos.

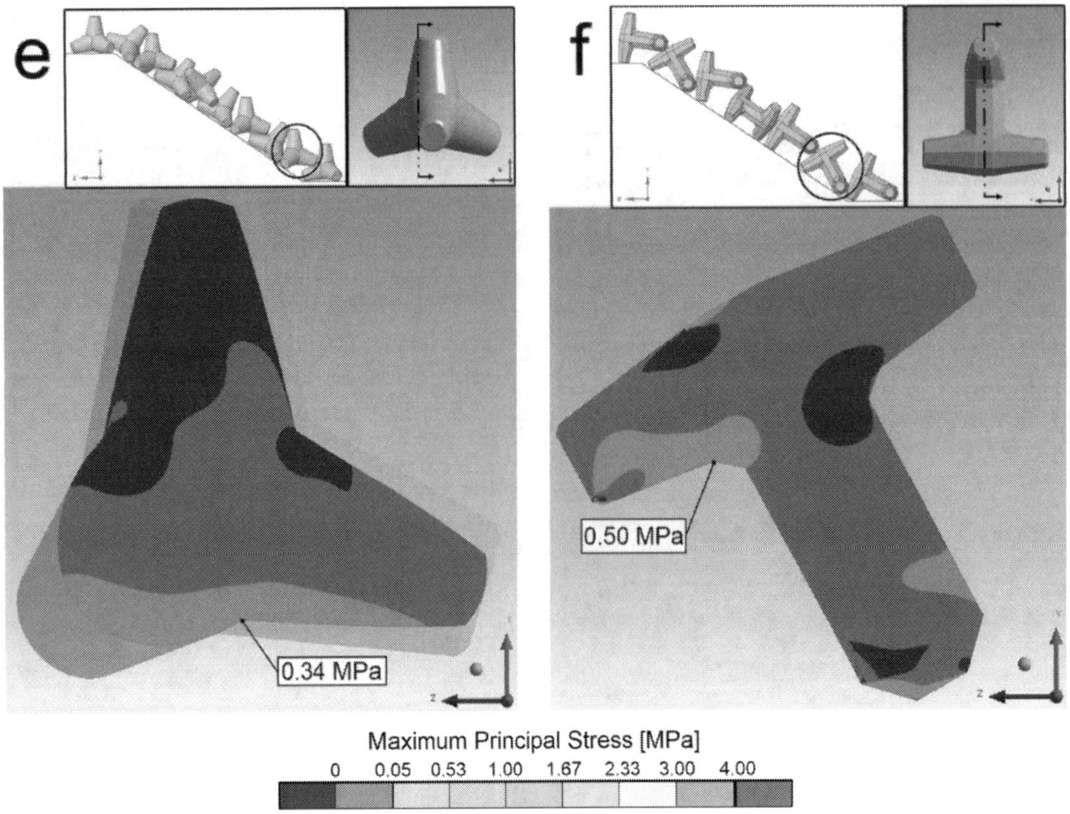

Figure 4. (Continued)

Table 2 shows the maximum hydrostatic pressure acting on the seven armor unit positions, whose value is affected by the armors' vertical height.

$$p_h = \rho g h_s \tag{2}$$

Table 2. Maximum hydrostatic pressure acting on armor units.

	Accropode		A-Jack	BPPT-Lock		Tetrapod	Dolos
	Pos. 1	Pos. 2		Pos. 1	Pos. 2		
Armor vertical height (m)	2.19	2.27	3.30	2.42	2.42	2.60	3.15
Max. hydrostatic pressure p_h (kPa)	22.0	22.8	33.1	24.4	24.4	26.1	31.7

Wave conditions were based on data from the coast of southern Java in Indonesia. Wave celerity C_b on Equation (1) for shallow water was determined by Equation (3), where g is gravitational acceleration and h is shallow water depth [8]. Table 3 shows the calculation results of wave attack pressure using Equation (1) and Equation (3).

Design wave height H for each armor unit was determined by Equation (4), adapted from [9]. W is the armor unit weight, K_D is the Hudson stability coefficient, w_r and w_w are the specific weight of armor and seawater respectively, and θ is the angle of breakwater slope as measured from the horizontal. Wave height calculation for breakwater trunk and head is shown in Table 4. All values of wave height in Table 4 are higher than the vertical height of any armor unit.

$$C_b = \sqrt{gh} \tag{3}$$

$$H = \left(\frac{W \cdot K_D \cdot \cot\theta}{w_r}\right)^{1/3} \left(\frac{w_r}{w_w} - 1\right) \quad (4)$$

Table 3. Wave attack pressure calculation using Equation (1) and Equation (3).

Variable	Symbol	Value
Wave impact pressure coeff.	λ	0.5 s^{-1}
Seawater density	ρ	1024.5 kg m^{-3}
Shallow water depth	h	4 m
Wave celerity	C_b	6.26 m s^{-1}
Wave period	T	6.5 s
Wave attack pressure	p_I	130.5 kPa

Table 4. Design wave height calculation using Equation (4)[a].

	Accropode		A-Jack		BPPT-Lock		Tetrapod		Dolos	
	K_D	H (m)	K_D	H (m)	K_D	H (m)	K_D	H (m)	K_D	H (m)
Breakwater trunk ($\cot\theta = 1.5$)	15	5.62	20	6.19	17	5.86	7	4.36	15.8	5.80
Breakwater head ($\cot\theta = 2$)	11.5	5.66	20	6.81	13	5.90	4.5	4.14	8	5.09

[a] For all calculations, $w_r = 21{,}575$ N m^{-3} and $w_w = 10{,}074$ N m^{-3}. Values of K_D are from [9-12]. Armor unit mass of 11,500 kg yields $W = 112{,}776$ N. For dolos, mass of 12,000 kg yields $W = 117{,}680$ N.

Wave attack pressure was directed along the positive z-axis. The wave pressure acted on all the armor unit faces above the fixed support that were visible from an observer looking along the positive z-axis. Figure 5 shows the armor unit positions of this loading case, along with the location of fixed support and the direction of wave attack.

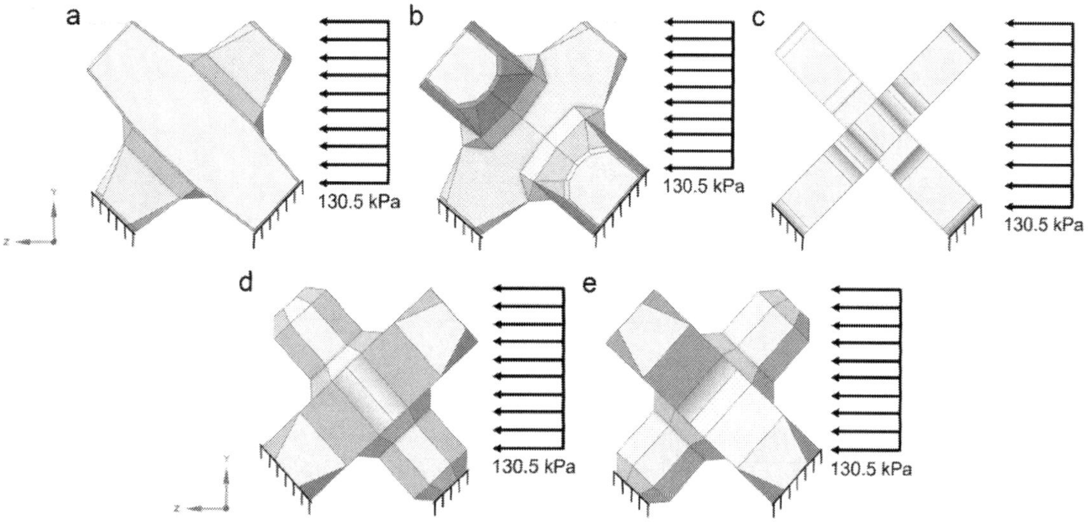

Figure 5. Armor unit positions for Case #2. Fixed support is marked by hatched lines below the armor units. Hydrostatic pressure is not shown. (a) Accropode Pos. 1, (b) Accropode Pos. 2, (c) A-Jack, (d) BPPT-Lock Pos. 1, (e) BPPT-Lock Pos. 2, (f) Tetrapod, (g) Dolos.

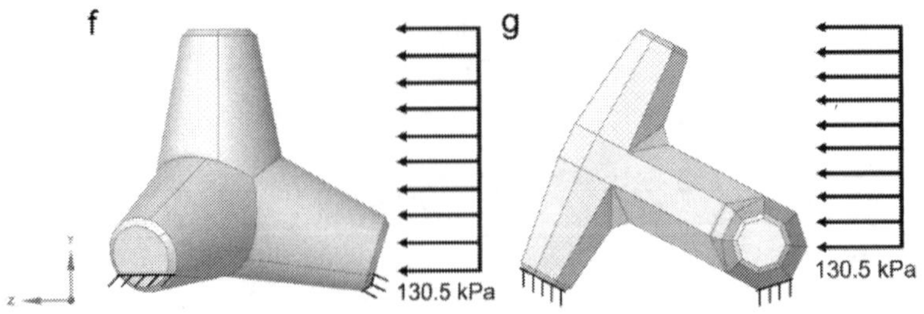

Figure 5. (Continued).

4.2. *Mesh convergence*

Mesh convergence test was performed on Accropode Position 2, and the stress probe results are shown in Figure 6. Similar to Case #1, Figure 6 shows that mesh independence was achieved for mesh elements above 3 million. Thus, 3.7 million mesh elements (with element size 0.023 m) were used for all simulations in this loading case.

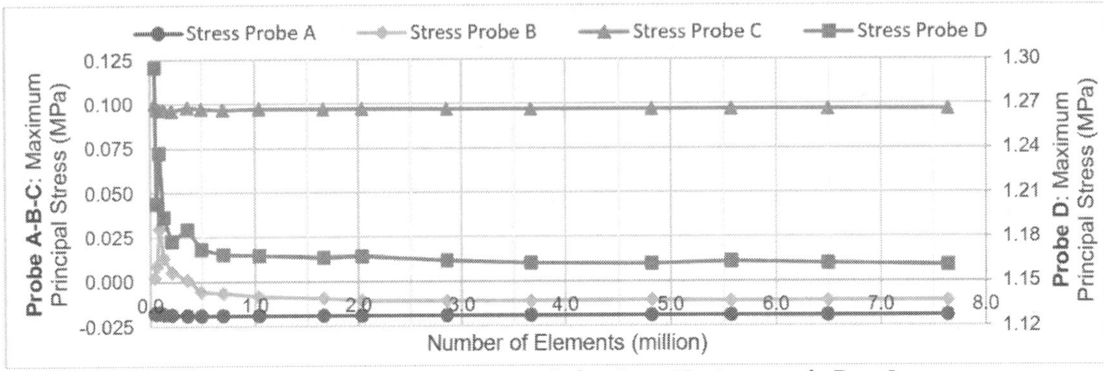

Figure 6. Convergence test result for Case #2, Accropode Pos. 2.

4.3. *Results*

Figure 7 shows the stress distribution of armor unit surface for all positions of Case #2. Shown also are the longitudinal sections of each armor unit, accompanied by the value and location of the maximum corner tensile stress.

5. Discussion

The order of armor units in Case #1, based on the value of maximum corner tensile stress as shown in Figure 4, is as follows: A-Jack (0.55 MPa), dolos (0.50 MPa), Accropode Conf. 2 (0.39 MPa), Accropode Conf. 1 (0.37 MPa), tetrapod (0.34 MPa), and BPPT-Lock (0.20 MPa). It is apparent that armor units with stocky shapes—BPPT-Lock, tetrapod, and Accropode—experience lower tensile stresses in their corner regions. On the other hand, due to their slender limbs, dolos and A-Jack experience higher tensile stresses.

However, it should be noted that the finite element model used in Case #1 has a *limited accuracy* in simulating multi-body interactions. In this study particularly, contacts between armor units were only modeled as bonded contacts, creating an unrealistic situation in which one armor unit surface could *pull* another armor unit's surface. This could affect the result accuracy of A-Jack's configuration, which involve the most armor-to-armor contacts. If a more realistic model had been used—which would have allowed armor unit surfaces to separate from each other—computer simulation time would have increased significantly due to the use of a *non-linear* mathematical model. Thus, for a simple comparison of armor unit stresses in a static loading case, the bonded contact model in Case #1 is acceptable.

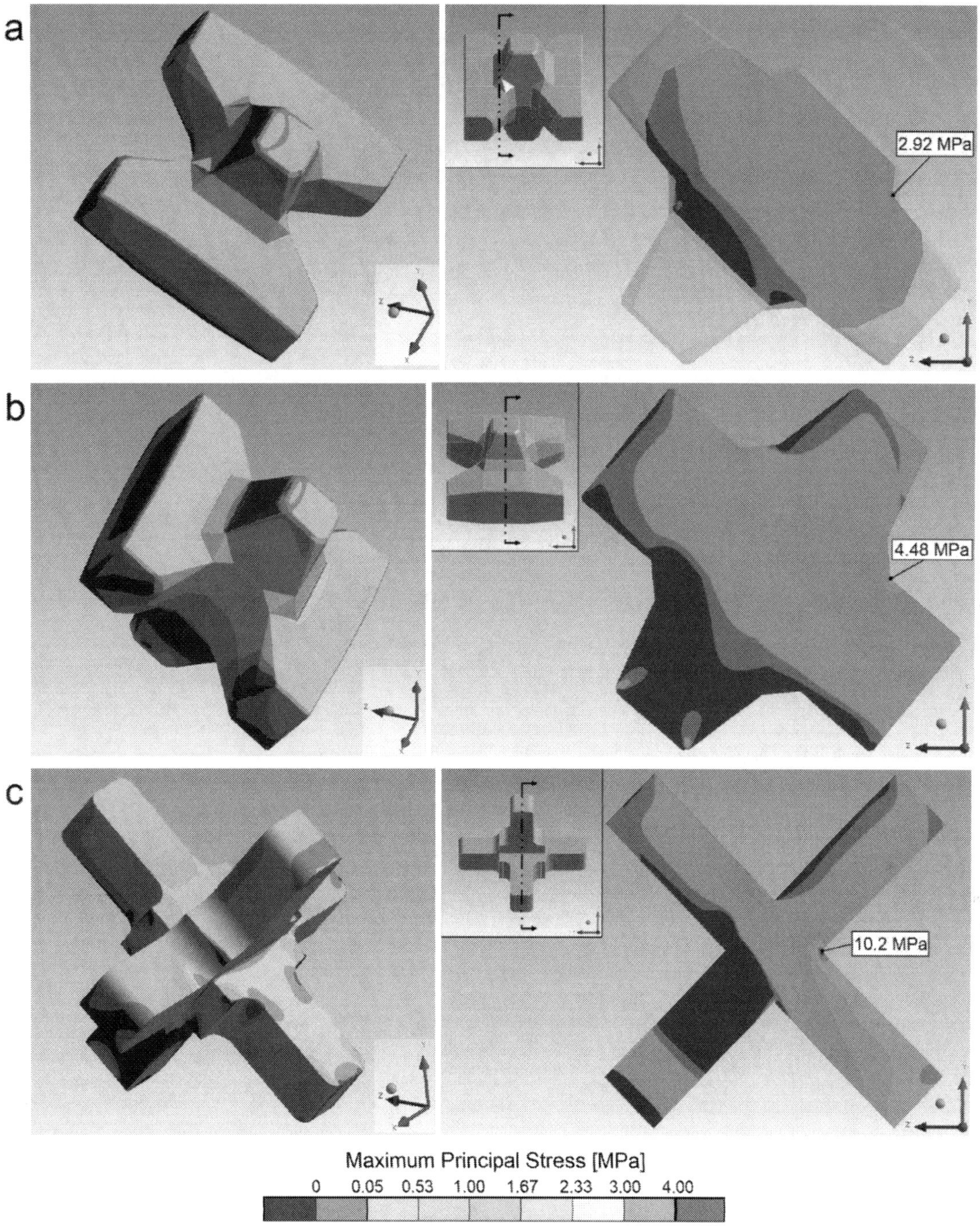

Figure 7. Case #2: Stress distributions of armor unit surfaces and longitudinal sections. For each section, the value and location of the maximum corner tensile stress are shown. (a) Accropode Pos. 1, (b) Accropode Pos. 2, (c) A-Jack, (d) BPPT-Lock Pos. 1, (e) BPPT-Lock Pos. 2, (f) Tetrapod, (g) Dolos.

Using the same criterion as Case #1, the order of armor units in Case #2, based on Figure 7, is as follows: A-Jack (10.2 MPa), BPPT-Lock Pos. 1 (6.21 MPa), dolos (5.91 MPa), Accropode Pos. 2 (4.48 MPa), BPPT-Lock Pos. 2 (4.20 MPa), tetrapod (3.91 MPa), and Accropode Pos. 1 (2.92 MPa). Except for Accropode in Position 1, the tensile stresses exceeded the strength of concrete (3 MPa). This is to be

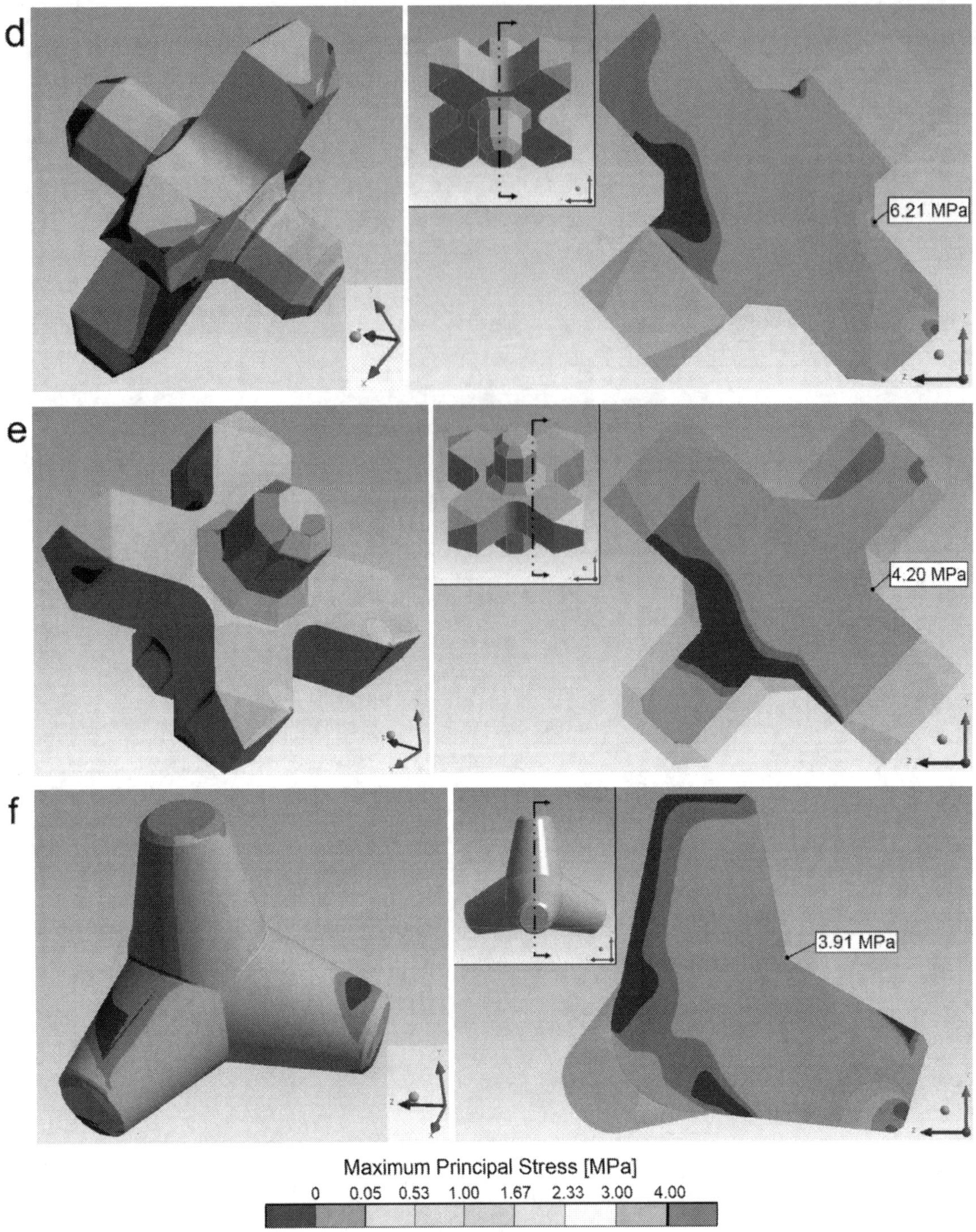

Figure 7. (Continued)

expected: In Case #2, the full pressure of sea wave was faced only by one armor unit. The magnitude of the pressure was also *overestimated*, as established in Section 1. Moreover, Case #2 was an idealization because it modelled wave attack as a *static* pressure. Due to these factors, the results of Case #2 can only serve as basis for a simple comparison of armor-unit stress responses under an external loading.

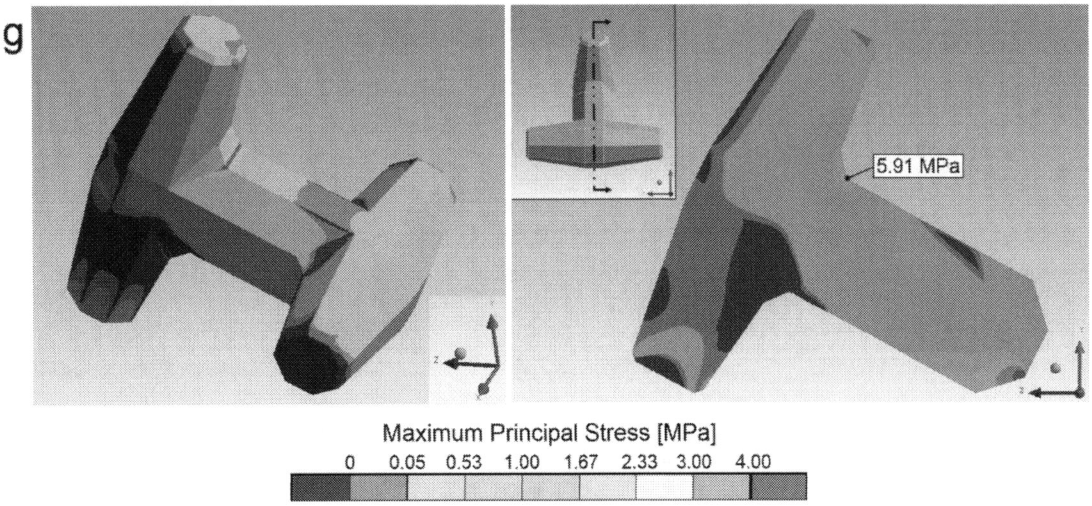

Figure 7. (Continued)

The armor that performed the best in Case #2 is Accropode in Position 1. As shown by its surface stress distribution (Figure 7(a)), Accropode's corner regions were safe from failure in this position. However, Accropode in Position 2 (Figure 7(b)) showed a significant concentration of tensile stress on the corner region facing the wave. A similar situation can be found on BPPT-Lock. A higher concentration of tensile stress happened on BPPT-Lock Position 1 (Figure 7(d)) compared to Position 2 (Figure 7(e)). These results suggests that the magnitude of tensile stress is affected by an armor's position. Accropode experience lower tensile stress when it rests on *three* legs; BPPT-Lock, when its octagonal limb faces *upward* toward the wave.

Tetrapod (Figure 7(f)) performed decently in Case #2, as it had the second lowest corner tensile stress overall. On the other hand, a significant concentration of tensile stress could be found on dolos (Figure 7(g)); its maximum tensile stress value exceeding tetrapod, Accropode, and BPPT-Lock Position 2. Finally, same as Case #1, A-Jack performed the worst in Case #2. Figure 7(c) shows that there are *two* significant regions of tensile stress concentrations: A-Jack's corners and fillets. This suggests that A-Jack is particularly vulnerable to external forces. Thus, the results of Case #2 are consistent with Case #1: stocky armor units experience lower tensile stress than the slender ones.

6. Conclusions

In this study, structural integrity has been treated as an important criterion in the selection of concrete armor units for use on breakwaters. Ansys was used to simulate five kinds of armor units (1) arranged on a breakwater slope and subjected to their own weight and (2) placed on a horizontal ground and subjected to idealized wave attacks. The results of both cases agree with each other: armor units with stocky shapes have better structural integrity compared to the slender ones, if that integrity is solely based on their internal tensile stress response. Particularly, Accropode, BPPT-Lock and tetrapod have better structural integrity compared to dolos and A-Jack.

For future research, the use of FEMDEM codes can be highly beneficial. FEMDEM can model realistic armor unit configurations on breakwater slopes, thereby providing a more accurate analysis of the armor-unit stress response. However, this is a challenge as currently there is no popular commercial software to implement FEMDEM for structural simulations. Another step that can be taken is to couple structural analysis with computational fluid dynamics (CFD) codes, which enables the study of armor units' dynamic stress response. CFD can be coupled with finite element structural analysis to simulate the stress response of *one* armor unit subjected to a wave attack. This simulation should be validated by experiment if possible.

CFD will become even more useful if coupled with FEMDEM to study *multiple* armor units under a wave attack. This also brings another important benefit, as CFD-FEMDEM couple can enable a simultaneous study of armor-unit structural integrity and interlocking ability. Thus, in addition to studying armor-unit dynamic tensile stress response, it will also be possible to know whether a wave attack will break apart armor interconnections. Combining these two considerations will greatly improve coastal protection practitioners' ability to choose the most suitable armor unit.

Acknowledgments
The authors would like to thank Iman Riyadi for providing armor unit drawings and 3D configurations.

References

[1] Allsop N W H, Cork R S and Verhagen H J 2009 A database of major breakwaters around the world *Proceedings of the 9th international conference organised by the Institution of Civil Engineers* (Edinburgh, UK, 16-18 September 2009) 10.1680/cmsb.41318.0064
[2] Scaravaglione G, Latham J and Xiang J 2021 *J. Mar. Sci. Eng.* **9** 460
[3] Guo L, Latham J and Xiang J 2015 *Comput. Struct.* **146** 117-142
[4] Anastasaki E 2015 Numerical Modelling of Coastal Structures Armoured with Concrete Units *PhD dissertation* Imperial College London
[5] Eden D 2019 Forces and Pressures on Core-Loc Armour Units in Rubble Mound Breakwaters Measured via Instrumented "Smart Units" *Master's thesis* University of Ottawa
[6] Blackmore P A and Hewson P J 1984 *Coastal Eng.* **8** 331-346
[7] Kisacik D, Van Bogaert P and Troch P 2010 Comparative Study on Breaking Wave Forces on Vertical Walls with Cantilever Surfaces *Proceedings of the 20th International Offshore and Polar Engineering Conference* (Beijing, China, 20-25 June 2010)
[8] Knauss J A 1997 *Introduction to Physical Oceanography* (Long Grove: Waveland Press) p 203
[9] US Army Corps of Engineers 1984 *Shore Protection Manual Volume II* (Washington, D.C.: U.S. Government Printing Office) p 7-205–7-206
[10] Concrete Layer Innovations 2015 *ACCROPODE™* (Échirolles: CLI) p 2
[11] K. Ketut, H. Tuah, A. Wurjanto and K. Idris 2005 *Jurnal Teknik Sipil ITB* **12** 1-12
[12] Badan Pengkajian dan Penerapan Teknologi 2021 *Katalog BPPT-Lock* (Yogyakarta: BPPT) p 7

Numerical modelling of land subsidence effect on coastal sediment transport in Pekalongan

F S Wijaya[1*], A Kurniawan[2], N Yuanita[2], E O Nugroho[3], R J Martin[1], I F Muhammad[1], Yola[1] and N Fauzi[1]

[1]Ocean Engineering Master Program, Faculty of Civil and Environmental Engineering, Bandung Institute of Technology, Bandung 40132, Republic of Indonesia
[2]Coastal Engineering Research Group, Faculty of Civil and Environmental Engineering, Bandung Institute of Technology, Bandung 40132, Republic of Indonesia
[3]Water Resources Engineering Research Group, Faculty of Civil and Environmental Engineering, Bandung Institute of Technology, Bandung 40132, Republic of Indonesia

*fauziseptianwijaya@gmail.com

Abstract. This study examines the condition of sediment transport on the Pekalongan coast by a numerical model computation using the Delft3D Flow module. The research was started by collecting and analysing field data both primary and secondary. Thus, the hydrodynamic and sediment transport model are then calculated in two scenarios which are representing the conditions before (2016) and after (2021) land subsidence by assuming a land subsidence of 10 cm/year. The numerical model is then calibrated and verified on the data of tidal water level elevation and current velocity, the best results are obtained at a chezy number of 60 $m^{0.5}/s$, and sensitivity tests are carried out on changes in sediment grain size, wave height and wave direction in order to determine the characteristics of the numerical model. The sensitivity analysis of sediment grain size is the most sensitive effect to sediment transport characteristic. The results of the numerical model are then compared based on the scenarios before (2016) and after the occurrence of land subsidence (2021). The results indicate that after land subsidence, the sedimentation rate is increasing by 56.34%, whereas the erosion rate is decreasing by 3.98% after one neap-spring tidal cycle (around 15 days) in the study area.
Keywords: *sediment transport, land subsidence, erosion, sedimentation, numerical modeling*

1. Introduction

Pekalongan Regency is one of the districts in Central Java that experienced severe beach erosion or abrasion problems within the last decade. Pekalongan City has a coastline of ± 6 km stretching from west to east facing each other directly with the Java Sea which is in north Pekalongan District. Morphologically the beach sloping shape is dominated by a stretch of sand, not rocky, the waters are open, not a bay and the beach waves are relatively low in strength [1]. The coast will restore its own form due to the following normal waves, or this will lead to erosion on one location and sedimentation on another side [2]. As a result of this abrasion, the distance between residential areas and the coastline is getting closer and becoming an area that is prone to tidal flooding.

Tidal flood or 'rob' is a serious problem in many coastal areas in Indonesia, including Pekalongan in the northern coast of Java island [3]. The current in Pekalongan Waters is dominated by tidal current with the percentage of 57.49%, the non-tidal current percentage in Pekalongan Waters is 42.51% [4]. This flow pattern causes sediment characteristics that exist in different research areas such as sand, silty sand, silt, sand-silt-clay, silt clay and silty clay [4]. With this tidal event, the ponds in the coastal area of Pekalongan, which are the main livelihoods of residents in the area, will be damaged due to being submerged in water. The worst possibility is that residents are forced to move from the area because it is no longer habitable. The coastal protection system right now at Pekalongan coast consisting of geobag and mangroves like proposed by [5].

Energy waves, currents and tides is the most important factor in erosion, especially waves at time storms and tsunamis. However, not only waves have an effect against erosion, but also the type factor and rock durability, rock structure, stability of the beach, whether or not the beach is open to influence of waves, water depth, and the size of the abrasive material being transported by wave [6].

Land subsidence has been an important subject of studies in the past decades [7]. The objective of this research is to obtain the condition of sediment transport on the coast of Pekalongan before and after the occurrence of land subsidence with numerical modelling and determine the affected area of the phenomenon. The model approach refers to the publication of sediment transport and morphological simulation using process-based models in estuary areas e.g. [8].

2. Research Method and Modelling Approach

The methodologies used for numerical modelling in this study include:

1. Hydrodynamic and wave model configuration also simulation
2. Calibration of hydrodynamic and wave model
3. Hydrodynamic, wave and sediment transport simulation
4. Interpretation of sediment transport model result

The research method includes field data collection and is supported by laboratory analysis [9]. Sediment transport and the shape of the beach morphology affect the shape and type of sediment on the beach, both bottom and floating sediments, cohesive or non-cohesive [10]. The calculation of sediment transport used in the Delft3D software uses the Van Rijn equation [11] which is shown in equation (1) to equation (5). Frame work of numerical modelling shown in Figure 1.

$$q_s = 0.012\bar{u}\frac{(\bar{u}-\bar{u}_{cr})^{2.4}d_{50}D_*^{-0.6}}{\left((s-1)gd_{50}\right)^{1.2}} \tag{1}$$

$$q_b = 0.005\bar{u}h\left(\frac{\bar{u}-\bar{u}_{cr}}{\left((s-1)gd_{50}\right)^{0.5}}\right)^{2.4}\left(\frac{d_{50}}{h}\right)^{1.2} \tag{2}$$

$$D_* = \left[\frac{g}{v^2}\right]^{\frac{1}{3}} D_{50} \tag{3}$$

$$\bar{u}_{cr} = 0.19(D_{50})^{0.1}\log\left(\frac{12h}{3D_{90}}\right) \text{ for } 0.0001\,m \le D_{50} < 0.0005\,m \tag{4}$$

$$\bar{u}_{cr} = 8.50(D_{50})^{0.6}\log\left(\frac{12h}{3D_{90}}\right) \text{ for } 0.0005\,m \le D_{50} < 0.002\,m \tag{5}$$

Description:

D_{50} : median sediment grain diameter (m)
D_{90} : 1.5 times D_{50}
D_* : grain diameter
$\bar{u}$: current velocity at mean depth
$\bar{u}_{cr}$: critical current speed

h : depth (m)
g : acceleration due to gravity (m/s^2)
v : speed in y direction
q_b : bottom sediment transport
q_s : float sediment transport

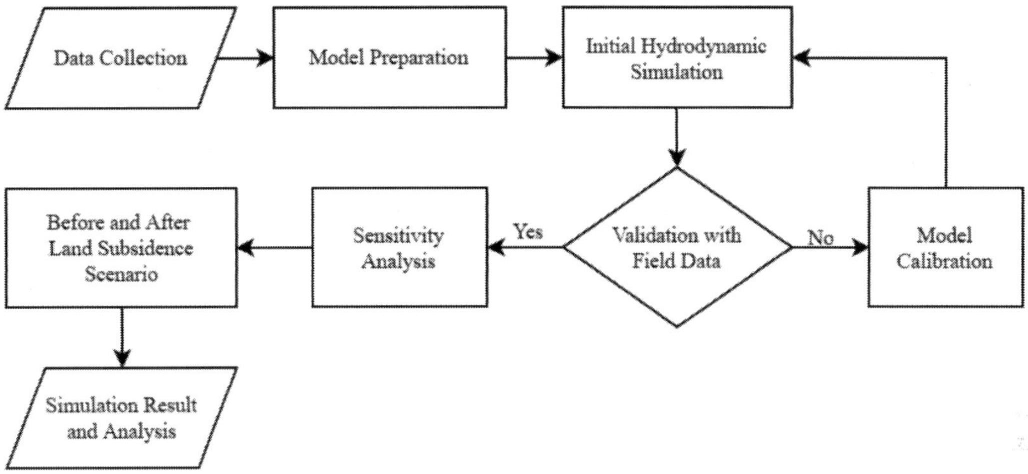

Figure 1. Frame work of numerical modelling.

Delft3D consists of various process modules and can operate in different modes, such as 2DH, Q3D or 3D [12]. Validation of these models requires specific data. To run model simulations detailed measurements of the boundary conditions are required. To validate the model results a dense spatial instrumental coverage within the modelling area is required, to give an accurate representation of the spatial flow patterns [10]. The numerical modelling takes the main role in the study which performed to construct a reliable simulation of tidal propagation at Java Sea and its intrusion to the urban residential area in the northern coast of Pekalongan City [7].

Waves that propagate towards the coast will undergo a transformation due to interactions with currents or changes in depth (refraction and shoaling). Wave transformation modelling is used to simulate wave propagation towards the coast. In the Delft3D software the wave components are propagated in geographic space with a wind direction scheme. In order to correctly account for the boundary conditions, the computation is performed iteratively at each time step.

3. Data Collection

Primary data collection was carried out by conducting a field survey directly in the coastal area of Pekalongan by the FCEE ITB (Faculty of Civil and Environmental Engineering Institute Technology of Bandung) team in 2021 as well as data from observation stations around the coastal area of Pekalongan. Primary data collected includes bathymetry data, tidal elevation data, wind speed data, wave data, current velocity data, and sediment data (Total Suspended Soil). The location of the data collection point is shown in Figure 2. Current and wave measurement using the Argonout SonTek's ADCP (Acoustic Doppler Current Profiler) tool XR [1], suspended sediment data collection using a Nansen bottle [1].

Figure 2. Data Collection points at the Pekalongan coastal area (Source: google earth pro at 6.853897°S; 109.669333°E).

3.1. Bathymetry Data

Bathymetric data from primary data that has been obtained includes data on river cross-sectional depths and bathymetric surveys for the Pekalongan waters area with the lowest low tide datum (LWS).

3.2. Current Velocity Data

In the field data collection process, the current velocity measurement process using ADCP was carried out in the range of April 6, 2021 at 03:00 pm GMT+7 to April 9, 2021 at 05:10 pm GMT+7 with ten minutes at interval. The coordinates are 109.689317° East Longitude and 6.828245° South Latitude covering five depths. From the data obtained, most of the u-direction currents speeds is around 0 – 0.4 m/s while the v-direction majority is around 0 – 0.1 m/s. It can also be concluded that the current at the measurement point leads to the southeast and northwest.

3.3. Sediment Data

After performing a sieve test for the bottom sediment, the results of the test are processed and analyzed to determine the grain size of the bottom sediment. In addition to the bottom sediment parameters, suspended sediment concentration testing is also carried out in the laboratory. From the field data, it was obtained that the type of sediment found in the waters of the prospective location of the Pekalongan mangrove park was of non-cohesive type with the type of grain in the form of sand. The results suggested that the largest average for Total Suspended Solid (TSS) was at the observation point MA-08 (407.5 ppm) and the lowest TSS content was at the observation point MA-07 (35.9 ppm).

3.4. Tidal Data

Tidal data was taken from April 06 to April 10, 2021 with an hour interval at five points on the river, namely MA02, MA07, MA08, MA10, and MA12 (Figure 2). It is noted that the tidal range is 0.75 m.

3.5. Wave Data

Field data collection also includes measurement of wave parameters consisting of measurements of wave height and period. The wave parameter data was taken from April 6 to April 9, 2021. The location is the same as current measurement. The measurements of wave height and period are carried out at 10-minute intervals. From the field data, it can be concluded that the wave height is in the range of 0.4 – 1.3 m with a period of 3.5 – 7 seconds. Based on the analysis of the wave parameters, it can be obtained that a significant wave height value is 0.89 m with a period of 5.57 seconds.

4. Result of Hydrodynamics Simulation

At this initial modelling stage, it will produce output in the form of a time series of water level elevation and current velocity. In addition to the output in the form of a time series, the output from this initial modelling stage is also in the form of a spatial area to show how the distribution pattern of the direction and velocity of the current in the modelling domain is made.

4.1. Hydrodynamic Model Calibration

Table 1 shows the comparison of the RMSE value of the tidal water level to the variation of the Chezy number parameter.

Table 1. Comparison of RMSE value of water level.

Chezy Number ($m^{0.5}/s$)	RMSE Water Level (%)					
	ADCP	MA02	MA07	MA10	MA12	Average
55	3.66	7,17	16.79	10.82	5.20	8.73
60	3.67	7.14	16.82	10.79	5.18	8.72
65	3.66	7.27	16.85	10.83	5.18	8.76
70	3.67	7.24	16.85	10.77	5.13	8.73
75	3.67	7.09	16.86	10.84	5.20	8.73

Based on the comparison of tidal water level elevations (Figure 3) and the result of calculating the error value represented using the RMSE method between the current velocity from field observations (Table 2), the hydrodynamic model is deemed valid for the subsequent sediment transport modelling. In addition, three scenarios of wave parameter are carried, namely the input of the wave parameter variation every time with the original data value from ERA5 at the open boundary point (A04), the second scenario, namely the time-based wave parameter which is multiplied by a multiplier factor of 1.35 which is obtained from the comparison of the parameter values (A05). The last scenario is to make a larger multiplier factor of 2.50 to see how it affects the simulation results (A06). As can be seen Figure 4, wave simulation results are also suggested that the hydrodynamic model is valid.

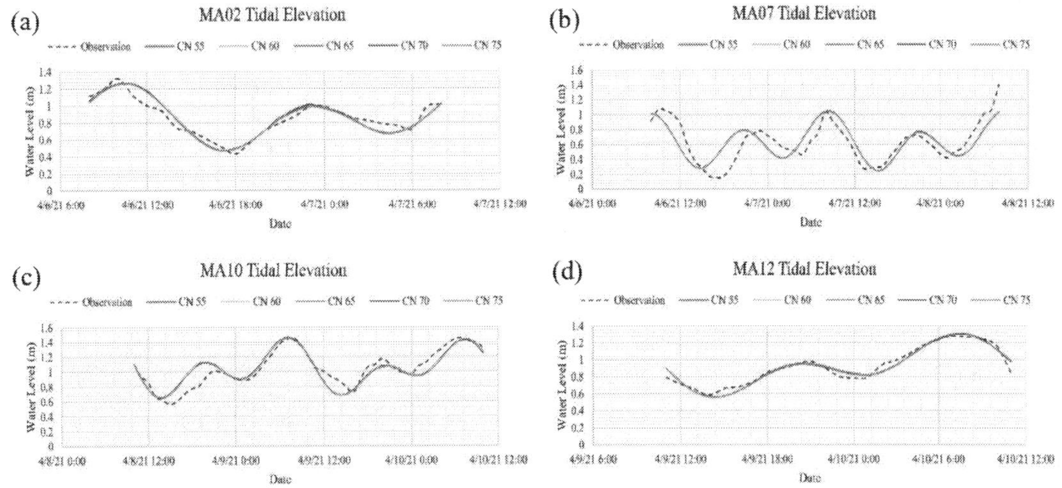

Figure 3. Comparison of tidal water level modeled with observations for (a) point MA02, (b) point MA07, (c) point MA10 and (d) point MA12.

Table 2. Current velocity RMSE value comparison.

Chezy Number ($m^{0.5}/s$)	RMSE Current Velocity (%)
55	11.66
60	11.60
65	11.40
70	11.54
75	11.54

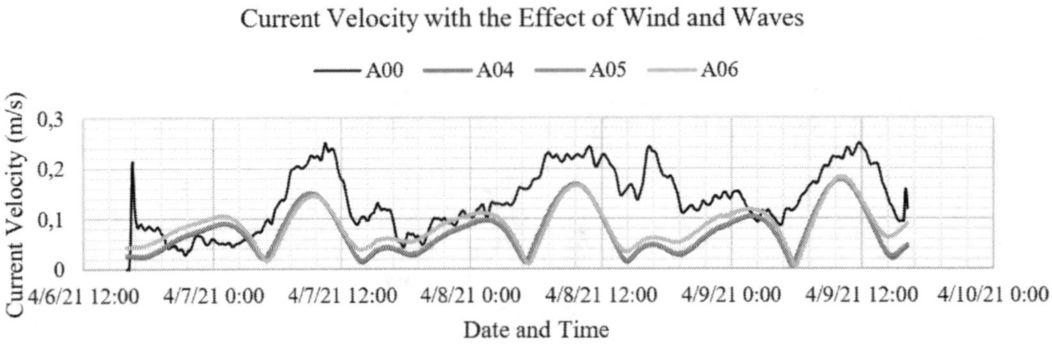

Figure 4. Comparison of current velocity with variation of wave time varying.

5. Result of Sediment Transport Simulation

The following is the condition of sedimentation and erosion that occurred after a simulation of a numerical model of sediment transport was carried out for 15 days in the scenario conditions before and after the land subsidence occurred with the scenario model shown in Table 3.

Table 3. Model scenario.

Scenario ID	Differences	
	Bathymetry	Forcing
Climate changes	Same as 2021	Tides and various number of wave parameters at open boundaries, wind forcing
After land subsidence	Same as 2021	Same as climate changes
Before land subsidence	Same as 2021 shallower 0.5 m on area of interest	Same as climate changes

5.1. Result of Sediment Transport Simulation

The scenario after the land subsidence occurs is the current existing condition, while the scenario before the land subsidence is the current condition where the bathymetry is increased 50 cm higher in the area of interest. The simulation results are generally shown the difference in Figure 5.

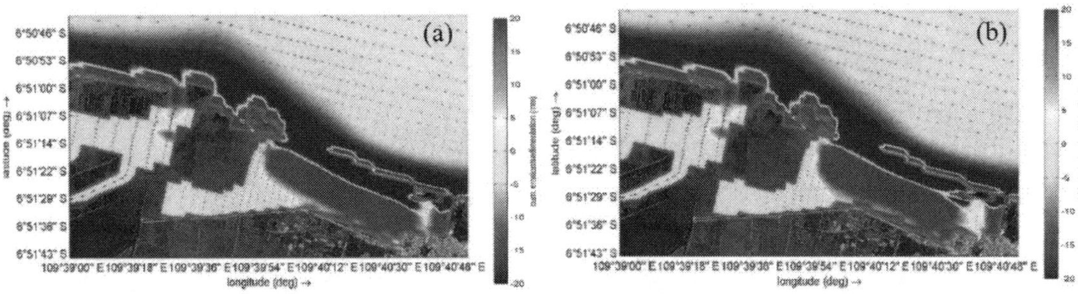

Figure 5. Sedimentation/erosion conditions at the end of day 15 simulation (a) before land subsidence and (b) after land subsidence.

Green to brown is an indicator of sedimentation and green to blue is an indicator of erosion. By the end of the 15th day, it can be seen that the erosion and sedimentation conditions are getting bigger in certain areas. For the concentration of sediment at high tide and low tide in conditions before the occurrence of land subsidence, it shows that the sediment concentration is still quite a lot when compared to conditions after the occurrence of land subsidence when viewed at the research location.

5.1.1. Sediment Concentration. Sediment is collected and trapped at the back of the geo bag with the greatest concentration in the open part of the gap because when viewed from the flow vector the part in the gap shows more turbulence so that the area is the only way for sediment to enter and leave so that the concentration will remain higher when compared to with other areas as seen in Figure 6 and Figure 7.

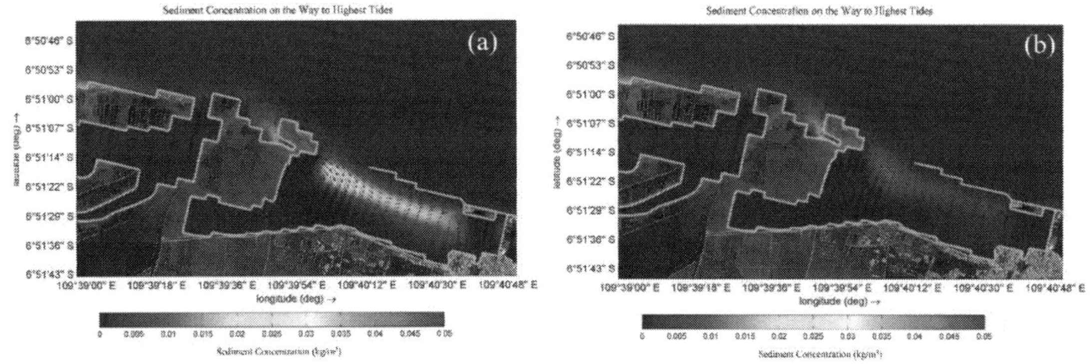

Figure 6. Sediment concentration at highest tidal period (a) before land subsidence and (b) after land subsidence.

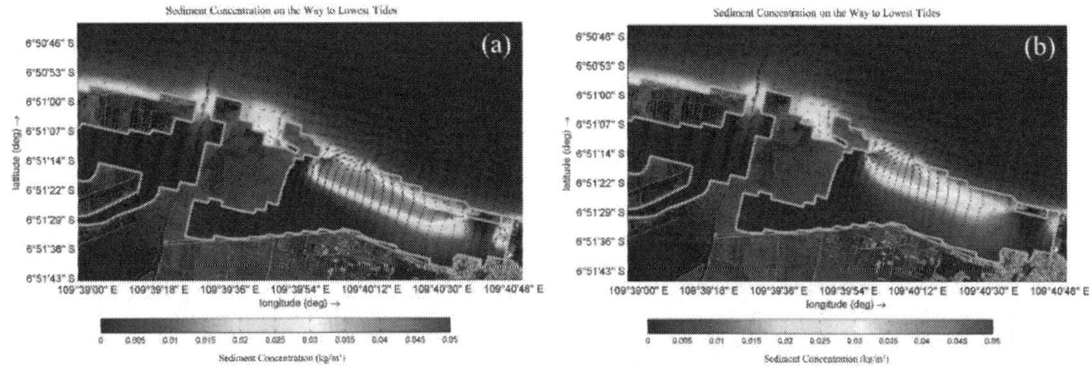

Figure 7. Sediment concentration at lowest tidal period (a) before land subsidence and (b) after land subsidence.

5.1.2. Mean Bed Load Transport. In fact, when viewed from the pattern of mean bed load transport that occurs in conditions before and after the occurrence of land subsidence, there is no significant difference, the difference is only in the magnitude of the bed load transport for the pattern and direction are still the same as seen in Figure 8.

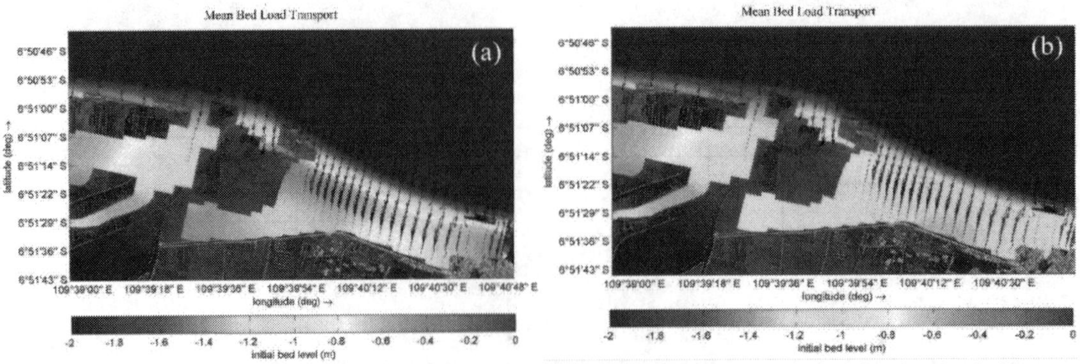

Figure 8. Mean bed load transport scenario (a) before land subsidence and (b) after land subsidence.

5.2. Sediment Transport Analysis

To facilitate the review of sedimentation and erosion patterns that occur in the study area, a cross section is made to see the sedimentation and erosion profiles along the cross section of the review. The location of the created cross section is shown in Figure 9. The results of the sedimentation and erosion profiles formed along the previously created reference lines are shown in Figure 10. There is a significant influence on erosion and sedimentation along the cross-section AB where from point A to the north about 400 m the difference in the effect of changes in depth causes the sediment transport process to also change, while from a distance of 400 m from point A to the north to point B there is no change in transportation significant sedimentation due to changes in water depth is also not significant.

The difference in depth from point C to the north as far as 500 m there is a significant difference in depth so that the rate of sedimentation and erosion is also significant, while from 500 m to the north to point D there is no significant difference in depth so that the sediment transport pattern does not change significantly. From point E to the north as far as 400 m there is erosion which is indicated on the graph that the cumulative erosion obtained is less than zero. And from a distance of 400 m to the north towards point F, sedimentation occurs with the marked cumulative sedimentation occurring more than zero. From a distance of 1000 m from point E to the north to point F there is a decrease in cumulative sedimentation but does not reach the point of erosion until at the end of the fifteenth day of simulation it goes to zero. From point G to the north as far as 250 m there is sedimentation then experiencing very drastic erosion from a distance of 250 m to 500 m from point G, then slowly sedimentation from a distance of 500 m from point G to point H. Along the cross section of BI there is no difference so that the sediment transport that occurs also does not change. From point B to the north as far as 300 m erosion continues to the north towards point I but with a decreasing level of erosion.

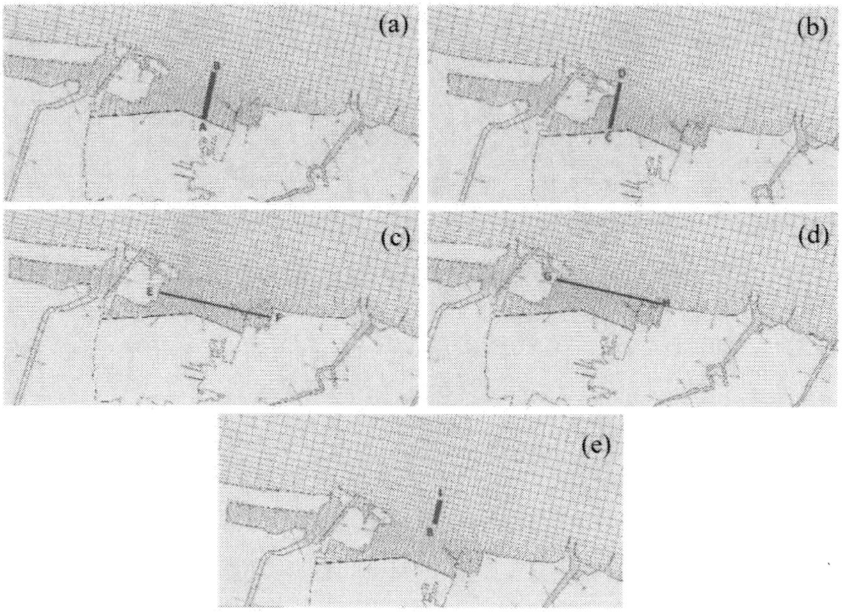

Figure 9. Cross sectional locations in the modeling domain for (a) section AB, (b) section CD, (c) section EF, (d) section GH and (e) section BI.

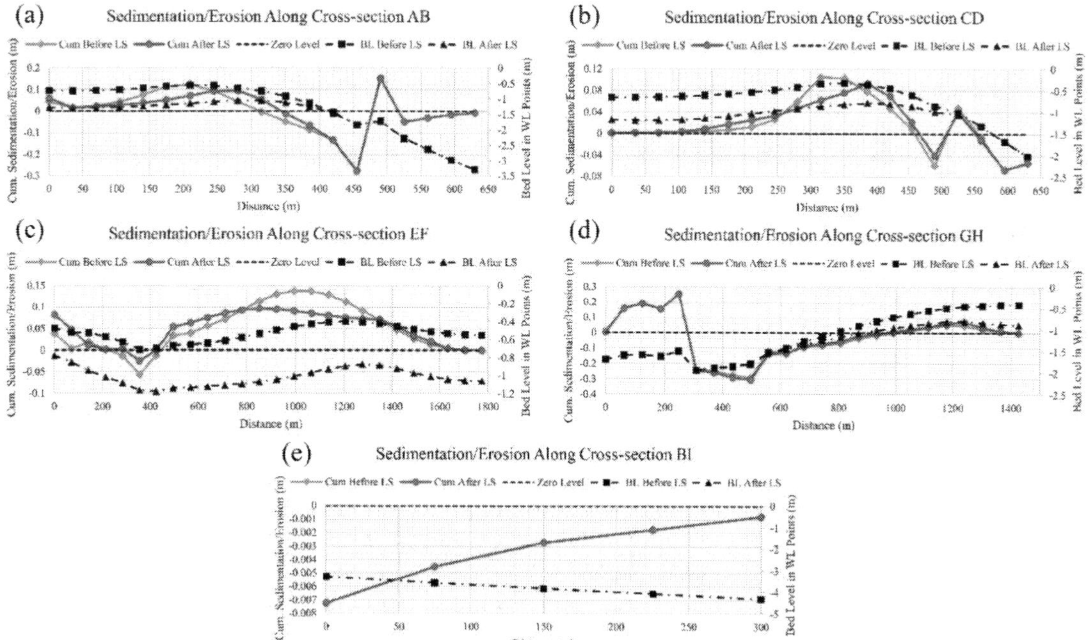

Figure 10. Sedimentation/erosion along the cross-section for (a) section AB, (b) section CD, (c) section EF, (d) section GH and (e) section BI.

In addition to using the cross-sectional line method, the sampling point method is also used in the research area that represents the embankment area, gap area, and the area around the coast as shown in Figure 11. The results of the analysis at the sample points are shown in Figure 12 and Figure 13.

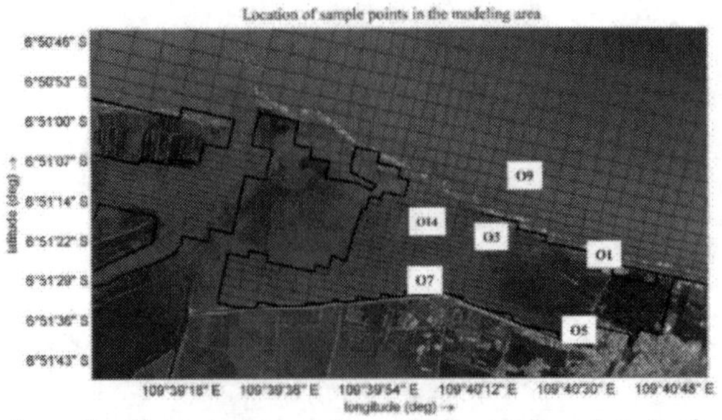

Figure 11. Location of sedimentation and erosion sample points in the area of interest.

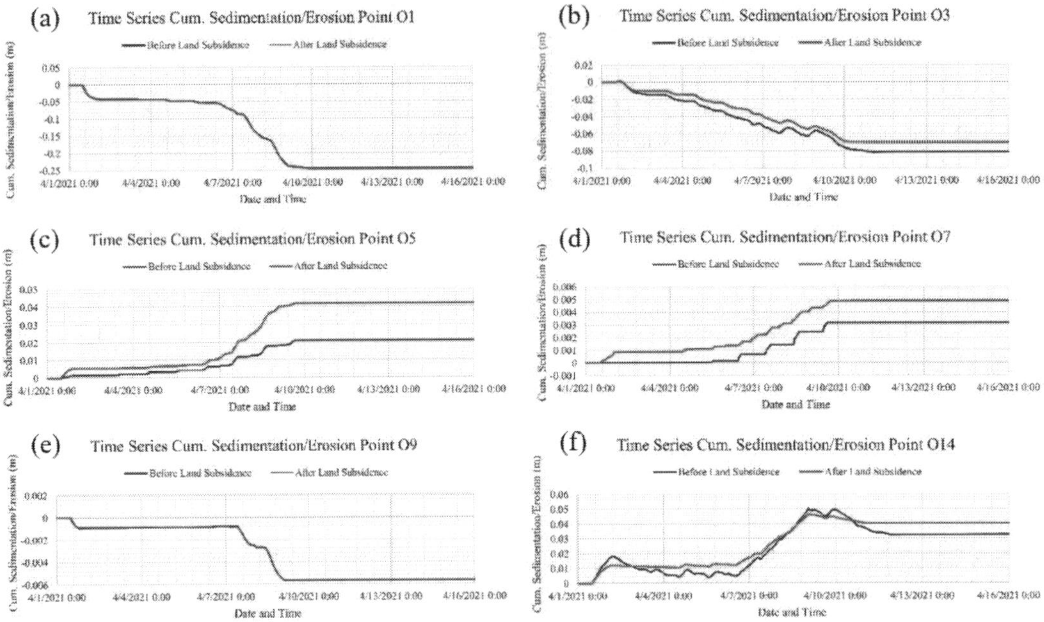

Figure 12. Time series cum. sedimentation/erosion for (a) point O1, (b) point O3, (c) point O5, (d) point O7, (e) point O9 and (f) point O14.

Based on Figure 12, it can be seen that for points O1 and O3 on the first day to the ninth day of the simulation erosion occurred then at a later time there was a relatively much smaller change than before until it reached a saturation condition. It can also be observed that there is a significant difference in the erosion pattern in the scenarios before and after the land subsidence where after the land subsidence the erosion rate becomes smaller. The O5 and O7 points generally occur in the sedimentation area near the embankment on the first to tenth day of the simulation, sedimentation occurs then the sedimentation rate decreases over time until it reaches a saturation condition. At point O9 there is no significant difference in the geo bag front area because there is no significant difference in depth changes from the two scenarios used so that the results are not different. For the review area, point O14 experienced erosion from the first to the seventh day then proceeded to sedimentation until the eleventh day of the simulation and then erosion occurred until the twelfth day of the simulation until it finally reached the saturation point again.

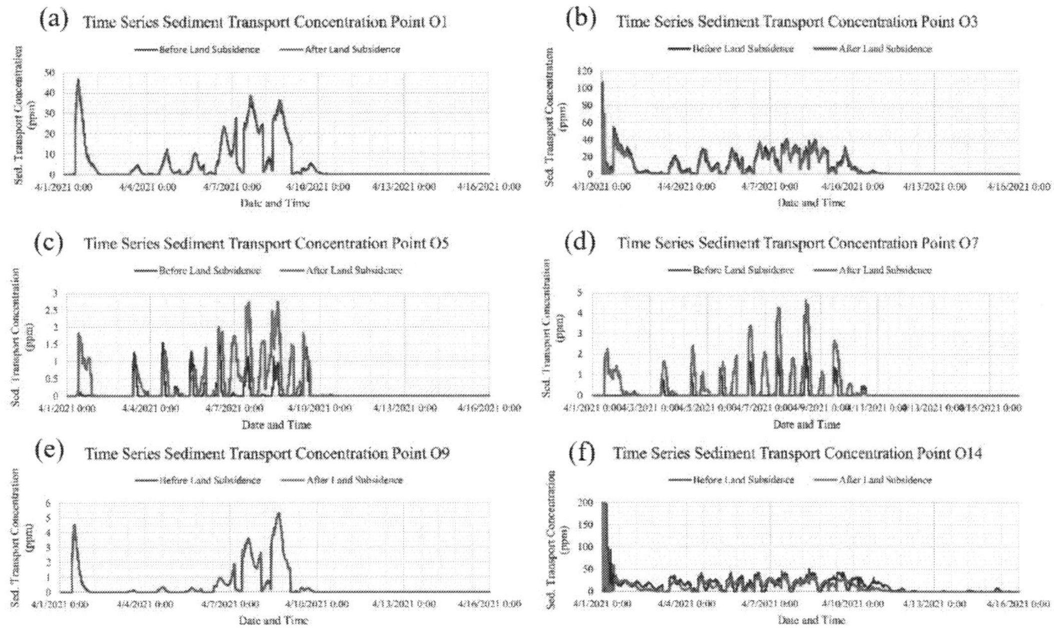

Figure 13. Time series concentration of sediment transport for (a) point O1, (b) point O3, (c) point O5, (d) point O7, (e) point O9 and (f) point O14.

From the graph on Figure 13, it can be seen that the concentration of sediment transport at points O1 and O3 decreased when land subsidence occurred. From the graph in general, it can be seen that the concentration of sediment transport at points O5 and O7 increases when land subsidence occurs. The concentration of sediment transport at point O9 did not change because the depth did not change. From the graph in general, it can be seen that the concentration of sediment transport at point O14 decreased after land subsidence occurred.

The accumulation of sedimentation and erosion in the study area during the fifteen simulation days is shown in Table 4.

Table 4. Accumulation of sedimentation and erosion in the study area for 15 days.

Observation Point	Sedimentation (m/15 days)		Erosion (m/15 days)	
	Before	After	Before	After
O1	-	-	0.246	0.247
O3	-	-	0.082	0.071
O5	0.021	0.042	-	-
O7	0.003	0.005	-	-
O9	-	-	0.006	0.006
O14	0.033	0.040	-	-

Based on the results in Table 4 it can be seen that sedimentation occurs in the area in front of the embankment represented by points O3 and O5 and the area in the geo bag gap represented by point O14, while erosion occurs in the area around the geo bag and in the geo bag gap area represented by points O1, O3, and O9.

6. Conclusion

During the fifteen days of the simulation, Sedimentation rates were initially 0.246 m and 0.247 m, then changed to 0.082 m and 0.071 m after land subsidence. Erosion occurred at the southern location of the geo bag-dike. Then for the area north of the dike, sedimentation occurred of 0.021 m and 0.003 to 0.042 m and 0.005 m. For the northern part of the geo bag-dike, an erosion of 0.006 m occurred. At the geo bag-dike gap area, sedimentation occurred from 0.033 m to 0.040 m.

Sedimentation occurs in the dike's northern area, and the impact of erosion occurs in part to the north and south of the geo bag. These results indicate a correlation between the hydrodynamic simulation results, which show the high current velocity in the area around the geo bag-dike that caused erosion.

The land subsidence affects the sediment transport pattern as indicated by the simulation results of study area.

Acknowledgments

This research is partially funded by the ITB Research and Community Service (P2MI) grant through the Coastal Engineering Research Group, Faculty of Civil and Environmental Engineering, and the Indonesia Collaboration Research Program grant from Bandung Institute of Technology initiated by the Ministry of Education and Culture Republic of Indonesia.

References

[1] Damayanti R, Hariadi and Atmodjo W 2013 The Effect of Currents on the Distribution of Suspended Solids at Slamaran Beach Pekalongan *Journal of Oceanography* **2(1)** 128 – 142

[2] Yudowaty S O, Atmodjo W and Wulandari S Y 2012 Sediment Transport Study at Slamaran Beach Pekalongan *Journal of Oceanography.* **1(2)** 197 – 204

[3] Nashrullah S, Aprijanto, Pasaribu J M, Hazarika M K and Samarakoon L 2013 Study on Flood Inundation in Pekalongan, Central Java *International Journal of Remote Sensing and Earth Sciences.* **10(2)** 76 – 83

[4] Ikhwan R, Saputro S and Hariadi 2015 Study of Base Sediment Distribution Around the Estuary of the Pekalongan River, Pekalongan City *Journal of Oceanography.* **4(3)** 617 – 624

[5] Yuanita N, Kurniawan A, Setiawan H, Hasan F, and Khasanah M 2019 Physical Model of Natural Coastal Protection System: Wave Transmission Over Mangrove Seedling Trees *The 3rd International Water Safety Symposium. Journal of Coastal Research, Special Issue.* **(91)** 176 – 180

[6] Satriadi A, Subardjo P, Saputra S, Hariadi and Haryadi 2013 Recent Geology of the Coastal Area Between the Pekalongan River to the Klidang River, Central Java *Marina Oceanographic Bulletin.* **(2)** 24 – 34

[7] Pratama M B 2019 Tidal Flood in Pekalongan: Utilizing and Operating Open Resources for Modelling *The 2nd EPI International Conference on Science and Engineering.* 1 – 10

[8] Yuanita N and Tingsanchali T 2008 Development of a River Delta: A Case Study of Cimanuk River Mouth, *Indonesia Hydrological Processes: An International Journal.* **22(18)** 3785 - 3801

[9] Martosuwito S and Fakhruddin R 2014 Tidal Sedimentary Rocks at the Keruh Creek, Lor Agung Village, Pekalongan Residents *Journal of Geology and Mineral Resources.* **15(1)** 41 – 53

[10] Novico F, Astjario P and Bachtiar H 2013 Erosion-Sedimentation and Tidal Current Conditions Around Coastline in the front of Steam Power Plant Tarahan Lampung Using Delft3D Version 3.28 *Journal of Marine Geology.* **11(1)** 39 – 54

[11] Li L and Huang Z 2013 Modelling the Change of Beach Profile Under Tsunami Wave: A Comparison of Selected Sediment Transport Models *Journal of Earthquake and Tsunami.* **7(1)** 1350001-1 – 1350001-29

[12] Elias R, Roelvink D J A, Walstra D J, Stive M and Klein M D 2000 Hydrodynamic Validation of Delft3D with Field Measurements at Egmond *27th International Conference on Coastal Engineering 'ICCE 2000' in Sydney.* 1 – 14

The eco enzyme application to reduce nitrite in wastewater as the sustainability alternative solution in garbage and wastewater problems

T Wikaningrum*, P L Anggraina
Environmental Engineering, Faculty of Engineering, President University, Indonesia

*Email: temmy@president.ac.id

Keywords: eco enzyme, nitrite, organic garbage, wastewater

Abstract. The increasing population growth has an impact on increasing problems in wastewater quality and garbage volume. It is a need to develop a sustainability solution for both problems. In this research, the papaya fruit and spinach vegetable garbage were used as raw organic materials and fermented for six months. This study objective was to determine whether eco enzymes can be used to reduce nitrite concentration in wastewater. The research method was conducted on a laboratory scale with the artificial samples in a batch system. The result showed that eco enzyme made from a mixture of papaya and spinach can reduce the nitrite concentration in the water sample. The t-test showed that the effect of eco enzyme application in reducing nitrite in water samples was significant. The application of eco enzyme 0%, 2%, 4%, and 6% respectively during 10 hours exposure showed the nitrite removal were 0.3 %, 20%, 29%, and 35 %. It can be seen that the reduction of nitrite concentration in samples added by eco enzyme was caused by eco enzyme activity. The application at a higher concentration of 10%, 15%, and 20% showed that nitrite removal efficiencies were 35.7%, 36.7%, 46.7%, 49.4% respectively during 7 hours exposure. It can be concluded that a longer time exposure, and also the more eco enzyme concentration, showed the more result in the effect of nitrite reduction concentration in water. 4. By this study it is proven that application of eco enzyme made from organic garbage can be an alternative solution for garbage and wastewater quality problem.

1. Introduction

It was estimated that one-third of the world's food is wasted every year in the form of food waste. the value of the discarded food waste is USD 936 million or greater than the gross domestic product of the Netherlands as a whole [1]. Most of the food waste is dumped in the waste landfill which will eventually produce greenhouse gases from the degradation process. The Intergovernmental Panel on Climate Change (IPCC, 2010)) estimated that about 3 percent of the world's greenhouse gas emissions were the result of transporting and treating solid waste. the largest source of these emissions was generated from landfill activities.

It is estimated that 190 thousand waste is generated in Indonesia every day, and most of it is in the form of organic waste. In Indonesia, the major urban centers generate about 10 million ton of waste every year. The waste amount is estimated to increase by 2 to 4% annually (Ministry of Environment, 2008). With the continuous increase in waste along with population growth with the limited area for final landfill, it becomes a big challenge to the government. However, the waste problem is not only the duty of the government but also the responsibility of society as a whole. For this reason, the role of the community is needed to overcome the problem of waste, especially organic waste that comes from food scraps. The main sources of municipal solid waste in Indonesia generates from households andtraditional markets. The community role that makes it possible to process organic waste is the composting system, but the survey results showed that the community does not accept that their place of residence is close to the composting facility [2].

With this condition, the use of organic waste in other useful materials can be a solution, especially if it can be used as a material agent in wastewater treatment. Dr.Rosukon Poompanvong introduced eco Enzyme intending to process enzymes from organic waste. Eco enzyme is the result of fermentation of organic waste consisting of vegetable and fruit waste, brown sugar, and water. Eco enzyme Has a characteristic dark brown to yellowish colour and has a strong sweet and sour aroma due to the fermentation process [3]

Eco Enzyme can be a potential solution to the issue of global warming due to the high production of methane gas produced by the decomposition process of organic vegetables and fruits. There are many benefits obtained from the manufacture of eco-enzymes, one of which is as a disinfectant due to the presence of organic acids and alcohol. The use of eco-enzyme as a biochemical catalyst for reducing pollutants in rivers has also begun to be studied because its ability to decompose waste is proven from previous studies [4]. From several previous studies, it was found that eco-enzyme can be used as an alternative material to reduce various pollutants in wastewater. For example, reducing COD concentrations, lowering the pH of the solution, and reducing ammonia concentrations. Organic waste that was previously only waste can be processed into eco-enzymes that have high economic value [3].

In this wastewater quality study, the parameter studied was the nitrite concentration in the water. The selection of nitrite parameter is based on the importance of this parameter in water quality standards, both drinking water, sanitation water, and wastewater. The presence of nitrite in water can occur due to contamination of underground water, wastewater from livestock, excessive use of fertilizers from agricultural land, it can also be from domestic activities [5].

High concentrations of nitrite can cause a decrease in the immunity of the organism and this makes the organism susceptible to disease. Nitrite also has the potential to react with haemoglobin and trigger the formation of methaemoglobin and can be fatal because the blood cannot bind and transport oxygen to all body tissues [6].

2. Literature review

2.1 Organic waste

Household waste contributes to polluting the environment and causes aesthetic problems if there is no serious commitment to handling it. Organic waste management is still not implemented efficiently and is even dominated by irresponsible waste disposal. Meanwhile, organic waste is very useful if it can be processed into products that have economic value such as compost [7]. Improper disposal of waste can cause environmental pollution and is a serious threat to living things, including humans. Environmental problems arise as a result of ineffective waste management [8].

The increase in population and human activities in an area is in line with the increase in the amount of waste produced, therefore efficient and effective waste management is needed so that it does not cause environmental pollution. This problem cannot be handled by conventional waste management methods, namely disposal at the final waste disposal site. The weakness of this method in addition to requiring a high cost also does not reduce the volume of waste disposed of, even though the landfill hasa service life and maximum capacity [9].

2.2 Eco enzyme

Eco-enzyme solutions have many benefits, including as a pollutant-reducing agent in river water, providing nutrients for the soil and as an antiseptic [10]

Manufacturing eco enzyme can be done by mixing organic waste, brown sugar, and water in a ratio of 3:1:10 then fermented for up to three months. After fermentation is complete, the solution is filtered and stored at room temperature. The application of the eco-enzyme can also be applied to treat sludge and wastewater [11].

Fermentation is a process of chemical change in organic substrates that can survive due to the action of biochemical catalysts, namely enzymes produced by certain living microbes, such as organic acids, single-cell proteins, antibiotics, and biopolymers [10].

Enzymes are the most abundant types of proteins in living cells and have an important role as biocatalysts in biochemical reactions. One of the enzymes that play an important role is the protease enzyme. Protease enzymes function to hydrolyze peptide bonds in proteins into amino acids. Proteases are complex enzymes that have varied chemical, physical and catalytic properties. This enzyme is produced by microorganisms extracellularly and has an important function in cell metabolism [12]. Types of protease enzymes produced from plants include papain and bromelain enzymes. One of the compounds contained in the eco enzyme is acetic acid which has bactericidal properties. The types of enzymes contained in this eco enzyme are trypsin and amylase which can prevent pathogenic bacteria. [12]. The presence of NO_3 (nitrate) and CO3 (carbon trioxide) content is useful for the soil as nutrients [13].

2.3 Nitrite in water

Nitrate and nitrite ions are naturally as the are part of the nitrogen cycle. The nitrate ion (NO_3^-) is the stable form of combined nitrogen for oxygenated systems. Although chemically unreactive, it can be reduced by microbial action. The nitrite ion (NO_2^-) contains nitrogen in a relatively unstable oxidation state [5] The organic nitrogen wastes are first decomposed to give ammonia, which is followed by an oxidation process to nitrite and nitrate (WHO).

The presence of nitrite indicates the ongoing process of biological degradation of organic matter. The degradation process of nitrogen from ammonium ions to nitrate ions is called nitrification. This process is an enzymatic oxidation process by microorganisms that occur in two continuous stages [14] The first stage involves obligate autotrophic nitrite bacteria known as Nitrosomonas and Nitrococcus with the final product nitrite ion as follows [15] :

$$2NH_4^+ + 3O_2 \rightarrow 2NO_2^- + 2H_2O + 4H^+$$

The second stage is the conversion of nitrite ions to soluble nitrate ions involving obligate autotrophic nitrate bacteria known as Nitrobacter and Nitrocystis as follows [15]:

$$2NO_2^- + O_2 \rightarrow 2NO_3^-$$

The guideline for nitrite of 3 mg/l as nitrite is based on human data showing that doses of nitrite that cause methemoglobinemia in infants range from 0.4 to more than 200 mg/kg of body weight [5]. In the United States of America (USA), naturally occurring levels do not exceed 4–9 mg/l for nitrate and 0.3 mg/l for nitrite (USEPA, 1987). Nitrite levels in drinking water are usually below 0.1 mg/l [5]. In Indonesia, based on the Minister of Health Regulation No. 32 of 2017 concerning clean water for environmental health sanitation, the maximum level of nitrite is 1 mg/L and nitrate is 3 mg/L. Meanwhile, according to the wastewater quality standard No. 5 of 2014 issued by the Minister of the Environment, the maximum level of nitrite is 1 mg/L. According to the Indonesian National Standard (SNI) of 3554:2006 regarding bottled drinking water quality standards, the maximum concentration of nitrite is 0.005 mg/L and nitrate is 45 mg/L.

3. Research method

3.1Population and Sample

The samples in this study were artificial samples that were prepared by adding a standard solution of nitrite into distilled water so that a solution with a concentration of about 20 mg/L. The water sample population is homogeneous because the sample is artificial. The sampling method used in this study was a random sampling method.

3.2 Data Collection Method

The data used as a source of data in this study was through primary data generated from observations in the laboratory. The data collection technique was through an experimental method to determine the effect of the eco enzyme on the reduction of nitrite. This research was conducted from September 2020 – April 2021

3.3 Materials and tools
The main equipment that used in the research were laboratory glassware, analytical balance, and UV-Vis double beam spectrophotometer UNICO 4082.

The materials used were eco enzyme made from mixing of papaya and spinach waste, and chemicals according to analysis procedure for nitrite measurement in water sample of SNI 06-6989.9-2004. The main chemicals are N-(1-naphthyl)-ethylenediamine dihydrochloride, sulphanilamide ($H_2NC_6H_4SO_2NH_2$).

3.4 Research framework

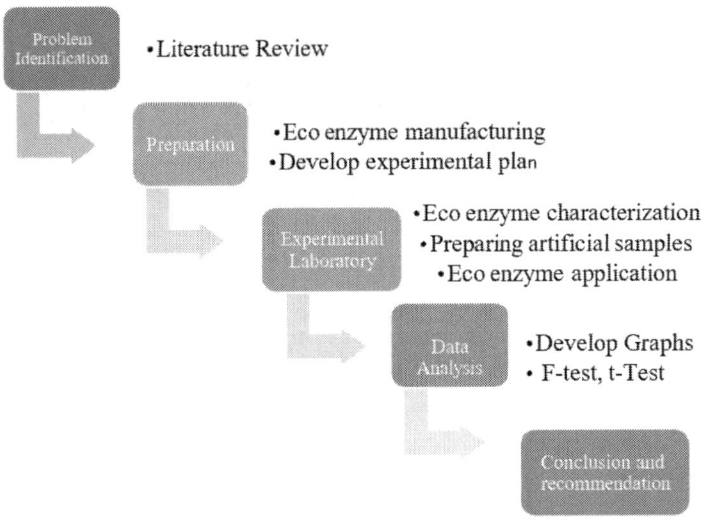

Figure 1. Research framework

3.5 Experimental methods
The steps in the research experiments were as follow:

- Prepare an artificial sample of NO_2 with a concentration range of 20 ppm which is prepared from $NaNO_2$ standard solution in 8 beaker glasses.
- Add eco enzyme with concentrations of 0 %, 2%, 4%, 6% into each artificial sample and mix well.
- Duplicate the experiments.
- The samples and control were then allowed to stand exposure for observation time.
- The nitrite concentration of the samples was measured every hour until there was a significant decrease.

3.6 Data analysis method
The hypothesis in this study was that the application of eco enzyme can reduce nitrite concentrations in the water. Hypothesis Ho, if $\mu A = \mu B$ it means the nitrite concentrations in the water sample added by the eco enzyme are the same or not lower than the control sample. Ha or Ho is rejected, if $\mu A \neq \mu B$, it means the nitrite concentration of the sample added by the eco enzyme is lower than the control sample. Statistical analysis for this study was F-test to know the variance whether it is equal or an-equal and

followed by t-Test for testing its significant concentration of nitrite reduction with a value of P =0.05, which was performed with Microsoft Excel software.

4. Result and discussion

4.1 Eco enzyme characterization

Eco enzyme was prepared by anaerobic fermentation for six months made from papaya and spinach mixing garbage waste. The composition of raw materials was water: organic garbage: brown sugar was 10:3:1. The eco enzyme result was filtered before applying for removing solids materials. The laboratory analysis showed that the eco enzyme has a pH = 3.48. The pH value of eco enzyme manufactured from organic mix garbage was 3.6 [4] The pH of eco enzyme made from pineapple was reported at 3.4-3.7; made from oranges was reported at 3.2-3.3; made from tomatoes was reported at 3.1-3.4 [4], and made from *Carica papaya* L was reported at 3.29 [16]In nature, the pH of garbage enzyme solution is acidic because of its acetic acid contents [17].

In this study, there was no nitrite concentration detected in the eco enzyme by laboratory analysis procedure refer to SNI 06-6989.9-2004. The test results of the pure garbage enzyme, it was reported that it was acidic and did not contain ammonia nitrogen, and phosphorus [18].

4.2 The effect of eco enzyme application on the water samples

The effect of applying eco enzyme with concentrations of 0% (as a control), 2%, 4%, and 6% on nitrite concentration in water samples for 10 hours can be seen in Table 1. These observations were made by analyzing the concentration of nitrite concentration every hour.

Table 1. The effect of eco enzyme and time exposure on nitrite concentration

Exposure Time (hrs)	Control (0 % eco enzyme)	2% -Eco enzyme	4% -Eco enzyme	6% -Eco enzyme	Control (0 % eco enzyme)	2% - Eco enzyme	4% -Eco enzyme	6% -Eco enzyme
	NO_2 concentration (mg/L)				**NO_2 reduction (%)**			
0	21.53	21.53	21.53	21.53		0.0%	0.0%	0.0%
1	21.51	19.15	17.81	18.37	0.1%	11.0%	17.2%	14.6%
2	21.22	20.30	20.07	19.26	1.4%	4.3%	5.4%	9.2%
3	21.53	20.20	19.90	18.70	0.0%	6.2%	7.6%	13.2%
4	21.56	18.50	17.06	17.28	-0.1%	14.2%	20.9%	19.9%
5	21.42	18.67	17.39	15.83	0.5%	12.8%	18.8%	26.1%
6	21.53	17.31	15.02	16.61	0.0%	19.6%	30.2%	22.9%
7	21.48	17.16	15.16	13.98	0.3%	**20.1%**	**29.4%**	**34.9%**
8	21.48	17.31	15.25	14.82	0.3%	19.4%	29.0%	31.0%
9	21.48	17.67	16.17	14.32	0.3%	17.7%	24.7%	33.3%
10	21.28	17.27	15.94	14.93	1.2%	18.8%	25.1%	29.8%

Table 1 showed that there was a decrease in nitrite concentration in the samples that were added by eco enzyme compared to the control. The nitrite concentration of control was relatively stable within 10 hours of exposure. The stability of nitrite concentration in the control samples is proven statistically by its result regression analysis which showed equation: nitrite (mg/L) = -0.0007 time (hr) + 21.49, and value of significance F (P) = 0.528 > 0.05. A very low linear coefficient and P > 0.05 showed that there was no significant time exposure effect on the control samples during 10 hrs. It can be seen that the reduction of nitrite concentration in samples added by eco enzyme was caused by eco enzyme activity.

The key enzymes in nitrification are located in the complex internal membrane system in many species of nitrifying bacteria, are ammonia monooxygenase (which oxidizes ammonia to hydroxylamine), hydroxylamine oxidoreductase (which oxidizes hydroxylamine to nitric oxide - which is oxidized to nitrite by a currently unidentified enzyme), and nitrite oxidoreductase (which oxidizes nitrite to nitrate) [15]. The enzyme of oxidoreductase is used by the microorganism to oxidize nitrite to nitrate. The nitrite reductases are present in Nitrobacter spp., Nitrococcus Marinus, Nitrospira spp. The microorganism of Nitrospina could reduce nitrite to nitric oxide. The oxidation of nitric oxide has been observed in Nitrobacter, but it was still unclear whether this reaction is biotic or abiotic, what kind of enzyme that responsible for this mechanism remains unknown [15]. The nitrogen sources such as NO2 is degraded by microorganism with secretes their enzyme [19]. The observations confirm that the activity of the enzymes, which are amylase, protease, and lipase, is present in the eco enzyme [3].

The reduction of nitrite is a part of the nitrification process in reducing ammonia. The reduction of ammonia by hydrolytic reaction of eco enzyme in converting ammonia, nitrogen, and ammonium into more soluble were reported [20]. Some evidence also supported the suggestion that nitrite may use by DNPAO (denitrifying polyp-P-accumulating organisms) in P uptake [21]. The application of garbage enzyme was showed remove ammonia in waste wastewater effectively [22]. Many studies have reported that the application of the eco enzyme to treat organic pollutant in domestic wastewater and greywater can remove ammonia nitrogen [17].

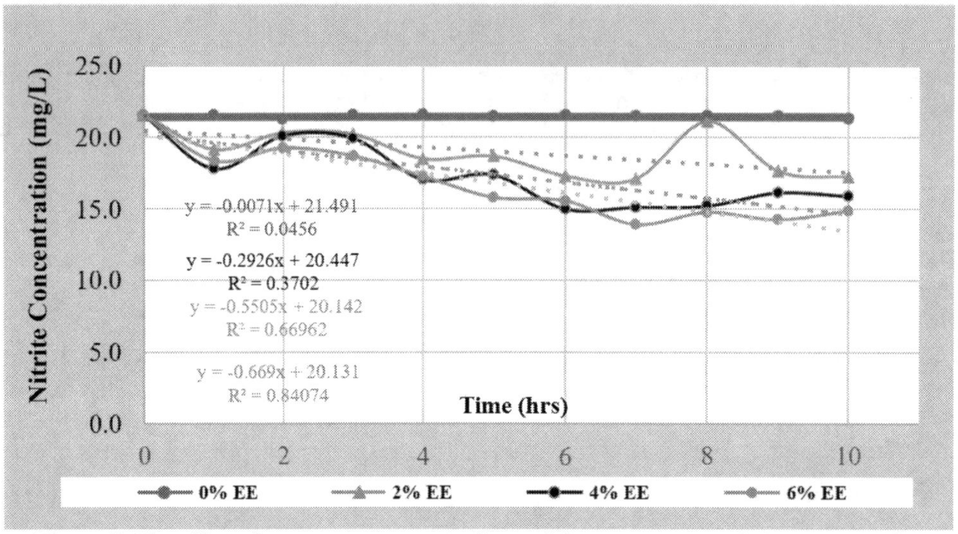

Figure 2. The effect of eco enzyme concentration and time exposure on nitrite concentration

The linear regression between nitrite concentration and exposure time on various concentrations of eco enzyme application is shown in Figure 2. The linear model was getting better with increasing the eco enzyme that was shown on a higher value of R^2 (the coefficient of multiple determination) [23]. The value of R^2 is the fit indicator of the regression equation; that is to give the proportion or percentage of the total variation in the dependent variable (time) that is explained by the independent variable (nitrite concentration).

The coefficient of the linear equation in Figure 2 showed that the more concentration of eco enzyme application has a more negative value, they are -0.29 for 2%-eco enzyme, -0.55 for the 4%-eco enzyme, and -0.67 for 6%-eco enzyme. This analysis showed that the addition of eco enzyme has an effect on reduction of nitrite concentration in the water samples, and the effect getting more on the higher eco enzyme concentration application. It showed that adding eco enzyme with a concentration of 2%, 4%, and 6% can reduce the nitrite concentration successively by 20.1%, 29.4%, dan 34.9% respectively.

It has been reported that the application of the higher concentration of eco enzyme and the longer the time exposure, given the greater results in the reduction of ammonia concentration in wastewater [17] The nitrates reduction level from 5.54 to 3.39 mg/L was reported in the 0.5%-eco enzyme testing application in the drain water [24].

The F-test for variance analysis of the nitrite concentration of samples compared to the control samples showed the results are unequal for all the samples. The unequal variances can be seen by P-value are < 0.05, which are 3 x 10^{-10} for 2%-eco enzyme, 1.1 x 10^{-11} for the 4%-eco enzyme, and 4.7 x 10^{-12} for 6%-eco enzyme application. Based on the results of these F-tests, then a t-test is carried out to know the significance of different concentrations of nitrite concentration of samples and control samples. The t-test result showed that the P-value for all the samples is $P<0.05$, which showed there is a significant difference in each sample's nitrite concentration compared to their controls. The result of t-Test: Two-sample assuming unequal variances, P(T<=t) two-tail are 0.00045 for 2%-eco enzyme, 0.00013 for the 4%-eco enzyme, dan 7.9 x 10^{-5} for 6%-eco enzyme. These t-tests indicate that eco enzyme application on water samples can reduce the nitrite concentration significantly. ANOVA test for comparing treatment without eco enzyme (control) with eco enzyme also in line with t-Test that theresults were P = 5.16 x 10^{-5} for 2%-eco enzyme, P=6.7 x 10^{-6} for 4 % eco enzyme, and P = 3.09 x 10^{-6} for 6%-eco enzyme.

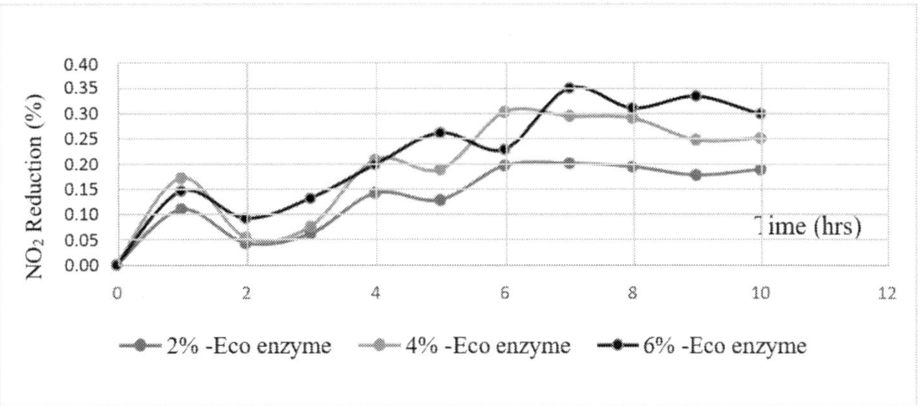

Figure 3. The effect of eco enzyme concentration and time exposure on nitrite reduction

From the data in Table 1, it can be developed a graph in Figure 3 that showed the nitrite reduction of samples at the 7th hour the highest results were achieved, while at the 8th to 10th hours there was tend to decrease.

4.3 The effect of eco enzyme concentration on the nitrite reduction

Based on the result that showed there was a significant reduction in nitrite concentration in water samples by eco enzyme addition, the next experiments were application higher concentration of eco enzyme to find more nitrite reduction. The experiments were done with eco enzyme concentration addition of 6%, 10%, 15%, and 20% application in 7 hours exposure time. The selection of 7 hours based on the previous results that nitrite reduction showed the best value at the 7th exposure time.

Table 2. The effect of eco enzyme concentration on nitrite reduction

Exposure Time (hrs)	Control (0 % eco enzyme)	6 % -Eco enzyme	10% -Eco enzyme	15% -Eco enzyme	20% -Eco enzyme
	NO_2 (mg/L)				
0	21.74	21.74	21.74	21.74	21.74
7	21.69	13.98	13.77	11.58	11.01

The experiments results can be seen in Table 2 that indicated the nitrite concentration in control samples were stable, while the samples added eco enzyme has a nitrite lower concentration after 7 hours. Figure 4 shows the effect of eco enzyme applied concentration (%) on nitrite concentration (mg/L). The graph shows that there was a negative linear relationship between the eco enzymes applied concentration and nitrite concentration with a linear coefficient of -23.79 and coefficient determination (R^2) was 91.72%.

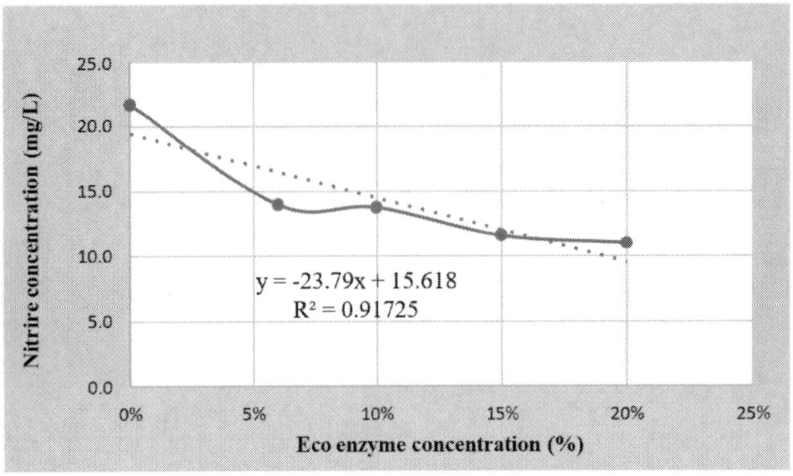

Figure 4. Nitrite reduction with various concentration eco enzyme application

Figure 5 showed more clearly that after 7 hours of exposure, the nitrite concentration in the control sample was relatively stable, while the sample with the eco enzyme 2%, 4%, and 6% have a negative coefficient or there was a decrease in nitrite concentration.

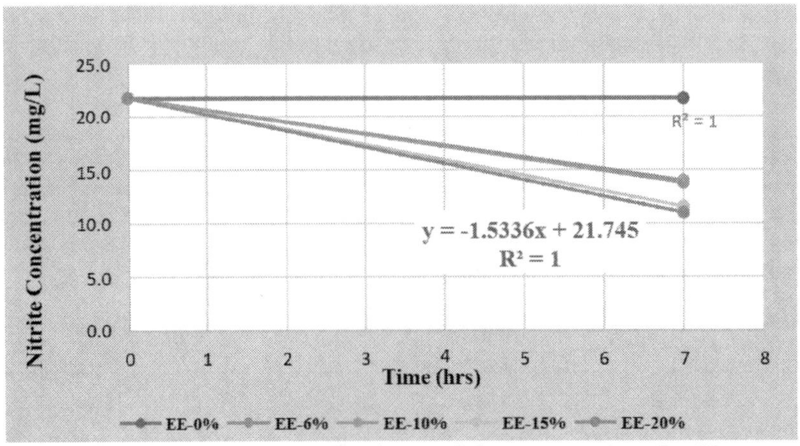

Figure 5. The Effect of eco enzyme concentration on nitrite concentration after 7 hours application

The effect of eco enzyme application concentration on nitrite can be seen in Figure 6, which showed a positive linear relationship. The data is shown in Table 3. The effect showed that the higher the concentration of added eco-enzyme, the greater the decrease in the concentration of nitrite.

The application for 5 days of 10% eco enzyme in artificial greywater samples reported can reduce the ammonia levels from 9.6 mg/L to 0 mg/L [17].

Table 3. The effect of eco enzyme concentration on nitrite reduction

Exposure Time (hrs)	Control (0 % eco enzyme)	6 % -Eco enzyme	10% -Eco enzyme	15% -Eco enzyme	20% -Eco enzyme
	NO_2 reduction (%)				
7	0.3	35.7	36.7	46.7	49.4

The positive linear correlation has a coefficient of + 2.954 which reaches a 49.4% reduction with 20% -eco enzyme addition. In this study, the initial nitrite of 21.74 mg/L can be reduced to 11.01 mg/L or a 49.4% reduction in applying 20% eco enzyme. The achievement of the nitrite concentration from the original 21.74 mg/L to 11.07 mg/L still does not meet the quality standard of 1 mg/L according to the Minister of the Environment No. 5 of 2014 concerning the wastewater quality standard. However, the result of a 49.4% decrease in nitrite concentration has significantly reduced the level of water pollution.

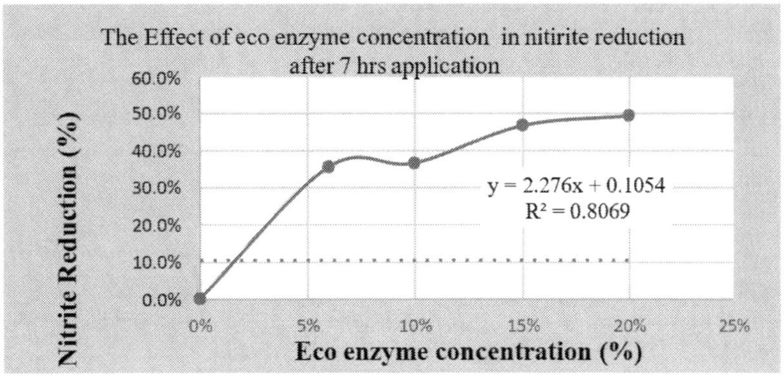

Figure 6. The Effect of eco enzyme concentration on nitrite reduction after 7 hours application

The reduction of TAN (Total ammonia Nitrogen) was sharply more effective with the exposure time. The study result showed that the reduction rates of TAN with 15 % eco enzyme were significantly higher than 10 % and 5 % [3]. It also found that eco enzyme not only very efficient in removal TAN, but also very efficient in removal Total Suspended Solid (TSS), Volatile Suspended Solids (VSS), Total Phosphorus (TP), and stabilization of the Chemical Oxygen Demand [3]. The effectiveness of eco enzyme solution application for removing the concentration of ammonia, nitrogen, and phosphorus also reported that the result is highly effective [24]. The effectivity in removing Total Phosphate (TP) and TAN that achieved up to 90 by eco enzyme was initiated by its characteristic and biocatalytic property [3] The optimum concentration of eco enzyme application was found 10% in removing TSS, VSS, TP, TAN, and COD by the reduction efficiency of 87%, 67%, 99%, 91%, and 77% respectively [3].

In this study, the decrease in nitrite concentration continued to occur up to 49% even in an acidic environment. This will be an advantage of the nitrite removal by eco enzyme because nitrification occurs naturally at the pH range of 6.45-8.95 and lower than 6.45 for complete inhibition nitrification to take place.

5. Conclusion and recommendation

5.1 Conclusion

Based on the study, conclusions and recommendations can be summarized as follows:

The study on the application of eco-enzyme made from a mixture of papaya and spinach in artificial water samples to reduce nitrite levels showed the following results:

1. Eco enzyme made from a mixture of papaya and spinach can reduce the nitrite concentration in the water sample. The t-Test showed that the effect of eco enzyme application in reducing nitrite in water samples was significant.

2. The application of eco enzyme 0%, 2%, 4%, and 6% respectively during 10 hours exposure showed the nitrite removal were 0.3 %, 20%, 29%, and 35 %. It can be seen that the reduction of nitrite concentration in samples added by eco enzyme was caused by eco enzyme activity. The application at a higher concentration of 10%, 15%, and 20% showed that nitrite removal efficiencies were 35.7%, 36.7%, 46.7%, 49.4% respectively during 7 hours exposure.
3. The result showed that longer time exposure, and also the more eco enzyme concentration, showed the more result in the effect of nitrite reduction concentration in water samples.
4. By this study it is proven that application of eco enzyme made from organic garbage can be an alternative solution for garbage and wastewater quality problem.

5.2 Recommendation

Further research should be carried out by measuring the concentration of nitrate to determine whether a decrease in nitrite indicates an increase in nitrate in the water sample. In addition, further research should use the wastewater samples to determine whether the reduction of nitrite is still effective in the presence of other materials contained in the real wastewater samples.

6. References

[1] FAO, "Food wastage footprint & Climate Change," *Fao*, no. 1, pp. 1–4, 2015, [Online]. Available: http://www.fao.org/3/a-bb144e.pdf.

[2] A. Aprilia, T. Tezuka, and G. Spaargare, "Household Solid Waste Management in Jakarta, Indonesia: A Socio-Economic Evaluation," *Waste Manag. - An Integr. Vis.*, 2012, doi: 10.5772/51464.

[3] O. Galintin, N. Rasit, and S. Hamzah, "Production and characterization of eco enzyme produced from fruit and vegetable wastes and its influence on the aquaculture sludge," *Biointerface Res. Appl. Chem.*, vol. 11, no. 3, pp. 10205–10214, 2021, doi: 10.33263/BRIAC113.1020510214.

[4] C. Arun and P. Sivashanmugam, "Investigation of biocatalytic potential of garbage enzyme and its influence on stabilization of industrial waste activated sludge," *Process Saf. Environ. Prot.*, vol. 94, no. C, pp. 471–478, 2015, doi: 10.1016/j.psep.2014.10.008.

[5] California Department of Health Services, "Health Concerns Related to Nitrate and Nitrite in Private Well Water," no. February, p. 4, 2000, [Online]. Available: http://www.tularehhsa.org/?LinkServID=2D7CA07C-FED0-6A01-19858ADA512AE780&showMeta=0.

[6] Kiswanto, Wintah, and N. L. Rahayu, "Analisis logam berat (mn, fe , cd), sianida dan nitrit pada air asam tambang batu bara," *J. Litbang Kota Pekalongan*, vol. 18, pp. 20–26, 2020.

[7] A. Rosmala, D. Mirantika, and W. Rabbani, "Takakura Sebagai Solusi Penanganan Sampah Organik Rumah Tangga," *Abdimas Galuh*, vol. 2, no. 2, p. 165, 2020, doi: 10.25157/ag.v2i2.4088.

[8] I. G. Ngurah, A. Suryaputra, and I. W. Mudianta, "Pengelolaan Sampah Organik Di Sd Negeri 5 Panji," *Proceding Senadimas Undiksha*, pp. 1082–1085, 2020.

[9] R. W. Sekarsari *et al.*, "Pemanfaatan Sampah Organik Untuk Pengolahan Kompos," *J. Pembelajaran Pemberdaya. Masy.*, vol. 1, no. 3, p. 200, 2020, doi: 10.33474/jp2m.v1i3.6510.

[10] B. Dwi M., Desmintari, and Yuhanijaya, "Pemberdayaan Masyarakat Desa Citeras Rangkasbitung Melalui Pengelolaan Sampah dengan Konsep Eco-enzyme dan Produk Kreatif Yang Bernilai Ekonomi," *Pros. Pengabdi. Kpd. Masy. 2017*, vol. 2, no. 1, pp. 1–6, 2017.

[11] M. Hemalatha and P. Visantini, "Potential use of eco-enzyme for the treatment of metal based effluent," *IOP Conf. Ser. Mater. Sci. Eng.*, vol. 716, no. 1, 2020, doi: 10.1088/1757-

899X/716/1/012016.

[12] A. Akhdiya, "Isolasi Bakteri Penghasil enzim protease alkalin termostabil," *Bul. Plasma Nutfah*, vol. 9, no. 2, pp. 38–44, 2003, [Online]. Available: http://indoplasma.or.id/publikasi/buletin_pn/pdf/buletin_pn_9_2_2003_38-44_alina.pdf.

[13] Y. Tong and B. Liu, "Test research of different material made garbage enzyme's effect to soil total nitrogen and organic matter," *IOP Conf. Ser. Earth Environ. Sci.*, vol. 510, no. 4, 2020, doi: 10.1088/1755-1315/510/4/042015.

[14] "17. Brady 1990.pdf." .

[15] G. Ruiz, D. Jeison, and R. Chamy, "Nitrification with high nitrite accumulation for the treatment of wastewater with high ammonia concentration," *Water Res.*, vol. 37, no. 6, pp. 1371–1377, 2003, doi: 10.1016/S0043-1354(02)00475-X.

[16] N.- Rochyani, R. L. Utpalasari, and I. Dahliana, "ANALISIS HASIL KONVERSI ECO ENZYME MENGGUNAKAN NENAS (Ananas comosus) DAN PEPAYA (Carica papaya L.)," *J. Redoks*, vol. 5, no. 2, p. 135, 2020, doi: 10.31851/redoks.v5i2.5060.

[17] F. Nazim, "Treatment of Synthetic Greywater Using 5% and 10% Garbage Enzyme Solution," *Bonfring Int. J. Ind. Eng. Manag. Sci.*, vol. 3, no. 4, pp. 111–117, 2013, doi: 10.9756/bijiems.4733.

[18] V. Deepak, A. N. Singh, and P. S. A.K, "Use of Garbage Enzyme," *Int. J. Sci. Resarch Rev.*, vol. 07, no. September No.07, pp. 210–205, 2019, [Online]. Available: https://www.researchgate.net/publication/335528212%0AUSE.

[19] V. C. Renge, S. V Khedkar, and N. R. Nandurkar, "Enzyme Synthesis By Fermentation Method : a Review," vol. 2, no. 6, pp. 585–590, 2012.

[20] M. F. Syahirah and R. Nazaitulshila, "The Utilization of Pineapples Waste Enzyme for the Improvement of Hydrolysis Solubility in Aquaculture Sludge," *J. Energy Saf. Technol.*, vol. 1, no. 2–2, pp. 35–41, 2019, doi: 10.11113/jest.v1n2-2.29.

[21] R. J. Seviour, T. Mino, and M. Onuki, "The microbiology of biological phosphorus removal in activated sludge systems," *FEMS Microbiol. Rev.*, vol. 27, no. 1, pp. 99–127, 2003, doi: 10.1016/S0168-6445(03)00021-4.

[22] F. Tang and C. Tong, "Garbage-Enzyme-University-Sarawak," *World Acad. Sci. Eng. Technol.*, vol. 60, pp. 1143–1148, 2011.

[23] C. Arun and P. Sivashanmugam, "Study on optimization of process parameters for enhancing the multi-hydrolytic enzyme activity in garbage enzyme produced from preconsumer organic waste," *Bioresour. Technol.*, vol. 226, pp. 200–210, 2017, doi: 10.1016/j.biortech.2016.12.029.

[24] M. Kumar, D. N. Rao, S. Mohanty, A. Selvi, and S. Bhoi, "International Journal of Advanced Research in Biological Sciences," *Int. J. Adv. Res. Biol.Sci*, vol. 2, no. 7, pp. 12–20, 2015, [Online]. Available: www.ijarbs.com.

Numerical modeling: the effect of land subsidence on hydrodynamic conditions and coastal morphology in Pekalongan

I F Muhammad[1], N Yuanita[2], E O Nugroho[3], F S Wijaya[1], R J Martin[1], Yola[1] and N Fauzi[1]

[1]Ocean Engineering Master Program, Faculty of Civil and Environmental Engineering, Bandung Institute of Technology, Bandung 40132, Indonesia
[2]Coastal Engineering Research Group, Faculty of Civil and Environmental Engineering, Bandung Institute of Technology, Bandung 40132, Indonesia
[3]Water Resources Engineering Research Group, Faculty of Civil and Environmental Engineering, Bandung Institute of Technology, Bandung 40132, Indonesia

iqbalfmuhammad@gmail.com

Abstract. This study examines the effect of land subsidence on hydrodynamic conditions and coastal morphology through numerical modeling using MIKE21 with 30-day simulation time. Land subsidence is assumed to be 10 cm/year, which affects bathymetry. There are 2 modeling scenarios, scenario 1 before (2016) and scenario 2 after land subsidence (2021). River discharge is assumed to be constant at 10 m^3/s. The model calibration result on water level elevation and current velocity of the model simulation output and field data comparison based on the variation of Chezy Number, the smallest error was obtained at Chezy Number 50 $m^{0.5}/s$. The sediment transport model sensitivity test was conducted on parameters of model type, sediment grain size, and river discharge. The result of sensitivity test demonstrated that the sediment grain size is the most sensitive parameter. By comparing the two model scenario results show that the current velocity in scenario 1 is greater than scenario 2 by 20%, significant wave height in scenario 1 is smaller than scenario 2 up to 0.1 m, and sedimentation in Mangrove Park increases up to ± 0.01m. Land subsidence reduces the current velocity, increases the significant wave height and sedimentation.

Keywords: Numerical model, hydrodynamics, sediment transport, land subsidence

1. Introduction

In the past 10 years, Pekalongan is one of the cities in Northern coast of Java that are suffering from tidal flood due to the combine effect of sea level rise and land subsidence [1]. As a result of the land subsidence, Pekalongan coastline experienced severe coastal erosion, which made the coastline closer to the urban areas and sea water is able to flow further into residential areas [2]. This combination contributes to the tidal flood in Pekalongan [3]. There are economic, social, and infrastructure damage due to the flood. Local authorities have tried to slow down and stop the coastal erosion by building revetments made of concrete, stones, and geobags start from 2014, but it is discovered that some of the revetments are heavily damaged and could not protect the coast anymore [4]. It is predicted that permanently flooded areas will be larger as a result of sea level rise and land subsidence if there are no solution [5].

The objective of this study is to examine the effect of land subsidence on hydrodynamic conditions and coastal morphology through numerical modeling using MIKE21 with 30-day simulation time through Hydrodynamic (HD) [6], Spectral Wave (SW) [7], and Sand Transport (ST) [8] modules. The effect of land subsidence is determined by comparing two modeling scenarios, scenario 1 before land subsidence (2016) and scenario 2 after land subsidence (2021). Numerical modeling uses primary and secondary data, the primary data are bathymetry, tides, wave data, and sediment samplings which collected by field survey while the secondary data are wind, current, and wave data from open data.

2. Methodology

The study location is in the Pekalongan coastal area, northern part of Java Island where the main area is the Mangrove Park shown in **Figure 1**.

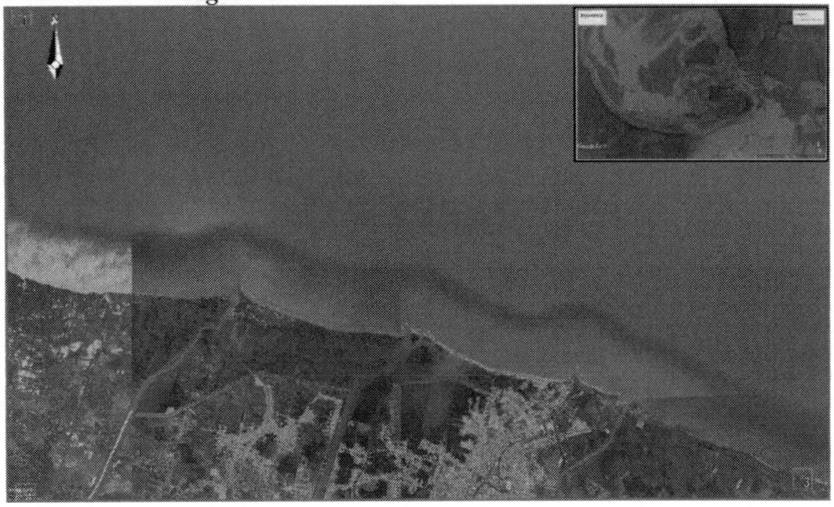

Figure 1. Study location.

The numerical modeling is performed by systematic sequences starting form data collecting, model preparation, and model validation to result analysis which will answer the study objectives [9]. The study work flow is shown in **Figure 2**.

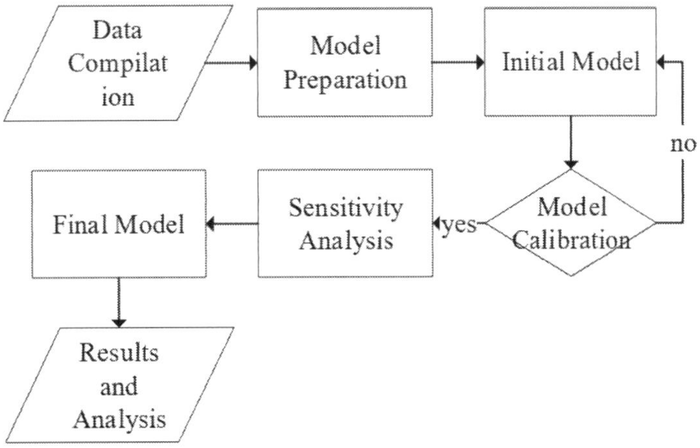

Figure 2. Study work flow.

The numerical model is set up several times before obtaining the reliable model. The modeling stages are initial model, calibrated model, sensitivity analyzed model, and the final model. The initial model is the rough simulation using the collected data, the output of this model is the general pattern of current circulation and surface water elevation. The calibrated model is the model that performed with several Chezy Number which is compared to obtain the smallest simulation error based on the field data. The sensitivity analyzed model is the model that performed with different parameters value which is model type, sediment grain size, and river discharge to obtain the most sensitive parameter. The final model is composed from the previous model that will be used in determining the effect of land subsidence on hydrodynamic condition and coastal morphology.

2.1. Data Compilation

There are primary and secondary environmental data used in the model. The primary data was obtained from field survey by Faculty of Civil and Environmental Engineering (FTSL ITB) team in 2021 that includes tides, bathymetry, waves, current, sediment samples, and Total Suspended Solid (TSS). The secondary data was obtain from open data that includes wind, shoreline, bathymetry, wave, and current data. The primary data area is shown in **Figure 3**.

Figure 3. Field survey, observation points, and cross section location.

2.1.1. Tide

Tide measurement was conducted in 6th to 10th April 2021 hourly at four location namely MA02, MA07, MA10, and MA12. The average tidal range in four locations is 0.75 m, the Formzahl number is 1.104 which indicated that tidal type in Pekalongan is semi diurnal.

In addition to tidal measurement, tidal data was also obtained from Indonesian Geospatial Information Agency (BIG) that is used as tidal boundary [10] in three different areas which is west, north, and east with longer duration (30 days).

2.1.2. Current

Current measurement was conducted using Argonaut ADCP in 6th to 9th April 2021 with ten minutes interval on five depths from surface to bottom. Current measurement was conducted with wave observation concurrently. Current velocity at surface has the most value and it gradually decrease toward bottom ranges from 0 m/s to 0.7 m/s.

2.1.3. Wave

Wave observation was conducted using Argonaut ADCP in 6th to 9th April 2021 with ten minutes interval. Wave observation was conducted with current measurement concurrently. Based on the collected data, the wave height is around 0.4 m to 1.3 m with 3.5 s to 7 s and the significant wave height is 0.89 m with 5.57 seconds period. There were also secondary wave data from ERA5 mainly used for scenario 1 model input.

2.1.4. Bathymetry

Bathymetric survey was conducted in 6th to 9th April 2021 from Mangrove Park to nine meters depth with Lowest Water Spring (LWS) as vertical datum. To accomodate deeper water in model domain,

GEBCO data was used to complete the bathymetry from ten meters to cover the whole model domain, the GEBCO data first refers to LWS before use as input.

2.1.5. Sediment Samples and TSS

The sediment samples was analyzed in laboratory to obtain its grain size. Based on the laboratory testing, the type of sediment in Pekalongan coastal area was non-cohesive sediment mainly sand with 0.187 mm d50. There were also TSS analysis at MA02, MA07, MA08, MA10, and MA12 with MA08 had the highest concentration at 407.5 ppm and MA07 had the lowest concentration at 35.9 ppm.

3. Numerical Model

There are two modeling scenarios, scenario 1 before land subsidence (2016) and scenario 2 after land subsidence (2021) both used MIKE 21 software with tide and wind as main input. Both models use Hydrodynamic (HD), Spectral Wave (SW), and Sand Transport (ST) modules. Model calibration is performed by comparing the water level elevation and current velocity of the model simulation output with field data based on the five different Chezy Number through Root Mean Square Error (RMSE) methods.

3.1. Model Description

There are three open boundaries that use tide as boundary condition, there are Eastern, Northern, and Western as shown in **Figure 4**. There are at least 10 points tide input in each boundary condition generated from BIG tidal model. There is also land boundary in the Southern part, as well as four river boundaries in which its discharge is assumed to be constant at 10 m^3/s. In scenario 1, the land subsidence rate is assumed to be 10 cm/year in which the bathymetry in scenario 1 is 0.5 m shallower than in scenario 2 indicating that there is 0.5 m land subsidence from 2016 to 2021 (5 years). Both models use 30-days simulation time with the sediment transport model was generated by wave and current interaction where the former was obtained from SW module while the latter was from HD module.

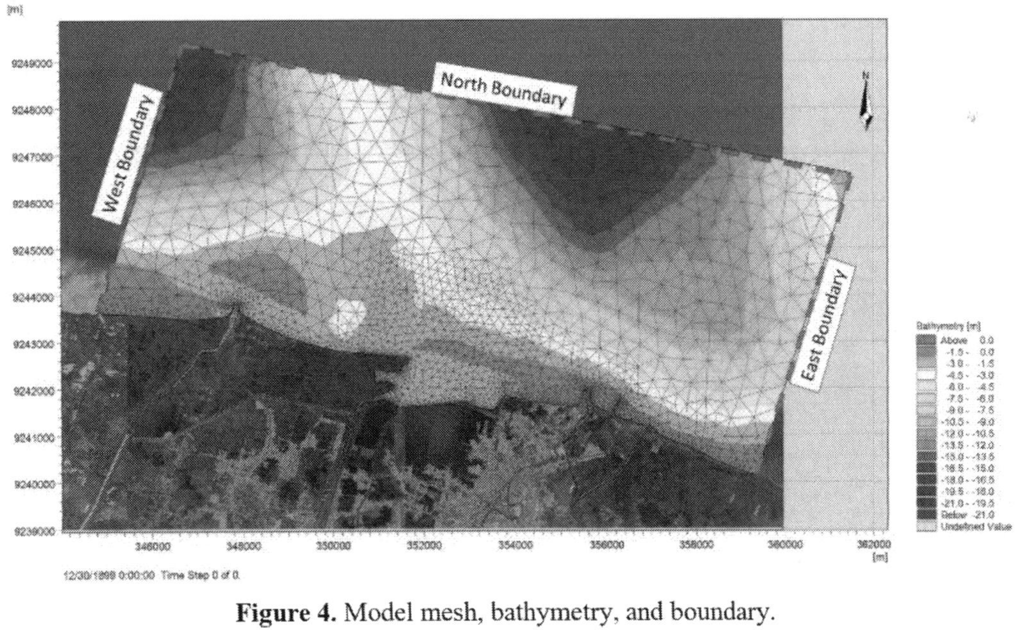

Figure 4. Model mesh, bathymetry, and boundary.

3.2. Model Calibration

Model calibration is performed by comparing the model results (water surface elevation and current velocity) with the collected field data and calculating its error value using RMSE method with variations of 5 different Chezy Number namely 50 $m^{0.5}/s$, 55 $m^{0.5}/s$, 60 $m^{0.5}/s$, 65 $m^{0.5}/s$, and 70 $m^{0.5}/s$ as bottom roughness. By comparing the water surface elevation between the model and field data, it was found that the 50 $m^{0.5}/s$ Chezy Number has the smallest RMSE average value of 9.57% as shown in **Table 1**.

Table 1. RMSE water level comparison.

Chezy Number ($m^{0.5}/s$)	RMSE WATER LEVEL (%)				
	MA02	MA07	MA10	MA12	Average
50	**7.392**	**14.825**	**10.905**	**5.151**	**9.568**
55	7.447	14.858	10.908	5.148	9.591
60	7.489	14.887	10.912	5.147	9.610
65	7.523	14.914	10.916	5.147	9.625
70	7.561	14.941	10.927	5.147	9.644

In the current velocity case, Chezy Number 50 $m^{0.5}/s$ also has the smallest RMSE average value of 10.43% as shown in **Table 2**. With Chezy Number 50 $m^{0.5}/s$ has the smallest RMSE average value in both parameters, this Chezy Number will be used in further models.

Table 2. RMSE current velocity comparison.

Chezy Number ($m^{0.5}/s$)	RMSE Current Velocity
50	**10.434**
55	10.439
60	10.517
65	10.702
70	11.075

3.3. Sensitivity Analysis

The ST model sensitivity test was conducted on parameters of model type, sediment grain size, and river discharge and their effect on the TSS concentration and bed level change [11]. For model type there are pure current and wave-current interaction model type. The sensitivity analysis was carried in order to obtain the most sensitive parameter in ST model. There are five locations for sensitivity analysis that is shown in **Figure 5**, namely CR1, CR3, CR7, ST1, and ST2, while the sensitivity analysis scenario can be seen in **Table 3.**

Figure 5. Observation points location.

Table 3. Sensitivity test scenarios.

Sensitivity Test Scheme			
Scheme No.	**Discharge (m^3/s)**	**Model Type**	**Sediment Grain Size (mm)**
1	10	*current-wave*	0.187
2	10	*pure current*	0.187
3	10	*pure current*	0.0025
4	10	*current-wave*	0.0025
5	20	*pure current*	0.187
6	20	*pure current*	0.0025

In the sediment transport model, sediment grain size has the most significant effect than the two other parameters based on the TSS concentration and the bed level changes. For the discharge parameter sensitivity test, the comparison is on scheme 2, 5, and 6 (**Figure 6**), it is shown that the greater the discharge, the more it influenced sediment transport.

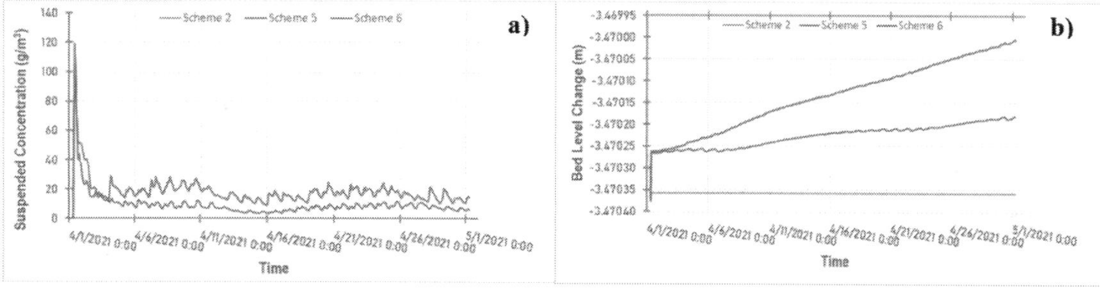

Figure 6. Discharge amplification sensitivity test on **a)** suspended concentration and **b)** bed level change at CR3 with scheme 2, 5, and 6.

For closed area, in this case in the Mangrove Park, model type wave-current interaction is the most significant parameter than river discharge and sediment grain size parameter both in TSS concentration and bed level changes **Figure 7**.

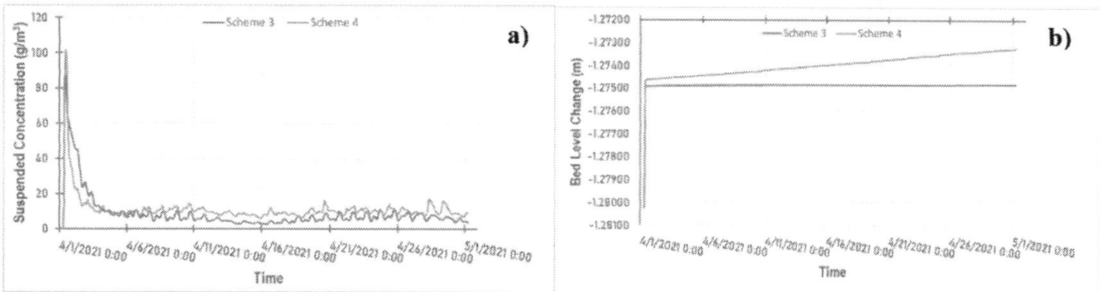

Figure 7. Model type sensitivity test on **a)** suspended concentration and **b)** bed level change at CR7 with scheme 3 and 4.

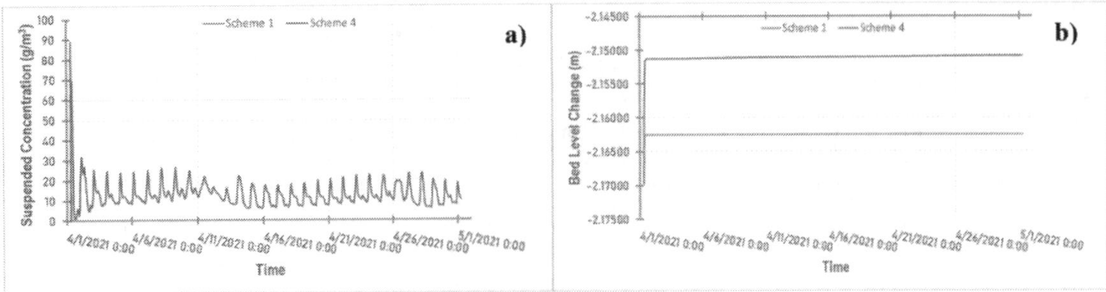

Figure 8. Sedimen grain size amplification sensitivity test on **a)** suspended concentration and **b)** bed level change at CR1 with scheme 1 and 4.

In all sensitivity analysis model, the results shown that the most sensitive parameter is sedimen grain size (**Figure 8**) for the sediment transport models.

4. Model Result and Analysis

The effect of land subsidence on hydrodynamic conditions and coastal morphology [12] could be studied through numerical modeling by comparing the simulation before and after land subsidence. In order to get the more detail comparison, there are cross section comparison spatially and temporally. The tidal condition in scenario 1 and scenario 2 is different, thus there is no neap and spring comparison.

For both models, it is assumed that the river discharge is constant at 10 m^3/s, the land subsidence is 0.1 m/year, 0.5 m for five years. The scenario 1 hydrodynamic model result taken on 14th April 2016 in maximum current can be seen in **Figure 9**.

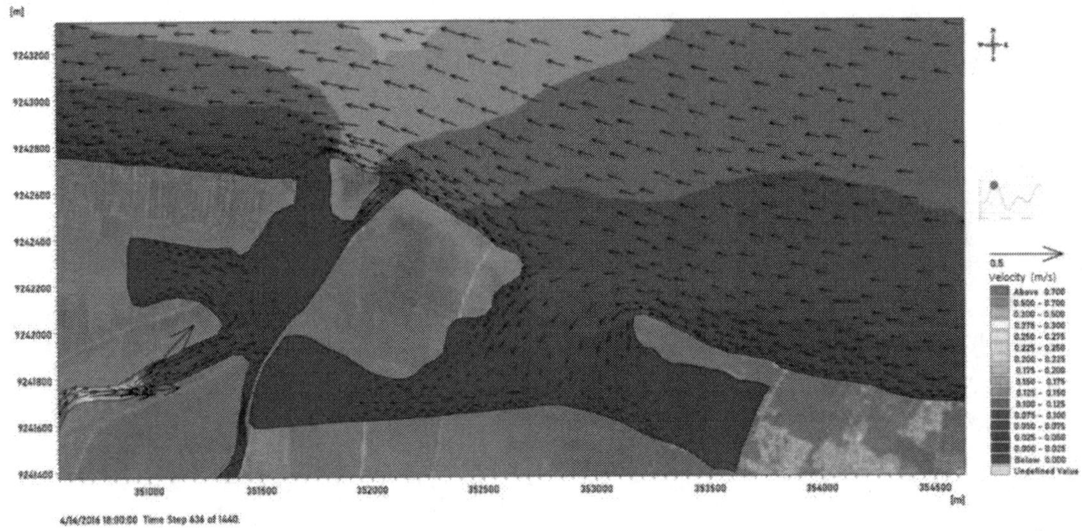

Figure 9. Scenario 1 maximum current velocity at 14th April 2016, 06:00 PM.

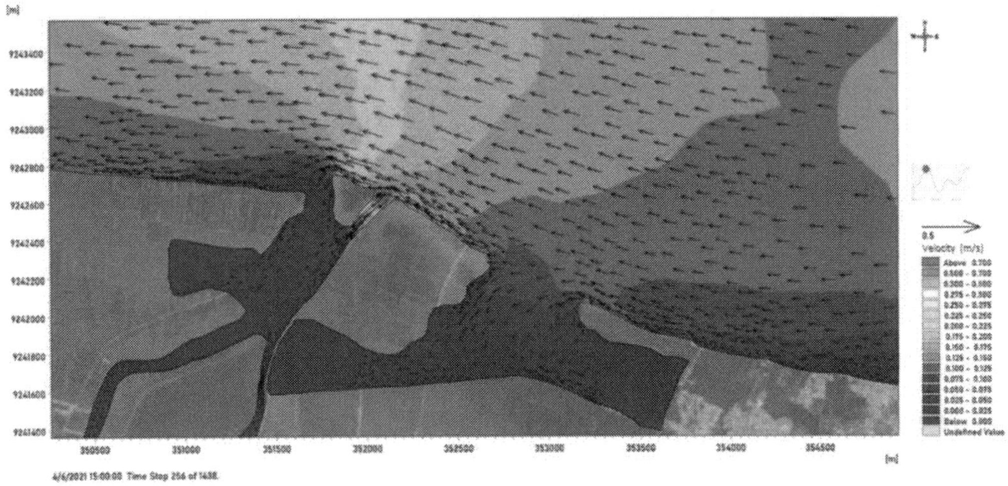

Figure 10. Scenario 2 maximum current velocity at 6th April 2021, 03:00 PM.

The scenario 2 hydrodynamic model result taken at 6 April[th] 2021 in maximum current can be seen in **Figure 10**. There are no significant differences between the two model overall, but in the Mangrove park area, where the water depth is 0.5 m deeper, the current velocity is lower as we can see that the purple color is spread wider indicating that is the area with low velocity. It is known that the deeper the water, the lower the surface current velocity will be.

4.1. Current Velocity Comparison

Current velocity comparison between scenario 1 and scenario 2 is taken at long cross section CR1, CR3, CR5, and cross section ST1 and ST2 at high tide (flood). In the long cross section case, the comparison is based on time series (**Figure 11**) and for the cross section it based on distance (**Figure 12**).

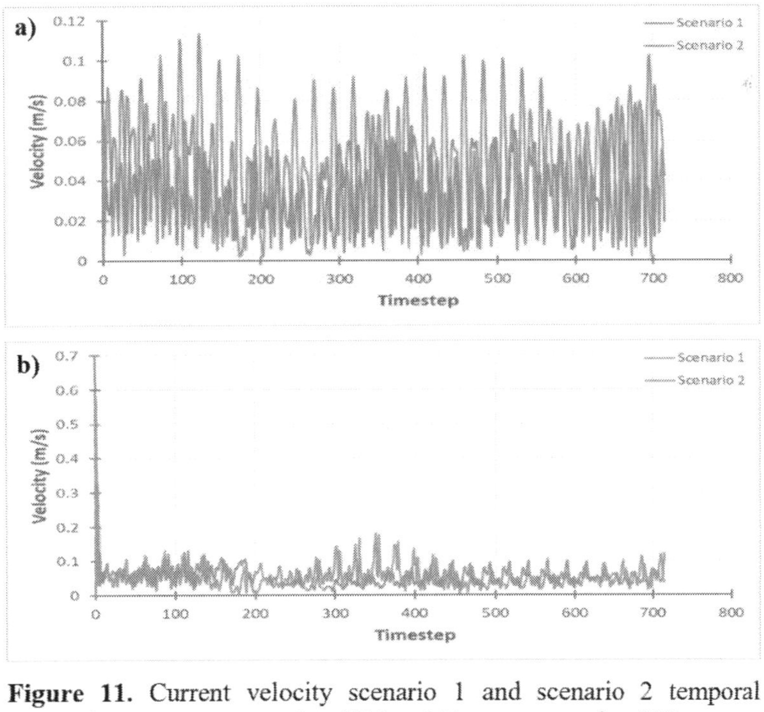

Figure 11. Current velocity scenario 1 and scenario 2 temporal comparison **a)** at cross section CR3 and **b)** at cross section ST1.

Current velocity in scenario 1 is greater than scenario 2 both temporally and spatially due to the water depth as it is stated that water depth in scenario 1 is shallower than scenario 2 hence the surface velocity will be greater. In this case the land subsidence affects the current velocity due to it deepen the water depth, the deeper the depth the slower the current will be.

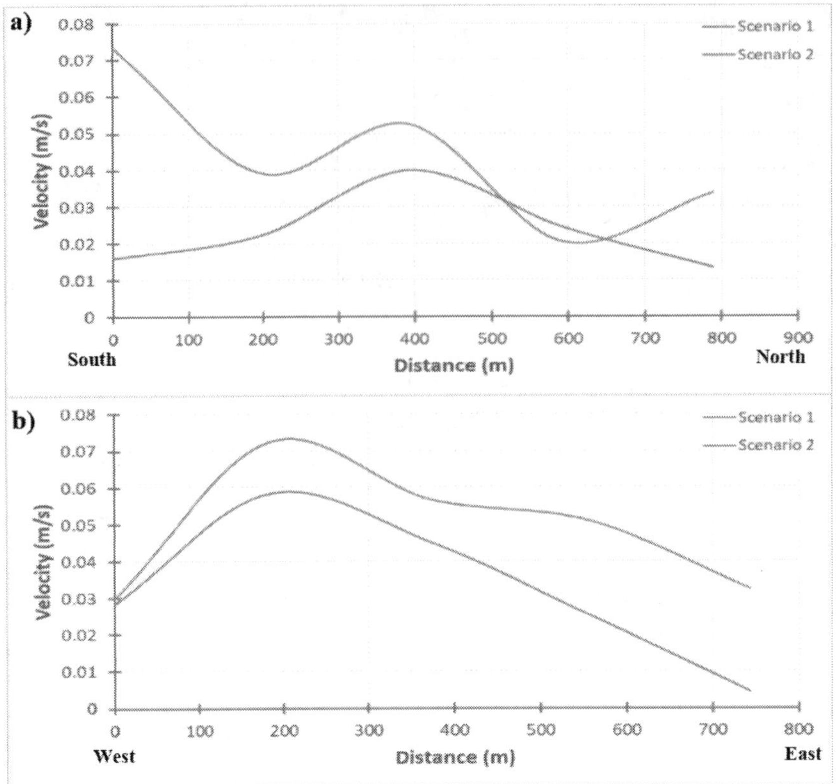

Figure 12. Current velocity scenario 1 and scenario 2 spatial comparison **a)** at cross section CR3 and **b)** at cross section ST2

4.2. Significant Wave Height Comparison

As it stated that land subsidence made the water getting deeper, different with the current case, in significant wave height the deeper the water the more it propagated will be. Due to the waves propagate further, its significant wave height will also greater. The significant wave height in scenario 1 is lower than scenario 2 due to the wave propagating the water depth. The difference range is around 0.05 m to 0.2 m at the five locations. It can be seen that the significant wave height in scenario 2 is greater than scenario 1 both spatially (**Figure 13**) and temporally (**Figure 14**).

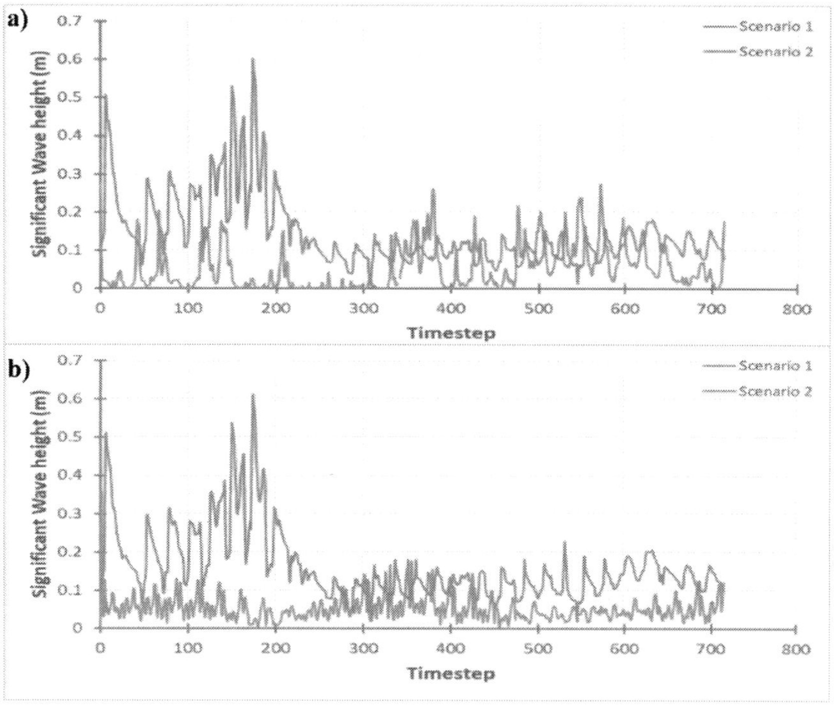

Figure 13. Significant wave height scenario 1 and scenario 2 temporal comparison **a)** at cross section CR3 and **b)** at cross section ST1.

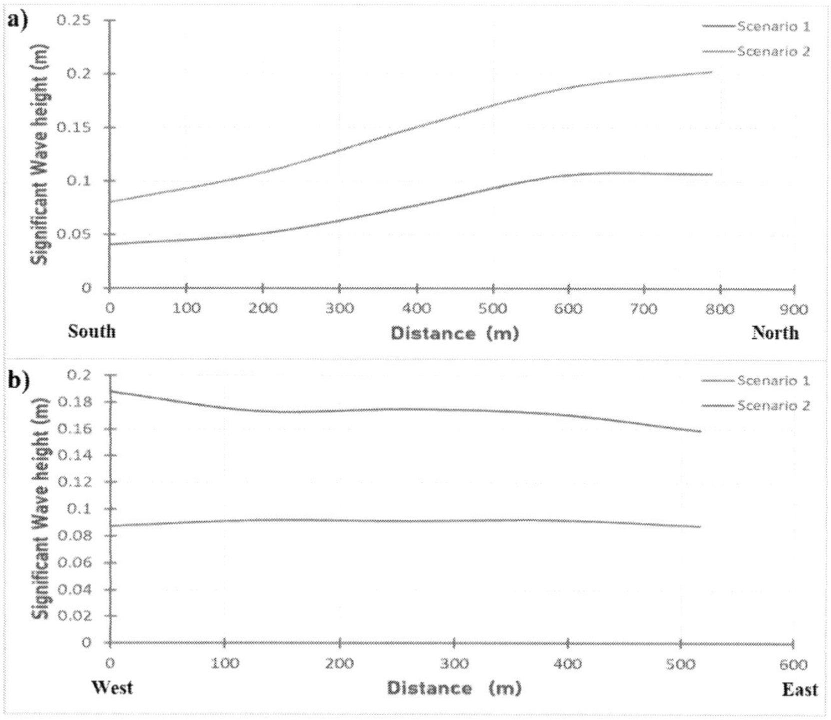

Figure 14. Significant wave height scenario 1 and scenario 2 spatial comparison **a)** at cross section CR3 and **b)** at cross section ST1.

It is not just the significant wave height that greater, the average wave height as well due to the deeper water depth. The greater the wave height, the more likely it carried the sediment transport, especially in an enclosed area as it was identified in the sensitivity analysis that greater wave height has more influence in closed area sediment transport. Wave height different between scenario 1 and scenario 2 is about 0.005 to 0.01 meters.

4.3. Bed Level Changes Comparison

Bed level changes could be positive (+) that indicate sedimentation. It also could be negative (-) which indicate erosion. Mangrove Park is a generally closed area where the current effect on sediment transport is not significant while the wave is. There is sedimentation at Mangrove Park area, both in scenario 1 and scenario 2, but the sedimentation in scenario 2 is greater than scenario 1 up to 10 fold. (**Figure 15**).

Sediment transport in an enclosed area is greatly affected by wave rather than current hence sedimentation is greater than erosion due to sediment cumulation. The sediment carried by incoming waves to Mangrove Park will be trapped and not brought back to the sea by current due to the influence of current is smaller than waves in sediment transport. In scenario 2, the significant wave is greater and so is the sedimentation. The sedimentation in scenario 1 was around 0.001 m and in the scenario 2 it was 0.01 m, almost ten folding increment. The increment is due to the increasing significant wave height after land subsidence case that carried the sediment toward the Mangrove Park area. The carried sediment could not be carried back toward open water by current due to lower current velocity, hence the sediment was trapped and it increases the sedimentation.

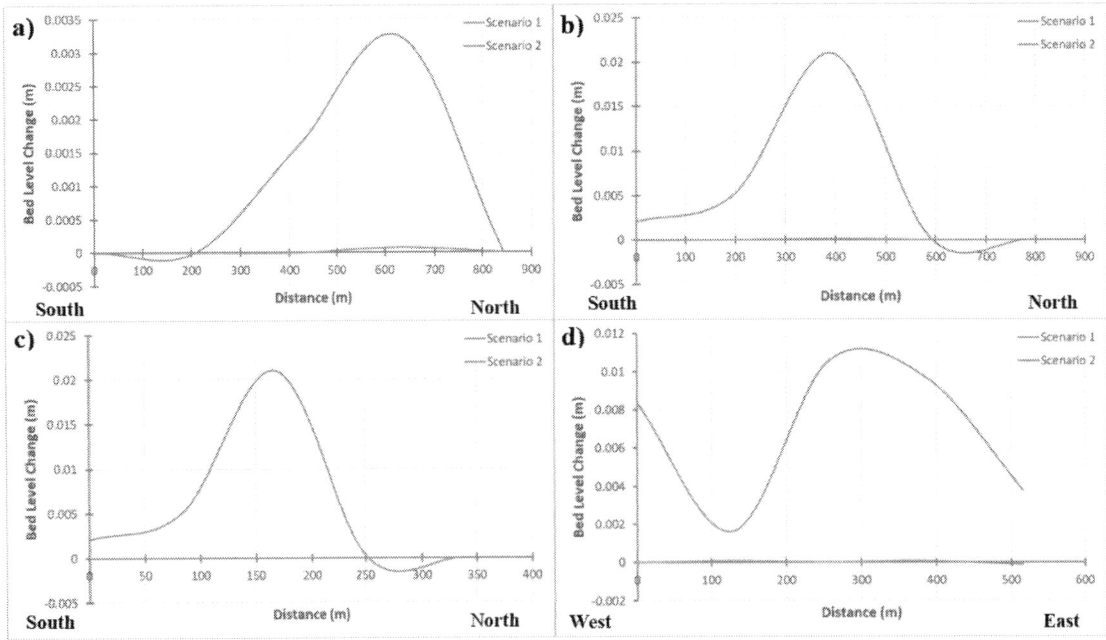

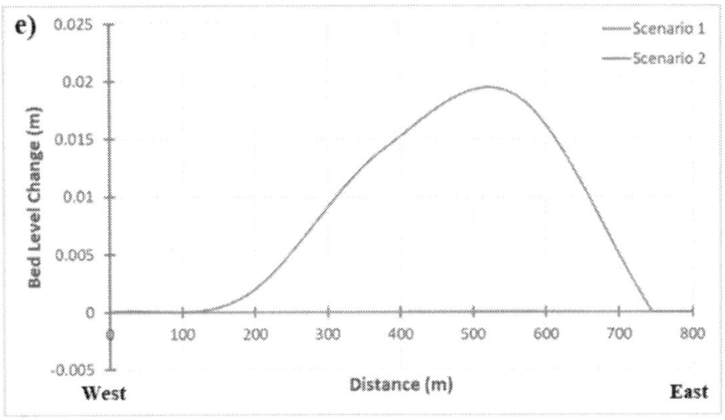

Figure 15. Bed level change scenario 1 and scenario 2 comparison at cross section **a)** CR1, **b)** CR3, **c)** CR7, **d)** ST1, and **e)** ST2 observation points.

5. Conclusion

The effect of land subsidence on hydrodynamic conditions and coastal morphology could be studied through numerical modeling by comparing the simulation before and after land subsidence using DHI MIKE 21 software with Hydrodynamic, Spectral Wave, and Sand Transport modules. There are primary and secondary data used in the model with the former was collected by FTSL ITB team in April 2021 including bathymetry, tide, current, and wave data, while the latter is collected from open source data like BIG forecast for tide, ERA5 for wind and waves, and TIPS NOVELTIS for current.

Model calibration is performed by comparing the water level elevation and current velocity of the model simulation result with field data based on five different Chezy Number, 50 $m^{0.5}/s$, 55 $m^{0.5}/s$, 60 $m^{0.5}/s$, 65 $m^{0.5}/s$, and 70 $m^{0.5}/s$, through Root Mean Square Error (RMSE) methods. The sediment transport in Mangrove Park, Pekalongan, is greatly influenced by wave, especially in an enclosed area as it was identified in sensitivity analysis based on the model type parameter between pure current and current – wave interaction sediment transport. It is also concluded that sediment grain size is the most sensitive parameter in the sediment transport model.

Land subsidence made the water depth deeper, thus will affects the other parameter. It is known that the deeper the water, the lower surface current velocity will be, with land subsidence the current velocity in scenario 2 is lower than scenario 1 with around 0.05 m/s to 0.1 m/s disparity. The lower the velocity the more unlikely it could transport the sediment.

The land subsidence also affects the wave height. It is identified that wave height in scenario 2 is greater than scenario 1 due to the wave could propagate further because of the water is deeper. The wave increase is around 0.005 to 0.01 meters. The greater the wave height, the more sediment transport it carried. Thus, in the scenario 2, the sedimentation is greater than scenario 1 the differences are up to 0.01 meter or ten folding increment. The increment in sedimentation is not just due to wave, but the decrease in current velocity as well.

Land subsidence cause changes in hydrodynamic condition and coastal morphology. It causes the water deeper, the surface current velocity lower, the greater wave height due to the waves propagates further, and sedimentation increase in Mangrove Park or enclosed area. It is predicted that the Pekalongan flood inundation area will be larger [13] and its damage will be more frightening if there is no solution applied either from local authorities or others [14].

Acknowledgments

This study was funded by Indonesia Collaboration Research Program grant from Bandung Institute of Technology. The MIKE 21 licensed was provided by Faculty of Civil and Environmental Engineering, Bandung Institute of Technology.

References

[1] E Chaussard, F Amelung, H Abidin, and S-H Hong. 2013. *Sinking cities in Indonesia: ALOS PALSAR detects rapid subsidence due to groundwater and gas extraction*. Remote Sensing of Environment, **128** 150-161.

[2] M A Marfai. 2014. *Impact of sea level rise to coastal ecology: a case study on the northern part of Java Island, Indonesia*. Quasestiones Geographicae **33(1)** 107 – 114.

[3] S Nashrrullah, Aprijanto, J M Pasaribu, M K Hazarika, and L Samarakoon. 2013. S*tudy on flood inundation in Pekalongan, Central Java*. International Journal of Remote Sensing and Earth Sciences **10(2)** 76-83.

[4] T de Bruijn, and T de Waele. 2016. *Water system research Pekalongan. Mapping and analysing the water infrastructures and drainage systems in Tirto and Kota Pekalongan.* Rotterdam: Waterboard Schieland and the Krimpeneerwaard, 47.

[5] M B Pratama. 2019. *Tidal flood in Pekalongan: Utilizing and operating open resources for modelling*. IOP Conf. Ser.: Mater. Sci. Eng. **676** 012029

[6] DHI, Water & Environment. 2014. *MIKE 21 FM Hydrodynamic: scientific documentation. New York.*

[7] DHI, Water & Environment. 2014. *MIKE 21 Spectral wave module: scientific documentation. New York.*

[8] DHI, Water & Environment. 2014. *MIKE 21 FM Sand transport: scientific documentation. New York.*

[9] H Bachtiar, H E Setiantoro, A Prasetyo, L Eliasta, and A E Putriasri. 2020. *Pendekatan model komputasi untuk penanggulangan sedimentasi pelabuhan akibat pengaruh muara sungai: studi kasus pelabuhan tanjung mas Semarang*. Jurnal Teknik Hidraulik **11(2)** 119-134.

[10] M Mangopo. 2019. *Pemodelan numerik hidrodinamika dan transportasi sedimen paska kegiatan reklamasi di perairan pltu Suralaya.* Tesis Program Magister, Bandung: Institut Teknologi Bandung.

[11] G I Prabowo. 2019. *Studi transportasi sedimen area pltu 1 NTT Ropa, Kabupaten Ende.* Tesis Program Magister, Bandung: Institut Teknologi Bandung.

[12] N Sharmila, R Venkatachalapathy, and M Mugilarasan. 2018. *Studies on the morphological changes by numerical modeling along Kakinada Coasts*. Proceedings of the Fourth International Conference in Ocean Engineering (ICOE2018) **23**.

[13] D N Febriani, M Helmi, and Hariyadi. 2017. *Kajian genangan banjir Pekalongan di Kecamatan Pekalongan Utara, Kota Pekalongan.* Jurnal Oseanografi **6(4)** 579-587.

[14] D Sarah and E Soebowo. 2018. *Land subsidence threats and its management in The North Coast of Java.* IOP Conf. Ser.: Earth Environ. Sci. **118** 012042.

Constructed wetlands with Cyperus alternifolius as a sustainable solution for household greywater treatment

S Qomariyah[1], **B Utomo**[1] **and A H Wahyudi**[1]

[1]Department of Civil Engineering, Faculty of Engineering, Sebelas Maret University, Jalan Ir. Sutami 36A, Surakarta, Indonesia

E-mail: sitiqomariyah@staff.uns.ac.id

Abstract. Compared to the conventional wastewater treatment plants, constructed wetlands (CWs) are low-cost and affordable green technologies for treating many types of wastewater. This study aimed to evaluate the sustainability of an old-single CW, after four years of operation, and compare its performance with a new-smaller CW, using *Cyperus alternifolius* plant and applying Hydraulic Retention Time (HRT) for one day. Two artificial wetlands, namely CW-B and CW-S, were used with dimensions (length x width x height) of 170 x 70 x 70cm and 71 x 42.5 x 52.5cm, respectively. The first operation of the CW-B was in September 2015 and the CW-S was in April 2019. Several concentrations of household greywater influent were loaded into the CWs to analyze their performance. Two aeration pipes were installed in the wetland system to increase the oxygen level in the systems. Water quality parameters namely Biological Oxygen Demand (BOD), Total Suspended Solid (TSS), detergent, and Dissolved Oxygen (DO) were laboratory tested before and after treatment. Research carried out during April to October 2019 resulted in the mean removal efficiencies for CW-B and CW-S being 94.13% and 96.84% for BOD; 95.04% and 95.62% for TSS; 97.11% and 94.61% for detergent; and 99.5% and 79.7% for DO increase, respectively. These findings indicate that the removal efficiencies of those parameters in both wetlands were not significantly different. Hence, the system of the old CW-B wetland still keeps a good working process. Referring to the national standards of water quality, the greywater effluents were safely discharged into water bodies and could be used for irrigation needs. A further study needs to be developed to maximize the detergent removal to meet the existing standard for hygiene sanitation. These results suggest that CWs are becoming a sustainable solution for the treatment of greywater at the household level and finally contributing to minimizing the pollution of water.

Keywords: constructed wetland, cyperus alternifolius, greywater, household

1. Introduction

Water pollution is becoming a serious concern due to the large population and the worsening environmentally unfriendly activity [1]. Globally, 80% of all wastewater is discharged into the environment without treatment [2]. In Indonesia, in larger cities with a high population density, about two-thirds of the raw water supply comes from polluted surface water [3]. Rivers flowing in and around the large cities in Indonesia contain 70% of organic loading originating from domestic wastewater, in which most of which are in the type of greywater [4]. Greywater, defined as wastewater without any input from toilet water, generally includes water originating from sources such as kitchen sinks and dishwashers, hand washbasins, baths, showers, and washing machines, and accounts for 50 - 80% of the total volume of household wastewater [5]. The greywater is commonly discharged into the drainage system without any treatment. As households regularly dispose of greywater, it has become a significant contributor to water pollution.

Several conventional centralized wastewater treatment technologies have been developed and successfully applied in many regions to regulate water pollution [6, 7]. However, the suitability of the conventional wastewater treatment plants for developing countries was questionable because the costs for implementing the plants are exorbitant. They need high costs for construction and operation, require vast amounts of energy and trained personnel to operate and maintain them, and have large mechanical parts made of non-renewable materials (e.g., concrete and steel) [8, 10]. In Indonesia, 13 conventional centralized wastewater treatment plants have been built in 13 cities, but only 10% of the Indonesian population has been served by the systems [11]. In this context, it is hard to think about cost-effective, easy-to-operate, and sustainable solutions to the present day's water pollution crisis.

Wastewater treatment facilities in developing countries can only be accessed under the criteria of low cost, applicable technology, and natural water cycle [1]. Man-made wetlands or construction wetlands (CWs) are such systems of wastewater treatment that meet those criteria [7, 12]. Constructed wetland technology is a flexible system for improving water quality and can be developed in a decentralized manner, namely in individual households or communally from several houses [13, 14]. Compared to the conventional wastewater treatment systems, constructed wetlands (CWs) offer an environment-friendly green approach for managing various types of wastewater [7, 15]. They are cost-effective and energy-efficient in the process of wastewater treatment [16]. Hence, they have a high potential for being applied in developing countries or regions with limited funds.

CWs are a type of wastewater treatment technology based on nature. They are designed and constructed to utilize the natural purification processes involving wetland plants, substrates, and the associated microbes [1]. According to the water flow path through the system, the CWs can be categorized into two different CWs, i.e., free water surface and subsurface flow CWs, in which for the subsurface flow CWs, they can be created using either vertical or horizontal flow [6, 17]. Initially, CWs were applied to treat municipal wastewaters only. In the last few decades, the CWs have been widely used to treat several types of wastewater such as domestic sewage, industrial effluent, agricultural wastewater, landfill leachate, polluted river water, urban runoff, pharmaceutical wastewater, heavy metal contaminated water, aquaculture wastewaters, sludge treatment effluent, and winery wastewaters [13, 15]. However, there are limited studies on the successful application and sustainable operation of these systems [14]. The limitation of acquiring such information represents a significant knowledge gap in sustainable operation. This paper presents the sustainable operation of a single CW after four years of operation.

Many studies have demonstrated that wetland plants, directly or indirectly, contribute to the process of wastewater treatment in CWs [18, 19]. Plants are important elements in natural and artificial systems. Plants in CWs mainly release oxygen through the roots into the root zone. As a consequence of this, an area around the roots, called the rhizosphere, facilitates the growth and survival of diverse microorganisms which play an important role in the degradation of organic pollutants [13, 20]. The presence of these plants often defines wetlands as green technology. There are numerous aquatic plant species, but it was recommended that locally available aquatic plant species need to be chosen [7, 14]. A previous study found that among the existing plants in the study area, *Cyperus alternifolius* was the most suitable for use in CWs for greywater treatment in tropical areas [21]. Because of that, the *Cyperus alternifolius* is used in this study.

HRT is the time given to wastewater residing CWs. The effect of HRT may differ between CWs depending on the dominant plant species and temperature [22]. Even though much research has been undertaken using different HRTs [7], limited information is available on using an HRT of one day. The limitation of getting such information represents a significant knowledge gap in one-day treatment. As households regularly discharge greywater each day, this study applied the HRT for one day. Accordingly, the objective of the present study was to evaluate the sustainability of an old-single CW, after four years of operation, and compare its performance with a new-smaller CW, using *Cyperus alternifolius* and applying HRT for one day.

2. Experimental methods

2.1. Descriptions of wetland models

The old-single CW was constructed in July 2015 and first operated in Sep 2015. The CW was made of polyvinyl chloride (PVC), having a length of 170 cm, a width of 70 cm, and a depth of 70 cm. The new-single CW was an identical plastic tank, built-in March 2019, with the dimensions of 71 cm long, 42.4 cm wide, and 52.5 cm deep. The dimensions of the old CW are bigger than the new one. Therefore, this paper uses a term CW-B and CW-S, representing the big-old and the small-new CW. The type of both CWs was horizontal subsurface flow. The total working volume-area of the CWs were 833 litres-1900 cm^2 and 158.42 liters-3017.5 cm^2 for the CW-B and the CW-S, respectively. The slope of both CWs was 1%, needed to obtain regular flow. Two pipes 20 mm in diameter were installed for aeration purposes. It aimed to find higher removal efficiency of organic pollutants as confirmed by [23] who improved bed aeration and observed high efficiency of organic removal in their wetland systems. The CWs were located outdoors and the influent was greywater from a single household in Surakarta City, Central Java, Indonesia. The location is categorized as a humid tropical climate with temperatures ranging from 15.1 and 33.5°C, and humidity varied between 69 and 87% on average. Figure 1 illustrates the CWs employed.

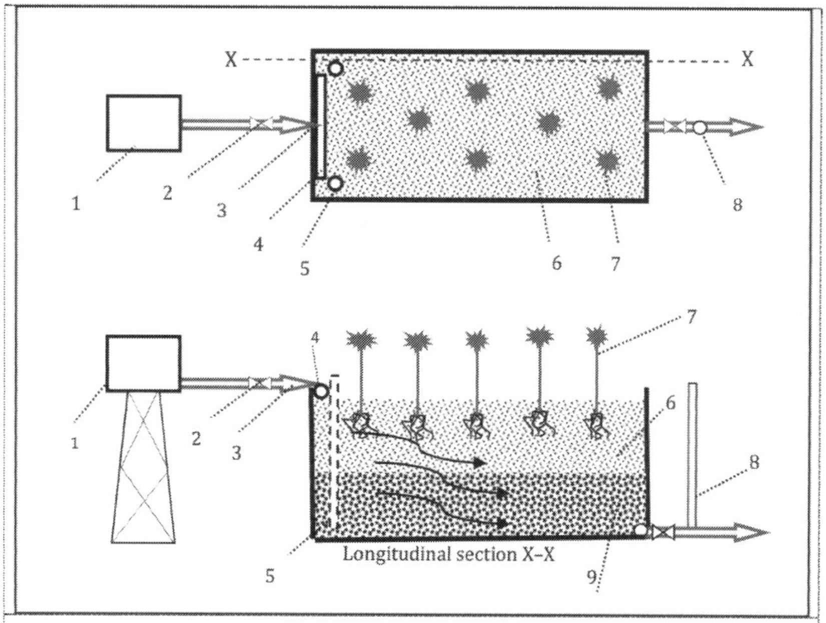

Figure 1. Schematic layout of horizontal subsurface flow CW system: 1. Greywater tank; 2. Valve; 3. Inlet pipe; 4. Feeding pipe; 5. Aeration pipe; 6. Fine sand layer (1 - 5 mm diameter); 7. Plants; 8. Adjusting pipe; 9. Coarse sand layer (5 - 10 mm diameter)

2.2. Substrate

Substrates, as media filters, can eliminate contaminants from wastewater by exchange, adsorption, precipitation together with complex formation [1]. The media filter supports the growth of wetland plants and microorganisms, enhancing biodegradation of wastewater pollution [15]. Gravel and sand, known as conventional-natural components of the substrate, were used in this study. The CWs were filled with two different layers of a washed substrate of river sand-gravel of which the top layer was 1 – 5 mm and the bottom layer was 5 – 10 mm in size. The top and the bottom layers' depth was 30 cm and 20 cm, respectively. Thus, the total depth of the substrate was 50 cm. Before placing the substrate into the CWs, they were cleaned first to remove any original soil from its sources. The porosity of the substrate was about 35%. The water depth in both CWs was maintained at 2 cm below the surface.

2.3. *Hydraulic retention time (HRT)*

This study applied HRT for 1 day as households practically discharge each day. In order to characterize the influent and a steady-state system, the following parameters were calculated [24].

$$HRT = \frac{\rho V}{Q} \tag{1}$$

$$HLR = \frac{Q}{A} \tag{2}$$

where *HRT* is hydraulic retention time (d), *HLR* is hydraulic loading rate (m d^{-1}), V is the volume of the media (m^3), Q is the greywater flow rate (m^3 d^{-1}), A is the area of the bed (m^2), ρ is the porosity of the filter media. *HRT* of 1 day and the porosity value of 0.35 are respectively to the Q and *HLR* of 199.92 liter d^{-1} and 168 mm d^{-1} for the CW-B; and 50.70 liter d^{-1} and 168 mm d^{-1} for the CW-S.

2.4. *Aquatic plants*

Plants are responsible for transferring oxygen from their roots to the rhizosphere, providing aerobic conditions to enhance the contaminant degradation in the system. The plants can also absorb pollutants from the wastewater and accumulate them in their tissue and provide microorganisms in the system [18]. Plants with uniform sizes (cut down to about 500 mm plant height, 150 mm root long, and 50 grams biomass weight) were planted in the CWs. Before planting them, the roots of the plants needed to be carefully washed to remove any remaining soils. The plants were then watered daily for acclimatization for which the concentration of the greywater was set up to be gradually increased from 25 to 100%. This was required in order to allow the plants and their bio-film to develop in the new environment [20]. In the CW-B, the old *Cyperus alternifolius* plants were discarded. The young plants with around 500 mm height were allowed to remain in the CW-B and used in the present study.

Figure 2. The condition of the CW-S (left) and the CW-B (right) at the end of the monitoring period. The height of the plants in both CWs was almost the same, reaching 1.2 – 2.0 meters.

2.5. *Feeding and sampling of greywater*

Batch feeding mode generally showed the best performance compared to continuous mode as the batch mode affects the oxygen to be transferred and diffused in the wetland system resulting in treatment efficiency [22]. Therefore, the batch feeding mode was applied in this study. To make the greywater flow gravitationally into the CWs, a greywater tank was placed higher than the surface of the CW. The influent samples were taken before entering the CWs in which the greywater was distributed across the width of the system at the inlet through a diameter of 20 mm feeding pipe. The greywater then flew slowly through the porous substrate under the surface of the CW in more or less horizontal paths until it reached the outlet. The water level in the CWs was maintained at 2 cm below the surface of the substrate by an adjusting pipe at the outlet, while the valve at the outlet pipe was kept close. After remaining in the CWs for one day, the water was drained completely by opening the outlet valve, the treated effluent was then collected for laboratory test.

Monitoring on both CWs started on April 15 and ended on October 21, 2019. The Indonesian National Standards no. 6989.72-2009 and 06-6989.51-2005 were used to analyze the BOD (Biological Oxygen Demand) and detergent, respectively; whereas In-House Method and APHA, 2012-Section 4500-OG were applied for TSS (Total Suspended Solid) and DO (Dissolved Oxygen) analysis, respectively. Parameters of pH and temperature were measured in the field, using a standard digital pH and temperature meter. The efficiency of removal was calculated by comparing the concentration level of the parameter at both inlet and outlet of the CWs as percent removal (%R).

$$\%R = \frac{(Ci - Ce)}{Ci} x100\% \quad (3)$$

where Ci represents the influent and Ce is the effluent concentration (mg/L).

Tests to determine significant differences between treatments in the CW-B and CW-S were performed by comparing the critical value through one-way ANOVA (t-test) with the aid of Microsoft software calculation. A significant level of $p < 0.05$ was applied for all tests.

3. Results and discussion

The influent and effluent BOD, TSS, detergent and DO concentrations as well as the treatment efficiency (%) of the measured parameters are presented in Table 1. Results pointed out that the tested constructed wetlands effectively removed BOD, TSS, and detergent, as well as increased DO. Additionally, there were no significant statistical differences in the effluents in the CW-B and CW-S.

Table 1. The average removal of BOD, TSS, Detergent, and the increase of DO

Parameter	n	Influent	Effluent: CW-B		Effluent: CW-S	
		mg/L	mg/L	%	mg/L	%
BOD	12	304.9 ± 202.1	12.1 ± 8.1	94.1 ± 4.7	6.6 ± 3.4	96.8 ± 2.6
TSS	12	178.5 ± 110.0	7.6 ± 3.2	94.6 ± 3.2	5.9 ± 2.1	95.6 ± 2.6
Detergent	9	45.3 ± 56.7	1.6 ± 3.0	97.1 ± 5.9	3.8 ± 7.5	94.6 ± 4.8
DO	12	2.6 ± 1.6	3.5 ± 1.3	99.5 ± 111.6	2.9 ± 1.2	79.7 ± 99.3

3.1. BOD

The removal of BOD depends on the combination of physical and microbial mechanisms, and the porosity of the media filter allows the organic solids to be filtered and trapped in the media spaces where the process of the decomposition of the organic solids occurs [13]. BOD concentrations of greywater in the city of Bandung and Surabaya varied between the range of 111 - 690 mg/L and 200 - 490 mg/L, respectively [25], while in low-income countries such as India and Nigeria, the BOD concentrations of greywater varied from 100 and 518 mg/L [5]. Referring to those data, this study was carried out with the highest BOD concentration of 702 mg/L in the influent. To examine the CWs' performance for low levels of BOD concentration, the lowest influent concentration of 61 mg/L was applied [Figure 3].

Table 1 showed that the influent concentration of BOD_5 reduced from an average of 304.9 mg/L to 12.1 and 6.6 mg/L at the effluent, representing an average removal rate of about 94.1% and 96.8% for CW-B and CW-S, respectively. Figure 3a indicated that along the study period, the various concentrations of BOD influent resulted in the low BOD effluents (lower than 30 mg/L), representing the high removal rates of the BOD as presented in Figure 3b.

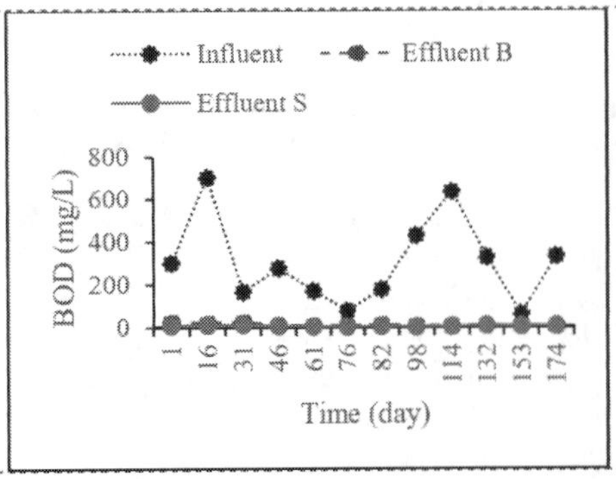

Figure 3a. Influent and effluent of BOD in the CW-B and the CW-S.

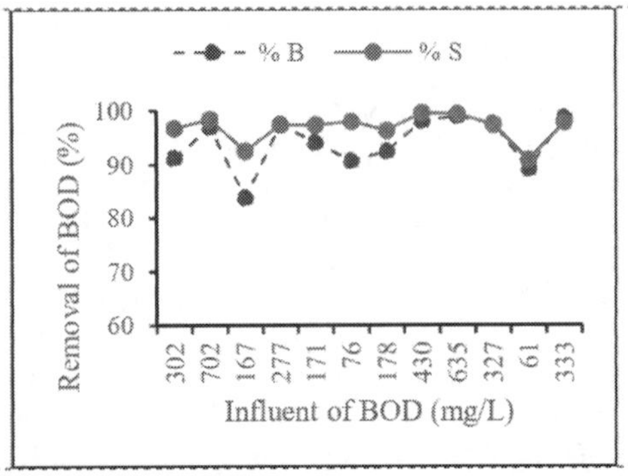

Figure 3b. The percentage removal of BOD in the CW-B and the CW-S.

The removal of BOD in CW-S are mostly higher than in the CW-B. It indicated that the *Cyperus alternifolius* in the CW-S was more capable to oxygenize to support aerobic degradation of organic substances in wastewater. It might be due to the newly-formed system in the CW-S compared to the CW-B. In the CW-B, the 94.1% average BOD removal value of this current study was slightly lower than that of its first operation in 2015, using the same aquatic plant, applying the same HRT of one day, and resulting in 98.55% BOD_5 removal [26]. However, after four years of operation, the percentage of BOD removals in the CW-B was not significantly different compared to the new CW-S. It indicated that the CW-B still performed significantly high. Moreover, the average BOD removal of 94.1 ± 4.7 was higher than that of the [27] study, resulting in the average BOD removal of 85% ± 4.4%. The performance of the CW-B in removing BOD could be enhanced by adding such sand layers on the top of the existing surface of the substrate. It was confirmed by the [23] study which indicated that organic matter (BOD or COD) breakdown mainly occurs in the top layers of a wetland system, predominantly in the upper layer (10–20 cm) due to the high availability of oxygen and microbial density in these layers.

3.2. TSS

In this study, the influent concentration of TSS reduced from an average of 178.5 to 7.6 and 5.9 mg/L in the CW-B and the CW-S respectively (Table 1).

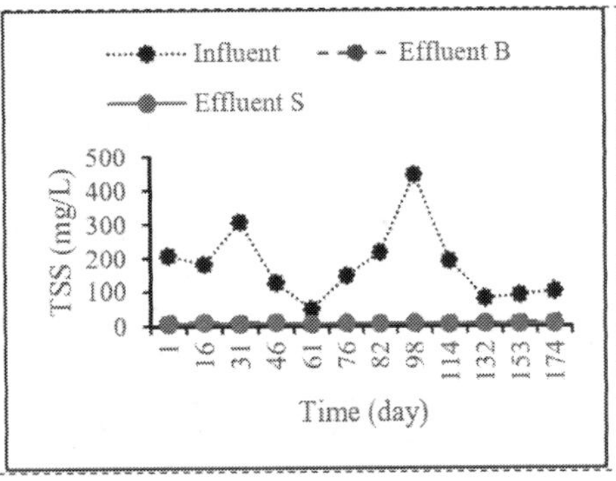

Figure 4a. Influent and effluent of TSS in the CW-B and the CW-S.

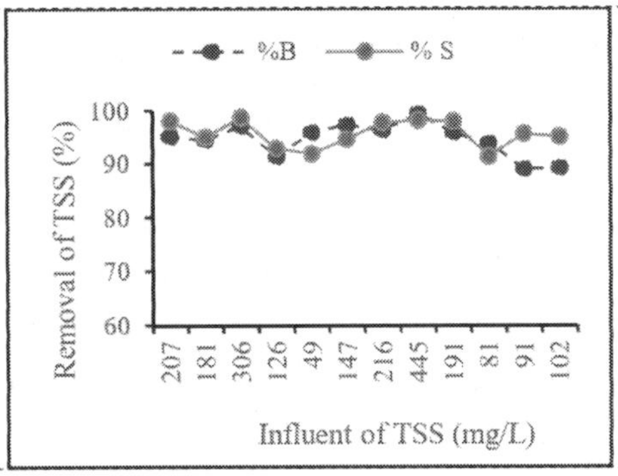

Figure 4b. The percentage removal of TSS in the CW-B and the CW-S.

Figure 4a shows various differences in TSS influents, however, the TSS effluents are almost the same, not significantly different between both CWs. As presented in Figure 4b, the trend of the TSS removals is also similar, not significantly different between both CWs.

It was confirmed that the TSS removal in wetlands is mainly due to physical processes such as sedimentation and filtration [13], which are not influent-dependent processes. Furthermore, the mean efficiency of TSS removal in the CW-B and the CW-S reached 94.6% and 95.6%, respectively. Even though the CW-B in the current study resulted in the TSS removal (94.6%) lower than that of its first operation (98.06%) [26], the removal TSS values are still higher than the range of existing TSS removal data (73% - 83%) in most single horizontal flow types of CW from around the world [17]. Again, the old CW-B still indicated high performance in removing the TSS concentration.

3.3. Detergents

Detergents are household chemical cleaning compounds required in a wide range of daily human life for diverse purposes. The draining out of detergents into the water bodies will adversely affect its aquatic fauna leading to cultural eutrophication [29]. Moreover, detergents greatly contribute to increasing wastewater pollution that comes from household activities or certain industries. In developing countries, the problem of detergents is growing rapidly [30]. This study resulted in detergent percentage removal as presented in Table 1 and Figure 5.

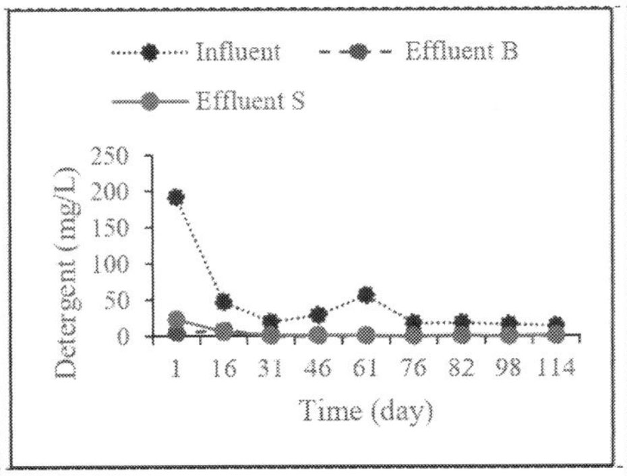

Figure 5a. Influent and effluent of detergent in the CW-B and the CW-S.

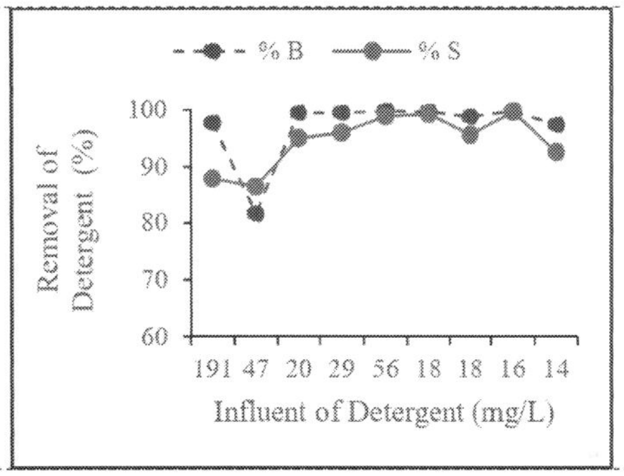

Figure 5b. The percentage removal of detergent in the CW-B and the CW-S.

Figure 5a shows that all experiments could decrease the detergent concentrations, revealing all effluents in both CWs are lower than their influents. The mean detergent effluent significantly reduced from 45.3 to 1.6 mg/L in the CW-B and 45.3 to 3.8 mg/L in the CW-S (Table 1). Figure 5b shows that the CW-B system resulted in more detergent removal (97.1% on average) than the CW-S (94.6%). It might be due to the surface area of the CW-B being larger than the CW-S, providing more oxygen entering the system through the diffusion processes on the CW surface. The high detergent removal in this study, either in the CW-B or in the CW-S, may be caused by aerobic processes in the CWs as the effect of batch feeding mode that delivers oxygen to enter the system. It was confirmed by researchers that aerobic processes are able to degrade most of the detergents [29]. Compared to the first operation of the CW-B in 2015, resulting in 99.86% of detergent removal [26], there was a 2.76% decrease in detergent removal. However, both values of detergent removals in the CW-B and CW-S are higher than the result of the previous study by [31] which obtained 90% detergent removal, using *S. americanus* plants, and applying HRT for 15 days. Once again, the old CW-B exhibited high performance in removing such pollutants, in terms of detergents.

3.4. DO

Figure 6a shows that the effluents of DO in both CW-B and CW-S systems are higher than their influents throughout the study period, except in the effluents of 3rd and the last experiments which were no DO increase. However, the mean DO concentration increased significantly from 2.6 to 3.5 mg/L in the CW-B, and from 2.6 to 2.9 mg/L in the CW-S (Table 1). The DO increases might be due to the batch feeding mode which affected oxygen to introduce into the CW systems. It was affirmed by [1] that the batch operation can affect oxidation along with oxygen transfer and diffusion in wetland schemes. The average increase in DO of 99.5% in the CW-B was higher than that of 79.7% in the CW-S. Figure 6b clearly shows that the increase in DO in the CW-B occurred in most experiments compared to the CW-S. It is reasonable because the surface area of the CW-B is larger than the CW-S, allowing more oxygen to come into the systems through the process of diffusion on the surface of the substrate. The average DO effluent of 3.5 mg/L in the CW-B is larger than 3mg/L, the minimum DO condition for fishery water needs in general. Repetitively, the CW-B indicated high performance after four years of operation. In contrast, the average DO effluent in the CW-S is slightly less than 3 mg/L. This deficiency of DO could be overcome by adding aeration pipes to the CW system as suggested by [32] who stated that aeration of sub-surface flow wetlands using air distribution pipes installed at the bottom of the wetlands could be utilized to increase oxygen transfer in wetland treatment systems.

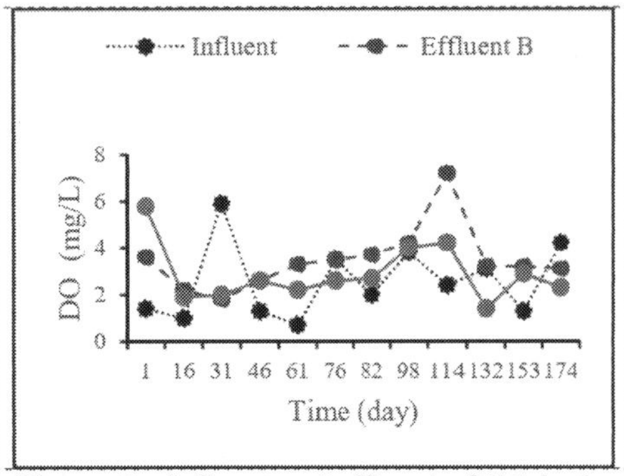

Figure 6a. Influent and effluent of DO in the CW-B and the CW-S.

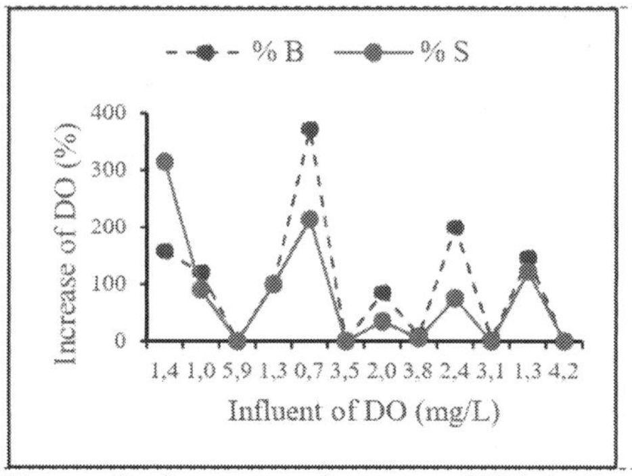

Figure 6b. Percentage increase of DO in the CW-B and the CW-S.

3.5. pH and Temperature

In wetlands, the pH of wastewater greatly affects the removal of organic matter [13]. Temperature is also one of the main factors affecting pollutant removal efficiency in constructed wetlands [1]. Table 2 presents the pH and temperature along with the study.

Table 2. pH and temperature

Parameter	n	Influent (mg/L)	Effluent (mg/L)	
			CW-B	CW-S
pH (pH unit)	12	7.2 ± 0.3	7.1 ± 0.5	6.9 ± 0.8
Temperature (°C)	12	26.6 ± 0.5	27.1 ± 0.3	27.2 ± 0.4

Temperature and pH influence the removal mechanisms for organic matter and nitrogen in constructed wetlands [27]. This is because organisms present in biological wastewater treatments are sensitive to these parameters. Table 2 shows that the values of pH are within the range of pH ($4.0 < pH < 9.5$) recommended by [22], in which the pH values allow a good activity of bacteria. These results suggest that the CW-B and CW-S systems could maintain stable pH levels. Thus, the pH stability along this study provides support to microorganisms in the contaminant degradation. Table 2 also presents the water temperature in both CWs, ranging from 26.1 to 27.6 °C in the influent and the effluent. This is accordant with previous findings [5] in which the greywater temperature ranged from 18 to 38°C. The study was conducted in a tropical region where is more exposed to the sun and hence remains warm throughout the year. Associated with temperature, there is a direct link between microbial activity and temperature in the wetland and the subsequent impact on pollutant removal efficiency [7]. Indeed, the treatment efficiency of constructed wetlands is supposed to be higher in tropical countries because most of the biological activities are enhanced with rising temperatures [15].

3.6. Water quality standards

Results of the household greywater treatment in this study in terms of the average effluent concentrations of BOD, TSS, detergent, and DO are compared to the existing water quality standards as presented in Table 3.

Table 3. The average effluent concentrations and water quality standards

Parameter	CW-B mg/L	CW-S mg/L	Standard[a] mg/L	Standard[b] mg/L	Standard[c] mg/L
BOD	12.1 ± 8.1	6.6 ± 3.4	30	-	12
TSS	7.6 ± 3.2	5.9 ± 2.1	30	-	400
Detergent	1.6 ± 3.0	3.8 ± 7.5	-	0,05	NR
DO	3.5 ± 1.3	2.9 ± 1.2	-	-	NR

[a] the Regulation of the Minister of Environment and Forestry No.68 the year 2016
[b] the Minister of Health No. 32 the year 2017 for hygiene sanitation purposes
[c] the Indonesian Government Regulation No.82 the year 2001 for irrigation (class 4)
NR: not required

Table 3 shows that throughout the monitoring period, the treated greywater (BOD and TSS effluent concentration is lower than 30mg/L) meets the water quality standards. Therefore, the treated greywater is safe to be released into water bodies. Additionally, they can be reused for irrigation purposes. Even though the removal efficiencies of both CWs systems in treating detergent were comparatively high, reaching more than 94% removal on average (Table 1), the detergent effluent concentrations did not meet the maximum limit allowed by the regulation from the Minister of Health number 32 the year 2017 for water media for sanitation hygiene needs. Appropriately, to increase the detergent removal, this study can be further developed by utilizing other adsorbents such as zeolite or granular activated carbon which are known to have higher adsorption capacity.

4. Conclusion

The performance of single constructed wetlands to treat household greywater, applying the HRT of one day, planting with the *Cyperus alternifolius,* had been demonstrated. Results of the present work highlight that the CW-B, after four years in operation, still effectively removes the BOD, TSS, and detergents, as well as increases in DO concentration. Effluent concentrations in the old CW-B and the new CW-S systems were not significantly different. The treated greywater is safe to be discharged into water bodies and can be reused for irrigation needs. The filter media and the aquatic plant used in this study were all available locally and easily found. It makes this work adaptable to other similar tropical areas to build cost-effective treatment of greywater for individual or even communal households. It can be concluded that the CW represents a feasible solution for household greywater treatment and subsequently minimizes pollution in rivers. This work can be further developed by utilizing other higher adsorbents to increase detergent removal and considering other significant parameters related to water quality.

Acknowledgments
The authors are very grateful to the University of Sebelas Maret for providing some financial supports for this research.

References

[1] Rahman ME, Halim MIE, Samad MYA, Udin MK, Mahmud K, Shukor MYA, Abdullah SRS and Shamsuzman SM 2020 *Int. J. Environ. Res. Public Health* **17** 8339 https://doi.org/10.3390/ijerph17228339
[2] United Nation 2017 *Executive summary - Wastewater The Untapped Resource*
[3] Fulazzaky MA 2014 *Water* **6** 2000 - 20 https://doi.org/10.3390/w6072000
[4] Sisnayati, Said M, Nasir S and Priadi DP 2018 *Int. J. Adv. Sci. Eng. Inform. Tech.* **8** 5 2114 - 21 https://doi.org/10.18517/ijaseit.8.5.4724
[5] Oteng-Peprah M, Acheampong MA and deVries NK 2018 *Water Air Soil Pollut.* **229** 255 https://doi.org/10.1007/s11270-018-3909-8
[6] Stefanakis AI 2019 *Sustainability* **11** 6981 https://doi.org/10.3390/su11246981
[7] Almuktar SAAAN, Abed SN and Scholz M 2018 *Environ. Sci. Pollut. Res.* **25** 23595 - 625 https://doi.org/10.1007/s11356-018-2629-3
[8] Wang M, Zhang DQ, Dong JW and Tan SK 2017 *J. Environ. Sci. (China)* **57** 293 - 311 https://doi.org/10.1016/j.jes.2016.12.019
[9] Mustafa A 2013 *Int. J. Environ. Sci. Dev.* **4** 1 20 - 24 https://doi.org/10.7763/IJESD.2013.V4.296
[10] Konnerap D, Kottatep T and Brix H 2009 *Ecol. Eng.* **35** 2 248 - 257 https://doi.org/10.1016/j.ecoleng.2008.04.018
[11] Hendrawan D, Widanarko S, Moersidik SS and Triweko RW 2013 *European Sci. J.* **9** 17 229 - 39
[12] Flores L, García J, Pena R and Garfí M 2019 *Sci. Total Environ.* **659** 1567 - 1576 https://doi.org/10.1016/j.scitotenv.2018.12.348
[13] Stefanakis A, Akratos CS and Tsihrintzis VA 2014 *Vertical Flow Constructed Wetlands: Eco-engineering Systems for Wastewater and Sludge Treatment*, 1st ed. Elsevier, Netherland https://doi.org/10.1016/B978-0-12-404612-2.00001-5
[14] Gorgoglione A and Torretta V 2018 *Sustainability* **10** 3910 https://doi.org/10.3390/su10113910
[15] Varma M, Gupta AK, Ghosal PS and Majumder A 2021 *Sci. of the Total Environ.* **755** 142540 https://doi.org/10.1016/j.scitotenv.2020.142540
[16] Rodríguez-Dominguez MA, Konnerup D, Brix H and Arias CA 2020 *Water* **12** 6 1744 https://doi.org/10.3390/w12061744
[17] Vymazal J 2010 *Water* **2** 530 – 49 https://doi.org/10.3390/w2030530
[18] Vymazal J 2011 *Hydrobiologia* **674** 133 – 56 https://doi.org/10.1007/s10750-011-0738-9
[19] Vymazal J 2013 *Ecol. Eng.* **61** 582 – 592 https://doi.org/10.1016/j.ecoleng.2013.06.023
[20] Sandoval L, Zamora-Castro SA, Vidal-Álvarez M and Marín-Muñiz JL 2019 *Applied Sci.* **9** 685 1 - 17 https://doi.org/10.3390/app9040685
[21] Qomaryah S, Ramelan AH, Sobriyah and Setyono P 2021 *Int. J. Adv. Sci. Eng. and Inform. Tech.* **11** 3 1014 – 21 https://doi.org/10.18517/ijaseit.11.3.12343
[22] Wu S, Carvalho PN, Müller JA, Manoj VR and Dong R 2016 *Sci. of the Total Environ.* **541** 8 - 22 https://doi.org/10.1016/j.scitotenv.2015.09.047
[23] Stefanakis AI and Tsihrintzis VA 2012 *Chemical Eng. J.* **181** 416 - 30 https://doi.org/10.1016/j.cej.2011.11.108
[24] Frazer-William RAD 2010 *J. of Chemical Eng.* **25** 1 29 - 42 https://doi.org/10.3329/jce.v25i0.7237
[25] Firdayanti M, Indriyani A, Prihandrijanti M and Otterpohl R 2015 *J. Tek. Ling.* **21** 2 98 - 114 https://doi.org/10.5614/jtl.2015.21.2.1

[26] Qomaryah S, Ramelan AH, Sobriyah and Setyono P 2017 *IOP Conf. Series: Material Sci. and Eng.* **176** 012018 https://doi.org/10.1088/1757-899X/176/1/012018
[27] Merino-Solis ML, Villegas E, deAnda J and Lopez A 2015 *Water* **7** 1149 - 63 https://doi.org/10.3390/w7031149
[28] Tsihrintzis VA, Akratos CS, Gikas GD, Karamouzis D and Angelakis AN 2007 *Environ. Technol.* **28** 621 - 628 https://doi.org/10.1080/09593332808618820
[29] Mousavi SA and Khodadoost F 2019 *Environ. Sci. Pollut. Res. Int.* **26** 26 26439 - 48 https://doi.org/10.1007/s11356-019-05802-x
[30] El-Gawad HAS 2014 *Water Sci.* **28** 51 – 64 https://doi.org/10.1016/j.wsj.2014.09.001
[31] Perez-Lopez ME, Arreola-Ortiz AE and Zamora PM 2018 *Ecological Eng.* **122** 135 - 42 https://doi.org/10.1016/j.ecoleng.2018.07.036
[32] Nivala J, Wallace S, Headley T, Kassa K, Brix H, Afferden M and Muller R 2013 *Ecological Eng.* **61** 544 – 54 https://doi.org/10.1016/j.ecoleng.2012.08.028

Influence of random variable on reliability index of braced monopod platform based on fatigue analysis in Madura North Java Sea

Paramashanti[1], **Rildova**[1], **V N B Sucipto**[2] **and A Kurniawan**[1]

[1] Ocean Engineering Department, Bandung Institute of Technology, 10 Ganesha Street, Bandung
[2] Ocean Engineering Student, Bandung Institute of Technology, 10 Ganesha Street, Bandung

virtynobelia@gmail.com

Abstract. This paper carried out fatigue reliability analysis on braced monopod structure located in Madura North Java Sea to find the most suitable fatigue performance function so better service life prediction of each structure's components can be done by testing the influence of some load random variable. Fatigue reliability analysis is a local analysis therefore analysis will be conducted only on four critical joints. Reliability analysis is done on several performance function using First Order Reliability Method (FORM). The load random variables which will be tested are yearly fatigue damage, local and global stress analysis, and S-N curve parameter. Randomness of fatigue damage is obtained by running fatigue analysis using only one-year wave data for 59-year SEAFINE wave data. Fatigue damage from analysis will be divided by safety factor and service life to obtain yearly fatigue damage. Reliability index from several performance functions show S-N curve parameter hardly influence the result of reliability analysis. Therefore, fatigue reliability analysis will only include yearly fatigue damage and local and global stress analysis as load random variable. Reliability index with 2 as the value of safety factor and 20 years of service life for four critical joint which are 301L, 303L, 203L, and 201L respectively are 1.24, 1.30, 1.30, and 1.30.

Keywords: Braced monopod, fatigue analysis, reliability analysis, reliability index

1. Introduction

Structural design using API RP2A WSD will give a structure that fulfill safety and functional requirements. That with the addition of economic factor will be the main purpose of design [10]. Differences in local environmental conditions and those used in the standard calibration may cause an unfulfilled economic factor requirement. API RP2A WSD is using Gulf of Mexico calibrations. Therefore, analysis using local environment must be conducted to test whether the standard's factor is overdesign or under design or suitable for local environment.

Structure design must fulfill four limit state categories, one of it is Fatigue Limit State (FLS). FLS is a limit state category to test the capability of element's joint under repeated load. Fatigue is a kind of deterioration that can occur in any metallic structure and mechanical part [7]. Structure's capacity and load that are used to express limit state depend on some variable whether it is deterministic or random.

A variable may be considered random if the value it contains is uncertain which may affect the capability of the structure to hold during its serviceable period. This uncertainty caused by lack of offshore structure data or varied variable value on structure. Variable with varied value tend to be given by environmental condition.

The influence of random variable to structure's capacity and deterioration of structure capability in time can be tested using reliability analysis. Reliability analysis needs performance function to determine the relationship between resistance and load. Structure resistance to fatigue influenced by tubular thickness and material. There have been several performance functions used by [3], [7], [8], [10], and [11] to find fatigue reliability index of structure but without the randomness of yearly fatigue damage. Randomness of yearly fatigue damage must be taken to include the behavior of local wave. Reliability index can be calculated using FORM II. Structure design can be said to meet the capacity requirement if the value of index reliability reach the reliability index target (**Subsection 2.5**).

2. Methodology

2.1. Structure Modelling

Study case takes place at Madura North Java Sea (**Figure 1**) with 45-meter depth. Braced monopod structure with one level deck will be designed using SACS software. Structure design will follow criteria given by API RP2A WSD for static in-place analysis and fatigue analysis.

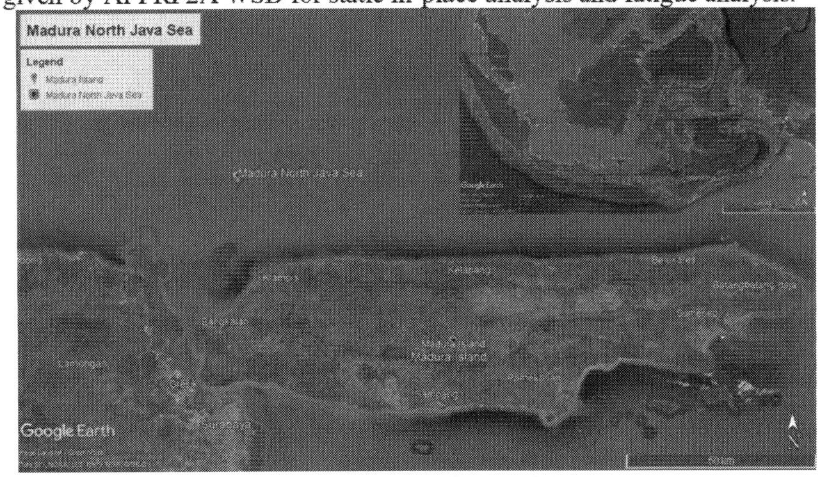

Figure 1. Structure location.
(Source: Google Earth)

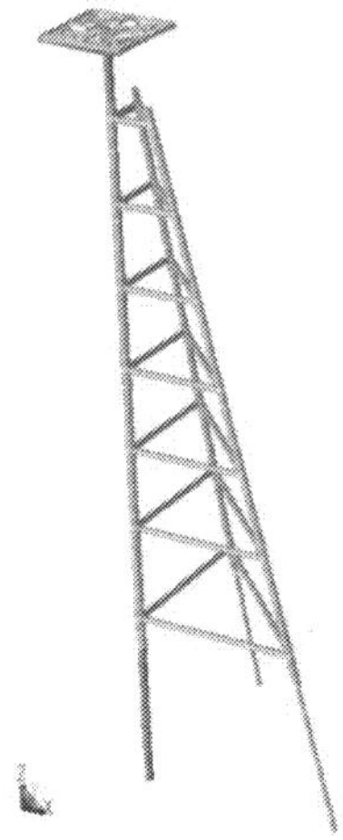

Figure 2. Structure design.

Structure's legs consist of one caisson and two raked piles and their sleeves. The deck will be located at 6.6 meter above MSL (Mean Sea Level) and modeled as a simple beam with a weight of 1333.33 kN and 6 x 6 meter dimension (**Figure 2**). The structure is assumed to have zero marine growth. The whole structure will be made of steel and have 248 MPa of yield stress with 7.849 ton/m^3 of density. Wave design in static in-place analysis will use storm wave with 100 years return period. This study will focus on wave as environmental load and neglecting wind and current.

2.2. Extreme Wave Analysis

Wave data is obtained from SEAFINE with duration from June 30th,1956 to July 1st,2015. SEAFINE is an hourly significant wave height, peak period, and dominant direction. Extreme wave analysis must be done to obtain maximum theoretical distribution for calculating storm wave with 100 years return period. Theoretical distribution that will be tested are Gumbel, Normal, Lognormal, Frechét, and Weibull. Data sample that will be used in analysis is chosen using annual maxima method (yearly maximum wave height). Sample data fitting on theoretical distribution function will be done using least square method. These distributions are tested using goodness of fit test with MIR (Minimum Ratio of Residual Correlation Coefficient) method.

Maximum hourly significant wave height of every year data, which are provided by SEAFINE, will be converted to its maximum wave height value using the conversion formula proposed by Chakrabarti [2]. The equation to convert it can be seen below. The 100-year return period wave height is calculated using extreme analysis equation from the selected distribution.

$$H_{max} = 1.86H_s \tag{1}$$

Designed wave period is calculated in accordance with the designed wave height. Relationship between wave period and height is made from making scattered relationship of hourly maximum wave height as x-axis and peak period as y-axis. According to Goda [4], scattered diagram can be approached using power equation. Designed wave will be used in designing the model structure.

2.3. Static In-Place Analysis

Static in-place analysis is performed to obtain the optimum structure of the overall design. Additionally, the joint can analysis which is also part of the static in-place analysis will be used together with the fatigue analysis to acquire the optimal tubular joint parameter and brace member size. Structure is considered to fulfill in-place requirement if Unity Check Ratio (UCR) value is less than or equal to one and joint deflections are below allowable deflection. Optimum structure has caisson pile's UCR above 0.9. Members and joint cans' UCR are calculated using equation from API RP2A WSD. Allowable horizontal deflection is recommended by ISO 9001, while allowable vertical deflection is recommended by SNI 03-1729-2002.

2.4. Fatigue Analysis

The fatigue analysis in this paper will use 2 as the value of safety factor and 20-years for its service life. Joint design in fatigue will only use the first 1-year duration of SEAFINE wave data. This analysis must be done to check structure's fatigue caused by cyclic load which result in deterioration of structure's capacity. Structure's capability of enduring cyclic load from wave can be checked using fatigue damage value which has criterion less than or equal to one. Analysis will be done using spectral fatigue method.

Fatigue damage is calculated using Palmgren Miner's rule which written is equation (2) where n_i is i-th number of wave cycle and N_i is permissible cycles of load. The variable can be derived to equation (3) for n_i and equation (4) for N_i.

$$D = \sum_{i=1}^{I} \frac{n_i}{N_i} \tag{2}$$

$$n_i = \frac{u_i \cdot U}{T_{z,i}} = U \cdot n_i' \tag{3}$$

$$N_i = K.s^{-m} = K \cdot \left(s' \cdot B_{scf} \cdot B_{glob}\right)^{-m} \tag{4}$$

Where:

- n_i' = i-th number of wave cycle in 1-year
- u_i = Probability of i-th spectrum occurrence
- T_z = Mean zero-crossing period
- U = Structure life
- B_{scf} = Random variable of local stress analysis on joint
- B_{glob} = Random variable of global stress analysis on structure
- K and m = S-N curve parameter
- s = Stress range
- s' = Stress range without the randomness of stress analysis

Substituting equation (3) and (4) to equation (2) will result on equation (5) where $\overline{K}$ is average and deterministic value of S-N curve parameter.

$$D = \sum_{i=1}^{I} \frac{U.n_i'}{K}\left(s \cdot B_{scf} \cdot B_{glob}\right)^m$$
$$D = B_{scf}^{m_g} \cdot B_{glob}^{m_g} \cdot \frac{\overline{K}}{K} \cdot U \sum_{i=1}^{I} \frac{n_i'}{\overline{K}} s^m \tag{5}$$

Equation (5) can be simplified to equation (6) by taking out uncertainty factor and service life, therefore the sigma is yearly fatigue damage.

$$D = B_{scf}^m \cdot B_{glob}^m \cdot \frac{1.794 \times 10^{10}}{K} \cdot U \cdot d_j \tag{6}$$

After optimum structure is obtained, fatigue analysis will be performed for the rest of 58-years wave data which analyzed yearly. The purpose of using only 1-year period of wave data is to obtain the randomness of yearly fatigue damage. Reliability analysis will be done for the four joints with the most fatigue damage based on fatigue analysis. Fatigue damages from analysis are processed to gain yearly fatigue damage by dividing it with safety factor and service life. Yearly fatigue damages of each joint then processed to find its theoretical distribution and statistic parameter for reliability analysis.

2.5. Reliability Analysis

Reliability analysis is done to find out structure's reliability on resisting acting load with evaluating structure's failure based on resistant load system probability. Relation between resistance and load is indicated by fatigue limit state equation as follows (shown in equation 7):

$$g = R - L \tag{7}$$

Structure will undergo failure if g less than zero. Fatigue damage calculated from fatigue analysis acted as load so by substituting equation (6) to variable L of equation (7) and change resistance symbol (R) to fatigue damage capacity (Δ), the equation will be as follows (shown in equation 8):

$$g = \Delta - B_{scf}^m \cdot B_{glob}^m \cdot \frac{1.794 \times 10^{10}}{K} \cdot U \cdot d_j \tag{8}$$

This paper will check the influence of load random variable therefore there will be five performance function used in analysis (**Table 1**).

Table 1. Performance function.

No.	Random Variable	Performance Function
1	Δ	$g(X) = \Delta - U \cdot d_j$
2	Δ and d_j	$g(X) = \Delta - U \cdot d_j$
3	$\Delta, B_{scf}, B_{glob}$ and d_j	$g(X) = \Delta - B_{scf}^m \cdot B_{glob}^m \cdot U \cdot d_j$
4	Δ, K and d_j	$g(X) = \Delta - U \cdot \frac{1.794\times10^{10}}{K} d_j$
5	$\Delta, B_{scf}, B_{glob}, K$ and d_j	$g(X) = \Delta - B_{scf}^{m_g} \cdot B_{glob}^{m_g} \cdot U \cdot \frac{1.794\times10^{10}}{K} d_j$

Variable description for performance function can be seen in **Table 2**. Reliability analysis uses First Order Reliability Method (FORM) II to determine reliability index. FORM II calculate reliability index by the finding the closest distance between origin point and performance function in normal standard room. Reliability index will be calculated from 1-year old structure to $(SF \times SL)$-years old. Where SF is safety factor and SL is service life. Reliability index when structure reach its service life is going to be used to find out the influence of load random variable. The service life's reliability index needs to be compared with reliability target to establish a reliability-based design. Reliability target is defined as the acceptable minimum safety level to represent the probability of failure of a structure. According to Onoufriou [9], reliability index target for newly designed structure is 4.19 (Probability of failure 1.4×10^{-5}) for main leg and 3.69 (Probability of failure 1.1×10^{-4}) for bracing.

Table 2. Performance function's variable information.

Variable	Description	Parameter and Distribution	Information
Δ	Fatigue capacity of joint component	Lognormal Distribution Mean = 1.0 Standard Deviation = 0.3	Recommendation from DNVGL RP C203: *Fatigue design of offshore steel structures* (2014)
B_{scf}	Uncertainty of local stress analysis on joint	Lognormal Distribution Mean = 1.0 Standard Deviation = 0.2	Recommendation from JCSS (*Joint Committee for Structural Safety*) *probabilistic model code* (2011)
B_{glob}	Uncertainty of global stress analysis on structure	Lognormal Distribution Mean = 1.0 Standard Deviation = 0.1	Recommendation from JCSS (*Joint Committee for Structural Safety*) *probabilistic model code* (2011)
K	API X' S-N curve material parameter	Lognormal Distribution Mean = 1.794×10^{10} CoV = 0.3	Recommendation from Wirsching, 1984
m	API X' S-N curve gradient	Deterministic = 3.74	API RP2A WSD
U	Structure life	Deterministic = 3.74	Reliability index will be calculated for every 1-year of structure age (1 to Safety Factor x Service Life)
d_j	Yearly fatigue damage	Random	Obtained from fatigue analysis for every 1-year of SEAFINE wave data

3. Result and Discussion

3.1. Design Wave

Result from goodness of fit test using MIR is shown in **Table 3**.

Table 3. MIR test result.

		Minimum Ratio of Residual Correlation Coefficient (MIR)			
Distribution	k	r	Δr	Δrmean	Δr/Δrmean
Gumbel		0.9916	0.0084	0.0215	0.3928
Weibull	2	0.9958	0.0042	0.0092	0.4592
Lognormal		0.9954	0.0046	0.0098	0.4680
Weibull	1.4	0.9874	0.0126	0.0116	1.0938
Frechet	10	0.9764	0.0236	0.0180	1.3126
Frechet	5	0.9480	0.0520	0.0247	2.0996
Weibull	1	0.9573	0.0427	0.0158	2.6963
Frechet	3.3	0.9024	0.0976	0.0335	2.9099
Frechet	2.5	0.8430	0.1570	0.0449	3.4937
Weibull	0.75	0.9059	0.0941	0.0223	4.2219
Normal		0.9857	0.0143		

The result shows that Gumbel distribution is the fittest distribution to the SEAFINE wave data at structure's location because the value of $\Delta r/\Delta r_{mean}$ is the smallest among the other distribution. From the analysis using Gumbel distribution, 100-years return period wave height is 7.11 meter.

Relation between wave height and period of hourly SEAFINE data is being approached using power trend line which shown in equation (9). The designed wave period is 8.06 second. Wave design parameter will be used in static in-place analysis.

$$T = 4.3227 \times H^{0.3177} \tag{9}$$

3.2. Structure Design

Structure design as can be seen in **Figure 2** have 1:4 single batter. Static in-place analysis gives optimized structure with 54cm diameter and 1.4cm thickness for caisson pile while raked pile have 30cm diameter and 0.75cm thickness. Design wave applied in twelve directions according to API recommendation.

Spectral fatigue analysis use probability of wave occurrence table for every direction as an input. Structural design use only 1-year wave data as an input for fatigue analysis. The result of fatigue analysis on 10 critical joints can be seen in **Table 4**.

Table 4. Fatigue analysis result.

Joint	Member Type	Joint Type	Fatigue Damage	Service Life (years)
301L	Chord	T	0.8922	22.42
303L	Chord	Y	0.8433	23.71
203L	Chord	Y	0.8396	23.82
201L	Chord	T	0.8378	23.87
401L	Chord	T	0.5841	34.24
501L	Chord	T	0.4006	49.93
403L	Chord	Y	0.3844	52.03
202L	Chord	Y	0.3813	52.46
503L	Chord	Y	0.2629	76.06
302L	Chord	Y	0.1888	105.9

3.3. *Fatigue Analysis*

After obtaining optimum structure, fatigue analysis is done for the rest of 58-years wave data using the same structure. Four joints with the highest fatigue damage which will be analyzed are 301L, 303L, 203L, and 201L (see **Figure 3**). Yearly fatigue damage obtained by dividing fatigue damage with 2 for eliminating safety factor and 20 for eliminating service life.

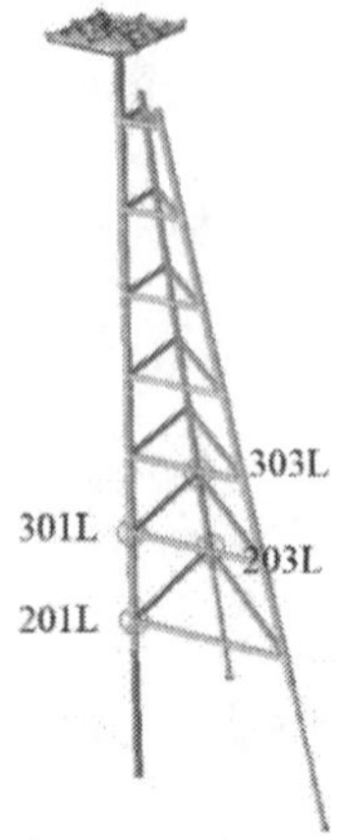

Figure 3. Critical joints location.

Every joint has 59 data of yearly fatigue damage which being processed to obtain theoretical distribution. The same theoretical distribution and sample data fitting method as wave data is used, but these distributions are tested using goodness of fit test with K-S (Kolmogorov-Smirnov) method. The chosen theoretical distribution and its statistic parameter for every joint can be seen in **Table 5**.

Table 5. Yearly fatigue damage distribution and statistic parameter.

Joint	Distribution	Mean	Standard Deviation	COV
303L	Weibull, k = 1	0.0155	0.0106	0.686
301L		0.0165	0.0112	0.681
203L		0.0154	0.0104	0.672
201L		0.0155	0.0103	0.664

3.4. *Reliability Index*

Reliability index of every joint is calculated every year until structure reach 40 years old which is safety factor multiplied by service life. Graph of reliability index and structure life can be seen in **Figure 4** until **Figure 7** (**Figure 4** for joint 301L, **Figure 5** for joint 303L, **Figure 6** for joint 203L, and **Figure 7** for joint 201L), the number indicate reliability index when structure reach its service life which is 20 years old. Performance function with random yearly fatigue damage is more reliable than with deterministic yearly fatigue damage (performance function 1) when structure's reach 40 years old. Up until structure's service life (20 years), deterministic fatigue damage still gives much higher reliability index with a rate of decrease higher than those with random fatigue. This may lead to under-design of structure with low service life.

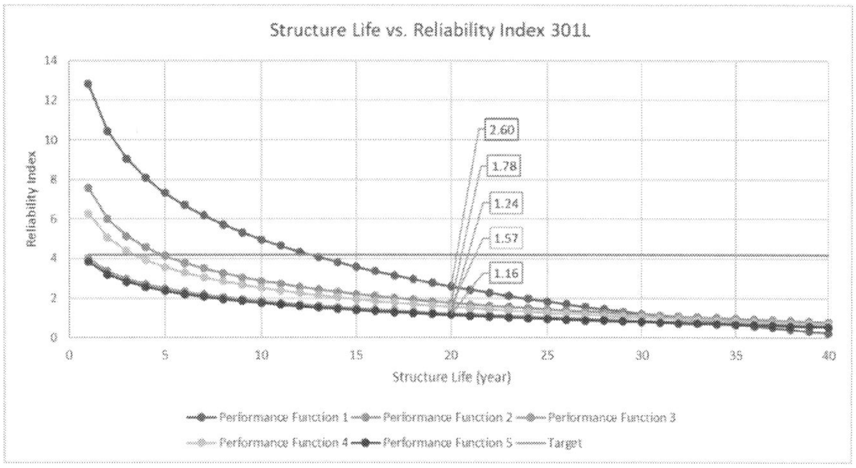

Figure 4. Relation between reliability index and structure life joint 301L.

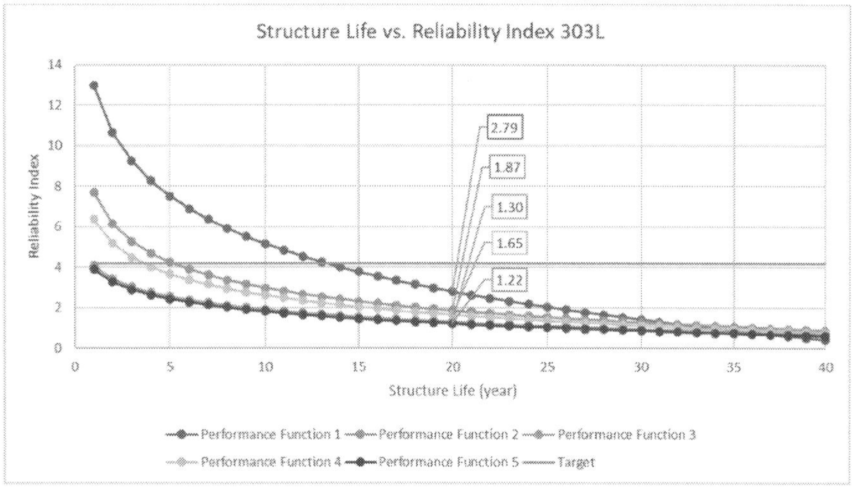

Figure 5. Relation between reliability index and structure life joint 303L.

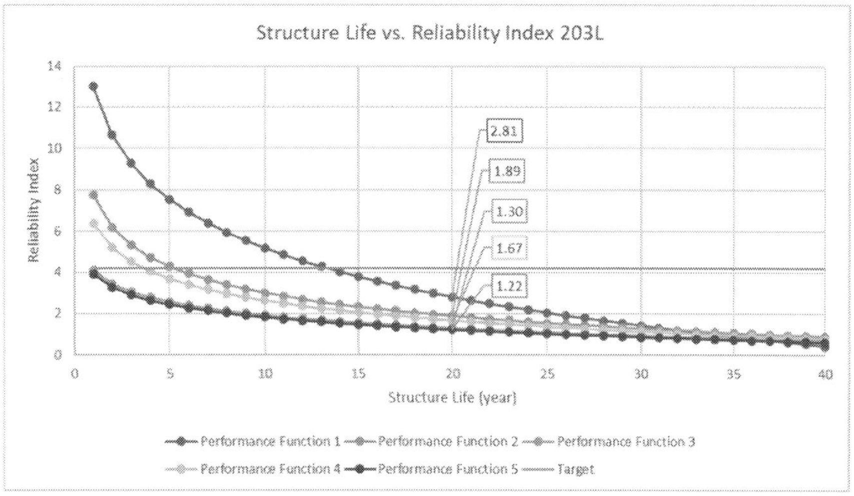

Figure 6. Relation between reliability index and structure life joint 203L.

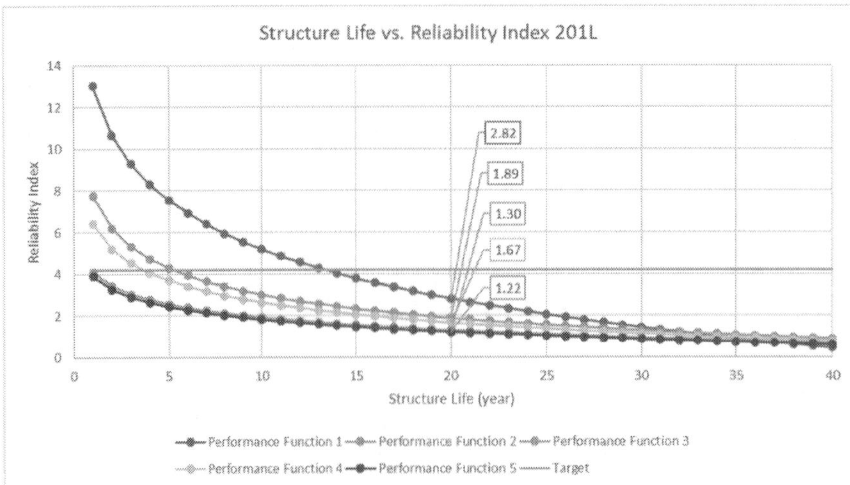

Figure 7. Relation between reliability index and structure life joint 201L.

All joints have the same comparison among performance function. Yearly fatigue damage has the biggest influence because it has the biggest value of Coefficient of Variation (CoV). By comparing performance function 3 and 4 to performance function 2, randomness of stress analysis has bigger influence than S-N curve parameter even though it has smaller value of CoV. This happen because randomness of stress analysis consists of two variable and powered by 3.74 (value of m). Addition of S-N curve parameter in performance function 5 does not give much difference reliability index than performance function 3. It means that influence from S-N curve parameter can be neglected.

Reliability index from every performance function when structure at its service life is below the reliability target or can be said structure does not meet design criterion. This indicate that when structure at its service life, there is a probability that structure is no longer able to receive the same amount of load as designed load or structure encounter failure. Failure is still uncertain because structure's reliability index has not reached zero which means structure still have capacity to withstand specific load. Zero reliability index means the failure probability is 1.0.

Further analysis using higher safety factor is done because using 2 as the value of safety factor is not enough to meet the target reliability. According to recommendation from API RP2A WSD, structure design using 5 as the value of safety factor is done to get a new optimum structure that meets the requirement of in-place and fatigue analysis. Dimension change of critical joints from design using 2 as the value of safety factor to design using 5 as the value of safety factor can be seen in **Table 6.**

Table 6. Dimension change.

Joint	Dimension (cm)	
	SF 2	SF 5
301L	Ø62.2 × 2.6	Ø62.8 × 2.9
303L	Ø52.3 × 2.4	Ø52.7 × 2.6
203L	Ø52.1 × 2.3	Ø52.5 × 2.5
201L	Ø62.8 × 2.9	Ø64.0 × 3.2

The same analysis is done to find yearly fatigue damage from the new structure. Reliability analysis is just done on the most critical joint which is 301L. Statistic parameter of yearly fatigue damage joint 301L can be seen in **Table 7**.

Table 7. Yearly fatigue damage distribution and statistic parameter, SF = 5.

Joint	Distribution	Mean	Standard Deviation	COV
301L	Weibull, k = 1	0.0024	0.0020	0.8087

Reliability index is calculated until 100 years (service life × safety factor) as the safety factor change to 5. Relation of reliability index and structure life in a graph can be seen in **Figure 8**. Number in graph indicate the reliability index when structure reach its service life.

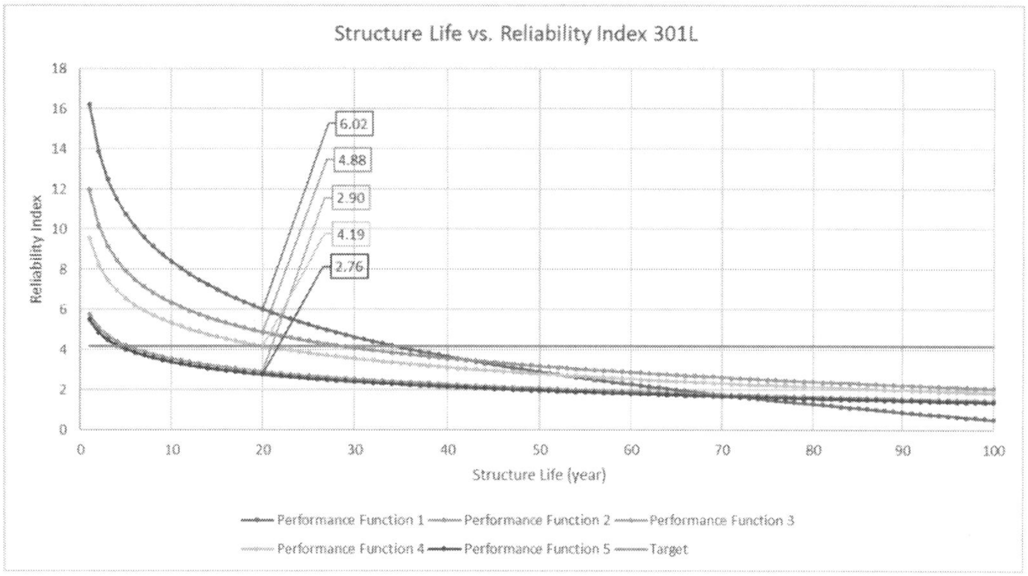

Figure 8. Relation between reliability index and structure life joint 301L, SF = 5.

Comparison of reliability index using performance function 3 between design using 2 and 5 as the value of safety factor can be seen in **Figure 9**. Reliability index using higher safety factor which is five is still not enough to meet the reliability target but still have lower probability of structure encounter failure.

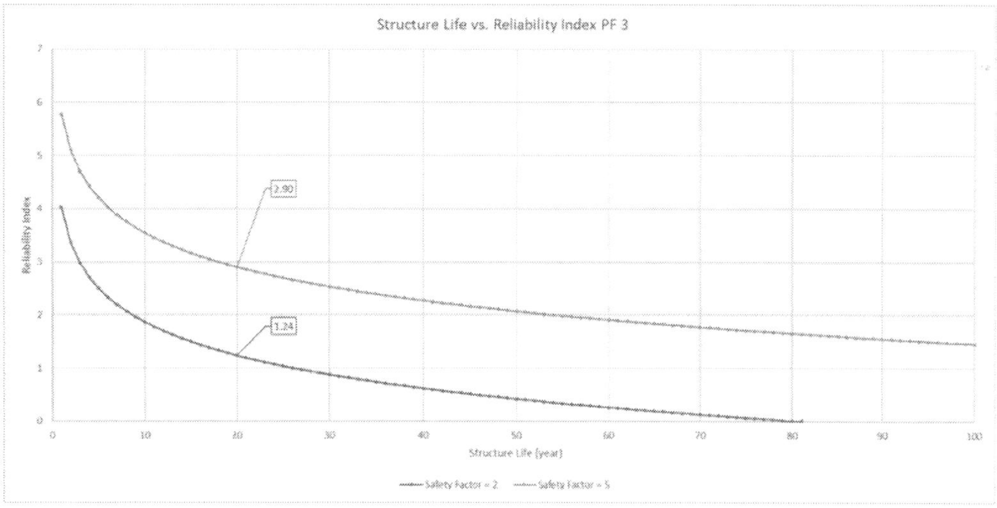

Figure 9. Reliability index comparison between SF = 2 and SF = 5.

4. Conclusion

This research can be concluded as follows:

1. The deterministic fatigue damage lead to higher reliability index at the start of structure life. However, its reliability index rate of decrease is higher than those with random fatigue damage,

leading to lower reliability index over long period of time. This may lead to under-design of structure with low service life.

2. The order of load random variable influence from the biggest is yearly fatigue damage, stress analysis, and S-N curve material parameter.
3. The performance function recommendation for fatigue reliability analysis is performance function 3 which has randomness of yearly fatigue damage and stress analysis.
4. Reliability index with 2 as the value of safety factor using performance function recommendation for four critical joints which are 301L, 303L, 203L, and 201L respectively are 1.24, 1.30, 1.30, and 1.30.
5. As structure life increases, structure's capacity degrades.
6. Statistic parameter of random variable is taken based on standard recommendation and another research. Structure's stress calculation using finite element is considered quite accurate so standard deviation of B_{glob} can be further reviewed. Stress concentration factor's calculation in this case is using Efthimiou theory which gives the smallest fatigue life compared to other theory see, Khedmati et al. [7]. Based on that consideration, standard deviation of B_{scf} can be further reviewed if calculation used theory which gives conservative value.
7. Structure design using 5 as the value of safety factor is still not enough to meet reliability index target, therefore, further research must be done to obtain fatigue safety factor recommendation for Madura North Java Sea using the recommended performance function in this paper.
8. For typical offshore platform operation, reliability index may be increased after a certain period of service by performing appropriate inspection and / or maintenance to reduce the uncertainty and increase the reliability of the structure.

References

[1] American Petroleum Institute 2007 *Recommended Practice for Planning, Designing and Constructing Fixed Offshore Platform* (Washington D.C.: API Publishing Service)

[2] Chakrabarti S 1987 *Hydrodynamics of Offshore Structures* (Illinois: CBI Industries, Inc)

[3] Falahi F 2020 *Penentuan Interval Inspeksi Struktur Anjungan Lepas Pantai Dengan Metode Risk Reliability Based Underwater Inspection* (Bandung: Institut Teknologi Bandung)

[4] Goda Y 2000 *Advanced Series on Ocean Engineering 15: Random Seas and Design of Maritime Structures* (Singapore: World Scientific Publishing Co. Pte. Ltd)

[5] Goda Y 2000 *Random Seas and Design of Maritime Structures* (Singapore: World Scientific)

[6] Haldar A and Mahadevan S 2000 *Probability, Reliability and Statistical Methods in Engineering Design* (Danvers: John Wiley & Sons, Inc)

[7] Khedmati M R, Rigo P, Amrane A and Nazari M 2013 Assessment of Fatigue Reliability for Jacket-Type Offshore Platforms Considering Dynamic Behavior *OMAE2013-11568* 1-8 (Nantes: ASME)

[8] Kirkegaard P H, Sørensen J and Brincker R 1988 Fatigue Analysis of a Mono-Tower Platform *Dept. of Building Technology and Structural Engineering, Aalborg University, Fracture and Dynamics* **R8840** No. 9

[9] Onoufriou T 1999 Reliability Based Inspection Planning of Offshore Structure *Marine Structure* **12** 521-39

[10] Tawekal R L and Heriana R 2007 Analisis Kehandalan Anjungan Lepas Pantai Tipe Jacket Berdasarkan Kapasitas Fatigue pada Sambungan *Jurnal Teknik Sipil* **14**, *No. 2*, 93-104

[11] Wouts M A, Van der Graaf J, Vrouwenvelder A, Heerema Engineering Service BV and TNO-IBCC 1988 Fatigue Reliability Analyses of Fixed Offshore Structures. *SPE European Petroleum Conference* 137-148 (London: Society of Pertoleum Engineers)

Numerical study of hydrodynamic and sedimentation for sustainable marine floating cage aquaculture in Indonesia: a case study in Situbondo, East Java.

Marsha Sabrina[1,2] and Entin A. Karjadi[1,3]

[1]Ocean Engineering Program, Faculty of Civil and Environmental Engineering, Institut Teknologi Bandung, Indonesia
[2]marsha.sabrina.s@gmail.com
[3]entinkarjadi@gmail.com

Abstract. In this study, the floating cage aquaculture in Indonesia, specifically the effect on sedimentation in surrounding areas will be studied. In addition to waves and currents, sedimentation is a considerable environmental concern. This result could help establish additional criteria for 1) selecting suitable aquaculture locations, 2) assessing the effect of aquaculture on the environment, and 3) developing sustainable marine aquaculture industry in Indonesia. This is important because, as an archipelago, Indonesia is expected to play a significant role in providing food supplies from marine aquaculture. The results are from a case study in Situbondo, East Java in which the hydrodynamics and sedimentation are simulated using the Delft3D model with an energy-loss coefficient included in the model to account for floating cages.
Keywords: *Aquaculture, Numerical Model, Hydrodynamic, Sedimentation, Floating Cage*

1. Introduction

In response to the United Nations 2nd and 14th Sustainable Development Goals (SGD): "Zero Hunger" and "Life Below Water", the Government of Indonesia has set targets and indicators for each of these goals to achieve food security and a sustainable marine ecosystem. The targets of the 14th SGD include reducing sea pollution, conserving coastal areas, and sustainably managing fishery resources, while one of the targets of the 2nd SGD is to increase the productivity of food production, including aquaculture products [1] [2]. With the growing popularity of cage aquaculture, sustainability becomes its biggest challenge. Fish cages contribute to water quality degradation due to their effluents. Furthermore, their presence affects the hydrodynamic and sedimentation of the surrounding area. The focus of this study is to examine the effect of floating cage aquaculture on the hydrodynamic and sedimentation in the surrounding area, which can be used as additional criteria for the development of sustainable marine aquaculture in Indonesia.

Several studies have been carried out to investigate the effect of fish cages on currents and sedimentation near the aquaculture area. One of the key studies was conducted in the Bay of Fundy [3] which found that the presence of fish cages with a certain height and drag coefficient decreased tidal currents in the upper layers while increasing the currents in the bottom layer, thereby increasing the rate of erosion of the bottom sediment beneath the cages. This increase in erosion rate will reduce excessive sediment accumulation due to waste generated by aquaculture operations. A study of mussel aquaculture in the coast of East China Sea [4] in an area with a water depth of 15 m to 25 m found that the surface friction induced by the aquaculture was much stronger than the bottom friction, which caused a reduction in the tidal current of 75-90% and 45% in the upper and the bottom layers, respectively.

A numerical study of the flow field around fish net cages was conducted using the porous-media fluid model to establish the relationship between the porous coefficient and the net features. The porous coefficient was determined based on field measurements and empirical formula [5].

Most studies on marine aquaculture in Indonesia are conducted to find suitable locations for aquaculture development using currents and waves as criteria, such as studies in Situbondo in Java Sea [6] and Prigi Bay in the southern part of Java [7]. A study in Bali [8] provides a simple hydrodynamic method to estimate the production carrying capacity for sustainable cage aquaculture. The study also includes modeling waste from the fish cages and their impact on water and sediment quality. However, the presence of the fish cage structure itself is not taken into account in the sedimentation simulation.

In this study, simulations of tidal and wave-induced currents as well as sedimentation and morphology for a case study in Situbondo, East Java are carried out to determine the impact of fish cage structures on the surrounding area. Several fish cage configurations and energy-loss coefficient values are considered for sensitivity assessment. The model results are compared with available water levels and current data.

2. Method

The numerical model used in this study is the Delft3D-FLOW [9] to simulate the water level and tidal current and coupled with the Delft3D-WAVE [10] to simulate wave-induced currents and sedimentation. Simulations are carried out for two cases: 1) Bay of Fundy, Canada, and 2) Situbondo, Indonesia. The results of the Bay of Fundy model are compared with the measured data and modeling results provided by [3] and used to determine the value of the energy-loss coefficient of the fish cage structure. This energy-loss coefficient is then used for simulation in Situbondo, East Java, Indonesia.

2.1. Model Setup and Validation

2.1.1. Bay of Fundy Model Setup and Validation

The first case is located in the Bliss Harbor fish farm, Bay of Fundy, Canada with a stand-alone hydrodynamic simulation of tidal forcing using the FLOW module. A nesting method is used to reduce computational time for a more detailed simulation at the aquaculture site. The overall model covers the area of Scotian Shelf, Georges Bank, Gulf of Maine, and the Bay of Fundy and the nested model is placed near the Bliss Harbor fish farm. The overall model domain and the nested model domain are shown in **Figure 1** and **Figure 2**, respectively.

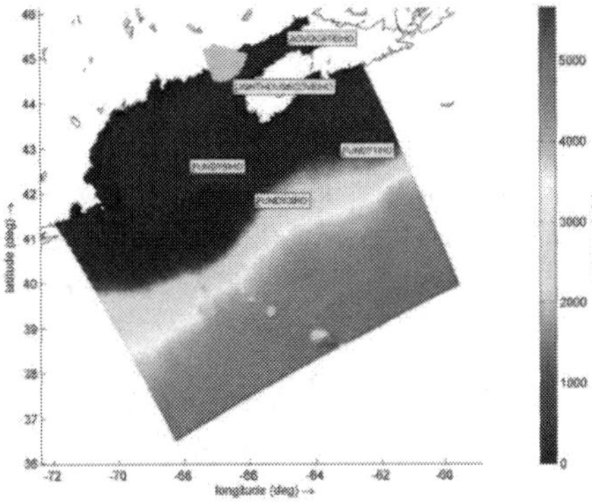

Figure 1. Overall model domain, tidal elevation stations, and bathymetry of Bay of Fundy. The nested grid is shown with grey areas.

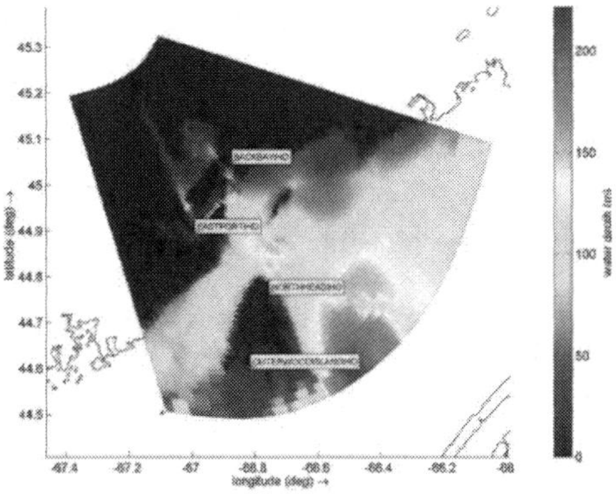

Figure 2. Nested model domain, tidal elevation stations, and bathymetry of Bay of Fundy

Table 1 shows the domain description of the Bay of Fundy model. The overall and nested models are simulated from 1 January 2021 to 8 February 2021. The 7-day spin-up interval is taken into account in the model results. The Manning coefficient of 0.032 gives the smallest error for water level and current so that it is then used for the nested model simulations.

Figure 3. shows the comparisons of model results with data. The water level compares well with data from the International Hydrographic Organization (IHO) [11] with an average error of 2.8%. The current M2 amplitudes also compare well with measured data at 13 locations (C1-C13) as well as the previously modelled result using FVCOM by Wu et al. [3].

Table 1. Bay of Fundy Domain Description

Domain Description	Bay of Fundy - Overall	Bay of Fundy - Nested
Grid type	Rectilinear	Curvilinear
Grid resolution	3.7 km x 4 km	50 x 60 meters - 2 x 1.7 km
Layers	5	5
Bathymetry	GEBCO [12]	CHS NONNA-100 [13]
Boundary Condition	TPXO 7.0 [14] water level constituent	Water level time series (curved boundary), current time series (straight boundaries)
Manning Coefficient	0.02, 0.03, 0.032	0.032

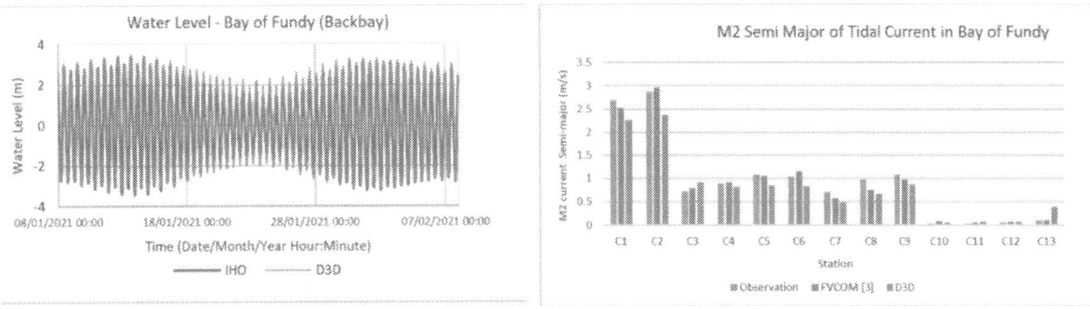

Figure 3. Bay of Fundy model comparison for water level (left) and current magnitude (right)

2.1.2. Situbondo Model Setup and Validation

The second case is located in Klatakan, Situbondo, East Java, Indonesia with the tide, wave, and sedimentation simulations. This case uses two levels of nesting as shown in **Figure 4 – Figure 6**. The first domain (the largest domain) shown in **Figure 4** includes the area of the north side of East Java; the second domain shown in **Figure 5** includes the Situbondo coastal area; and the third domain (the smallest domain) shown in **Figure 6** includes the area of the aquaculture site in Situbondo. The second domain will be nested in the first domain and the third domain will be nested in the second domain as shown in **Figures 4, 5,** and **6. Table 2** shows the domain description of the Situbondo model.

Table 2 Situbondo Domain Descriptions

Domain descriptions	Situbondo - Overall	Situbondo - 1st Nested	Situbondo - 2nd Nested
Grid Type	Curvilinear	Curvilinear	Curvilinear
Grid resolution	900 x 600 m - 1500 x 1500 m	55 x 40 m - 290 m x 160 m	3 x 3 m - 38 x 38 m
Layers	5	5	5
Bathymetry	BATNAS [15]	BATNAS [15]	Survey [6]
Boundary Condition	TPXO 7.0 water level constituent	West: Water Level North & East : Current	North: Water Level West & East: Current
Manning Coefficient	0.02, 0.02, 0.025, 0.027	0.027	0.027

In the smallest domain, simulations of waves, sediment transport, and morphology are also carried out. For the morphology simulation, a morphological scale factor of 365 is used to simulate the long-term effect of sedimentation. The type of sediment in this location is sand (non-cohesive sediment) [16] [17]. The morphology simulation will not include the accumulation of waste from the fish cage. To take into account the interaction of wave and tidal current, the simulations are carried out using an online coupling of FLOW and WAVE modules. The WAVE module requires a wave spectrum input as the boundary condition. The wave input in this model is a JONSWAP spectrum with a significant wave height (H_s) of 3 m and a peak period (T_p) of 4.1 s [6].

All three simulations of Situbondo are carried out from 25 January 2021 – 15 February 2021 with a spin-up interval of 7 days. The morphological simulation will represent the change of morphology in 15 years.

The Situbondo model results are compared with current data from HYCOM [18] and water level data from *Badan Informasi Geospatial* (BIG) [19] as shown in **Figure 7**. The smallest error is obtained using the Manning coefficient of 0.027, which is then used for the nested simulations. The average water level error is 3.3% and the average current magnitude error is 29.3%. From **Figure 7**, it can be seen that although the error percentage of the current is large, the current from the model results and HYCOM data are in the same order of magnitude.

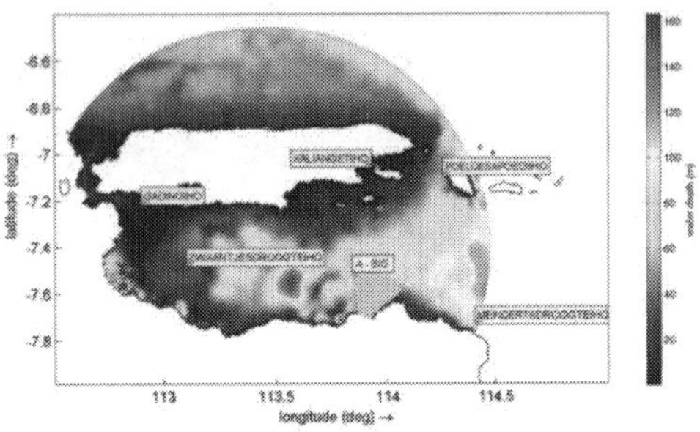

Figure 4. Overall model domain, tidal elevation stations, and bathymetry of Situbondo. The nested grid is shown with grey areas.

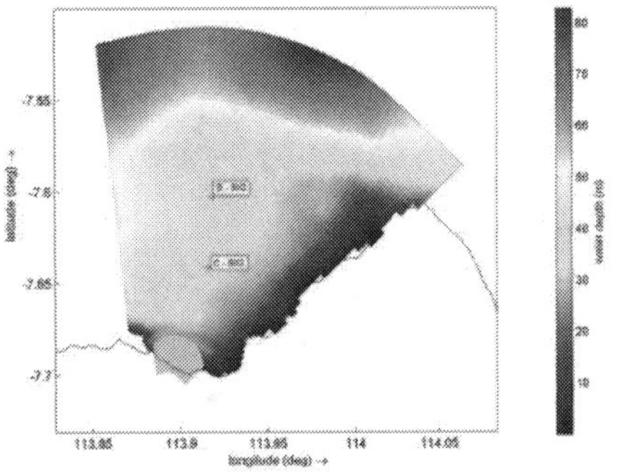

Figure 5. First nested model domain, tidal elevation stations, and bathymetry of Situbondo. The second nested grid is shown with grey areas.

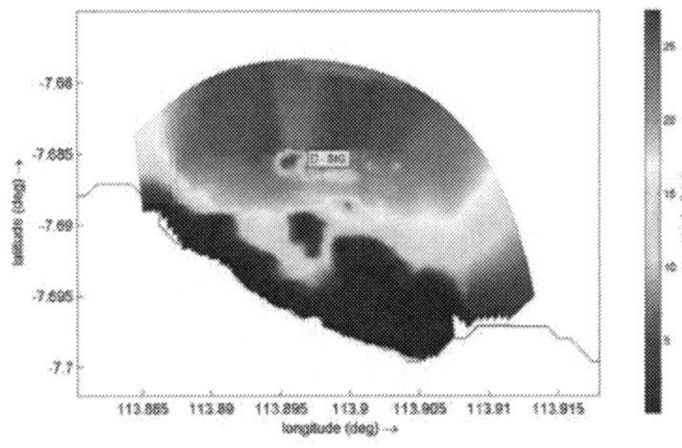

Figure 6. Second nested model domain, tidal elevation stations, and bathymetry of Situbondo

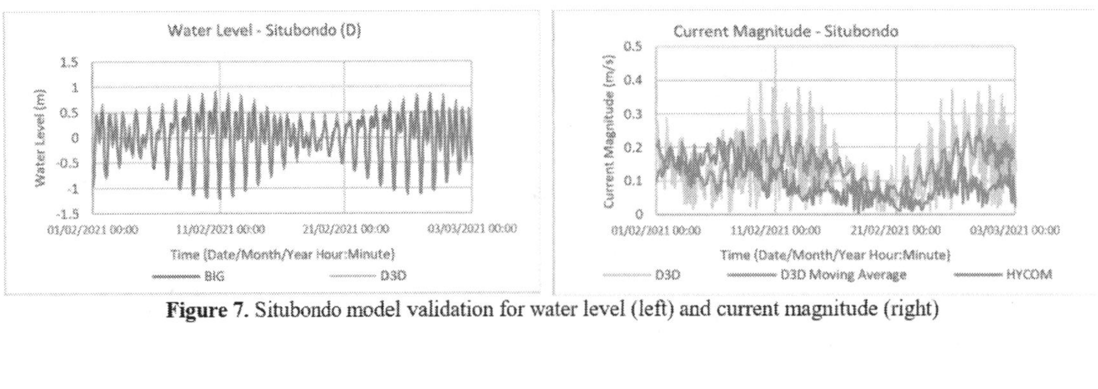

Figure 7. Situbondo model validation for water level (left) and current magnitude (right)

2.2. Fish Cage Simulation

The fish cage in the Bay of Fundy and Situbondo is modeled using a porous plate feature from Delft3D-FLOW and an obstacle feature from Delft3D-WAVE. The default setting of the obstacle in the WAVE module will be used due to the lack of data regarding the effect of the fish cage on the waves.

The porous plate is governed by a quadratic friction equation. A coefficient of energy loss is included in the quadratic friction equation to take into account the effect of the porous plate on the current through the plate.

The coefficient of energy loss is a non-dimensional coefficient depending on the drag coefficient, which can be calculated by Equation 1 [9]:

$$c_{loss-u} = \frac{N C_{drag} d_{pile}}{2\Delta y}\left(\frac{A_{tot}}{A_{eff}}\right)^2 \quad (1)$$

where

A_{tot} = total cross-sectional area ($A_{tot} = \zeta \Delta y$)
A_{eff} = effective wet cross-sectional area ($A_{eff} = A_{tot} - \zeta N d_{pile}$)
C_{drag} = drag coefficient of a pile
d_{pile} = diameter of pile
N = number of piles in the grid cells
ζ = water level
Δy = grid width

To calculate the c-loss, each basin of a cage is considered as a pile, and one unit of cage consists of nine (3 x 3) basins. The basin dimension of 3 m x 3 m is used as the pile diameter. The distance between each basin is 0.5 meters. This results in Atot of 10η and Aeff of η. A drag coefficient of 0.599 used in the model is based on a study by Bi and Xu (2018) [5]. **Figure 8** shows the location of fish cage units in the Bay of Fundy. Based on this configuration and the value of the drag coefficient, the value of c-loss can be calculated and the result is in the order of 10^1. Therefore, a variation of c-loss values of 10, 20, and 30, is used in this study. These values of c-loss are first applied in the Bay of Fundy simulation to examine their validity by comparing the model results of current velocities with the ADCP data [3]. Subsequently, the values of c-loss will be applied in the Situbondo case.

For the Situbondo case, two sensitivity studies are carried out, i.e., a sensitivity study of c-loss values on the hydrodynamics and sedimentation, and a sensitivity study of the fish cage configuration. For the study of c-loss values, simulation is carried out using c-loss values of 10, 20, 30. For the fish cage configuration, three units of cages will be placed in the Situbondo model as shown in **Figure 9**. The first configuration will only include cage Number 2, the second configuration will include cage Number 1 and Number 2, and the third configuration will include all three units of the cages.

In the 3-D model, the effect of the cage can be applied at arbitrary levels in the vertical direction. In this paper, the porous plate will be placed on the top three layers of the water column for both the Bay of Fundy and the Situbondo models. For the Bay of Fundy model, the fish cage is placed in a water depth of around 20 m, resulting in a fish cage height of 7-10 m. On the other hand, the fish cage in the Situbondo model is placed in a water depth of around 10 m, resulting in a fish cage height of 4-5 m. To simplify the problem, it is assumed that the fish cage in the Bay of Fundy and Situbondo is of the same type so that the same value of c-loss can be applied in both cases.

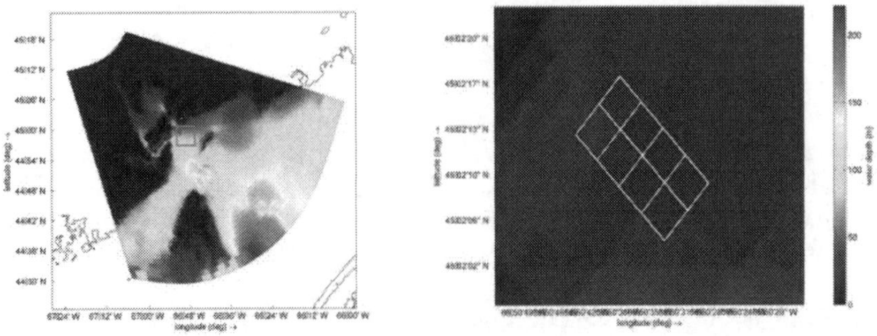

Figure 8. Location of the fish cage units in the Bay of Fundy marked by the orange square (left) and the zoomed-in map of fish cage configuration (right). The fish cage is drawn with white lines.

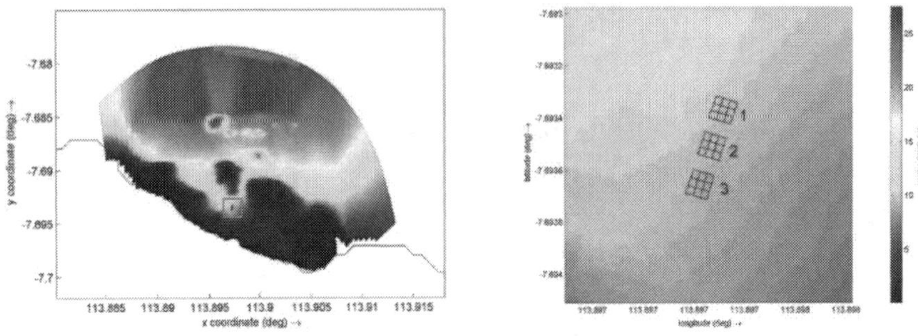

Figure 9. Location of the fish cage units in Situbondo marked by the orange square (left), and the zoomed-in map of fish cage configuration in Situbondo (right). The fish cages are drawn with black lines

3. Results

3.1. Hydrodynamic Simulation Results

3.1.1. Bay of Fundy Hydrodynamic Simulation Results

In this section, the tidal current from the simulation using the FLOW module in the Bay of Fundy and the effect of the fish cage will be discussed.

To examine the c-loss value of the cage, the current velocities from the simulations using the c-loss of 20 are compared with the current velocity components in M2 amplitude data from Wu et al. [3]. The data provided in this paper includes measured current data from various ADCP that is moored in several locations, namely mooring 5 and 6 that is placed outside of the fish cage and mooring 7 and 8 that is placed inside of the fish cage. With these measured data, the drag coefficient of the fish cage net can be calibrated. From the comparison shown in that the current amplitudes at Mooring 6 which is located outside the cage show a significant difference with data. The difference may be due to Mooring 6 being near the corner of the modeled fish cage and insufficient resolution of the model grid.

A sensitivity study is also conducted to investigate the variation of the c-loss values to the current magnitude. **Figure 11** shows model results using the c-loss values of 10, 20, and 30. The result shows that the increase of c-loss causes weaker currents inside the cage and at the edge of the cage. However, the change of current magnitude is insignificant. The difference of current reduction percentage between c-loss of 10 and 20 is about 2-4%, while the difference of current reduction percentage between c-loss of 20 and 30 is only about 0.5-2%.

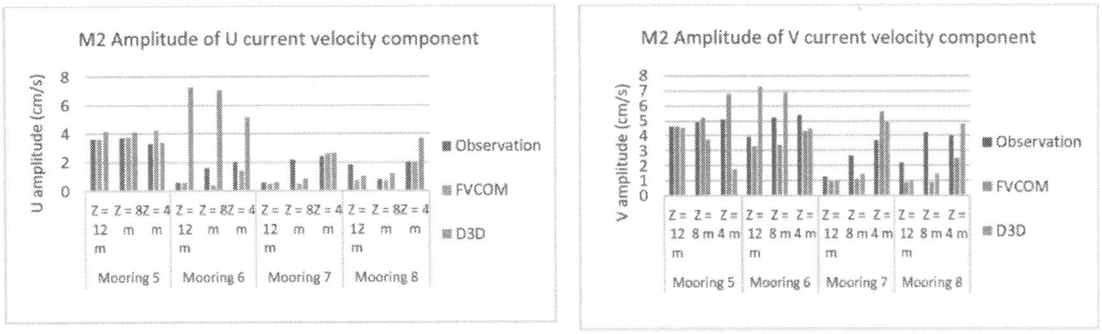

Figure 10. Comparison of eastward (left) and northward (right) tidal current M2 amplitude from observation, FVCOM modeling, and Delft3D modeling with the presence of the fish cage in the Bay of Fundy. Observation and FVCOM data are provided by Wu, et al. [3]. Z indicates vertical position measured from the seabed.

Figure 11 shows that the presence of the fish cage reduces the current velocity inside the cage significantly by about 60%. Below the cages (depth larger than 5 m), the current reduction due to the presence of the cages ranges from 25-40%. Meanwhile, near the seabed, the reduction of current velocity is much smaller which is only about 4%. This large difference of current reduction percentage creates a different current velocity profile compared to that without cages. It is noted that below the cages, the current velocity increases with increasing depth. However, the current velocity below the cages does not exceed the current velocity without cages. At the edge of the cage, the current velocity is also reduced by the presence of the cage. The percentage reduction is mostly uniform at all depths, ranging between 15-20%, thus maintaining the shape of the current velocity profile.

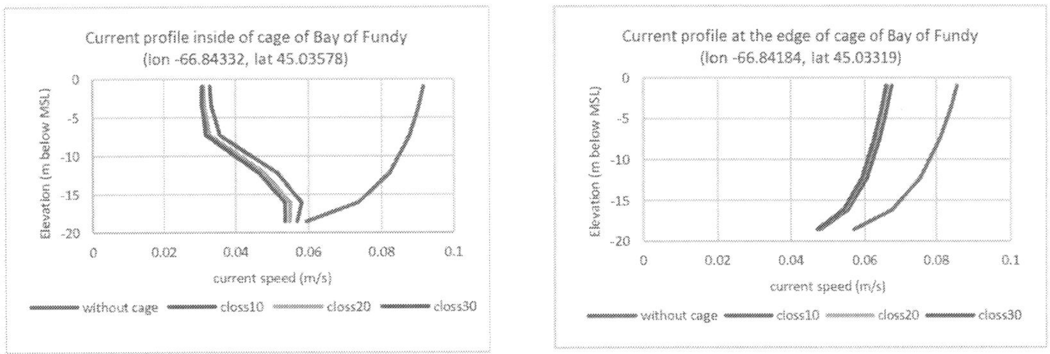

Figure 11. Tide-averaged current magnitude profile of the Bay of Fundy model inside of the cage (left) and outside of the cage (right) with varied values of c-loss.

Figure 12 shows the model results of the flow field around the Bay of Fundy fish cage during ebb tide (8 January 2021 00:00 UTC). **Figures 12(a)** and **(c)** show the simulation results without cages in the upper and bottom layers, respectively, whereas **Figures 12(b)** and **(d)** show the simulation results with a porous plate and c-loss value of 20 in the upper and bottom layers, respectively. It can be seen that the reduction of current velocity due to the presence of the cages is larger on the upper layer than that on the bottom layer, while the current direction changes in both layers.

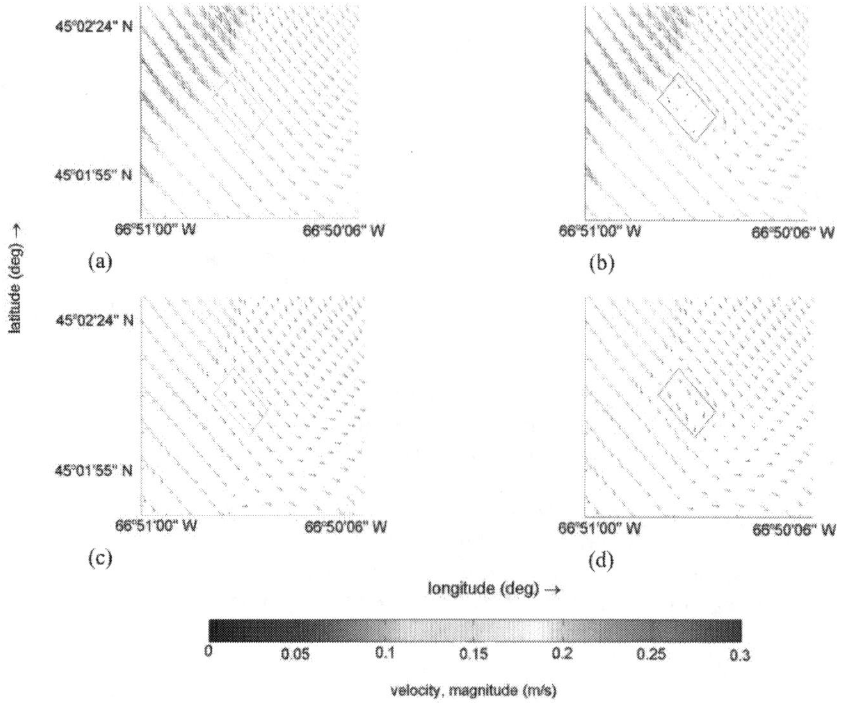

Figure 12. Flow fields around the Bay of Fundy fish cage from Delft3D Simulations during ebb tide (8 January 2021 00:00 UTC). (a) and (c) shows results without a cage. (b) and (d) shows results including a porous plate with a C-loss of 20. (a) and (b) shows flow fields in the upper layer and (c) and (d) shows flow fields in the bottom layer. The cage unit area is shown with black lines.

3.1.2. Situbondo Hydrodynamic Simulation Results

In this section, the current and wave results from the coupled simulation of the Situbondo model will be discussed as well as the effect of the fish cage with three different configurations.

The presence of one unit of fish cage in Situbondo reduces the tidal current inside the cage up to 80% at the water depth of the cage height as shown in **Figure 13**. This decrease of the tidal current is much more significant than that of the Bay of Fundy. Near the bottom layer, the current increases 15-30%. Outside the cage, a similar current profile shape occurred. The current within the cage's height is decreased to around 60%, while the current near the bottom layer is increased to 10-20%. Similar to the Bay of Fundy case, an increased value of c-loss leads to a slightly weaker current near the fish cage area. The decrease of current due to a fish cage with a c-loss value of 30 is 1-3% more than that of a fish cage with a c-loss value of 20.

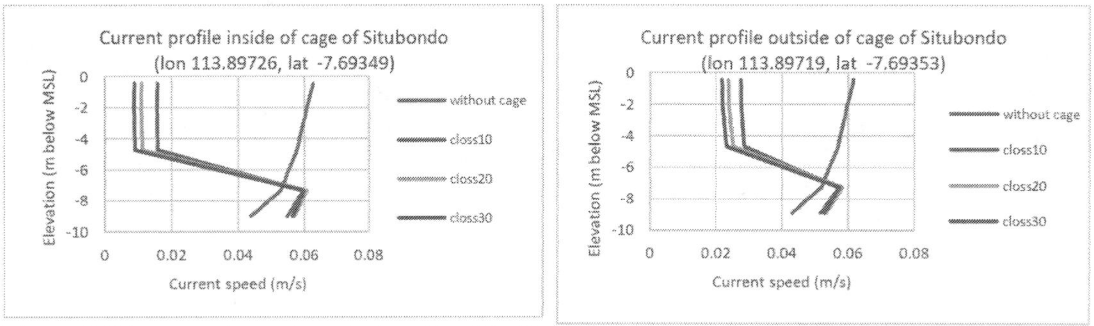

Figure 13. Tide-averaged current magnitude profile of the Situbondo model inside of the cage (left) and outside of the cage (right) with varied values of c-loss.

Figure 14 shows flow fields around Situbondo fish cage from simulations during flood tide (2 February 2021 at 02:00 UTC) for three fish cage configurations using a c-loss of 20.

Without the fish cage, the tidal current ranges between 0.1-0.2 m/s and 0.05-0.1 m/s at the top and bottom layers, respectively (**Figures 14a** and **14c**). With the presence of one fish cage, the current inside the cage at the top layer is reduced to below 0.05 m/s. Weaker current flows also occurred behind the fish cage structure. On the other hand, the magnitude is increased at the bottom layer. The direction of the flow is to the west on the top layer and the south on the bottom layer. These effects are more apparent with the increasing number of cages.

The presence of the fish cage in the WAVE module is represented by the obstacle feature. A default setting is used as the transmission and reflection coefficients of the obstacle. Due to the lack of data to confirm the validity of the obstacle parameters, the obstacle effect on the wave is analyzed qualitatively. **Figure 15** shows the simulation results of wave height without fish cage and with fish cage for three configurations. The presence of one fish cage reduces the wave height inside and behind the cage significantly. It also changes the direction of the wave propagation around it. With the increase in the number of cages, the wave heights inside cages Number 2 and 3 are almost zero. **Figure 15** clearly shows that behind the cages, the wave direction changes from southeast to south on the 2nd and 3rd cage configurations.

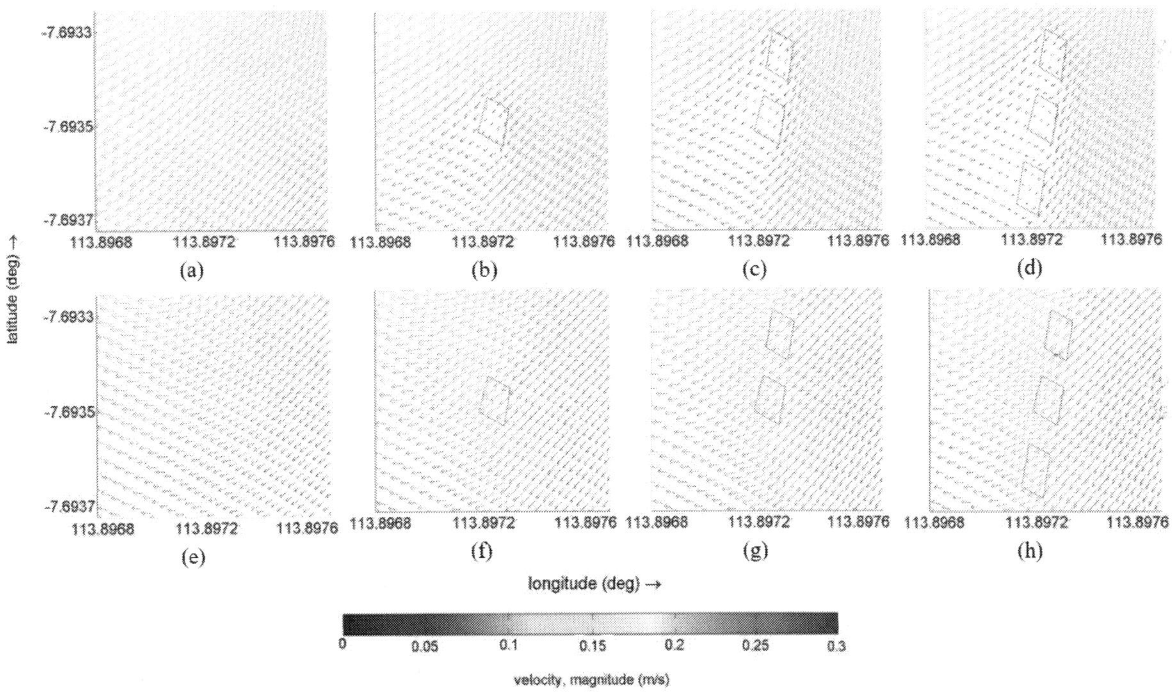

Figure 14. Flow fields around Situbondo fish cage from Delft3D simulations during flood tide (2 February 2021 at 02:00 UTC). The top row shows the surface-level current and the bottom row shows the current on the sea-bed level. (a) and (e) shows results without a cage. (b) and (f) shows results of the first cage configuration, (c) and (g) the second cage configuration. (d) and (h) the third cage configuration. The c-loss value used in all of the fish cages is 20. Cage units are shown with black lines.

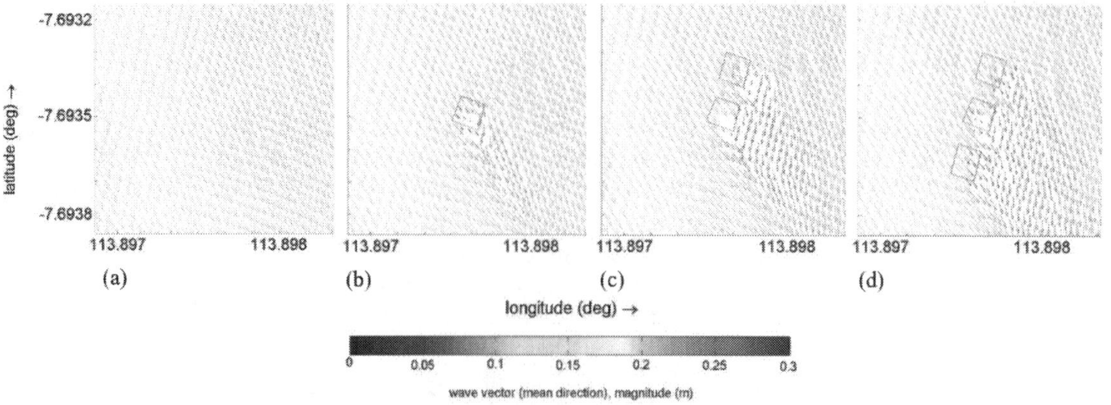

Figure 15 Wave height vector around the Situbondo fish cage from Delft3D simulations (during flood tide). (a) shows result without a cage. (b) shows result for the first cage configuration, (c) for the second cage configuration, and (d) for the third cage configuration. The cage units are shown with black lines.

3.2. Sediment and Morphology Simulation Results

In this section, the results of the sedimentation simulation of the Situbondo model will be discussed. The total deposition and erosion thickness in 15 years without cage is shown in **Figure 16(a)**. The areas where erosion and deposition occur are shown in blue and red, respectively. The average erosion and deposition thickness in these areas range from 1 m to 2 m. In some areas, significant morphology changes up to 5 m occur. However, at the location of the fish cage itself, the morphology only changes slightly. It is noted that the simulation does not include the fish cage waste. The cross-section of the sedimentation thickness near the shoreline (see **Figure 16a**) with different c-loss and cage configurations is shown in **Figure 16(b)** (one unit of cage) and **Figure 16(c)** (C-loss 20). A higher c-loss value results in higher erosion up to 1.3 m and cage configurations affect the change in morphology considerably.

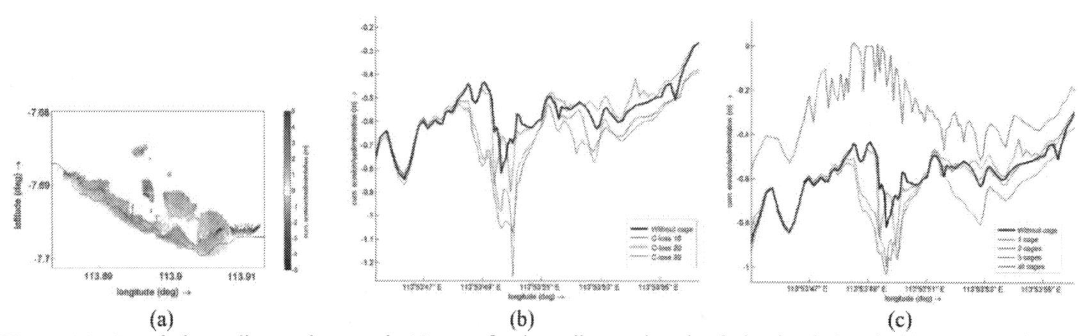

Figure 16. Cumulative sedimentation map in 15 years for the sedimentation simulation in Situbondo (a). Cross-section of the cumulative deposition and erosion thickness near the shoreline with variations of c-loss (b) and cage configurations (c). The cross section location in figure (a) is shown with the purple line.

Figure 17 shows the difference between the cumulative erosion and deposition simulated with the presence of fish cages and those without a fish cage. As shown in these figures, the presence of the fish cages causes a significant change of erosion and deposition on the area outside of the aquaculture site of Situbondo. However, at the aquaculture site itself, the sedimentation only changes slightly. The white color in this map shows sedimentation change that is less than 1 cm.

Figure 17(a) – **Figure 17(c)** show the sensitivity of the c-loss values on the change of sediment thickness. A cage with c-loss of 10 and 30 shows similar results, while the c-loss of 20 results in a smaller sediment reduction near the shoreline. **Figure 17(b), Figure 17(d),** and **Figure 17(e)** show the sensitivity of the number of cages unit to the sedimentation thickness. An apparent difference occurs in the areas near the shoreline where the reduction of the sedimentation is larger as the number of cages increases. This result can also be observed in the northern part of the model. **Figure 17(f)** shows the sedimentation changes with all the cage units present in the Situbondo aquaculture site. Unlike the previous sensitivity studies where the three cages are placed in a concentrated location, these cages are spread around the area of the aquaculture site. As a result, the sediment thickness near the shoreline increases considerably while the sediment thickness on the north part of the model decreases.

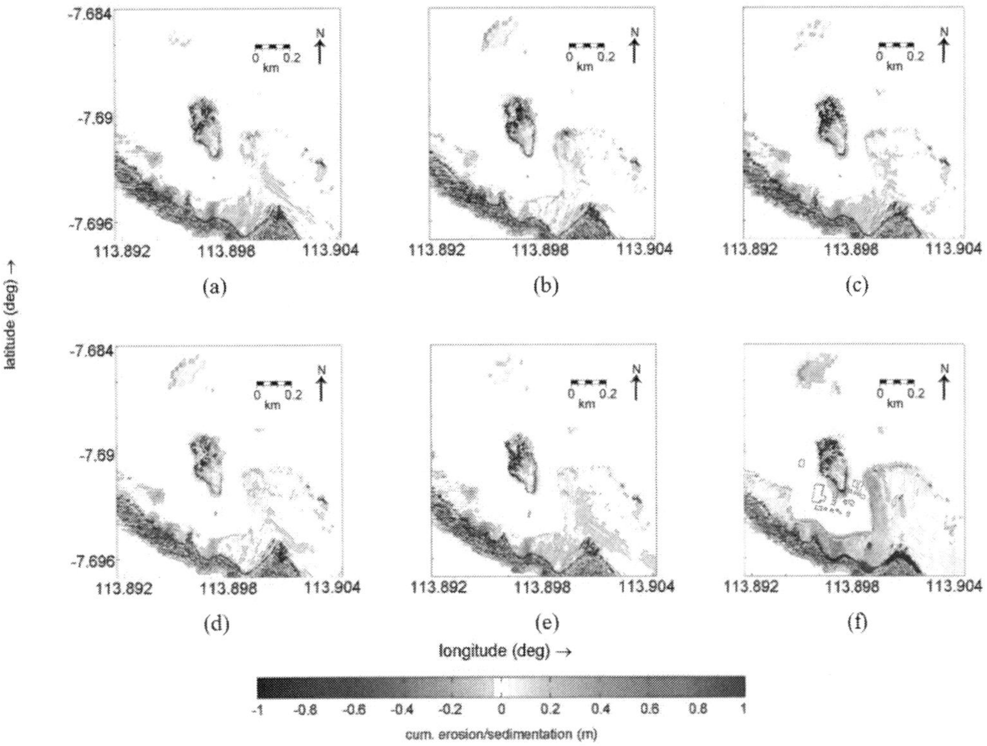

Figure 17 Changes of cumulative erosion and deposition with one cage unit (a, b and c) with c-loss of 10 (a), 20 (b), 30 (c), with two cage units (d), three cage units (e), and all cage units present in the Situbondo aquaculture site (f). C-loss of 20 is used in (d), (e), and (f).

4. Conclusion

The simulation results show that the presence of floating fish cages changes the hydrodynamic and sedimentation in the surrounding area. In the Bay of Fundy simulation, the presence of the cages causes a reduction of current velocity inside the cages but does not affect the current below the cages significantly. The insignificant effect on the current below the cages may be due to insufficient number of layers in water depth used in the simulation. In the Situbondo case, the current velocity decreases inside the cages and increases below the cages. The cages also affect the wave height and direction in the area. As for the sedimentation, the presence of fish cages in the Situbondo model does not cause a significant change of sedimentation in the fish cage area itself. However, its effect on the sedimentation in several locations near the cage area is noticeable. A 15 year-span simulation results in the average sea bottom changes of 1 m to 2 m in some areas and can be up to 5 m at a certain location. Therefore, the study of sedimentation is important to determine the impact of aquaculture on the environment of the surrounding area.

A sensitivity study of two cage variables, i.e., coefficient of energy loss (c-loss) and cage configuration, is conducted to see the impact of these variables on the current, waves, and sedimentation. The c-loss variation within the order of 10^1 does not change the current significantly. The increase of the c-loss value of 10 causes a reduction of the current inside the cage in Situbondo by only 1-3%. On the other hand, the current and wave fields are sensitive to the configuration of the cage. Sedimentation around the fish cage area is also sensitive to energy loss coefficient and cage configuration. However, the relationship between these cage parameters and the sedimentation is not as implicit as the relation between cage parameters and current speed. Generally, the presence of one to three small units of fish cage causes an increase in erosion near the shoreline. On the other hand, if all of the fish cage present in Situbondo is taken into account in the simulation, the erosion near the shoreline is reduced. The location of the fish cage and the dimension of the fish cage units will determine the sediment transport and morphological change in this area. Due to the lack of data in the Situbondo case, some inputs such as sediment and obstacle parameters are simplified to default values. These inputs may affect the reliability and accuracy of the model results. Future study with measured data needed for the simulation is expected to improve the results significantly.

The main sources of sedimentation from the aquaculture system include feed and fecal wastes. In the Situbondo case, the increase of erosion rate below the fish cage is not significant, which can cause problems such as over-accumulation of wastes at the aquaculture site. To keep a sustainable environment, the flushing of the fish cages has to be maintained by keeping good water circulation to avoid settling of the wastes in the seabed. For future works, a study of the effects of these wastes on the sedimentation and quality of water is needed. This study can be done by coupling the FLOW simulations with Delft3D-WAQ (Water Quality) module in the Delft 3D model.

5. References

[1] https://sdgs.un.org/goals (Accessed on October 30th, 2021).
[2] https://www.sdg2030indonesia.org/ (Accessed on October 30th, 2021).
[3] Wu, Y. Chaffey, J., Law, B., Greenberg, D., Drozdowski, A., Page, F., and Haigh, S., 2014. A three-dimensional hydrodynamic model for aquaculture: a case study in the Bay of Fundy, *Aquacult. Environ. Interact.* **5**, 235-48.
[4] Lin, J., Li, C., and Zhang, S., 2016. Hydrodynamic effect of a large offshore mussel suspended aquaculture farm, *Aquaculture* **451,** 147-155.
[5] Bi, C.W. and Xu, T.J., 2018. Numerical study on the flow field around a fish farm in tidal current, *Turkish J. Fish. Aqua. Sci.* **18,** 705-718.
[6] Risandi, J., Sagala, S., and Pranowo, W., 2015. The application of wave characteristic numerical model for site selection of marine aquaculture development in Situbondo, East Java, *Jurnal Kelautan Nasional* **10,** 21-31.
[7] Zikra, M., Armono, H., and Pratama, F., 2020. Wave modeling for the establishment potential area of offshore aquaculture in Indonesia, *Fluids* **5**, 229.
[8] Mayerle, R., Niederndorfer, K.R., Fernandez Jaramillo, J.M., and Runte K-Heinz, 2019. Hydrodynamic method for estimating production carrying capacity of coastal finfish cage aquaculture in Southeast Asia, *Aquacultural Engineering* **88** *Preprint*/102038.
[9] Deltares, 2014, Delft3D-FLOW user's manual, Deltares, Delft, The Netherlands.
[10] Deltares, 2014, Delft3D-WAVE user's manual, Deltares, Delft, The Netherlands.
[11] https://iho.int/en/twcwg (Accessed on August 27th, 2021).
[12] https://www.gebco.net/ (Accessed on October 21st, 2021).
[13] https://open.canada.ca/data/en/dataset/d3881c4c-650d-4070-bf9b-1e00aabf0a1d (Accessed on October 30th, 2021).
[14] https://www.tpxo.net/global (Accessed on August 4th, 2021).
[15] https://tanahair.indonesia.go.id/demnas/#/batnas (Accessed on March 1st, 2021)
[16] Bayhaqi, A. and Dungga, C.M.A., 2015. Distribusi butiran sedimen di pantai Dalegan, Gresik, Jawa Timur, *Depik* **4(3),** 153-159.
[17] Anggraini, R.R., Yanuhar, U., and Risjani, Y., 2020. Karakteristik Sedimen di Perairan Pesisir Lekok, Kabupaten Pasuruan, Jawa Timur, *Jurnal Ilmu dan Teknologi Kelautan Tropis* **12,** 235-46.
[18] https://www.hycom.org/ (Accessed on 5th August 2021).
[19] http://tides.big.go.id/pasut/index.html (Accessed on 6th August 2021).

6. Acknowledgment

This study is supported by P3MI-ITB. We thank A. Kurniawan and E. R. Gandanegara for their inputs and comments. We also thank M. Bintang and M. Satyadi. for their assistance in the Delft3D model.

Economic Valuation and Potential Pollution Load Analysis of Domestic Wastewater in Greater Bandung

Adhitya Rakhmadi(1), Suharyanto(2), Anindrya Nastiti(2)*

1Clean Water Supply and Sanitation Infrastructure Management Study Program, Institut Teknologi Bandung, Jalan Let. Jen. Purn. Dr. (HC). Mashudi No. 1, Sumedang, Indonesia
2Environmental Management Technology Research Group, Institut Teknologi Bandung, Indonesia, Jalan Ganeca 10 Bandung, Indonesia

Corresponding author: anindrya@tl.itb.ac.id

Abstract. Sanitation is still a significant problem in most developing countries. The lack of awareness and low education level are the main factors causing 30% of people to still practice open defecation. The problems are getting worse because of the population growth. They are not followed by the availability of sanitation infrastructure and services, leading to people mainly still discharging their domestic wastewater into the river. According to the Ministry of Environment and Forestry (2017), domestic wastewater contributes significantly to Citarum River's pollution by 68. From the calculation of domestic waste pollution estimation potential and analysis using SFD tools, the people in Greater Bandung produced 292 tons of BOD/day (106.580,25 ton/year), 401,5 tons of COD/day (131.101 ton/year), and 277,4 tons of TSS/day (101.251 ton/year). Ironically only 10,54% of the waste generated are safely managed, and 89,46% of the rest are unsafely managed. Pollution can cause several problems, including health and some more economic loss. The financial loss of the people of MBR is IDR 4.234.059.114.465/year or IDR 461.543/person/year. The economic loss from the health sector is responsible for about 29% of the total cost, it costs IDR 1.217.720.640.356/year or Rp 148.016/person/year. This is followed by clean water access cost for IDR 3.054.566.638.642/year or IDR 371.289/person/year (71%). This cost is equivalence to 1,3% of the Gross Domestic Product (GDP) of the people in the Metropolitan Bandung Area.
Keywords: Greater Bandung Area, domestic wastewater, economic loss.

1. Introduction

The sanitation problem in developing countries is still unresolved. The low public awareness and education cause the number of residents who still practice open defecation is relatively high, about 30% of the population [1]. The BOD load discharged to Citarum River from domestic sources is the highest compared to industrial, agriculture and animal husbandry [2]. The low priority of the sanitation sector for the central and local governments, as indicated by the amount of investment in the sanitation sector of IDR 200/KK/year [3] (ESI, 2008). Indonesia experienced a loss of IDR 56 trillion per year due to poor sanitation or around 2.3% of gross domestic product [3]. Economic loss due to poor sanitation is around IDR 252,000/person/year. The high population and economic growth are not accompanied by an increase in the quality and quantity of waste processing infrastructure, which causes domestic waste to reach water bodies without prior processing [4].

The Greater Bandung Metropolitan Area (MBR) is a regency/city area in the Bandung Basin consisting of Bandung City, Cimahi City, Bandung Regency, West Bandung Regency, and a small part of Sumedang Regency. Dense human activities in the Greater Bandung Metropolitan area, either domestic or industrial, can potentially cause contamination due to human activity waste, leading to environmental degradation if not appropriately managed. To prevent river pollution from becoming more severe, periodic monitoring is required, and the commitment from stakeholders to control the environment, in this case, regional heads, industrial actors, and the community. Therefore, to encourage stakeholder awareness, a calculation for the impact of economic losses and a management strategy for domestic wastewater pollution in the Greater Bandung area are needed.

2. Methodology

This research is quantitative research. The research flow is as shown in **Figure 1**.

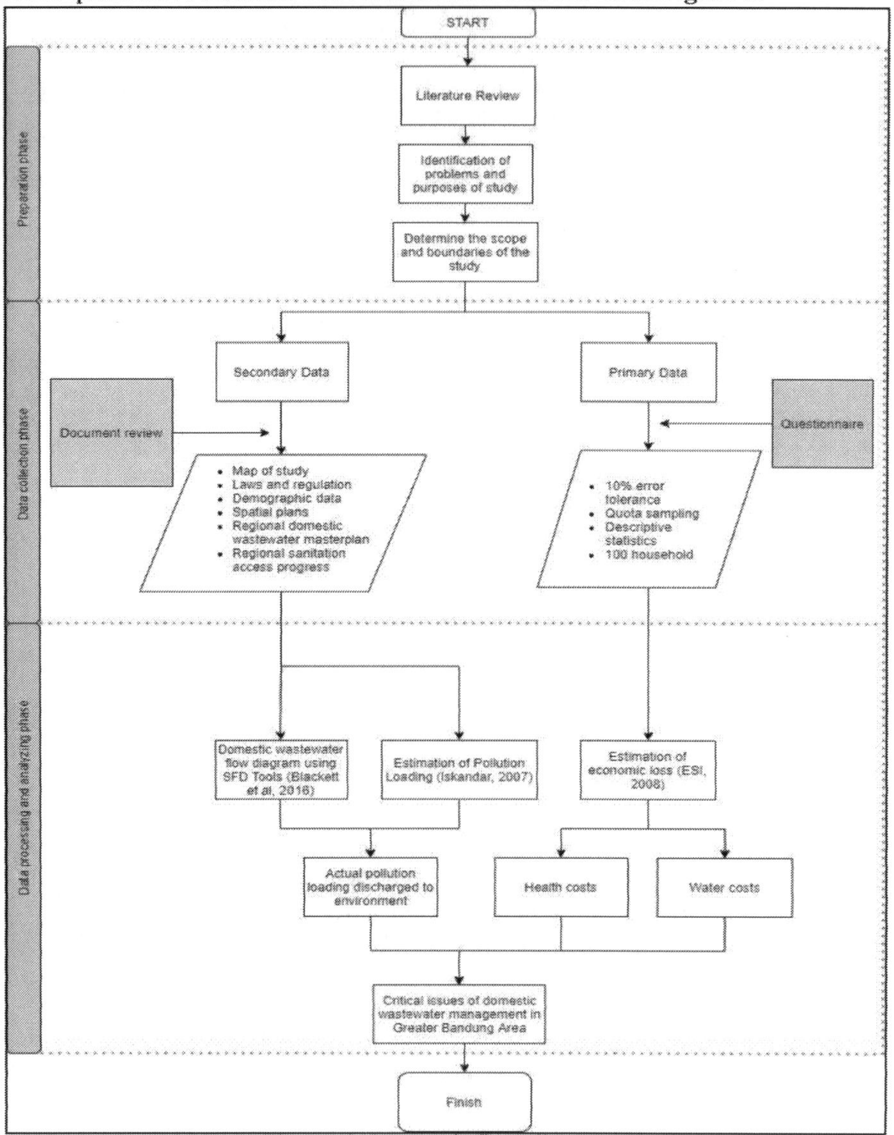

Figure 1. Research flow diagram

2.1. *Number of Samples and Sampling Technique*

In this study, data collection was conducted through a literature study to get secondary data and questionnaires to the community for the primary data to find out the amount of economic loss experienced by the community in MBR, which is caused by domestic liquid waste pollution. The number of sample sizes can be calculated using the number of households in the Greater Bandung area. According to data from the Central Bureau of Statistics Republic of Indonesia (2020), households living in the Greater Bandung area are 2,157,199. In this study, the level of confidence sought is 90%, so by Taro Yamane formula in Ridwan [4]:

$$n = \frac{N}{1 + Ne^2}$$

$$n = \frac{2.157.199}{1 + (2.157.199 \times 0{,}1^2)}$$

$$n = \frac{2.157.199}{21.572{,}99} = 99{,}995 \approx 100$$

Where :
n = Number of samples required
N = Number of households living in Greater Bandung Area
e = 10% error tolerance level

The method used was a non-probability sampling method. This sampling technique does not provide equal opportunities or chances for each element or member to be selected as a sample. The sampling technique used was purposive sampling.

2.2. *Validity Test and Reliability Test*

The validity test is the ability level of the research instrument to reveal data according to the problem to be uncovered. Meanwhile, the reliability test is related to the reliability problem toward the instrument test. This test is used to test the consistency of the data obtained. The validity and reliability tests of the questionnaires were conducted on 30 respondents, and the analysis used the Karl Pearson product-moment formula and the Cronbach Alpha test software using SPSS 26 software.

2.3. *Type of Data*

Data collection was conducted to support the analysis to be undertaken. The data requirements needed in this study were divided into two, primary data and secondary data. The detail of the primary and secondary data required can be seen in **Table 1** and **Table 2**:

Table 1. The primary data

Type of Data	Data Source	Number of Samples	Method
Economic impact of wastewater pollution	Respondents	114	Questionnaire
SWOT analysis	Respondents	10	Interview and Online Interview

Table 2. The detail of secondary data

Type of Data	Data Source
Base map of the research area	Development Planning Agency at Sub-National Level (Bappeda) of District/City
Land use map	Development Planning Agency at Sub-National Level (Bappeda) of West Java
Demographic data	Central Bureau of Statistics Republic of Indonesia (BPS) of District/City
Community sanitation access	Health Department and Settlement Department of District/City
Data of disease incidence	District Health Department and BPJS Kesehatan
Rules and Regulations	District Government/City Government/Provicial Government of West Java

2.4. *Calculation of Domestic Waste Pollution Load*

The data obtained from the results of questionnaires about sanitation conditions in the research area were then analyzed and shown in the form of a wastewater flow diagram, a modification of the shit flow diagram (SFD), a method for making the assessments for urban scale waste management.

2.4.1. *Water Quality Monitoring with the Daily Pollution Load*

The estimation method of pollutants generated from uncertain sources in this study was limited to 3 parameters, BOD, COD, and TSS, using the following calculations [5]:

$$\mathrm{PBPD} = \mathrm{F}\left(\frac{\mathrm{gr}}{\mathrm{person-day}}\right) \times \mathrm{Population} \times \mathrm{REK} \times \alpha \quad (1)$$

Where:

PBPD	=	Calculation of Domestic waste pollution load
F	=	Domestic emission factor (BOD = 40 gr/p/d ; COD = 55 gr/p/d ; TSS = 0,625 gr/p/d)
REK	=	City equivalence ratio (Urban = 1; Peri-urban = 0,8215 ; Rural = 0,6250)
α	=	Weight transfer ratio [0-100 m = 1,0 ; 100 m – 500 m = 0,85 ; >500 m = 0,30]

According to [6], the correction of weight transfer ratio (α) in metropolitan cities (>1 million population) and big cities (>500 million people) is equal to 1. The area with the lowest population in this research is Cimahi City, with a population of 553.755. Therefore, the weight transfer ratio in all areas is equal to 1.

2.4.2. *Mapping of Potential Pollution Loads with SFD tools*

After we calculated the estimated value of the domestic pollutant load in the Greater Bandung area, the next step was to assess the sanitary conditions of the community. This needs to be done to find out the factual situations in the field. Data of community sanitation conditions was obtained through secondary data collection in performance data or community sanitation access obtained from related government agencies. The conditions of community sanitation were described using a tool, the faecal waste flow diagram (SFD), to determine what percentage of domestic waste was handled safely and what percentage of domestic waste was discharged into the environment in the form of a mass balance. Therefore it was then calculated with the results of the calculation of the potential pollutant load obtained previously

2.4.3. *Sanitation Impact Value on Financial and Economic Aspects*

According to ESI [3], the calculation in the value of financial and economic impacts from domestic wastewater pollution is essential to be conducted. This is used to see whether the poor management of domestic waste in the Greater Bandung area can affect revenue and expenditure, either from the community or regional revenue. Calculation of the total cost of financial and economic impacts from poor wastewater management in the Greater Bandung area can be done by the estimate of the health cost and water cost [3], as shown in **Table 3.**

Table 3. Economic Impact Cost Calculation Methods [3]

Equation		Notes
$\boldsymbol{C = CH + CW}$	(2)	C : Total economic cost CH : Health cost CW : Water related cost
$\mathbf{CH = CH\_HC + CH\_P}$	(3)	$\sum CH_{HC}$: Total health cost $\sum CH_{P}$: Productivity loss cost
$CH\_HC = \sum_i CH\_HCi$	(4)	CH_HC: Total health cost CH_HC_i: Medical cost of type-i disease
$CH\_HCi = \alpha_i \times pop \times \sum_h \times phealth_{ih}$	(5)	CH_HCi: Total medical cost of type-i disease α_i : Incidence rate of type-i disease pop : Population $phealth_{ih}$: Average of medical cost of type-i disease
$CH\_P = \sum_i CH\_P_i$	(6)	CH_P: Total *productivity cost* CH_P_i:*Productivity cost* of type-i disease
$CH\_P_i = \alpha_i \times pop \times dh_i \times ptime$	(7)	CH_Pi: Productivity cost of type-i disease α_i : Incidence rate of type-i disease dh_i: Days of absence from work/school due to i-type disease ptime: Average income per day
$\mathrm{CW} = \mathrm{CW}_{domestic} + \mathrm{CW}_{\mathrm{treatment}} + \mathrm{CW}_{\mathrm{access}}$	(8)	$CW_{domestik}$: Cost incurred for getting a replacement of water source (Rp/year) $CW_{treatment}$: Cost incurred to treat/process water (Rp/year) CW_{access}: Cost incurred for accessing clean water (Rp/year)
$\mathrm{CW}_{\mathrm{treatment}} = (\bar{\mathrm{X}}_{\mathrm{construction}} + \bar{\mathrm{X}}_{\mathrm{om}}) \times \mathrm{pop}$	(9)	$CW_{treatment}$: Cost incurred for processing water $X_{construction}$: Water treatment unit construction cost X_{om}: Annual maintenance cost Pop: Number of households
$\mathrm{CW}_{\mathrm{access}} = (\mathrm{t}_{\mathrm{access}} \times p_{time}) \times \mathrm{pop}$	(10)	CW_{access}: Productivity loss cost to access clean water source t_{access} : Average time needed to access clean water source (day) p_{time} : Average income per day Pop : Number of Household

2.4.4. *Strategic Issues of Domestic Wastewater Management in Greater Bandung Area*

After knowing the amount of pollutant load wasted into the environment and the impact of its economic loss, a descriptive analysis was conducted to map the critical issues that occurred due to domestic wastewater pollution in the Greater Bandung area. Mapping the existing problems certainly eases the process of diagnosis and formulating strategies for managing domestic wastewater in Greater Bandung so that it can be managed safely.

3. Results

3.1 *Calculation of the potential MBR domestic Pollutant Load and SFD Analysis*

Overall, the people in the Greater Bandung produced 292 tons of BOD per day or 106,580 tons of BOD per year, 401.5 tons of COD per day, or 146,647 tons of COD per year, and 277.4 tons of TSS per day or 101,251 tons of TSS per year. **Table 4** shows the calculation of the potential domestic pollution load in the Greater Bandung area.

Table 4. Potential Pollution Load in MBR

District/City	Number of population[1]	REK	BOD[2] (ton/day)	COD [2] (ton/day)	TSS[2] (ton/day)
Bandung City	2.480.464	1	99,22	136,43	94,26
Bandung Regency	3.524.960	0,8215	115,83	159,27	110,04
West Bandung Regency	1.667.734	0,8215	54,80	75,35	52,06
Cimahi City	553.755	1	22,15	30,46	21,04
Total	8.226.913		292,00	401,5	277,40

Source: [1][8], [2]Data Processing

Access to sanitation services in the Greater Bandung Metropolitan area was divided into four types of services: using an individual septic tank, communal, regional on-site, and *cubluk* (traditional septic tanks). **Table 5** shows the percentage of sanitation access based on the service:

Table 5. Sanitation access in MBR based on the type of service [9]

No	Types of services	Number of households	Per cent of access (%)
1.	Individual septic tank	1.434631	56,92
2.	Communal septic tank	159.005	6,88
3.	Off-site system	118.236	5,11
4.	Unsealed pit latrine (*Cubluk*)	208.442	9,01
5.	Open defecation	510.600	22,08
	Total	2.157.199	100,00

It can be concluded that the sanitation conditions in the Greater Bandung area are still in poor condition. The percentage of domestic wastewater safely managed in the Greater Bandung area was only 10.54%, or only 32,24 tons of BOD per day, while those that were not safely managed were 259,76 tons of BOD per day. According to the maximum daily load (MDL) analysis, 76,87% of BOD reduction needs to be achieved to meet the MDL condition. From the SFD graphic above, it can be seen that if the local government in the Greater Bandung is committed to focusing on building a sludge treatment plant and providing faecal trucks, more than 50% of domestic wastewater pollution in the Greater Bandung area can be managed.

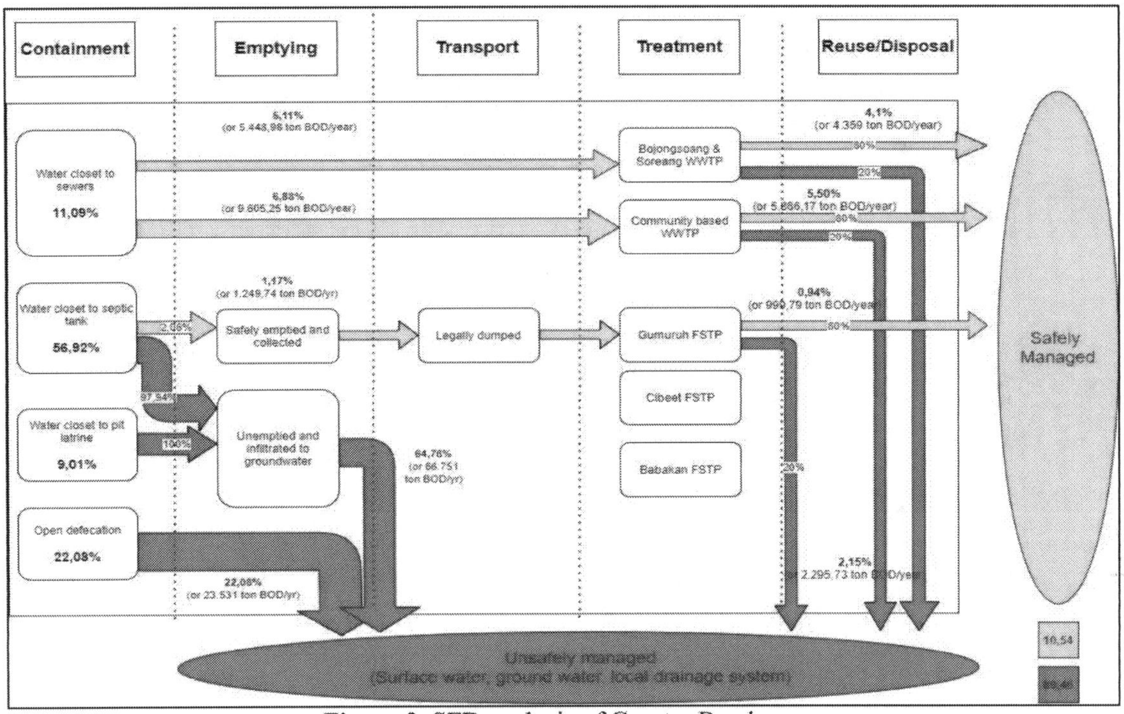

Figure 2. SFD analysis of Greater Bandung

Following the SFD analysis, we can compare the actual pollution load that goes into the watershed to the self-purification capability of the Citarum River. According to the Ministry of Environment and Forestry, the total maximum daily load (TMDL) of BOD that can be self-purified in the segment of Citarum River is only 60 tons BOD/day. The daily load of BOD that the people in Greater Bandung generate is 292 tons BOD/day, and after the small portion of it got treated, the actual BOD load that flows to the environment is 259,75 tons BOD/day or more than 400% of the TMDL of Citarum River. The following comparison is as shown in **Table 6** below:

Table 6. Existing BOD Load compared to Total Maximum Daily BOD Load

District	[1]BOD load (ton/day)	[1] BOD load discharged to the watershed (ton/day)	[2]Total maximum daily BOD load (ton/day)	Reduce of BOD load needed (%)
Bandung City	99,22	79,39	25,53	67,84
Bandung Regency	115,83	106,89	12,65	88,16
West Bandung Regency	54,80	21,41	10,50	78,13
Cimahi City	22,15	52,03	11,38	50,97
Total	292,00	259,75	60,07	76,87

Source: [1]Data Processing, [2][5]

We still need to reduce 76,87% of the BOD generated according to the calculation above. **Figure 3** shows that the wastewater treatment facilities in Greater Bandung Area are inadequate, especially in Bandung Regency and West Bandung Regency. The BOD load from Cimahi City and Bandung City are better than the two Regencies. However, the number is still too high, that means there are still a lot of room for improvement in domestic wastewater management in Greater Bandung Area.

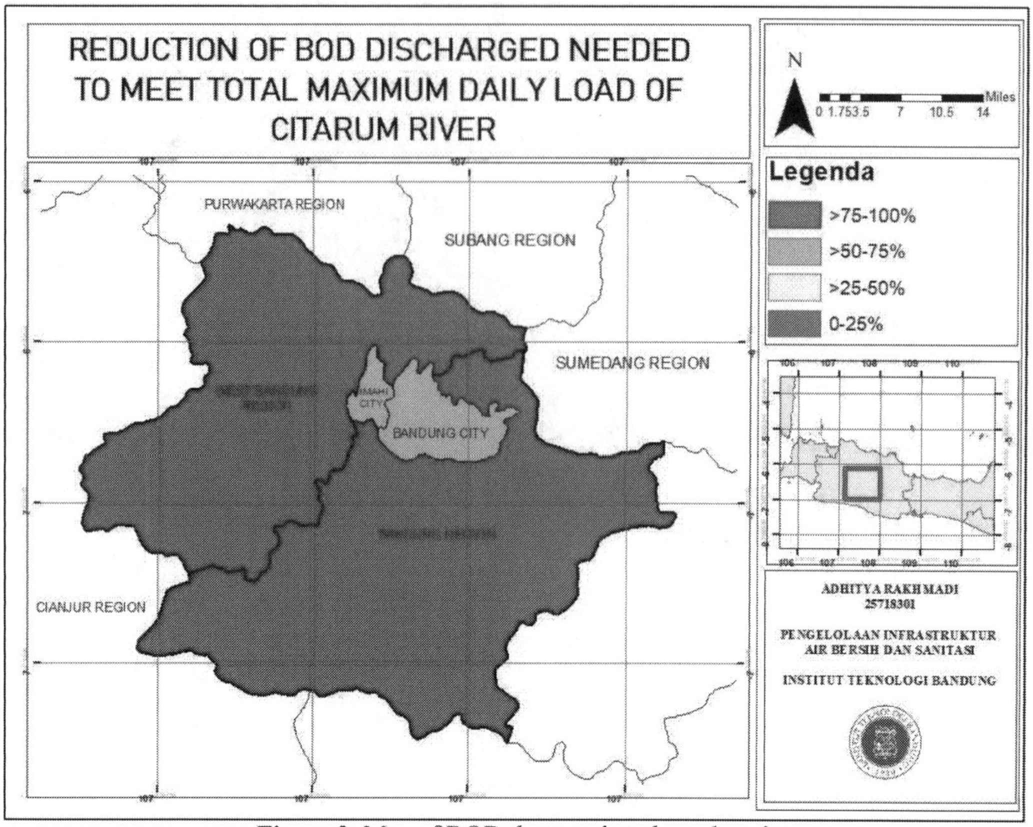

Figure 3. Map of BOD that needs to be reduced

3.2 The impact of Economic Losses

The calculation found that the community's costs for health service costs for diseases caused by domestic wastewater pollution were IDR 310,226,505,747/year or IDR 37,708/person/year. Furthermore, it can be seen that the cost of productivity loss was IDR 907,494,134,608/year or IDR 110,307/person/year as illustrated in **Figure 4** below.

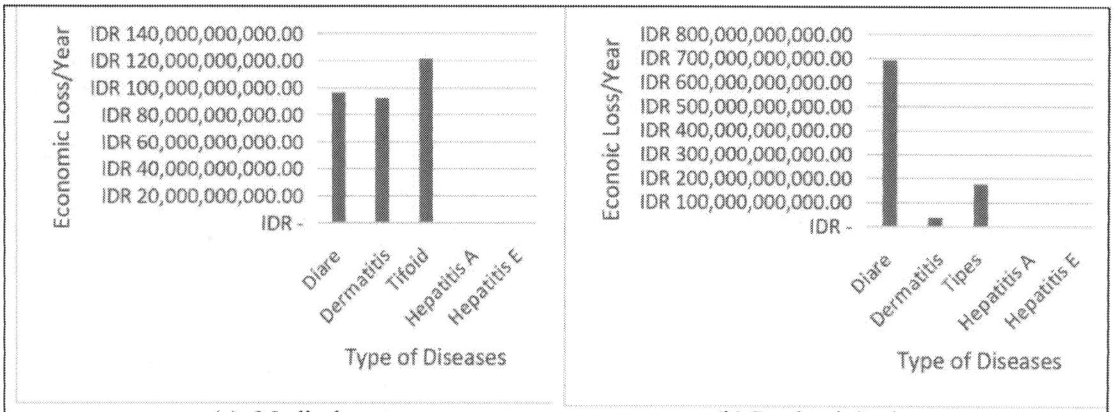

(a) Medical cost (b) Productivity loss cost

Figure 4. The cost of economic losses from the health sector

After the calculation, it can be seen that the total cost of economic losses for people in the Greater Bandung area from the health sector was IDR 1,217,720,640,356/year or IDR 148,016/person/year (32% of total economic loss), consisting of medical costs of IDR 37,708/person/year and cost of productivity loss of IDR 110,307/person/year.

Table 7 shows the economic losses from the water sector are as follows.

Table 7. Total economic losses from the water sector

Types of losses	Total losses (IDR/year)	Economic loss/ person (IDR/people/year)
Water sources replacement	2.510.967.097.467	260.997
Water processing	355.119.137.368	36.912
Productivity loss	150.252.239.273	15.617
PDAM Marginal cost	38.228.164.533	4.646
Total	3.054.566.638.642	371.289

From the table above, the water sector produces the most impact on the cost of economic loss for the low-income community with a loss of more than 3 trillion rupiahs/year or 71,23% of total financial losses of the people in the Greater Bandung Area. The poor water quality causes people to have to treat water before it can be consumed or used for their daily needs. Also, it adds up to the PDAM operational costs because PDAM will need to use more chemicals like chlorine and coagulant agent to treat the low-quality raw water into clean water. **Table 8** shows the total economic loss cost of MBR.

Table 8. Total Economic Loss Cost of MBR

Type of Cost	Total cost /person/year (IDR)	Total cost/year (IDR)	%
Health cost	148.016	1.217.720.640.356	28,77
Water cost	371.289	3.054.566.638.642	71,23
Total	519.306	4.272.287.278.998	100,00

3.3 SWOT Analysis

After we analyse the condition of domestic wastewater management in Greater Bandung, we need to make a strategy alongside the action plan that we have to do to make a better domestic wastewater management system, that way we need the SWOT analysis. The SWOT analysis data was collected using interview and an online interview with a guideline prepared beforehand in this research. The interview's respondence is limited to domestic wastewater management stakeholders, such as practitioner/professional, researcher, and operator (PDAM and Department of Public Works and Housing). The chosen strategy is Weaknesses and Opportunities (W-O), which consists of developing the offsite system in the selective area and the aggressive development of an onsite system that is out of service of the offsite system.

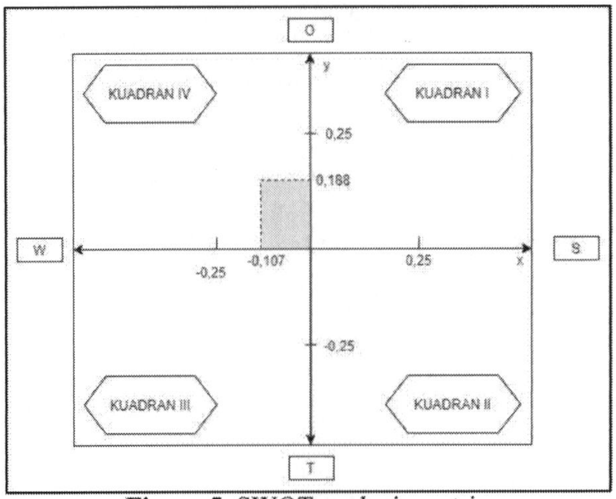

Figure 5. SWOT analysis matrix

4. Conclusion

The open defecation rate in the MBR is still very high; around 501.600 households still practise open defecation (22,08%) although about 56,92% of the household already had an individual septic tanks, but there are only 5,11% of them that are connected to the piping system and flows to Bojongsoang and Soreang wastewater treatment plant. This causes a low amount of domestic wastewater safely managed in the MBR of only 10,54%, while the remaining 89,46% are still unsafely discharged into the environment. This has the necessary role as a cause of economic loss for the low-income community, with an average loss of Rp 519.306/person/year, or equivalence to 1,3% of Greater Bandung Area's gross domestic product (GDP).

References

[1] Soedjono. Setiadi E, Fitriani and Nurina 2016 *Jurnal Sains dan Teknologi Lingkungan* **8**

[2] Suharyanto, Matsushita J 2011 *Lakes and Reservoirs: Research and Management* **16** 185-193 Doi:10.111/j.1440-1770.2011.00473.x

[3] Economic of Sanitation Initiatives 2008 Economic Impacts of Sanitation in Indonesia

[4] Marganingrum D, Roosmini D, Pradono, Sabar A 2008 *Jurnal Riset Geologi dan Pertambangan* **23** 37-48

[5] Ridwan 2005 Metode dan Teknik Penyusunan Tesis

[6] Ministry of Environment and Forestry 2017 Buku Kajian Daya Tampung dan Alokasi Beban Pencemaran Sungai Citarum

[7] Kersten S M, Hutton G, Firmansyah I, Leusbrock I and Zeeman G 2016 *Journal of Environmental Protection* **7** 1439-1465 http://dx.doi.org/10.4236/jep.2016.711122

[8] District Central Bureau of Statistic 2020 Kabupaten/Kota dalam Angka Tahun 2020

[9] District Department of Public Works and Housing 2020 Laporan Kinerja Tahun 2020

Acknowledgement

This research is supported by the "Global Sanitation Graduate School" funded by the Bill & Melinda Gates Foundation (OP1192599) and managed by the IHE Delft Institute for Water Education.

The properties of rolled compacted concrete as a preservation material

A B Utomo[1], **G Primaswari**[1]

[1]Departement of Highway Construction Technology, Polytechnic of Public Work, Prof Soedarto, SH, Tembalang, Semarang, Central Java, Indonesia
adityobudiutomo@pu.go.id, gita.primaswari@pu.go.id

Abstract. In Indonesia the frequency of road damage's repairing (like pothole, unsmooth, uneven, and crack) is increasing every year and takes precedence when the damage is still early in small size. In rainy conditions, practically the activity of patching the hole can not be done while usually many holes occur during the rainy season. The study was conducted to evaluate the optimal composition of Rolled Compacted Concrete (RCC) materials as a road-repaired material on a laboratory scale. RCC is a type of concrete that is compacted with mechanic equipment (like vibrator roller and plate tamper). This compacting method is like the application of cold mix asphalt on road preservations. It was found that using RCC with cement content minimum of 18%, adding 2,0% of viscocrete to the weight of cement, and w/c ratio 0,3 will produce compressive strength 10 MPa in a day. The average value of RCC density is 2270 kg/m^3 so its classified as normal concrete. From economical analysis, RCC production cost more efficien 5% compared to ordinary concrete at the sama quality and age. The application of RCC can increase the efficiency of worker and equipment component in range 35-85%.

Keywords : rolled compacted concrete, road damage, efficiency

1. Introduction

The types of road pavement in Indonesia consist of 4 types are flexible, rigid, composite, and granular pavement. Common road pavement damage can be ruts, uneven, weaving, cracked blocks, crocodile cracks, irregular cracks, transverse cracks, longitudinal cracks, hair cracks, dan edge cracks. Damage to road pavement or asphalt layers should be prioritized for repair because in high rainfall areas such as Indonesia, pavement can be damaged faster than its service life. The implementation of hole patching or road preservation in high traffic volume locations should:

a. Preparation for procedures and schedules appropriately
b. Equipment for road preservation must be in good condition
c. Field preparation, dispose of damaged material and immediately replace it with class an aggregate that meets the requirement or Cold Mix Asphalt mixture (CMA)
d. Do not leave digging holes or potholes on the road surface until night
e. The entire thickness of the patch CMA should be compacted in one layer at a time so that the final surface of the layer after compacted to a flat or slightly higher than the existing pavement surface. If the thickness of the layer is too thin, there is a tendency for the layer to peel off.
f. Road patching activities should be carried out in the shortest possible time so that traffic can be reopened

The material used for patching or repairing of pothole surface must be of equal or higher quality than existing materials. The material that is often used for patching is cold mix asphalt (CMA) which is available in 25 kilograms packaging. Hole patching should be carried out only on dry surfaces and should not be carried out when winds are strong, rainy, or will rain [1]. Patching with asphalt adhesive that is done when it will rain will make the patch material easily detached and lost when mixed with water.

RCC is a type of rigid pavement made of concrete with a slump value of 0 cm which has a lower cement content than ordinary concrete [2]. Advantages of using RCC include [3] :

a. Cost depends on the level of difficulty of the structure. The cost of making RCC is 25-50% cheaper than regular concrete.
b. Faster construction. For large-scale work, RCC dams can be completed 1-2 years faster than when using mass concrete.
c. Less cement is used due to simpler concrete mixtures
d. The cost of formwork becomes lower because the casting method is done gradually layer by layer.
e. Cooling pipes are not needed due to lower concrete temperature
f. The cost of transportation, casting, and compaction is lower because it can be transported by a dump truck, scattered with a bulldozer, and compacted with a vibratory roller.
g. The portion of equipment and labor use becomes larger due to the higher rate of casting speed

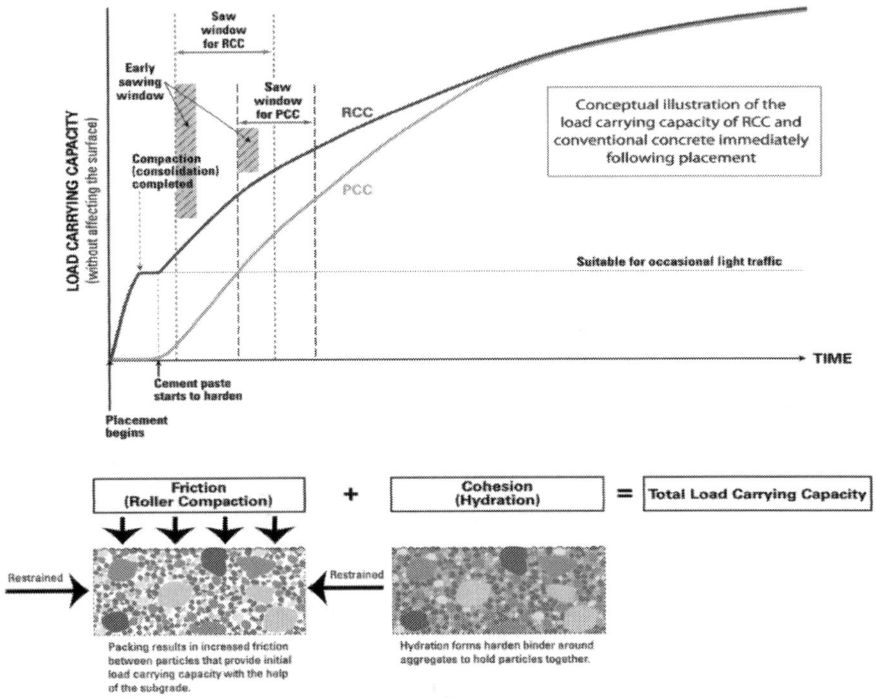

Figure 1. Load distribution capacity on conventional concrete compared to RCC.

In figure 1, we can see the illustration load distribution on conventional concrete compared to RCC in early pouring time. Immediately after pouring, conventional concrete became hardly in plastic conditions starting from the initial setting time until the final setting time. Concrete or rigid pavement usually does not have enough capacity to carry the light traffic load until the time for grooving on the surface. Instead, RCC has enough capacity to carry the light traffic loads (like cars) immediately after pouring. The capacity of carrying traffic loads came when the compaction process cause friction and cohesion interlocking between fine and coarse aggregate which support the movement of vehicles without damage or ravelling on the material. Despite theoretically explaining it, it is not suggested to open traffic above RCC or conventional concrete until the value of compression strength is between 13,8 – 17,2 MPa [4]. Based on the advantages of RCC compared to CMA, especially application in the rainy season, the purpose of this study is to know the properties of RCC material as a road-repairing material.

2. Methodology

The stage of research implementation begins with the collection and acquisition of previous research results, the selection of related previous research, and consecutively until analysis and reports. The complete research flow chart is in Figure 2 below.

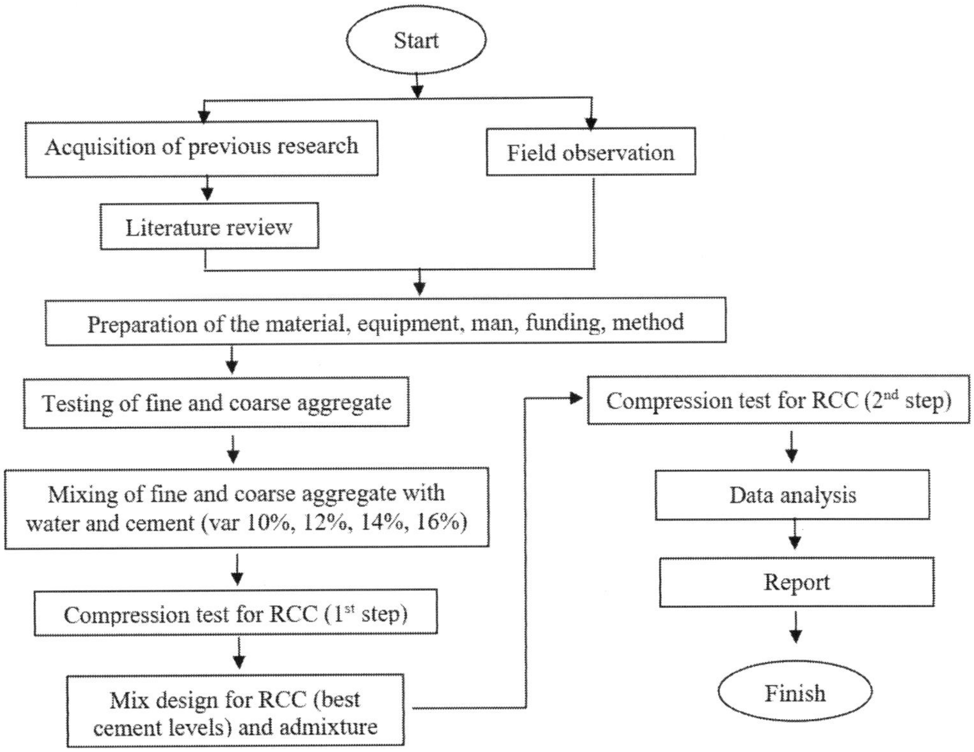

Figure 2. Research flow chart.

There are 2 approaches to making RCC mixtures [3] are:

a. The first approach uses the principle of soil compaction to produce a solid RCC where the optimum moisture content in concrete will produce concrete with the highest dry type mass. This method is different from conventional concrete which minimizes the use of cement water ratio (FAS) to produce the highest concrete press strength. The best compaction will provide the best power and the best compaction will occur when the entire aggregate is wet which will support the performance of the vibrating roller.

b. The second approach uses conventional concrete methods to produce a high cement paste on the RCC mixture. The acceptance criteria in this method are low sliding capacity between aggregates and permeability.

The working method of RCC are :

a. Material and equipment preparation
b. Testing of fine and coarse aggregate include test of materials Finer than 75-µm (No. 200) Sieve in mineral aggregates by washing, test of aggregate sieve analysis, test of specific gravity and absorption, test of weight density, and test of water content
c. Determination of compact gradation of coarse and fine aggregates based on sieve analysis
d. Determination of the amount of water content based on the amount of water needed to create Saturated Surface Dry (SSD) condition aggregates (based on water absorption tests) and amount water to hydrate cement is 30% of the weight (based on the chemical reaction of calcium hydrate)
e. Mixing aggregate with various cement contents of 10%, 12%, 14%, 16%, and 18%, then tests of compression after 7 days. Material composition for 1 m^3 RCC on 1st step can be seen in table 1
f. Formulation of the amount of cement that produces optimal compression test based on 1st step
g. RCC mix design with various in admixture levels of 0%, 0.8%, 1.2%, 1.6%, and 2%. The admixture that is used in this study is high-performance superplasticizer Sika ViscoCrete-8055. Material composition for 1 m^3 RCC on the 2nd step can be seen in table 2.
h. Compression test of RCC after 3 days, 4 days, 28 days, and 42 days (2nd step)
i. Data analysis
j. Report

Table 1. RCC composition with various cement levels for 1m^3.

Code	Cement levels	Fine aggregate (kg)	Coarse Aggregate (kg)	Cement (kg)	Water (kg)
RCC-10%	10%	770,76	1408,95	242,19	84,77
RCC-12%	12%	743,45	1359,02	286,7	100,34
RCC-14%	14%	716,86	1310,42	330,02	115,51
RCC-16%	16%	690,98	1263,10	372,21	130,27

Table 2. RCC composition with various admixture levels for 1m^3.

Code	Admixture levels	Fine aggregate (kg)	Coarse Aggregate (kg)	Cement (kg)	Water (kg)	Admix (kg)
RCC-14%-0.8%	0.8%	716,86	1310,42	330,02	115,51	2,64
RCC-14%-1,2%	1.2%	716,86	1310,42	330,02	115,51	3,96
RCC-14%-1.6%	1.6%	716,86	1310,42	330,02	115,51	5,28
RCC-14%-2%	2%	716,86	1310,42	330,02	115,51	6,60

3. Result and Discussion

3.1. Aggregates Characteristic

Tests conducted on coarse and fine aggregates derived from Muntilan include Test of Material finer than 75 µm (no. 200) sieve in mineral aggregates by washing, specific gravity testing, water absorption testing, fine modulus aggregate testing, weight density, void ratio, and aggregate abrasion with Los Angeles machine. The result of the test aggregate characteristic can be seen in table 3.

Table 3. Characteristic aggregates from Muntilan that use for RCC.

No	Item Test	Standard Test	Result	
			Fine Aggregate	Coarse Aggregate
1.	Test of Material finner than 75 µm (No. 200) Sieve in Mineral Aggregates by Washing	ASTM C117:2012	2,08 %	1,5 %
2	Specific Gravity	SNI 1970-2008 and SNI 1969-2008	2,7	2,46
3	Water Absorption		2,19 %	3,73 %
4	Fine Modulus Aggregate	ASTM C136-2012	2,51	5,28
5	Weight Density	SNI 03-4804-1998	1665,46 Kg/m3	1494,86 kg/m3
6	Void Ratio		37,49 %	36,93 %
7	Aggregate Abrasion with Los Angeles Machine	SNI 2417:2008	-	59,16 %

Acording to Bina Marga technical specifications in 2018 [5], Fine aggregate samples from Muntilan have the value of Material finer than 75 µm (No. 200) <5% dan coarse aggregate sample <1%, so it qualified for RCC mixture. According to SNI 03-2834-2000, fine aggregate samples have the value of fine modulus between 2,2 – 2,6, so it is categorized as fine sand. While coarse aggregate samples have the value of fine modulus >5, it's qualified for RCC mixture. According to Bina Marga technical specifications in 2018, Fine aggregate samples from Muntilan have the value of water absorption < 5%, specific gravity > 2,1, weight gravity >1200 kg/m3. Even though coarse aggregate samples from muntilan have specific gravity >2,1, weight gravity > 1200 kg/m3 that is qualified for RCC but the value of water absorption >2%, the value of abrasion > 40% that is mean it has a lot of pores inside.

3.2. Density of RCC

The addition of viscocrete material in the RCC mixture does not indicate a significant change in density value. The correlation between density and variance of admixtures can be seen in figure 3. That chart shows the value density of RCC > 1800 kg/cm3 so it is categorized as normal concrete.

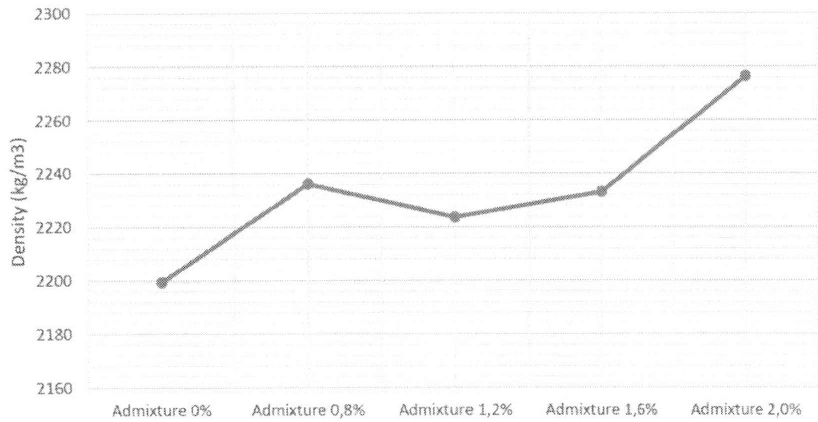

Figure 3. Correlation between the variance of admixture and density.

3.3. *Compression Strength of RCC*

In general, increasing the content of cement in concrete will increase the compressive strength. In the first stage, we make concrete samples on a standard cylinder of 15x30 cm then carried out a compressive test to determine the optimum level of cement. The optimum level of cement on RCC sufficiently reached the initial strength for light vehicle traffic. The result of compressive test RCC with various cement content can be seen in figure 4.

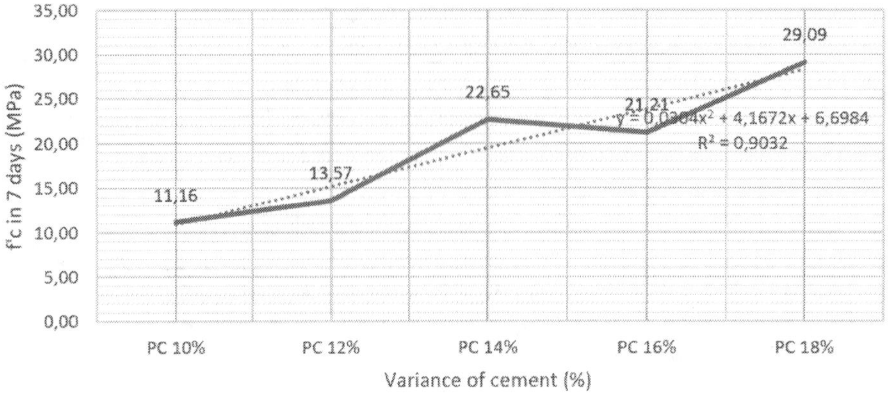

Figure 4. Correlation between the variance of cement and RCC's compression strength.

In figure 4, the increasing number of cement content in the RCC will cause an increase in compressive strength value. The correlation between compression strength and cement content can be described in the polynomial equation. The compression strength at 7 days has a conversion value of 0.65 when compared to 28 days [6]. To accelerate the opening of traffic in 1 day, it is needed admixtures to provide high initial strength and we use viscocrete from SIKA. In the second stage, we test variations of admixture levels that can provide a compression strength of min 10 MPa in 1 day. The result of the admixture variation test against the compressive strength can be seen in figure 5.

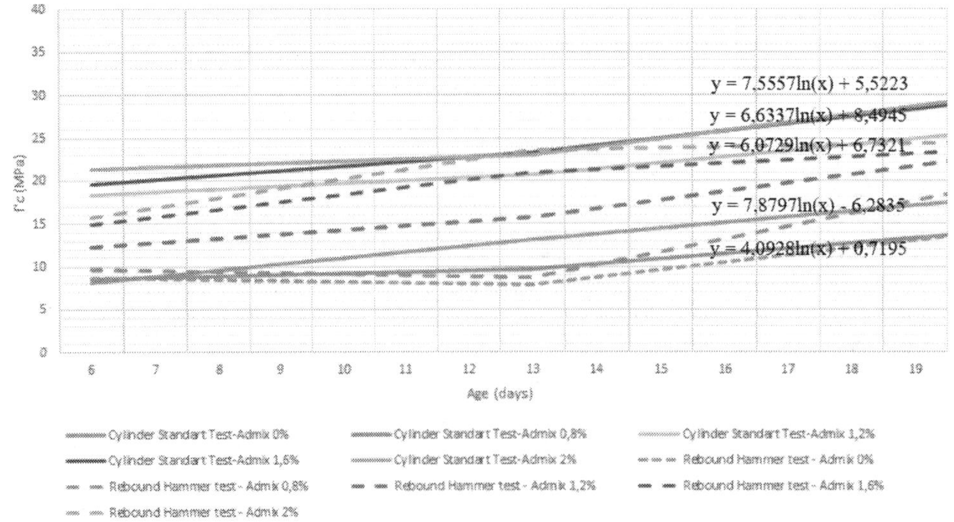

Figure 5. Correlation between age and RCC's compression strength.

The addition of viscocrete material has the impact of providing a high initial compressive strength where the increasing admixture content on RCC will be higher than the value of compressive strength. The addition of 0.8%, 1.2%, 16.6%, 2.0% viscocrete will impact on additional compressive strength of 6%, 28%, 42%, and 63%. For the next analysis to get compressive strength (f'c) 10 MPa, the characteristic of the tradeline can be seen in table 4.

Table 4. Predictions of RCC hardening age based on logarithmic functions.

Admixture Content	Logarithmic Equation	Y (MPa)	X (days)
0%	y = 7,8797ln(x) - 6,2835	10	7,89
0,8%	y = 4,0928ln(x) + 0,7195	10	9,65
1,2%	y = 6,0729ln(x) + 6,7321	10	1,71
1,6%	y = 7,5557ln(x) + 5,5223	10	1,8
2,0%	y = 6,6337ln(x) + 8,4945	10	1,25

* X = age (days)
Y = compression strength (MPa)

The result in table 4 shows that the compressive strength of RCC minimal 10 MPa can be achieved in 2 (two) days with viscocrete addiction 1,2 % - 2,0 % of cement mass. This characteristic is not enough to be used as road preservation material. The material should have 10 MPa of compression strength in a day so the traffic can be opened immediately. The solution to solve this problem is by increasing cement content based on figure 4. The increasing of cement content will increase the initial compressive strength and it means the time to hardening is more shortly. The predictions of days for hardening RCC based on cement content can be seen in table 5.

Table 5. Predictions of RCC hardening age based on cement content correlation.

Cement content	f'c (MPa)	Correlation	Age to 10 MPa
14%	22,65	1	1,25 days
16%	21,21	0,93	1,34 days
18%	29,09	1,28	0,97 days

Road preservation material should have characteristics easy to use, as soon as possible in traffic opening, strong enough to carry the load traffic, and have strong bonds with the existing pavement. The result of the vehicle trajectory test on road pavement showed concrete pavement with compressive strength of 10,8 MPa [7] or 13,8 MPa [4] can be crossed with a maximum axle load of 8 tons. Compressive strength of 10,8 MPa on conventional concrete at the age of 28 days can be achieved with RCC concrete at 7 days with a cement content of 10% or 2 days with a cement content of 14%. The opening time of traffic is attempted to take place within a few hours to 1 day after the material is applied. Table 5 shows RCC with a cement content of at least 18% and the addition of viscocrete 2.0% to the weight of the cement will produce 10 MPa in a day so that it can be used as a road preservation material.

3.4. Economical Analysis of RCC

Economical analysis for produce RCC dan conventional concrete consist of 3 components are material, labor, and equipment. Cost analysis is carried out using local prices [8] and standart unit price analysis [9]. Comparison of the cost of producing 1 m3 for each type concrete can be seen in table 6 below.

Table 6. Cost analysis to produce 1 m3 RCC compare to ordinary concrete.

No	Description	RCC (f'c 40 MPa – 28 days)				Ordinary Concrete (f'c 40 MPa – 28 days)			
		Qty	Unit	Unit Cost (IDR)	Total Cost (IDR)	Qty	Unit	Unit Cost (IDR)	Total Cost (IDR)
A	**Material**								
1	Cement	413,29	kg	1.175	495.948	512,5	kg	1.175	615.000
2	Fine aggregate	665,76	kg	280.000	116.508	0,4622	kg	280.000	129.442
3	Coarse Aggregate	1217,01	kg	350.000	266.221	0,610	kg	350.000	213.728
4	Viscocrete	8,27	ltr	40.000	330.800	1,3013	ltr	40.000	52.052
5	Water	123,98	ltr	100	12.398	162,26	ltr	100	16.226
6	Formwork		m2			0,05	m2	3.400.000	170.000
7	Nail		kg			0,6	kg	12.000	7.200
8	Minyak Bekisting		ltr			0,3	ltr	16.800	5.040
B	**Labor**								
1	Worker	0,2008	hour	16.500	3.313	1,5065	hour	16.500	24.857
2	Foreman	0,0201	hour	17.500	352	0,0502	hour	17.500	879
C	**Equipment**								
1	Vibro plate tamper	0,26	hour	35.000	5.600		hour		
2	Concrete vibrator		hour			0,3012	hour	42.000	12.650
3	Dump truck	0,2509	hour	300.000	75.270		hour		
4	Agitator truck		hour			0,1709	hour	576.100	98.455
5	Asphalt finisher	0,0125	hour	925.000	11.563		hour		
6	Concrete paver		hour			0,0502	hour	582.000	29.216
				Total Cost	**1.317.973**				**1.374.746**

The application of RCC and ordinary concrete is so different that it affects the total production cost. From table 6, it can be seen that the production cost of RCC is 5% cheaper than ordinary than ordinary concrete at the same quality and age of concrete. When analized in more detail the material component for RCC is 2% more expensive, worker component is 85% cheaper, equipment component is 35% cheaper when compared to ordinary concrete cost. The different application method in RCC which uses a mechanical compaction system can increase the efficiency of worker and equipment component. The increase of RCC in material component is not significant when compared to ordinary concrete.

4. Conclusion

Damage to road pavement or asphalt layers should be prioritized for repair because in high rainfall areas such as Indonesia, pavement can be damaged faster than its service life. Based on the analysis, it was found that using RCC with a cement content minimum of 18% and adding 2% of viscocrete to the weight of cement will produce compressive strength of 10 MPa in a day. The density of RCC is 2270 kg/m3 so that it classified as normal concrete. From economical analysis, RCC production cost more efficien 5% compared to ordinary concrete at the sama quality and age. The application of RCC can increase the efficiency of worker and eqipmwent component in range 35-85%. Characteristics of these materials can be used as a material to repair road holes (potholes). The performance of RCC material can be used for road preservation material if combined with cutback asphalt material or instant mortar as an adhesive material existing material. To easily apply this material in fields, RCC must be packed in a 25 kg package. We recommend conducting a trajectory test review or prototype field applications of RCC material with cutback asphalt or instant mortar to determine the measure of bonding and durability when used in light traffic. Further research also needs to be done to find out the level of public acceptance of instant RCC in 25 kilograms packaging.

References

[1] Murjanto D. Standart Repair For Routine Maintenance. Jakarta: Directorate General Of Highway; 2011.

[2] Hashemi M, Shafigh P, Karim M. The effect of coarse to fine aggregate ratio on the fresh and hardened properties of roller. Constr. Build. Mater. J. [Internet] 2018;169:553–66. Available from: https://doi.org/10.1016/j.conbuildmat.2018.02.216

[3] Mehta PK, Monteiro PJM. concrete microstructure, properties, and materials. 3rd ed. California: McGraw-Hill COmpanies, Inc; 2006.

[4] Delatte N. Concrete pavement design, construction, and performance. 1st ed. New York: Taylor & Francis; 2008.

[5] Ministry Of Public Work And Housing; Directorat General of Highway. General Specifications for Road Works and Bridges. Jakarta: 2018.

[6] PBI 1971 N.I. - 2. Indonesian Reinforced Concrete Regulation 1971 N.I. - 2. 7th ed. Jakarta: Departement of Public Work and Power Plants; 1971.

[7] Utomo AB, Hidayatiningrum LF, Dhanardono B, Konstruksi T, Umum PP. Vehicle Track Test On No Fine Aggregate Concrete Interlocking Pavement. Journal of Media Komunikasi Tek. Sipil [Internet] 2021;27:143–50. Available from: https://ejournal.undip.ac.id/index.php/mkts/article/view/27387

[8] Central Java Agency of Highway and Setlement; Information of Basic unit price construction services [Internet]; 2021;2–4. Available from: http://maspetruk.dpubinmarcipka.jatengprov.go.id/harga_satuan/hsd

[9] SNI no 28/PRT/M/2016, Guidelines for the Analysis of Unit Prices For Public Work [Internet]. Indonesia: 2016. Available from: http://maspetruk.dpubinmarcipka.jatengprov.go.id/foto/Permen PUPR No 28 Tahun 2016 Tentang Pedoman Analisis Harga Satuan Pekerjaan Bidang Pekerjaan Umum.pdf

Acknowledgments

Authors wishing to acknowledge assistance or encouragement from colleagues, special work by technical staff, or financial support from Polytechnic of Public Work.

Modular incinerator with pre-treatment plant for municipal solid waste treatment in the super-priority tourism destination of Labuan Bajo Indonesia

Wahyu Purwanta[1*], Pandji Prawisudha[2], Firman Bagja Juangsa[2], Ayudia Mutiara Fani[1], Evan Philander[2]

[1]Environmental Research Centre, National Research and Innovation Agency
[2]Faculty of Mechanical and Aerospace Engineering, Bandung Institute of Technology

*E-mail: wahyu.purwanta@brin.go.id

Abstract. One of the competitiveness parameters of tourist destinations is the cleanliness index. Labuan Bajo as a super-priority tourist destination has a very serious municipal solid waste (MSW) problem. Waste generation in Labuan Bajo and Komodo National Park in 2020 reached 14.06 tons/day and is projected to be 15.51 tons/day in 2023. The main sources of MSW come from settlements, hotels, traditional markets, and seaports. The current MSW management system is that waste is collected from the source and then transported to a temporary storage tank to be transported by truck to the Warloka landfill. This system poses a threat in the form of shortening the service life of the landfill. Management through recycling is still a very low percentage because the recycling ecosystem has not yet developed. One alternative for MSW management that is currently being carried out by the central government in Labuan Bajo is to implement a modular incinerator. This paper describes the process of designing and operating a test incinerator with a combustion capacity of 20 tons/day consisting of 2 lines each with a capacity of 10 tons/day. The incinerator is equipped with a pre-treatment plant consisting of a sorting conveyor, thermal dryer, and storage conveyor. The incinerator furnace consists of two chambers and each is equipped with a burner. This furnace is designed to be able to burn waste with a calorific value of 1300 kcal/kg with a maximum moisture content (MC) of 60%. The test results show that the temperature in the primary chamber can reach a minimum of 800^{0}C and a maximum of 1000^{0}C in the secondary chamber. The burning rate has obtained a maximum of 527 kg/h which means it has passed the initial design of 417 kg/h. To meet emission-quality standards, exhaust gas is processed through flue gas treatment consisting of a cyclone, quencher scrubber, and chimney. Emissions measured by CEMS include parameters of particulates, CO, SO_2, and NOx, all of which are below the national emission-quality standards. The heat generated from partial combustion will be channeled to the thermal dryer to reduce MC so that it can reach 40%. The results of burning waste with this incinerator produce fly ash and bottom ash about 4% of the weight of the input waste.

Keywords: MSW, pre-treatment, thermal dryring, incinerator, performance

1. Introduction
1.1 Background
There are at least three sectors that need to be improved in order to boost the competitiveness of Indonesia tourism on the world stage. These sectors include environmental sustainability, health and hygiene, as well as tourist service infrastructure. Based on data from Indonesia Travel and Tourism Competitiveness, Indonesia's position in these three fields is still quite low. In environmental

sustainability, for example, Indonesia is still ranked 135th out of 140 countries. Meanwhile, in health and hygiene, Indonesia ranked 102 and in tourist service infrastructure is ranked 98 [1]. One of the problems that still stands out in tourist areas is domestic solid waste that is not managed properly so that is interferes with the aesthetics and health of tourist. Labuan Bajo as one of the premium tourist areas is the entrance gate for the Komodo National Park tourism area, it should be able to manage its waste effectively and sustainability. However, the waste problem in Labuan Bajo is at an alarming stage, even an emergency. The main sources of waste come from settlements, hotels, traditional markets, and ports. The current solid waste management system is that waste is collected from the source and then transported to a temporary storage tank to be transported by truck to the Warloka landfill. This system poses a threat in the form of short landfill life. Waste management through recycling is still slow in growth because the recycling ecosystem has not yet grown. Recyclable waste must be taken outside the island if it is to be sold so that it becomes economically unfeasible. The Ministry of Environment and Forestry (MoEF) in collaboration with global companies has built a Waste Recycling Center (WRC) in Labuan Bajo[2]. Currently, the type of waste collected in the WRC has only used bottles of mineral water (Polyethylene Terephthalate/PET). Considering the waste reduction target in 2025 is 30% in terms of reduction and 70% in terms of handling so that the target of 100% waste service is achieved, a breakthrough is needed that can reduce waste quickly and significantly. The method that can destroy waste with these criteria is known as a thermal process [3]. Efforts made by the Ministry of Public Works and National Research and Innovation Agency are implementing modular incinerators in Labuan Bajo. The incinerator was built at the Warloka landfill to reduce the volume of waste before being dumped in the landfill. This incinerator has a combustion capacity of 20 tons/day consisting of 2 lines each with a capacity of 10 tons/day. The incinerator is equipped with a pre-treatment unit consisting of a sorting conveyor, thermal dryer, and storage conveyor. The incinerator furnace consists of two chambers and each is equipped with a burner. The primary chamber is set to burn at a minimum temperature of 800^{0}C while the secondary chamber is set at a minimum temperature of 1000^{0}C. This ensures complete combustion and prevents the formation of dioxins in the flue gas. This modular incinerator which is integrated with a pre-treatment plant and utilizes hot air for the thermal drying process in Labuan Bajo is the first in Indonesia. This makes the performance data of the designed system the novelty in this study.

1.2 Objective

The purpose of implementing a modular incinerator is to reduce the generation of MSW that will be dumped in the Warloka landfill, starting with a survey of the composition and characteristics of the MSW, designing and testing plant performance. This paper will describe the characteristics of waste in Labuan Bajo, the parameters and the incinerator design process as well as the results of the incinerator system performance test during the observation period.

2. MSW Incinerator in Labuan Bajo

2.1 MSW Characteristic

The composition and characteristics of MSW are important data in the planning of waste processing facilities to achieve effective and efficient principles [4]. Data on the composition and characteristics of Labuan Bajo waste in this study were taken from secondary data and also from field surveys. The sampling and testing were carried out in August 2020 using the SNI 19-3964-1994 method. The samples taken were 1-2 m^3 of waste, each of which came from the traditional market, settlements, and hotels.

The results of the measurement of solid waste density obtained for waste from the market are 0.08 – 0.16 tons/m^3. For waste from residential areas is 0.08 – 0.15 ton/m^3 while hotel waste is in the range of 0.08 – 0.15 ton/m^3. From the determination of waste composition, the average waste from markets, settlements, and hotels is 53% biowaste, 17% plastic, 14% paper, and less than 10% other types of waste [5]. This result is reasonable considering that Labuan Bajo is a tourist city so that the main source of waste other than settlements is from tourists in the form of food waste and packaging materials (seen in Table 1).

Table 1. Municipal Solid Waste Composition

Component	Housing 1 (% w/w)	Housing 2 (% w/w)	Market (% w/w)	Hotels (% w/w)	Offices (% w/w)	Average (% w/w)
Biowaste	54	50	51	56	50	53
Paper	19	22	14	9	22	14
Cloth	0	0	1	0	0	1
Plastic	14	19	19	18	19	17
Rubber/leather	0	0	6	3	0	2
Metal	1	5	1	3	5	2
Glass	0	0	0	2	0	1
E-waste	0	0	0	0	0	0
Hazardous waste non e-waste	0	0	0	0	0	0
Others	12	4	8	9	4	9
	100	100	100	100	100	100

Proximate analysis was carried out to determine the moisture content (MC), volatile matter, ash content, and fixed carbon while the ultimate analysis was carried out to determine the content of chemical elements such as carbon, oxygen, hydrogen, nitrogen, sulfur in the waste [6]. For proximate and ultimate analysis carried out by the Solid Waste Laboratory and Hazardous Waste, Bandung Institute of Technology. The results of the MC measurement for all samples obtained an average value of 46.85%. However, when this measurement is carried out, it is in the dry season, when the rainy season the MC value can reach 60%.

The ash content of waste from various samples is in the range of 14.03% to 28.42% or an average of 22.07%. From the ultimate analysis, the carbon content in the waste is 41.95% (representation of organic matter). The results of the analysis for the characteristics of the waste can be seen in Table 2. From the three values of MC, ash content, and carbon content based on the Tanner diagram, it can be concluded that the waste in Labuan Bajo can be self-combusting or can be treated thermally. According to the Tanner Diagram, waste is included in the self-combustion criteria if MC $\leq$ 50%, ash content $\leq$ 60%, and organic matter $\geq$ 25% [7].

Table 2. Characteristics of MSW in Labuan Bajo.

Parameter	Method	Quantity	Unit
Water content (as received)	ASTM D2216-80	46.85	% ww
Water content (air dried)	ASTM D2216-80	17.52	% dw
Ash (air dried)	ASTM D2216-80	22.07	% dw
Volatile Matter (air dried)	ASTM D2216-80	58.33	% dw
Fixed Carbon (air dried)	ASTM D2216-80	2.08	% dw
High Heating Value (HHV)	Bomb Calorimeter	4254.22	kcal/kg
Low Heating Value (LHV)	Bomb Calorimeter	1480.11	kcal/kg

Carbon	ASTM 3172-3175	41.95	% dw
Hydrogen	ASTM 3172-3175	6.54	% dw
Oxygen	ASTM 3172-3175	30.94	% dw
Nitrogen	ASTM 3172-3175	1.24	% dw
Sulfide	ASTM 3172-3175	0.21	% dw
Chloride	ASTM 3172-3175	0.49	% dw

Although the results of the characteristic analysis show that the condition of waste can be burned in the incinerator, to overcome fluctuations in MC levels, a drying system is needed before the waste is fed to the incinerator. The drying system is integrated into the pre-treatment plant which also includes waste sorting. The selection of the drying system was carried out based on a literature study by comparing the methods of bio drying, bio stabilization, thermal drying, and solar drying. The choice of drying method was based on the results of studies from Ragazzi (2007), He (2009), Toshniwal (2013), Bukhmirov (2015), and Tun & Juchelkova (2019). The comparison of the four methods is based on the variable drying time, drying temperature, moisture reduction, waste weight reduction, waste volume reduction, and humidity. One of the disadvantages of bio-drying is that the process quite long, so it requires a large area oh land [8]. While the bio-stabilization method was able to reduce the weight of the waste up to 85% after 59 days [9]. The solar drying method is actually fuel efficient but still requires a long drying process and is weather dependent [10]. Solar drying also has a relatively small environmental impact. While another study explained that the application of a thermal dryer was able to reduce the relative humidity along with heating temperature [11]. In the application of bio-drying, between developed and developing countries there are also differences in results. In developed countries at the same level of moisture reduction, the reduction in waste weight is smaller then in developing countries [12]. The results of the comparison of several variables are then adjusted to the conditions on the site such as land area, electricity availability, and also ease of operation. From the study conducted, it was decided that the drying system was chosen by the thermal dryer by utilizing the exhaust gas from the combustion in the furnace which was still at a high temperature.

2.2 Incinerator Design

Based on the results of the waste generation survey in August 2020, it was found that the number of trucks carrying waste to the Warloka landfill is 12 trucks a day with a capacity of 6 m^3. If the density of waste in the truck is 250 kg/m^3, then 18 tons of waste per day will be disposed of at the landfill. If the waste is assumed to enter the plant as much as 18 tons/day or theoretically as much as 750 kg/h which will be spilled into the receiving hopper before being sorted on a sorting conveyor. The process of sorting waste and drying in a thermal dryer will produce waste as much as 417 kg/h (10 tons/day) of waste that is ready to be burned. For this reason, the incinerator is designed with a capacity of 417 kg/hour per line and there are 2 lines so that the total capacity is 834 kg/hour (20 tons/day). The incinerator plant series consists of a sorting conveyor, thermal dryer, storage conveyor, combustion furnace, air pollution control, and ash handling (seen in Figure 1).

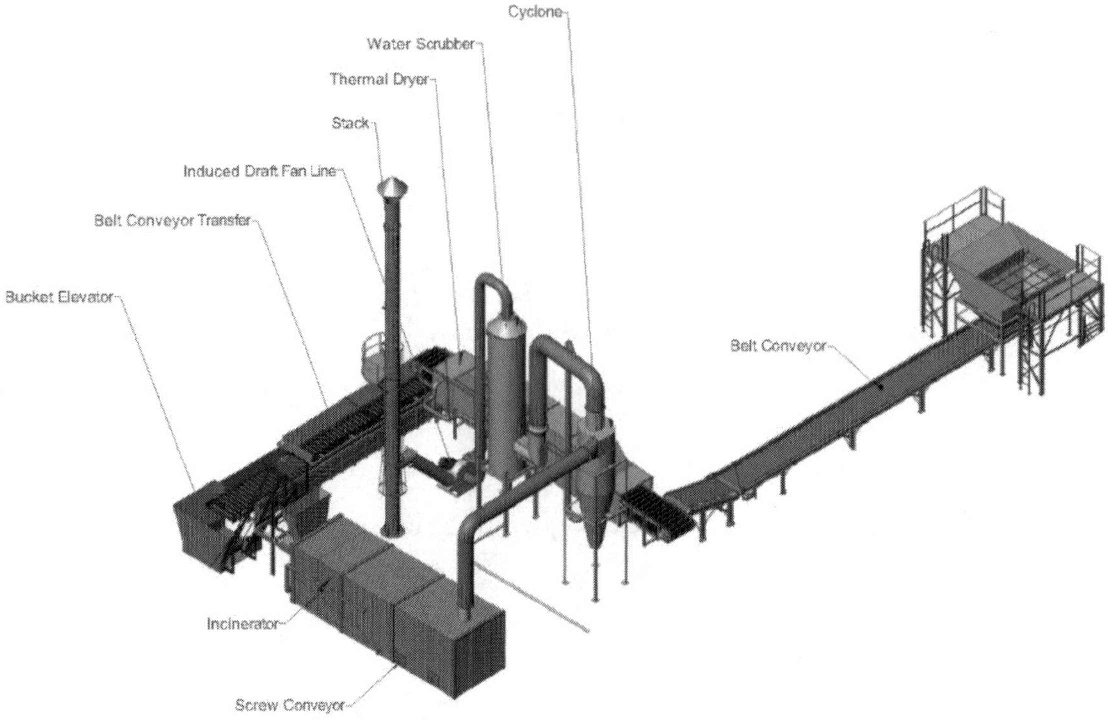

Figure 1. Isometric of incinerators and pre-treatment plant in 1 line of 2 lines.

a. Sorting conveyor
Waste from the truck will be poured into a sorting conveyor through a hopper. If bulky waste is found in the truck, the waste will be poured into the receiving area for sorting first. A sorting conveyor is a pre-treatment unit that functions as a place for sorting between combustible and noncombustible materials. In addition, hazardous waste, metal, glass, PVC, and aluminum foil are also removed. The sorting conveyor is designed with a length of 16.6 m and a belt width of 1 m and a waste height of 10 cm.

b. Thermal dryer
After passing through the sorting conveyor, the waste will then enter the thermal dryer. A thermal dryer is a unit that functions to reduce the water content of the waste that passes through it by utilizing the hot airflow from the combustion in the furnace. The thermal dryer technology used is a conveyor dryer that is placed before burning waste in the incinerator. In general, conveyor dryers use the heat of the flue gases to heat the air which is then used to dry the waste.

Thermal dryers must be able to dry waste from a water content of 52% to about 40% with a water temperature of 30^{0}C inlet and 45^{0}C outlet. The exhaust gas feed rate is designed to be 6,500 kg/h with a pressure of 1.05 bar and the hot air rate to the waste is 3,100 kg/h. The length of the thermal dryer conveyor is 12 m with a belt width of 1 m and the height of the waste is 20 cm with a conveyor speed of 8.5 m/hour.

c. Storage conveyor
The main function of the temporary storage conveyor is to accommodate dry waste so that it can be used during the incinerator start-up process. In addition, in this system, there is also a conveyor system that is used to move the dry waste from the thermal dryer to the waste feeder. The storage conveyor is

designed to have a minimum volume of storage that can accommodate dry waste with the need for ignition of the incinerator for two hours. The conveyor speed is set at a certain speed so that waste does not accumulate on the input side and waste can be evenly distributed along the conveyor. The length of the storage is 10 m, the conveyor width is 0.75 m and the top section width is 1.1 m. The conveyor speed is 4.62 m/h.

d. Incinerator
The incinerator is the main component used to reduce solid materials in the MSW treatment process. Therefore, the main performance of an incinerator is the ability of the device to burn all the volatile matter content and fix carbon in the waste. The performance of an incinerator furnace is generally based on the parameters of temperature, residence time, and turbulence (3T) [13]. In general, there is an incinerator combustion chamber that can be divided into two, namely primary and secondary combustion furnaces.

To obtain optimum furnace dimensions, gas residence time, and heat distribution pattern, FLIC and ANSYS FLUENT software were used in the design process. The solid waste characteristics measured in table 1 will be used as input parameters in the FLIC software. The incinerator design process is also defined as the following design parameters;
(a) Waste feeding rate is 417 kg/h
(b) Waste input temperature is 45^{0}C
(c) MC of waste in the range of 30%-60% wt
(d) Minimum combustion temperature of the primary chamber is 800^{0}C
(e) Minimum combustion temperature of secondary chamber is 1000^{0}C
(f) Furnace operating pressure is 0.89 bar A
(g) Exhaust gas residence time > 2 s
(h) Solid waste residence time is 80 minute
(i) 80% excess air
The dimensions of the furnace are interior length (L) 7.1 m, interior width (L) 2 m, primary chamber height (T1) 2 m, and grate length (P_{grate}) 4 m, and grate width 1.8 m as seen in Figure 2.

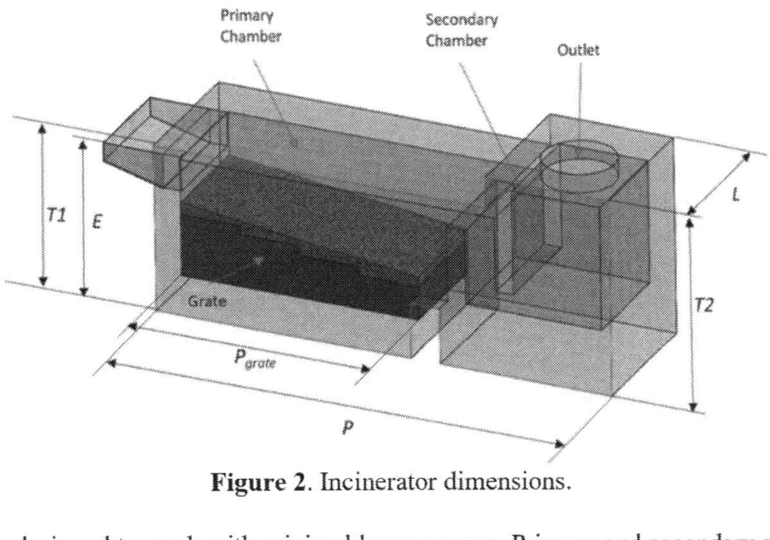

Figure 2. Incinerator dimensions.

Incinerators are designed to work with minimal burner usage. Primary and secondary air flow must be regulated so that it can adjust to the conditions of the waste input and the humidity conditions of the surrounding air. The design of the secondary airflow system must ensure that C_xH_y, CO, and H_2 gases can be burned out. The excess air delivery system for both primary and secondary air must be able to work in the specified excess air range.

e. Air pollution control
The flue gas treatment system for this incinerator consists of a cyclone, a quencher scrubber, and a chimney. Cyclone is a means of controlling exhaust emissions, especially large particulates with the principle of gravity [14]. The main function of the cyclone is the initial separation of medium and large size particulates (coarse particles). Quencher-Scrubber is one of the exhaust emission control units, especially particulates in small sizes with the principle of decreasing temperature and at the same time adsorption using water media [15]. The main function of the quencher scrubber is as an operating unit for the cooling process of exhaust gases resulting from burning waste. Its additional function is to trap small particulates in the flue gas. For the additional function to work properly, the given mass of water spray is added from the theoretical calculation. Water from the cooling process will be recycled in a wastewater treatment plant (WWTP) equipped with a sludge drying bed. This means that the quencher-scrubber and WWTP are closed systems.

Air pollution control is expected to reduce pollutant emissions such as particulates matter (PM), NO_x, SO_2, and CO to meet quality standards. All pollutant parameters in flue gases are measured in real-time with a continuous emission monitoring system (CEMS). Dioxin parameters will be measured manually but only once every five years.

f. Ash handling
The process of burning waste in addition to generating heat and flue gas also produces solid residues in the form of fly ash and bottom ash. Fly ash is captured by a cyclone and a quencher scrubber so that what is discharged into the atmosphere through the chimney is minimal. Meanwhile, bottom ash is formed at the bottom of the furnace and will be carried out by a screw conveyor to then be accommodated in a tub and dumped periodically to the landfill. In this incinerator design, the maximum amount of fly ash is 3% and bottom ash is 20% of the weight of the burned waste.

3. Results and discussion

3.1 Pre-treatment plant

The results of the incinerator performance test at the site were carried out in the period November – January 2022. The pre-treatment plant performance test aims to determine the performance of the pre-treatment unit and the level of material segregation. In addition, it also aims to measure the skills of the operators, what weaknesses are still emerging, and what additional equipment is needed. At the beginning of the trial, it was found that the sorting rate was still slow. This is because the operators are still adapting and looking for the right work rhythm. Sorting on the conveyor aims to take the saleable materials and pass the combustible materials. Some materials that should not be burned are also set aside such as household hazardous waste, metals, glasses, PVC, and aluminum foil by the Regulation of the Minister of Environment and Forestry No. 70/2016. The sorting activity on the conveyor is shown in Figure 4.

The first day of the sorting test only obtained 40% combustible materials but over time it continued to increase as the operator's sorting speed increased which reached a maximum of 95%. From observations during the trial period, it was found that the average combustible material obtained and fuel waste reached 62.4% as seen in table 2. The sorting rate generally depends on the skill of the operator, the composition of the waste, and the availability of supporting tools for the operator's work [16]. While some equipment problems that arise are that there are still gaps between the conveyors or between equipment units so that there is scattered garbage or not a centric conveyor belt, all problems can be overcome.

3.2 Incinerator combustion performance

An incinerator performance test was carried out to determine the temperature level in the combustion chamber, waste burning rate, flue gas residence time, and complete combustion associated with turbulence, CO formation, and smoke opacity in the chimney. The temperature setting for the primary chamber (High Flame) is 800^{0}C and the secondary chamber is 850^{0}C (Low Flame). The time to reach the desired temperature in the furnace varies from tens of minutes to hundreds of minutes depending on the type, composition, and MC of the fuel waste. This is why an incinerator is better if it is operated 24 hours continuously for several days. This saves fuel by eliminating the need for preheating the furnace and preventing excessive dioxin build-up at the start and end of the process.

The recorded temperature in the primary chamber is in the range of 800 – 900^{0}C while in the secondary chamber it is between 850 – 1000^{0}C. The high temperature achieved is expected to occur complete combustion and avoid the formation of dioxins. Modeling with FLIC and ANSYS FLUENT resulted in residence time above 2 seconds for all conditions. Field observations based on gas volume rate and furnace size obtained residence time above 2 seconds. Meanwhile, turbulence occurs in the secondary air supply system as well as the reversal of waste with a moving grate. The results of turbulence will be seen from the results of measurements of CO and the opacity of the thin white smoke in the chimney.

Table 3. Incinerator performance test result [17].

Date	**#Line**	**Primary chamber**		**Secondary chamber**		**Burning rate (kg/h)**	**Combustible (% weight)**
		Temp.(^{0}C)	After	Temp.(^{0}C)	After		
01/11/2021	1	800	10’	850	15’	200	40
04/11/2021	1	800	25’	850	40’	332.3	60
06/11/2021	1	800	30’	850	180’	264.4	88
08/11/2021	1	800	30’	850	125’	286.7	51
10/11/2021	1	850	50’	850	115’	262.5	65
13/11/2021	1	900	15’	850	160’	366.7	61
15/11/2021	1	900	15’	850	180’	375	78
17/11/2021	1	900	20’	800	100’	337.5	56
18/11/2021	1	900	15’	742	80’	244.5	52
19/11/2021	2	900	30’	800	140’	295	44
09/12/2021	2	900	60’	1000	120’	477	61
10/12/2021	1	900	25’	950	120’	413	95
11/12/2021	2	850	15’	900	135’	442	72
14/12/2021	2	800	30’	900	105’	381	69
15/12/2021	2	900	15’	1000	130’	527	72
16/12/2021	2	850	20’	900	120’	485	51
17/12/2021	2	900	30’	1000	110’	500	41
20/12/2021	2	850	25’	900	70’	515	83
30/12/2021	2	900	20’	850	20’	335	47
05/01/2022	2	950	15’	1000	35’	484	67
06/01/2022	2	950	25’	1000	40’	445	75
07/01/2022	2	950	30’	1000	180’	484	68
08/01/2022	2	950	30’	1000	125’	337	76
10/01/2022	2	850	50’	1000	115’	453	77
11/01/2022	2	900	15’	1000	160’	244	64
12/01/2022	1	900	25’	1000	180’	425	70
13/01/2022	1	900	20’	1000	100’	304	59

17/01/2022	1	950	15’	1000	80’	406	74
18/01/2022	2	950	30’	1000	140’	439	68
19/01/2022	2	950	30’	1000	60’	376	67
20/01/2022	1	900	25’	1000	120’	381	64
21/01/2022	1	900	15’	1000	135’	442	72
22/01/2022	1	950	30’	1000	105’	437	79
23/01/2022	2	950	25’	900	130’	324	71
24/01/2022	2	950	20’	900	120’	405	80

This incinerator is designed to burn 10 tons/day MSW or 417 kg/h at 40% MC. From the results of the combustion test during the observation period, it was found that the combustion rate varied (seen in Table 3). This is due to the unstable operation of the incinerator, both regarding the incoming waste supply, the operator's work rhythm, and the continuity of combustion due to frequent technical problems along the way. The lowest combustion rate is 200 kg/h and the highest is 527 kg/h with an average MC of 58.06% (seen in Figure 3). The rate of combustion will depend on the MC content of the waste which also depends on the season. This happens because waste in Indonesia is generally stored open so that it can be infiltrated by rainwater. Incinerator combustion furnace is shown in Figure 5.

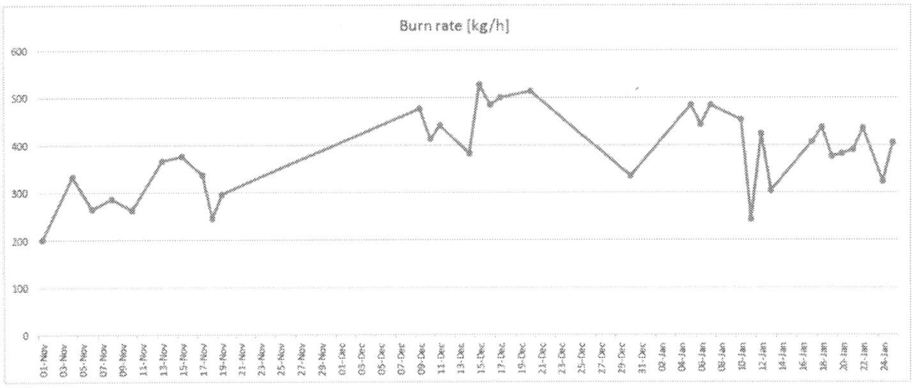

(a)

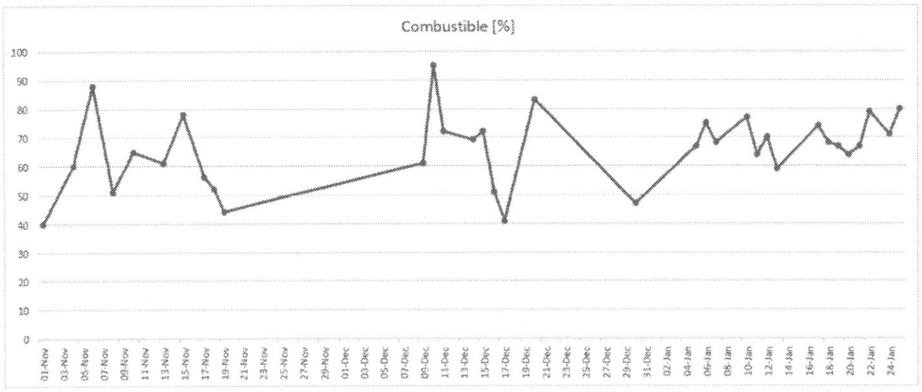

(b)

Figure 3. (a) Burning rate and (b) percentage of combustible material.

3.3 Exhaust gas emission

The rest of the burning of waste that must be given attention is the emission of gaseous pollutants. To prevent air pollution, the Emission Quality Standard has been set in Regulation of MoEF No. 70 of 2016. The measurement of gas and particulate emissions from incinerators is measured sensory and read through CEMS. The results of CEMS monitoring during the performance test period are shown in table 3. CEMS only contains four sensors measuring the levels of particulate, SO_2, CO, and NOx. The provisions in the emission-quality standards for thermal processing waste also regulate the levels of Hg, HCl, HF, and dioxins which were not measured in this trial due to equipment limitations. From the measured parameters, all are below the emission-quality standard values (seen in Table 4).

Table 4. Flue gas emission parameter measurement result.

Parameter	Unit	Measurement with CEMS	Emission Quality Standard (Reg. MoEF No.70/2016)
#Line 1			
Particulate	mg/Nm^3	60	120
SO_2	mg/Nm^3	31	210
CO	mg/Nm^3	30	625
NOx	mg/Nm^3	11	470
#Line 2			
Particulate	mg/Nm^3	103	120
SO_2	mg/Nm^3	22,5	210
CO	mg/Nm^3	37,5	625
NOx	mg/Nm^3	11	470

In addition to the four parameters that have been measured by CEMS, there are four other parameters regulated in the national emission quality standards namely HF, HCl, Hg and PCDDs/PCDFs which will be measured manually. Reports from the incinerator plant in Bantargebang with the same air pollution control devices as the incinerator in Warloka plant, obtained HF values of 0.8 – 2 mg/Nm^3, HCl 5 – 8 mg/Nm^3, Hg close to zero and PCDDs/PCDFs of 0.0088 ng/Nm^3. This shows that the four parameters are below the national emission quality standard values [18].

3.4 Fly ash and bottom ash

In general, burning waste with an incinerator will produce bottom ash ± 20% and fly ash ± 3% of the weight of the burned waste [19]. During the trial period of this incinerator, the weight of the bottom ash at the bottom of the furnace was measured and an average of 130 kg or 2.44% of the weight of the burned waste was obtained. The weight of fly ash was measured from the material collected in the cyclone and obtained an average of 50 kg or 1.52%. Currently, it is not possible to measure the amount of fly ash that enters the WWTP. These results show that the reduction rate of waste burning in the furnace reaches ± 96%. From the weight percentage, bottom ash also shows that the level of sorting, especially metals and glasses, has been going well.

In MoEF Regulation No.26/2020 concerning fly ash and bottom ash, it is stated that the solid residue from the combustion process of domestic solid waste is not considered hazardous waste. However, considering its size, the fly ash still needs to be solidified/stabilized before being dumped in landfills. According to the regulation, bottom ash can be directly dumped into landfills. Fly ash from the combustion residue at the Warloka plant is solidified in the form of paving blocks which will be used as construction material. The bottom ash is directly dumped into a landfill near the plant.

Figure 4. Pre-treatment plant.

Figure 5. The furnace of incinerator.

4. Conclusions

The modular incinerator in Labuan Bajo is an MSW incineration system that is integrated with a pre-treatment plant to obtain fuel waste that is more homogeneous in type and size and has a minimum MC of the waste. The average combustible material sorting rate in a pre-treatment plant is 62.4%. The combustion temperature in the primary chamber has been able to reach a minimum of 800^{0}C and in the secondary chamber 850^{0}C with a residence time for exhaust gas of more than 2 seconds. The waste burning rate in the furnace reaches 527 kg/h at 58.4% MC, which is above the design figure of 417 kg/h at 40% MC. Exhaust gas emissions as measured by CEMS both particulate, SO_2, CO, and NOx are below the national emission-quality standards. The total fly ash and bottom ash produced is ±4% of the weight of the waste, which means it is better than the initial design which was 23%.

Acknowledgment

We would like to thank the Directorate of Sanitation, Ministry of Public Works and Public Housing, East Nusa Tenggara Regional Settlement Infrastructure Center and West Manggarai Regency Environmental Office who have collaborated with the National Research and Innovation Agency - Environmental Technology Center since the planning, advisory for construction and operation of the Incinerator Plant so that they can obtain performance data that is very useful for science.

References

[1] Fariz, M., 2017, "Studi Penyusunan Panduan Pengelolaan Sampah Kawasan Taman Nasional Komodo dan Kota Labuan Bajo."

[2] 2018, "KLHK Optimalkan Pengelolaan Sampah di Labuan Bajo," hal. 1–2 [Online]. Available: https://mediaindonesia.com/humaniora/205023/klhk-optimalkan-pengelolaan-sampah-di-labuan-bajo. [Accessed: 15-Jan-2022].

[3] Helsen, L., dan Bosmans, A., 2010, "Waste-to-Energy through thermochemical processes : matching waste with process Thermochemical conversion technologies," Symp. A Q. J. Mod. Foreign Lit., (January 2010).

[4] Ozcan, H. K., Guvenc, S. Y., Guvenc, L., dan Demir, G., 2016, "Municipal solid waste characterization according to different income levels: A case study," Sustain., **8**(10).

[5] BPPT, 2020, *Laporan Akhir Penyusunan Desain Awal dan Pendampingan Rancang Bangun Pilot Proyek Pengolahan Sampah Proses Termal.*

[6] Zhou, H., Meng, A., Long, Y., Li, Q., dan Zhang, Y., 2014, "Classification and comparison of municipal solid waste based on thermochemical characteristics," J. Air Waste Manag. Assoc., **64**(5), hal. 597–616.

[7] Pasek, A. D., Gultom, K. W., dan Suwono, A., 2013, "Feasibility of recovering energy from

municipal solid waste to generate electricity," J. Eng. Technol. Sci., **45**(3), hal. 241–256.

[8] Ragazzi, M., Rada, E. C., Panaitescu, V., dan Aposto, T., 2007, "Municipal solid waste pre-treatment: A comparison between two dewatering options," WIT Trans. Ecol. Environ., **102**(April), hal. 943–949.

[9] He, P., Shao, Z., Zhang, D., dan Shao, L., 2009, "Bio-stabilization of municipal solid waste prior to landfill : Environmental and Economic Assessment," Int. Symp. MBT MRF, hal. 1–11.

[10] Toshniwal, U., dan Karale, S. ., 2013, "A Review Paper on Solar Greenhouse Dryer," Int. J. Mech. Appl., **03**(02), hal. 896–902.

[11] Bukhmirov, V. V., Kolibaba, O. B., dan Gabitov, R. N., 2015, "Experimental research of solid waste drying in the process of thermal processing," IOP Conf. Ser. Mater. Sci. Eng., **93**(1).

[12] Tun, M. M., dan Juchelková, D., 2019, "Drying methods for municipal solid waste quality improvement in the developed and developing countries: A review," Environ. Eng. Res., **24**(4), hal. 529–542.

[13] Yuliani, M., 2016, "Insinerasi untuk Pengolahan Sampah Kota," J. Rekayasa Lingkung., **9**(No.2), hal. 89–96.

[14] Funk, P. A., dan Baker, K. D., 2013, "Dust cyclone technology- A literature review," J. Cotton Sci., **17**(1), hal. 40–51.

[15] Cooper, C. D., dan Alley, F. ., 2011, *Air Pollution Control : A Design Approach*, Waveland Press, Inc. IL.

[16] Sin-Yee, T., dan Sheau-Ting, L., 2016, "Attributes in Fostering Waste Segregation Behaviour," Int. J. Environ. Sci. Dev., **7**(9), hal. 672–675.

[17] BRIN, 2021, *Laporan Akhir Advisory untuk Konstruksi dan Operasional Pengolahan Sampah Proses Termal.*

[18] BPPT, 2020, "Laporan Akhir Pemeliharaan dan Operasional PLTSA Bantargebang."

[19] Ornebjerg, H., Morin, R., Brua, J.-F., Friege, H., dan Bader, C., 2008, "Management of APC Residues from W-T-E Plants," Int. Solid Waste Assoc., **44**(10), hal. 1180–1182.

Spatial Analysis of Potential Greenhouse Gas (GHG) Emissions from Household-Scale LPG Consumption in Urban and Sub-Urban Areas of Medan City during the COVID-19 Pandemic

I Suryati[1*], N Hasibuan[1], M Hanisah[1], L Nadapdap[1], M Ruadelia[1] , J Sigalingging[1]
[1]Environmental Engineering Department, Universitas Sumatera Utara, Medan, Sumatera Utara, 20155,Indonesia

*isratl@usu.ac.id

Abstract. Household activities such as the use of electricity, LPG, and fuel for vehicles are sources of greenhouse gas emissions. Energy consumption at the household scale continues to increase, especially during the COVID-19 pandemic due to the policy of working from home. The global COVID-19 pandemic also affects household-scale energy consumption for urban and suburban areas. The increase in energy consumption is in line with the increase ingreenhouse gases so to reduce greenhouse gas emissions, spatial mapping is necessary. This mapping aims to determine the potential distribution of greenhouse gas emissions from household-scale LPG consumption so that programs, plans, or policies can be determined in strategies for reducing greenhouse gas emissions. The stages of this research begin with calculating GHG emissions using the IPPC method and making a GHG potential map. The research results from the calculation of greenhouse gas emissions in the household-scale LPG sector in the study area ranged from 110 – 2792 tons CO_2eq/year. GHG emissions in sub-urbanareas are greater than in urban areas because they are influenced by population, LPG usage patterns, and income per capita.

Keywords: climate change, COVID-19, GHG emission, LPG consumption and SDGs

1. Introduction

Handling climate change is one of the goals of the SDGs, with the occurrence of the COVID-19 pandemic, which limits human activities, thereby affecting greenhouse gas emissions as the cause of climate change. The Energy and Clean Air Research Center (CREA) has announced that the COVID-19 quarantine conducted in various countries has reduced global CO_2 emissions by up to 17%. Almost half (43%) of global emission reductions at the peak of the blockade came from the transportation and industrial sectors, especially automobiles and commercial production. [1].

Indonesia is one of the countries that has also been affected by the COVID-19 pandemic, the development of this disease outbreak has prompted the government to make policies ranging from calls for social distancing, work, study, and worship from home to the enactment of large-scale social restrictions. This condition certainly affects the activities of the community. This change in activity affects greenhouse gas emissions from several sectors, especially the energy sector.

According to Rudiyanto, carbon dioxide emissions fell by about 25% when China implemented lockdown. However, emissions have decreased since the pandemic, but there is no measurable change in time, long and significant scale [2].

The COVID-19 pandemic caused changes in production and energy consumption habits, 80% of workplaces were closed, household electricity demand increased, people's mobility decreased by about 50%, some air transportation between countries was stopped, so the International Energy Agency forecasts a reduction in emissions energy-related CO_2 is about 7% [3].

Based on data published by the Directorate General of Climate Change, Ministry of Environment and Forestry during the COVID-19 pandemic, National GHG emissions from the household scale have increased by 12.96% - 28.50%, and an increase in GHG emissions from household LPG use is 6.02% in May 2020 [4].

At the COP 26 forum, Indonesia has shown the lowest deforestation rates in the last 20 years. Forest fires reduced by 82% in 2020, restoring 600,000 hectares of mangrove forest by 2024 and successfully restoring 3 million critical lands from 2010-2019. Indonesia will achieve a carbon net sink by 2030, the use of renewable energy, mobilization of climate finance and innovative financing, and the carbon market and carbon price are part of national efforts to tackle climate change.

Research [5] related to the inventory of the energy sector obtained emissions from the energy sector in North Sumatra of 8.3 million tons of CO_2eq or 4.39% of the total emissions. From this research, no spatial analysis related to the distribution of GHG emissions has been carried out in North Sumatra. The energy sector has the highest emission growth, followed by the waste sector. Economic and population growth is directly proportional to the increase in energy demand.

The inventory of GHG emissions in Medan City for the household energy sector, especially electricity and LPG consumption, was carried out in 2019 [6]; the inventory was carried out before the pandemic so that GHG emissions in the energy sector, especially electricity and LPG consumption, amounted to 1.16 million tons CO_2eq/year. This study also shows that the energy sector in the form of electricity consumption and household-scale LPG is the most dominant in contributing GHG emissions of ± 83.11% of the total household-scale GHG emission sector.

Carbon emission space is a characteristic of spatial distribution, such as the strength of the planned structure, the direction of roads, the direction of public facilities, and the connection of industries. The spatial distribution of CO_2 emissions around rural and urban areas of East Asia shows the same characteristics. According to industry classification, the industrial community produces the highest CO_2 emissions, far higher than other communities such as modern logistics, housing, and tourism. The physical energy use aspect contributes to the driving force of the spatial distribution of CO_2 emissions, followed by the spatial planning, social life, and industrial economy aspects. Fuel consumption per unit of land is the most important factor affecting CO_2 emissions [7].

Research related to the intensity of GHG emissions from urban dwellings in North and South China obtained a spatial autocorrelation from 2000 – 2016. Climate, consumption level, and building area are significant factors in increasing the intensity of GHG emissions in urban dwellings [8].

Handling climate change to protect the environment, which is expected to be achieved by 2030, is one of the Sustainable Development Goals (SDGs) goals. This research is fundamental to be carried out because it is in line with international and Indonesian programs to reduce GHG emissions. The existence of spatial analysis or map of the distribution of GHG emissions will identify the distribution area of GHG emissions; can assist in regional spatial planning, especially for increasing green open space; can provide input for health risk studies; can determine the level of compliance with emission-quality standard threshold regulations and establish ambient air monitoring locations.

In this research, spatial mapping analysis with the concept of a Geographic Information System will be carried out. This spatial analysis can help visualize data on the distribution of GHG emissions from household activities consuming energy (electricity and LPG) and the use of fuel for transportation in each study area using ArcGIS software version 10.5. The result of the study is a map of the distribution of CO_2 gas emissions obtained that can be used to support GHG mitigation reduction strategies to overcome global warming in the city of Medan.

2. Methodology

2.1. Location of study

The study area researched in this research is the urban and sub-urban areas of Medan City. The study area selection is adjusted to the designation and regional zones in the Medan City Regional Regulation No. 13 of 2011 concerning the Medan City Spatial Plan 2011-2031. Urban areas represented by the Districts of Medan Barat, Medan Timur and Medan Perjuangan, while thesuburbs were represented by Medan Deli, Medan Sunggal and Medan Amplas. The distribution of the population in the study area and respondents can be seen in Table 1.

Table 1. Number of Population and Respondents in the Study Area

Study Area	Number of Population (People)	Number of Respondent (Household)
Urban Area		
Medan Barat	90551	168
Medan Timur	117871	257
Medan Perjuangan	104893	228
Sub Urban Area		
Medan Sunggal	122743	182
Medan Amplas	134733	194
Medan Deli	193899	280

Source: [9]

2.2. Data collection

Secondary data needed for research are (1) baseline data on GHG emissions in North Sumatra Province. This data was obtained from the results of previous studies [5]; (2) Baseline GHG emission data from the residential sector in Medan City is obtained from the results of [6] research;
(3) Data on population, number of households, area of research are obtained from the Central Statistics Agency [9]; (4) The land use map of Medan City is obtained from BIG (Geospatial Information Agency); (5) The map of the spatial and regional planning of the city of Medan is obtained from the Bappeda of the City of Medan;

The primary data to be taken are (1) fuel consumption for cooking (LPG, kerosene, or wood) and collecting data through interviews (questionnaires) to randomly selected respondents. Determination of the number of respondents using the Kricjie and Morgan method with a sampling error of ±1% ; (2) the coordinates of the study area to input the spatial mapping process.

2.3. Data analysis method

2.3.1. Test the reliability and validity of the questionnaire data

Questionnaire data obtained need to be validated and the reliability of 10%. Validation and reliability tests were carried out on several questions. Questionnaire questions related to the respondent's personal data such as age, occupation, number of family members, income, and others. Other questions such as fuel for cooking, one month's fuel requirement, cooking habits, etc. In this study, the validation and reliability tests were considered to represent the total number of respondents: validation and reliability testing using SPSS 25.

According to [10], the research questionnaire is declared valid if it is seen that the r count is greater than r table. Where r count is obtained from the results of operations using SPSS 25 and r table is obtained from the table of critical value values, with a degree of freedom df = n-2. Thus, the value of r table is pdf = 130 - 2 = 128 at the level of = 0.1 is 0.2058.

Meanwhile, for the reliability test according to [10], if the value of Cronbach's alpha is more significant than 0.6, then the data is declared reliable, meaning that the questionnaire can be trusted and can be used in this study. The validation test for this survey will help you determine if the questionnaire distributed is valid or if the correlation values obtained are valid. Confidence tests were performed to assess the consistency of respondents' responses.

2.3.2. Calculation of GHG emissions from household-scale LPG use

Data analysis was carried out to obtain data on the value of GHG emissions at each sampling point in the research area. The GHG emission values would be obtained from each sub-district per sub-district. Meanwhile, secondary data is used to support primary data processing, such as emission factor data used to calculate emissions. The emission calculation is based on IPCC [11].

In this study, the formula for calculating GHG emissions from LPG use is given in Equation 1 for CO_2, CH_4 and N_2O.

$$Emission\ CO_2, CH_4, N_2O = EF\ (CO_2, CH_4, N_2O) \times F_{cy} \times NCV \quad (1)$$

Where is the CO_2, CH_4, and N_2O emission of LPG use in tons of CO_2eq/month; Fcy is fuel consumed (Kg/month); EF CO_2 is emission factor for CO_2 (63,100 kg/TJ), EF CH_4 is the emission factor for CH_4 (5 kg/TJ), EF N_2O is the emission factor for N_2O (0.1 kg/TJ), and NCV is Net CalorificVolume (47.3 x 10^{-6} TJ/kg).

In this study, the unit used is Ton CO_2eq. The calculation of CO_2, N_2O, and CH_4 emissions are multiplied by the value of GWP (Global Warming Potential) to convert the units of tons of CO_2eq. The GWP value of CO_2 is 1, the GWP of CH_4 is 25 and the GWP of N_2O is 298. To get Tons of CO_2eq, use equation 2.

$$CO_{2eq} = \{(CO_2 \times GWP\ CO_2) + (CH_4 \times GWP\ CH_4) + (N_2O \times GWP\ N_2O)\} \quad (2)$$

2.3.3. Method of spatial analysis of GHG emissions household-scale LPG use

Spatial data is an accurate image of an area of the earth's surface. It is typically represented digitally in the form of graphics, maps, and images, and is stored in the form of x, y coordinates (vectors) or images (raster's) with specific values [12].

Making a map of the potential distribution of GHG emissions from the use of household-scale LPG in the study area using Quantum GIS (QGIS) which aims to convert the data in the form of measurement results into data points that can be read by QGIS. Data on the distribution of GHG emissions using LPG from the measurement results will be represented in one coordinate for each village so that in the process of creating data points, X and Y coordinates will be needed as representatives from each village.

The uploaded data points are data X, Y, the name of the village, and the results of the calculation of GHG emissions using LPG which have been converted into CSV (comma delimited) format. After the data is entered in QGIS, a joint of data is carried out between the map of the suburban and urban areas with the data points that have been previously entered through the map properties. The data that has been joined with the map will produce a new map that contains data on the potential results of the distribution of GHG emissions using LPG. Maps that already contain data on the distribution of GHG emissions from the use of LPG will be classified into 3 classifications with the same classification level, so that uniform color differences are obtained on the sub-urban map and the urban map of the city of Medan. The color difference on this map is divided into green, yellow, and red with red indications representing the highest potential distribution of GHG emissions using household-scale LPG.

Spatial analysis of GHG emissions is the final process to achieve the objectives of this research. Data visualization uses different coloring for each level to be seen and read clearly. Based on research [12], the darker the color, the higher the emission concentration level. On the contrary, the lighter the color, the lower the emission concentration.

3. Results and discussion

3.1 GHG emissions using household-scale LPG in the study area

GHG emissions in the household-scale energy sector, which are taken into account in this study, are from LPG consumption. The data entered into the calculation was obtained from a questionnaire to 1,309 respondents. Based on the results of the questionnaire sampling, respondents generally use LPG as fuel for cooking with varying sizes, namely 3 kg and 12 kg. CO_2 emissions in this study were calculated based on the respondent's total LPG consumption in a month. The total consumption of LPG is then multiplied by the emission factor and NCV (Net Calorific Volume) for the CO_2 emissionfactor of 63100 kg/TJ based on the 1996 IPCC for the N_2O emission factor of 0.1 kg/TJ and CH_4: 5 kg /TJ. In this study, the unit used is ton CO_2eq, multiply the calculation result of CO_2, N_2O, CH_4 emissions by the value of GWP (Global Warming Potential) related to $CO_2 = 1$, $CH_4 = 25$, $N_2O = 298$. The results of the calculation of GHG emissions in the household-scale LPG for each sub-district in the study area are as shown in Table 2.

Table 2. GHG emissions using household-scale LPG in the study area

Study Area	Population (people)	GHG Emission			Total GHG Emission ton
		CO_2 (ton/year)	CH_4 (ton/year)	N_2O (ton/year)	CO_2eq/year
Urban					
Medan Barat	90551	2707.297	0.376	0.008	2709.530
Medan Timur	117817	3910.536	0.542	0.011	3927.340
Medan Perjuangan	104893	3701.268	0.514	0.010	3717.180
Sub Urban					
Medan Sunggal	130768	4290.332	0.594	0.012	4308.729
Medan Amplas	131451	8935.629	0.664	0.013	8395.629
Medan Deli	191650	6679.509	0.917	0.018	6707.923

The household-scale GHG emissions inventory results during the COVID-19 pandemic in the study area, namely urban and sub-urban areas of Medan City in 2021, ranged from 2709.530 – 8395.629 tons CO_2eq/year. Based on Table 2, the site that contributes the most GHG emissions is in the suburban area (± 65.21%) of the total GHG emissions in the study area. Table 2 also shows that the sub-district with the most significant GHG emission using household-scale LPG is Medan AmplasSub-district at 8395.629 tons CO_2eq/year or ± 28.20% of the total GHG emissions in the study area. The high GHG emissions in this sub-district are proportional to the number of residents in the place. This can be seen from the correlation between the total population and total GHG emissions showingthe value of R^2 = 0.43. It means the more the population, the more household activities, the greater theGHG emissions produced.

3.2 Spatial analysis of GHG emissions from household-scale LPG use in the study area

Geographic Information System (GIS) is designed for paintings with spatially referenced facts and geographic coordinates. Mapping is the grouping of regions associated with numerous geographical locations, together with highlands, mountains, resources, and populace potential.

The prerequisite for the realization of ArcGIS is spatial data. Spatial data is geographic objects or elements that can be interpreted and have a reference location based on specified coordinates. Spatial data includes graphic data and attribute data. This data has a coordinate system that is unique to the reference base and has the necessary data: location information (space) and descriptive information (attributes). With the existence of a Geographic Information System (GIS) or mapping of distribution potential with spatial analysis, it will be easier to determine areas with high GHG emission distribution so that climate change mitigation actions can be planned. The potential distribution of GHG emissions from household-scale LPG use in urban and sub-urban areas of Medan City can be seen in Figures 1 and 2.

IOP Conf. Series: Earth and Environmental Science 1065 (2022) 012031 doi:10.1088/1755-1315/1065/1/012031

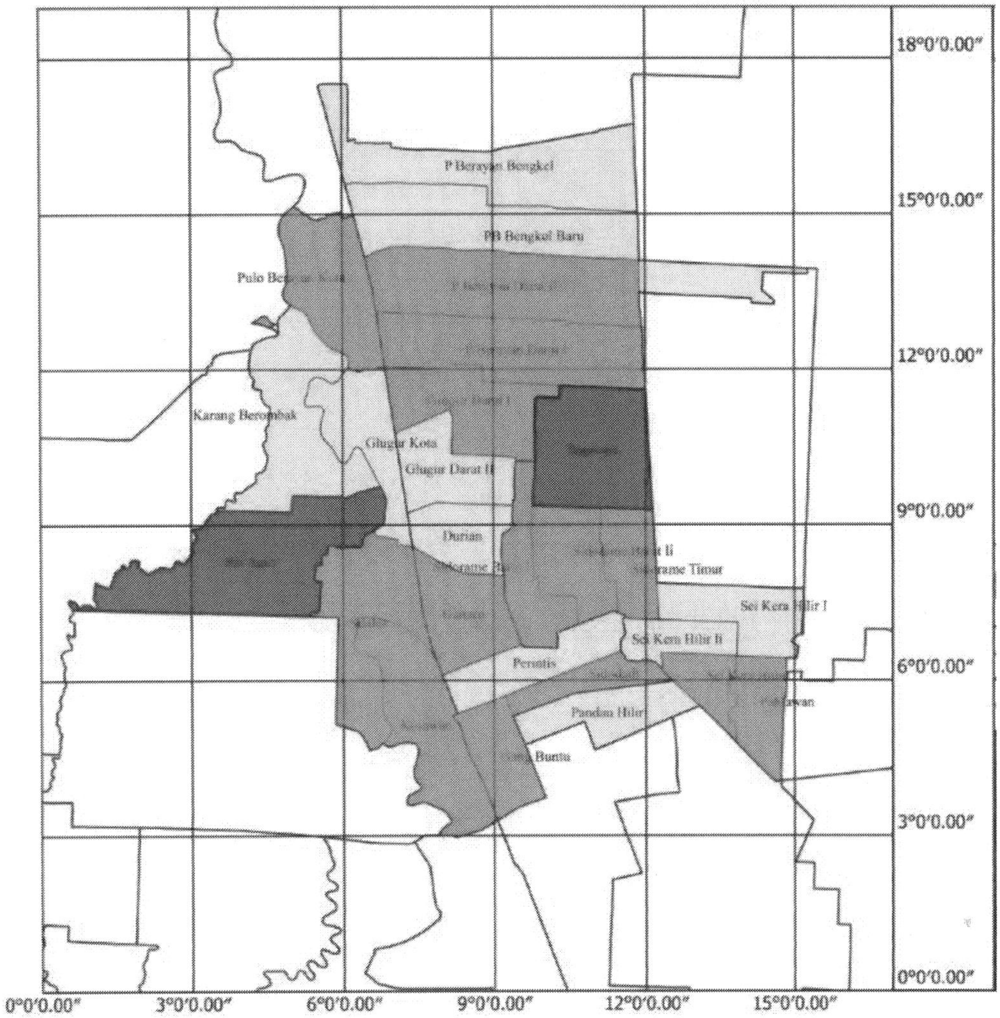

Fig 1. Map of the potential distribution of GHG emissions from household-scale LPG use in urban areas Medan city

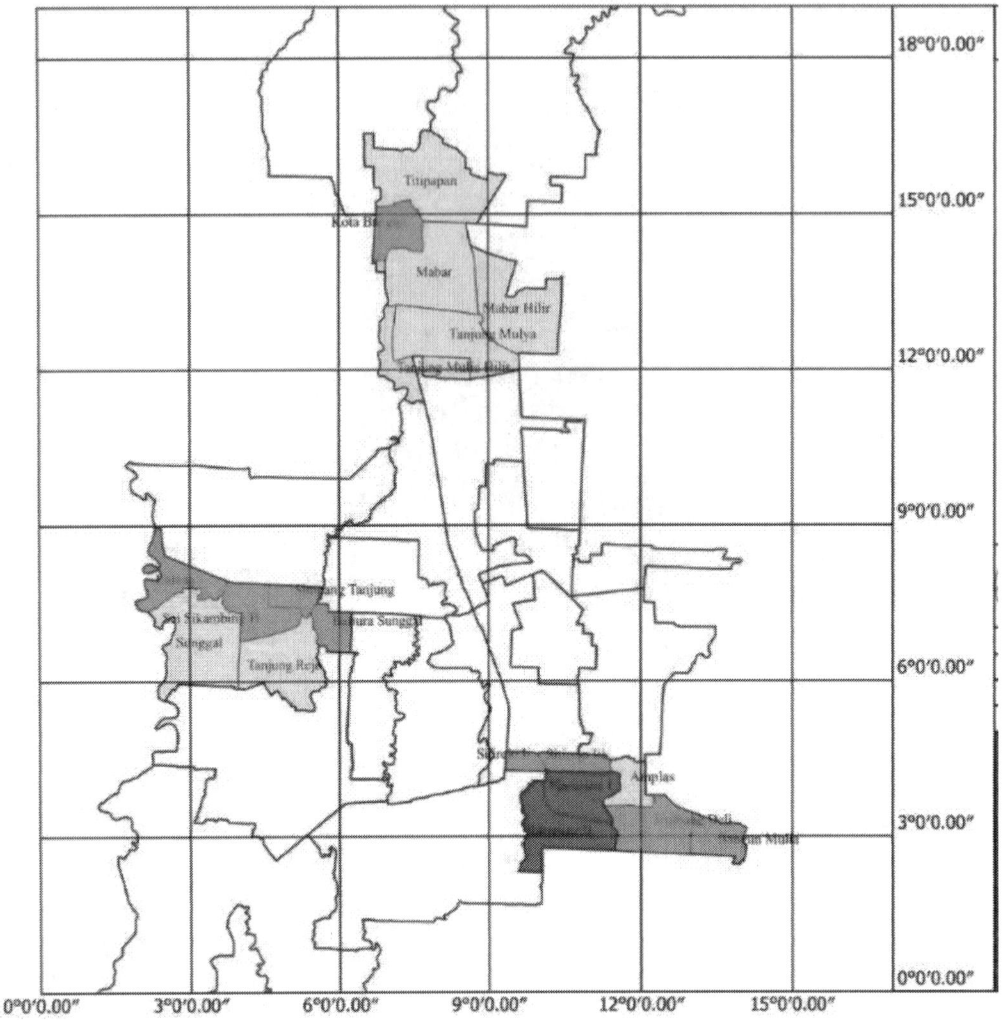

Fig 2. Map of the potential distribution of GHG emissions from household-scale LPG use in suburban areas Medan city

The color gradation shows the range of GHG emission figures obtained from the calculation of the emission inventory. The green color represents the lowest number and the red color represents the highest number. Determination of the range based on GHG emission inventory calculation data by making class intervals from the lowest value to the highest value.

Figures 1 and 2 show the distribution of GHG emissions from household-scale LPG use varies between urban and suburban areas. In urban areas (Figure 1) it can be seen that the dominant color is green for the study area with GHG emission ranges ranging from 110 – 388 tons CO_2eq/year, yellow is indicated in 10 (ten) urban villages in the study area with GHG emissions ranging from 388 – 666 tons CO_2eq/year. In this urban area, there are only 2 (two) kelurahan that are indicated as red, namely Kelurahan Tegalrejo and Kelurahan Sei Agul with a GHG emission range of 666 – 943 tons CO_2eq/year, so it can be concluded that broadly speaking, GHG emissions from household-scale LPG use in the area are: urban areas are still in the low to moderate range.

Meanwhile in Figure 2, it can be seen that the color yellow dominates the study area. It can be seen in Medan Deli District, which is dominated by yellow color with a range of GHG emissions from 952

– 1872 CO_2eq/year in each village. In the sub-urban area, there are 2 villages that are colored red, namely Harjosari I and Harjosari II villages which are located in Medan Amplas District with GHG emission ranges ranging from 1872 - 2792 tons CO_2eq/year. According to [13], the amount of CO_2 emissions resulting from energy utilization is influenced by the number of households in the research location, resulting in CO_2 emissions for each household.

This is in line with research conducted by [14] in Waru Subdistrict, Tambak Oso Village which stated that the number of household heads is relatively small so the emissions generated from the use of LPG and electricity are also low. According to [15], the main factors that influence the largest total CO_2 emissions are the total consumption of each energy and the number of households in the study area.

3.3 The driving factors for increasing GHG emissions using household-scale LPG during the COVID-19 pandemic

One of the factors driving the increase in GHG emissions using household-scale LPG during the COVID-19 pandemic is the population. The result of the correlation between the population and GHG emissions using household-scale LPG in the study area during the COVID-19 pandemic was 0.43. This figure shows a strong correlation between the population and GHG emissions, which means thatthe larger the population, the higher the GHG emissions from the use of household-scale LPG produced. This is in line with the spatial analysis of GHG emissions in Figures 1 and 2 which show that the yellow and red colors are found in suburban areas which have more population than urban areas. Meanwhile, based on previous research in Medan City [6], the correlation between GHG emissions from the use of LPG and the total population is 0.68.

The use of LPG in Ecuador for cooking is quite high where LPG is imported and subsidized by the government. The government of Ecuador has also started to promote the use of electric induction cookers for sustainability, but this must be balanced with environmentally friendly power generation [16]. Residential energy use accounts for about 20% of greenhouse gas (GHG) emissions in the United States, where suburban areas with higher levels of social affluence will produce GHG emissions 15 times higher than surrounding areas [17].

Some developing countries like Nepal, India, Pakistan, Sri Lanka, etc. use various fuels for cooking such as kerosene, biogas, traditional wood stove, natural gas, LPG, etc. The oil stove produces the highest emission of 350 g CO_2eq/MJ. To mitigate GHG emissions from fuel for household cooking, it is necessary to increase the use of modern biomass such as increasing the use of biomass and biogas [18].

Another factor that influences the size of CO_2 emissions resulting from the use of cooking fuel is the consumption pattern of the cooking fuel, the number of households as well as the economic growth and income per capita of a region [19]. During the COVID-19 pandemic, the use of LPG increased, especially 3 kg LPG, but the increase in the use of LPG by the community in the study area was not very significant, only around 1 – 5% so that it did not have much impact on GHG emissions resulting from the use of LPG.

Based on the results of this study, it can be determined that GHG emissions from the use of LPG per household for urban areas range from 0.150 – 0.177 tons CO_2eq/household and suburban areas range from 0.165 – 0.319 tons CO_2eq/household. The use of LPG for cooking during the COVID-19 pandemic has also increased. Respondents who used 3 kg LPG cylinders before the COVID-19 pandemic spent 2-3 cylinders per month and after the COVID-19 pandemic it became 4-5 cylinders per month. Meanwhile, respondents who used 12 kg LPG cylinders, before COVID-19 spent 1 12 kg LPG cylinder for 45 days and after the COVID-19 pandemic it was only enough for 1 month, so this also affects the monthly expenditure of each household.

4. Conclusion

Based on a spatial analysis of the potential distribution of GHG emissions from household-scale LPG use in urban and sub-urban areas of Medan City, it was found that GHG emissions in sub-urban areas were higher than in urban areas with indications of the dominant color being yellow and red on the map. The results showed that GHG emissions from the use of LPG in the study area ranged from

0.150 – 0.319 tons CO_2eq/household. One of the most influencing factors is that the population of sub-urban areas is more than urban areas. The larger the population, the higher the GHG emissions produced. In addition to population, other factors that affect GHG emissions are the pattern of LPG consumption for cooking and per capita income

Acknowledgments
The authors appreciate the cooperation of the University of Sumatera Utara in this study. The University of Sumatera Utara supported this project through the TALENTA CEST USU 2021 No 191/UN5.2.3.1/PPM/SPP-TALENTA USU/2021.

References
[1] Anih S S. 2020. Kajian singkat terhadao isu aktual dan strategis Vol XII, No. 13/Puslit/Juli/2020
[2] Rudiyanto, A. 2020. Pengaruh COVID-19 terhadap tujuan pembangunan berkelanjutana, Kementrian PPN/Bappenas
[3] Siksnelyte-Butkiene, I.2021. Sustainability 2021, 13, 12973
[4] KLHK. 2020. Emisi gas rumah kaca selama masa pandemic covid-19 di Indonesia
[5] Ambarita H, et al 2018 IOP Conf. Ser.: Earth Environ. Sci. 126 012114 1- 8
[6] Suryati I, Herlina N, Hasibuan N H, Siregar R L, Husna A, Silalahi A, Audina M and Sari S 2021 IOP Conf Ser: Earth and Environ Sci 802 012034 1 - 8
[7] Zhu L, et al. 2020. De-carbonization of global energy use during the COVID-19 pandemic
[8] Qingwei S , Jingxin G , Xia W, Hong R, Weiguang C and Haifeng W. 2020. Sustainability 2020, 12, 2695; 1 -23
[9] BPS Kota Medan. 2021. Kota Medan dalam Angka.
[10] Sugiyono. 2016. Metode Penelitian Kuantitatif, Kualitatif dan R&D. Bandung: Alfabeta
[11] Intergovernmental Panel on Climate Change (IPCC) 2006 Guidelines for National Greenhouse Gas Inventories (Institute for global environmental strategies (IGS) Hayama Japan)
[12] Somba, S. 2017. Analisis Karakteristik Spasial Kota Pare-Pare Berbasis GIS dan Remote Sensing Menggunakan Citra Landsat 8.
[13] Subkhan, A., Setyowati, D. L., & Setyaningsih, W. (2017). Kajian Emisi CO_2 dari Pemanfaatan Energi Rumah Tangga di Kelurahan Candi Kota Semarang. GeoImage, 6(2),147-157.
[14] Akmalina, D. (2020). Analisis jejak karbon pada aktivitas permukiman di Kecamatan Waru Kabupaten Sidoarjo (Doctoral dissertation, UIN Sunan Ampel Surabaya).
[15] Indrawati, E D. 2015. Masters thesis, Program Pascasarjana UNDIP, Semarang.
[16] Angel D. Ramirez, Andrea Boero, Edgar F. Perez, Daniel Salas 2017 ASME Conference Paper International Mechanical Engineering Congress and Exposition
[17] Benjamin G, Dimitrios G, and Joshua P. N. 2020. PNAS, vol. 117. (32); 19122 – 19130
[18] S.C. Bhattacharya and P. Abdul Salam 2002 Biomas and Bioenergy 22 (305 – 317)
[19] Q Nugrahayu et al. 2017 Journal of Environmental Science and Technology 9(1) 25 - 36

Role of Non-Orthogonal Hydrodynamic and Sedimentation Model for Port Facilities Analysis on Port of Kuala Tanjung

M Muin[1*], R Sihombing[1], M W Hermawan[1], A B Muslim[1], H N Alam[2]

[1]Ocean Engineering Research Group, Faculty of Civil and Enviromental Engineering, Institut Teknologi Bandung, Jalan Ganesha No 10, Indonesia
[2]Lembaga Afilisasi Penelitian dan Industri, Institut Teknologi Bandung, Jalan Dayang Sumbi No 7, Indonesia
*Corresponding author: muslimmuin@gmail.com

Abstract. The port-channel and the availability of piers are strongly influenced by physical environmental conditions such as sea current, sea tide, sediment transport, and wave. The best size of vessel and pier length depends on the location's demand and sea environment. The uncertainty of these environmental factors results in the low efficiency of port infrastructure, the load factor for back cargo, and the reliability level of port users. This paper presents the role of MuHydro3D and MuSed3D software, a three-dimensional hydrodynamic, and sedimentation models to determine the best vessel size. Using a non-orthogonal curvilinear coordinate technique, also known as the boundary-fitted technique, MuHydro3D and MuSed3D can accurately and quickly simulate the environmental conditions at the port. Knowledge of environmental conditions is expected to be an accurate basis in determining the depth and dredging duration, pier length, ship size, or the need of breakwater. The Port of Kuala Tanjung is one of Indonesia's designated international hubs and was chosen as the case study in this paper. The result shows that the sedimentation rate of dredging for a vessel with up to 3,000 TEUs capacity is not too sensitive. Moreover, a vessel design with an approximate capacity of 3,000 TEUs is the optimum size from the pier length aspect.
Keywords: Sediment Dispersion Modeling, Kuala Tanjung, Optimum Vessel Size, Non-Orthogonal, Boundary Fitted, Port Infrastructure

1. Introduction

Port operations are affected by physical environmental conditions, such as current conditions, water level, sedimentation, and wave conditions [1,2]. The wave condition could impact the vessel manoeuvring in the port-channel area [3]. Tidal conditions in the port area influenced the vessel arrival schedule and berthing allocations [4]. The uncertainty of these physical environmental conditions may result in the low efficiency of port infrastructure.

One of the best ways to identify these physical environmental conditions is by hydrodynamic model. Then the model is validated using observation data where the location observation data based on the model. MuHydro3D and MuSed3D are three-dimensional hydrodynamic models using boundary-fitted technique model [5,6]. The model was developed in spherical coordinate and was transformed to a non curvilinear coordinate to accurately and quickly simulate the environmental conditions in the port sea area.

The result of the model can further inform the environmental conditions of the port sea area. This information can be used to determine the requirements of port facilities such as pier design or the needs of breakwater structures. Furthermore, the information of physical conditions can be used to determine the best vessel size used to berth in the port.

Port of Kuala Tanjung is one of Indonesia's designated international hub ports and strategically located in Malacca Strait. The construction of the Port of Kuala Tanjung is expected to attract international cargo in the Malacca Strait [7], as well as being utilized as a hub port in *Tol Laut* (Marine Highway) Program [8]. Marine Highway is a containerized sea freight program to reduce the disparity of logistic prices in Indonesia. In addition, this program is aimed to enhance the logistic distribution within Indonesia's developing region.

Besides the international cargo, the Port of Kuala Tanjung is also planned to handle domestic transhipment. In 2028, the expected amount of container demand in Port of Kuala Tanjung will be more or less 1,000,000 TEUs/year [9]. This paper is expected to find the optimum vessel size for domestic container activity in the Port of Kuala Tanjung.

2. Model Description

2.1. MuHydro3D Description

The MuHydro3D model is a boundary-fitted ocean hydrodynamic model [5] using a non-orthogonal curvilinear coordinate technique in a spherical coordinate that can be used to compute tidal elevations, velocities, salinity, and temperature distributions in the computational domain at the same time. The non-orthogonal grid system matches the shoreline and bathymetric feature boundaries for accurately representing areas with complex coastal geometry. The grid system on the model can be adjusted in high resolution for the area of interest and lower resolution at other locations for computational efficiency. MuHydro3D can also be run in either two or three dimensions depending on the complexity of the area of study. A detailed description of the model with associated test cases is also described in Muin and Spaulding [10]. The model has undergone extensive testing against analytical solutions. Anticipating the importance of mangroves as a driving force, in 1999, Muin included the water exchange between wet/dry area and main channel in the computation [11].

The mathematical basis for the model begins with continuity and momentum equations. Both equations were developed with the underlying assumption that the flow is incompressible, variation of water mass density negligible except when multiplied with the earth's gravity (Boussinesq approximation), and vertical acceleration negligible compared to gravitational acceleration (hydrostatic approximation).

2.2. MuSed3D Description

MuSed3D integrates the three-dimensional hydrodynamics model and suspended sediment transport [6]. The sediment is divided into several grain size classes, and the sediment can be cohesive or non-cohesive sediment. The properties of cohesive sediment are quite different from those of non-cohesive sediment.

For very fine sediments, the sediments are considered cohesive, where the settling velocity is a function of sediment concentration and salinity [12]. The empirical formulas of settling velocity are derived by Muin from laboratory experiments of fine sediment in Papua Indonesia as shown in Equations 1 and 2 [12]. The empirical formulas used based on particular salinity and the settling velocity is set to zero or very small when its value is negative

For salinity 10 ppt

$$W_s = 0.023c^3 - 0.304c^2 + 0.895c - 0.131 \quad (1)$$

For salinity 32 ppt

$$W_s = 0.044c^3 - 0.457c^2 + 1.173c - 0.155 \quad (2)$$

Where W_s is settling velocity in mm/sec, and c is the concentration of fine suspended sediment in gr/l. The settling velocity for cohesive sediment is shown in Figure 1.

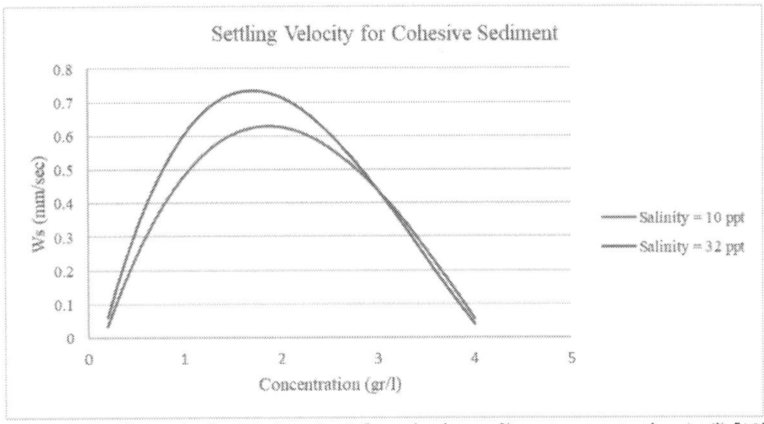

Figure 1. Settling velocity Ws (mm/sec) for cohesive sediment concentration (gr/l) [12].

3. Previous Application of MuSed3D to Support Port Design

MuSed3D has been widely used in Indonesia; in addition, to support oil company operations, this software is also used for Navigational Channel dredging plan for several ports such as, Pomako (Papua Province) and Dumai (Riau Province) [13]. The dredging plan at Pomako is very challenging because the sediment source comes from the mine tailings of PT Freeport Indonesia [14]. The model successfully simulates the change in bathymetry in the dredged plan area accurately.

MuSed3D also successfully simulates the distribution of sediments in the Mahakam Delta [15]. The sedimentation process in the Mahakam Delta is very complicated because of its complex geometry. Changes in water depth are suitable for several observation locations.

These previous studies prove that MuSed3D can accurately simulate existing sedimentation conditions in the port area and sedimentation due to dredging works.

4. Description of Study Area

Port of Kuala Tanjung is one of Indonesia's designated international hub ports. This port is located in Batubara Regency, North Sumatera, Indonesia. Existing berthing facilities in Port of Kuala Tanjung are multipurpose pier (Kuala Tanjung Multipurpose, KTMP) with dimensions 500 x 60 m^2 and connected by a 2,800 x 18.5 m^2 trestle. Apart from this pier, there are also two piers for the operations of PT. Inalum and PT. MNA. Figure 2 shows the satellite view of the location and general layout of Port of Kuala Tanjung.

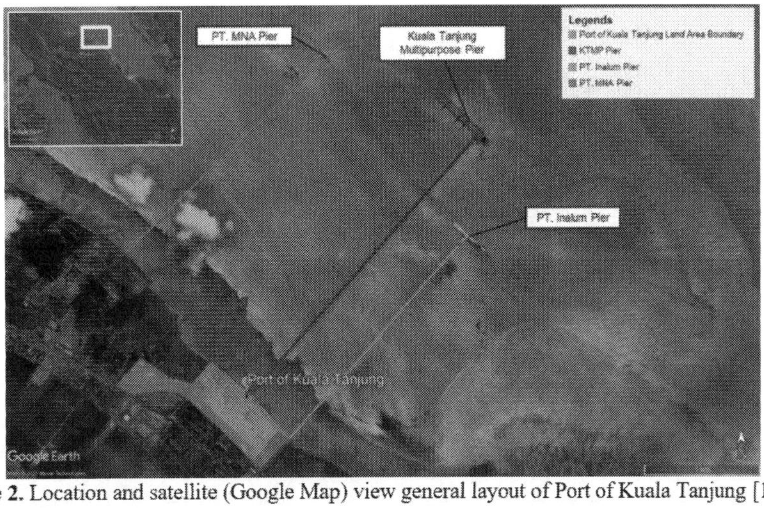

Figure 2. Location and satellite (Google Map) view general layout of Port of Kuala Tanjung [16].

In the area around Port of Kuala Tanjung, there are four rivers with downstream areas located around the Port of Kuala Tanjung coastline area. These rivers are Siparepare River and Bahapal River in the West, Tanjung River, and Gambus River. The location of these rivers to the Port of Kuala Tanjung can be seen from the map in Figure 3. The rivers will be used as input for the sedimentation model.

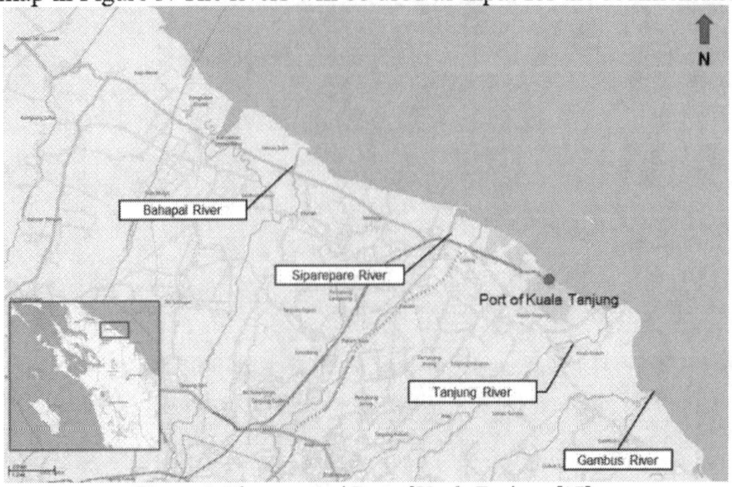

Figure 3. Rivers around Port of Kuala Tanjung [17].

5. Optimum Vessel Size of Port

A ship that can dock at a pier depends on the vessel's draft. The minimum depth is about 1,1 of the draft. The objective is to determine the optimum vessel size. Reducing the size of the ship will certainly reduce the cost of dredging in the port and navigational channel. As a result, the Berth Occupation Ratio will increase and require longer berthing structure. The deepening of the navigational channel will also increase the sedimentation rate because the current speed will be lowered.

The design length of the berthing structure (L_B) is heavily influenced by the design number of berthing vessels required to accommodate the total number of ship call and to keep the value of Berth Occupancy Ratio below the limit value. The equation of design length of berthing structure can be given as follows.

$$L_B = nLOA \times 1.2 \quad (3)$$

Where:
n = Number of berthing vessels
LOA = Length Overall of design vessel

The number of berthing vessels will be obtained from the number of demands compared to the capacity of the designed vessel.

5.1. *Design Vessels*

In this analysis, several design vessels are used as a reference to calculate the length of berthing structure requirement. Table 1 presents the vessel specification for several container vessels with the capacity ranging from 500 TEUs to 18,000 TEUs.

Table 1. Dimensions of Design Vessels

Vessel Capacity (TEUs)	Vessel Specification (m)		
	Beam	Length	Draft
500	17.00	137.00	9.00
1,000	20.00	215.00	10.00
3,000	32.00	250.00	12.50

Vessel Capacity (TEUs)	Vessel Specification (m)		
	Beam	Length	Draft
4,000	40.00	300.00	13.00
6,000	43.00	340.00	14.50
12,500	49.00	366.00	15.20
15,000	56.00	397.00	15.50
18,000	59.00	400.00	16.00

5.2. Required Berthing Structure Length

The required length of berthing structure for container vessels is affected by the container demand in the port. Table 2 presents the required number of berthing vessels and length of berthing structure for container demand 1,000,000 TEUs/year. Figure 4 presents the comparison of required length of berthing structure for each design vessels.

Table 2 Required Length of Berthing Structure for Each Design Vessels

Vessel Capacity (TEUs)	Ship Call (Call / year)	Estimated Berth Time (day)	Number of Berth	Required Length of Berthing Structure (m)
500	1200	0.20	6	986.4
1,000	600	0.24	3	774.0
3,000	200	0.71	1	300.0
4,000	150	0.79	1	360.0
6,000	100	1.17	1	408.0
12,500	48	2.13	1	439.2
15,000	40	2.55	1	476.4
18,000	34	3.06	1	480.0

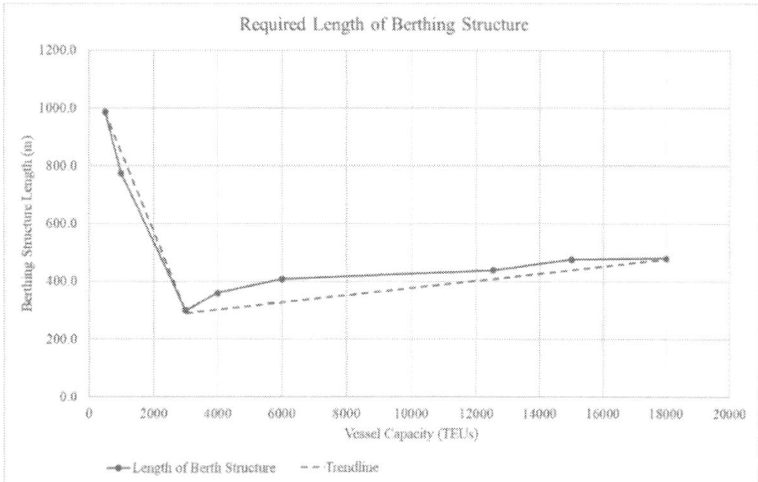

Figure 4. Relationship between the required Length of Berthing Structure and Vessel Size.

The Table 2 and Figure 4 show the optimum vessel size from the berthing structure aspect is 3,000 TEUs.

6. Model Application in Kuala Tanjung Domestic Transhipment Port

6.1. Model Setup

6.1.1. Grid System

For modeling purposes, the area of study is divided into computational grids. The open boundary on the model is as far from the study site as possible to reduce the effect of the open boundary on the simulation model results. It is also recommended to extend the domain model for a wider area to eliminate the possibility of boundary effects due to complex flow patterns. The model grid system is shown in Figure 5. The main water bodies are represented by 5,826 boundary fitted water cells.

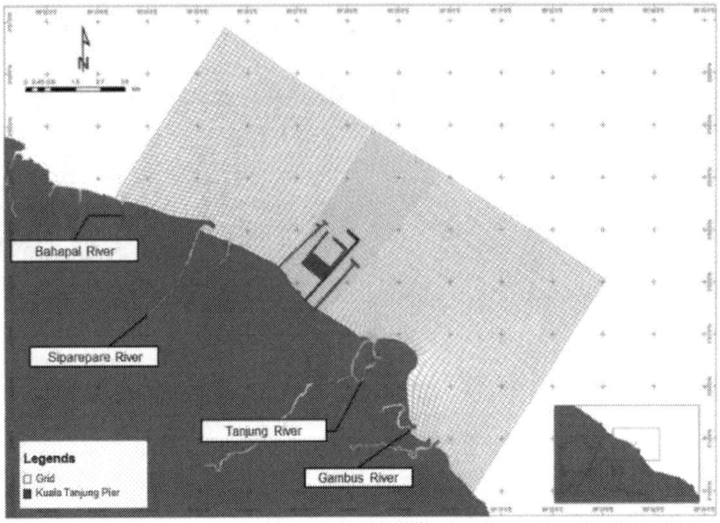

Figure 5. Model grid system.

6.1.2. Bathymetry

The water depth in the computing area is obtained from the results of field measurements carried out by PT Multi Purpose Terminal. The water depth is very shallow along the coast consisting of fine sediment originating from four rivers. The waters start to drastically deepen within 1,500 meters distance from the shore. The bathymetry of the study area is presented in Figure 6.

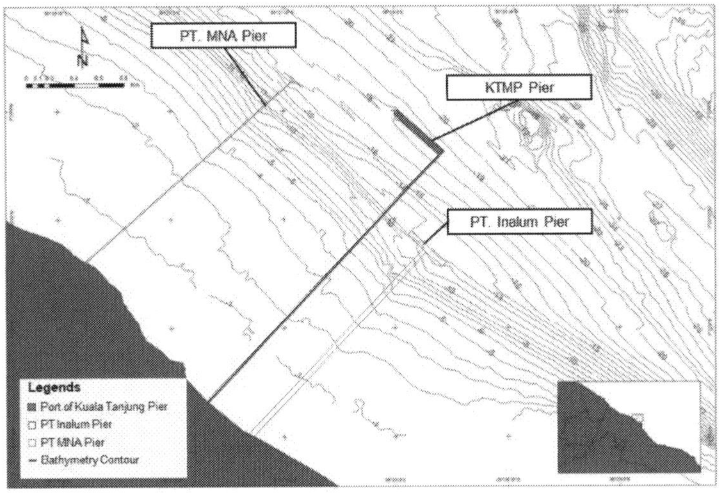

Figure 6. Bathymetric map of Port of Kuala Tanjung area.

6.1.3. Open Boundary

Seven main tidal components (M2, S2, N2, O1, K1, P1, and Q1) are used to calculate the tidal elevation along the open boundary. The tidal components are obtained from larger computational domain in Malacca Strait.

6.1.4. River Boundary

There are four rivers that are used as input in the computing area, namely Siparepare River and Bahapal River in the West, Tanjung River and Gambus River in the East as can be seen from Figure 5. The rivers that are included in the computing area are a source of sediment that poses a risk of sedimentation in the Port Area.

The river flow was calculated using rational method with the rainfall data from Meteorology Climatology and Geophysical Agency Indonesia (BMKG) at Kualanamu Station. The time series of flow and sediment concentration for each river can be seen in Figure 7 and Figure 8.

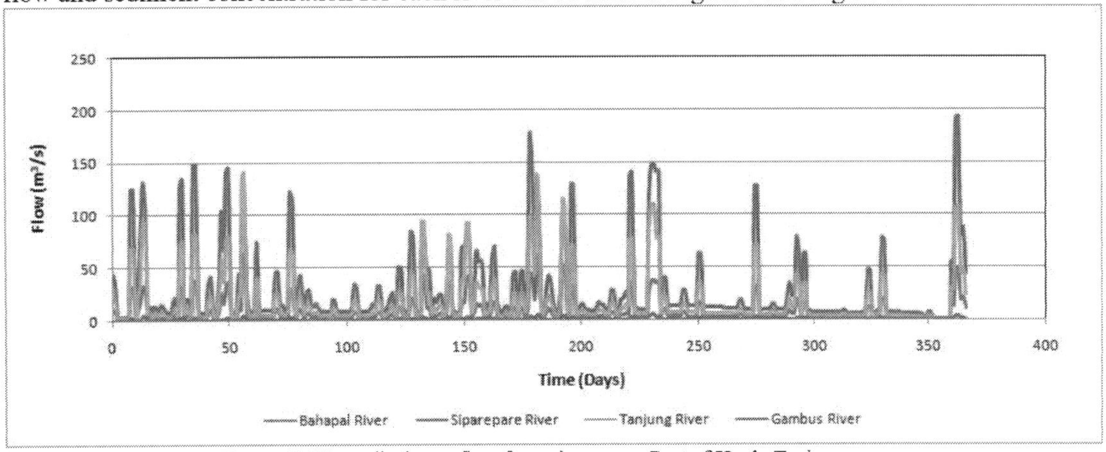

Figure 7. Water discharge flow from rivers near Port of Kuala Tanjung area.

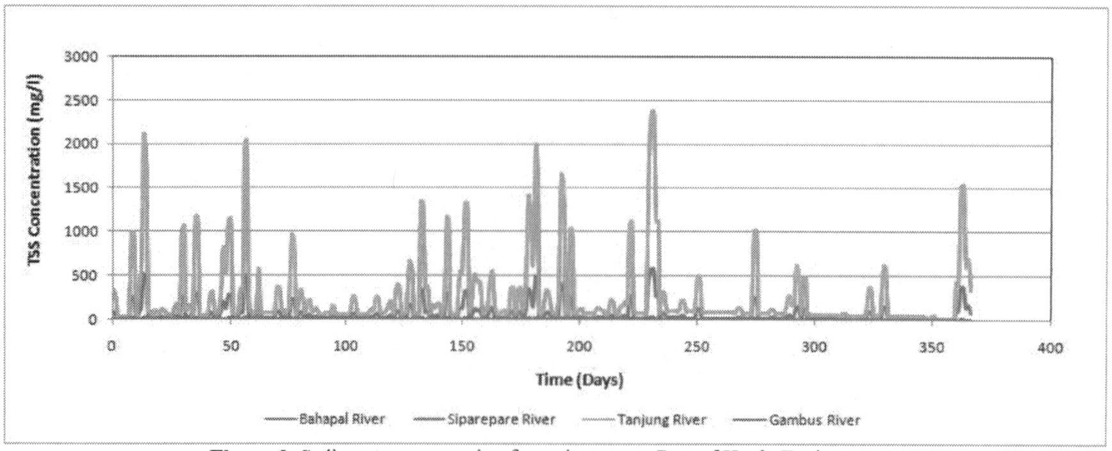

Figure 8. Sediment concentration from rivers near Port of Kuala Tanjung area.

6.2. Field Survey

Measurements were carried out to understand the characteristics of sediment and ocean currents at the study location. The field survey was conducted from 24th April to 14th May 2019. The measurement of water velocity and direction was carried out using ADCP and Total Suspended Solid (TSS) concentration were obtained from Laboratory Analysis of water samples. The measurement locations are presented in Figure 9. The measurement results are also used to calibrate the model used.

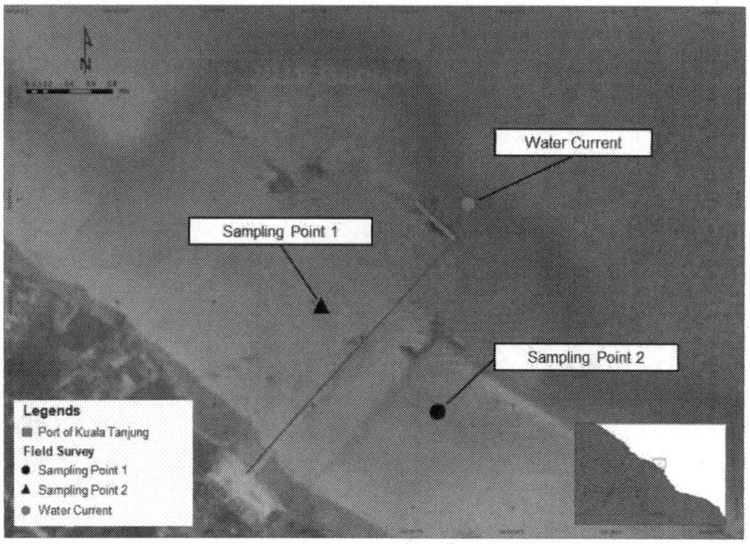

Figure 9. Locations of Observation.

6.3. Model Calibration

The success of the model application depends on the credibility of the model results. This credibility is obtained by comparing the model results with observation data. The purpose of calibration is to select the model coefficients to match the observed data. The calibrated model parameters are seabed roughness and critical shear stress for erosion and deposition.

Figure 10 and Figure 11 show the comparison of current speed between the measurement results and the model on 7th – 12th May 2019. The comparison between the observation data and simulation results for water current in the surface and bottom shows a good agreement both for amplitude and phase. The model also calibrates the concentration at the sampling of points 1 and 2, which is shown in Figure 12 and Figure 13.

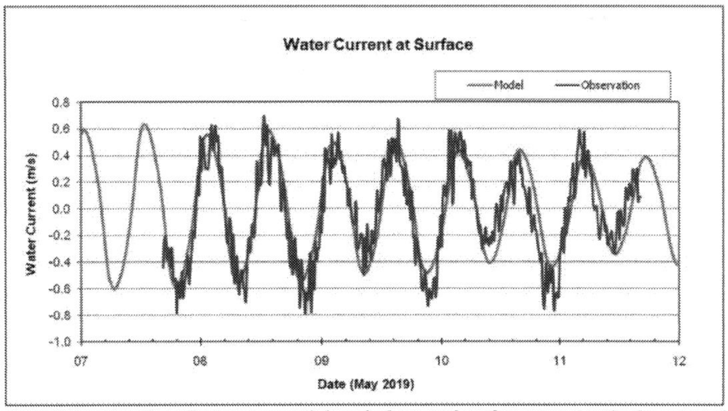

Figure 10. Comparison between model and observation for current velocity at surface.

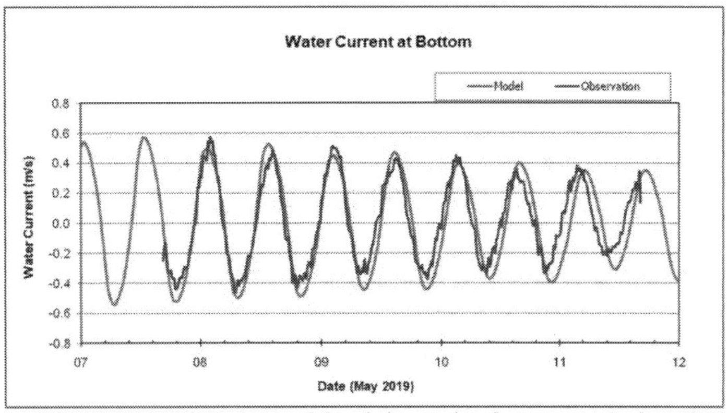

Figure 11. Comparison between model and observation for current velocity at bottom.

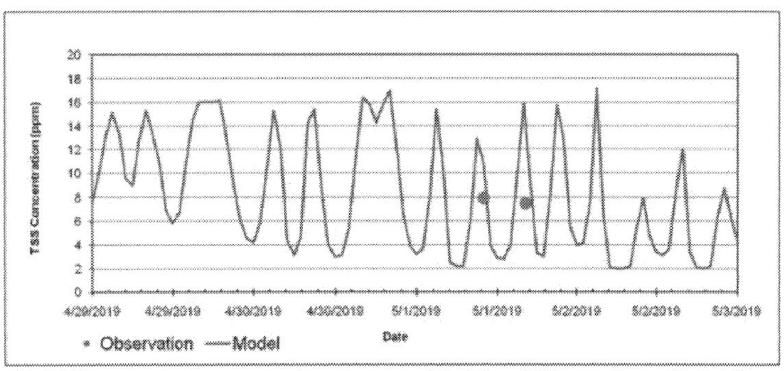

Figure 12. TSS concentration comparison between Model and Observation at Point Sampling-1.

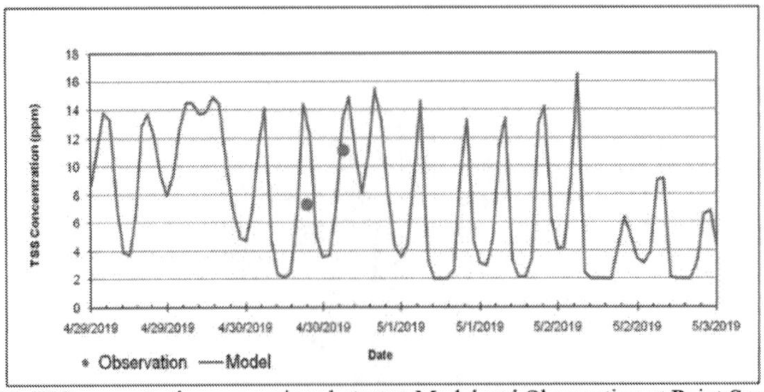

Figure 13. TSS concentration comparison between Model and Observation at Point Sampling 2.

6.4. Modelling Scenario

The existing Kuala Tanjung pier is equipped with a curtain breakwater to protect container ships from waves. Therefore, the port will be built in the south of the existing pier for the next development for domestic vessel. Dredging process needs to be done because most of the development area has a depth of 5 meters.

Two simulation scenarios are used to find the best selection of ship size, namely dredging for small 500 TEUs vessels (Scenario-1) and dredging for 3,000 TEUs vessels (Scenario-2). Simulations were carried out to compare the sedimentation rates for these scenarios.

7. Result and Discussions

Ocean currents in the computational area are tidal dominant. The simulation results show that the Port of Kuala Tanjung pier is outside the plume of high concentration sediment. The water depth change in the pier area is very small because strong current on the area prevents the suspended sediments from settling.

The simulation results of hydrodynamics and sedimentation models have been compared to the results of field observations with satisfactory results. The water velocity at the surface is about 10% greater than near the seabed.

Figure 14 shows the sediment deposition in the study area for Scenario-1 simulation, while Figure 15 shows the sediment deposition for Scenario-2 simulation. The time series of sediment deposition for Scenario-1 and Scenario-2 is presented in Figure 16.

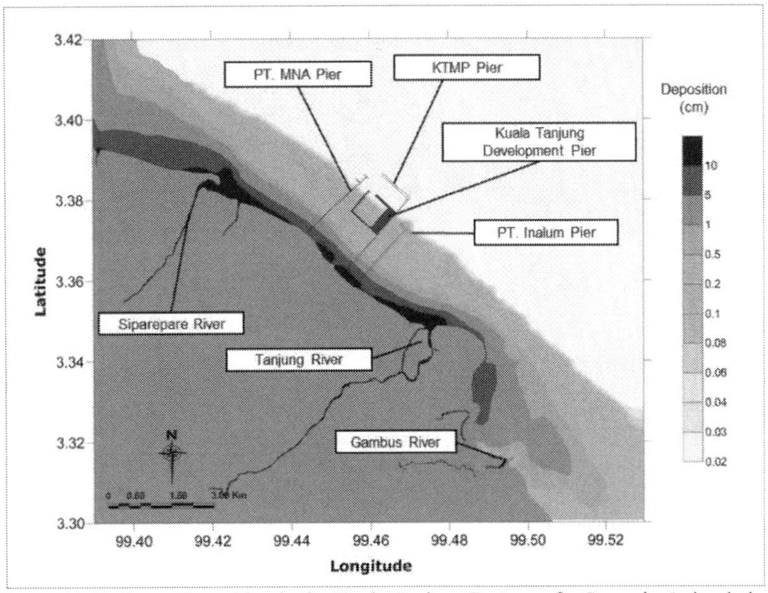

Figure 14. Sediment deposition in the Kuala Tanjung Port area for Scenario-1 simulation.

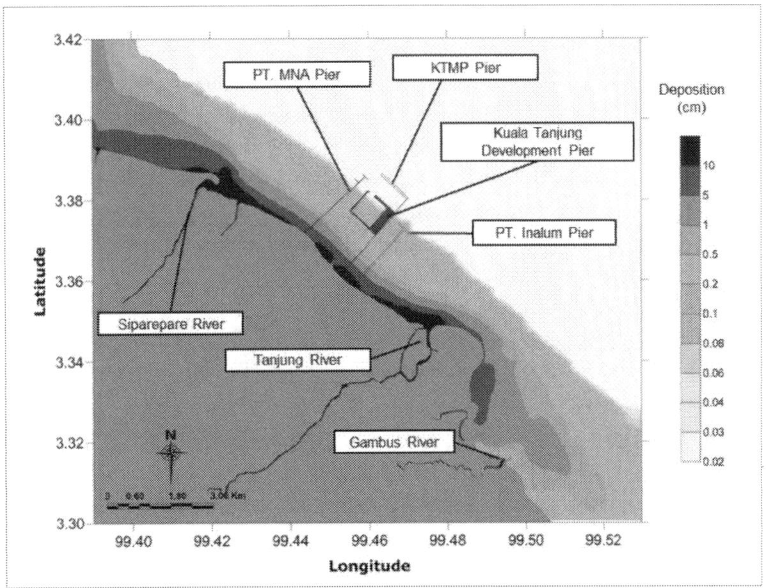

Figure 15. Sediment deposition in the Kuala Tanjung Port area for Scenario-2 simulation.

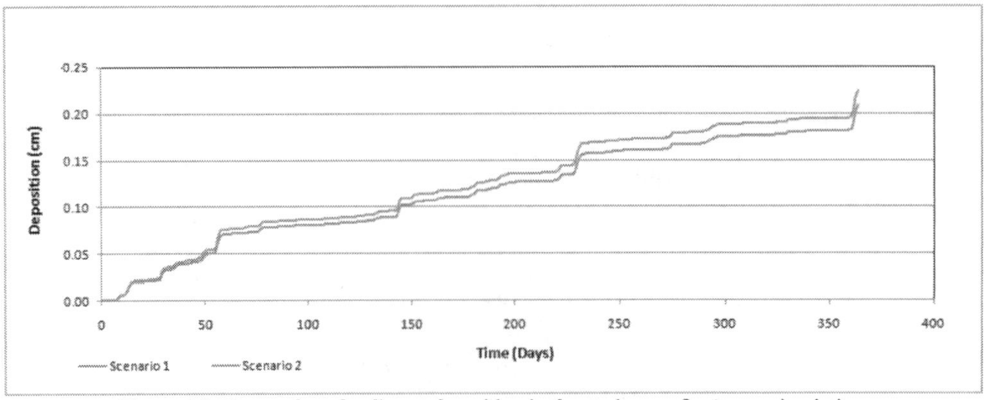

Figure 16. Time series of sediment deposition in the study area for 1 year simulation.

From the results of Scenario-1 and Scenario-2, changes in dredging depth impact the sedimentation rate in the study area. Dredging for small container ships of 500 TEUs will affect the deposition rate in the study area for around 0.21 cm per year, while dredging for Scenario-2 results in a deposition rate of approximately 0.23 cm per year.

8. Conclusion

This paper introduces the relationship between berth length requirements and container ship size. The result shows that the optimum ship size from the pier length perspective is 3,000 TEUs. However, the size of the vessel will depend on the network of the transportation system. The hydrodynamic and sedimentation simulations play an important role in port design. Moreover, the simulation results are useful in determining the optimum size vessel for a port.

MuHydro3D and MuSed3D software had been previously applied to support the Indonesian Government Marine Highway in Dumai (Riau Province), Pomako (Papua Province), and Delta Mahakam (East Kalimantan Province). The simulation results can support the decision-makers in planning the development of the navigational channel.

The model results show that the sedimentation rate for the proposed domestic jetty in Kuala Tanjung is not too sensitive for the dredging depth of a vessel with up to 3,000 TEUs capacity since it is not built in the sediment plume area. The result of this study can be used for further research related to the space of port turning basin area, the availability of ships, and the networking of the sea transportation itself, whether the 3,000 TEUs vessel is the optimum size for the domestic container activity in the Port of Kuala Tanjung.

References

[1] Yang Z, Ng A K Y, Lee P T W, Wang T, Qu Z, Sanchez Rodrigues V, Pettit S, Harris I, Zhang D and Lau Y yip 2018 Risk and cost evaluation of port adaptation measures to climate change impacts *Transportation Research Part D: Transport and Environment* **61** 444–58

[2] Saengsupavanich C 2011 A current Environmental Impact Assessment of a port in Thailand: Marine physical aspects *Ocean and Coastal Management* **54** 101–9

[3] Sprenger F, Maron A, Delefortrie G, Zwijnsvoorde T van, Cura-Hochbaum A, Lengwinat A and Papanikolaou A 2017 Experimental Studies on Seakeeping and Maneuverability of Ships in Adverse Weather Conditions *Journal of Ship Research* **61** 131–52

[4] Dadashi A, Dulebenets M A, Golias M M and Sheikholeslami A 2017 A novel continuous berth scheduling model at multiple marine container terminals with tidal considerations *Maritime Business Review* **2** 142–57

[5] Muin M 1993 *Three-Dimensional Boundary-Fitted Circulation Model in Spherical Coordinates, Dissertation, University of Rhode Island.*

[6] Muin M, Idris K and Yuanita N 2016 Application of large-scale 3D non-orthogonal boundary fitted sediment transport model and small-scale approach for offshore structure in Cimanuk Delta North Java Sea *Journal of Engineering and Technological Sciences* **48** 301–19
[7] Kementerian Perhubungan Republik Indonesia 2017 *Lampiran Rencana Induk Pelabuhan Nasional KP 432 Tahun 2017*
[8] Biro Komunikasi dan Informasi Publik Kementerian Perhubungan Republik Indonesia 2015 Pembangunan Tol Laut Libatkan 24 Pelabuhan
[9] Kementerian Perhubungan Republik Indonesia 2020 *Rencana Induk Pelabuhan Kuala Tanjung* (Kementerian Perhubungan)
[10] Muin M and Spaulding M 1997 Three-Dimensional Boundary-Fitted Circulation Model *J. Hydraulic Engineering* **123** 2–12
[11] Muin Muslim 1999 *Application of 3-D Boundary Fitted Hydrodynamics and Sediment Transport Model in ADA Estuary*
[12] Muin M 2013 Integration Lagrangian Sediment Transport into Non-Orthogonal Ocean Hydrodynamics Model to Simulate Drilling Cutting and Mud Dispersion in Indonesia Sea Water *Proceedings of the Twenty-third International Offshore and Polar Engineering* (Anchorage: International Society of Offshore and Polar Engineers) pp 656–62
[13] PT LAPI ITB 2016 *Jasa Konsultansi Studi Coastal Untuk Optimasi Pengembangan Alur Pelayaran Alternatif (Shortcut) Pelabuhan Dumai* (Bandung)
[14] PT LAPI ITB 2015 *Pemodelan Arus dan Sedimentasi Alur Pelayaran Pelabuhan Pomako* (Bandung)
[15] PT LAPI ITB 2019 *Provision of Sedimentation Study Services in Delta Mahakam* (Bandung)
[16] Google 2020 Google Earth
[17] OpenStreetMap Open Street Map

Development of seismic map for offshore platform design in Indonesia, case study of the North Coast of Java

Hendriyawan[1,2]*, D D Aditramulyadi[2], R L Tawekal[1,2] and Rildova[1,2]

[1] Offshore Engineering Research Group, Faculty of Civil and Environmental Engineering, Institut Teknologi Bandung, Ganesha 10 Street, Bandung 40132, Indonesia
[2] Ocean Engineering Program, Faculty of Civil and Environmental Engineering, Institut Teknologi Bandung, Ganesha 10 Street, Bandung 40132, Indonesia

*Corresponding author's e-mail: hendriyawan_nln@yahoo.com

Abstract. Indonesia's design of earthquake-resistant offshore platforms does not yet have specific regulations. The commonly used standard in Indonesia, API RP 2A WSD, does not explicitly state the earthquake return period for offshore structures. A more detailed procedure is described in ISO 19901-2:2004. Two methods are introduced in the standard. First, the simplified method allows the use of earthquake maps with a return period of 1000 years. Second, the detailed method requires a site-specific seismic hazard study to determine the return period of earthquakes. This paper presents the development of seismic hazard maps using a simplified method following ISO 19901-2:2004. The maps are prepared using the probabilistic method with a case study on the North Coast of Java. The results show that the study area falls into seismic zones 1 and 2. The seismic hazard maps can be used for earthquake resistance design for structures with exposure level L2 and L3 based on the seismic risk category following the ISO.

Keywords: return period, seismic hazard, seismic maps, simplified method, exposure level

1. Introduction

Indonesian region is located in the Circum-Pacific belt, which contributes about 90% of the world's earthquakes and nearly half of them are world great earthquakes [1]. Over the last decades, some great earthquakes repeatedly occurred in Indonesia have increased experts' awareness regarding the seismic activities in Indonesia. Unfortunately, the research in seismic design of offshore structures in Indonesia is relatively not as many as that for other structures. For instance, earthquake-resistant design for Indonesia's offshore structures does not yet have specific regulations.

Indonesian National Standard (SNI) 1726-2019 contains the latest regulations related to seismic design for building and non-building structures [2]. The standard adopts the seismic hazard maps published by the National Center for Earthquake Studies (PuSgeN) in 2017. The maps were developed based on the probability of exceedance 2% in 50 yr (earthquake return period of 2475 years) [3]. However, the maps and SNI 1726-2019 provision are specified for building structures and exclude offshore platform structures. On the other hand, the commonly used standard for offshore structures in Indonesia, API RP 2A WSD, does not explicitly state the earthquake return period for offshore structures [4].

IOP Conf. Series: Earth and Environmental Science 1065 (2022) 012033 doi:10.1088/1755-1315/1065/1/012033

As described above, the selection of return periods for the strength level earthquake (SLE) and the ductility level earthquake (DLE) of offshore structures is not determined explicitly by API but decided by the owner's discretion [5]. The seismic hazard map in SNI 1726-2002 is commonly used if no detailed seismic hazard assessment is conducted for the offshore structures. The standard is used because it contains the seismic hazard maps for offshore areas [6]. It should be noted that the earthquake load in the SNI 2002 is only for the peak ground acceleration (PGA) (T~0 second), as shown in Figure 1. There is no other spectral acceleration for other periods in the standard.

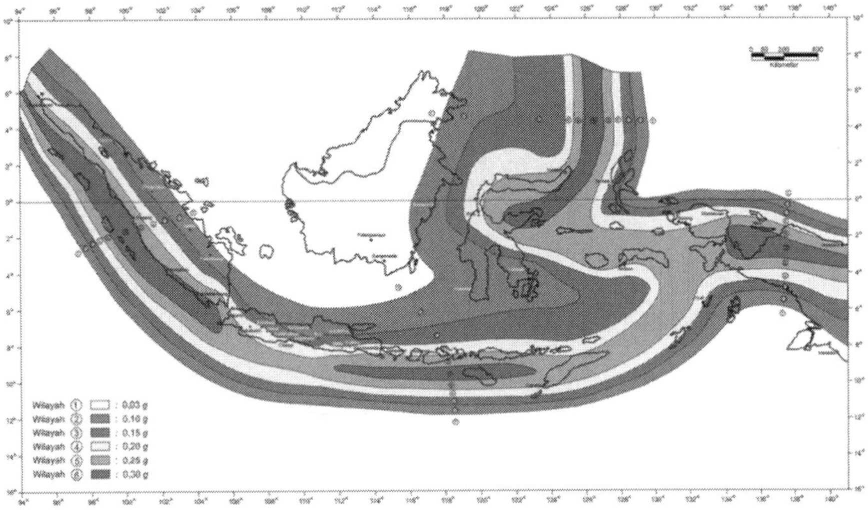

Figure 1. Indonesia Hazard Map of PGA as in SNI 1726-2002 [6].

The most recent standard containing regulations for seismicity is ISO 19901-2:2004, which was later adopted by the API as API RP 2EQ [7][8]. In this standard, the seismic loading design is divided into two (2) levels, namely Abnormal Level Earthquake (ALE) and Extreme Level Earthquake (ELE). Both are based on the results of mapping earthquake loads that are specific for each location. In addition, ISO provides more detail and specific recommended return periods for offshore structures in seismic design. The return period is determined based on the seismicity characteristics of the region surrounding each particular site [5].

According to ISO 19901-2:2004, the selection of the procedure for determining earthquake loads depends on the exposure level of the structure and the characteristics of the seismic event, defined by the seismic acceleration zone of a 1000-year return at 1.0 s oscillator period. The simplified method can be used when seismic considerations are unlikely to govern the structural design. In contrast, the detailed method can be used when seismic considerations significantly impact the design [7].

ISO 19901-2:2004 contains seismic hazard maps for offshore Indonesian, as shown in Figure 2. The maps can be used to determine the site's seismic loads (actions), particularly for those sites where the simplified seismic design is acceptable [9]. However, the map's resolution is very rough, with a relatively large interval of 0.1 and 0.25g for 1.0 s and 0.2 s oscillator periods, respectively. Therefore, instead, the use of spectral hazard maps with 500 (SNI 1726-2002) or 2500 years return period (SNI 1726-2012, 2019), this study presents the development of more detailed seismic hazard maps for a 1000-year return period on the northern coast of Java Island, Indonesia. PuSgeN [3] in 2017 also developed spectral hazard maps for 1000-year; unfortunately, the maps exclude information on offshore locations, which is required for offshore platform design. This study will use a probabilistic approach to develop two spectral hazard maps for the 1.0s and the 0.2s oscillator periods.

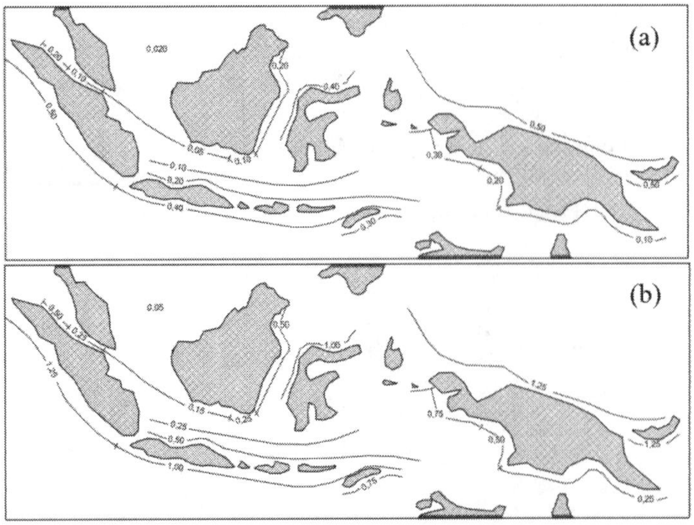

Figure 2. The 5% spectral response accelerations for offshore Indonesia in ISO 19901-2:2004, (a) for 1.0 s oscillator period and (b) for 0.2 s oscillator period [7].

2. Methodology

2.1. Seismic source model

This study begins by developing the seismic source model. The model is developed based on Indonesia's latest data and information compiled by PuSGeN and the Center for Housing and Settlement Research and Development – Ministry of Public Works of Indonesia (PUSKIM-PUPR). The model consists of three different source types. Two of them are developed based on geodetic and geophysics surveys of visible earthquake sources that have been well identified, categorized as fault and subduction sources. The fault sources required the earthquake parameters such as slip-rates, fault mechanisms, dip angles, and moment magnitudes. While for subduction sources, the seismic source parameters represent by *a-b* values and moment magnitude. Both source types required trace coordinates of the source. The earthquake sources parameters for fault and subduction are presented in Table 1 and Table 2. The map of fault and subduction sources in the area around the study location is presented in Figure 3.

If the historical earthquake record mechanism cannot be identified clearly, the earthquake data are classified as background sources. Then, the sources are calculated as gridded seismicity. The grid sources models are accommodated for low to medium scale earthquakes with depth from 0 to 50km, represented by moment magnitude (Mw) 5.0-6.5 for earthquakes. This model is also called a shallow background model. Meanwhile, for earthquakes in the Benioff zone (depth between 50-300km), the magnitude ranges between Mw 5.0 and 7.8.

The historical earthquake records used to estimate future seismic activity must be free of dependent events such as foreshocks and aftershocks. The inclusion of dependent events in the seismic hazard analysis has resulted in a minor increase in the hazard estimation [10]. In this study, dependent events from the catalog were removed by Gardner and Knopoff's method [11]. The second problem for this kind of model is the completeness of the data. The historical earthquake records are usually more complete for large earthquakes than small earthquakes [1]. This is because of seismograph station density or, in the early record, population density was not well distributed. Therefore, the method proposed by Stepp was performed to solve this problem [12]. The earthquake catalog map that has been analyzed for dependency and completeness is presented in Figure 4.

Table 1. Fault source parameters [3].

ID	Fault Segment	M_{max}	Mechanism	Slip-rate (mm/yr)	Dip (°)
1	Sumatera Fault - Siulak	7.2	Strike-slip	14	90
2	Sumatera Fault - Dikit	7.1	Strike-slip	12	90
3	Sumatera Fault - Ketaun	7.3	Strike-slip	12	90
4	Sumatera Fault - Musi	7.2	Strike-slip	13.5	90
5	Sumatera Fault - Manna	7.3	Strike-slip	13.5	90
6	Sumatera Fault - Kumering North	7.5	Strike-slip	12.5	90
7	Sumatera Fault - Kumering South	7.2	Strike-slip	12.5	90
8	Sumatera Fault - Semangko Barat A	7.4	Strike-slip	8	90
9	Sumatera Fault - Semangko Barat B	7.3	Strike-slip	8	90
10	Sumatera Fault - Semangko Timur A	6.5	Strike-slip	5	90
11	Sumatera Fault - Semangko Timur B	6.9	Strike-slip	3	90
12	Sumatera Fault - Semangko Graben	6.5	Normal	3	60
13	Sumatera Fault - Ujung Kulon A	7.3	Strike-slip	10	90
14	Sumatera Fault - Ujung Kulon B	7.6	Strike-slip	10	90
15	Mentawai Fault - Mentawai	8.2	Reverse	5	45
16	Mentawai Fault - Enggano	7.6	Reverse	5	45
17	Cimandiri Fault - Cimandiri	6.7	Reverse	0.55	45
18	Cimandiri Fault - Nyalindung Cibeber	6.5	Reverse	0.4	45
19	Cimandiri Fault - Rajamandala	6.6	Strike-slip	0.1	90
20	Lembang Fault - Lembang	6.8	Strike-slip	2	90
21	Baribis - Kendeng Fold - Subang	6.6	Reverse	0.1	45
22	Baribis - Kendeng Fold - Cirebon 1	6.5	Reverse	0.1	45
23	Baribis - Kendeng Fold - Cirebon 2	6.5	Reverse	0.1	45
24	Baribis - Kendeng Fold - Brebes	6.5	Reverse	0.1	45
25	Baribis - Kendeng Fold - Tegal	6.5	Reverse	0.1	45
26	Baribis - Kendeng Fold - Pekalongan	6.6	Reverse	0.1	45
27	Baribis - Kendeng Fold - Weleri	6.5	Reverse	0.1	45
28	Baribis - Kendeng Fold - Semarang	6.5	Reverse	0.1	45
29	Baribis - Kendeng Fold - Rawapening	6.5	Reverse	0.1	45
30	Baribis - Kendeng Fold - Demak	6.5	Reverse	0.1	45
31	Baribis - Kendeng Fold - Purwodadi	6.5	Reverse	0.1	45
32	Baribis - Kendeng Fold - Cepu	6.5	Reverse	0.1	45
33	Baribis - Kendeng Fold - Waru	6.5	Reverse	0.05	45
34	Baribis - Kendeng Fold - Surabaya	6.5	Reverse	0.05	45
35	Baribis - Kendeng Fold - Blumbang	6.6	Reverse	0.05	45
36	Ciremai	6.5	Strike-slip	0.1	90
37	Ajibarang	6.5	Strike-slip	0.1	90
38	Opak	6.6	Strike-slip	0.75	60
39	Merapi Merbabu	6.6	Strike-slip	0.1	90
40	Pati Thrust	6.5	Strike-slip	0.1	90
41	Bawean Fault	7.6	Strike-slip	0.5	90
42	RMKS Fault - East	7.8	Strike-slip	1.5	90

Table 2. Subduction source parameters [3].

ID	Segmentation	M_{max}	a	b
M1	Mentawai-Pagai	8.90	3.02	0.63
M2	Enggano	8.40	5.57	1.05
M3	Selat Sunda	8.70	5.99	1.15
M4	West-Central Java	8.70	5.55	1.08
M5	East Java	8.70	5.63	1.08

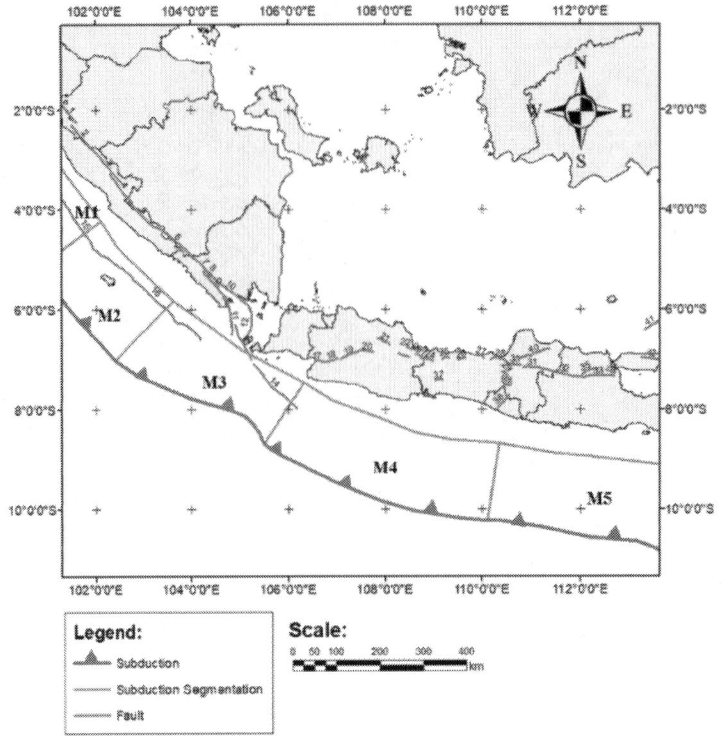

Figure 3. Map of fault and subduction sources in the area around the study location [3].

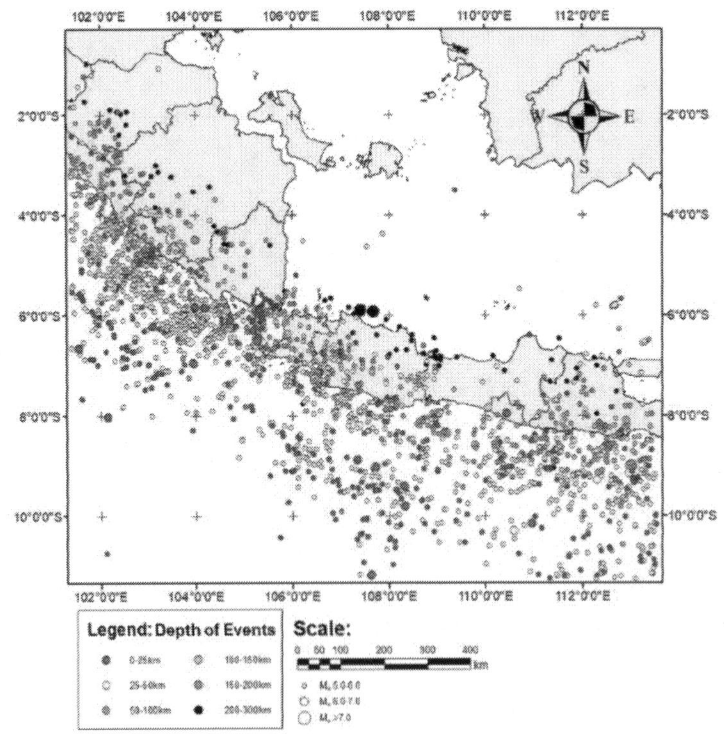

Figure 4. Earthquake catalogue map that has been declustered and analyzed for completeness [3].

2.2. Seismic hazard analysis

The development of seismic maps is conducted by using Probabilistic Seismic Hazard Analysis (PSHA). The method was developed by McGuire [13] based on the total probability theorem initiated by Cornell [14]. The theorem of total probability assumes earthquake size (magnitude, M) and location (defined by range from the hypocenter, R) as continuous independent random variables [15]. It formulated as examples:

$$P[I > i] = \iint P[I > i \mid M, R] f_M(m) f_R(r)\, dm\, dr \quad (1)$$

The probability of intensity (I) will exceed a value of i at a certain site will depend on the integration of the conditional probability of $I > i$ at a certain site from earthquake source with magnitude M and hypocentral distance R, multiply by probability density functions for M and R [15][16][17]. The analysis is performed by USGS PSHA [18]. The program was developed by the United States Geological Survey (USGS).

2.3. Ground Motion Prediction Equations (GMPEs)

The relation between the earthquake source parameters (such as the location of the sources, the magnitude, slip-rate) and the ground motion parameters (such as acceleration, velocity, and displacement) at site location is represented by the attenuation function [15]. The attenuation function, also called ground motion prediction equations (GMPEs), must be used based on the seismotectonic conditions at specific locations. However, there are currently no GMPEs developed specifically for the Indonesian region [19]. The only way is to adapt the existing attenuation functions derived in other regions that are similar to Indonesia's region tectonically and geologically [17]. The GMPEs used in this study are shown in Table 3.

Table 3. Ground Motion Prediction Equations (GMPEs) for this study.

Source Model	Ground Motion Prediction Equations (GMPEs)
a) Subduction sources	1) BC Hydro (2012) [20]
	2) Atkinson - Boore (2003) [21]
	3) Zhao et al. (2006) [22]
b) Shallow crustal (fault and shallow background)	1) Boore - Atkinson NGA. (2008) [23]
	2) Campbell - Bozorgnia NGA. (2008) [24]
	3) Chiou - Youngs NGA. (2008) [25]
c) Benioff / intraslab	1) Atkinson - Boore intraslab seismicity Cascadia (2003) [21]
	2) Youngs et al. (1997) [26]
	3) Atkinson - Boore intraslab seismicity orldwide data (2003) [21]

3. Results and discussion

Based on this study, three earthquake maps for a 1000-year return period have been developed. Compared to PuSGeN's maps, the results are slightly higher for the PGA and spectral accelerations at 0.2 seconds and relatively smaller for the 1.0 second period. This is probably due to the different attenuation functions at fault and subduction sources. However, the values are relatively close to ISO 19901-2:2004. For example, for the acceleration values in the 0.2 second period, the USGS PSHA results range from 0.4-0.5g at the northern coast of Indramayu, while the ISO 19901-2:2004 values range from 0.5g or higher. Meanwhile, for the 1.0 second period, the acceleration value ranges from 0.1-0.2g compared to the range of 0.2g or higher, given by ISO 19901-2:2004. Unfortunately, the PGA map cannot be compared with ISO 19901-2:2004 since the code does not give the acceleration map for this period. Results and comparisons for the 1000-year return period earthquake maps for PGA, periods 0.2s and 1.0s are shown in Figure 5, Figure 6, and Figure 7, respectively.

The seismic zone map is developed based on a seismic hazard map at a 1.0 second period for a 1000-year return period following ISO 19901-2:2004. Site zoning refers to the range of earthquake

IOP Conf. Series: Earth and Environmental Science 1065 (2022) 012033 doi:10.1088/1755-1315/1065/1/012033

acceleration values shown in Table 4. Based on the seismic zone map of the study area, the site are can be classified as seismic zone 1 (*Sa,map(1.0)* = 0.03g – 0.10g) and 2 (*Sa,map(1.0)* = 0.11g – 0.25g). Based on the seismic risk category following ISO, the seismic hazard maps can be used for earthquake resistance design for structures with L2 and L3 exposure levels (medium and low consequence category for unmanned or manned evacuated life safety category). The seismic zone map from this study is shown in Figure 8.

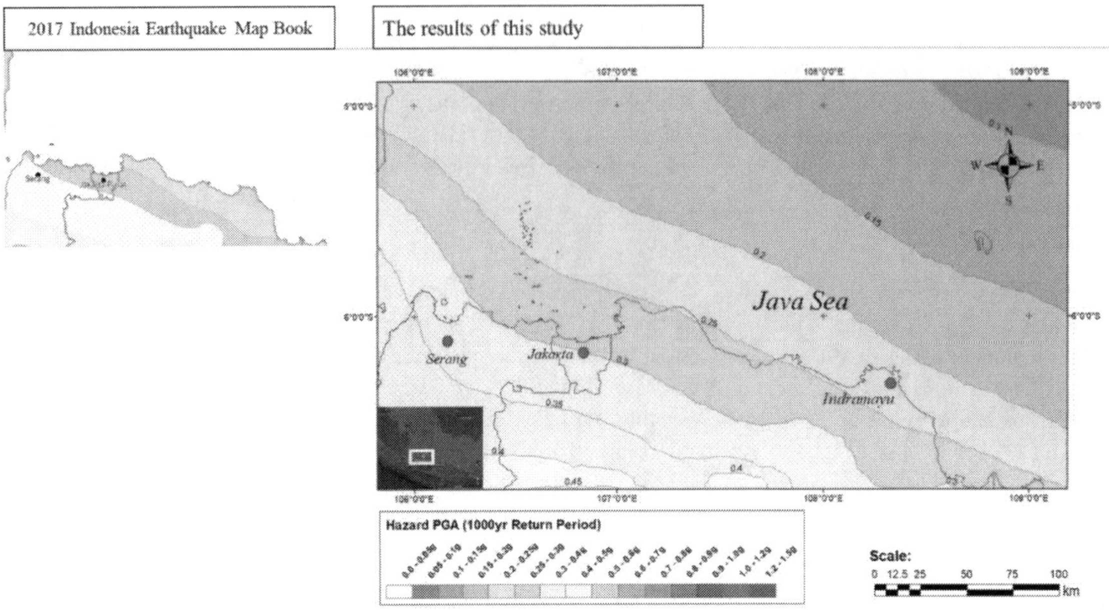

Figure 5. Earthquake map comparison for PGA on return period of 1000 years.

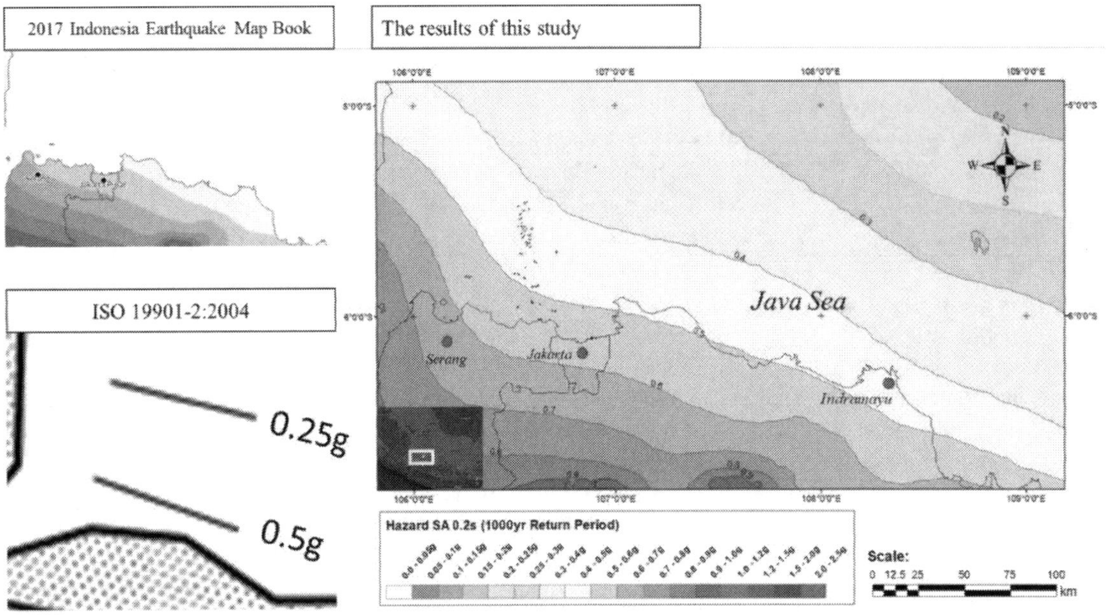

Figure 6. Earthquake map comparison for 0.2s oscillator period on return period of 1000 years.

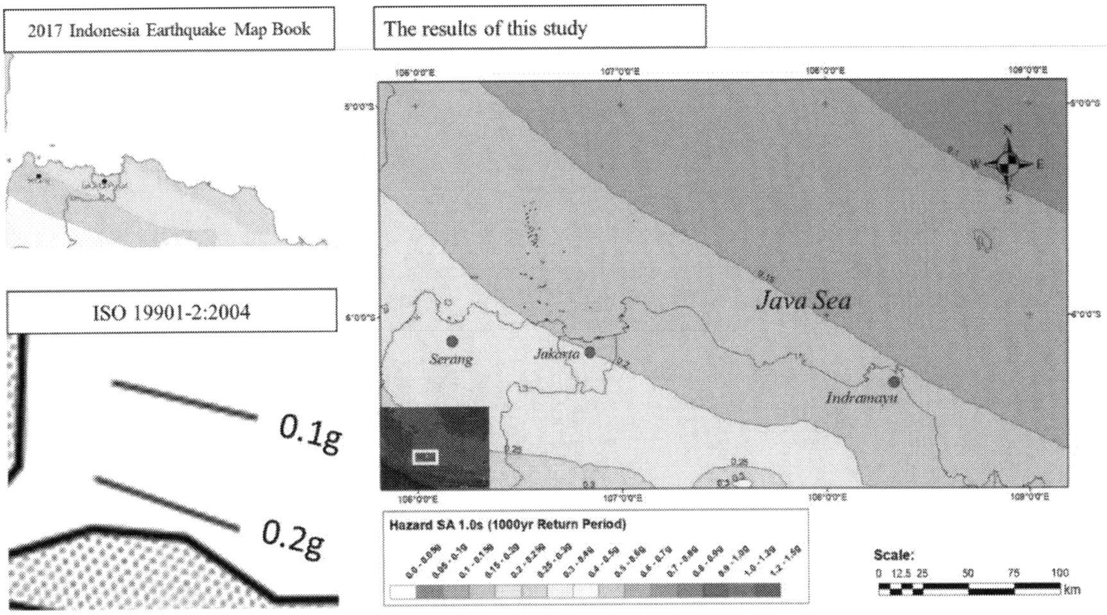

Figure 7. Earthquake map comparison for 1.0s oscillator period on return period of 1000 years.

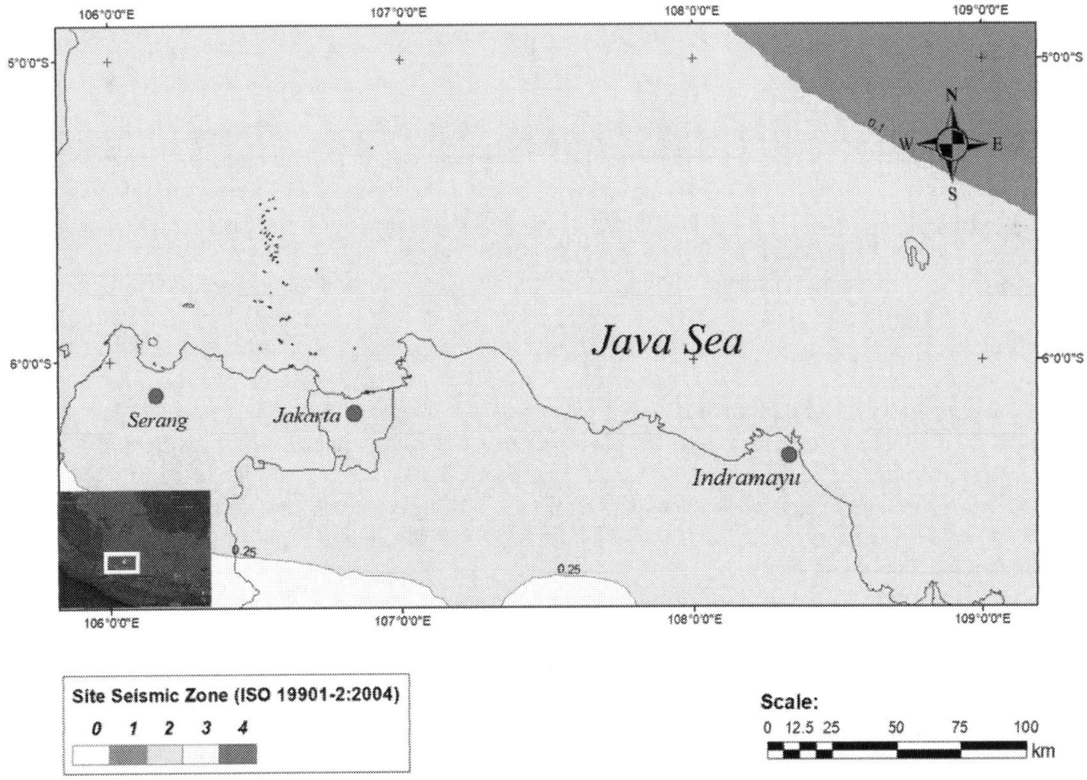

Figure 8. Site seismic zone map for the North Coast of Java based on ISO 19901-2:2004.

Table 4. ISO 19901-2:2004 seismic zone classification [7].

S_a (1.0)	Site Seismic Zone
< 0.03g	0
0.03g – 0.10g	1
0.11g – 0.25g	2
0.26g – 0.45g	3
> 0.45g	4

4. Conclusion

This study presents the development of seismic maps of the North Coast of Java, Indonesia. It comprised maps of PGA, 0.2s, 1.0s oscillator period, and seismic zone map that complies with ISO 19901-2:2004 standard. Compared to the currently published maps in ISO 19901-2:2004, this study's results are only slightly different. The results show that the study area falls into seismic zones 1 and 2. Based on the seismic risk category in ISO 19901-2:2004, the maps can be used to design earthquake resistance structures with L2 and L3.

Acknowledgments

This research was successfully completed with the fund by ITB Research, Community Service and Innovation Program (P3MI-ITB). The authors would also like to thank the National Earthquake Study Center (PuSGeN) and the Center for Housing and Settlement Research and Development (PUSKIM) for their support in providing the data. The authors' special thanks are extended to all supporting people who were involed both directly and indirectly in this project.

References

[1] Kramer S L 1996 *Geotechnical Earthquake Engineering* (New Jersey: Prentice Hall)

[2] SNI 1726:2019 2019 *SNI 1726:2019 - Tata cara perencanaan ketahanan gempa untuk struktur bangunan gedung dan nongedung* (Jakarta: Badan Standardisasi Nasional) (in Indonesian)

[3] PuSGeN 2017 *Peta Sumber dan Bahaya Gempa Indonesia Tahun 2017 1st ed.* (Bandung: Puslitbang Perumahan dan Pemukiman) (in Indonesian)

[4] American Petroleum Institute 2004 *API RP-2A-WSD - Recommended practice of planning, designing and constructing fixed offshore platform – working stress design 21st edition.*

[5] Lubkowski Z A, Alarcon, J E and Razak, Z A 2011 Comparison of ISO 19901-2 and API RP 2A seismic design criteria for a site in the Caspian Sea, Turkmenistan *Frontiers in Offshore Geotechnics II*, ed S Gourvenec & D White p 909

[6] SNI 03-1726-2002 2002 *Tata Cara Perencanaan Ketahanan Gempa untuk Bangunan Gedung* (Jakarta: Badan Standardisasi Nasional) (in Indonesian)

[7] ISO 19901-2 2004 *International Standard ISO 19901-2, Petroleum and natural gas industries - Specific requirements for offshore structures - Part 2: Seismic design procedures and criteria*

[8] American Petroleum Institute 2014 *ANSI/API Recommended Practice 2EQ – Seismic Design Procedures and Criteria for Offshore Structures 1st ed*

[9] Chang B, Abraham M and Peng B F 2005 Comparison of ISO and API Seismic Design Guidelines Using Three Existing Offshore Platforms *Offshore. Tech. Conf.* **17285**

[10] Pacheco J F and Sykes L R 1992 Seismic Moment Catalog of Large Shallow Earthquakes, 1900 to 1989 *Bull. Seis. Soc. Ame.* **82** 3 pp 1306-1349

[11] Gardner J K and Knopoff L 1974 Is the Sequence of Earthquakes in Southern California, with Aftershocks removed, Poissonian? *Bull. Seis. Soc. Ame.* **64** 5 pp 1363-1367

[12] Stepp J C 1972 *Analysis of completeness of the earthquake sample in the Puget Sound Area and its effect on statistical estimates of earthquake hazard* (Boulder Colorado: National Oceanic and Atmospheric Administration Environmental Research Laboratories)

[13] McGuire R 1995 Probabilistic Seismic Hazard Analysis and Design Earthquakes: Closing the Loop *Bull. Seis. Soc. Ame.* **85** 5 pp 1275–1284
[14] Cornel C A 1968 Engineering seismic rick analysis *Bull. Seis. Soc. Ame.* **58** pp 1583–1606
[15] Hendriyawan H, Aditramulyadi D D and Syahbana A J 2021 Seismic hazard analyses for North Java Offshore Indonesia using ISO 19901-2:2004 *IOP Conf. Ser.: Earth Environ. Sci.* **729** 012006
[16] Asrurifak M, Irsyam M, Budiono B, Triyoso W and Hendriyawan 2010 Development of Spectral Hazard Map for Indonesia with a Return Period of 2500 Years using Probabilistic Method *Civ. Eng. Dim.* **12** 1 52–62
[17] Irsyam M, Asrurifak M, Hendriyawan, Budiono B, Triyoso W and Firmanti A 2010 Development of spectral hazard maps for a proposed revision of the Indonesian Seismic Building Code *Geom. and Geoe. Int. J.* **4** 1 35–47
[18] Harmsen S 2008 *USGS Software for Probabilistic Seismic Hazard Analysis (PSHA)* (unpublished)
[19] Irsyam M *et al.* 2020 Development of the 2017 national seismic hazard maps of Indonesia *Earth. Spec.* **36** 1 pp 112-136
[20] BC Hydro 2012 Probabilistic Seismic Hazard Analysis (PSHA) Model Volumes 1, 2, 3 and 4. BC Hydro Engineering Report E658 (Vancouver: BC Hydro)
[21] Atkinson G and Boore D 2003 Empirical ground-motion relations for subduction-zone earthquakes and their application to Cascadia and other regions *Bull. Seis. Soc. Ame.* **93** 4 pp. 1703–1729
[22] Zhao J et al. 2006 Attenuation relations of strong motion in Japan using site classification based on predominant period *Bull. Seis. Soc. Ame.* **96** 898
[23] Boore D and Atkinson G 2008 Ground-motion prediction equations for the average horizontal component of PGA, PGV, and 5%-damped PSA at spectral periods between 0.01 s and 10.0 s *Eart. Spec.* **24** 1 99–138
[24] Campbell K and Bozorgnia Y 2008 Ground motion model for the geometric mean horizontal component of PGA, PGV, PGD and 5% damped linear elastic response spectra for periods ranging from 0.01 to 10.0 s *Eart. Spec.* **24** 1 139–171
[25] Chiou B and Youngs R 2008 A NGA model for the average horizontal component of peak ground motion and response spectra *Eart. Spec.* **24** 1 173–215.
[26] Youngs R, Chiou S, Silva W and Humphrey J 1997 Strong ground motion attenuation relationships for subduction zone earthquakes *Seis. Re. Let.* **68** 58–73

Removal of phthalocyanine reactive dye using plasma corona discharge

B Handoko[1,2], S Notodarmojo[1], and R Iqbal[1,*]

[1]Environmental Engineering Department, Institut Teknologi Bandung, West Java, Indonesia
[2]Textile Chemistry Department, Politeknik STTT Bandung, West Java, Indonesia

*Corresponding authors: iqbalrofiq@gmail.com

Abstract. Phthalocyanine reactive dye is typically difficult to biodegrade in water because of its big molecular size coming from the complex metal structure of the dye. Corona plasma discharge which can produce active radical species and molecules with high oxidation potentials, shockwaves, and ultraviolet light is the potential to efficiently degrade the compounds. The experiments were carried out in a batch reactor with a capacity of 400 ml. The plasma generator used in this study had a point-plane electrode configuration with a voltage of 40 kV. The concentration of dye used as a target for degradation was 25 ppm. The reactor was operated with the addition of pure oxygen gas flowing into the reactor. The distance between electrodes and the time of exposure were varied and their removal efficiency (color and COD value) were tested. The results showed that the highest removal efficiency of the dye (83.15% and 52.32% for color removal and COD value respectively) was achieved from the experiment with an electrode distance of 4.0 cm and exposure time of 60 minutes.

Keywords: phthalocyanine, plasma corona discharge, degradation

1. Introduction

The textile industry is considered a major source of pollutants for the aquatic environment in terms of quality and quantity [1], is colorful, has a high content of organic matter, and is difficult to biodegrade. Wastewater generated from textile industry is generally colored and non-biodegradable with high BOD and COD due to the high concentration of dyes. Most of dyes are toxic, mutagenic, and carcinogenic, and hence they are very dangerous to the environment [2]. Treatment of the textile wastewater by any conventional methods is therefore difficult because it is non-biodegradable and soluble in water. Those wet textile processing requires very high-water consumption that then generates a high amount of liquid waste [3]. Most of the textile production processes consume large amounts of water and dispose it in the form of waste with high color intensity and high concentration of organic substances with varied contents [4]. For example, on the one hand, the dyeing process with reactive dyes has the potential to produce liquid waste above 10% because it cannot be fixed into the fiber. Previous research has stated that about 10-15% of reactive dyes has wasted into waste because it was not absorbed by fabrics or textile materials [5]. On the other hand, the consumption of dyes is more than 700,000 tons per year, putting pressure on the textile industry which is considered one of the main polluters [6].

The dye content in textile wastewater is high, non-biodegradable, and highly soluble, so it is difficult to be processed through coagulation, adsorption, or biological methods. Several AOPs, such as Fenton, photo-Fenton, photo-catalysis, and ultrasound, have been carried out with high efficiency of dye decolorization but they have disadvantages such as the addition of chemical substances, certain pH ranges, and large energy usage. Plasma, alternatively, is a partially ionized gas under a high electrical field, and hence it is also defined as the quasi-neutral mixture of electrons, radicals, positive and negative ions [7]. The energetic electrons produced by plasma can collide with other molecules in the air such as N_2, O_2, H_2, and others to produce secondary electrons, photons, ions, and radicals [8]. Species produced by plasma such as ozone (O_3) and hydroxyl radicals ($OH\bullet$) have very high oxidation potentials that can degrade organic substances in textile waste. Thus, it is significant to find alternative methods for degrading organic substances like dyes contained in the textile wastewater in a more environmentally benign process without adding some more harmful chemical substances.

Several recent studies stated that plasma technology had been the one potential method in producing oxidation agents useful for degraded organic substances through the AOPs method. Several studies showed that the textile dyes can be decolorized or even further mineralized with high efficiency by the plasma method [2,9–13], and the addition of other substances besides dyes can also be degraded with high efficiency [12]. Several types of plasma such as corona [13], pulse plasma [14], Dielectric Barrier Discharge [15], and plasma arch [10] had been reported able to effectively degrade various organic substances. However, research to degrade dyes with phthalocyanine chromophore which has a very complex molecule is rarely studied.

In this study, a corona discharge plasma reactor with point-field electrodes was used to degrade a very complex type of dye, namely a reactive dye with a phthalocyanine chromophore. The distance between the electrodes and the exposure time and their effect on the degradation of the dye have been investigated.

2. Experiment

The experiment was carried out on a laboratory scale at the Laboratory of Analytical Chemistry in the Polytechnic of STTT Bandung. The dye used is Galaxyl Cyanine XLN which has a phthalocyanine chromophore with a turquoise blue color. The type of plasma used is corona discharge (non-thermal plasma). To produce a corona discharge, a point-plane electrode configuration made of copper was used. The point electrodes were mounted on a copper plate with a point electrode length of 3.0 cm. The plane electrode was mounted on the bottom of the reactor which was submerged in the wastewater. A high voltage of 40 kV was attached to the point electrode, while the ground was attached to the plane electrode. The reactor was made of acrylic material with a thickness of 3 mm with a cover. The reactor was operated on a batch system with samples drawn at specific times. The height of the point electrodes can be adjusted according to the plane electrodes. The reactor system can be seen in Figure 1.

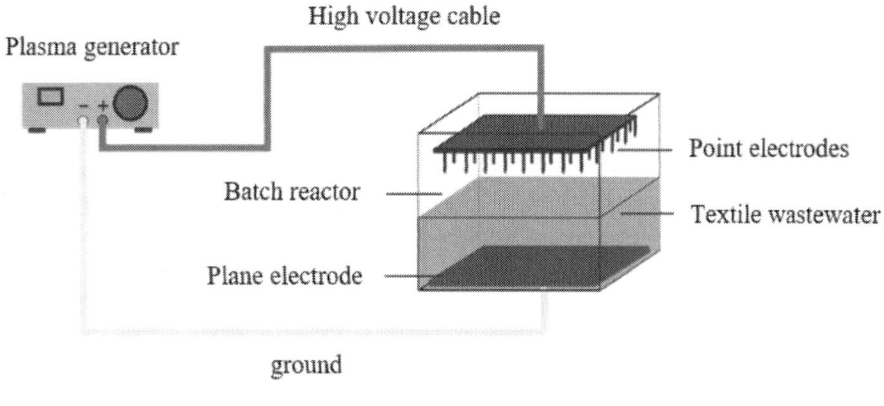

Figure 1. Non-thermal plasma reactor system.

2.1. Instrument and Material

The corona discharge plasma generator with a DC generator was provided by analysis chemical laboratory of the Polytechnic of STTT Bandung. The chemicals for COD testing were H_2SO_4 98%, $K_2Cr_2O_7$, $AgSO_4$, $HgSO_4$, and ferro ammonium sulfate of analytical grade that were purchased from Merck. The dye used in this study is Galaxy Cyanine XLN provided by a dye supplier in Indonesia. The instruments used for testing and analysis were Spectrophotometer UV-vis (Thermo Scientific Genesys 20), COD HACH reactor 20 cuvette, and COD tube.

2.2. Experimental procedures and analysis

A series of experiments were carried out in a plasma corona discharge reactor with a needle-shaped point-plane electrode made of copper with a pointed tip placed on the surface of the textile wastewater at a specified distance (Figure 1) [16]. The number of needles attached to the point electrode was 60 points with a needle length of 3.0 cm. The dimensions used were a rectangle with 12 dots and 5 lines each. The point electrode was designated as the positive electrode (anode). The plane electrode was a metal plate made of copper with a length of 20 cm, and a plate edge of 10 cm wide. This field electrode acted as the cathode. The two electrodes were installed parallel and connected to a high voltage DC source. The corona formed in the reactor for this study were positive. This plasma reactor is a cold plasma with plasma discharge under atmospheric conditions. At the cathode, an aerator functional as a mixer for wastewater was installed. The experiment in this study used 400 ml of wastewater that was put into the reactor, while the distance between the electrodes was adjusted to the distance of 4, 4.5, and 5 cm. The generator was run at 40 kV and the exposure was carried out for 10, 20, 30, 40, 50, and 60 minutes. Exposure time was limited to 60 minutes for easy expansion to scale up reactor sizes. For each unit of exposure time, samples were taken for color analysis, UV Spectrum, and COD removal. To determine the concentration of reactive dye with phthalocyanine chromophore, measurement was conducted at a maximum wavelength of 667 nm by measuring the absorbance of the solution using a UV-Vis spectrophotometer. UV spectrum analysis were done by a UV-Vis spectrophotometer. COD analysis was carried out according to Standard Methods for the Examination of Water and Wastewater 23th Edition (2017), Methods 5220 C: Closed Reflux, Titrimetric Methods.

The efficiency of color removal and COD was calculated using the following equations:

$$\textit{Color removal efficiency (\%)} = \left(1 - \frac{A}{A_0}\right) \times 100\% \tag{1}$$

$$\textit{Degradation efficiency (\%)} = \left(1 - \frac{C}{C_0}\right) \times 100\% \tag{2}$$

Where A is absorbance in exposure time t, A_0 is initial absorbance of dye color, C is value of COD at time of exposure t (mg/L) and C_0 is initial COD value (mg/L)

3. Result and discussion

3.1. Effect of exposure time on COD and Color removal efficiency

To produce active species, experiment was conducted to treat liquid waste by corona plasma using 40 kV energy. To study the effect of exposure time on COD and color removal efficiency, the processing time was set at 10, 20, 30, 40, 50, and 60 minutes. The results of the study are shown in Figures 2 and 3. Based on the figure, it can be inferred that the textile wastewater treatment using plasma corona discharge was able to significantly reduce COD and color values. This experiment indicates that the efficiency of COD removal increased as the exposure time reached 52,32 % for a contact time of 60 minutes.

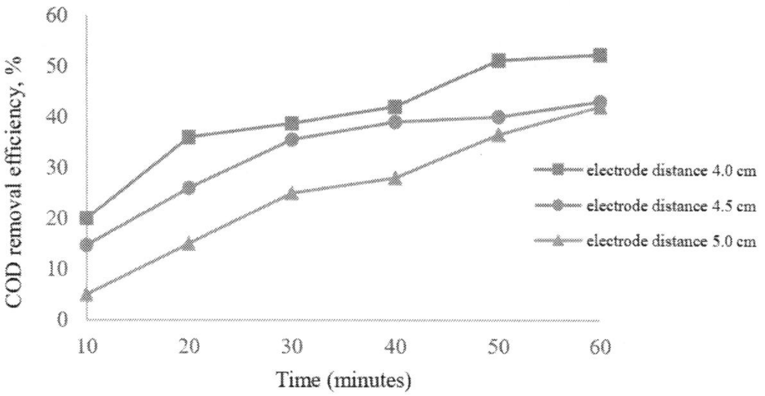

Figure 2. Effect of time exposure on efficiency of COD removal.

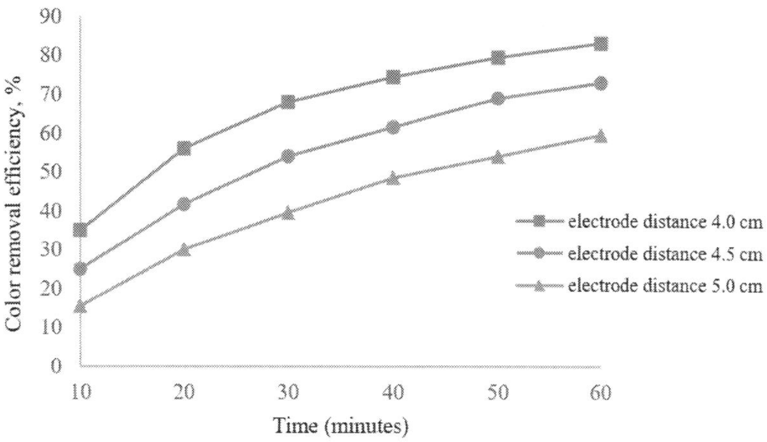

Figure 3. Effect of time exposure on efficiency of color removal.

This is in accordance with the results of the research performed by Tao [17] in which plasma technology was able to remove up to 94.1% methyl orange. Another research supported this research was conducted by [13] that was able to obtain TOC removal efficiency by 65.5%. In this case, the organic substances were randomly degraded by radical species such as hydroxyl radicals. The reduced COD value indicated that there has been a mineralization process of organic substances in textile wastewater. The mechanism of degradation of organic substances by hydroxyl radical species in chemical processes can be explained according to the hypothesis made by previous researchers [18,19] as follows:

$$e^- + H_2O \rightarrow e^- + H\bullet + OH\bullet \quad (3)$$

$$e^- + H_2O \rightarrow H_2O^+ + 2e^- \quad (4)$$

$$H_2O^+ + H_2O \rightarrow H_3O^+ + OH\bullet \quad (5)$$

$$RH + OH\bullet \rightarrow R\bullet + H_2O \quad (6)$$

Based on equation (3), hydroxyl radical species which had a high oxidation potential (2.8 eV) can be formed from the ionization of water. The reaction can continue according to equation (3) - (5) which produced more hydroxyl radicals. Equation (6) showed that organic substances were degraded by hydroxyl radical substances into smaller substances. This explains the phenomena that occurred in Figures 2 and 3 that the degradation and color removal efficiency increased as the longer exposure time got longer. More free electrons, free ions, and free atoms would be generated at a longer exposure time.

In accordance with the experiments conducted by Tao [17], the efficiency of degradation of methyl orange increased as the plasma residence time increased. Figures 2 and 3 shows the degradation of organic matter occurred since the 10 minutes and continued to increase until 60 minutes. The maximum degradation efficiency value was obtained at 52.32% and 83.15% color removal at 60 minutes. This is due to the higher formation of hydroxyl radicals (2.8 eV). These species will degrade reactive dye with phthalocyanine chromophore in water. The mechanism of organic substances degradation by hydroxyl radicals is described as follows [20]:

$$OH\bullet + RH \rightarrow R\bullet + H_2O \quad (7)$$

$$O\bullet^{-} + RH \rightarrow R\bullet + OH^{-} \quad (8)$$

$$OH\bullet + OH\bullet \rightarrow H_2O_2 \quad (9)$$

Equation (7) indicates the organic matter was gradually and continuously degraded until mineralization. The hydroxyl radical can oxidize organic substances into water and new organic substances. Equation (8) reveals further degradation reactions of organic substances when the pH of the solution is alkaline. Equation (9) shows that hydrogen peroxide can be formed from 2 moles of hydroxyl radicals even though the reaction rate is slower. Furthermore, the color removal at the optimum distance was measured using a UV-Vis spectrophotometer. The results are shown in Figure 4 below.

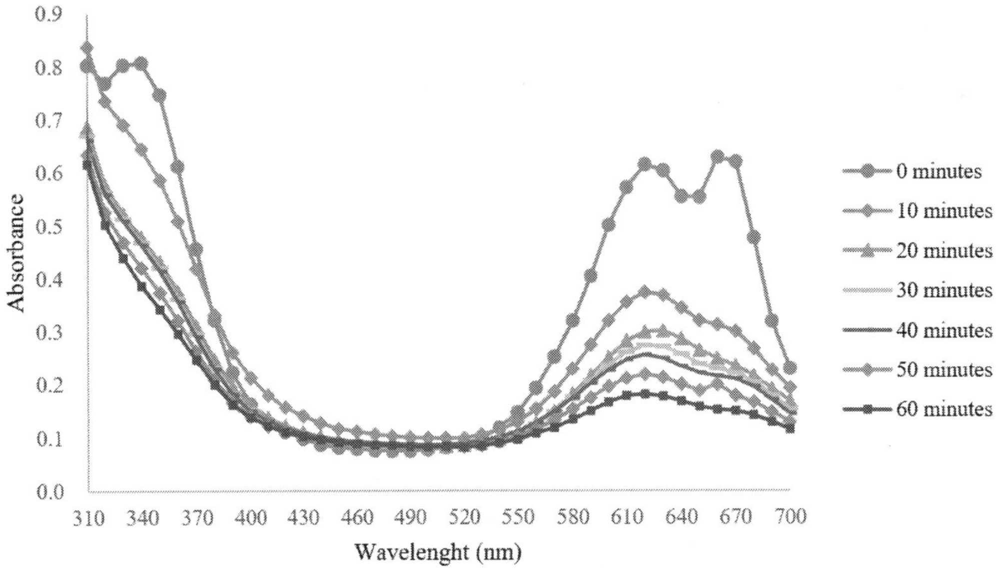

Figure 4. Galaxyl Cyanine XLN dye removal curve at an electrode distance of 4.0 cm.

In Figure 4 the chromophore removal reaction of phthalocyanine proceeds rapidly in the first 10 minutes and is relatively sloping up to 60 minutes. This could be the further reaction between hydroxyl radicals to form hydrogen peroxide with a lower oxidation potential value (eq. 9). Hydrogen peroxide has an oxidation potential of 1.78 eV which is lower than the hydroxyl radical. Therefore, the degradation becomes slower at a longer exposure time. However, it can be seen from Figure 4 that there was a decrease in the absorbance value of the dye with a longer exposure time. Although we see that the removal is not up to 100%, it means that the mineralization process is still ongoing. This indicates that the degradation of the dyestuff increases with a function of time. The longer irradiation time, the more optimum the dye degradation process. The dye curve at the initial condition had two peaks at the wavelengths of 620 and 667, but after 60 minutes of irradiation, both peaks sloped, especially for the peak at the wavelength of 667 nm. From equations (1) - (9) then, if the radiation time is extended beyond 60 minutes, it can be concluded that the reactive dye removal reaction with the phthalocyanine chromophore will continue, and the mineralization will be thorough. However, with the formation of species with lower oxidation potential such as hydrogen peroxide (eq. 9), the rate of the dye removal reaction will be slower. This phenomenon also occurs in the removal of Acid Orange 7 dye which runs fast at the beginning of the process and slows down as the processing time increases [21]. However, the effect of exposure time on dyes with phthalocyanine chromophore needs to be investigated further for a longer time.

3.2. Effect of electrodes distance on COD and color removal efficiency

Plasma corona discharge is formed from a non-uniform electrical field, due to differences in the geometry of electrode (point-plane) which creates a strong magnetic field around the tip of point electrode. The strong electrical field between the electrodes is a function of the distance (x) between the electrodes [7]. The equation used to calculate the electric field based on the distance between the electrodes is as follows:

$$E\,(x) = \frac{2V}{\left(r+2x-\frac{x^2}{d}\right)ln\,(1+\frac{2d}{r})} \qquad (10)$$

V = voltage on the electrodes
r = the radius of an electrode tip
d = distance between electrodes
x = distance between the point electrode tip and a point between the two electrodes

From equation (10), it can be concluded that the electrical field was affected by the distance between the electrodes. In this study, the voltage used was 40 kV. The corona discharge plasma can be created at high voltages between 25-40 kV, and it has very short pulse densities (approximately $2\,x\,10^{-12} - 10 -^{-10}$ seconds) [22]. To get a high electrical field, the distance between the electrodes should be as close as possible. However, if the electrodes are too close there will be a breakdown. The results can be seen in Figures 5 and 6.

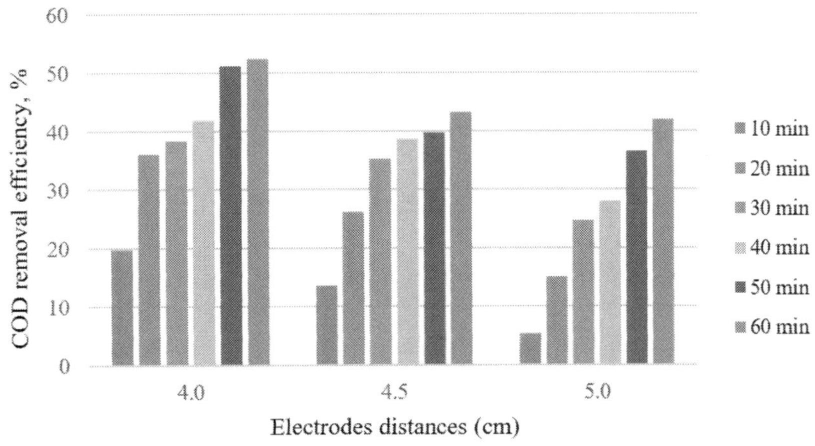

Figure 5. Effect of electrodes distance on efficiency of COD removal.

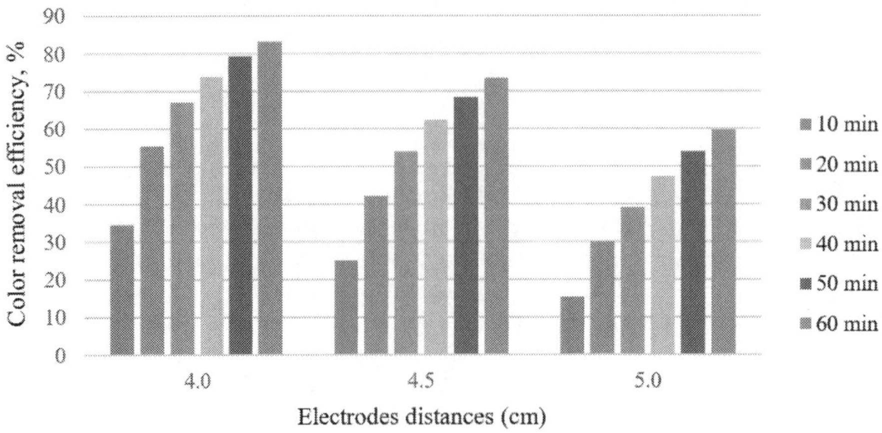

Figure 6. Effect of electrodes distance on efficiency of color removal.

At a distance between the electrodes of 5 cm, the efficiency of degradation and color removal obtained was low, then it increased at distances of 4.5 and 4 cm. If the distance between the electrodes is too far (5 cm), there would be a low-energy electrical field and fewer species are formed. This explains Figure 4 and 5 that the low efficiency of waste degradation and color removal occurred at 5.0 cm distance between the electrodes. These results are in accordance with the research carried out by [16] which showed a similar effect on the corona plasma experiment to modify the absorbency of cotton and polyester fabrics. Based on the experimental results, the optimum distance between the electrodes for treating textile wastewater with plasma corona discharge was at a distance of 4.0 cm.

4. Conclusion

Reactive dyes with chromophore of phthalocyanine which has complex molecules can be degraded effectively and efficiently using plasma corona discharge. The exposure time and the distance between the electrodes affected the efficiency of degradation. The longer exposure time resulted in a higher degradation efficiency. The best result with COD and color removal of 83.15% and 52.32% respectively was achieved by the process at 60 minutes exposure time with a distance between the electrodes of 4 cm.

Acknowledgments

Polytechnic of STTT Bandung for the opportunity of this research collaboration and Nadya Amelia for contributing to this research.

References

[1] Aditya W, Notodarmodjo S and Helmy Q Decolorization of Reactive black-5 Dye by UV based photocatalytic with immobilized ZnO nanoparticles onto ceramic plate surface Recent citations Decolorization of Reactive black-5 Dye by UV based photocatalytic with immobilized ZnO nanoparticles onto cera

[2] Iervolino G, Vaiano V, Pepe G, Campiglia P and Palma V 2020 Degradation of acid orange 7 azo dye in aqueous solution by a catalytic-assisted, non-thermal plasma process *Catalysts* **10** 1–18

[3] Hussain T and Wahab A 2018 A critical review of the current water conservation practices in textile wet processing *J. Clean. Prod.* **198** 806–19

[4] Chen L, Wang L, Wu X and Ding X 2017 A process-level water conservation and pollution control performance evaluation tool of cleaner production technology in textile industry *J. Clean. Prod.* **143** 1137–43

[5] Pratiwi R, Notodarmojo S and Helmy Q 2018 Decolourization of remazol black-5 textile dyes using moving bed bio-film reactor *IOP Conf. Ser. Earth Environ. Sci.* **106**

[6] Samsami S, Mohamadi M, Sarrafzadeh M H, Rene E R and Firoozbahr M 2020 Recent advances in the treatment of dye-containing wastewater from textile industries: Overview and perspectives *Process Saf. Environ. Prot.* **143** 138–63
[7] Nur M 2011 *Plasma Physics and Aplications*
[8] Jiang B, Zheng J, Qiu S, Wu M, Zhang Q, Yan Z and Xue Q 2014 Review on electrical discharge plasma technology for wastewater remediation *Chem. Eng. J.* **236** 348–68
[9] Dojčinović B P, Roglić G M, Obradović B M, Kuraica M M, Kostić M M, Nešić J and Manojlović D D 2011 Decolorization of reactive textile dyes using water falling film dielectric barrier discharge *J. Hazard. Mater.* **192** 763–71
[10] García M C, Mora M, Esquivel D, Foster J E, Rodero A, Jiménez-Sanchidrián C and Romero-Salguero F J 2017 Microwave atmospheric pressure plasma jets for wastewater treatment: Degradation of methylene blue as a model dye *Chemosphere* **180** 239–46
[11] Hama Aziz K H, Mahyar A, Miessner H, Mueller S, Kalass D, Moeller D, Khorshid I and Rashid M A M 2018 Application of a planar falling film reactor for decomposition and mineralization of methylene blue in the aqueous media via ozonation, Fenton, photocatalysis and non-thermal plasma: A comparative study *Process Saf. Environ. Prot.* **113** 319–29
[12] Zhang C, Sun Y, Yu Z, Zhang G and Feng J 2018 Simultaneous removal of Cr(VI) and acid orange 7 from water solution by dielectric barrier discharge plasma *Chemosphere* **191** 527–36
[13] Nippatla N and Philip L 2019 Electrocoagulation-floatation assisted pulsed power plasma technology for the complete mineralization of potentially toxic dyes and real textile wastewater *Process Saf. Environ. Prot.* **125** 143–56
[14] Zhang Y, Zhou M, Hao X and Lei L 2007 Degradation mechanisms of 4-chlorophenol in a novel gas-liquid hybrid discharge reactor by pulsed high voltage system with oxygen or nitrogen bubbling *Chemosphere* **67** 702–11
[15] Marotta E, Ceriani E, Schiorlin M, Ceretta C and Paradisi C 2012 Comparison of the rates of phenol advanced oxidation in deionized and tap water within a dielectric barrier discharge reactor *Water Res.* **46** 6239–46
[16] Susan A I, Tayibnapis A S, Widodo M and Nur M 2016 Kajian Kelistrikan Plasma Pijar Korona Menggunakan Elektroda Multi Titik-Bidang Dalam Perlakuan Tekstil *Arena Tekst.* **31** 11–6
[17] Tao X, Wang G, Huang L, Ye Q and Xu D 2016 A novel two-level dielectric barrier discharge reactor for methyl orange degradation *J. Environ. Manage.* **184** 480–6
[18] Liu X, Zhang H, Qin D, Yang Y, Kang Y, Zou F and Wu Z 2015 Radical-Initiated Decoloration of Methylene Blue in a Gas–Liquid Multiphase System Via DC Corona Plasma *Plasma Chem. Plasma Process.* **35** 321–37
[19] Ghezzar M R, Abdelmalek F, Belhadj M, Benderdouche N and Addou A 2007 Gliding arc plasma assisted photocatalytic degradation of anthraquinonic acid green 25 in solution with TiO2 *Appl. Catal. B Environ.* **72** 304–13
[20] Joshi R P and Thagard S M 2013 Streamer-like electrical discharges in water: Part II. environmental applications *Plasma Chem. Plasma Process.* **33** 17–49
[21] Iervolino G, Vaiano V and Palma V 2020 Enhanced azo dye removal in aqueous solution by H2O2 assisted non-thermal plasma technology *Environ. Technol. Innov.* **19** 100969
[22] He Z, Liu J and Cai W 2005 The important role of the hydroxy ion in phenol removal using pulsed corona discharge *J. Electrostat.* **63** 371–86

Study of Hybrid Annuity Model on Indonesia Toll Road

D T Ramadhanti, R H Karsaman, and S S Wibowo

Transportation Engineering Research Group, Civil Engineering department, Institut Teknologi Bandung, Ganesha Street number 10, Bandung 40132, Indonesia

E-mail: dizkatriarsaniramadhanti@gmail.com

Abstract. Indonesia strives for toll road connectivity to push the high logistics costs and lead to economic growth. However, the expansion of toll road construction on a large-scale has cost much money. So, an innovative funding scheme must be developed to attract the private to invest in those projects. Hybrid Annuity Model is a new public-private partnership developed by the Government of India in 2016 to reduce concessionaires' risk on toll road project funding. This study aimed to examine the compatibility of the Hybrid Annuity Model implementation on Indonesia toll road. The analysis was carried out through financial feasibility and sensitivity analysis on Serang – Panimbang toll road as the object study. Moreover, a survey was conducted to know the stakeholders' perception of this funding scheme implemented in Indonesia. The results showed that Hybrid Annuity Model could give financial feasibility on Serang – Panimbang toll road under any changes of traffic volumes, toll fares, discount rates, investment costs, investment return period, and the investment proportion between the Government and the concessionaires. From stakeholders' perspective, Hybrid Annuity Model could be implemented on Indonesia toll roads, especially those which were not financially feasible and located in rural areas. Furthermore, it needed some improvements and supporting regulations to be more suitable applied on Indonesia toll road.
Keywords: toll road, Hybrid Annuity Model, feasibility analysis, sensitivity analysis, stakeholders' perception

1. Introduction

Logistic costs in Indonesia are relatively high, especially in rural areas. The poor connectivity and accessibility in rural areas become the main cause of those high logistic costs. The Indonesian government currently strives for toll road construction on a large scale. Toll roads can reduce travel time as a benefit, which leads to lower transportation costs. From 2014 to 2020, Indonesia had operated a total of 1,348-kilometers new toll roads. In the future, the government tends to build more new toll roads for increasing connectivity and accessibility to push down the logistic costs and accelerate the nation's economic growth.

Serang - Panimbang toll road becomes one of Indonesia's priority projects. It will connect Serang as the capital city of Banten Province to Tanjung Lesung as a special economic zone in the Panimbang regency. Hopefully, this toll road will increase the number of visitations in Tanjung Lesung and Ujung Kulon as promoted tourism destinations. Moreover, it will also support the connectivity between the industrial estate in Pandeglang and ports in Tanjung Priok and Merak. Hence, the transportation cost will decrease. Furthermore, it also increases accessibility to the south area of Banten Province, whose economic growth has been left behind by other areas of Banten Province such as Tangerang. Serang – Panimbang toll road consists of three sections with a total length of 83.6 kilometers.

As a consequence of expanding toll roads in Indonesia, lack of funding becomes a challenge to the government. The construction of toll roads on a large scale requires a huge amount of money, while the government's budget is limited. However, it is hard to get funds for financially unfeasible toll road projects. The concessionaires may not be attracted to invest in those toll road projects. Then, the urgency of developing an innovative toll road funding scheme has arisen. So that, the concessionaires may be interested in participating in the government's program of expanding toll roads in Indonesia.

The Hybrid Annuity Model is a new public-private partnership developed by the government of India in 2016 to reduce the concessionaires' risk on toll road project funding. This model has combined three existing funding models - EPC (Engineering, Procurement, and Construction), BOT (Build - Operate - Transfer) toll model, and BOT annuity model [1]. By adopting the EPC model, the government becomes a part who has the authority to hold the tender as the project owner. As the adaptation of the BOT toll model, the concessionaires, who have participated and won the bidding process, will be responsible to construct and operate the toll road. In this scheme, the government offers to pay forty percent of the total project cost while the concessionaire pays the rest. This feature will reduce the capital outflow for the government and equity investments for the concessionaires [2]. By adopting the BOT annuity model, the government will pay for the maintenance and operational cost annually and return the concessionaires' investment with biannual payment during the concession period. These features are beneficial to the concessionaire because they will ease their cash flow pressure. Furthermore, the government will be responsible for managing toll revenues from road users to handle the traffic risk. Besides the allocation risk between the government and the concessionaires, the calculation of NPV will use the prespecified rate as the bank rate + 3% [3].

The impacts of the Hybrid Annuity Model (HAM) implementation need to be analyzed, especially in Indonesia toll roads. As an adopted model from other countries, the suitability of this model also needs to be identified through the perspectives of the stakeholders on toll road financing in Indonesia.

2. Research objectives

The research aimed to investigate the impacts and the suitability of the Hybrid Annuity Model implementation in Indonesia toll roads. The following research objectives were addressed:

1. Identify the financial feasibility of Indonesia toll road projects through HAM.
2. Identify the changes of financial feasibility indicator under several uncertain conditions in the implementation of HAM.
3. Investigate the Hybrid Annuity Model's suitability on toll road funding in Indonesia through the stakeholders' point of view

3. Methodology

3.1. Financial feasibility analysis

This study was conducted on Serang – Panimbang toll road as the object study. It had run two scenarios on the object study. First scenario was done under the PPP scheme with government's funding to the third section of this toll road as the existing scheme. Second scenario was run under the Hybrid Annuity Model. It used secondary data for analysis, including the investment cost, traffic volume estimation, bank rate, and inflation. The financial feasibility was assessed by using Net Present Value (NPV), Internal Rate of Return (IRR), Benefit Cost Ratio (BCR), and Payback Period (PBP). The assumptions used in this study were:

- The basic toll fare was Rp1,000/km for the first-class vehicles. The ratio of toll fare to the vehicle classification I : II : III : IV : V = 1 : 1.5 : 1.5 : 2 : 2
- The concession period of the toll road was 40 years, started from a year of 2020. The construction of the toll road started from 2017 for 3 years.
- Toll road's additional revenues = 1.5% from toll revenues
- Toll fare would be rose by 3% per two years due to inflation

- The Debt Equity Ratio (DER) used for this project was 70:30 with 20 years of loan term and the bank rate of 10%

3.2. Sensitivity analysis

Sensitivity analysis was conducted under the Hybrid Annuity Model. The impact of changes in the financial feasibility (NPV, IRR, BCR, and PBP) was observed under several changes as follows.

- Traffic volume with changes -40% to +40% of the initial estimation with 10% interval
- Discounted rate with changes -4% to +4% of the initial used discount rate with 1% interval.
- Investment percentage between the Government and the concessionaires. Both the investment portion of the Government and the concessionaires varied from 10% to 90% with 10% interval
- Investment return period varied from 10 years to 40 years with 5 years interval
- Toll fare with changes -Rp150/km and +Rp150/km of the initial toll fare
- Investment cost with changes -20% to+20% of the initial investment cost with 5% interval.

3.3. Stakeholders' perception analysis

This study surveyed through online questionnaire that targets respondents who have a minimum of one-year experience in Indonesia toll road financing. The questionnaire was developed by conducting thorough literature review to identify the characteristics of the Hybrid Annuity Model.

Table 1: The characteristics of the Hybrid Annuity Model

No	Characteristics	No	Characteristics
1	The government and the concessionaire share the investment funding 40% and 60%, respectively [3]	11	Operational and maintenance of toll road is managed by the concessionaire [5]
2	The investment return will be given as an annual payment during the concession period [3]	12	Operational and maintenance risk is borne by the concessionaire [5]
3	The investment return will be given biannually during the concession period [3]	13	The government will pay operational and maintenance costs during the concession period [3]
4	The used discount rate will add 3% from the bank rate [3]	14	Operational and maintenance costs will be given along with the annual return [3]
5	Financial risk caused by a change of discount rate is borne by the concessionaire [5]	15	Operational and maintenance expense is inflation adjusted [2]
6	The annual payment is inflation adjusted [3]	16	Government is responsible to collect income from user fees [5]
7	Toll road construction is managed by the concessionaire [5]	17	Revenue risk is borne by the Government [5]
8	Construction risk is borne by the concessionaire [5]	18	The government will give compensation to the concessionaire as the consequences of policy and politic situation changes [4]
9	The concessionaire will receive bonus 0.5% of its investment cost if the project has been finished on or before to the scheduled completion date [4]	19	The government will give compensation to the concessionaire as the consequences of the project scope change [5]
10	The concessionaire will receive a penalty for delay in the construction phase [4]		

On this survey, the respondents were asked to choose their agreement on each characteristic of the Hybrid Annuity Model by using A 5-point Likert scale (1= strongly disagree, 2= disagree, 3=neutral, 4=agree, 5= strongly agree). Moreover, the respondents were also asked for their opinion on this model's implementation through one open-ended question. The questionnaire was distributed randomly to toll road funding stakeholders i.e., the government, the concessionaires, and academics.

4. Data Analysis

4.1. Financial feasibility analysis

At first, Serang – Panimbang toll road was not financially feasible. It had the Net Present Value (NPV) of -Rp3,064,661,915,041 the Internal Rate of Return (IRR) of 7.07%, and the Benefit Cost Ratio (BCR) of 0.80. So that, the government offered public – private partnership (PPP) scheme to make this toll road financially feasible and attract the concessionaires' interest for investing on this project.

In the first scenario (PPP scheme), the construction cost for the first and second section became the concessionaires' responsibility while the government would fund the construction cost of the third section of Rp4,467,234,268,992. On this scheme, the concessionaire would be responsible to pay the operational and maintenance cost. Moreover, they also received toll revenue as their investment return during the concession period. For financial feasibility calculation, it would use the discount rate of 9% based on data from Bank Indonesia.

In the second scenario (HAM scheme), the government and the concessionaire shared the investment funding 40% and 60%, respectively. On this scheme, the government would return the concessionaires' investment by paying annuities during the concession period. Moreover, the government would also be responsible to pay the operational and maintenance cost of the toll road. Since the concessionaire had received annuities as their investment return, toll revenue would be received and managed by the government. For financial feasibility calculation, it would use the discount rate of 12%. This value had been added by 3% from the used discount rate on the first scenario as the characteristics of the Hybrid Annuity Model.

The result of financial feasibility analysis from those scenarios was compared, shown on Table 2.

Table 2: Comparison of Financial Feasibility Result

PPP scheme (1st scenario)		HAM scheme (2nd scenario)	
Discount rate	9%	Discount rate	12%
NPV	Rp2,209,759,391,460	NPV	Rp2,363,134,991,237
IRR	9.95%	IRR	21.42%
BCR	1.21	BCR	1.26
PBP	28th year	PBP	7th year
Government's NPV	-Rp4,467,234,268,992	Government's NPV	-Rp8,786,132,312,201

Both schemes had given the financial feasibility for the toll road project. HAM scheme was more beneficial to the concessionaires, shown by the higher result of NPV, IRR, and BCR and faster payback period. However, HAM scheme caused the bigger financial burden for the government because of a high annuity that must be paid during concession period (40 years).

4.2. Sensitivity analysis

The changes in financial feasibility indicators were observed under several conditions through sensitivity analysis. The variables used in this study were:

4.2.1. Traffic volume

Traffic volume changes did not affect all assessment indicators of financial feasibility. As the traffic risk had been given to the government, it affected the government's financial loss. The result of the sensitivity analysis due to traffic volume changes is shown in Table 3.

Table 3 : Sensitivity analysis result due to traffic volume changes

Changes	Annuity (billion)	NPV (billion)	BCR	IRR	PBP	Government's NPV (billion)
-40%	Rp1,098	Rp2,363	1.26	21.42%	8th year	-Rp12,121
-30%	Rp1,098	Rp2,363	1.26	21.42%	8th year	-Rp11,287
-20%	Rp1,098	Rp2,363	1.26	21.42%	8th year	-Rp10,454
-10%	Rp1,098	Rp2,363	1.26	21.42%	8th year	-Rp9,620
0%	Rp1,098	Rp2,363	1.26	21.42%	8th year	-Rp8,786
+10%	Rp1,098	Rp2,363	1.26	21.42%	8th year	-Rp7,952
+20%	Rp1,098	Rp2,363	1.26	21.42%	8th year	-Rp7,119
+30%	Rp1,098	Rp2,363	1.26	21.42%	8th year	-Rp6,285
+40%	Rp1,098	Rp2,363	1.26	21.42%	8th year	-Rp5,451

4.2.2. Discount rate

The toll road project was still financially feasible due to changes in the discount rate, although all assessment indicators had also changed. An increase in the discount rate resulted in a higher discount factor as this factor was equal to one plus discount rate raised by an exponent. This condition had caused the value of NPV and BCR to drop. Yet, it escalated the payback period and the value of IRR as the annuity had also increased. As the annuity had increased, the government experienced greater financial loss due to changes in the discounted rate. The result of the sensitivity analysis due to discounted rate changes is shown in Table 4.

Table 4 : Sensitivity analysis result due to discounted rate changes

Changes	Annuity (billion)	NPV (billion)	BCR	IRR	PBP	Government's NPV (billion)
-4%	Rp924	Rp3,010	1.27	14.70%	16th year	-Rp4,491
-3%	Rp967	Rp2,787	1.27	16.28%	13th year	-Rp5,961
-2%	Rp1,011	Rp2,611	1.26	17.94%	12th year	-Rp7,116
-1%	Rp1,055	Rp2,472	1.26	19.65%	11th year	-Rp8,037
0%	Rp1,098	Rp2,363	1.26	21.42%	8th year	-Rp8,786
+1%	Rp1,142	Rp2,279	1.26	23.23%	8th year	-Rp9,406
+2%	Rp1,186	Rp2,215	1.26	25.09%	7th year	-Rp9,930
+3%	Rp1,231	Rp2,167	1.27	26.98%	7th year	-Rp10,382
+4%	Rp1,276	Rp2,133	1.27	28.90%	6th year	-Rp10,779

4.2.3. Investment portion between the government and the concessionaires

Investment portion changes on the Hybrid Annuity Model still gave financial feasibility to the concessionaires. The bigger portion of the concessionaires' investment would rise the annuity. Consequently, it also would increase the value of NPV. But it did not change the value of BCR as the annuity increased proportionally to the changes in investment cost. However, it escalated the payback period and the value of IRR as the annuity had also increased. As the annuity had increased, the government experienced greater financial loss. The result of the sensitivity analysis due to the changes of investment portion is shown in Table 5.

Table 5 : Sensitivity analysis result due to changes in investment portion

Concessionaires - Government	Annuity (billion)	NPV (million)	BCR	IRR	PBP	Government's NPV (billion)
10% - 90%	Rp373	Rp802	1.26	50.89%	3rd year	-Rp8,377
20% - 80%	Rp518	Rp1,115	1.26	29.26%	6th year	-Rp8,459
30% - 70%	Rp663	Rp1,427	1.26	24.84%	7th year	-Rp8,541
40% - 60%	Rp808	Rp1,739	1.26	23.03%	7th year	-Rp8,622
50% - 50%	Rp953	Rp2,051	1.26	22.04%	8th year	-Rp8,704
60% - 40%	Rp1,098	Rp2,363	1.26	21.42%	8th year	-Rp8,786
70% - 30%	Rp1,244	Rp2,675	1.26	20.99%	9th year	-Rp8,868
80% - 20%	Rp1,389	Rp2,987	1.26	20.68%	9th year	-Rp8,950
90% - 10%	Rp1,534	Rp3,300	1.26	20.45%	9th year	-Rp9,032

4.2.4. Investment return period

The toll road project was still financially feasible due to changes in investment return period, although all assessment indicators had also changed. A reduction to investment return period caused the NPV and BCR to drop. However, it escalated the payback period and the value of IRR as the annuity had increased. Furthermore, it eased financial burden on the government. The result of the sensitivity analysis due to changes of investment return period is shown in Table 6.

Table 6 : Sensitivity analysis result due to changes in investment return period

Changes	Annuity (billion)	NPV (million)	BCR	IRR	PBP	Government's NPV (billion)
10 years	Rp1,603	Rp2,171	1.24	56.70%	3rd year	-Rp8,594
15 years	Rp1,330	Rp2,106	1.23	34.36%	4th year	-Rp8,529
20 years	Rp1,212	Rp2,154	1.24	27.10%	6th year	-Rp8,577
25 years	Rp1,155	Rp2,219	1.25	24.07%	7th year	-Rp8,642
30 years	Rp1,124	Rp2,279	1.25	22.61%	8th year	-Rp8,702
35 years	Rp1,108	Rp2,327	1.26	21.84%	8th year	-Rp8,750
40 years	Rp1,098	Rp2,363	1.26	21.42%	8th year	-Rp8,786

4.2.5. Toll fare

Toll fare changes did not affect all assessment indicators of financial feasibility. As the traffic risk had given to the government, it affected to the government's financial loss. The result of the sensitivity analysis due to change of toll fare is shown on Table 7.

Table 7 : Sensitivity analysis result due to basic toll fare changes

Changes	Annuity (billion)	NPV (million)	BCR	IRR	PBP	Government's NPV (billion)
-Rp150	Rp1,098	Rp2,363	1.26	21.42%	8th year	-Rp9,290
0	Rp1,098	Rp2,363	1.26	21.42%	8th year	-Rp8,786
+Rp150	Rp1,098	Rp2,363	1.26	21.42%	8th year	-Rp8,463

4.2.6. Investment cost

The changes in investment cost still gave the financial feasibility to the concessionaires. The annuity was set up constant, although the investment cost had changed from the previous estimation. The concessionaires' investment had become more efficient due to a decrease in the investment cost, shown by a faster payback period and a greater value of IRR. Furthermore, it raised the concessionaires' profit, shown by an increase in NPV and BCR. Moreover, it also cut the government's financial loss. The result of the sensitivity analysis due to changes of investment cost is shown on Table 8.

Table 8 : Sensitivity analysis result due to investment cost changes

Changes	Annuity (billion)	NPV (million)	BCR	IRR	PBP	Government's NPV (billion)
-20%	Rp1,098	Rp3,798	1.50	35.78%	4th year	-Rp7,645
-15%	Rp1,098	Rp3,440	1.43	31.01%	5th year	-Rp7,930
-10%	Rp1,098	Rp3,081	1.37	27.16%	6th year	-Rp8,216
-5%	Rp1,098	Rp2,722	1.31	24.01%	7th year	-Rp8,501
0%	Rp1,098	Rp2,363	1.26	21.42%	8th year	-Rp8,786
+5%	Rp1,098	Rp2,004	1.21	19.27%	12th year	-Rp9,071
+10%	Rp1,098	Rp1,645	1.17	17.46%	14th year	-Rp9,357
+15%	Rp1,098	Rp1,287	1.13	15.94%	18th year	-Rp9,642
+20%	Rp1,098	Rp928	1.09	14.63%	22nd year	-Rp9,927

4.3. Stakeholders' perception analysis

The survey had obtained data from 32 respondents. Using descriptive analysis, the average experience years of respondents on toll road funding was 9.66 years (SD = 9.35). 46.9% of respondents had 1 to 5 experience years on toll road funding, 15.6% of respondents had 6 to 10 experience years on toll road funding, and 37.5% of respondents had more than 10 experience years on toll road funding. In terms of their institution, respondents consisted of those who were from the government (31.3%), the concessionaires (46.9%), and academics (21.9%).

The respondents were asked to choose their agreement on 19 questions about the characteristics of the Hybrid Annuity Model through a 5-point Likert scale. The calculation of mean on each question was required to assess stakeholders' agreement generally on each characteristic given on the questionnaire. Pimentel (2019) had stated that people tend to show their agreement on the 5-point Likert scale question when the mean value was equal or more than 3.40. On this research, those questions which had mean score less than 3.40 were considered to decrease stakeholders' interest on the Hybrid Annuity Model implementation on Indonesia toll road. Meanwhile, those questions which had mean score more than 3.40 were considered to increase stakeholders' interest on the Hybrid Annuity Model implementation on Indonesia toll road. Moreover, the respondents were also asked their opinion about the characteristics of Hybrid Annuity Model implemented on Indonesia toll road in the future through one open-ended question.

The result of stakeholders' perception survey on the characteristics of the Hybrid Annuity Model through a 5-point Likert scale is shown on Table 9.

Table 9 : Statistical test of stakeholders' perception survey

Question	Scale					Mean	Impact on stakeholders' interest on HAM
	1	2	3	4	5		
1	3.13%	18.75%	40.63%	37.50%	0.00%	3.13	Decrease
2	3.13%	15.63%	12.50%	62.50%	6.25%	3.53	Increase
3	3.13%	37.50%	31.25%	28.13%	0.00%	2.84	Decrease
4	3.13%	25.00%	50.00%	21.88%	0.00%	2.91	Decrease
5	6.25%	34.38%	40.63%	15.63%	3.13%	2.75	Decrease
6	3.13%	6.5%	9.38%	75.00%	6.25%	3.75	Increase
7	0.00%	3.13%	12.50%	65.63%	18.75%	4.00	Increase
8	0.00%	3.13%	18.75%	65.63%	12.50%	3.88	Increase
9	0.00%	21.88%	15.63%	53.13%	9.38%	3.50	Increase
10	6.25%	12.50%	21.88%	56.25%	3.13%	3.38	Decrease
11	0.00%	3.13%	12.50%	71.88%	12.50%	3.94	Increase
12	0.00%	3.13%	15.63%	68.75%	12.50%	3.91	Increase
13	9.38%	25.00%	18.75%	46.88%	0.00%	3.03	Decrease
14	6.25%	15.63%	15.63%	62.50%	0.00%	3.34	Decrease
15	0.00%	3.13%	15.63%	81.25%	0.00%	3.78	Increase
16	12.50%	18.75%	18.75%	43.75%	6.25%	3.13	Decrease
17	3.13%	9.38%	31.25%	43.75%	12.50%	3.53	Increase
18	0.00%	0.00%	6.25%	75.00%	18.75%	4.13	Increase
19	0.00%	0.00%	12.50%	65.63%	21.88%	4.09	Increase

From these results, we could conclude several characteristics of the Hybrid Annuity Model which were gaining disagreements from stakeholders as follows.

- **The government and the concessionaire share the investment funding 40% and 60%, respectively**. Some respondents thought that the Hybrid Annuity Model should have flexibility on investment portion between the government and the concessionaire based on their agreements. Since the characteristic of toll roads in Indonesia were diverse, this recommendation could become win-win solution between the government and the concessionaire to adjust the investment portion based on their own financial condition.
- **The investment return will be given biannually during the concession period**. A long period of investment return would overburden the governance budget. Moreover, it might get the concessionaires losing interest to invest in toll road project. A recommendation for reducing the period of investment return might ease the government's fiscal burden and attract the concessionaires for investing.
- **The used discount rate will add 3% from the bank rate**. From sensitivity analysis, additional percentage on the used discount rate could reduce the NPV and BCR since an increase in the discount rate resulted in a higher discount factor.
- **Financial risk caused by a change of discount rate is borne by the concessionaire**. Since the investment funding was shared between the government and the concessionaire, it should be fair to share the financial risk to both the concessionaire and the government.
- **The concessionaire will receive a penalty for delay in the construction phase**. This characteristic was mostly opposed by respondents from the concessionaires as there was a lot of uncertainties during the construction phase. However, this characteristic should not be a

problem to the concessionaire since most respondents had agreed that construction risk was borne by the concessionaire. So, the concessionaire, as the risk taker, should be responsible to deliver the project on time and take the consequence for a delay in the construction phase.

- **The government will pay operational and maintenance costs during the concession period along with the annual return**. Respondents from the concessionaire considered the uncertainties on operational and maintenance costs. Since it had not been stated on the previous study whether these costs were constant or not, it was necessary to put this issue on the regulation.
- **Government is responsible to collect income from user fees**. This characteristic was mostly opposed by respondents from the concessionaires since the government was inexperienced to manage public fund, such as toll revenue. Some respondents suggested to establish an instrument or institution to manage toll revenue. However, it was considered inefficient since India had NHAI, an Indian governmental institution that had a similarity to BPJT in Indonesia, to collect and manage toll revenue through the Hybrid Annuity Model.

5. Conclusion

The Hybrid Annuity Model could give financial feasibility to toll road project since the concessionaires would receive a certain amount of annuity. However, the annuity itself, as one feature on this model, caused more burden to the government. Under changes of several conditions (traffic volume estimation, discounted rate, investment portion between the government and the concessionaire, investment return period, toll fare, and investment cost), Hybrid Annuity Model still gave financial feasibility to the concessionaires. The tenacity of this model from changes in traffic volume and toll fare could be used to attract the concessionaires' interest. On the other hand, those changes still could not give profit to the government. The Hybrid Annuity Model was possible to be implemented on Indonesia toll road. Though it should be applied to toll roads that either financially unfeasible or located in rural areas. Since there were several features received a lot of disagreements from stakeholders, it needed adjustment to make this model more suitable to be implemented in Indonesia. Moreover, the government should create regulation about the Hybrid Annuity Model implementation once this model was ready to be applied on Indonesia toll road funding.

References

[1] Singh, V., & Madurwar, D. (2019). Study of HAM in Highway Construction Projects-Owner and Contractor Prospective. SSRN Electronic Journal.

[2] Taneja, A., & Karla, D. (2019). Analysis of Hybrid Annuity Model in Construction. *Comprehensive Advanced Specific Summarised Studies*, 61-68.

[3] Garg, S. (2019). Hybrid Annuity Model: Hamming risk allocations in highway. Journal of Public Affairs, 1-28.

[4] Ghayal, M. S., & Salgude, R. R. (2019). Effect of Hybrid Annuity Model on Road Project. *International Journal of Engineering and Advanced Technology (IJEAT)*, 1525-1530.

[5] Patwardhan, A. V. (2016). *Hybrid Annuity Model (HAM) for PPP Projects.* New Delhi: CUTS Institute for Regulation & Competition

[6] Pimentel, Jonald L. (2019). *Some Biases in Likert Scaling Usage and its Correction.* International Journal of Sciences: Basic and Applied Research (IJSBAR) Volume 45 No. 1, 183-191.

Estimation of Extreme Rainfall over Kalimantan Island based on GPM IMERG Daily Data

A A Kuntoro[1,2], R K Hapsari[2], M B Adityawan[1,3], M Farid[1,3], Widyaningtias[1,2] and Radhika[4]

[1]Faculty of Civil Engineering and Environment, ITB
[2]Center for Water Resources Development, ITB
[3]Center for Coastal and Marine Development, ITB
[4]Directorate of Engineering Affairs of Water Resources, Ministry of Public Works

E-mail: arnoak@itb.ac.id

Abstract. Rainfall is one of the critical data for water resources infrastructure planning. In many cases in developing countries such as Indonesia, rainfall stations are not evenly distributed. In many cases, regional development occurs much faster than the improvement of hydrological measurement instruments. The plan to move the capital city of Indonesia to Kalimantan is one example. Satellites rainfall products can be utilized, especially for areas with a limited number of rainfall stations. This study examines the potential use of Global Precipitation Measurement (GPM) satellite products to estimate the spatial distribution of rainfall in the Kalimantan region. Twenty years data of daily maximum rainfall from GPM satellite rainfall products in 2001-2020 were compared to twenty years data of daily maximum rainfall from 16 rainfall stations under the Meteorology, Climatology, and Geophysical Agency (BMKG), with data time spanning from the 1970s to 2020. The analysis results show a significant difference between extreme rainfall analysis computed by using station data and the satellite. The use of the correction function can increase the accuracy of the GPM rainfall product. It can be used as an alternative data source for a region with limited rainfall stations.

Keywords: Rainfall, Satellite, GPM, Kalimantan

1. Introduction

Kalimantan is the largest island in Indonesia the territory. The island has attracted worldwide attention due to the high rate of deforestation in recent decades. The biggest drivers of deforestation in Kalimantan are expanding oil palm plantations and converting forests to open area which then overgrown by bush/grass [1]. The expansion of oil palm land in Kalimantan reached 4.5 million ha in 1995-2015 [2]. Deforestation has a strong correlation with an increase in the frequency of forest fires in Kalimantan [3], [4], [5], and become more frequent with annual occurrence [6]. On the other side, the frequency of floods in Kalimantan has also increased in the last few decades. There were several flood events in Kalimantan in 2021, e.g., January 2021 in South Kalimantan, May 2021 in East and North Kalimantan, August 2021 in East Kalimantan, and November 2021 in West Kalimantan. The flood incident in South Kalimantan in January 2021 was considered as the biggest flood in recent decades, which caused huge loss [7], [8]. The Indonesian Government has planned to move the state capital to Kalimantan. The new capital city needs more infrastructures such as roads, bridges, office areas, commerce, settlements, agriculture, plantations, and others supporting infrastructure. Those infrastructures must be planned and appropriately designed to deal with the potential natural disasters, including flood. A common problem in developing countries, including Indonesia, is the limited availability of the data, including rainfall data. BMKG is one of the agencies in Indonesia which provides climatological data. However, the availability of rainfall data in the BMKG online websites (http://dataonline.bmkg.go.id) is quiet limited. During analysis of this study, only 16 stations available in Kalimatan with long time series of more than 20 years. Compared to the area of Kalimantan Island, the availability of rainfall data which can be accessed by the public is quite limited. In this case, rainfall satellite data becomes one of the potential alternatives. TRMM (Tropical Rainfall Measurement Mission) is one of the most widely used rainfall satellites. Several studies evaluated the accuracy of TRMM data in Indonesia. In case study of Central Java, TRMM 3B42 product tends to be underestimates in rainy season and less accurate in mountainous area [9]. In case study of Makassar Straits, TRMM rainfall product shows weak correlation with ground station data in the flat terrain area in Kalimantan but strong correlation in the east coast of South Sulawesi [10]. The above results shows that accuracy of TRMM product varies over different topographical features. TRMM operation ended in 2015, and continued by the GPM (Global Precipitation Measurement) satellite. In case study of Surabaya, GPM IMERG (The Integrated Multi-satellitE Retrievals for GPM) shows strong correlation with ground station data in monthly basis, but weak correlation in daily basis [11]. In case study of Brantas Watershed, correction factors are applied to improve GPM IMERG accuracy in monthly rainfall [12]. The above research results show evaluation of TRMM and GPM product in monthly dan daily series, but not in annual daily maximum rainfall. Meanwhile, extreme rainfall analysis is used in most of the water resources infrastructure designs. This research focussed on evaluation of GPM daily rainfall product to estimate extreme rainfall in case study of Kalimantan Island.

2. Materials and Method

2.1. Rainfall Data

GPM IMERG provides time-series data daily from 2001-2020 with a grid size of 0.1 degrees square. Statistically, the 20-year time series is adequate for extreme rain analysis. Evaluation of GPM IMERG products for extreme rain analysis was conducted by comparing 16 rain stations operated under BMKG, which have data series of more than 20-years. Location of the rainfall stations which has more than 20-years series data is shown in Figure 1.

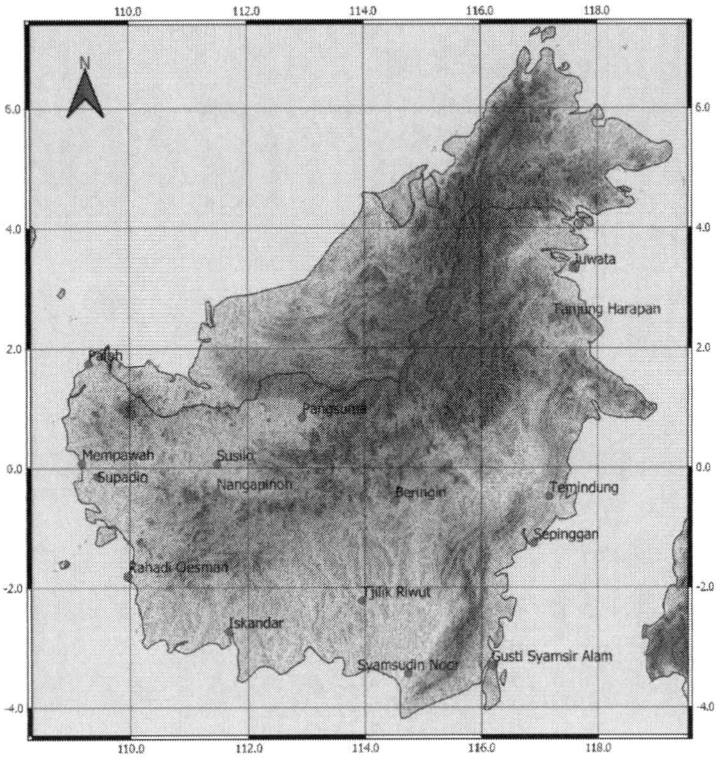

Figure 1. Rainfall stations in Kalimantan Island

Each stations contains blank data. Data completeness of the sixteen rainfall stations is shown in Figure 2.

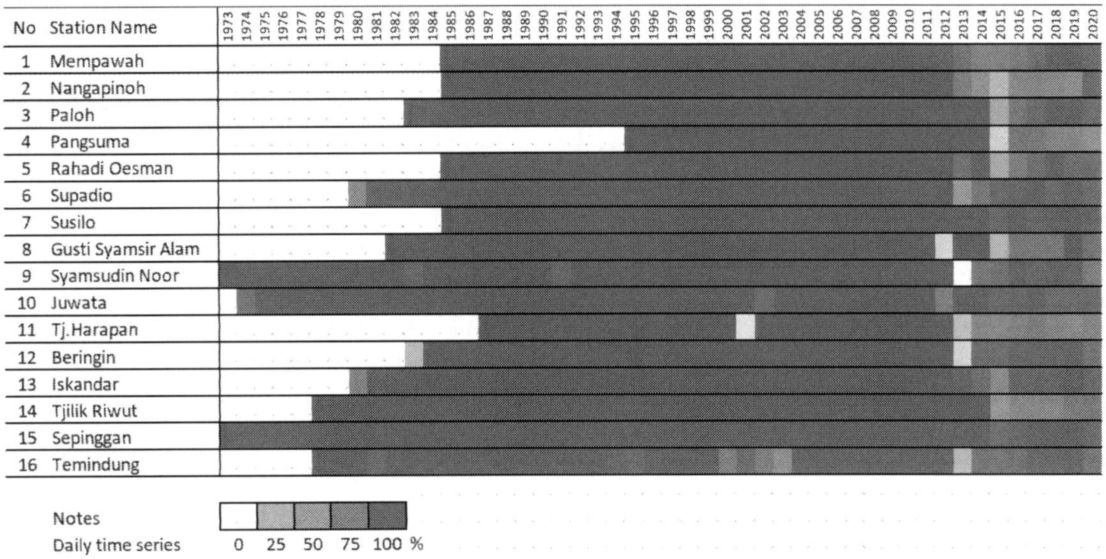

Figure 2. Rainfall data completeness

Figure 2 shows relatively significant blank data from 2013 to 2020 which shows in light blue color. Therefore, in this study, the 20-year time series comparison only use year with no blank data by assuming the trend of extreme rainfall has not increased significantly during 1980s to 2020s. This assumption was taken to avoid blank data filling which might result statistical properties changes of the extreme rainfall series. Table 1 shows rainfall series which was used for GPM data comparison.

Table 1. Selected time series from 16 rainfall stations

No	Station Name	Time Series	No	Station Name	Time Series
1	Mempawah	1992-2011	9	Syamsudin Noor	1993-2012
2	Nangapinoh	1993-2012	10	Juwata	1982-2001
3	Paloh	1995-2014	11	Tj.Harapan	1993-2012
4	Pangsuma	1995-2014	12	Beringin	1993-2012
5	Rahadi Oesman	1993-2012	13	Iskandar	1993-2012
6	Supadio	1992-2011	14	Tjilik Riwut	1995-2014
7	Susilo	1995-2014	15	Sepinggan	1995-2014
8	Gusti Syamsir Alam	1992-2011	16	Temindung	1980-1999

2.2. Correction Factor of Satellite Data

2.2.1. Rainfall data checking

One of the Indonesian Technical Instruction for flood analysis contains a chapter about method for TRMM data correction. The document is "*Petunjuk Teknis Perhitungan Debit Banjir Pada Bendungan* (Technical Instruction of Flood Discharge Design for Dam)," published by the Ministry of Public Works, 2017. According to the above document, TRMM can be used as comparison to the ground station data. However, rainfall data obtained from TRMM needs to be evaluated/corrected. The evaluation is carried out on a monthly basis by using two statistical tests: coefficient correlation and root-mean-square (RMSE). If ground station rainfall data quality is good enough, then the correlation coefficient between TRMM and post-rain is higher or equal to 0.6. For the monsoon rainfall pattern, the typical RMSE is between 97-158, while for the equatorial rainfall pattern, the typical RMSE is between 58-84. Procedures for extension of rainfall data using TRMM for dam flood discharge calculation are as follow:1) Select the ground rainfall station with minimum 3 years series of data; 2) Check the availability of daily rain data period at the selected station; 3) Select the TRMM grid closest to the coordinates of the selected rainfall station; 4) Plot of monthly rainfall from stations vs TRMM rainfall; 5) Calculate the correlation coefficient value of the monthly rainfall from stations and TRMM rainfall (value correlation coefficient > 0.6; 6) If the correlation coefficient > 0.6 then the data can be used for flood discharge calculation.

2.2.2. Time series rainfall data correction

According to Technical Instruction of Flood Discharge Design for Dam, TRMM Daily Data Correction, the procedure of TRMM data correction is as follow: 1) The two rainfall data series (TRMM and station) with the same period are compared into two columns; 2) plot non-exceedance probability for each rainfall data with an interval of 10-20 mm; 3) TRMM rainfall data less than the minimum rainfall station data value is considered 0. The value usually ranges between 0-10 mm; 4) TRMM rainfall data with value less rainfall station is multiplied by a constant. The typical rainfall ranges between 50–100mm and the constant ranges between 0.8-1; 5) TRMM rainfall data with value greater than point 4) is multiplied by a different constant, typically ranges between 1–1.3.

2.2.3. Extreme rainfall data correction

This research proposes simple correction procedure for GPM IMERG only for extreme rainfall data. The proposed procedure is as follow: 1) Select rainfall stations with minimum time series data of 20 years; 2) Retrieve GPM IMERG data in the same area as the station location from 2001 to 2020 (20 years); 3) Perform probability analysis of daily maximum rainfall (e.g.: using Weilbull probability plot equation) and calculate daily rainfall for return period, e.g.: 2, 5, 10, and 20 years; 4) Compare extreme rainfall with certain return periods, e.g.: 2, 5, 10, and 20 years calculated based on station data, and extreme rainfall calculated based on satellite data. The correction factor is defined as the ratio between rainfall with certain return period calculated from station data and from GPM IMERG; 5) Correction factor for other area can be interpolated from the available stations. In this study, GPM IMERG correction factors only calculated for 5 and 10 years return period as follow:

$$C_{5y} = \frac{R_{5\,station}}{R_{5\,GMP\,IMERG}} \tag{1}$$

$$C_{10y} = \frac{R_{10\,station}}{R_{10\,GMP\,IMERG}} \tag{2}$$

3. Result and Discussion

Example result of extreme rainfall data correction from procedure 3 to 5 in sections 2.2.3 is shown in Figure 3. Figure 3 (above-left) shows the data plots for Mempawah and Supadio Stations, representing the area of West Kalimantan which influenced by the climate of Karimata Strait. Figure 3 (below-left) shows the data plots for the Temindung and Sepinggan Stations, representing the area of East Kalimantan, which is influenced the climate of Sulawesi Strait. Figure 3 (above-right) shows the data plots for Syamsudin Noor and Iskandar Stations, representing Central and South Kalimantan which influenced by the climate of Java Sea. Figure 3 (below-right) shows the plots for Pangsuma and Beringin Stations, representing the central part of the mainland of Kalimantan Island. In general, the above data shows that in most stations in Kalimantan, the higher the rainfall (the higher the return period), the higher the deviation of the extreme rain value given by GPM IMERG (compared to the results obtained from the station data). In other words, the higher the return period, the more GPM IMERG tend to be overestimated, and the more significant the correction factor must be applied (with the correction factor further away from the value 1). For example, the data plot for the Beringin Station located in the central part of Kalimantan Island shows the IMERG GPM correction values for 5-year and 10-year return periods are 0.80 and 0.54, respectively.

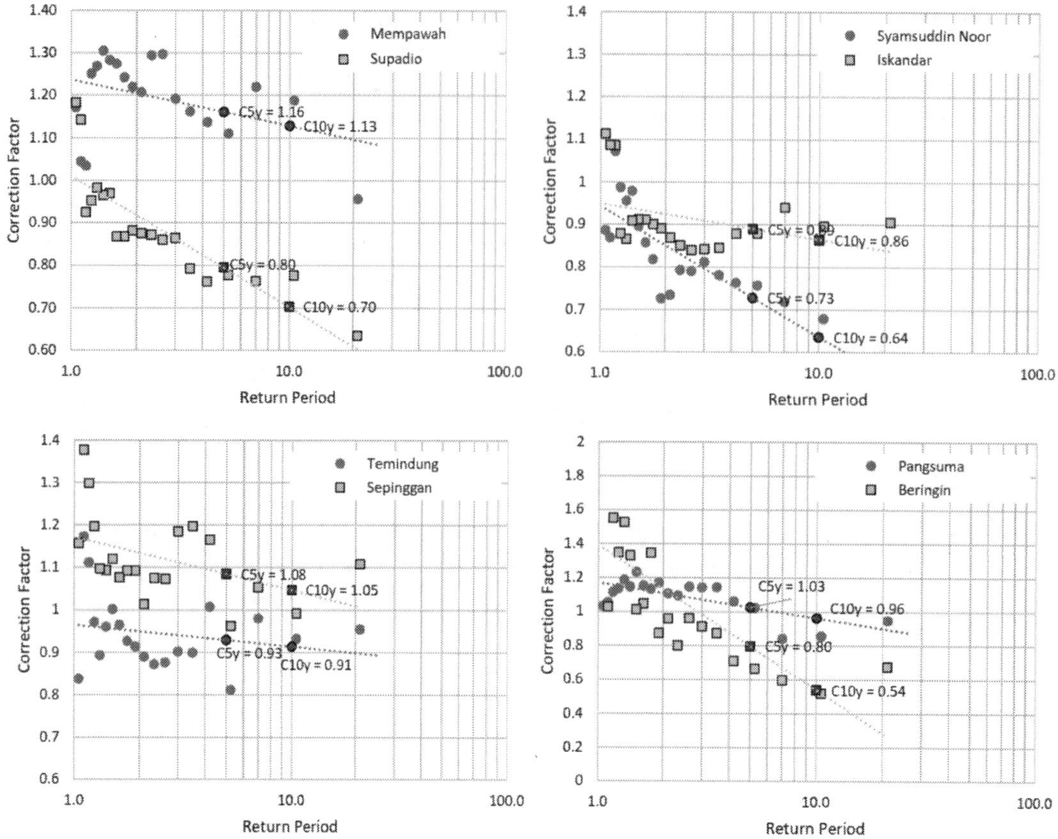

Figure 3. Example of correction factor for GPM IMERG rainfall based on return period

Correction factor of GPM IMERG for 5-years and 10-years rainfall is shown in Table 2.

Table 2. Correction factor of GPM IMERG for 5-years and 10-years rainfall

No	Station Name	Correction Factor for 5 years Rainfall	Correction Factor for 10 years Rainfall
1	Mempawah	1.09	1.07
2	Nangapinoh	1.13	1.13
3	Paloh	1.20	1.24
4	Pangsuma	1.03	0.96
5	Rahadi Oesman	1.08	1.11
6	Supadio	0.80	0.70
7	Susilo	0.97	0.93
8	Gusti Syamsir Alam	0.86	0.80
9	Syamsudin Noor	0.73	0.64
10	Juwata	0.78	0.68
11	Tj.Harapan	0.99	1.01
12	Beringin	0.80	0.54
13	Iskandar	0.89	0.86
14	Tjilik Riwut	1.11	1.06
15	Sepinggan	1.08	1.05
16	Temindung	0.93	0.91

From the 16-stations, three stations show the opposite trend, i.e. Paloh, Rahadi Oesman, and Tj.Harapan. In these three stations, extreme rainfall given by GPM IMERG tend to be underestimated. Nine from 16-stations shows a trend of overestimated GPM IMERG in high rainfall. These stations are Pangsuma, Supadio, Susilo, Gusti Syamsir Alam, Syamsudin Noor, Juwata, Beringin, Iskandar, and Temindung. This trend can be seen in correction factor of 10-years rainfall which are less than one. In the four other stations, i.e.: Mempawah, Tjilik Riwut, Sepinggan, and Nangapinoh, rainfall with lower return period is tend to be more overestimated than rainfall with higher return period. If correction factor in Table 2 is plotted spatially, then correction factor map of GPM IMERG for 5-years and 10-years rainfall can be generated. The map is shown in Figure 4.

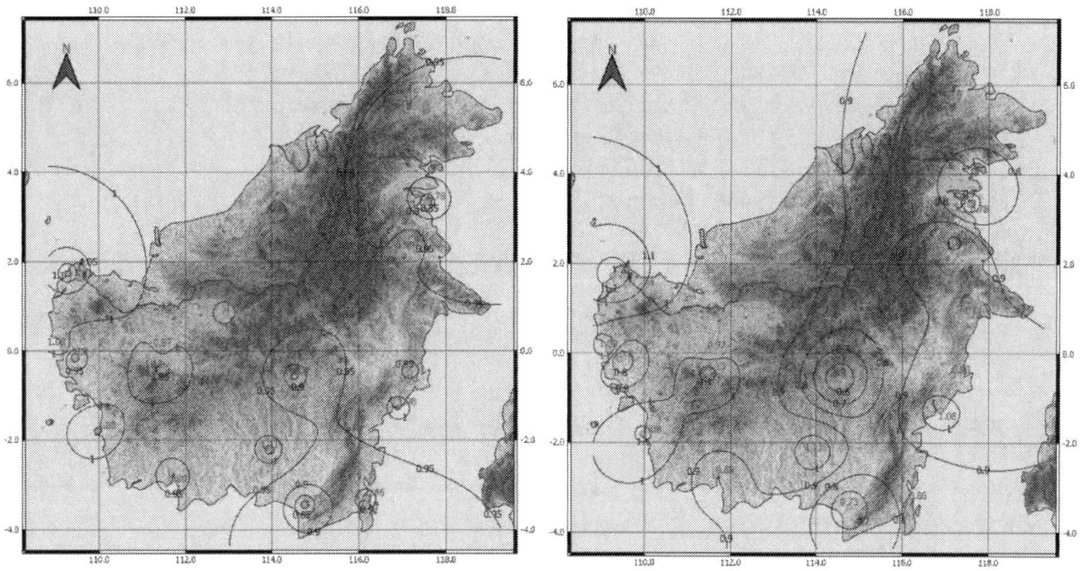

Figure 4. GPM IMERG Correction factor for 5-years (left) and 10-years (right) rainfall

The above result shows that analysis of extreme rainfall in area with less (or no) rainfall stations can be conducted using the corrected GPM IMERG data. However, more data is needed to generate more comprehensive map than is shown in Figure 4. This study can be used as a comparison to the other correction procedure of satellite rainfall data.

4. Conclusion

GPM IMERG data is potential to be used for extreme rainfall calculations in area with limited or no rainfall stations. One of the advantages of GPM IMERG is available in long time-series (20 years). Statistically, this long time series meet the minimum data requirements for water resources infrastructure planning in Indonesia. However, the IMERG GPM data does not always give the similar extreme rainfall value as the station data. In this study, a comparison of extreme rainfall analysis for 5-year and 10-year return periods by using station data and GPM IMERG was carried out. Analysis of the 16 stations showed that more than 50% (9 out of 16 locations) showed an overestimated trend of GPM IMERG in higher return period rainfall (e.g., in 10-year return period). A correction factor can be applied to reduce the deviation between of the extreme rainfall value calculated by GPM IMERG and by station data. For case study of Kalimantan Island, GPM IMERG correction factor varies spatially, and shown in a contour map. Further analysis using more rainfall data is needed to produce more comprehensive correction factors of the GPM IMERG.

5. References

[1] Austin K G, Schwantes A, Gu Y, and Kasibhatla P S, 2019 What causes deforestation in Indonesia? *Environ.Res.Lett.* **14** 024007

[2] Austin K G, Mosnierb A, Pirkerb J, McCallumb I, Fritzb S and Kasibhatlaet P S, 2017 Shifting patterns of oil palm driven deforestation in Indonesia and implications for zero-deforestation commitments *Land Use Policy* **69** 41–48

[3] Kuntoro A A, Yamashita T and Wahyu A, 2010 Modification of Fire Module In LPJ-DGVM For Application In Tropical Area: A Case Study Of West Kalimantan *Journal of Global Environment Engineering* **15** 25–38

[4] Farahnaz A, Kuntoro A A and Kusuma M S B, 2020 Statistical Downscaling for the Projection of the Keetch Byram Drought Index in the Barito Basin *International Journal on Advanced Science, Engineering and Information Technology* **10** 873-879

[5] Singh M and Siheng Y, 2021 Spatial–temporal variations in deforestation hotspots in Sumatra and Kalimantan from 2001–2018 *Ecology and Evolution* **11** 7302–7314

[6] Thoha A S, Saharjo B H, Boer R and Ardiansyah M, 2019 Characteristics and causes of forest and land fires in Kapuas District *Biodiversitas* **20** 110-117

[7] Pratama M B, Withono R M, and Azkiarizqi I N, 2021 Hydro-Meteorological Aspects of The 2021 South Kalimantan Flood: Topography, Tides, And Precipitation *International Journal of Remote Sensing and Earth Sciences* **18** 73-90

[8] Oktovin, Tutpai G and Rachman A, 2021 Bantuan Kesehatan Bagi Warga Terdampak Banjir di Wilayah Sungai Lulut Banjarmasin Kalimantan Selatan *Jurnal Suaka Insan Mengabdi* **3** 18-26

[9] Sekaranom A B, Nurjani E, Hadi M P and Margai M A, 2018 Comparsion of TRMM Precipitation Satellite Data Over Central Java Region – Indonesia *Quaestiones Geographicae* **37**(3) 97-114

[10] Giarno, Hadi M P, Suprayogi S and Murti S H, 2018 Distribution of Accuracy of TRMM Daily Rainfall in Makassar Strait *Forum Geografi* **32**(1) 38-52

[11] Azka M A, Sugianto P A, Silitonga A K and Nugraheni I R, 2018 Accuracy Test of GPM IMERG Rainfall Estimation Product over Surabaya, Indonesia, *Jurnal Sains & Teknologi Modifikasi Cuaca* **19** (2) 83 – 88

[12] Hapsari R K, Kuntoro A A, Widyaningtias, Riawan E, Farid M, and Adityawan M B, 2021 A Correction Factor for GPM IMERG Precipitation Product Over Brantas Watershe, *Proc.the 7th HATHI International Seminar* (Indonesian Association of Hydraulic Engineers) Theme 1 pp 145-160.

Estimating household water consumptions in the Bandung Metropolitan area

Yuniati Zevi[1*], *Wika Maulany Fatimah*[2], *Yusep Ramdani*[1], *Muhammad Yusuf Habibbullah*[1], *Nurdianti Mursyida*[1]

[1]Department of Environmental Engineering, Institut Teknologi Bandung, Bandung, Indonesia
[2]Department of Anthropology and Development Studies, Radboud University, Nijmegen, The Netherlands
*Email:yz59@itb.ac.id, yz59@cornell.edu

Abstract:

The aim of this study is to determine relevant factors contributing to the water consumption in the Bandung Metropolitan area or Greater Bandung, which covered Bandung City, Cimahi City, Bandung Regency and West Bandung Regency. Towards achieving this aim, questionnaires were developed and served on the households in various communities to collect relevant data on the physical and socio-economic factors. Online questionnaires and direct survey were conducted with the number of samples proportional to the population of the Bandung Metropolitan area which is 1219 samples. Results of the study revealed that the daily average household water consumption in Bandung and Cimahi cities is 163.6 liters per person per day (lpd), Bandung Regency is 168.6 lpd, and West Bandung is 151.7 lpd, and in general was 161.26 lpd for Bandung Metropolitan. These values are higher than the standard commonly used for planning water supply systems which is 120 lpd. The factor affecting water consumptions are the type of water sources (public and private) and based on the Spearman Rho correlation test socio-economic conditions (economic level, education level, gender, and age) have a significant effect on water consumption as well. Comprehensively understanding household water consumption is necessary to design efficient and effective water demand management.

Keywords: Bandung, household, water consumption

1. Introduction

Water is a natural resource that is very important in the continuity of human life. Water is used for various human activities such as cooking, washing, bathing and others. The quantity, continuity and quality of water resources are factors that need to be considered in meeting water needs (Durán-Sánchez et al., 2018). The increasing demand for water resources is caused by increasing population, drought, and unpredictable weather patterns due to climate change. Many things have been done to deal with this problem, one of which is the effort to improve in terms of water demand management (WDM) which is used to help change consumers towards sustainable water consumption behavior. The water demand management (WDM) approach relies heavily on consumers to understand how to reduce their water consumption and to apply this understanding to their daily activities to consume sustainably (Willis et al., 2011).

Increased water demand without proper management can lead to water scarcity. In 2040, Java is predicted to experience total water scarcity with a water supply of 476 m^3/person/year. According to Walton (2019), water scarcity in Java is entirely caused by humans. Increasing urbanization and environmental degradation make it difficult to capture and store rainwater so that more rain is run off causing groundwater supplies to decrease. The prospect of a water crisis that may occur is a challenge to find effective ways to ensure the availability of sufficient water for the future without disrupting the sustainability of the current water supply system. Water demand management (WDM) can be an alternative to maintain the supply of water sources. Appropriate WDM policies can improve the supply–demand balance in areas that are vulnerable to water scarcity. An important aspect of WDM is an understanding of the need for water to support population and economic development without compromising the sustainability of water resources (Dziegielewski, 2003). Therefore, this study aims to provide an understanding of the pattern of water consumption for household purposes and the influencing factors which are expected to be the basis for WDM.

Based on related studies, it is known that water consumption in a household depends on many factors. These factors include the number of people in the house, age, education level, property size, income, efficiency of water consuming devices (e.g. washing machine, type of tap, type of toilet, etc.) and consumer attitudes and behavior (Fan et al., 2013). The importance of estimating water consumption which is more relevant for the purpose of planning water demand is needed to solve water resource problems.

Futhermore, based on BPS-Statistics of Jawa Barat Province, Indonesia's population from 2015 to 2019 have increased by 1.15%. The province that has a major influence on this population increase is West Java Province which has the largest population in 2019 with a total population of 49.316 million. This research was conducted in the Bandung Metropolitan Area, including Bandung and Cimahi city, West Bandung Regency and Bandung regency. Regional development in these cities/regencies continue to increase every year linear with the increasing population, so that the intensity of water consumption will continue to increase from year to year. Related to this, it is important to know the average household consumption of clean water for urban communities that are experiencing rapid regional development and population growth, in Metropolitan Bandung Area. This research is expected to be useful in assisting the maker decision in determining the planning of water supply systems in the future.

2. Methodology

This research includes preliminary studies, primary and secondary data collection, and data processing and analysis.

2.1. Research design

This research was quantitative and descriptive type with the research instrument in the form of a questionnaire. 50 questions were developed and divided into six themes: 1) attributes of respondents, 2) water source, 3) method of water collection for each activity, 4) personal water use, 5) household water use, dan 5) pattern of water use. A preliminary test was carried out ro re-examine the completeness, readability, clarity, consistency, suitability, relevance, and uniformity of the answers and how they corresponded to the questions. Later on, a minimum sample size was determined using Slovin's equation considering a 95% confidence level. The minimum sample required was 400 respondents for each city; thefore the total number of samples was 1219 for Bandung Metropolitan area.

A simple random sampling was conducted for three months by distributing the questionnaire through different social media platforms and in persons.

2.2 Statistical data analysis

The primary data that have been obtained are then systematically documented and calculated on MS. Excel. The plotting between clean water consumption and water sources as well as socio-economic conditions is carried out in the box plot graph on MS. Excel. Then the analysis of the plotting results is carried out using descriptive and inferential analysis. Based on the normality test, the data is not normally distributed, so the inferential analysis is using a non-parametric statistical test. For the average test of two independent samples, the Mann-Whitney test is performed, while for the average test of three or more independent samples, the Kruskal Wallis test is used. To determine the relationship between variables, a non-parametric correlation analysis is performed using the Spearman Rho Correlation test.

3. Results and Discussion

3.1. Characteristics of Respondents

Of the total 1219 respondents, 63.17% of them are female. As many as 72% of the total respondents are younger than 25 years old, 17% are from 25 to 40 years of age, and 9 % are older than 40 years old. The most widely used water source by respondents are private water sources consisting of dug wells and drilled wells with pumps, which is 74%. A total of 11% of respondents belong to the lower economic class, 52% are lower middle class, 26% upper middle class, and 9% are upper class. The distribution of respondents by education level is 33% for middle level and 63% for high level. The distribution of respondents' characteristics is show on **Table 1**.

3.2. Average Household Water Consumption in Metropolitan Bandung

The calculation of household water consumption resulted in an average of 161.26 lpd with the proportions for each household activity shown in **Figure 1**. The three activities that use the most water are bathing (32%), ablution (27%) and flushing (18%), Household activities consisting of food preparation and cooking, washing dishes, washing clothes, mopping floors, watering plants, and washing vehicles contributed 22% to the total daily drinking water needs. The three activities that use drinking water the least are drinking, mopping floors, and washing vehicles, each of which contributes 1%.

Compared to Indonesian Standard (SNI 6728.1-2015), the calculated average drinking water demand are still within the range of metropolitan household drinking water demand, which is 150–200 lpd. According to WHO (2011), household drinking water demand in the range of 100–200 lpd are generally found in households that have optimal access to drinking water source where drinking water is provided in residential buildings through several taps.

Table 1. Characteristics of respondents

Criteria	Total Respondents	Percentage (%)
Age		
- Younger (<25 y.o.)	887	72,76
- Middle (25–40 y.o.)	213	17,47
- Older (>40 y.o.)	119	9,76
Gender		
- Male	449	36,83
- Female	770	63,17
Education Level		
- Basic	29	2,38
- Middle	410	33,63
- High	780	63,99
Economic Group		
- Lower	143	11,73
- Lower-middle	635	52,09
- Upper-middle	328	26,91
- Upper	113	9,27
Water Source		
- Public	317	26,00
- Private	902	74,00

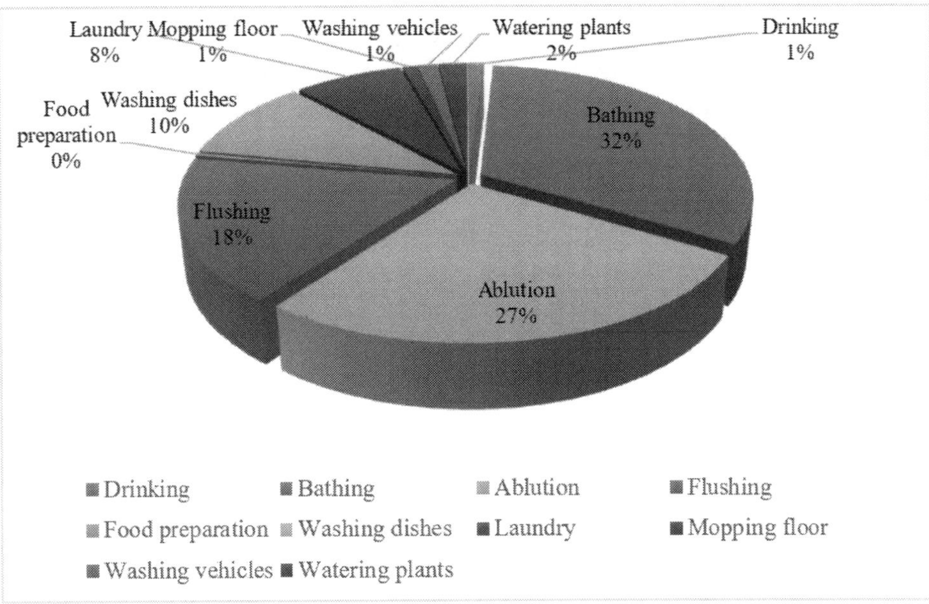

Figure 1. Household drinking water demand percentage for each activity

The factors influencing water consumption are population, environmental conservation, water conservation, technological advances, climate, price of water and how to get it (water sources) and socio-economic conditions (Tedjakusuma et al.,2001). However, in this study, the factors discussed are water sources and socio-economic conditions (economic level, education level, gender, and age) of the community.

3.3 Clean Water Consumption by Water Source

The sources of clean water used by the respondents are divided into two, such as a private water and a public water. Public water sources come from water supply company (PDAM) while private water sources come from bucket wells, drilled wells, and springs. Based on **Figure 2,** the average consumption of clean water for the community with different public and private water sources, the average water consumption for public water sources is 156.10 lpd and the average water consumption for private water sources is 163.07 lpd. In other words, the consumption of water from private sources is higher than public water sources. One possible reason is because people with private water sources can get the water freely, which cause them to have less awareness of the value of water (Jorgensen et al., 2009). In addition, residents who have responsibility for household water bills are more aware of their water consumption (Kelly and Fong, 2015).

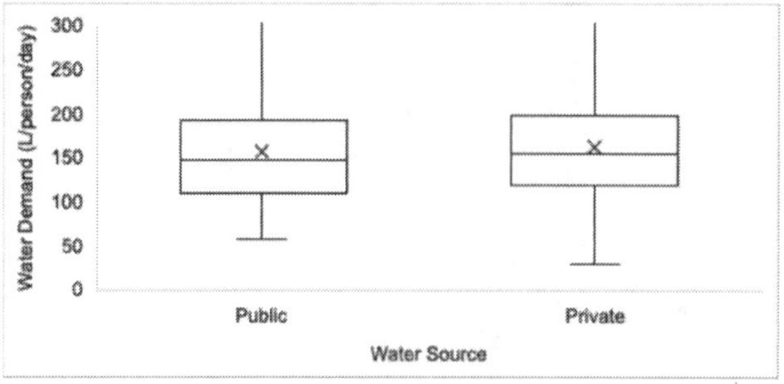

Figure 2. Water Consumption by Water Source

The results of the significance test for the mean difference by using Mann Whitney test show a significance value of 0 which is less than 0.05 (sig <0.05). It is concluded that there is a significant difference in the average water consumption between public and private water sources. The results of the Spearman Rho correlation test show a significance value of 0, which is less than 0.05 (sig <0.05), meaning that there is a significant relationship between public and private water source variables. In addition, a correlation coefficient of 0.281 is obtained (this value is in the range 0.26-0.50), which means that there is a strong relationship between variables. Thus, it can be said that the type of water source affects the amount of water consumption significantly.

3.4 Clean Water Consumption Based on Socio-Economic Characteristics

Socio-demographic characteristics include household size, education level, income, age, and gender (Shan et al., 2015). According to Tedjakusuma et al. (2001), the use of clean water is an absolute element in modern life. Apart from being an industrial facility and infrastructure, clean water is also able to meet the needs of life in everyday life in the household sector. Therefore, socio-economic changes will affect the behavior of clean water consumption, both for primary and secondary needs. In addition, the increase in the amount of income per capita of the population from year to year has caused the need for clean water to increase.

3.4.1 Water Consumption Based on Economic Level

Based on **Figure 3,** the average consumption of clean water for low-income economies (less than IDR 1,250,000), low-middle economy (IDR 1,250,001 - IDR 5,000,000), upper middle economy (IDR 5,000,001 - IDR 15,000,000), and upper economy (more than IDR 15,000,001) respectively 148.74 lpd, 157.17 lpd, 168.04 lpd, 180.42 lpd. It can be seen that the average water consumption shows an increase along with the increment of economic status. The upper economy has the largest average water consumption, while the lower economy has the smallest average water consumption. The possible reasons are linked to the relatively low price of the water which caused anti-social behaviors, wasteful water behaviors, and less awareness of the value of water for the higher income people (Jorgensen et al., 2009). High incomes will facilitate the purchase of more sophisticated and efficient equipment, but this can be correlated with reluctance to change habits, for example, by not delaying the usage of washing machine until the laundry is fully accumulated (Shan et al., 2015). In contrast, low-income households are less likely to take efficiency measures (purchasing high-tech equipment) but may have money-saving habits, such as only using the washing machine when the laundry is full.

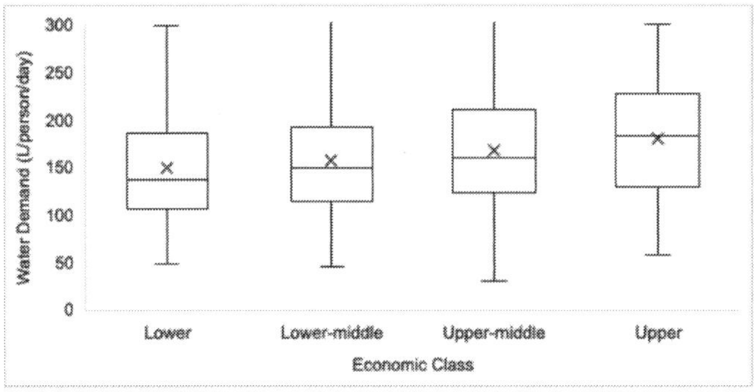

Figure 3. Water Consumption Based on Economic Level

The results of the mean difference significance test by using Kruskal Wallis test show a significance value of 0.005 which is less than 0.05 (sig <0.05). It is concluded that there is a significant difference in the average water consumption based on the economic level. Then the results of the Spearman Rho correlation test show a significance value of 0.027, which is less than 0.05 (sig <0.05), meaning that there is a significant relationship between differences in economic levels and clean water consumption. In addition, a correlation coefficient of 0.95 was obtained (this value is in the range 0.76-0.99), which means that there is a very strong relationship between variables. The correlation coefficient is positive, so the relationship between variables is unidirectional, which means the higher the economic level, the more water consumption will increase. It can be said that economic level affects the amount of water consumption significantly.

3.4.2 Clean Water Consumption Based on Education Level

Basic education, namely elementary schools (SD), secondary education, namely SMP / equivalent and SMA / equivalent and higher education, which can be in the form of universities, institutes, polytechnics, high schools, and academies. Based on **Figure 4**, the highest water consumption are respondents with high education level, namely 163.58 lpd, while the lowest water consumption are respondents with basic education level with average water consumption of 146.74 lpd. Water consumption based on education level increases with higher education. This is in accordance with the human capital theory which explains that the educational process has a positive influence on economic growth (Becker,2006). The argument put forward by proponents of this theory is that people with higher education level will have better jobs and wages (i.e., higher levels of economy) than those with lower education level, so water consumption will also increase because of their ability to buy or pay for water (Purwanto, 2006).

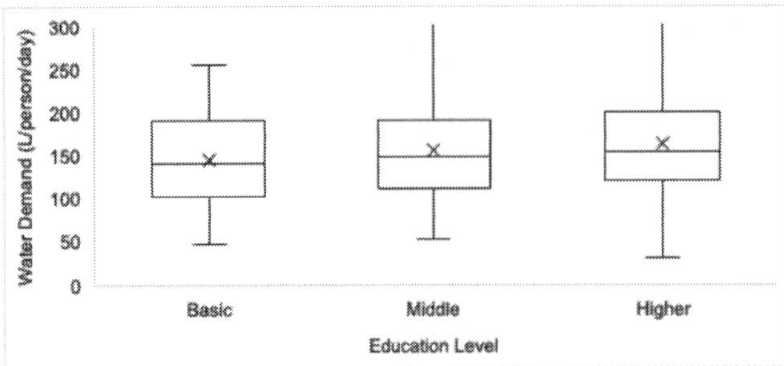

Figure 4. Clean Water Consumption Based on Education Level

The results of the mean difference significance test by using Kruskal Wallis test show a significance value of 0.000 which is less than 0.05 (sig <0.05). It is concluded that there is a significant difference in the average water consumption based on the education level. Then the results of the Spearman Rho correlation test show a significance value of 0.000, which is less than 0.05 (sig <0.05), meaning that there is a significant relationship between differences education levels and clean water consumption. In addition, a correlation coefficient of 0.202 is obtained (this value is in the range 0.00-0.25), which means that there is a weak relationship between variables. The correlation coefficient is positive, so the relationship between variables is unidirectional, it can be interpreted that the higher education level, the water consumption will also increase. Thus, it can be said that the level of education affects the amount of water consumption significantly.

3.4.3 Clean Water Consumption Based on Gender

Based on **Figure 5,** the average consumption of clean water between male and female is different, the average water consumption for male is 150.26 and the average water consumption for female is 167.67 lpd. We could say that, in this study female consumed more water than male. One of the reasons for this is that male and children need less time to shower than female (Shan et al., 2015). This is also accordance with the data obtained, which shows that the results of the largest consumption of water are for the needs of bathing, urinating and defecating (53%). In addition, females are more associated with activities that use water every day, because they are the ones who take care of the availability of drinks, food, water for bathing, house cleaning and others (Slamet, 2014).

The results of mean difference significance test by using Mann Whitney test show a significance value of 0.005 which is less than 0.05 (sig <0.05). It is concluded that there is a significant difference in the average water consumption based on gender Then the results of the Spearman Rho correlation test show a significance value of 0.002 which is less than 0.05 (sig <0.05), meaning that there is a significant relationship between the difference in gender and clean water consumption. In addition, a correlation coefficient of 0.14 is obtained (this value is in the range 0.00-0.25), which means that there is a weak relationship between variables. The correlation coefficient is positive, so the relationship between variables is unidirectional. Thus, it can be said that gender affects the amount of water consumption significantly.

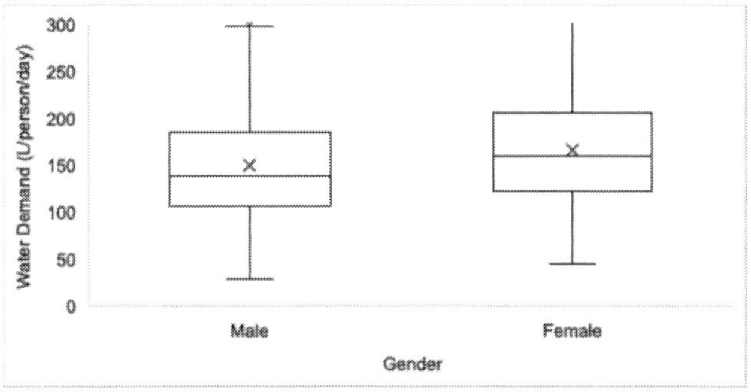

Figure 5. Clean Water Consumption Based on Gender

3.4.4 Clean Water Consumption by Age

Based on **Figure 6**, the average consumption of clean water between young, middle, and old ages are different, the average water consumption for young age is 164.38 lpd, the average water consumption for middle age is 147.15 lpd, and water consumption for the old age is 162.44 lpd. We could say that this study found that young people consume more water than middle age and old age. This happens because older people know themselves as people who are aware of water and tend to pay water bills. In other words, older people will have the awareness to save in consuming water (Kelly and Fong, 2015).

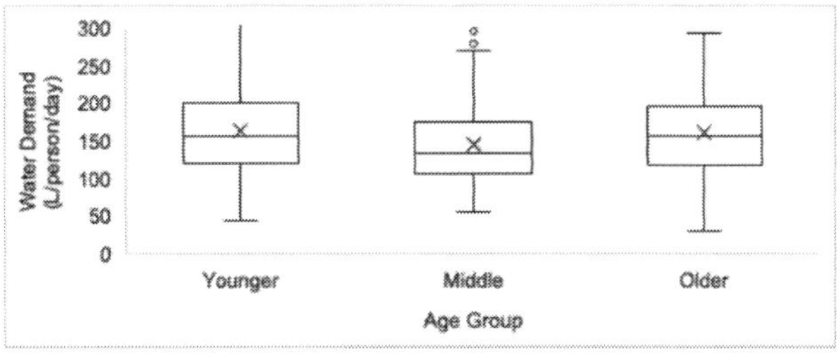

Figure 6. Water Consumption by Age Group

The results of mean difference significance test by using Kruskal Wallis test show a significance value of 0.000 which is less than 0.05 (sig <0.05). It is concluded that there is a significant difference in the average water consumption based on age. Then the results of the Spearman Rho correlation test show a significance value of 0.002 which is less than 0.05 (sig <0.05), it means that there is a significant relationship between the difference of age and clean water consumption. In addition, the correlation coefficient is -0.146 (this value is in the range 0.00-0.25), which means that there is a weak relationship between variables. The correlation coefficient is negative, so the relationship between variables is unidirectional, it can be interpreted that the higher the age the water consumption will decrease. Thus, it can be said that age affects the amount of water consumption significantly.

4. Conclusion

1. The calculation result of the average household water consumption in Bandung Metropolitan area is 161.26 liters/person/day. This proves that the basic water needs in Indonesia (120 lpd) which is usually used for planning the water supply system is no longer relevant.

2. Respondents with private water sources consume more clean water than respondents with public water sources. Based on Mann Whitney test results, it is known that there is a significant difference by the average water consumption between public and private water sources. Based on the Sperman rho correlation test, differences clean water sources are proven to have a significant positive effect on water consumption.

3. Based on the results of Kruskal-Wallis and Mann Whitney test, it is known that there are significant differences by average water consumption between economic levels, education levels, gender and age. Based on the Spearman Rho correlation test, the socio-economic aspects of the respondents (economic level, education level, gender, and age) have a significant effect on water consumption.

REFERENCES

Badan Pusat Statistik Kabupaten Bandung Barat. (2019). Kabupaten Bandung Barat dalam Angka 2019. Bandung Barat: BPS Kabupaten Bandung Barat.

Badan Pusat Statistik Provinsi Jawa Barat. (2020). Provinsi Jawa Barat dalam Angka. Jawa Barat: BPS Provinsi Jawa Barat

Becker, Gary S.,2006. "Health and Human Capital: The Inaugural T.W. Schultz
Lecture," Review of Agricultural Economics, Agricultural and Applied Economics Association, vol. 28(3), pages 323-325.

DJCK. PUPR. (1996). Modul Proyeksi Kebutuhan Air Dan Identifikasi Pola Perenc. Jar. Pipa Transm. Dan Distrib. Air Minum. pp. 1–16.

Durán-Sánchez, A., Álvarez-García, J., and del Río-Rama, M. de la C. (2018): Sustainable water resources management: A bibliometric overview, *Water (Switzerland)*, **10**(9), 1–19. https://doi.org/10.3390/w10091191 ok

Dziegielewski, B. (2003). Strategi for Managing Water Demand, Universities Council on Water Resources, Water Resources Update, 126, pp 29-39.

Fan, L., Liu, G., Wang, F., Geissen, V., and Ritsema, C. J. (2013). Factors Affecting Domestic Water Consumption in Rural Households upon Access to Improved Water Supply:Insights from the Wei River Basin, China, *PLoS ONE*, **8**(8). https://doi.org/10.1371/journal.pone.0071977

Jorgensen, B., Graymore, M., and O'Toole, K. (2009): Household water use behavior: An integrated model, *Journal of Environmental Management*, **91**(1), pp. 227–236. ok

Kelly, D., & Fong, D. (2015). *Water conservation: the implications of user awareness, attitude, and behaviour*. Paper presented at 41st International Symposium of CIB W062 Water Supply and Drainage for Buildings 2015, Beijing, China.

Purwanto, N. (2006): Kontribusi Pendidikan Bagi Pembangunan Ekonomi Negara, *Jurnal Manajemen Pendidikan*, **2**(2), 1–7.

Shan, Y., Yang, L., Perren, K., and Zhang, Y. (2015): Household water consumption: Insight from a survey in Greece and Poland, *Procedia Engineering*, **119**(1), 1409–1418.

Slamet, Juli Soemirat. (2014). *Kesehatan Lingkungan.* Gadjah Mada University Press.Yogyakarta

The World Bank. (2016). Annual Report 2016

Tedjakusuma, R., S. Hartini, and Muryani. 2001. Analysis of Factors Affecting Consumer Behavior in Purchasing Mineral Water in the Municipality of Surabaya. Journal of Social Dynamics Research, Volume 2, No. 3 Pages 48-58.

Walton, K. (2019). In Java, the water is running out, from website: https://www.lowyinstitute.org/the-interpreter/in-java-water-is-running-out.

Willis, R. M., Stewart, R. A., Panuwatwanich, K., Williams, P. R., and Hollingsworth, A. L. (2011): Quantifying the influence of environmental and water conservation attitudes on household end use water consumption, *Journal of Environmental Management*, **92**(8), 1996–2009. https://doi.org/10.1016/j.jenvman.2011.03.023

Properties of recycled hot asphalt mixture added with mastic asphalt component

I N A Thanaya[1], **I M A Ariawan**[1], and **I D G R Yudistira**[1], **I M Widiyarta**[2]

[1]Civil Engineering Magister Study Program, Udayana University, Bali, Indonesia.
[2]Mechanical Engineering Magister Study Program, Udayana University, Bali, Indonesia.

The corresponding author's email: aryathanaya@unud.ac.id

Abstract. The use of natural aggregates can be reduced by utilizing Reclaimed Asphalt Pavement (RAP). In field, there have been many simple recycling applications of RAP with addition of new asphalt and stone ash as road pavement on community housing areas. The objective of this experiment is to provide engineering properties of the mixture. Asphalt pen 60/70 was softened or rejuvenated with 11% SAE 10 engine oil, to obtain asphalt pen 180/200. The mixtures were produced by adding the RAP with 10% and 20% rock ash and soften asphalt, in hot mix process. The added asphalt content was varied by 2%, 2.5%, 3%, 3.5% and 4% by weight of the RAP, and compacted at 2x50 Marshall blows. The mixture of RAP with 10% stone ash at 3% added asphalt gave better performances compared to the mixture with 20% stone ash, in which the stability value was 3380.5 kg, flow 3.08 mm, Marshall Quotient of 1097.65 kg/mm, VIM 3.077%, VMA 18.95%, VFB 83.84%, Cantabro 9.02% , ITSM 5422 MPa, creep slope 0.267, and fatigue with load of 700; 900; 1100 KPa failed at 6101; 1471; 831 load repetitions. The mixture well meet the specification for typical mixtures for low to medium trafficked roads.

Keywords: recycle, RAP, mastic component.

1. Introduction

Road pavement recycling need to be promoted in most pat of the world as, it can reduce demand on natural aggregates. In many places road maintenance had been carried out by doing repeated overlays on top of the existing deteriorated road pavement. This had resulted in the raised up of road pavement elevation in many areas, hence had caused problems on drainage, site surface elevation disharmony especially in heritage sites that is located on road sides.

RAP is an attractive material to be used as either as recycled surface layer or road base layer, blended with natural materials, stabilized with cement and/or fly ash. To improve strength and reduce deformation, geo cell can be used to confine the blended RAP materials [1].

RAP recycling had been done since 1970's. RAP is most reused material into new road pavement in the USA. RAP can and had been recycled with limited pollution and it is recommended to be applied in less acidic environment (pH≤4). Acidic environment can be due to high sulfur content and environment with decomposed organic matters [2].

The use of RAP would reduce demands on natural aggregates. RAP can be used up to more than 50% of the mixture. Higher RAP content would increase the stiffness of the compacted mixtures and improve the resistant to deformation in higher temperature. However, higher stiffness would cause the mixture more susceptible to fatigue and thermal cracking [3].

RAP recycling had been implemented more widely in developed countries as it is supported by government policies, the availability of equipment and material technology particularly in rejuvenating asphalt binder. Meanwhile in developing countries, previously there had been less implementation on RAP recycling due to constrain in economic and technology, and it is being encouraged [4].

In line with a researcher [4], hot asphalt mixture recycling has not yet widely applied in many parts of developing countries, including in Bali-Indonesia especially for official government projects for surface pavement. Road milling is only carried out in limited areas that heavily damaged, before it is overlaid using new asphalt mixture for the whole road section. As only limited road milling carried out there is less Reclaimed Asphalt Pavement (RAP) stockpiles available. However in some privately run projects for local (communities housing) access road with light traffic, hot asphalt mixture recycling had been done. These works were done semi manually using simple hot mixing method, where the Reclaimed Asphalt Pavement (RAP) added with some amount of stone dust are heated on a metal sheet base container using wood fires, then added with hot liquefied asphalt. It is manually mixed, spread and compacted using a simple light road roller. The properties of the mixture is not known, as there is only visual observation carried out on a privately run projects. So far the road pavement recycling carried out in such way, can last for quite long and satisfied customers.

The proportion of RAP and added stone dust were done by experienced with no support form laboratory test results. Typical stone dust content were in a range of 10-40% by weight of the RAP. What had been done in this way is basically adding mastic component (fine aggregates, filler and asphalt) onto the RAP. This would somewhat change the aggregate grading with an increase of mastic component on the system. The aggregate gradation of the system would become finer or become more gap (gap grading) then the RAP itself. This experiment is aimed at evaluating the properties of the recycled hot asphalt mixture using reclaimed asphalt pavement added with mastic asphalt components.

2. Literature review

2.1 Hot recycling asphalt mixture

Hot recycling asphalt mixture had been done by researchers in laboratory scale. The RAP content incorporated may vary from low percentage up to using maximum even all RAP materials with satisfactory results. Various type of rejuvenators had been used and available in developed countries, but may not yet sufficiently available in developing countries.

RAP management and stockpiles system is an important aspect to promote RAP recycling, in addition to the policy of the government. The availability of specification is essential to support wider application of RAP recycling. Aggregate modification is one aspects of RAP recycling to meet the aggregate gradation specification. It had been known that aggregate gradation of asphalt mixture is either continuous or gap grading. In lab scale recycling, more accurate aggregate modification to meet aggregate grading specification can be done. Addition of a particular fraction of aggregate can be carefully done into the RAP [5]. However this works may not be practicable on site application.

Full scale site hot mix recycling had been tried and heavy and sophisticated equipment are available in developed countries. This may not yet available in developing countries. Therefore simpler method needs to be done. For recycling, can basically be done by changing the aggregate gradation of the RAP to become more gap. This can be done by adding mastic component of the mixture, by adding fines component (fine aggregates and filler) and the added asphalt binder. The amount of mastic component added into the mixture need to be tried in order to obtain a balance between engineering properties and efficiency [6].

2.2 Mastic Component of Asphalt Mixture

Mastic asphalt is a term that is commonly used in Europe to figure out an aggregate asphalt mixture that consists of filler and fine aggregates with asphalt as the binder. The asphalt content usually high that cause the mastic asphalt mixture has low porosity. Mastic asphalt had been applied in buildings (as screed, roof covering, flooring, ect.), in bridges for absorbing dynamics stresses, and in pavement and paths [7]. The mastic component of aggregate asphalt mixture is illustrated in figure 1.

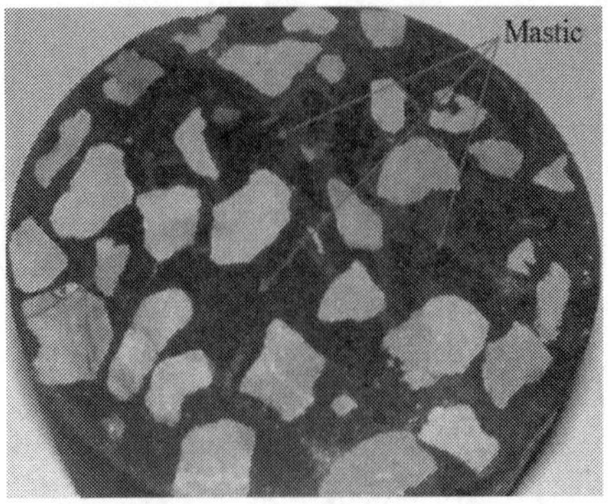

Figure 1. Mastic component of asphalt aggregate mixture.

2.3 Cantabro abrasion test

This test is used to evaluate the the resistance of the compacted specimens particle losses due to impact. The test was done using the Los Angeles rattler without the steel balls. The results are given in terms of mass loss after 300 drum revolutions using Equation 1 [8].

$$P = \frac{P1 - P2}{P1} \times 100\% \quad (1)$$

where: P = Abrasion Loss (%), P_1 = initial sample mass (g), P_2 = final sample mass (g).

2.4 Indirect Tensile Stiffness Modulus (ITSM), Fatigue and Dynamic Creep

On the ITSM test the samples were subjected to repeated diametrical line loading along the vertical diameter to induce indirect tensile stress. Calculation of the parameters on ITSM, Fatigue and Dynamic Creep has been available on the software of the Dynapave UTM30 machine used. The main equation used for ITSM calculation are Equation 2 and 3. The test is done at 20°C [9, 10]:

$$E = \frac{F \times (v + 0.27)}{(z \times h)} \quad (2)$$

$$E' = E \times (1 - 0.322 \times (log(E) - 1.82) \times (0.60 - k)) \quad (3)$$

where, E is the measured (indirect tensile) stiffness (ITSM) modulus in MPa; $F$is the peak value of the applied vertical load in Newton (N); z is the horizontal deformation in mm; v is the poisson's ratio; h is the thickness of the specimen in mm; $E'$is the stiffness modulus (ITSM) in MPa, adjusted to load area factor of 0.60; k is the measured load area factor.

For creep test the formulae used are shown in Equation 4 [11].

$$\varepsilon_n = 100 \times \left(\frac{h_o - h_n}{h_o}\right) \quad (4)$$

where, ε_n (microns) is the cumulative axial strain of the specimen after n stress (in kPa) application;h_o is the average height as measured by both displacement transducer after pre load of the specimen, in millimeters (mm); $h_n$is the average height as measured by both displacement transducer after n stress (of 100 kPa) application, in millimetres (mm). The tests are carried out at 40°C. The stress and strain data after 3600 repeated loading can be potted on a graph for further analysis. The creep slope is taken from the linear portion of the graph (from 3-4 last stress-strain data). The slope is the constant of x on the linear trend line. This constant is categorized as shown in Table 1.

Table 1. The slope of the dynamic creep.

Temp (°C)	Load $>10^6$ ESA (Heavy traffic)	Load 5×10^5 - 10^6 ESA (Medium traffic)	Load $<5 \times 10^5$ESA (light traffic)
> 30	< 0.5	0.5	> 3 – 6
20 – 30	< 1	1 – 6	> 6 – 10
10 – 20	< 2	2 – 10	-

Source : [12]

For fatigue loading, the formulae used are shown in Equations 5 and 6 [13].

$$\sigma = \frac{2F}{\pi \times t \times \Omega} \tag{5}$$

$$\varepsilon = \left(\frac{2\Delta H}{\Omega}\right) \times \left[\frac{1 + 3\nu}{4 + \pi \times \nu - \pi}\right] \tag{6}$$

where, σ is the horizontal tensile stress at the centre of the specimen, in KPa; $F$is the measured force, in Newton (N); t is te specimen thickness in mm; Ω is the specimen diameter in mm; $\Delta H$is the horizontal deformation in mm; ε is the horizontal strain in microns per meter(μm /m); $\nu$is the poisson's ratio. The tests are done out at 20°C. The F is calculated based on the applied stress, where minimum 3 levels of stresses are applied. The stress and number of load cycle to failure are plot into a graft with log-log scale for the analysis.

2.5 Materials and method

2.5.1 The Materials used. The materials used were RAP form local contractor stockpiles in Bali. This material was from cold milling from some road projects. The age of the RAP was unknown and can consist of RAP of different ages as it was stockpiles from different period of projects. The stone dust was taken from a locally available stone crusher in eastern part of Bali-Indonesia, where its particles size less than 5mm (contains 92,79% passing 5mm retained on 0.075mm) and 7.21% filler component (passing 0.075mm).

For the asphalt, ideally used softer asphalt penetration for rejuvenating the old asphalt in the RAP. However there is no softer asphalt penetration available in market, so the asphalt used was asphalt penetration of 60/70 that had been added with 11% by weight of SAE 10 motor cycle engine oil, to soften the asphalt until achieve penetration value close to 180-200. This experiment had been done in Indonesia [14].

2.5.2 Method. The experiment was started by doing extraction test on the RAP and sieve analysis on the stone dust. Then asphalt content of the RAP and the aggregate gradation of the combination of the materials obtained.

The experiment was done in a lab scale and in a simple way to represent the works that had been carried out on site. After trials it was chosen for better performance and efficiency to use RAP with 10% and 20% stone dust. This range of stone dust percentage had represented the amount used on site. The proportion of the materials is shown in Table 2.

The RAP and stone dust were heated on a metal pan. The softened asphalt was heated differently. After the RAP and stone dust achieve 160-170°C, the melted asphalt was added in a range of: 2.0; 2.5; 3.0; 3.5, and 4% of the RAP weight, then manually mixed. After that the mixture were compacted with 2x50blows of Marshall Hummer, for low to medium trafficked road. The asphalt content by weight of total mix was then calculated for each proportion of materials. The samples were tested for their Marshall properties, and Cantabro abrasion test at room temperature around 30°C. The Marshall properties of the mixtures were compared to the specification of asphalt mixtures commonly used to accommodate light to medium trafficked, i.e. Sand Sheet, Hot Rolled Asphalt (HRS), and Split Mastic Asphalt (SMA) mixtures [15].

Table 2. Proportion of materials.

Stone dust	RAP weight	RAP Asphalt content		Aggregate of RAP		Added stone dust by weight of RAP		Added asphalt by weight of RAP		Total weight of agg.	Total weight of mix
(%)	(gr)	(%)	(gr)	(gr)	(%)	(%)	(gr)	(%)	(gr)	(gr)	(gr)
10	1000	6.16	61.6	938.4	93.84	10	100	2.0	20	1038.4	1120
	1000	6.16	61.6	938.4	93.84	10	100	2.5	25	1038.4	1125
	1000	6.16	61.6	938.4	93.84	10	100	3.0	30	1038.4	1130
	1000	6.16	61.6	938.4	93.84	10	100	3.5	35	1038.4	1135
	1000	6.16	61.6	938.4	93.84	10	100	4.0	40	1038.4	1140
20	1000	6.16	61.6	938.4	93.84	20	200	2.0	20	1138.4	1220
	1000	6.16	61.6	938.4	93.84	20	200	2.5	25	1138.4	1225
	1000	6.16	61.6	938.4	93.84	20	200	3.5	30	1138.4	1230
	1000	6.16	61.6	938.4	93.84	20	200	3.5	35	1138.4	1235
	1000	6.16	61.6	938.4	93.84	20	200	4.0	40	1138.4	1240

The samples were also tested for their stiffness (Indirect Tensile Stiffness Modulus-ITSM) at 20°C, dynamic creep at 40°C and fatigue at 20°C (using Dynapave Universal Testing Machine-UTM 30). On fatigue test, the stresses were applied at three different stresses. The coordinate of the stress applied and the number cycles to failure were plotted into a graph. It is common to use log-log axis scale, and power trend line. The tests were carried out in line with British Standards [10, 11, 13]

Direct comparisons to other samples can be done using the same machine setting. For ITSM test setting: temperature 20°C; loading pulse width 250ms; pulse repetition period 3000ms; conditioning pulse count 5; estimated poisson's ratio 0.35; estimated modulus 3000MPa; target strain 5microns; seating force 20N; havershine loading shape. For Dynamic Creep: testing temperature 40°C; haversine loading shape; 100 KPa loading stress; 20 KPa seating stress for 600 sec; loading cycle width 1000ms; load cycle repeat time 1000ms; cycle count 3600. For ITFT: three level of stress Kpa: 700, 900, and 1100 KPa; 10 N seating force; loading cycle width 100ms; load cycle repeat time 500ms; and temperature of 20°C.

3 Results and analysis

3.1 The asphalt content of the RAP and the aggregate gradation of the mixture

The asphalt content of the RAP was found 6.16%. The aggregate gradation of the RAP, and the mixture with RAP plus 10% and 20% stone dust is shown in figure 2. It can be seen that the aggregate grading of the RAP closely following the upper limit of Concrete-Wearing Course (AW-WC) in line with the Indonesian Specification [15]. This showed that the RAP used was from AC-WC mixture in which some of the aggregate had been broken during the milling process, hence it became finer. It shown that the aggregate gradation of the milled RAP plus either 10% or 20% stone dust did not affect the aggregate gradation significantly and were also closely following the upper limit of the AC-WC gradation.

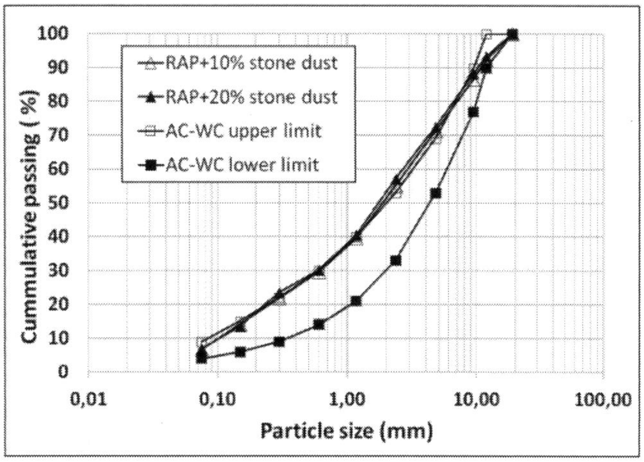

Figure 2. The aggregate grading of the mixture.

3.2 The Specific Gravity (SG) of the Materials

The SG of the aggregates as shown in Table 3, were very standard and met the specification.

Table 3. The aggregate specific gravity and water absorption.

Specific gravity of	SG value				Specification *)
	Bulk	SSD	Apparent	Effective	
The stone dust	2.55	2.63	2.75	2.63	
The RAP:					
Coarse aggregates	2.56	2.59	2.64	2.60	Max
Fine aggregates of	2.50	2.53	2.57	2.54	difference 0.2
Water abs of the coarse aggregates			1.16 %		Maks 3%
Water abs of the fine aggregates			1.02 %		Maks 3%
Filler component of the mixture			2.05 %		

*) in line with Indonesian specification [15]

3.3 The properties of the asphalt

The properties of the asphalt used is given in Table 4.

Table 4. Test result of asphalt.

No	Test	Result	Specification
1	Penetration of 60/70 pen asphalt (0.1mm)	63,86	60-70
2	SG of 60/70 pen asphalt	1,08	≥ 1,0
3	Penetration of 60/70 pen asphalt + 11 % oil SAE 10 (0.1mm)	184.74	Aimed at 180-200
4	SG of asphalt pen.60/70 + 11 % oil SAE 10	0,986	-
5	Softening point of asphalt pen 60/70	52°C	≥ 48 °C
6	Flash point of asphalt pen 60/70	305°C	≥ 232 °C
7	Ductility of asphalt pen 60/70	245 cm	≥ 100

3.4 Density and marshall stability of the mixtures

To be concise, in this part only figures of the density (figure 3) and the Marshall stability (figure 4) of the mixture are shown.

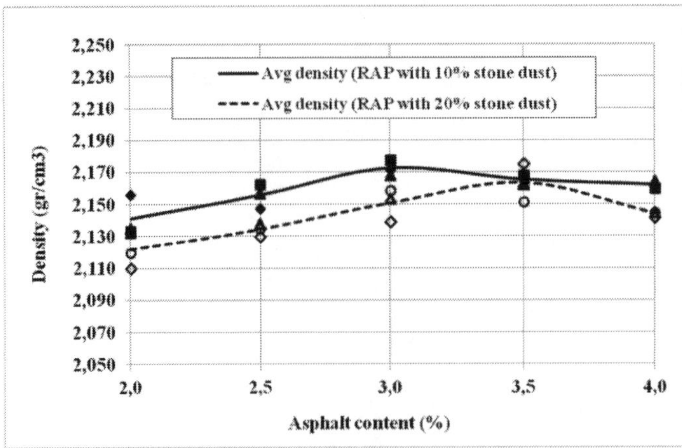

Figure 3. The average density of the mixtures.

The rest of the Marshall properties of the mixture at the recommended added asphalt content are shown in Table 5, with comparison to the specification of some mixture for light to medium trafficed road (Sand Sheet, Hot Rolled Sheet-HRS, and Stone Matrix Asphalt-SMA) in line with Indonesia Specification 2018 [15].

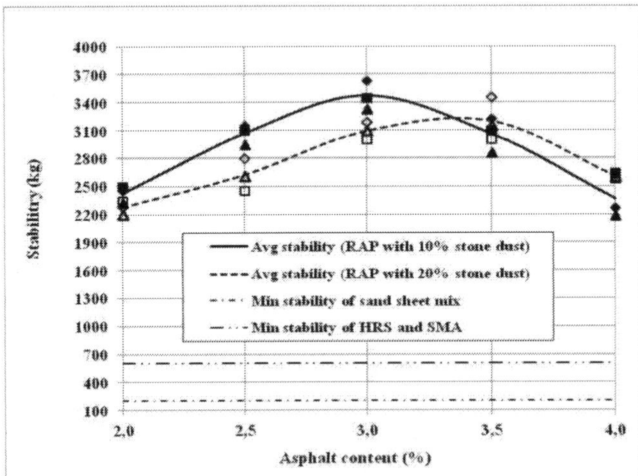

Figure 4. The average Marshall stability of the mixtures.

Table 5. Summary oh Marshall test results and comparisons with othe mixtures for medium to light trafficed road.

Marshall Properties		Mixture with RAP		Specification of mix for comparison		
		10% stone dust	20% stone dust	Sand sheet	HRS	SMA
		3% asphalt	3.5% asphalt			
Stability	(Kg)	3469.7	3200.5	Min.200	Min.600	Min.600
Flow	(mm)	3.08	2.99	Min.2-3	Min.3	Min.2-4.5
MQ	(kg/mm)	1097.6	1022.5	Min.80	Min.250	-
VIM	(%)	3.077	4.705	Min.3-6	Min.3-6	Min.4-5
VMA	(%)	18.95	20.08	Min.20	-	Min.17
VFB	(%)	83.84	76.60	Min.75	Min.68	-

Figures 3 and 4 showed that the trend of density the mixture is in line with the trend of stability. The mixture of RAP+10% stone dust gave better performance than the mixture of RAP+20% stone dust.

This means that the mixture with 10% stone dust gave sufficient amount of mastic in which the aggregate interlock within the system are enough and gave better strength of the mixture. Coupled with data in Table 3, it is revealed that the mixture of RAP+10% stone dust gave optimum asphalt content at 3% added asphalt by weight of total mixture, meanwhile the mixture of RAP+20% stone dust gave 3.5% optimum asphalt content. The stabilities of the mixtures well above the specification of Sand sheet, HRS and SMA mixtures (more than 600 kg). The rest of the Marshall properties are also met the specification of those mixtures for light and medium trafficked road, even more than minimum 800 kg stability for higher traffic loads [15].

The above data and analysis indicated that recycling RAP can be done by simply adding 10-20 % stone dust into the RAP. This simple method easily can be implemented on site. Some road paving works in community areas had been done semi manually using this method with satisfactory results in Bali.

3.5 The cantabro test results

The cantabro test results of the RAP+10% stone dust mixture was 9.02%, meanwhile the mixture with 20% stone dust gave 10.14%. This values are less then the recommended 16% at tropical temperature around 30°C [16]. Referring to figure 3 that is in line with figure 4, the mixture with 10% stone dust was denser, with added optimum asphalt content of 3%, has better resistance to aggregate raveling due to impact action.

3.6 Indirect tensile stiffness modulus (ITSM), dynamic creep, and fatigue

On these tests, different researchers may use slightly different method and/or test setting. These tests were done using testing machine (UTM30 Dynapave - by Control Italy). Direct comparison can be done with test results using the same machine and the same setting. The ITSM test results at 20°C are shown in Table 6, where it is shown that the mixture of RAP+10% stone dust gave higher stiffness. This is due to the mixture contain less mastic so the aggregate interlock is better.

Table 6. The ITSM test results.

Mixture	Avg. ITSM (MPa)
RAP+10% stone dust	5394,5
RAP+20% stone dust	4214,7

The ITSM value above are lower than the ITSM of mix done [17] tested using the same machine, UTM30 Dynapave. On that experiment the RAP was recycled with aggregate grading carefully proportioned and asphalt 80/100 pen were used. The ITSM was 7933 MPa [17]. This high stiffness was due to the aggregate grading was carefully proportioned in lab in line with the mid point grading of the Asphalt Concrete (AC) mixture, therefore the aggregate interlock were on its peak performance. Higher stiffness was also obtained when using recycled of old asphalt pavement as the bitumen had become harder. However this would give less fatigue performance [18].

3.7 The dynamic creep test results

The dynamic creep test results are shown in figure 5 at 40 °C. This test temperature was also used by other researcher [19]. The mixture with 10% stone dust gave lower deformation then the two other mixtures as it contains less mastic component. The dynamic creep line of AC-WC 60/70 pen asphalt with the RAP+20% stone dust intersect at about 1000 load cycles. At higher load cycles the later mixture deform higher. The test were done using the same machine and setting [20]. This is logical as the binder for the AC-WC mixture was 60/70 pen that was sufficiently hard, and designed with mid point aggregate grading specification, so the aggregate has proper aggregate interlock.

For creep slope analysis, the portion of the strain shown in figure 5 that gave linear shape was taken, i.e. the last four point from the results in figure 5. This is the secondary phase of typical repeated load creep test results [19]. Then a linear trend line was applied as shown in figure 6. The slope is the constant of x from the trend line equations of each mixture. It is obtained that the slope of the mixture with 10%

stone dust gave slope 0.2043 that is smaller than the slope of the mixture with 20% stone dust, i.e. 0.5962, but far higher than the slope of AC-WC 60/70 pen (i.e. 0.2673). Lower the slope indicates the better resistance of the mixture under dynamic creep loading (figure 6). As the slopes are < 0.5, and the test were done at 40 °C, the mixture can be classified suitable for heavy loads as shown in Table 1 [12].

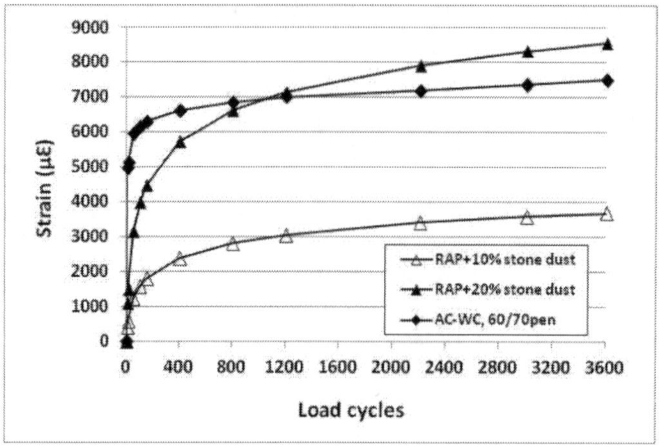

Figure 5. Creep test results.

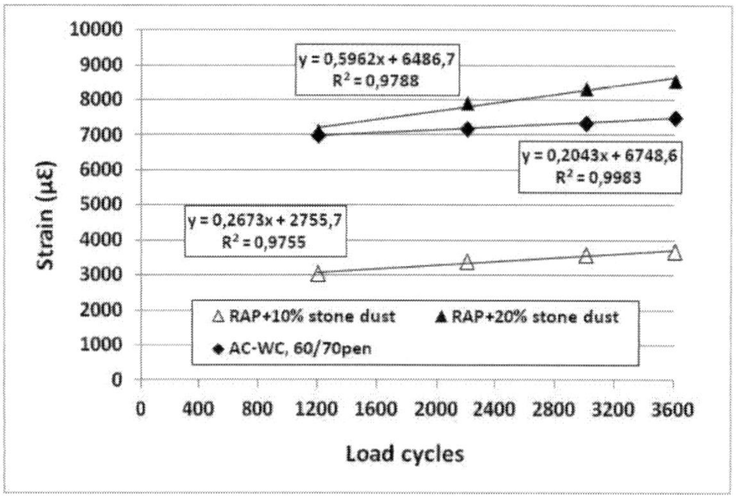

Figure 6. Dynamic creep slopes of the mixtures.

Fatigue test results meanwhile, are shown in figure 7, where there is no clear intersection among the fatigue lines. It is shown (if the lines extrapolated) that on any stress level applied the mixture with 10% stone dust gave slightly higher load cycles to failure then the mixture with 20% stone dust. The fatigue lines of the AC-WC 60/70 pen mixture lies between the recycled mixtures. These results gave similar trend when it is compared to other test results, where the mixture of RAP+10% stone dust had given better performances. Faster failure on fatigue repeated loading was also experienced [18], where the fatigue life of recycled asphalt mixture can be up to a half of those using natural materials. This can be affected by the type of asphalt used, asphalt content and the aggregate grading. For improving the fatigue life of recycled asphalt mixture, the use recycling agent may necessary [21].

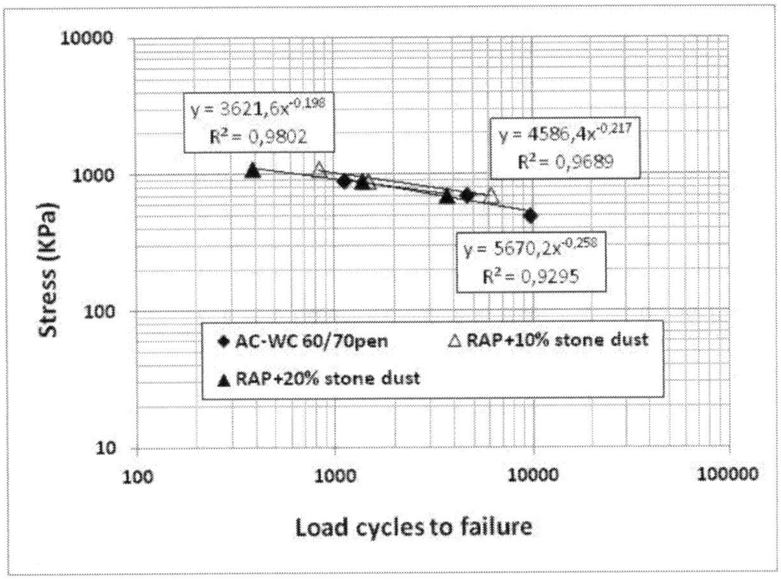

Figure7. Fatigue test results of the mixtures.

4 Conclusions

From the data and analysis, it can be concluded that RAP recycling can be done by simply adding 10-20% stone dust with properties met the minimum performances in line with the Indonesia specification. The stability of the mixture are well above the minimum 600 kg, for low to medium trafficked road. The stabilities even more than 800 kg which is designated for heavy trafficked load. The Marshall properties had also been sufficiently met. This method is simple and easier to apply. The aggregate grading of the final mixtures become slightly finer due to the present of additional mastic components.

References

[1] Thakur J K, Han J 2015 Recent Development of Recycled Asphalt Pavement (RAP) Bases Treated for Roadway Applications *Transp. Infrastruct. Geotech.* **2** pp 68–86. https://doi.org/10.1007/s40515-015-0018-7. https://link.springer.com/article/10.1007/s40515-015-0018-7 visited on 2nd April 2021.

[2] Rashid M, Mehta Y Y, Ali A, Yan B, Elroy Mc, Yin A E 2017 Environmental Impacts of Reclaimed Asphalt Pavement (RAP) *Final Report* May 2017 New Jersey Department of Transportation Bureau of Research And U. S. Department of Transportation Federal Highway Administration. https://www.njapa.com/wp-content/uploads/2017/05/EIRAP-Final-Report_FHWA_NJ-2017-008.pdf Visited on 29th March 2021.

[3] Abdo A M A 2016 Utilizing Reclaimed Asphalt Pavement (RAP) Materials in New Pavements - A Review. *The IJTEE* **12** (1): pp 61-66. DOI: 10.5383/ijtee.12.01.008. https://www.researchgate.net/publication/287800595_Utilizing_Reclaimed_Asphalt_Pavement_RAP_Materials_in_New_Pavements_-_A_Review visited on 30th March 2021.

[4] Pawar R, Patted A, Gowda K 2017 Recycling of Bituminous Pavement by RAP Method, in *Proceedings of International conference ICGCSC* ISBN: 978-93-5267-355-1 Mangalore Institute of Technology & Engineering, Moodbidri, India. https://www.researchgate.net/publication/315397515_Recycling_of_Bituminous_Pavement_by_RAP_Method visited on 30-03-2021.

[5] Thanaya I N A, Suweda I W, Putra G K 2017 Performance of Asphalt Concrete Wearing Course (AC-WC) Utilizing Reclaimed Asphalt Pavement from Cold Milling Bound with 80/100 Pen Asphalt, *Proceedings 15th International Conference on Quality in Research* (QiR), The Westin Resort Hotel, Nusa Dua, Bali, 24-27 July 2017.

[6] Yudistira I D G R 2019 Study on the Characteristics of Hot Asphalt Mixture Using Reclaimed Asphalt Pavement (RAP) Added With Mastic Component. *Master Thesis Report*. Master of Civil Engineering Program, Udayana University, Bali-Indonesia.

[7] International Mastic Asphalt Association (IMAA) 2020 *Mastic Asphalt.* https://mastic-asphalt.eu/mastic-asphalt#:~:text=Mastic%20Asphalt.%20Mastic%20asphalt%20is%20a%20mixture%20of,targeted%20adjustment%20of%20the%20composition%20of%20the%20mixture , visited on 10 August 2021.

[8] Nielsen C B 2006 Durability of Porous Asphalt-International Experience *Technical Note* **41**, Road Directorate, Danish Road Institute (DRI), Denmark.

[9] Strategic Highway Research Program (SHRP) 1994 *Stiffness of Asphalt-Aggregate Mixes* SHRP-A-388. National Research Council Washington DC.

[10] British Standard 2012 *Bituminous mixtures – Test methods for hot mix asphalt* Part **26**:Stiffness, BS EN 12697-26:2012.

[11] British Standard 2005 *Bituminous mixtures – Test methods for hot mix asphalt* Part **25**: Cyclic compression test, BS EN 12697-25:2005.

[12] Alderson A 1995 *Gyropac and Matta Training Course*. Australian: Australian Road Research Board Ltd.

[13] British Standard 2012 *Bituminous mixtures – Test methods for hot mix asphalt,* Part **24**: Resistance to fatigue, BS EN 12697-24 :2012 (2012).

[14] Indonesian Road and Bridge Research Center (IRBRC) 2016 The Influence of Rejuvenator on Continuous Graded Hot Mixed Asphalt Performance Using Reclaimed Asphalt Pavement. *R and B Journal*, Volume **33** No. 1 January-June 2016: pp 27-42. http://jurnal.pusjatan.pu.go.id/index.php/jurnaljalanjembatan/article/view/27/3, visited on 24th December 2020.

[15] Indonesian Ministry of Public Works and Community Housing (IMPWCH) 2018 *Specification for Road and Bridge Works.*

[16] Hamzah M O, Hasan M R M, Wan C N, and Abdullah N H A 2010 A Comparative Study on Performance of Malaysian Porous Asphalt Mixes Incorporating Conventional and Modified Binders, *Journal of App. Sci*. https://www.researchgate.net/publication/49591431_A_Comparative_Study_on_Performance_of_Malaysian_Porous_Asphalt_Mixes_Incorporating_Conventional_and_Modified_Binders , visited on 20th December 2020.

[17] Putra G K 2017 Performance of Recycled Concrete Wearing Course (AC-WC) From Cold Milling Bound With 80/100 Pen Asphalt *Master Thesis Report* Master of Civil Engineering Program Udayana University Bali-Indonesia.

[18] Sramek J 2018 Stiffness and Fatigue of Asphalt Mixtures for Pavement Construction *Slovak J of CE*, Vol. **26**, 2018, No. 2, 24 – 29, DOI: 10.2478/sjce-2018-0010. https://www.researchgate.net/publication/326155021_Stiffness_and_Fatigue_of_Asphalt_Mixtures_for_Pavement_Construction visited on 11th August 2021 http://onlinepubs.trb.org/onlinepubs/shrp/SHRP-A-388.pdf visited on 11th August 2021.

[19] Katman H Y, Ibrahim M R, Karim M R, Mashaan N S, and Koting S 2014 Evaluation of Permanent Deformation of Unmodified and Rubber-Reinforced SMA Asphalt Mixtures Using Dynamic Creep Test *Journal of Adv. in Mat. Sci. and Eng.* https://www.hindawi.com/journals/amse/2015/247149/ visited on 13th August 2021.

[20] Thanaya I N A, Purbanto I G, Nugraha I N S 2016 Study on the Characteristics of Asphalt Concrete-Wearing Course (AC-WC) Mixture Using Asphalt pen 60/70 added with Latex *Journal of CEMC* Volume **22** No 2 December 2016 ppp 77-86, ISSN 0954-1809.

[21] Kaseer F, Yin E A, Mercado, Martin A E 2017 Stiffness Characterization of Asphalt Mixtures with High Recycled Material Content and Recycling Agents, *TRR: JTRB* visited 15th December 2021. https://journals.sagepub.com/doi/10.3141/2633-08

Study of Clean Water and Sanitation Access and Its Relationship to Waterborne and Stunting Prevalence in Bandung Regency

Putri S Kamila[1], **Indah R S Salami**[1,*]

[1]Environmental Engineering Program, Faculty of Civil and Environmental Engineering, Bandung Institute of Technology, Jalan Ganesha no. 10 Bandung 40132, Indonesia

shafakam@students.itb.ac.id, indahrss@tl.itb.ac.id

Abstract. Water and proper sanitation access in Bandung Regency has not covered all population. Lack of availability of water and sanitation is certainly related to public health. This research was conducted to identify clean water access in terms of quality, quantity, and continuity as well as sanitation access in Bandung Regency, and then to determine its relationship to waterborne diseases and stunting prevalence. The study was conducted in three locations representing domestic, agricultural, and industrial areas. The study collected 33 water samples for physical and microbial analysis, and questionnaires to 75 household respondents about the waterborne disease and stunting prevalence, household income levels, hygiene behaviour, mother's education, and birth weight. The results showed clean water quality differences in those areas for parameters Total Coliform (p=0,051), temperature (p=0,044), and pH (p=0,002) as well as differences of wastewater system types (p=0,002). Factors namely clean water quantity and continuity, personal hygiene, and household income level have an association with waterborne diseases prevalence (PR>1). Whereas stunting is associated with sanitation access, personal hygiene, household income level, mother's education, and children's birth weight (PR>1). However, only household income level has significant relationship with stunting (p=0,005; PR=4.0; 95% CI=1.33-12). These factors should be improved to prevent waterborne disease and stunting children in the community.

Keywords: Clean water, sanitation, public health, stunting, waterborne diseases

1. Background

The rights of the Indonesian people to have access on clean water are guaranteed and protected by the state in terms of quality, quantity and continuity as stated in Article 8 Law No. 122 of 2015 and Article 4 Law No. 17 of 2019 [1, 2]. Indonesia has committed to achieving the Sustainable Development Goals (SDGs) in which the 6th Goal is related to clean water and sanitation, so all Indonesian must have access to safe and affordable sanitation and drinking water.

Bandung Regency is one of the areas that still have difficulty with clean water. Access to adequate drinking water has reached 83% in 2020, but water continuity and quality are still not guaranteed because 26 of 31 sub-districts are prone to drought, while in rainy season, some areas also have difficulty to get clean water due to flooding [3]. Access to adequate sanitation in Bandung Regency is considered good and above the national average, however there are some areas that have inadequate sanitation facility. Lack availability of safe water and sanitation have strong link to public health, including waterborne diseases whose impact can even cause death [4,5]. In addition to waterborne diseases, unimproved toilets and untreated drinking water can increase the risk of stunting in children [5,6] and the impact of stunting lasts for the next life or generation [7].

Based on those problems, research is needed to evaluate the clean water and sanitation access and its relationship to public health specifically in Bandung Regency. The objectives of this research are: (i) To determine the differences of clean water and sanitation access in studied locations; and (ii) To determine the relationship and risk of clean water and sanitation access, hygiene behaviours, and household economic levels with the prevalence of waterborne diseases and stunting in children. This information or results can be useful in overcoming the problem of stunting and waterborne diseases.

The research was conducted in 3 districts in Bandung Regency, West Java as shown in Figure 1. Those studied area are Margaasih District as a domestic area with predominantly used for human settlements, Dayeuhkolot District represents as an industrial area occupied by numerous industries, and Ciparay District as an agricultural area [8,9,10].

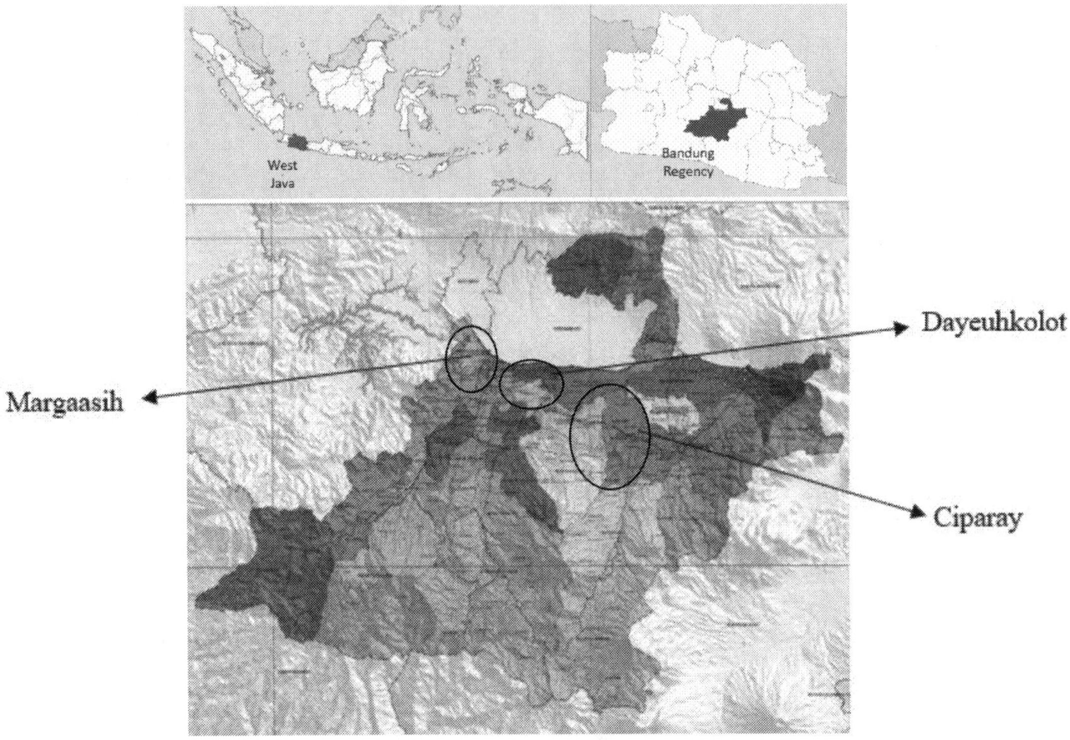

Figure 1. Study locations in Bandung Regency, West Java Province, Indonesia [21, 22, 23]

2. Materials and methods

The study collected primary data of water and sanitation condition and socioeconomic status with questionnaire to 75 respondents selected by simple random sampling method as calculated by Cochran equation with margin of error 10% [11] and distributed evenly in each area of 25 respondents. Moreover, clean water sampling was conducted to test its quality, specifically in-situ parameters (DO, temperature, pH, and turbidity) as well as microbiological parameters of *Total Coliform* and *E. coli* that were analysed according to EPA [12], The total water sample size was 33 water sources as calculated by Lead and Copper Rules (LCR) requirements [13]. The water samples were collected from clean water source in the respondents' house, that used water well or pump from individual well. In-situ parameters are analysed by Lovibond TB 211 IR (turbidity meter), Lutron PDO-520 (DO meter), and Lutron PH-222 (pH meter), while the microbiological test is conducted in the laboratory using membrane filter method [15]. Water samples for microbial analyses is collected in sterilised bottles and procedures. The quality of water then is compared with water quality standard for hygiene and sanitation stated in the Minister of Health Regulation No. 32 of 2017 [16]. Stunting status is determined by measuring the weight and height of 78 children's respondents. The children respondent is obtained from the household sampled, with the age below 5 years. There are houses with more than one child in the family. WHO defines the stunting condition by the category as follows:

Height-for-age z-score < -2SD as stunting condition, and

Height-for-age z-score < -3 SD as severe stunting.

Data from the questionnaire and laboratory are then analysed statistically using the Chi-square test, one-way Anova, or Kruskal-Wallis test. The PR (prevalence ratio) value is calculated based on 2x2 contingency table comparing prevalence of waterborne disease that is represented by diarrheal symptoms of the respondents during the last 3 months—Table 1 shows the contingency table for PR calculation.

Table 1. Contingency table 2 x 2 for Prevalence Ratio calculation

	With outcome	Without outcome
Exposed	a	b
Not exposed	c	d

$$PR = \frac{a/(a+b)}{c/(c+d)} \quad (1)$$

The outcomes are respondent with either waterborne disease or stunting children, and the exposures are from the risk factors tested in the study such as clean water access (quality, quantity, and continuity); sanitation access; hygiene behaviour (hand washing with soap); household income; mother's education and birth weight.

3. Results and discussions

3.1. Difference of clean water condition

The results showed that all areas have access to clean water with quantity and continuity that fulfil almost all respondents daily needs and is available at any time. However, from 33 sources of clean water samples, only 3 water sources that meet the water quality for hygiene and sanitation standard.

The water quality measurement showed that parameter temperature is in the range of 25.7°C - 33°C (27.4±1.7°C); pH 6.4-7.5 (7±0.4); DO 2.1-5.5 mg/ (3.5±0.8 mg/L); *Total Coliform* 10-4000 CFU/100 mL (1209±1331 CFU/100 mL); *E coli* 0-900 CFU/100 mL (225.2±217.3 CFU/100 mL); and turbidity 2.7-298 NTU (23.5±61.5 NTU). These results show some parameters exceeding the water quality standard for hygiene and sanitation, namely pH in one sample in Margaasih; turbidity in one sample in Dayeuhkolot, and two samples in Ciparay. *Total Coliform* and *E. Coli* are found in most of samples. Clean water sources in agricultural area tends to have higher turbidity and contain higher *Total Coliform* than other areas, while domestic area tends to have lower turbidity than other area as shown in Figure 2 and Figure 3. Moreover, Table 2 shows statistical comparison of water quality parameters in agricultural, industrial, and domestic areas and the result shows that the temperature, pH, and Total Coliform are the parameters with significant differences.

Based on the secondary data obtained, it is evaluated that piped drinking water access for Bandung Regency as government's target is only 4.1% for Margaasih District, 33.5% for Ciparay District, and 42.7% for Dayeuhkolot District. This means the access to piped drinking water in the three areas has not reached 50% with the lowest percentage in Margaasih District.

3.2. Difference of sanitation condition

The results showed that all areas had access to sanitation with improved latrine and almost all of them already had their own latrine facilities. Based on the type of the wastewater treatment system in each area, it was found, there was a significant difference in sanitation access condition (p=0.002). Figure 4 shows the percentage of households with a certain type of sanitation system (SPAL) in each sub-district.

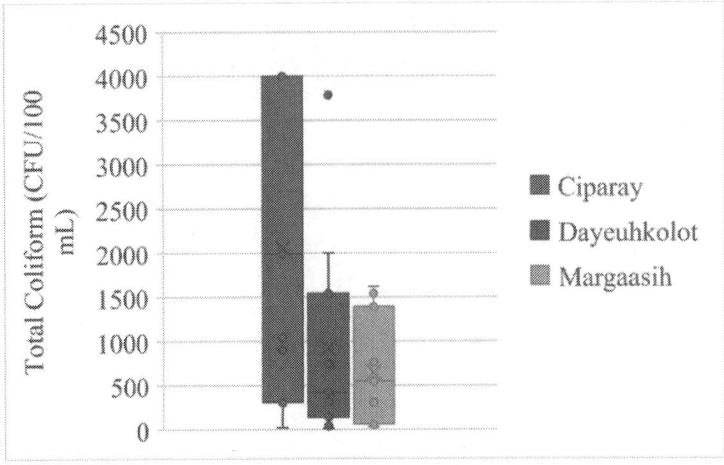

Figure 2. Total Coliform in water samples in each studied area

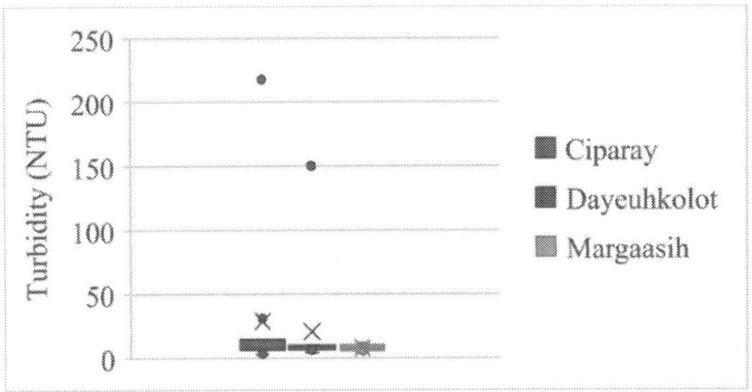

Figure 3. Turbidity of water samples in each studied area

Table 2. The statistic result of water quality parameters differences in agricultural, industrial, and domestic area

Water quality parameters	p value
E. coli	0.128
Total Coliform	0.051
pH	0.002
Temperature	0.044
DO	0.187
Turbidity	0.707

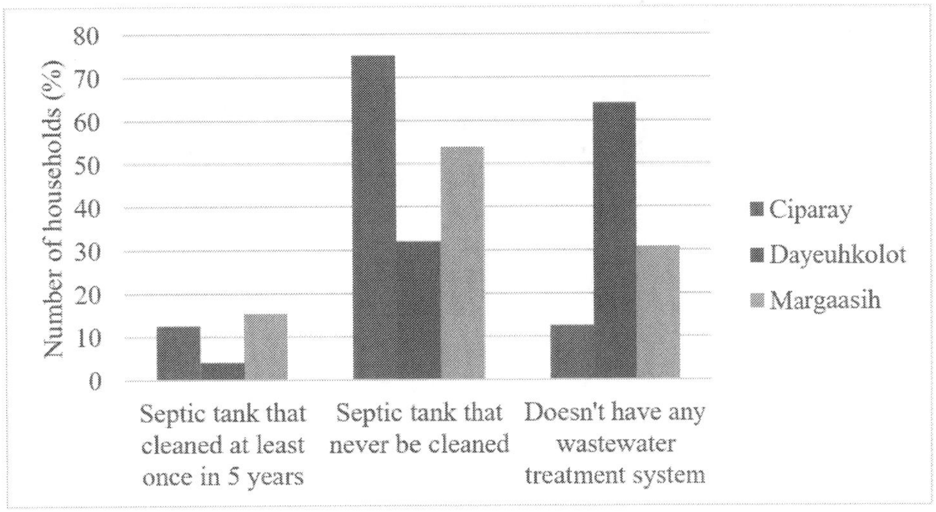

Figure 4. Percentage of households with certain wastewater system type (SPAL) in each area studied

Based on the secondary data obtained, it is obtained that the sanitation infrastructure built by the government in the three sub-districts are developed every year. However, when compared to the number of existing households, the total infrastructure of communal toilet (MCK), wastewater system (SPAL), and the existing system in the three sub-districts are still very small. However, the data of sanitation infrastructure target which will be built by the government in Bandung Regency is not available.

3.3. Waterborne diseases and stunting prevalence

The results showed that around 6.67% of the population had suffered from waterborne diseases in March-May 2021, while the other 93.33% had no complain of diarrhoea. There were 5 respondents who had diarrhoea where 2 of them had also skin infection disease. In addition, it was found that 13 children with severe stunting, 15 children were considered stunted, and 51 others were normal. Figure 5 and Figure 6 show the prevalence of stunting and waterborne diseases in each area.

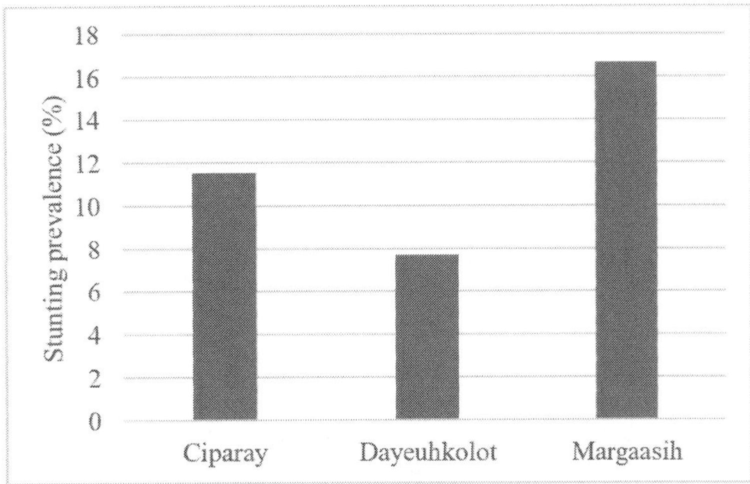

Figure 5. Stunting prevalence in each area

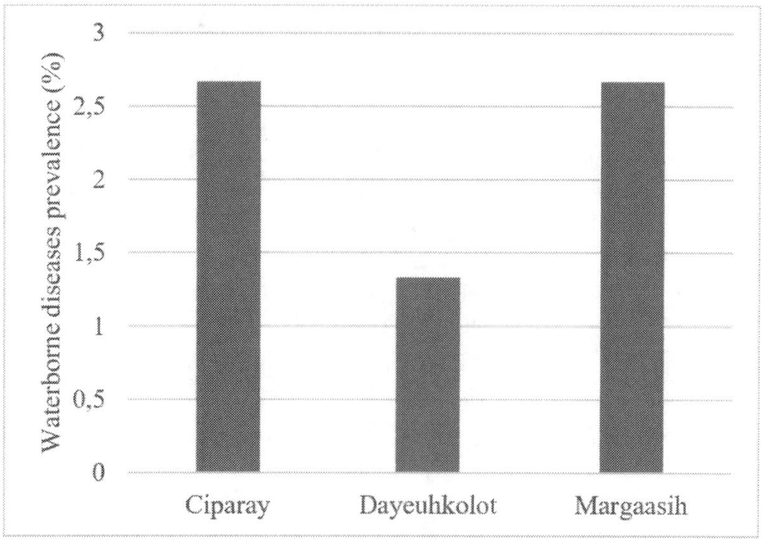

Figure 6. Waterborne diseases prevalence in each area

3.4. Clean water access and its relationship to waterborne diseases and stunting

The results showed that all the respondents already had access to clean water and its sources are categorised as protected water sources. For the quantity, 96% of clean water has met the daily needs of the community whereas for continuity, 8% of clean water sources are not available 24 hours continuously, both in rainy and dry seasons. In the shortage of water, the respondents need to collect and retain water when clean water is available. Based on the water quality data, only 6.67% of clean water sources meet the water quality standard. The statistical relationships of clean water to waterborne diseases and stunting are described in Table 3 and Table 4.

Table 3. Relationship of clean water condition and waterborne diseases

Independent variables	p value	PR	95% CI
Quality	1.000	0.9	0.87-0.99
Quantity	1.000	1.1	1.01-1.15
Continuity	1.000	1.08	1.01-1.15

Table 4. Relationship of clean water condition and stunting

Independent variables	p value	PR	95% CI
Quality	0.153	0.6	0.51-0.74
Quantity	1.000	0.9	0.18-4.7
Continuity	1.000	0.9	0.28-2.9

Based on Table 3 and Table 4, access to clean water in terms of quality, quantity and continuity do not show significant relationship with the prevalence of waterborne diseases and stunting. The relationship of water quantity and continuity to waterborne disease, also water quality for stunting have PR > 1 that may have association but not statistically significant. These variables are the risk factors of waterborne diseases and stunting occurrence in the studied areas. PR < 1, on the contrary such as the quality of clean water are included as protective factors of waterborne diseases occurrence. Similarly, the quantity, quality, and continuity of clean water are included as protective factors for stunting. This may occur because the water samples are limited and not representative. Clean water itself is included as an indirect factor for stunting. There are other direct factors which are related to stunting but are not included in this study. Firmanu (2016) states that environmental factors such as water and sanitation do not stand alone, but infectious disease factors and parenting patterns influence the incidence of stunting simultaneously [17]. Vilcin et al. (2018) reviewed 71 reports globally about stunting, and the findings reported factors associated with increased risk of childhood stunting are foodborne mycotoxins, a lack of adequate sanitation, poor quality cooking fuels, dirt floors in the home, and inadequate local waste disposal. The review also stated that a large number of research have studied the association of stunting and the access to safe water sources, but the results remain inconclusive due to inconsistent study findings [20].

3.5. Sanitation access and its relationship to waterborne diseases and stunting

The entire respondents had access to latrine facilities with improved latrine. Around 93.33% is self-owned latrine and the other 6.67% is jointly owned latrine. The wastewater disposal system facilities possessed by the community are 52% septic tanks, 12% are septic tanks which are vacuumed regularly at least every 5 years, and the other 36% do not have wastewater disposal facilities. Table 5 shows the result the relationship of sanitation access to waterborne diseases and stunting.

Table 5. Relationship of sanitation access to waterborne and stunting

Dependent variables	p value	PR	95% CI
Waterborne diseases	0.648	0.4	0.05-3.78
Stunting	0.372	1.4	0.78-2.58

There is no statistically significant relationship between access to sanitation and the prevalence of waterborne diseases and stunting. Access to sanitation is not a risk factor for stunting, but it is a protective factor for waterborne diseases. About 74% of the respondents already had access to proper sanitation, so the prevalence of diarrhoea is perhaps low. It may involve other factors that are not considered in the study such as food security or nutrition in each household. However, for people with lower access to sanitation or unimproved sanitation, they would have 1.4 times greater risk of stunting children than those who already have good access to improved sanitation.

3.6. Hygiene behavior and its relationship to waterborne diseases and stunting

According to questionnaires, 97.33% of respondent wash their hands using soap and running water often, especially before eating and after defecating and other 2.67% washing their hands with soap occasionally. This means that respondents are used to have good personal hygiene. This may also trigger by the COVID-19 pandemic which makes people pay more attention to health and personal hygiene at home. The statistical relationships of personal hygiene behaviour to waterborne diseases and stunting are shown in Table 6. Although the p-values are high, but the PR > 1, that there are possible association between personal hygiene, in this case is washing hand by soap, and waterborne disease and stunting. Public awareness of good handwashing is already high, so the incidence of waterborne diseases, especially diarrhoea is low. Better significant statistical results can be obtained if the samples are larger. In addition, other risk factors such as nutrition, other hygiene behaviours may be considered in further study. This study showed that the respondents who occasionally wash their hands with soap are having 1.1 times greater risk of getting waterborne diseases and 1.4 times higher risk of stunted children than those who often wash their hands with soap.

Table 6. Relationship of personal hygiene behaviour to waterborne disease and stunting

Dependent variables	p value	PR	95% CI
Waterborne Diseases	1.000	1.1	1.01-1.14
Stunting	1.000	1.4	0.34-5.8

3.7. Household income level and its relationship to waterborne diseases and stunting

The percentage level of household income in all locations of study is depicted in Figure 7 and the relationships of household income level to waterborne diseases and stunting is available in Table 7. Majority of respondents have income in the range of IDR 1 million to 3 million. Only 13% have income less than IDR 1 million. It is shown in Table 7 that household income level has a significant relationship with the prevalence of stunting in children under five and it is considered as a risk factor for stunting occurrence. For waterborne diseases, the PR value is also higher than 1, and showing an association with the waterborne disease, but not statistically significant.

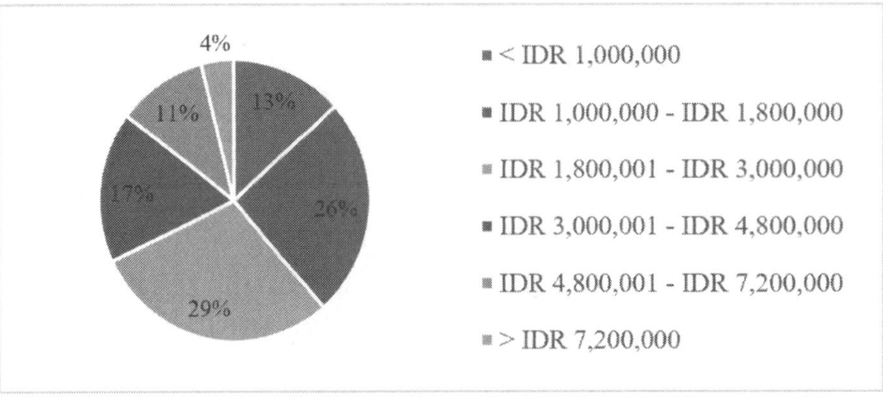

Figure 7. Percentage of household income level in all locations of the study

Table 7. Relationship of household income level to waterborne disease and stunting

Dependent variables	p value	PR	95% CI
Waterborne diseases	0.659	2.0	0.2-16.9
Stunting	0.005	4.0	1.33-12

In general, the socioeconomic level of the household is related to the purchasing ability of a family. The ability of a family to support their daily needs depends on the number of family's income, including the ability to buy food ingredients that can affect the fulfilment of nutrition in the children's growths, so this is certainly related to the health of children, especially the incidence of stunting in children [18]. For respondents whose household income level is below the minimum average, it is estimated to have 2.0 greater risks of getting waterborne diseases and 4.0 greater risk for stunting in children under five in their homes, as compared to those whose household income levels are above the minimum average. Argaw et al. (2019) studied the health data in 14 low- and middle-income countries and concluded that the stunting condition in under 5 years children showing a declining trend in these countries by 1.04 percentage points. The reduction of stunting prevalence should consider the economic development and nutrition-sensitive interventions, beside nutrition-specific programs [19].

3.8. Mother's education, birth weight of children under five, and its relationship to stunting

The percentage of education that was taken by mothers of children under five in the research area is in Figure 8. The education level of the mothers is dominantly senior high schools. Other factor considered in this study is the birth weight of the children.

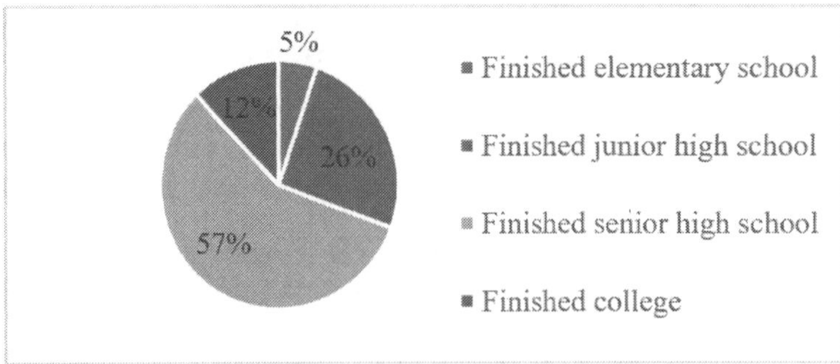

Figure 8. Percentage of mother's education

Lower birth weight can increase the risk of children being stunted. The results showed that 11% of children under five had a birth weight of less than 2.5 kg (classification of low birth weight), and another 89% of children under five had a birth weight more than or equal to 2.5 kg. The relationships of mother's education and birth weight of the children under five to stunting were explained in Table 8.

Table 8. Relationship of mother's education and birth weight of children under five to stunting

Independent variables	p value	PR	95% CI
Mother's education	0.130	2.2	1.15-4.26
Birth weight of children under five	0.128	1.9	1.01-3.6

The results showed that there is association between the mother's last education and the child's birth weight with the prevalence of stunting, but not statistically significant. It is possible that mothers already have sufficient knowledge about children nutrition so that nutritional intake in the first 100 days, for example by providing exclusive breastfeeding, adequate complementary feeding (MPASI) and complete immunization. However, it is calculated that mothers whose educational background has not finished junior high school have 2.2 times greater risk of stunting child compared to mothers whose educational background has graduated from junior high school or higher. In addition, children with low birth weight, they have 1.9 times greater risk of stunting than those babies with normal birth weights.

4. Conclusion

In all studied area the respondents already had access to clean water with protected water sources. However, the quality, quantity, and continuity are affected by seasons. Most water samples comply the water quality standard for hygiene and sanitation, except for pH, turbidity in one to two samples from Margaasih, Dayeuhkolot, and Ciparay. The *Total Coliform* and *E. Coli* are detected in most of water samples. All three location respondents had access to sanitation with improved latrine and almost all of them already had their own latrine facilities. There are differences of clean water quality condition (especially for temperature, pH, and *Total Coliform* parameters), and access to sanitation condition (especially the type of wastewater system) in agricultural, domestic, and industrial studied areas. The waterborne disease prevalence is found in 5 respondents (6.67%), and stunting children are detected in 36% children, with 17% are considered severe stunting. Areas of agricultural and residential have higher prevalence of waterborne and stunting children.

Relationship of studied factors to waterborne and stunting is represented by PR higher than 1. Waterborne disease has the association with limited water quantity (PR=1.1); water continuity (PR=1.08); personal hygiene (PR=1.1); and household income level (PR=2.0). Stunting prevalence is associated with sanitation access (PR=1.4); personal hygiene (PR=1.4); household income level (PR=4.0); mother's education (PR=2.2); and birth weight (PR=1.9). However, only household income level that significantly associated with stunting (PR=4.0, p=0.005, 95% CI of 1.33 – 12). To prevent any prevalence of waterborne disease and stunting children in studied locations, it is suggested to consider improving these related factors condition in the community.

Acknowledgement
The research has been supported by P2MI ITB 2021 research grant.

References

[1] Government Regulation No. 122 of 2015, *Drinking Water Supply System*
[2] Republic Indonesian Law No. 17 of 2019, *Water Resource*
[3] Heru S N, Prihartanto, and Deliyanti G 2019 *Jurnal Teknologi Lingkungan* **20(1)** 1-13 *(in Bahasa Indonesia)*
[4] Nwabor O F, Nnamonu E I, Martins P E, Ani O C 2016 *International Journal of TROPICAL DISEASE & Health* **12(4)** 1-14
[5] Harriet T, Aidan A C, Susy K S, Robin N 2016 *BMC Public Health* **16-11** 1-11
[6] Richard D S, Regina M, Kai S 2011 *Food and Nutrition Bulletin* **32(4)** 347-353
[7] Tim Nasional Percepatan Penanggulangan Kemiskinan 2018 *Gerakan Nasional Pencegahan*

Stunting dan Kerjasama Kemitraan Multi Sektor (National Action on Stunting Prevention and Multisectiral Cooperation)
[8] Badan Pusat Statistik 2020 *Kecamatan Margaasih Dalam Angka 2020 (Statistic Indonesia Bureau)*
[9] Badan Pusat Statistik 2020 *Kecamatan Ciparay Dalam Angka 2020 (Statistic Indonesia Bureau)*
[10] Badan Pusat Statistik 2020 *Kecamatan Dayeuhkolot Dalam Angka 2020 (Statistic Indonesia Bureau)*
[11] Sugiyono 2019 *Metode Penelitian Kuantitatif Kualitatif dan R&D (Quantitative and Qualitative Research Methods and R & D).* Alfabeta Bandung, Indonesia
[12] U.S. Environmental Protection Agency 2016 *Quick Guide to Drinking Water Sample Collection*
[13] U.S. Environmental Protection Agency 2010 *Revised Lead and Copper Monitoring Reporting Guidance for Public Water Systems*
[14] U.S. Environmental Protection Agency 2018 *In Situ Water Quality Monitoring*
[15] Cara G, Nick G 1997 *The Coliform Index and Waterborne Disease.* London (UK: E & FN Spon)
[16] Minister of Health Regulation No. 32 of 2017, *Environmental Health Quality Standards and Water Health Requirements for Sanitation Hygiene Purposes, Swimming Pools, Solus Per Aqua, and Public Bath*
[16] Minister of Health Regulation No. 32 of 2017, *Environmental Health Quality Standards and Water Health Requirements for Sanitation Hygiene Purposes, Swimming Pools, Solus Per Aqua, and Public Bath*
[17] Firmanu C, Stefanus P M, Intje P 2016 *J Gizi Pangan* **11(1)** *9-18 (in Bahasa Indonesia)*
[18] Gladys A, Sandra F 2018 *Media Litbangkes* **28(4)** 247-256 *(in Bahasa Indonesia)*
[19] Alemayehu Argaw, Giles Hanley-Cook, Nathalie De Cock, Patrick Kolsteren, Lieven Huybregts and Carl Lachat 2019, *Nutrients*, **11**, 2485
[20] Dwan Vilcins, Peter D. Sly and Paul Jagals. 2019, *Annals of Global Health.* **84(4)**, 551–562
[21] https://commons.wikimedia.org/wiki/File:West_Java_in_Indonesia.svg
[22] https://commons.wikimedia.org/wiki/File:Locator_kabupaten_bandung.png
[23] https://petatematikindo.wordpress.com/2014/12/07/administrasi-kabupaten-bandung/

IOP Conf. Series: Earth and Environmental Science 1065 (2022) 012040 doi:10.1088/1755-1315/1065/1/012040

Numerical Study on Connections of Replaceable Shear Link with Inserted Pins

M Moestopo[1,2], **D Kusumastuti**[1,2], **S Wibowo**[3]

[1] Faculty of Civil and Environmental Engineering, Bandung Institute of Technology, Bandung, 40132, Indonesia
[2] Research Center for Industry, Bandung Institute of Technology, Bandung, 40132, Indonesia
[3] Graduate Student, Bandung Institute of Technology, Bandung, 40132, Indonesia

Corresponding author : moestopo@si.itb.ac.id

Abstract. The link end plate of replaceable shear link in eccentrically braced frames has been modified with the 'inserted-pins' to ensure the high seismic performance and to reduce the number of bolts in its slip-critical connections. The maximum shear force in the connection is resisted by the pins, while the bolts resist the bending moment at the link-ends. Numerical work has been conducted using a non-linear ABAQUS program to study the deformation as well as the stress distribution of the shear-link and its connection elements: end plates, bolts, pins, and beam-outside the link. The results showed an adequate performance of the proposed modified shear link connections, where the end plate of the beam outside the link remained elastic as the shear-link reached 0.08 rad plastic rotation due cyclic loading.

Key-words : shear-link, replaceable, connection, inserted-pin, earthquake resistant structure

1. Introduction

Eccentrically braced frames (EBF) have shown its good performance and have been widely used as earthquake resistant structure. Previous studies [1-4] have indicated that the use of short shear link in EBFs with proper detailing can provide a stable and good energy dissipation as well as the seismic performance of EBFs.

The replaceable link with bolted-connection was proposed by Mansour et al. [5] to improve the performance of the conventional EBFs with a welded-connections that are difficult to repair after an earthquake occurs. The use of replaceable link also provides an improvement in constructibility of EBFs since the link can be installed using bolted-connection to the pre-fabricated column-brace-beam assemblies of the multi-story buildings.

The replaceable link is often supported by a slip-critical bolted-connection to sustain high shear force, V_{link} as well as high bending moment, M_{link} at the link-ends due to cyclic loading (Figure 1). The shear capacity of the slip-critical connection is, in general, less than the shear capacity of bearing-type connection, but is required to provide a good dissipation energy without slip during a cyclic loading due to earthquake. The limited size required for the end-plate of the link could limit the use of the high number of bolt in the slip-critical connection, especially for shear link connection that should withstand high shear of the yielding-link with a significant strain hardening. This study proposes a modified bolted-connection to improve the performance of a replaceable shear link.

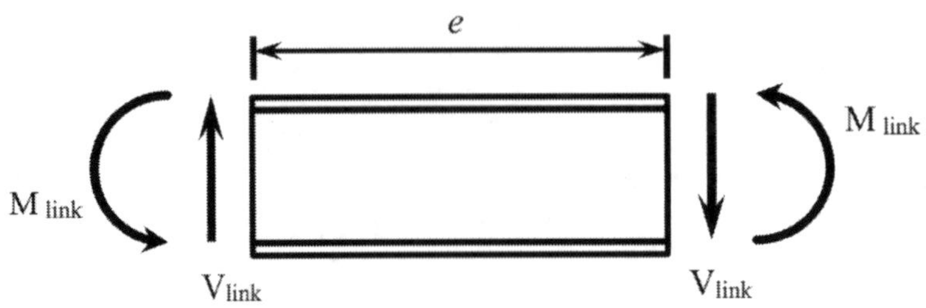

Figure 1. Link and Forces.

2. Modified Connection For Replaceable Shear Link

In this study, a modified bolted-connection is proposed for a replaceable shear link by inserting steel pins between the end-plate of the link and beam outside the link (Figure 2). Jie et al. [6] used a pin or shear key in a steel coupling beam of high-rise building and showed a good performance as a replaceable coupling beam. In this modified connection, there is no gap between pins and end-plate is assumed so that the pins is expected to resist most of the high shear force in the link connection. Meanwhile, the bolt is expected to resist the bending moment by the tensile force developed in bolt, that in general is less for shear link as compared to flexural link.

In a bearing-type bolted connection, the insertion of pins is expected to resist the total shear force in the shear link connection, so that the bolts only resist the tensile force due to bending moment at the link-end, and in turn could reduce the number of bolts in the end-plate. In a slip-critical bolted-connection, only some part of shear force is resisted by the pins and the other part of shear force is resisted by the friction between bolt-head and end-plate that in turn will reduce the capacity of the bolt to resist the tension force due to the bending moment.

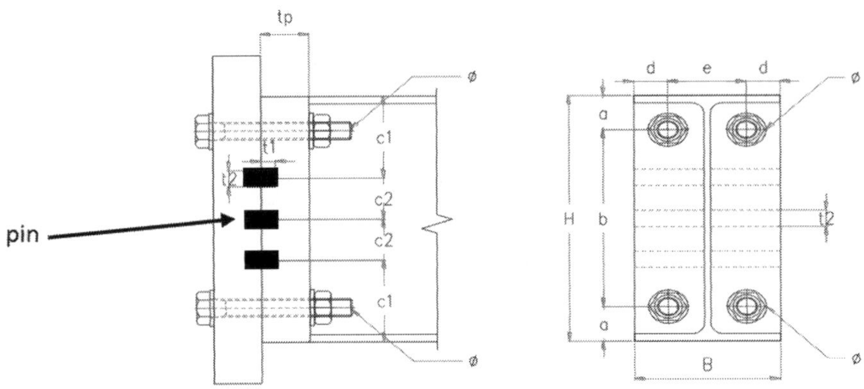

Figure 2. Modified Link Connection.

3. Numerical Study

This study investigated the performance of the replaceable shear link made of built-up I-section with the modified connection as shown in Figure 2 and Table 1. Link web is made of A36 steel with specified yield stress of 250 MPa, while the flange, the end-plate and the inserted pins are made of higher steel grade, i.e. A572 Gr.50 with specified yield stress of 345 MPa. According to the AISC 341-16, the value of Ry for steel plates are taken as 1.3 and 1.2 for A36 and A572 Gr.50, respectively. The 22-mm diameter of F568M 12.9 bolt is used in the modified connection. The yield strength, tensile strength and the elongation of the materials are shown in Table 2.

Table 1. Dimension of Modified Link Connection.

H (mm)	B (mm)	Ø (mm)	a (mm)	b (mm)	c1 (mm)	c2 (mm)	d (mm)	t1 (mm)	t2 (mm)	tp (mm)
300	150	22	42	216	100	50	35	15	20	50

Table 2. Material Specification.

Part	Specification	Yield strength (MPa)	Tensile Strength (MPa)	Elongation (%)
Web	A36	325	576	23
Flange	A572 Gr.50	380	540	20
End Plate	A572 Gr.50	380	540	20
Pin	A572 Gr.50	380	540	20
Beam	A572 Gr.50	380	540	20
Bolt	F658M 12.9	1,100	1,220	8

The shear link was modeled with fixed support at one end and the other end can move (with moment-restraint) due to transverse loading. Two models of modified connection were investigated to represent two fastening methods i.e. slip-critical model and bearing-type model. A non-linear finite element analysis (ABAQUS) using 8-nodes 3-D element with reduced integration point (Figure 3) and combined hardening was carried out to analyze the deformation and the stress distribution of the link and its modified connections subjected to a cyclic loading up to 0.09 radian as shown in Figure 4.

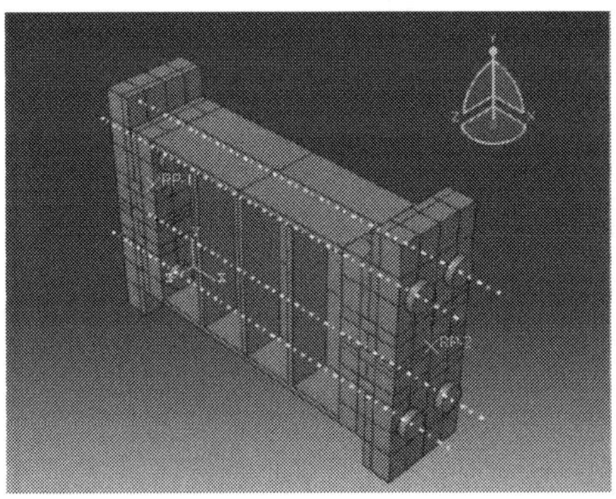

Figure 3. Model of Modified Connection.

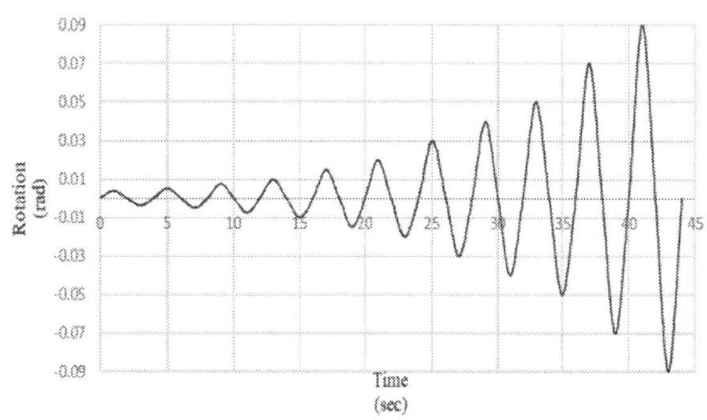

Figure 4. Cyclic Loading Pattern.

4. Result and Discussion

4.1 Load - displacement

The Figure 5 shows the shear force vs the vertical displacement difference between link-ends of modified link connections due to cyclic loading, for link with slip-critical connection (Figure 5a) and bearing-type connection (Figure 5b), respectively. The total link rotation of each model is shown in Figure 6. The plastic deformation and the plastic rotation of each model is shown in Figure 7 and Figure 8 respectively.

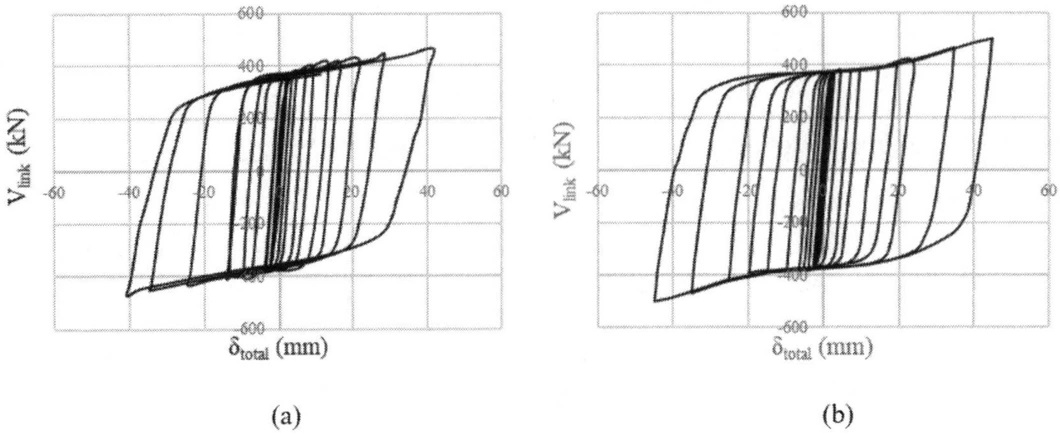

Figure 5. Link Shear and Link Vertical Displacement. (a) Slip-critical (b) Bearing-type

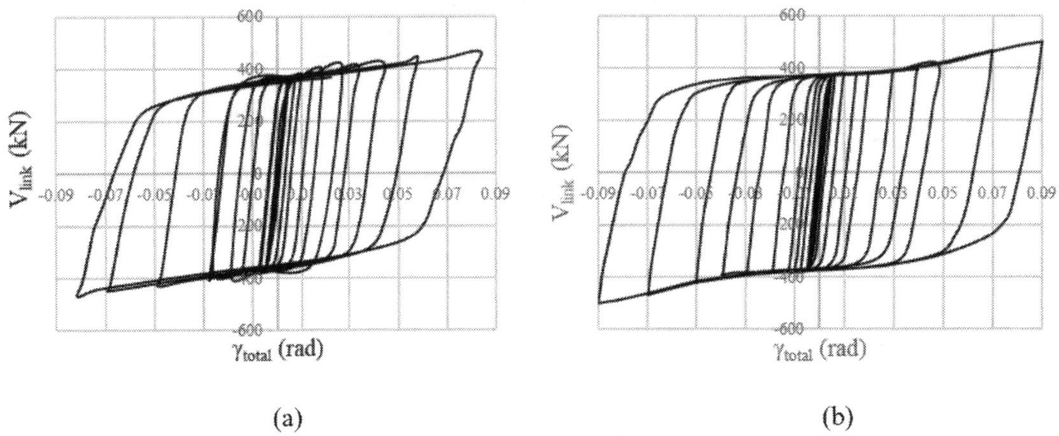

Figure 6. Link Shear and Total Link Rotation. a. Slip-critical b. Bearing-type

Table 3 and Table 4 show the displacement and the rotation of the link at the first yield of each component during the cyclic loading. The further observation and discussion is limited until the link plastic rotation reached 0.08 radian, as specified in the AISC Seismic Provision [7].

Table 3. Load and Displacement for Slip-critical Connection.

Component	Yield Stress (MPa)	V_{link} (kN)	δ_{total} (mm)	γ_{total} (rad)	$\delta_{plastic}$ (mm)	$\gamma_{plastic}$ (rad)
Web	325.30	335.93	1.87	0.00	0.00	0.00
Flange	379.70	415.59	16.99	0.03	14.68	0.03
End-plate	379.70	468.78	41.74	0.08	39.12	**0.08**
Bolt	1,100.00	469.35	40.81	0.08	38.20	**0.08**
Pin			Not yield			
Beam			Not yield			

Table 4. Load and Displacement for Bearing-type Connection.

Component	Yield Stress (MPa)	V_{link} (kN)	δ_{total} (mm)	γ_{total} (rad)	$\delta_{plastic}$ (mm)	$\gamma_{plastic}$ (rad)
Web	325.60	338.45	1.87	0.00	0.00	0.00
Flange	379.60	379.18	4.75	0.01	6.62	0.01
Bolt	1,105.00	462.63	33.91	0.07	32.04	0.06
End-plate	379.80	468.00	34.91	0.07	33.04	0.07
		487.35	40.63	0.08	38.76	**0.08**
Pin		Not yield				
Beam		Not yield				

The first yield of the web for both modified connections are very close, i.e 335.93 kN and 338.45 kN. The shear force of each modified connection at the plastic rotation of 0.08 radian reached 468.78 kN for slip-critical connection and 487.35 kN for bearing-type connection. This gives the overstrength values of 1.39 and 1.44 for shear link, respectively. Both modified connections show adequate ductility and strength.

4.2 Yielding sequence

Figure 7 and Figure 8 shows the plastic deformation and the plastic rotation of the shear link and also indicates the yielding of shear link and its components, i.e. web, flange, end-plate, bolt, and pin.

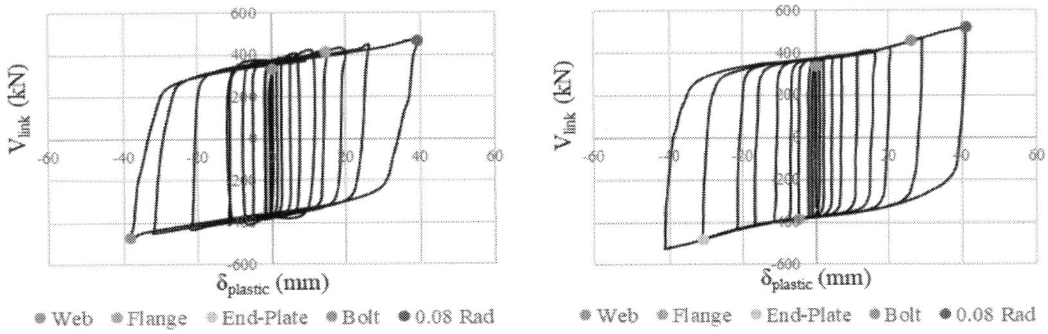

(a) (b)

Figure 7. Plastic Displacement and First-yield of Modified Connection. a. Slip-critical b. Bearing-type

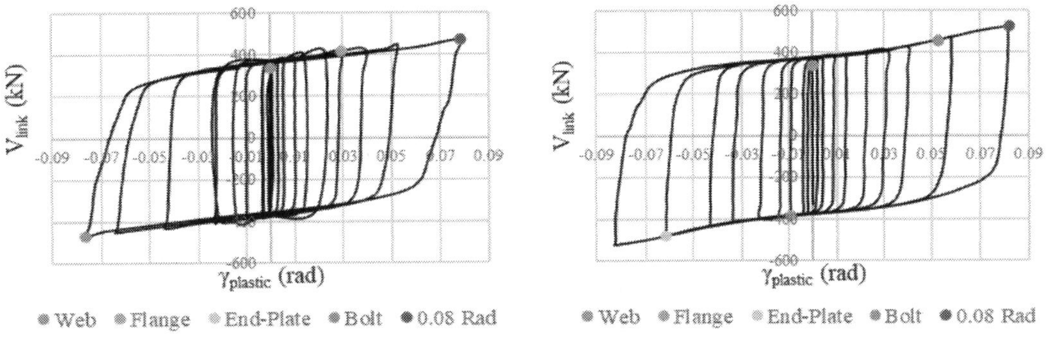

(a) (b)

Figure 8. Plastic Rotation and First-yield of Modified Connection. a. Slip-critical b. Bearing-type

The maximum effective stress occurred in each component during the increasing shear force, V_{link} is shown in Table 5 and Table 6, and is presented for each modified connection at the first yielding of each component, as well as when the plastic rotation reached 0.08 radian. The shaded numbers indicate the effective stress at the first yielding of each link component as reported by the ABAQUS.

Table 5. Stress at Link Components (Slip-criticial Connection).

Yielding Component	V_{link} kN	$\gamma_{plastic}$ (rad)	σ_e max (MPa)					
			Web	Flange	End-plate	Bolt	Pin	Beam
Web	335.93	0.00	325.30	159.30	297.10	993.50	66.13	92.51
Flange	415.59	0.03	369.30	379.70	314.50	985.90	104.60	105.90
End-plate	468.78	0.08	343.00	384.80	379.70	1,091.00	191.80	172.00
Bolt	469.35	**0.08**	431.80	384.80	379.80	1,100.00	183.50	161.80
Pin				Not yield				
Beam				Not yield				

Table 6. Stress at Link Components (Bearing-type Connection).

Yielding Component	V_{link} kN	$\gamma_{plastic}$ (rad)	σ_e max (MPa)					
			Web	Flange	End-plate	Bolt	Pin	Beam
Web	338.45	0.00	325.60	213.30	194.40	326.60	89.50	86.31
Flange	379.18	0.01	360.30	379.60	268.60	470.30	106.40	109.10
Bolt	462.63	0.06	382.90	382.90	375.80	1105.00	231.80	187.80
End-plate	468.00	0.07	382.90	382.90	379.80	1107.00	237.90	193.00
	487.35	**0.08**	389.40	383.80	380.40	1121.00	275.40	223.60
Pin					Not yield			
Beam					Not yield			

Table 7 shows that the yielding sequence was different between the two models of modified connections. Although the web of both connections yielded at the same shear force, the flange and bolts of the bearing-type connections yielded earlier than those of the slip-critical. However, the end-plate of the slip-critical connection yielded earlier than the end-plate of the bearing-type; and at the end of observation when the plastic rotation reached 0.08 radian, the maximum shear force at the link reached 468.48 kN at the slip-critical connection, less than 487.35 kN at the bearing-type connection.

4.3 Stress Distribution

The purpose of inserting pins in the modified connections is to reduce the shear force sustained by the bolts in order to minimize the number of bolts in the case of high shear force in limited size of end-plate. Figure 9 to 11 shows the stress distribution at link plastic rotation of 0.08 radian in bolts of slip-critical connection and bearing type connection.

Table 7. Yielding Sequence of Modified Connections.

Slip Critical			Bearing Type	
Pin σ_e max (MPa)	Yielding Component	V_{link} (kN)	Yielding Component	Pin σ_e max (MPa)
		338.45	Web	89.50
66.13	Web	335.93		
		379.18	Flange	106.40
104.60	Flange	415.00		
		462.63	Bolt	231.80
192.80	End plate	468.78		
192.80	0.08 rad	468.78		
183.50	Bolt	469.35		
		468.00	End plate	275.40
		487.35	0.08 rad	306.00

Figure 9 clearly shows that for both connections at link plastic rotation 0.08 radian, the shear stress was almost zero. The tensile stress of bolt in slip-critical connection (Figure 10a) varied with its height while a uniformly distributed tensile stress of bolt is shown in the bearing-type connection (Figure 10b). The friction developed between bolt heads and end plates in the slip-critical connection dictated the bolt to bend and reduced the tensile stress in half-part of the bolt-section. This, in turn, reduced the average tensile stress and the total tensile force of the bolts as well as the bending moment and the shear force of the link in the slip-critical connection. As a comparison, the average tensile stress was 804.20 MPa and 1099.37 MPa and the total tensile force was 305.70 kN and 380.19 kN, for slip-critical connection and bearing-type connection, respectively.

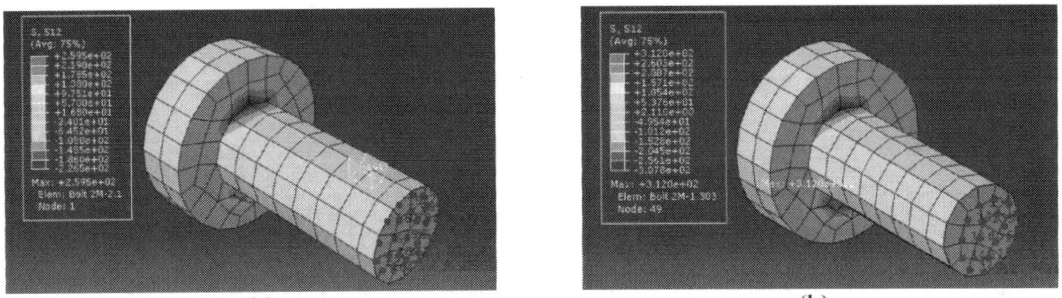

(a) (b)

Figure 9. Shear Stress Distribution at Bolt. a. Slip-critical b. Bearing-type

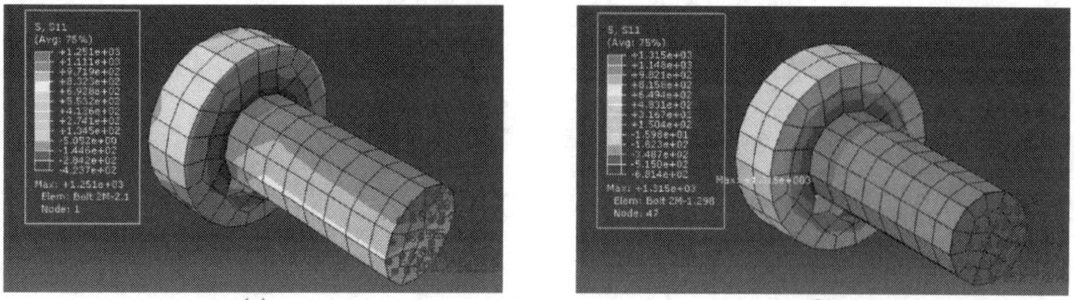

(a) (b)

Figure 10. Tensile Stress Distribution at Bolt. a. Slip-critical b. Bearing-type

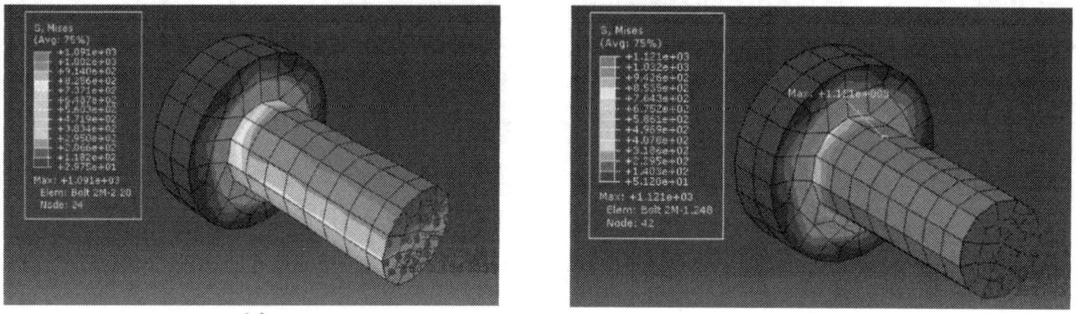

(a) (b)

Figure 11. Effective Stress Distribution at Bolt. a. Slip-critical b. Bearing-type

At the same time, the bearing-stress has been developed at the pin, much higher at the bolt in bearing-type connection since it sustained full shear-force of the link, as compared to only some part of shear force sustained by pins in the slip-critical connection. At the link plastic rotation of 0.08 radian, the maximum effective stresses of pins were 191.80 MPa and 275.40 MPa for slip-critical connection and bearing-type connection, respectively. This indicates that the pins in both models were still elastic. Figure 12 and Figure 13 shows the stress distribution at the pins and end plate of the link.

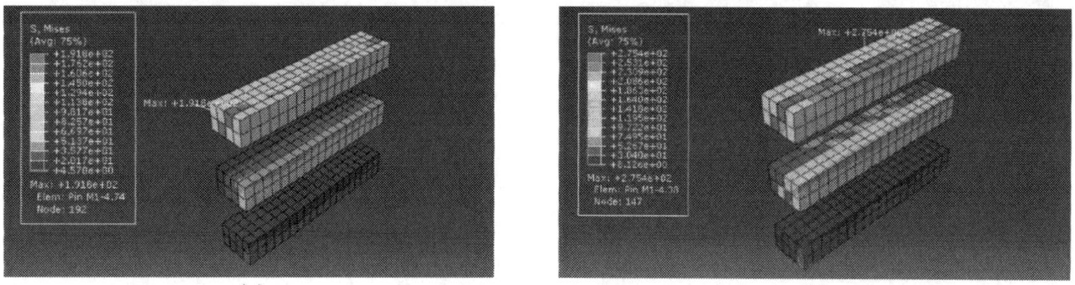

(a) (b)

Figure 12. Effective Stress Distribution at Pin. a. Slip-critical b. Bearing-type

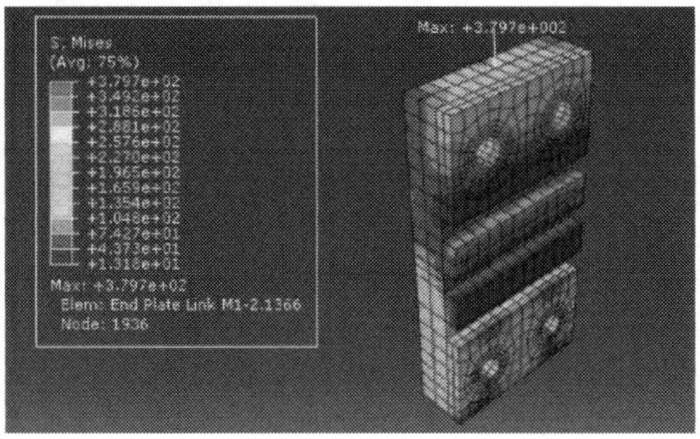

(a)

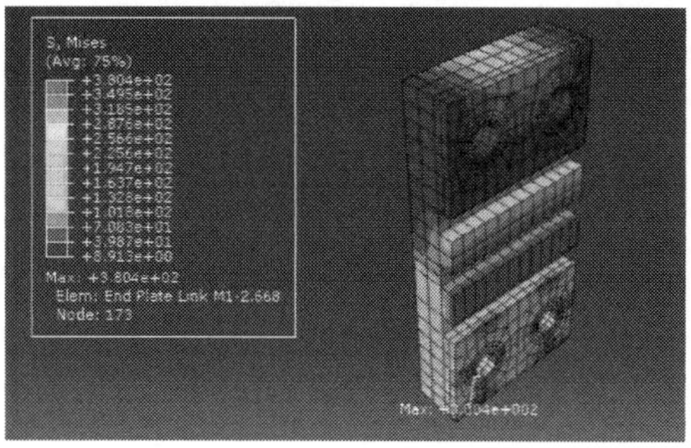

(b)

Figure 13. Effective Stress Distribution at End plate. a. Slip-critical b. Bearing-type

Meanwhile, it is also worth to note that up to the link plastic rotation of 0.08 radian, the endplate of beam outside link remained elastic and its maximum effective stress for slip-critical connection and bearing-type connection were 161.80 MPa and 223.60 MPa, respectively (Table 5 and Table 6), far less than its yield strength of 380 MPa. Based on this study results, further numerical work including the damage control as well as experimental work are expected to further investigate the performance of the modified connection for replaceable shear link.

5. Conclusion

A numerical work has been carried out to a modified shear link connection i.e. end plate with inserted pins, to reduce the number of bolts in slip-critical connection in eccentrically braced frames, especially when the size of link end-plate is limited. The study shows that hysteretic curves for the shear link with the modified connection indicate adequate strength and ductility for both slip-critical and bearing-type connections.

Furthermore, the bearing-type modified connection indicates better performance than the slip-critical connection due to the uniformly distributed tensile stress which causes more efficient tension bolt. This provides higher tensile force in the bearing-type connection as compared to less tensile force and less bending moment capacity of shear link due to reduce tensile stress in the slip-critical connection.

The study shows that up to the link plastic rotation of 0.08 radian, the beam outside the link of both models remained elastic and far from yielding. This indicates the replaceability of the shear link with the modified connection.

Acknowledgements

The authors acknowledge and express their sincere gratitude to Bandung Institute of Technology for supporting the work by the P3MI research fund.

References

[1] Malley J and Popov E 1984 Shear links in eccentrically braced frames *J.Struct.Div.* **110** 2275-95
[2] Kasai K and Popov E 1986 A Study of seismically resistant eccentrically braced frames *Rep.No. UCB/EERC-86-01* Earthquake Engineering Research Center, Univ. of California, Berkeley, CA.
[3] Okazaki T, Arce G, Ryu H and Engelhardt M 2005 Experimental study of local buckling, overstrength, and fracture of links in eccentrically braced frames *J. Struct. Eng.* **131** 1526-35
[4] Jie X, Wang Y, Ma Q and Okazaki T 2016 Cyclic behavior of very short steel shear links *J.Struct. Eng.* 04015114.
[5] Mansour N, Christopoulos C, and Tremblay R 2011 Experimental validation of replaceable shear links for eccentrically braced steel frames *J. Struct. Eng.* **137** 1141-52

[6] Jie X, Wang Y, Ma Q and Okazaki T 2016 Cyclic behavior of replaceable steel coupling beams *J.Struct.Eng.* 04016169
[7] American Institute of Steel Construction 2016 Seismic Provisions for Structural Steel Buildings
[8] Moestopo M, Kusumastuti D, Lim E, Akbar U and Ramadhita M 2018 Experimental study on the seismic behavior of replaceable shear links connected to coupling beam *IJASEIT* **8** 532-539
[9] Wibowo S 2021 Numerical study on bolted-connection of shear link in eccentrically braced frames (In Indonesian) *Master Thesis* Bandung Institute of Technology

Comparative Study in Bill of Quantity Estimates on Reinforcement Works of Pile Cap, Single Pier and Double Pier of Flyover Between Conventional Methods and BIM (Building Information Modelling)

Atanasius Agri Permadi Putra[1], Oei, Natanael Indrawan Wiharno[2] Hermawan[3] and Budi Hasiholan[4]

[1]Student of Civil Engineering, Departement of Engineering, Soegijapranata Catholic University
Email: 17b10100@student.unika.ac.id
[2]Student of Civil Engineering, Departement of Engineering, Soegijapranata Catholic University
Email: 17b10010@student.unika.ac.id
[3]Lecturers of Civil Engineering, Departement of Engineering, Soegijapranata Catholic University
Email: hermawan.mrk@gmail.com
[4]Lecturers of Civil Engineering, Faculty of Civil and Environmental Engineering, Institut Teknologi Bandung
Email: hasiholan.budi@si.itb.ac.id

Abstract. One of the construction materials with a high level of demand is steel rebar. Globally, the demand for steel rebar has quadrupled from 1960 to 2005—and this increase is predicted to double by 2050. In one project, steel rebar accounts for 16% to 60% of the total project budget. Therefore, if the volume of steel rebar is not estimated correctly it will have an impact on project budget wastage and swelling. The project budget wastage is indicated by the large volume of material remaining, while the budget swelling is indicated by the procurement of additional steel rebar because the material requirement is greater than the initial estimate. To minimize this, a comprehensive understanding of volume estimation is required. Currently, the method of estimating the reinforcement volume that is widely used is the conventional estimation method and the BIM (Building Information Modeling)-based estimation method. The conventional method is done manually based on the BBS (Bar Bending Schedule) as a reference. Meanwhile, the BIM-based method used in this study is the Allplan Engineering 2021 software and is a more modern estimation alternative because it is considered more practical, fast and efficient. Although these two methods are widely used, there are limitations and advantages that are taken into consideration when choosing the most appropriate method. Therefore, a study is needed to prove and compare the accuracy of the estimation of reinforcement volume between the conventional method and the BIM-based method. This study also highlights the advantages and limitations of the two methods for comparison. The conventional method is more transparent in terms of the estimation formula used because it is done manually. This can help the estimator track the estimation process to ensure the accuracy of the results. Whereas, the BIM-based estimation method is superior in terms of work automation and accuracy compared to the conventional method. Furthermore, the limitations of the conventional method are that it does not have a 3D visual model and requires a relatively long time to complete if there is a revision of the plan drawing. On the other hand, the BIM-based estimation method offers convenience in the form of data input using 3D drawings. The

practical data input process is proven to provide accurate results. However, the main limitation of the BIM-based method, especially for the software used in this study, is that the estimation formula cannot be traced because it relies more on three-dimensional modeling. This causes the estimator to be unable to ensure the accuracy of the estimation results. However, this limitation may differ from other BIM software products outside of this study.
Keywords : steel rebar, volume estimation, conventional-based method, BIM-based method.

1. Background

Infrastructure development is intensively carried out along with the increasing need for construction materials or engineering materials—the types of materials with a high level of demand are cement and steel rebar (Allwood, et al., 2010). The increasing global demand for construction materials is also accompanied by increased production activities for these materials, this can be seen through the number of carbon emissions produced during the production of materials (Allwood, et al., 2010). According to Allwood et al. (2010), the production of steel rebar covers 25% of the total carbon emissions produced by the industrial sector. This data indicates that steel rebar is a material with a dominant production level and has a higher level of demand and need than other construction materials so that steel rebar is produced more massively.

On the other hand, steel rebar produces a significant amount of waste material. According to Luangcharoenrat, et al., (2019), steel rebar is one of the most waste material categories produced by projects worldwide, especially in areas with rapid development such as the European Union, the United Kingdom and the United States. According to Luangcharoenrat, et al., (2019) in Indonesia and Malaysia, a number of factors that cause waste reinforcement material are reworking, delayed working, unskilled workers, lack of supervision, design mistakes and errors and lack of accuracy during the planning or estimation stage. Furthermore, Zheng, et al., (2019) stated that the process of estimating the volume of steel rebar is a crucial stage and has a significant influence on the BOQ. Therefore, if the volume of steel rebar is not estimated correctly it will have an impact on project budget wastage and swelling. The project budget wastage is indicated by the large volume of material remaining, while the budget swelling is indicated by the procurement of additional steel rebar because the material requirement is greater than the initial estimate.

One of the commonly used methods is the conventional estimation method which is done manually, but this method is considered inaccurate because it only considers the length and number of reinforcements (Datin, 2020). The limitations of the conventional method encourage the AEC (Architecture, Engineering, Construction) industry to adopt the use of BIM (Building Information Modeling) in their construction projects (Kim, et al., 2019). One of the BIM-based software that can optimize the steel rebar volume in the estimation process is Allplan Engineering. According to Xie (2018), the use of Allplan Engineering in residential projects is able to increase the efficiency of the design stage by up to 21% and reduce the design error rate to 100% (Excellent).

A number of previous studies discuss the calculation of the volume of the materials both conventional and based on BIM—the research objects are limited to residential structures and high-rise buildings. Therefore, a study with a different scope is needed to prove and compare the accuracy of the estimation of reinforcement volume between the conventional method and the BIM-based method. In addition, to accommodate the limitations of previous research, this study used a different type of structure, namely flyover pillars as the research object. Furthermore, this study also highlights the advantages and limitations of the two methods for comparison. Aspects or comparison parameters used are the evaluation aspect of the estimation methods and the volume aspect of the estimation results.

2. Literature review

2.1. Estimation

The estimation process is a predictive process to calculate the number of resources needed in a scope of work, investment, activity, or project (Dysert, 2004). According to Dysert (2004), the estimation methodology is divided into two categories, namely conceptual estimation and deterministic estimation. In deterministic estimation, the independent variable used in the estimation is part of the resource to be estimated such as item size multiplied by a known unit cost.

2.2. *Reinforcement work*
Reinforcement work is one of the important jobs in projects such as building, road, bridge and waterwork projects (McCormac, 2004). According to McCormac (2004), steel rebar used in reinforced concrete structures is in the form of bar or welded wire fabric. Steel rebar that has a bar shape can be divided into plain bar steel and deformed bar steel. According to Baskoro (2019), steel rebar is one of the materials that produces construction waste. Construction waste is the waste of construction materials that comes from a construction project or renovation of a building. In carrying out the project work, the steel rebar will be cut and then bent so that its configuration can match the planned work drawings. From the results of cutting the steel rebar, generally it produces *waste.*

2.3. *Estimated volume of steel rebar*
The estimation methods commonly used are conventional estimation and BIM-based estimation. The conventional estimation method, in general, is carried out manually by an estimator—what is meant by manual is grouping the shape of reinforcement based on work drawings and then calculating the volume. According to Angir, et al., (2017), estimators need to make groupings of the shape and diameter of the steel rebar used in the structural element in question. An example of the conventional steel rebar estimation method can be seen in Table 1. The following example is taken from the final project compiled by Damanik, 2019.

Table 1. An example of steel rebar volume estimation using the conventional method.

Bent Shape	Part	Diameter (mm)	Quantity (bar)	Bent Length (m)	One Bar Length (m)	All Bars Length (m)	Weight Per Meter (kg/m)	Work Volume (kg)
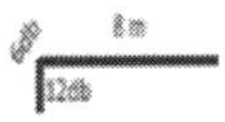	Top main reinforcement	16	2	0.288	8,288	16.576	1,578	26,157

The BIM-based estimation method is a method that has been used in recent years by the AEC (Architecture, Engineering, Construction) industry. According to Porwal and Hewage (2012), BIM can facilitate project teams to review architectural and structural designs simultaneously and make design changes quickly. BIM is a building modeling system that requires software to be implemented in a project. One of the software used by BIM modeler to estimate the volume of steel rebar is Allplan Engineering. Allplan Engineering can generate the results of the estimated volume of steel rebar in the form of a bar schedule. According to Rizqy, et al., (2021), the use of the BIM-based method at the project planning stage can increase time efficiency than the use of the conventional method.

2.4. *Bill of quantity*
According to Paikun, et al., (2018), bill of quantity (BOQ) is quantity and cost data as a result of identifying building component items and calculating volume with costs of building components contained in design drawings. The main function of the BOQ is as a document for project cost estimation, project cost determination (project costing), tender assessment benchmark, price negotiation—to payment completion according to the contract (Rashid, et al., 2006).

2.5. *Quantity take-off*
According to Pratoom and Tangwiboonpanich (2016), Quantity Take-Off or QTO is quantity data as the basis for cost estimation. The function of QTO is to provide a list of materials and their quantities required for construction projects (Apriansyah, 2021). One of the project work that requires QTO is structural work, especially steel rebar.

3. Research methods

This study focussed on the conventional estimation method and the BIM-based estimation method as comparison material. Aspects or comparison parameters used are the evaluation aspect of the estimation methods and the volume aspect of the estimation results. The evaluation aspect of the estimation methods includes the ease of the estimation process and the features that can be accommodated by each estimation method. The volume aspect is a parameter that compares the difference in reinforcement volume produced by the two methods. Thus, the most optimal estimation method can be determined.

3.1. Evaluation aspect of the estimation method

Table 2. contains nine features that are considered essential to describe how ideal the conventional estimation method and the BIM-based estimation method are. The nine essential features in Table 2. are compiled from a number of previous research literatures that have similar themes. This evaluation is carried out after the estimation of steel rebar using both methods has been completed.

Table 2. Essential features of the evaluation aspect of the estimation method.

No.	Essential Features	Literature Source		
		Baskoro, 2019	Anindya and Gondokusumo, 2020	Eroğlu, 2019
1	Classification and data input processes	●		
2	Object modeling process	●		
3	Have a 3D model as a representation	●	●	
4	Produce volume output of steel rebar	●		
5	The estimation results can adjust relatively quickly when there is a revision of the drawings	●	●	●
6	*Quantity* measurement or estimation formula can be displayed	●		
7	No additional costs for training and procurement of special equipment		●	●
8	Relatively little manpower requirement		●	
9	Have a marking system or object markers	●		

3.2. *Volume aspect of the estimation results*

The volume aspect is a parameter that compares the difference in reinforcement volume produced by the conventional estimation methods and the BIM-based estimation method. After the difference in the estimation results is found, further investigation and discussion of the factors causing the difference are carried out. The format and results of the volume aspect comparison are shown in Table 5. and Table 6.

The percentage difference in the estimated volume of steel rebar between the conventional estimation method and the BIM-based estimation method is determined using the following equation processed from Anindya and Gondokusumo (2020). In Equation 1., the total volume 1 refers to the largest volume between the conventional estimation method and the BIM-based method, while the total volume 2 refers to the smaller one between the two estimation methods:

$$\left(\frac{\text{(Total Volume 1-Total Volume 2)}}{\text{Total Volume 1}}\right) \times 100\ \% \tag{1}$$

Furthermore, the percentage difference per 1000 kg (1 Ton) of the estimated volume of steel rebar between the two methods is shown in Table 7. and Table 8. and determined using the following equation:

$$\left(\frac{\text{Percentage Difference of Estimated Volume}}{\text{Total Volume of Steel Rebar (kg)}}\right) \times 1000\ \text{kg} \tag{2}$$

The percentage difference per 1000 kg (1 Ton) of the known volume of steel rebar is then used to determine the value of the difference per 1000 kg in (kg) unit of weight using the following formula:

$$\left(\frac{\text{Equation (2)}}{100\ \%}\right) \times \text{Total Volume of Steel Rebar (kg)} \tag{3}$$

3.3. *Conventional estimation method*

Conventional estimation is done using Microsoft Excel and AutoCAD as tools. Dimensions, shapes, and specifications of steel rebar are obtained from the design drawings of single pillar and double pillar of the flyover. The final result of the conventional estimation is the volume of steel rebar (kg) which uses the estimation format processed from Damanik (2019) as shown in Table 3.

Table 3. Conventional method of steel rebar volume estimation format.

Bent Shape (A)	Reinforcement`s Type (B)	Diameter (C)	Number of Rebar Bars (D)	Bend Length (E)	One Rod Length (F)	Length of All Rods (G)	Weight Per Meter (H)	Working Volume (I)
-	-	(mm)	(pcs)	(m)	(m)	(D) × (F) (m)	(kg/m)	(G) × (H) (kg)

3.4. *BIM-based estimation method*

BIM-based estimation is done using Allplan Engineering software as a steel rebar modeling tool. The dimensions and shapes of the steel rebar are obtained from the design drawings of single pillar and double pillar of the flyover in dwg/CAD format which are imported to Allplan Engineering. The final result of the BIM-based estimation is the volume of steel rebar (kg) obtained after the modeling process in Allplan Engineering which processes using the Report feature. The specifications of the Allplan Engineering software used in this study are as follows:

- Program name: Allplan Engineering 2021
- Version: 2021-0-5
- Developer: Nemetschek Engineering GmbH (Austria)
- License type: Trial (30 days)

- Estimated standard used: Building Code Requirements for Structural Concrete (ACI 318-02) and Commentary (ACI 318R-02), ACI Detailing Manual 2004 Publication SP-66 (04)

4. Research discussion

This study discussed the advantages and limitations of the conventional estimation method and the BIM-based estimation method as comparison materials. In addition, the findings and results of the comparison of the two methods are discussed in detail. Aspects or comparison parameters used are the evaluation aspect of the estimation methods and the volume aspect of the estimation results.

4.1. Evaluation aspect of estimation method

Table 4. contains the evaluation of the two estimation methods on nine features that are considered essential for the researchers to describe how ideal an estimation method is. The ideal estimation method is a method that is able to increase the effectiveness of the work—and at the same time, it can be done with minimal operational personnel and procurement costs. This evaluation was carried out by the researchers after completing the estimation of steel rebar using the conventional estimation method and the BIM-based estimation method. The researchers consider an estimation method that has more essential features to be more ideal than other methods.

Table 4. Comparison of the evaluation aspect of the estimation methods.

No.	Essential Features	Conventional Method	BIM Based Method
1	Classification and data input processes	Yes	Yes
2	Object modeling process	Not	Yes
3	Have a 3D model as a representation	Not	Yes
4	Produce volume output of steel rebar	Yes	Yes
5	The estimation results can adjust relatively quickly when there is a revision of the drawings	Not	Yes
6	*Quantity* measurement or estimation formula can be displayed	Yes	Not
7	No additional costs for training and procurement of special equipment	Yes	Not
8	Relatively little manpower requirement	Yes	Yes
9	Have a marking system or object markers	Not	Yes

Based on the evaluation results in Table 4., it is known that the conventional estimation method fulfills five of the nine essential features. The features that the conventional method cannot fulfill are that it cannot model objects, cannot display objects in 3 dimensions for presentation needs, needs relatively long adjustments when there is a revision in the design drawings—and does not have a marking feature for objects that have been estimated. This finding is in line with the study by Eroğlu (2019) which states that the approach used by the conventional method to process QTO is still done manually and based on 2-dimensional images (As shown in Figure 1). Such an approach takes a lot of time and effort and has the potential to lead to erroneous measurements and miscalculations. In addition, this finding is also in line with the study by Baskoro (2019) that the main weakness of the conventional method is evaluation and clarification that often occurs, as well as long adjustments when there is a revision of the drawings. The reason is that the conventional method tends to be repetitive, gradual—and it requires a lot of studies.

Bent Shape	Reinforcement's Type	Diameter (mm)	Number of Reinforcing Bars	Bend Length (m)	One Rod Length (m)	Length of All Rods (m)	Weight Per Meter (kg/m)	Working Volume (kg)
a b a	F1	25	67	0,3	7,85	525,95	3,85	2024,9075
a b a	F2	25	45	0,3	10,55	474,75	3,85	1827,7875
a b a	F3	16	45	0,192	9,042	406,89	1,58	642,8862
a b a	F4	16	67	0,192	6,342	424,914	1,58	671,36412
a b a	F5	16	6	0,192	6,142	36,852	1,58	58,22616
a b a	F6	16	6	0,192	8,842	53,052	1,58	83,82216

Figure 1. Sample of QTO format of conventional estimation method arranged in the Microsoft Excel.

On the other hand, the BIM-based estimation method (Allplan Engineering 2021) fulfills seven of the nine essential features. The features that the BIM-based method cannot fulfill are that it cannot display the estimation formula and needs additional costs for training and procurement of special equipment in its operation. This finding is in line with the study by Eroğlu (2019) which states that each BIM program has a different method of modeling and processing QTO estimation data. As a result, each BIM program provides different benefits according to the features it has. Therefore, the limitations of the BIM features found in this study only apply to Allplan Engineering—and cannot be equated with other BIM programs.

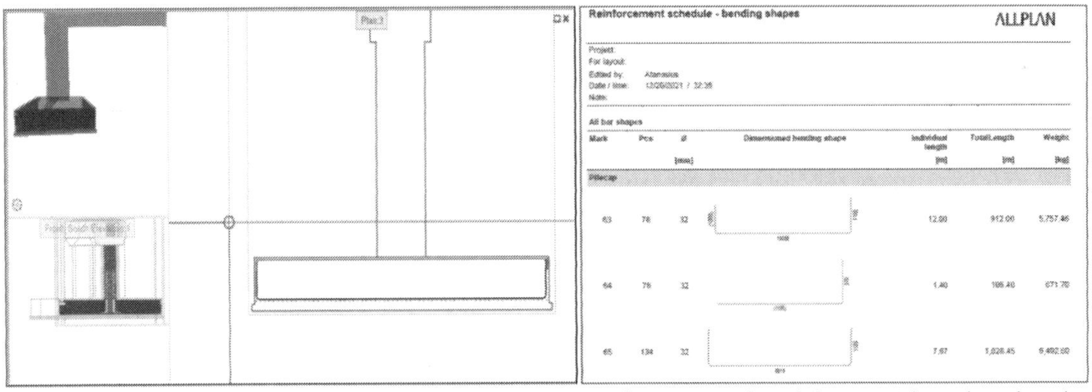

Figure 2. Sample of 3D drawing process and QTO format produced by BIM-based estimation method.

In addition, Anindya and Gondokusumo (2020) in their study state that the use of BIM in QTO work is proven to be able to reduce human resources or manpower while accelerating the completion of estimation. For example, BIM has an ability to produce QTO reports instantly from the completed 3D drawing process as shown in Figure 2. However, the implementation of the BIM-based method can be optimal if carried out by competent users. Therefore, an additional budget is needed to conduct training and procure special equipment, this is in line with the finding in this study. Based on the results of the evaluation and analysis of research findings, it can be concluded that the BIM-based estimation method is considered more ideal and more efficient than the conventional method.

4.1.1. Description of the advantages and limitations of the estimation methods.

The description of the advantages and limitations of the evaluation aspect of the BIM-based estimation method is as follows:

a. The advantages of the BIM-based estimation method (Allplan Engineering 2021):
   - BIM Allplan Engineering 2021 has a feature to distinguish structural objects and architectural objects. This feature is useful when estimating reinforcement, namely Allplan only calculates the volume of steel rebar which is a structural object even though it accidentally intersects with an architectural object.
   - BIM Allplan Engineering 2021 divides worksheets into three types, namely red, yellow and gray worksheets. The red worksheet contains active objects that can be changed, deleted and added with new objects. The yellow worksheet contains objects that can be viewed, deleted but cannot be added with new objects. Meanwhile, the gray worksheet contains objects that can only be seen but cannot be changed, deleted or added with new objects. With this feature, the BIM modeler can lock the finished steel rebar object from changing when creating a new steel rebar object. This feature prevents accidental modification of objects by the BIM modeler so that the volume estimation results are more accurate.
   - The estimation results in the form of quantity take off (QTO) of steel rebar are ready to use, this is because BIM Allplan Engineering 2021 has a feature to save or export QTO in PDF (Portable Document Format), Microsoft Word and Microsoft Excel. In addition, the table and QTO report formats have been provided by BIM Allplan Engineering 2021, with this feature the duration of preparing the QTO report is shorter.
   - BIM Allplan Engineering 2021 automatically ignores the accidental overlap condition, which is a condition when structural objects such as steel rebar accidentally intersect with other steel rebar objects that are located close together. With this feature, BIM modelers don't have to spend time tidying objects or checking the position of each steel rebar object.
   - BIM Allplan Engineering 2021 automatically performs rebar overlap if the length of the modeled steel rebar exceeds 12.0 m. Rebar overlap uses the split polygon feature which is equipped with markers for the position of the start and end of the overlap, while the overlap length is determined by the BIM modeler. This feature helps the BIM modeler to determine the position, length and type of rebar overlap quickly so that the duration of the work and the accuracy of the estimation results become more effective.

b. The limitations of the BIM-based estimation method (Allplan Engineering 2021):
   - The minimum hook length for BIM Allplan Engineering 2021 is only 56 mm. Thus, other hoop lengths shorter than 56 mm cannot be used. This can cause discrepancies in the results of the estimated volume of steel rebar when the length of the hoops used is less than 56 mm.
   - BIM Allplan Engineering 2021 has a less friendly appearance for some users, especially beginners and new users. This is because of the less strategic grouping and placement of menu toolbars and the existence of multiple toolbars that have the same function.
   - The steel rebar dimension input method modeled in BIM Allplan Engineering 2021 does not use the length of the steel rebar, but uses the size of the concrete cover. Thus, the steel rebar modeling process becomes more complicated, in addition if the size of the concrete cover is not correct—it has the potential to cause inaccurate estimation results.

The description of the advantages and limitations of the evaluation aspect of the conventional estimation method is as follows:

a. The advantages of the conventional estimation method:
   - The conventional estimation method does not require additional costs for training and procurement of electronic devices with special specifications. This is because the programs used, namely Microsoft Excel and AutoCAD, are lightweight programs that can still be run on devices with low system capacity. Furthermore, the conventional estimation method does

not require special expertise in its implementation because it is relatively simple and easy to understand. In addition, the need for estimator or manpower in making estimation is relatively small, this is because the work method is simple and the programs or tools used are quite familiar among estimators.

b. The limitations of the conventional estimation method:
- In the conventional estimation method, a three-dimensional model that displays the shape of the model made visually is not made, besides it also does not have a marking system or object marker. This limitation arises due to differences in work steps. Conventional estimation work steps are limited to the amount of reinforcement, specific gravity and length of steel rebar as input data during estimation, therefore this method does not provide a 3-dimensional visual and marking system.
- In the conventional estimation method, if there is a revision or change in the design drawings, the duration of the adjustment to the estimation results will be relatively long. This is because the tracking process for the revised part is still done manually.

4.2. *Volume aspect of the estimation results*

The results of this aspect are shown in Table 5. and Table 6. and were calculated using Equation 1. Both tables show that the volume of steel rebar based on the BIM method for single pillar and double-pillar structures produces a smaller volume than the conventional method. This finding is in line with the study by Pratoom and Tangwiboonpanich (2016) which states that estimation using the BIM-based method tends to have smaller results than the conventional method. The study found that one of the causes of this trend is that the conventional method uses the approximate dimensions of the steel rebar while the BIM-based method uses the actual dimensions of the steel rebar.

Table 5. Comparison of the volume aspects of the single pillar estimation results.

No. Steel Rebar Type	Conventional Method Weight (kg)	BIM-Based Method Weight (kg)	Percentage Difference of Reinforcement Volume (%)
I.	SINGLE PILLAR		
A. *Pile Cap*			
1. D16 (Waist reinforcement)	2170.31	2055,86	5.27
2. D25 (Main Reinforcement)	3852.70	3664.58	4.88
Total Volume	6023.01	5720.44	5.02
B. Column			
1. D13 (Brace Reinforcement)	203.67	145.57	28.53
2. D32 (Main Reinforcement)	3496.32	3362.94	3.81
Total Volume	3699.99	3508.51	5.18
C. *Pier Head*			
1. D13 (Brace Reinforcement)	709.17	689.67	2.75
2. D16 (Waist reinforcement)	101.45	110.29	8.02
3. D25 (Main Reinforcement)	2949.10	2936.75	0.42
Total Volume	3759.72	3736.71	0.61
Single Pillar Total Volume	13482.72	12965.66	3.83

Based on Table 5., it is known that the volume of single pillar steel rebar in the pile cap structure of the conventional method and the BIM-based method has a percentage difference of 5.02%. The difference is caused by the difference in the length of each rebar bar between the two methods. For example, the F1 reinforcement shape the length of each bar is 7.85 m in the conventional method while in the BIM-based method it is 7.42 m. This also occurs in all shapes of reinforcement in the pile cap structure with the same tendency, namely the length of each reinforcement bar in the conventional method is greater than the BIM-based method. This trend is caused by differences in input methods. The conventional estimation uses the length of steel rebar in the input process, while the BIM-based estimation (Allplan Engineering 2021) does not use the length of steel rebar, but uses the thikness of the pile cap concrete cover as the estimate reference.

In the column and pier head structure of the single pillar, several gaps in reinforcement`s volume were found due to the difference in the amount of reinforcement. The difference in column structure is shown in Table 5. Based on Table 5., the reinforcement`s volume between the two estimation methods has a percentage difference of 5.18%. Specifically on D13 (Brace reinforcement) has a percentage difference of 28.53%. The gap was caused by the conventional method having 47 stirrups more than the BIM-based estimate. This gap was caused by the BIM-based method that used the distance between reinforcement while the conventional method used the amount of reinforcement drawn in plan drawings. The difference in the pier head structure between conventional and BIM-based methods is relatively small, approximately 0.61%. Specifically on D16 (Waist reinforcement) produced by the conventional method has lower volume than BIM-based method. The difference was caused by the conventional method having 4 pieces of waist reinforcement less than the BIM-based estimation (Allplan Engineering 2021).

Table 6. Comparison of the volume aspects of the double pillar estimation results.

	Conventional Method	BIM Based Method	Percentage Difference of Reinforcement Volume
No. Steel Rebar Type	Weight (kg)	Weight (kg)	(%)
II.	DOUBLE PILLAR		
A. *Pile Cap*			
1. D16 (Waist reinforcement)	1526.81	1616.26	5.53
2. D19 (Main Reinforcement)	3749,48	3622.86	3.38
3. D32 (Main Reinforcement)	13172.25	12921.76	1.90
Total Volume	18448.54	18160.88	1.56
B. Column			
1. D16 (Brace Reinforcement)	1655.81	1522.95	8.02
2. D32 (Main Reinforcement)	11268,12	10857,73	3.64
Total Volume	12923.93	12380,68	4.20
C. *Pier Head*			
1. D16 (Waist reinforcement)	3106.26	3084.30	0.71
2. D32 (Main Reinforcement)	15348.97	14607.65	4.82
Total Volume	18455.23	17691.95	4.14
Total Volume of Double Pillar	49827.70	48233.51	3.20

Based on Table 6., it is known that the volume of double-pillar reinforcement steel in the pile cap structure between the two methods has a gap of 1.56%. There are several gaps in the volume of D16

(Waist reinforcement) produced by the conventional method that has lower volume than BIM-based method. This caused by the difference in length per steel bar between the two methods. The gap in length was caused by different input methods, the conventional estimation uses the length of reinforcement steel based on the plan drawings, while the BIM (Allplan Engineering 2021) based estimation uses the thickness of the pile cap concrete covers instead of the reinforcement steel`s length. Other detail of this gap is D19 reinforcement has a 10,878 m length of each rod in the conventional method while the BIM-based is 25,8 cm shorter (i.e. 10,620 m).

In the column structure of the double pillar, reinforcement`s volume between the two estimation methods has a percentage gap of 4.20%. This difference was caused by the conventional estimate having 1.25 m of stirrups length longer than the BIM-based estimate. Furthermore, the main bar length in the BIM-based method is 8.45 m longer than the conventional. In the pier head structure of the double pillar, reinforcement`s volume has a difference of 4.14% due to the length of the main reinforcement bar in BIM (Allplan Engineering 2021) which is 1.61 m longer than the conventional estimate.

In addition, the input method is not the only factor that caused the difference in the reinforcement`s volume, another factor is the difference in hook length. The size of the shortest hook length according to the reinforcement plan drawing is 50 mm, this size may use in the conventional estimation method. However, the size of the shortest hook length in the BIM (Allplan Engineering 2021) is only 56 mm. Thus, other hook lengths shorter than 56 mm cannot be used and cause a gap in the estimation results between both methods. The volume aspect of the estimation results also compares the difference in the reinforcement volume every 1000 kg (1 Ton) from the estimation results of the conventional method and the BIM-based method. The results of these comparisons are shown in Table 7. and Table 8. and were calculated using Equation 2. and Equation 3. Based on the comparison results in Table 7., the single pillar for every 1000 kg of reinforcement there is a gap in reinforcement`s volume of 51.8 kg in the column structure. In the pile cap structure for every 1000 kg of reinforcement there is a difference of 50.2 kg and in the pier head structure are 6.1 kg per 1000 kg for both estimation methods. The difference in the reinforcement`s volume of the column and pile cap structures is caused by the BIM-based method that uses the distance between reinforcement while the conventional method uses the amount of reinforcement required. This problem causes the gap in the amount of reinforcement to be significant.

Table 7. Comparison of the volume aspects of the single pillar estimation results.

No.	Structure Type	Percentage Difference of Reinforcement Volume per 1000 kg (%)		Difference of Reinforcement Volume per 1000 kg (kg)	
		Conventional Method	BIM Based Method	Conventional Method	BIM Based Method
I.	SINGLE PILLAR				
A.	*Pile Cap*	0.833	0.878	50.2	50.2
B.	Column	1.400	1.476	51.8	51.8
C.	*Pier Head*	0.162	0.163	6.1	6.1

Based on the comparison results in Table 8. for double pillar shows that for every 1000 kg of reinforcement there is a gap in the volume of reinforcement of 42.0 kg in the column structure. In the pier head structure, there is a difference of 41.4 kg for every 1000 kg of reinforcement and the pile cap structure of 15.6 kg for both estimation methods. The difference in the reinforcement volume in the column and pile cap structures is influenced by the difference in the length of each steel rebar produced by the conventional estimation method and the BIM-based estimation method. Furthermore, the difference in the reinforcement volume in the pier head structure is influenced by the complexity of the structure and the varied shapes of reinforcements.

Table 8. Comparison of the volume aspects of the double pillar estimation results.

No.	Structure Type	Percentage Difference of Reinforcement Volume per 1000 kg (%)		Difference of Reinforcement Volume per 1000 kg (kg)	
		Conventional Method	BIM Based Method	Conventional Method	BIM Based Method
II.	DOUBLE PILLAR				
A.	*Pile Cap*	0.085	0.086	15.6	15.6
B.	Column	0.325	0.339	42.0	42.0
C.	*Pier Head*	0.224	0.234	41.4	41.4

This research focus on the comparison between conventional method and BIM-based method to determine the most optimum estimation method when producing QTO of reinforcement steel. To achieve this purpose, researchers used two aspects as comparator parameters, there are the evaluation aspect of estimation method and the volume aspect of the estimation results. As a recommendation for future research and discussion, the coverage of the aspects could be expand and supported by adequate resources. Through this development, the comparison result between conventional method and BIM-based method become more objective. Another limitation in the current research is unable to make a comparison of the estimation result with the actual volume of steel rebar from the project. This limitation was caused by limited access to the real quantity data provided by the contractor. As an advice for future research, it is best if researchers could make comparison between estimation result and the actual volume from the project.

5. Research conclusion

Based on the results of the study on the estimation of steel rebar using the conventional method and the BIM-based method, the following conclusions can be formulated:

a. In the evaluation aspect of the estimation method, the conventional estimation method fulfils five of the nine essential features, while the BIM-based estimation method (Allplan Engineering 2021) fulfils seven of the nine essential features. Thus the BIM-based estimation method is considered more ideal than the conventional method.

b. In the volume aspect of the estimation results for single pillar, the total volume of steel rebar produced by the BIM-based method and the conventional method has a difference of 3.83%. For double pillar, the total volume of steel rebar produced by the two methods has a difference of 3.20%. Based on the results of the investigation, the gap in the reinforcement`s volume is caused by the difference in minimum hook length, the difference in the amount of reinforcement, and the difference in the input method.

6. References

[1] Allwood J M, Ashby M F, Gutowski T G dan Worrell E 2010 *Resoruces, Conservation, and Recycling* **55** 362–81

[2] Angir D C, Ekajaya D E, Santoso I dan Proboyo B 2017 *Cara Pendekatan Perhitungan Kuantitas Pembesian pada Kolom Struktur Beton Bertulang* 34–9

[3] Anindya A A, Gondokusumo O 2020 *Kajian Penggunaan Cubicost Untuk Pekerjaan Quantity Take Off pada Proses Tender* pp 90

[4] Apriansyah R 2021 *Implementasi Konsep Building Information Modeling (BIM) dalam Estimasi Quantity Take Off Material Pekerjaan Struktural* pp 7–32

[5] Badan Pusat Statistik 2020 *Konstruksi dalam Angka* 6301005 pp 1–2

[6] Baskoro I A 2019 *Penerapan Building Information Modeling Menggunakan Tekla Structures dalam Perhitungan Volume Besi Tulangan dan Bar Bending Schedule* pp 12–61

[7] Damanik Y S 2019 *Perhitungan Kebutuhan Tulangan, Sisa (Waste) Tulangan dan Biaya Tulangan pada Pekerjaan Struktur Balok dan Kolom Proyek Apartemen Wahid Hasyim Residence Medan* pp 12–14
[8] Datin I I 2020 *Jurnal Student Teknik Sipil* **1** 82–6
[9] Dysert L R 2004 *Skills & Knowledge of Cost Engineering 5th Edition* (AACE International) pp 102–134
[10] Eroğlu E 2019 *Evaluation Of The Reliability Of Bim-Based Quantity Take-Off Processes In Construction Projects* pp 4-5, 22, 26-27
[11] Kim S, Chin S dan Kwon S 2019 *a Discrepancy Analysis of BIM–Based Quantity Take–Off for Building Interior Components* **35** 1–12
[12] Luangchaoenrat C, Intrachooto S dan Sutthinarakorn W 2019 *Sustainability* **11** 1-17
[13] McCormac J C 2004 *Desain Beton Bertulang Edisi Kelima* jilid 1 (Jakarta: Penerbit Erlangga) pp 1–37
[14] Paikun, Sadikin D, Nova A I B, Kadarisman A dan Lestari M 2018 *Jurnal Rekayasa Teknologi Nusa Putra* **4** 1–13
[15] Porwal A dan Hewage K N 2012 *Building Information Modeling–Based Analysis to Minimize Waste Rate of Structural Reinforcement* **138** 943–54
[16] Pratoom W dan Tangwiboonpanich S 2016 *Suranaree J. Sci. Technol.* **23** 5–10
[17] Rashid R A, Mustapa M dan Wahid S N A 2006 *International Conference on Construction Industry* 1–11
[18] Rizqy R M, Martina N dan Purwanto H 2021 *Construction and Material Journal* **3** 15–24
[19] Xie J, Jiang D, Bao Z dan Zhou P 2018 *BIM Application Research of Assembly Building Design: Take Allplan as an Example* 138–47
[20] Zheng C, Yi C dan Lu M 2019 *Automation in Construction* **101** 32–47

Acknowledgements
The researchers would like to thank all the parties who have participated in the research and preparation of this study, especially to all staff of the Civil Engineering Study Program, Faculty of Engineering, Universitas Katolik Soegijapranata and the Civil Engineering Study Program, Faculty of Civil and Environmental Engineering, Institut Teknologi Bandung.

The Challenge Of Assisting Stakeholders In Data-Scarce Settings: Characterising Intermittent Water Systems Using A Citizen Science Approach

L Sioné, M R Templeton and C. Onof
Imperial College London, Department of Civil and Environmental Engineering, London, SW7 2AZ, United Kingdom

Abstract. The lack of relevant data on the failures of water infrastructure impedes efforts to remediate Intermittent Water Systems, and, by association, improve public health. This paper discusses implementing a citizen science approach that leverages the personal concern of citizens to gain information on the quantity and quality of services provided by IWS - data which would otherwise be intractable, laborious or costly, for developing countries to gather. In the first instance, this paper discusses the development of the citizen science approach and the supporting data-collection tool – a smartphone application. It then ascertains: (i) the feasibility and reproducibility of the method, and (ii) the reliability of the collected data by auto-validating the results using internal consistency tests. Finally, the paper reported on the acceptance of the method as a decision-support tool to develop evidence-based solutions to IWS. This paper demonstrates the first proof of concept that citizen science can be used in conjunction with personal communication technology to bridge the information gaps on the supply of municipal water in data-scarce and resource-constrained settings. It is a feasible and validated data collection method which requires a minimal time and resource investment.

1. Introduction

Intermittent Water Systems (IWS) are piped water distribution networks that do not deliver water to customers continuously 24 hours per day. Information available on these systems shows that it is present all over the world in varying degrees of severity, with a prevalence in developing countries. An estimated 1 billion people live with water intermittency[1][2], and yet there is no standard method for reporting or measuring piped water continuity or reliability[3][4]. As a result, here is an important lack of relevant data on the operation and failures of urban water supply infrastructure in low-income countries.

Furthermore, since there is little guidance provided to utilities on how to calculate their hours of supply, it makes sense that these may not match the hours of supply that customers receive water. It is therefore conceivable that household-reported outages may better capture variability than utility-supplied information, on provided hours of supply[5].

Intermittent Water Systems (IWS) are mostly supply-side driven, as a part of a reactive strategy to ration insufficient water resources[2]. This means that the operator is in charge of deciding when, where and how much water is delivered by the system. Ideally, water delivery should adhere to a schedule but typically an Intermittent Water System (IWS) is typically unreliable and the water often unsafe for direct consumption – but the lack of relevant data on the failures of water infrastructure impedes efforts to remediate IWS, and, by association, improve public health[6].

At governmental level, this knowledge gap prevents planners from tailoring water management policies[7]. At utility level, the regulatory agencies and water utilities managers are hindered from providing a more equitable supply[8][9][10]. Without accurate data, it is difficult to estimate real economic losses or the monetary investment offset by consumers' willingness to pay, and by extension, the budget

needed to improve the water supply[11]. This lack of information also trickles down to customer level, where it impedes effective communication with customers, especially surrounding water supply schedules and any service interruptions[12].
This multi-faceted knowledge gap prevents planners from fully understanding the water supply system, leading to an absence of evidence-based water management policies. It also prevents end-users from making informed decisions about their own water use.

To remediate this lack of knowledge, the research investigated implementing a citizen science approach to gain information on the quantity and quality of services provided by their IWS. Citizen science is the collection and analysis of scientific data by members of the community in collaboration with scientists. This presents itself as a low-cost means to collect data in large quantities over a large area and in a short space of time. This paper therefore seeks to establish a proof of concept demonstrating that citizen science can be coupled with personal communication technology in the form of mobile phones, to bridge the information gaps on the supply of municipal water in data-scarce and resource-constrained settings, where citizen science has a particularly high value.

2. Methods and Materials

The research employed a bottom-up approach to data collection, which started with conducting semi-structured interviews with stakeholders, to acquire pre-existing information on the local water supply intermittency, such as official water delivery schedules. Consultations with 15 stakeholders revealed that the water utility does not have the resources in terms of finance, intelligence and manpower, to appropriately monitor and manage the water distribution. Therefore there is little information and data on the water network, and not enough resources to adequately remediate this data gap.

The subsequent step consisted of designing a smartphone application, custom-built for the local situation in Kathmandu and through which data would be collected by citizen scientists. The app was designed for users to log information on the water arrival and timeliness. The app was made freely available on Android via the google play store, in English and Nepali, in order to be used by the majority of phone users in Nepal, regardless of education level. Responses were sent to the server and stored in an AWS database for future consultation.

Figure 1. Citizens gathered at a community meeting

The tool was seen as a chance for citizens to record where, when and what type of problem they had with their water. Recruited citizens were then trained to record quality data using the app and given free water quality test kits, to further encourage them to engage and to educate them on the quality of their water.

Importantly, before any involvement with potential citizen scientists began, the researchers carefully evaluated the ethical issues pertaining to enlisting members of the public to assist in the research. The main ethical issue raised concerned the collection of participants' personal details, which if leaked could be considered a breach in privacy. To address this, the data collected through the app was sent anonymously to a secure server, to which only the Imperial College researchers involved directly in the project had access. The Nepalese participants were asked to fill an informed consent form confirming that they were over 18 years old, and that they knowingly and voluntarily participated in the study with the understanding that their data was to be used by Imperial College London for research purposes. They were also asked to confirm that they understood that they could freely withdraw from the study at any time, without justification or impact on their legal rights. These ethical issues and solutions were presented to the Imperial College Research Ethics Committee for their review and approval.

3. Results and Discussion

So far citizen science has mostly been applied to the natural world, and this research sought to take it in a new direction by applying it to the monitoring the built environment. Therefore, the research was primarily interested in the ability of the method to provide data, with the main risk being that citizens do not engage and are not retained, resulting in insufficient data and a sparse data-set, making it difficult to validate - especially if there are no other sources of it available, as was the case according to stakeholders.

Analysis of the data collected showed that the methodology was feasible, being both practical to deploy and easily reproducible. It successfully recruited 110 citizens at the rate of 5 users per day, and ultimately resulted in the monitoring of a large geographical area of $82km^2$. The methodology also successfully retained citizens, with only 3.8% of citizen-scientists dropping out of the study prematurely.

A high level of citizen science reliability of 91% was observed starting with a sample size of only 24 participants.

Stakeholder interviews with water officials showed a wide acceptance of both the methodology and the validity of the data collected. It is an innovative approach which to data-generation which can complement existing data sources or replace them altogether, depending on the quantity of data generated. This approach has the potential to provide similarly useful operational data to installing flow meters, for a much reduced cost. This data could be used to develop strategies to improve the supply-side management, particularly the system operations. At the very least, acquiring this data would permit for a higher degree of control over the equity of the water delivery per household or neighbourhood, by knowing how much water is delivered and where.

4. Conclusions

The study showed for the first time that it is possible, with limited resources, time and personnel, to train a network of citizen scientists to provide water supply intermittency data that is validated and that is viewed as being operationally useful to key stakeholders, such as water utility managers and infrastructure funders. Importantly, the method was readily accepted by the targeted citizen population, and both the method and the data were trusted by the vast majority of decision-makers interviewed in this research. In the short term the citizen-app is a useful tool to simultaneously make water customers feel empowered while identifying low-effort high-reward remediation opportunities in the water supply network. In the long term, the approach could promote an inclusive approach to water management, where the water provision strategy is designed around the citizen's needs – needs that would be exposed by their data, and not the other way around.

References

[1] Bivins, A. W., Sumner, T., Kumpel, E., Howard, G., Cumming, O., Ross, I., Nelson, K., & Brown, J. (2017). Estimating Infection Risks and the Global Burden of Diarrheal Disease Attributable to Intermittent Water Supply Using QMRA. *Environmental Science and Technology*, *51*(13), 7542–7551. https://doi.org/10.1021/acs.est.7b01014

[2] Vairavamoorthy, K. (2021, February 22). Closing the SDG access gap – the challenge of intermittent supply. *The Source*. https://www.thesourcemagazine.org/closing-the-sdg-access-gap-the-challenge-of-intermittent-supply/
[3] Majuru, B., Suhrcke, M., & Hunter, P. R. (2018). Reliability of water supplies in low and middle-income countries: A structured review of definitions and assessment criteria. In *Journal of Water Sanitation and Hygiene for Development* (Vol. 8, Issue 2, pp. 142–164). IWA Publishing. https://doi.org/10.2166/washdev.2018.174
[4] WHO. (2017). *Progress on Drinking Water , Sanitation and Hygiene: 2017 Update and SDG Baselines* (Issue February). https://doi.org/10.1371/journal.pone.0164800
[5] Rawas, F., Bain, R., & Kumpel, E. (2020). Comparing utility-reported hours of piped water supply to households' experiences. *Npj Clean Water*, *3*(1). https://doi.org/10.1038/s41545-020-0053-y
[6] IWA Specialist Group on Intermittent Water Supply. (2020). *Intermittent Water Supply: Need for action, but what action?* https://iwa-connect.org/group/intermittent-water-supply-iws/timeline
[7] de Marchis, M., Milici, B., & Freni, G. (2015). Pressure-discharge law of local tanks connected to a water distribution network: Experimental and mathematical results. *Water (Switzerland)*, *7*(9), 4701–4723. https://doi.org/10.3390/w7094701
[8] Ilaya-Ayza, A. E., Martins, C., Campbell, E., & Izquierdo, J. (2017). Implementation of DMAs in Intermittentwater supply networks based on equity criteria. *Water (Switzerland)*, *9*(11). https://doi.org/10.3390/w9110851
[9] Gottipati, P. V. K. S. V., & Nanduri, U. v. (2014). Equity in water supply in intermittent water distribution networks. *Water and Environment Journal*, *28*(4), 509–515. https://doi.org/10.1111/wej.12065
[10] Ameyaw, E. E., Memon, F. A., & Bicik, J. (2013). Improving equity in intermittent water supply systems. *Journal of Water Supply: Research and Technology - AQUA*, *62*(8), 552–562. https://doi.org/10.2166/aqua.2013.065
[11] Vásquez, W. F., Mozumder, P., Hernández-Arce, J., & Berrens, R. P. (2009). Willingness to pay for safe drinking water: Evidence from Parral, Mexico. *Journal of Environmental Management*, *90*(11), 3391–3400. https://doi.org/10.1016/j.jenvman.2009.05.009
[12] Lutaaya, M., & Echelai, G. A. (2021, February 22). Intermittent supply in a rapidly growing city: the case of Kampala. *The Source*. https://www.thesourcemagazine.org/intermittent-supply-in-a-rapidly-growing-city-the-case-of-kampala/

Sustainability Analysis of The Upper Citarum Watershed Based on Water quality, Water quantity, and Landuse Indicators

M Marselina[1] **and N M Putri**[1]

[1]Faculty of Civil and Environmental Engineering, Institut Teknologi Bandung, Indonesia

Corresponding email: mariana.marselina@yahoo.com

Abstract. River is one of the surface water resources used to meet people's need for daily clean water. Along with the population growth, problems related to waste produced are also increasing and causing environmental pollution, including river water pollution. River sustainability is one of the issues that has attracted the attention of various institutions, one of which is the Citarum River which is one of the sources of raw water for people in West Java. In this study, analysis of the sustainability condition from the upstream of the Citarum watershed was carried out in terms of three indicators, namely water quality, water quantity, and land use. The study was conducted using water quality data at four monitoring points, river water discharge, and land use maps around the upstream of the Citarum watershed. The three types of data were processed separately where water quality data were used to determine the status of water quality using the CCME WQI method, river water discharge data were used to calculate the River Regime Coefficient (RRC) value, and land use data were used to calculate the Land Cover Index (LCI). The index values obtained from each calculation were then analyzed and combined into one value for the Upper Citarum Sustainability Index (UCSI) using the composite index method. The results of this study found that the values of UCSI in the period of April – September 2014 are 0.442; 0.419; 0.415; and 0.403. All these values are categorized in the sub-sustainability class with the interpretation that some of the river functions are slightly damaged. The UCSI value also tends to decrease from time to time towards the un-sustainability class with water quality being the indicator that plays the most role in decreasing the UCSI value in this study.

Keywords: indicators, landuse, sustainability index, Upper Citarum, water quality, water quantity

1. Introduction

Water is one of the most important resources in the life of living things. One of the raw water sources which commonly used to meet people's need is a river. One of the rivers that are generally used as a source of raw water in West Java to meet water needs and as a source of livelihood for the community is the Citarum River. Some of the uses of the Citarum River are as a source of clean water, domestic needs (such as washing, bathing, etc.), irrigation, supplying fishery water, industrial needs, and hydroelectric power plants. In recent years, Citarum River has attracted a lot of public attention because of its problems related to water quality. Pollutants in the Citarum River come from domestic activities, agriculture, livestock farming, fisheries, and industrial activities that dump their waste into the river. As a water body that plays an important role, it is necessary to monitor the sustainability of the river.

River sustainability is a concept that combines ecological and human aspects. The sustainability of the river depends on its ability to maintain its structure and function, to recover after a disturbance, to support local biota (including human communities), and to maintain key processes such as sediment transport, nutrient cycling, and energy exchange [1]. Sustainable rivers have the ability to provide clean water that can be safely used for daily activities. Assessment of the river sustainability using various indicators is generally used as the main method to obtain clearer picture of the river condition. However, the complex ecosystem of the river makes the use of a single indicator to assess river sustainability cannot fully describe the actual condition of the river. Several studies which used only a single indicator proved that it cannot be considered the only factor that affects the river environment [2]. Until now, many researchers have developed a combination of several indicators or indices into a single river sustainability index to be able to understand more about the complex dynamics between several variables related to the sustainability of a watershed.

This study aims to analyse the sustainability index of the Upper Citarum watershed by combining the three indicators consist of quality, quantity, and land use indicators. The assessment of water quality indicator was done by using the CCME WQI method, the assessment of water quantity indicator was done by using the calculation of the River Regime Coefficient (RRC), and the assessment of land use indicator was done by using the calculation of the Land Use Index (IPL). The combination of these indicators into one river sustainability index was done to be able to understand more about the condition.

2. Materials and methods

This study used secondary data obtained from several related institutions, such as river water quality data at four monitoring points from 2011 to 2019 conducted by Environmental Agency of West Java, river water discharge data at the Nanjung station from 2011 to 2019 obtained from Citarum River Basin Centre and digitized maps of land use around the Citarum watershed in 2009 and 2014 obtained from Citarum River Basin Centre and Water Resources of West Java.

2.1. Water quality

The method used to determine water quality status is the Canadian Council of the Minister of the Environment Water Quality Index (CCME WQI). This method is an objective based index that compares the water quality value with the quality standard to produce a score ranging from 0 (worst quality) to 100 (best quality) and flexible in the term of choosing the number and type of parameters, as well as the quality standards used according to their respective objectives. In this study, the reference quality standard was taken from the Regulation of the Government of the Republic of Indonesia Number 22 of 2021 about the Implementation of Environmental Protection and Management with Class II water quality classification whose designation can be used for water recreation infrastructure/facilities, freshwater fish cultivation, livestock, water for crops irrigation, and/or other designations that require the same water quality as the prior described user. The selected parameters consist of 14 parameters: total dissolved solids (TDS), total suspended solids (TSS), total detergent, phenol, total phosphate, biochemical oxygen demand (BOD), chemical oxygen demand (COD), oil and grease, nitrate, dissolved oxygen, pH, free chlorine, faecal coliform, and total coliform. The equation for the CCME WQI method is as follows:

$$CCME\ WQI = 100 - \frac{\sqrt{F_1^2 + F_2^2 + F_3^2}}{1{,}732} \quad (1)$$

with the parameter presence in equation (1) is described as follows:

1. F_1 (scope)

$$F_1 = \left(\frac{Number\ of\ failed\ variables}{Total\ number\ of\ variables}\right) \times 100 \quad (2)$$

F_1 represents the percentage of selected variables that do not meet their respective quality standards at least once during the period.

2. F_2 (frequency)

$$F_2 = \left(\frac{Number\ of\ failed\ samples}{Total\ number\ of\ samples}\right) \times 100 \tag{3}$$

F_2 represents the percentage of individual sample measurements (tests) that do not meet their respective quality standards within the period.

3. F_3 (amplitude)
When the measured variable does not meet its quality standard, it is considered an excursion and is calculated as follows:
When measured variable should not exceed the quality standard, then:

$$Excursion_i = \left(\frac{Failed\ test\ value_i}{Objective_i}\right) - 1 \tag{4}$$

When the measured variable should not be below the quality standard, then:

$$Excursion_i = \left(\frac{Objective_i}{Failed\ test\ value_i}\right) - 1 \tag{5}$$

Where j represents the various parameters included in the index calculation (from 1 to m). After the excursion value is obtained, the normalized sum of excursions (nse) value can be determined as follows:

$$nse = \frac{\sum_{i=1}^{n} Excursion_i}{Total\ number\ of\ samples} \tag{6}$$

Then, the value of F_3 can be determined as follows:

$$F_3 = \left(\frac{nse}{0{,}01nse + 0{,}01}\right) \times 100 \tag{7}$$

F_3 represents the number of measurements that do not meet their respective quality standards calculated in three steps and scaled to values between 0 – 100.

The water quality index value using the CCME WQI method is divided into five classes as shown in table 1 shown below [3].

Table 1. Class division of CCME WQI index scores.

Class	Score	Interpretation
Excellent	95.0 – 100.0	Water quality is good and protected from the presence of pollutants and the river conditions are very close to the natural quality of a body water
Good	85.0 – 94.9	There are low levels of pollutants and the river conditions have only slightly changed from the expected quality
Fair	65.0 – 79.9	There are pollutants and the river conditions sometimes change from the expected quality
Marginal	45.0 – 64.9	There are fairly high levels of pollutants and the river conditions changes quite often from the expected quality
Poor	0 – 44.9	There are high levels of pollutants and the river conditions almost always changed from the expected quality

2.2. Water quantity

The river discharge data were processed to determine the statistical values for the entire period including the maximum discharge value (Qmax), minimum discharge value (Qmin), and monthly average discharge value (Qavg), then plotted using a hydrograph plot and used to determine the River Regime Coefficient (RRC) value. The RRC value is one of the parameters used to monitor river discharge that can be used to determine the quantity of river flow over time, especially for the highest discharge in the rainy season and the lowest discharge in the dry season [4]. The RRC value can also show fluctuations of the river's flow that are in between the highest and the lowest. According to the Regulation of the Minister of Forestry of the Republic of Indonesia (P.61/Menhut-II/2014), the RRC value can be classified based on table 2.

Table 2. Classification of the River Regime Coefficient (RRC) value for wet areas.

Parameter	**Value**	**Class**	**Score**
	KRS ≤ 20	Very low	0.5
	20 < KRS ≤ 50	Low	0.75
RRC = Qmax/Qmin	50 < KRS ≤ 80	Moderate	1
	80 < KRS ≤ 110	High	1.25
	KRS > 110	Very High	1.5

The calculation of KRS value was carried out in wet areas because the Upper Citarum watershed itself is included in the wet area (there is water even at the dry season). A high KRS value indicates that the range of the runoff values in the rainy season that occurs is large, while in the dry season, the flow of water that occurs is very small or indicates a drought. Indirectly, this condition indicates that the infiltration capacity of land in the watershed lacks the ability to hold and store rainwater that falls and a lot of its runoff continues to enter the river and wasted into the sea so that the availability of the water in the watershed during the dry season is low.

2.3. Land use

The land use of Upper Citarum watershed based on the digitized map was reviewed to see the development of the composition of land use for a period of five years (2009 – 2014). The area of each type of land use was determined by classifying land use on a map using Geographic Information System (GIS) software. The land use classification is divided into seven classes, namely forest, garden/plantation, field/moor, paddy field, bush/shrub, freshwater, and built area. The analysis of the land-use changes was carried out by comparing the area of each type of land use each year. In addition, the calculation of the Land Cover Index (LCI) was also carried out which is a reference in obtaining information related to the land cover based on each class to maintain the sustainability of river functions. This was regulated in the Regulation of the Minister of Forestry of the Republic of Indonesia Number

P.32/Menhut-II/2009 about Procedures for the Preparation of a Technical Plan for the Rehabilitation of Forest and Watershed Lands. The equation used to determine the LCI value is as follows:

$$LCI = \left(\frac{Permanent\ vegetated\ area}{Watershed\ area}\right) \times 100, \quad (8)$$

where permanent vegetated area and watershed area are in hectares. The results of these calculations were then classified based on the class as shown in table 3 below.

Table 3. Classification of Land Cover Index.

No.	Score	Interpretation
1	> 75%	Good
2	30% – 75%	Moderate
3	< 30%	Poor

2.4. Sustainability index of the Upper Citarum

The three index values that have been obtained (CCME WQI, RRC, and LCI) were then used to determine the sustainability condition of the Upper Citarum watershed. The method used is the index combining method (composite index). Merging was done by a weighting system or scoring on each indicator. The index value was generated from the sum of each score on each indicator. The total value of all indicators is the value of the Upper Citarum Sustainability Index (UCSI), as the equation stated below:

$$UCHI = \sum_{i=1}^{n} Percentage_i \times Weight_i, \quad (9)$$

where UCSI is the Upper Citarum Sustainability Index, $Percentage_i$ is the percentage of indicator i, and $Weight_i$ is the weight assigned to indicator i.

The determination of the assessment standards is the main core of the assessment of river sustainability which directly impacts the rationality of the assessment results. However, there is currently no common understanding and uniform standards about the assessment standards for river ecosystems [4]. Considering the uncertainty and relativity of river sustainability, this study used the common river sustainability assessment standard which is divided into five classes with the interpretations listed in table 4.

Table 4. Interpretation of values on the river sustainability index.

No.	Class	Value	Description
1	Sustainable	0.8 – 1	River conditions are very good and able to maintain ecological function as well as socio-economic development needs for river resources
2	Basic sustainable	0.6 – 0.8	River conditions are between sustainable and sick, some river functions are in critical condition
3	Sub-sustainable	0.4 – 0.6	Some of the river functions were severely damaged
4	Unsustainable	0.2 – 0.4	Most of the river functions were severely damaged
5	Sick	≤ 0.2	The physical characteristics of the river, its ecological, and socio-economic functions are severely damaged

3. Results and discussion

Further discussion regarding the calculation results will be divided into several subchapters as follows.

3.1. Analysis of rainfall in Bandung City

Rainfall is one of the influential factors in measuring the quality and quantity of river water. Land use also indirectly becomes an important factor that influences the correlation between rainfall with the quality and quantity of river water. Therefore, further analysis related to rainfall is needed to be able to interpret further data better and improving the accuracy. The monthly average rainfall in Bandung City for the period of 2011 – 2019 is shown in figure 1.

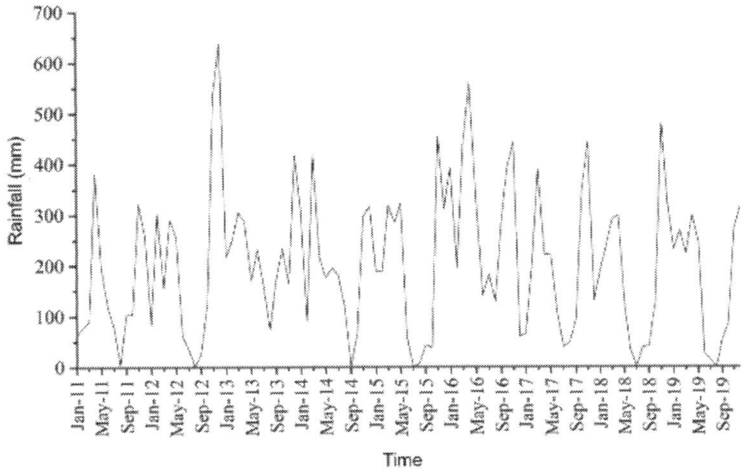

Figure 1. Bandung City's rainfall graph for the period of 2011 – 2019.

Based on the graph shown above, it can be seen that the monthly average rainfall is quite fluctuate. From this data, it is possible to classify months and years based on their seasons, namely wet month/year and dry month/year. The division was done by averaging the rainfall data each month and year, then dividing it into two classes (wet and dry) so that the following results were obtained.

1. Wet months: February, March, April, May, November, and December;
2. Dry months: January, June, July, August, September, and October;
3. Wet years: 2012, 2013, 2014, and 2016;
4. Dry years: 2011, 2015, 2017, 2018, and 2019.

3.2. Water quality analysis

3.2.1. Water quality parameter analysis. The calculation results of the average concentration value at each monitoring point during the period of 2011 – 2019 are shown in table 5.

Table 5. The average value of each parameter for the period of 2011 – 2019.

Parameter	QS	Monitoring Point			
		Wangisagara (1)	Koyod(2)	After Cisirung WWTP (3)	Nanjung (4)
TDS (mg/L)	1000	152.69	443.86	384.84	417.49
TSS (mg/L)	50	44.71	81.70*	65.72*	70.92*
Total Detergent (µg/L)	200	56.19	162.18	171.35	219.19*
Phenol (µg/L)	5	4.90	5.65*	7.73*	7.13*
Total Phosphate (mg/L)	0.2	0.12	0.26*	0.26*	0.33*
BOD (mg/L)	3	6.33*	25.81*	22.36*	25.71*

Parameter	QS	Monitoring Point			
		Wangisagara (1)	Koyod(2)	After Cisirung WWTP (3)	Nanjung (4)
COD (mg/L)	25	22.32	78.51*	77.39*	80.76*
Oil & Grease (µg/L)	1000	881.56	1271.78*	1246.27*	1281.57*
Nitrate (mg/L)	10	3.34	4.03	3.15	3.45
Dissolved Oxygen (mg/L)	4	5.74	2.78*	2.44*	2.07*
pH	6-9	7.75	8.02	7.50	7.42
Free Chlorine (mg/L)	0.03	0.09*	0.12*	0.14*	0.20*
Fecal Coliform (MPN/100 mL)	1000	71829.1*	373826.7*	643862.4*	1085453.2*
Total Coliform (MPN/100 mL)	5000	141073.1*	2490204.2*	5086219.6*	11318325.3*

* Does not meet the quality standard for class II based on PP 22/2021.

Based on the table above, the majority of water quality parameters have a concentration values that tend to get worse from monitoring point 1 to monitoring point 4. One of the factors that influenced this are changes in land use and population growth. Taken from the number of populations in the sub-watershed around the monitoring point [5], it can be seen that the population tends to increase from monitoring point 1 to monitoring point 4. This description has a good agreement with the statement that changes in land use and population growth have an impact on water quality vulnerability in various catchments areas [6]. In addition, changes in the designation of natural vegetation into built-up areas can increase runoff and sediment loads which result in the transfer of pollutants from the surface to water bodies. High concentrations of pollutants such as nutrients, sediment, and microbial contamination have been identified as water quality problems commonly found in rivers, especially in developing countries. This is ascribed to poor catchment management, the growth of human settlements without sewers, poor sanitation, population growth in rural areas, agricultural malpractice, dysfunctional sewerage networks, and use of powdered detergent [7].

3.2.2. CCME WQI calculation result analysis. The calculation of CCME WQI was carried out with four variations, namely the yearly average CCME WQI, seasonal monthly CCME WQI, seasonal yearly CCME WQI, and yearly CCME WQI. The calculation results are shown in figure 2.

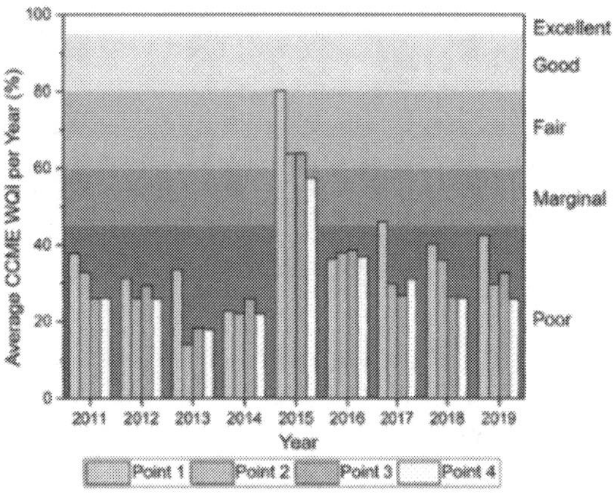

(a)

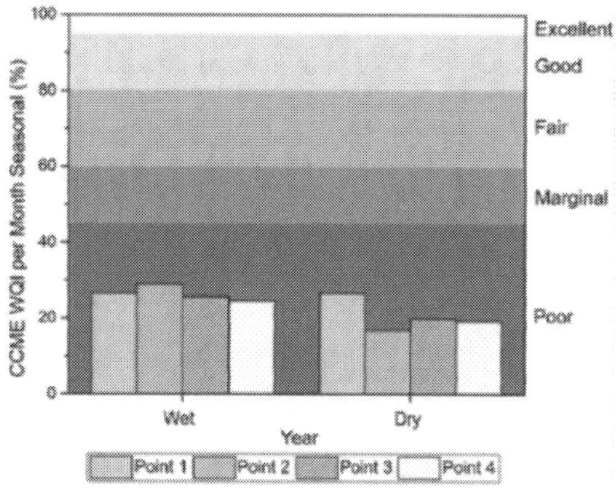

(b)

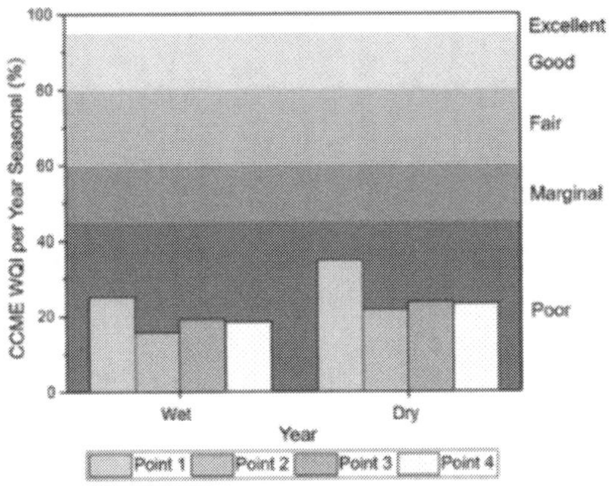

(c)

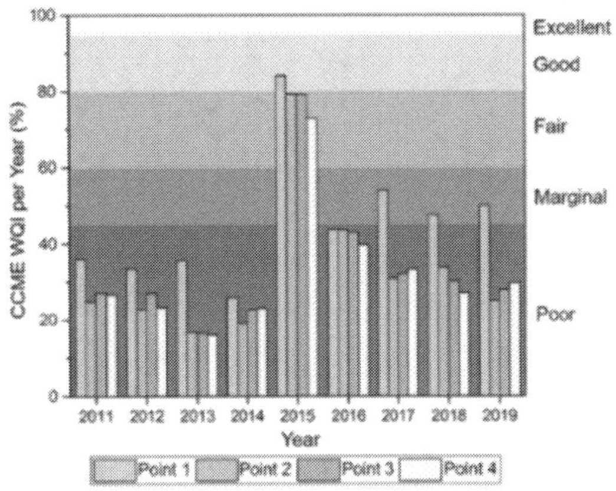

(d)

Figure 2. The calculation results of water quality using CCME WQI method: (a) yearly average CCME WQI; (b) seasonal monthly CCME WQI; (c) seasonal yearly CCME; (d) yearly CCME WQI.

The CCME WQI calculation show various results with quite significant differences. This is influenced by the amount of data used to determine the water quality index where the CCME WQI calculation per year used data with more frequency than the average CCME WQI per year as the input. Meanwhile, the CCME WQI calculation was done by grouping the data by season, showing relatively similar results because they are both in poor criteria. However, the grouping based on the seasonal month shows that the wet month has a water quality index value that tends to be worse than the dry month, while the grouping based on the seasonal year shows the opposite where the dry year has a water quality index value that tends to be worse than the wet year. These differences are also influenced by the amount of data inputted where grouping based on seasonal month produces an even distribution because it consists of 12 months with each division being 6 wet months and 6 dry months, while grouping based on seasonal year produces unequal distributions because it only consists of 9 years of monitoring with each division being 5 dry years and 4 wet years. This shows the sensitivity of the CCME WQI index to the number of samples included in the index calculation.

The selection of parameters, water quality standards/guidelines, and the amount of data that exceeds the quality standard are key factors that determine the sensitivity of WQI [7]. Based on several previous studies, the high sampling frequency is considered the best choice for water bodies with high water quality variation and can provide data for the estimation of pollutant flux. In addition, the WQI index itself depends on the availability of data where a decrease in frequency monitoring can cause the information being collected to be less representative. However, the high costs required for sampling and laboratory analysis costs are some of the many obstacles in water quality monitoring so that information on water quality becomes limited, especially in developing countries.

Based on the analysis presented above, it can be considered that to determine a representative quality status of a water body, it is necessary to carry out a long-term and sustainable water quality monitoring program. Things such as frequency of monitoring, sensitivity to input parameters, and quality standards that become references also need to be considered because they can affect the information generated from WQI calculations.

3.3. Water quality analysis

3.3.1. River water discharge analysis (Nanjung Point). The results of statistical calculation that include the maximum discharge value (Qmax) and the minimum discharge value (Qmin) are shown in figure 3 below.

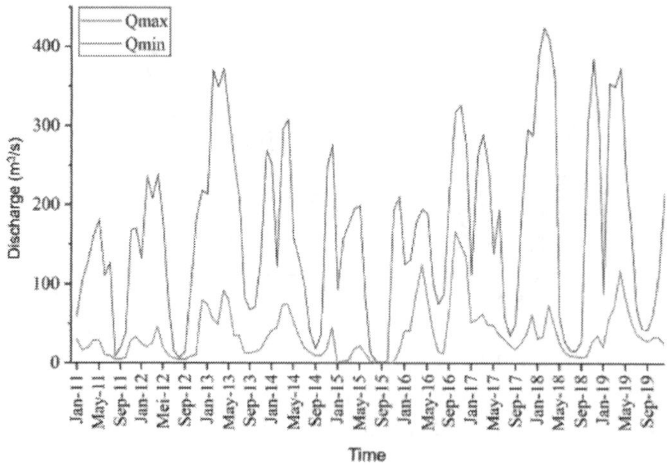

Figure 3. Chart of Citarum River (Nanjung point) discharge fluctuations for the period of 2011 – 2019.

The graph above shows the fluctuation of discharge which varies over the months. There is a similar pattern that can be seen in the graph that the increase in the maximum discharge value every month tends to be followed by the increase in the minimum discharge value and vice versa. Another similar pattern is also shown from the peak discharge which tends to be in the range of January – May each year, while the lowest discharge tends to be in the range of May – September each year. The peak discharge recorded during the 2011 – 2019 period was 425.15 m^3/second which occurred in February 2018. The graph also shows that the range of differences in values between the maximum and minimum discharges in each month is quite varied with the shortest range occurring in September 2015 with a maximum discharge value of 0.79 m^3/second and a minimum discharge value of 0.036 m^3/second.

3.3.2. River regime coefficient analysis (Nanjung Point). The calculation results of the RRC value in the Upper Citarum River (Nanjung point) are shown in figure 4.

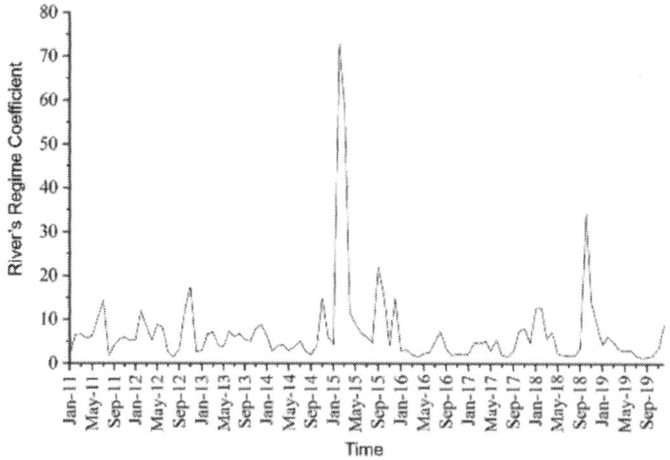

Figure 4. River regime coefficient of Upper Citarum River (Nanjung) for the period 2011 – 2019.

Based on the graph above, it can be seen that the majority of the RRC scores in the 2011 – 2019 period were in the very low class (RRC < 20). The low-grade RRC value occurred in September 2015 and

October 2018. Meanwhile, the moderate grade RRC value occurred in February 2015 and March 2015. The monthly fluctuations in the RRC value quite vary, but there was one measurement that has a quite significant difference compared to the others, shown in February 2015 with a RRC value of 72.9.

Referring to the rainfall data of Bandung City in figure 1, the rainfall in February 2015 with a value of 189.1 mm is not extreme in that period. This shows that the RRC value is not only influenced by rainfall factors, but there are several other factors such as the length of rainfall, land cover, soil type and structure, types of ground cover vegetation, and watershed management. In addition, a good ground cover vegetation can slow down the runoff rate so that water can be retained longer in the watershed [8].

3.4. Land use analysis

3.4.1. Land use area. The area of each type of land use obtained from processing map data using ArcMap (GIS) software can be seen in table 6.

Table 6. The area of each type of land use in the Upper Citarum watershed.

Type of Land Use	Area of Land Use (%)		Deviation
	2009	2014	
Forest	11.11	11.47	+0.37
Garden/Plantation	29.10	14.29	-14.81
Field/Moor	23.13	13.37	-9.76
Paddy Fields	25.01	24.08	-0.93
Shrubs	0.66	6.18	+5.52
Freshwater	0.88	2.22	+1.34
Built-up Area	10.11	28.38	+18.27
Total	100	100	0

Based on the data above, it can be seen that there are changes in the area of all types of land use. During the period, there was a decrease in the use of garden/plantation land, fields/moor, and paddy fields by 14.81%; 9.76%; and 0.93%. There was also an increase in the use of forest land, shrubs, freshwater (rivers/lakes/reservoirs/ponds), and built-up land by 0.37%; 5.52%; 1.34%; and 18.27%. Furthermore, the land use in 2009 was dominated by garden/plantation, while in 2014 was dominated by built-up land.

3.4.2. Land cover index analysis. The calculation of the LCI was done by dividing the area of permanent vegetation by the total area of the watershed. The area of permanent vegetation land consist of forest, shrubs, and garden/plantation. The results of the LCI calculation are shown in table 7.

Table 7. The results of land cover index calculation.

Year	Type of Land Use	Area of Land Use (%)	LCI (%)	Criteria
	Permanent vegetation	40.87		
2009	Others	59.13	40.87	Moderate
	Total	100		
	Permanent Vegetation	31.95		
2014	Others	68.05	31.95	Moderate
	Total	100		

The LCI value both in 2009 and 2014 was included in the moderate class category (35% – 75%). However, within five years, it can be seen that the value of IPL has decreased by 8.92%. This indicates that the area of permanent vegetation cover in the Upper Citarum watershed has changed its function. This decrement in vegetation cover can result in various negative impacts related to watershed sustainability, such as increasing the area of critical land, increasing runoff water, reducing land productivity, and affecting the hydrological characteristics of the watershed. This has a good agreement with the graph of maximum and minimum discharge fluctuations shown in figure 3 where the graph shows a tendency to increase in the maximum and minimum discharge values from 2011–2014.

Changes in land use in the watershed should also affect the RRC value where along with the reduced vegetation, the infiltration capacity of the land will also decrease so that the runoff will increase because the rainwater that falls cannot be held back and enters the ground but flows quickly into the sea. This has an impact on the amount of the measured discharge. The discharge during the rainy season will tend to be higher and the discharge during the dry season will tend to be lower. However, referring to the graph in figure 4, this statement does not apply because fluctuations in the RRC value do not increase from 2011 to 2014. This shows that the decline in the LCI value does not always affect the increase of the RRC value.

3.5. Upper Citarum sustainability index analysis

The Upper Citarum Sustainability Index (UCSI) compiled in this study is the average of the scores from each indicator (quality, quantity, and land use) so that the three indicators have the same percentage portion (33.33%) with weights adjusting the value obtained from the calculation of the index on each indicator. The approach can be taken by converting the index value into a percentage form (0% – 100%) to increase the sensitivity of the UCSI itself. In contrast to LCI, where the index value is already in percentage form, it is necessary to convert the CCME WQI score (table 1) and RRC score (table 2) into percentage form. The conversion of the CCME WQI score can be done directly because it is already in the range of 0 – 100, while the conversion of the RRC score can be done as follows:

$$\%RRC = \left(\frac{RRC\ score\ of\ each\ class}{Maximum\ RRC\ score}\right) \times 100\%, \tag{10}$$

where %RRC is the RRC value in the form of a percentage, RRC score of each class is the RRC score obtained from the calculation of the RRC value, and maximum RRC score is the highest score set by the Regulation of the Minister of Forestry (2014).

In addition, the interpretation of the RRC score is also different from the CCME WQI and LCI, where the higher the RRC score, the worse the hydrological condition of the river, and vice versa. As for CCME WQI and LCI, the higher the index value, the better the interpretation, and vice versa. Therefore, it is necessary to reverse the results of the calculation of %RRC as follows:

$$RRC\ weight = 100\% - \%RRC, \tag{11}$$

where the RRC weight is the value that will be included in the calculation of the river sustainability index, 100% is the maximum percentage in the RRC measurement, and %RRC is the RRC value in the form of a percentage obtained from the calculation results.

The calculation of UCSI was carried out using only 2014 data because that is the only year that complete data is available for the three indicators. UCSI comparisons were made monthly for the Nanjung monitoring point. In addition, in this UCSI calculation, it is assumed that the value of the LCI in each month is the same (31.95%, referring to table 7), considering the land use map which is updated once a year.

Based on these considerations, the calculation of CCME WQI and RRC at the Nanjung monitoring point in 2014 needs to be done. The calculation results can be seen in table 8 and table 9.

Table 8. Calculation results for CCME WQI at Nanjung point in 2014.

No.	Months	CCME WQI	Criteria
1	April	33.90	Poor
2	May	27.07	Poor
3	July	25.80	Poor
4	September	22.24	Poor

Table 9. Calculation results of RRC in 2014.

No.	Month	RRC Value	Class	Score	%KRS	RRC Weight (%)
1	April	4.14	Very low	0.5	33.3	66.67
2	May	2.95	Very low	0.5	33.3	66.67
3	July	5.14	Very low	0.5	33.3	66.67
4	September	2.02	Very low	0.5	33.3	66.67

Based on the calculation results above, it can be seen that the overall data, both the CCME WQI and RRC value, are in the same criteria or class. This affects the RRC weight that will be included in the UCSI calculation because even though the RRC value shows fluctuations, the RRC weight will be the same every month. The calculation of the Upper Citarum Sustainability Index can be determined with the results listed in table 10.

Table 10. Calculation results of Upper Citarum sustainability index in 2014.

No.	Month	Parameter	Percentage (%)	Weight	Value	UCSI
		Water Quality Index (CCME)	33.3	0.34	0.11	
1	April	River Regime Coefficient	33.3	0.67	0.22	0.442
		Land Cover Index	33.3	0.32	0.11	
		Water Quality Index (CCME)	33.3	0.27	0.09	
2	May	River Regime Coefficient	33.3	0.67	0.22	0.419
		Land Cover Index	33.3	0.32	0.11	
		Water Quality Index (CCME)	33.3	0.26	0.09	
3	July	River Regime Coefficient	33.3	0.67	0.22	0.415
		Land Cover Index	33.3	0.32	0.11	
		Water Quality Index (CCME)	33.3	0.22	0.07	
4	September	River Regime Coefficient	33.3	0.67	0.22	0.403
		Land Cover Index	33.3	0.32	0.11	

Based on the calculation results above, the monthly UCSI value are in the sub-sustainable class with the interpretation that some of the river functions are lightly damaged. Even though they are in the same class, it can be seen that the UCSI value tends to decrease every month with the UCSI value in September almost reach the unsustainable class with the interpretation of most river functions being heavily damaged. Based on these results, it can be said that from April to September 2014, there was a decline in river sustainability based on indicators of quality, quantity, and land use. The most important indicator in reducing the UCSI value, in this case, is the water quality determined by the CCME WQI method.

4. Conclusions

In this study, the CCME WQI calculation was carried out with 4 variations which show quite different results where it showed the sensitivity of the CCME WQI index to the number of samples included in the index calculation and the number of samples that exceeded the quality standard. The River Regime Coefficient (RRC) calculation was carried out on the discharge data at the Nanjung monitoring point. The calculation results show that the RRC value quite fluctuates with the majority of the RRC values being in the very low class (RRC < 20). The calculation of Land Cover Index (LCI) was carried out on land use maps in 2009 and 2014. The results of the calculation show a decrease in the value of LCI within 5 years by 8.92%, which means that there is a change in the function of land with permanent vegetation. The results of calculations from each indicator are then combined with the composite index method to determine the value of the Upper Citarum Sustainability Index (UCSI). The results of the UCSI calculation show a tendency to decrease from time to time in 2014. This shows a decline in river sustainability based on indicators of quality, quantity, and land use with quality indicators (CCME WQI) being the most important indicator in decreasing the value of UCSI.

5. References

[1] Bond N, Speed R, Gippel C, Catford J, Wei L, Shichuang W and Bunn S 2012 *Assessment of river health in the Gui River* (Brisbane, Australia: The International WaterCentre)
[2] Meng W, Zhang N and Zheng B 2009 *J. Environ. Sci.* **21**(8) p 1017-27
[3] Hurley T, Sadiq R, Mazumder A 2012 *Water Res.* **46**(11) p 2544-3552
[4] Tang X, Xie Y, Yang W and Jin F 2009 *Wuhan Univ. J. Sci.* **14**(6) p 537-42

[5] Rahardyan A 2020 Analisis ketidakpastian dan sensitivitas program pengendalian pencemaran terhadap kualitas air Sungai Citarum (Bandung, Indonesia: Institut Teknologi Bandung)
[6] Pullanikkatil D, Palamuleni L and Ruhiiga T 2015 *Afr. J. Aquat. Sci.* **40**(3) p 277-86
[7] Namugize J N and Jewitt G 2018 *Water SA* **44**(4) p 516-27
[8] Arsyad S 2010 Erosi dan Konservasi Tanah (Bogor, Indonesia: Institut Pertanian Bogor)

Development of Concrete Material Technology in the context of supply - demand based on geographical aspects

R G K Pradoto[1], B W Soemardi[1], A Gazali[1], A T Putri[1], R P Purba[1], I Mahardika[1]

[1]Faculty of Civil and Environmental Engineering, Bandung Institute of Technology, 10 Ganesha Street, Bandung, Indonesia
E-mail: ranipradoto@itb.ac.id

Abstract. The Indonesian construction industry is one of the leading sectors with significant contributions to economic growth. A current major issue in the industry is to strive for sustainable development focuses on three aspects, social, environmental, and economical. The use of technology, especially concrete material technology, is one of the supporting factors to achieve the goal. Concrete material usage, with cement as the main component, is quite prominent in the industry, supported by Indonesia's top 12th cement producer globally with a 2,05% contribution of the world's total cement production. However, the industry performance in each part of Indonesia also needs to be observed, considering that Indonesia has numerous islands. This study aims to evaluate the landscape of technological development of construction material regarding the supply-demand from geographical dimensions. This supply-demand analysis can consider policy strategies to support sustainable development programs. The data used in this study are derived from the production and consumption aspects of concrete material with an economic approach. The study shows the relationship between the development of concrete construction technology in terms of supply and market conditions. In addition, this study discusses the sustainable aspect of concrete material technology in the context of the trend of cement usage.
Keyword: construction technology, sustainable construction, supply-demand, concrete material

1. Introduction

The Indonesian construction industry is one of the leading sectors with significant contributions to economic growth. Currently, the construction industry accounts for 10-12% of Indonesia's total GDP, which currently has GDP growth in the (construction) sector, in line with Indonesia's total GDP growth [1]. The construction industry also has a reasonably high benefit because it has a multiplier effect that can contribute to the GDP of other sectors. The income multiplier output of construction is 2.3. Suppose the final demand of construction procurement is 1 million rupiahs. In that case, the income of another sector also increases by 2.3 million rupiahs [2]. This happens because the construction industry gets supply from other industrial sectors such as the cement and steel industries, which resulted in other industrial sectors also feeling a tremendous economic impact.

A current major issue in the industry is to strive for sustainable development focuses on three aspects, namely social, environmental, and economical. Construction performance in moving the economy is vital enough to support sustainable development. In addition, in supporting the achievement of the realization of sustainable development, three enablers are approaches in achieving sustainable development, namely technology, institutions, and value systems. One of the highlights is technology enabling sustainable construction[3].

Construction technology is currently developing rapidly and is helpful in its business processes. The use of technology, especially construction concrete material technology, is one of the supporting factors to achieve the goal. A survey conducted on 40 Indonesian contractors found that the priority of using concrete and steel materials ranked at the top during normal conditions and the covid-19 pandemic [4].

Concrete material usage, with cement as the main component, is quite prominent in the industry, supported by Indonesia's top 12th cement producer with a 2-3% contribution to the world's total cement production. However, the industry performance in each part of Indonesia also needs to be observed, considering that Indonesia is a vast country with numerous islands. Based on data given by datis export group - cement supplier (2019), Indonesia's cement production for export needs is relatively constant with an average of + 68,44 million metric tons per year in 2015 - 2019. Very small compared to China which can reach up to + 2.250 million metric tons per year. As a fellow ASEAN country, Vietnam has a relatively higher production rate with an average of + 70 million metric tons per year. This shows that Indonesia has potential in its market.

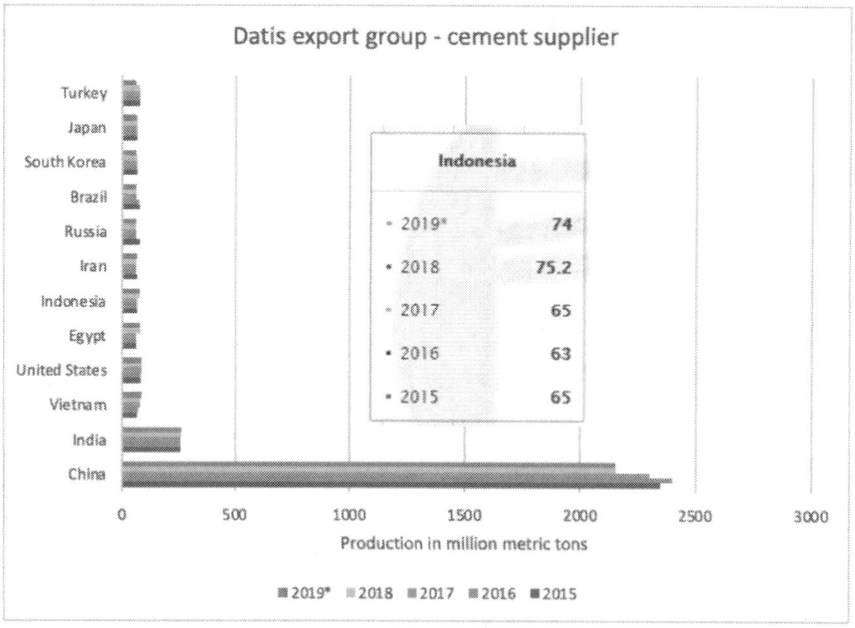

Figure 1. Datis export group-cement supplier (2019)

This study aims to evaluate the landscape of technological development of construction concrete material regarding the supply-demand from geographical dimensions. This supply-demand analysis can consider policy strategies to support sustainable development programs. In addition, this study aims to analyze which types of construction are most dominant in utilizing concrete material technology along with regional areas that have supply-demand developments that are identical to the correlation between technology use and the construction market in the area. In the end, the study is also expected to analyze the trend of the ratio of concrete material use to construction value that illustrates the sustainability aspect of using this technology.

2. The concrete material construction technology landscape

Construction Technology is defined as a collection of tools, equipment/machinery, materials, software, and construction methods/techniques [4]. Material technology plays an essential role in construction projects. The selection of materials in construction design will determine the overall project, from implementation methods to costs. As the demand for infrastructure development is on the rise, the use of concrete (including pre-cast concrete) has increased more than twice, followed by steel and other metal-base and bituminous material [5].

The use of these technologies can be mapped as a technological landscape. Technological landscapes help answer questions on how a particular technology is adopted and utilized, coping with growing problems, and predicting future needs [6]. Making this technology landscape is one of the efforts to map in realizing sustainable construction to get the correct route in initiating changes to develop the sustainability of this construction technology [3]. In a previous study, the use of concrete material technology is seen from the intensity of its use and the majority of the types of projects that use it. The intensity of the use of material technology is defined as how often the technology is used on projects that the respondent has done. This figure shows the questionnaire results regarding the intensity of the use of material technology.

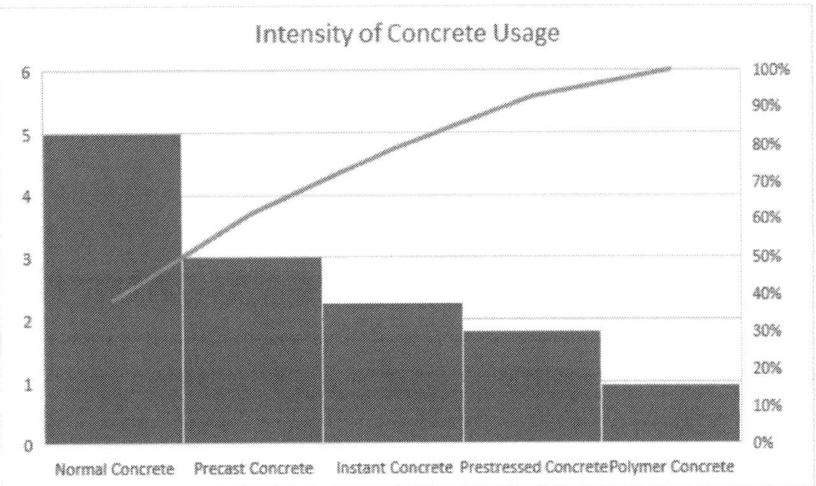

Figure 2. Intensity of Concrete Technology Usage [6]

The technological landscape reviewed in Indonesia can be seen in terms of geography because Indonesia consists of various regions and islands with different characteristics. According to ISO 3166-2, the country of Indonesia is divided into geographical units consisting of Java (ID-JW), Kalimantan (ID-KA), Maluku (ID-ML), Nusa Tenggara (ID-NU), Papua (ID-PP), Sulawesi (ID-SL), and Sumatra (ID-SM). Each region is described with an economic value called Gross Regional Domestic Product (PDRB). The amount of gross value added arises from all sectors of a region's economy, including the value of completed construction (BPS). The development of concrete material technology currently focuses on product innovation development and implementation methods.

3. Supply-demand aspects in technological development

Technological development comes from innovation, which can be traced to either a customer's initial idea or a supplier's suggestions that differentiate the two strategies of technological development: supply-push and demand-pull [8]. The supply-push model means that a technology holder or a supplier initiates the technology transfer and then pushes technology toward the market. A demand-pull model means that a technology transfer process is initiated by whatever entity is seeking a solution to satisfy demand [9]. Technology-push and demand-pull are interdependent because R&D objectives can be demand-selected, and market needs may arise in response to innovation [12]. This research focuses on concrete materials as construction technology, which development can't be separated from the usage of the technology itself. Market dynamics will influence technology development.

4. Methodology

This research is descriptive and statistical conducted with a focus on finding data on the use of concrete material construction technology and the construction market from the value of the completed contract. This data is then adjusted to the principles of supply and demand, which is reviewed through the study of literature related to construction technology and sustainable development. Then, this data was added from the results of focus group discussion (FGD) conducted to practitioners and experts who know the relationship of these data at once validated and followed up through data processing and analysis.

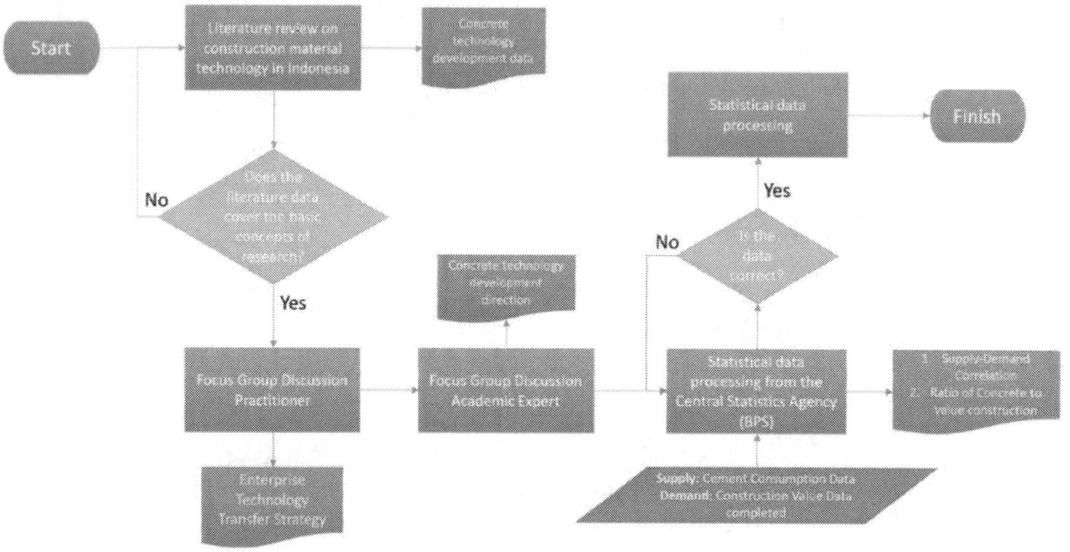

Figure 3. Methodology

This research method uses the term use of concrete construction technology as a supply illustrated by cement consumption data used in each region, obtained every year from Indonesian cement association data. Cement consumption is considered to transform cement into concrete technology by concrete suppliers to supply them to construction projects. Besides, demand data is illustrated by the construction market taken from construction value data completed in 2010 to 2018 from BPS (technology used in the construction completion process, so that the value of completed construction can describe technological market needs). This supply-demand analysis can consider policy strategies to support sustainable development programs. The data used in this study are derived from the production and consumption aspects of concrete material with an economic approach.

The approach taken in this study is to look at the demand factor of construction type according to KBLI (2020) as follows:

1. Building Construction, this group includes general construction activities of various buildings/buildings, including new buildings, building repairs, addition and renovation of buildings, the establishment of buildings or prefabricated structures on the site, and temporary construction.
2. Civil Building Construction, this type of principle includes general construction activities of civil buildings, new buildings, building improvements, building additions and building changes, and the establishment of prefabricated structures on project sites and temporary construction.
3. Specialized Construction, this type of principle includes specialized construction activities (those related to particular skills), usually specialized in one aspect common to different structures, which require specialized equipment or skills and are more done based on subcontracts.

In describing the landscape of concrete material construction technology with geographical dimensions, this study specifically discussed the region in regional Indonesia. The regional divisions to be discussed in this study are defined as follows:

1. Sumatra Regional (ID-SM)
2. Regional Java-Bali (ID-JW, ID-NU)
3. Kalimantan Regional (ID-KA)
4. Sulawesi Regional (ID-SL)
5. Regional Maluku-Papua (ID-ML, ID-PP)

The formula used in data processing in this study is as follows:

$$Growth\ of\ real\ GDP = \frac{Real\ GDP\ in\ current\ year - Real\ GDP\ in\ previous\ year}{Real\ GDP\ in\ previous\ year} \times 100$$

$$Ratio = Cement\ Consumption\ Value \div Value\ of\ Construction\ Works$$

5. Result and analysis

5.1 Development of supply-demand use of concrete material technology

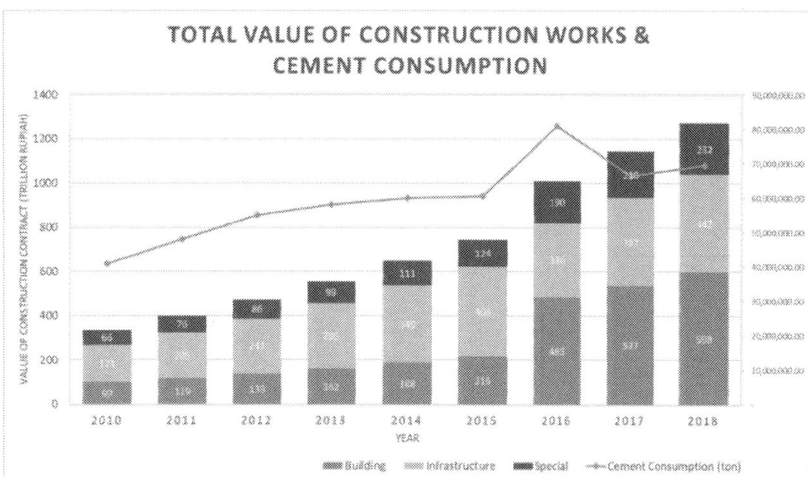

Figure 4. Graph of the development of cement material consumption (supply) & completed construction value (demand)

The data above shows the condition of the development of cement consumption which means that the number is recorded as a value for the supply that will be carried out on construction projects in applying concrete construction material technology. From 2010 to 2018, there was 170% or about 7.81% per year. There was an increase in cement consumption to fulfill concrete material construction technology in the form of ready mix and precast from 2015 to 2018 for the construction of strategic government infrastructure, and the peak occurred in 2016.

The data also shows the value of construction completed in the current year, connected to demand or demand for construction technology. This concrete material technology is quite dominant in its use.

In 2016 the real estate (property) sector in Indonesia had contributed 9.4% to national GDP, which caused housing development to reach the highest value with the highest selling price of 183.38 million / unit. 2016-2018 is quite significant because there is a difference in the definition of calculation of construction value completed, namely the addition of calculations to the cost of materials/materials provided by the work owner during the current year.

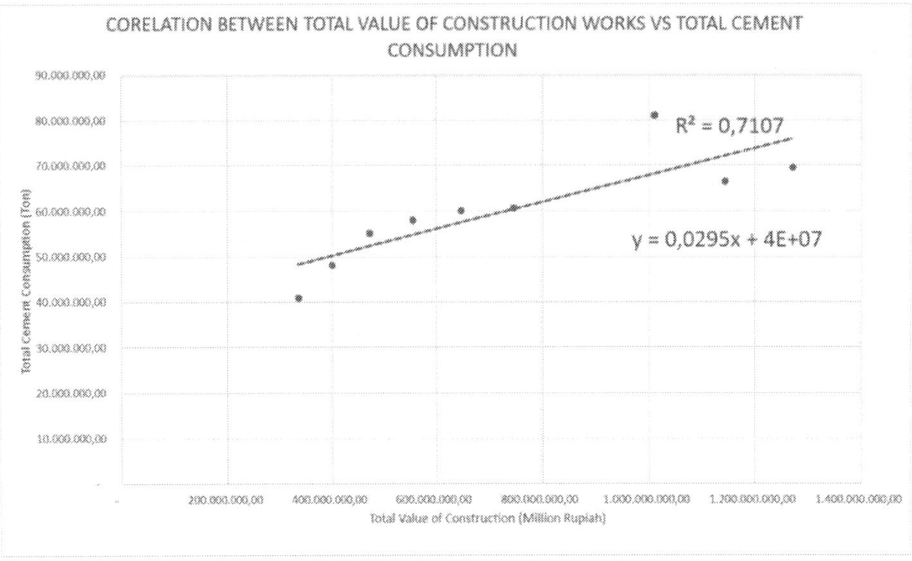

Figure 5. Supply-demand correlation

The two data above show a correlation analysis of the value of cement consumption as a supply and the value of contracts completed as demand from concrete material construction technology. From the correlation that occurs, it can be concluded that there is a relationship between supply-demand. It is proven by a regression value of 0.7107, which means it has a strong enough relationship to the addition of cement consumption if there is an increase in the value of completed construction or vice versa.

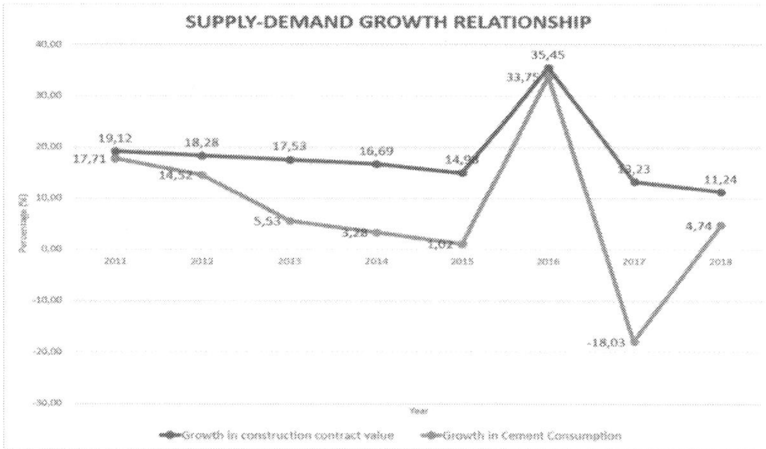

Figure 6. Supply-demand growth value

The chart above shows the supply-demand growth value. When viewed from the consistency of data, the value of demand (blue) has more consistent data in terms of growth. As for supply (orange), it is seen that the value of growth data is less consistent, as evidenced by the standard deviation of the value of contracts completed smaller (7.41) compared to the standard deviation of the value of cement consumption (14.95). However, these two supply-demand charts have the same tendency related to the increases and declines; even in 2016, both have a growth value close to each other.

Furthermore, the two mentioned variables will also be determined, which affect each other more. From the results of FGD conducted by researchers with experts in the concrete industry, it was found that the construction materials industry has its characteristics and, like other industrial developments, will rely heavily on government policies. The orientation of product development in the concrete industry and precast is demand-pull, where the ministry of public works and housing is the largest user. Considering Indonesia's market potential, the demand-pull technology approach can encourage construction technology innovation that utilizes concrete material, for example, prefabricated or precast concrete.

5.2 Relationship between supply-demand to geographic dimensions

Table 1. Growth correlation table per building and regional types

PARAMETER/REGIONAL	SUMATERA	JAWA - BALI	KALIMANTAN	SULAWESI	MALUKU - PAPUA
Mean Demand (Million Rupiah)	**96.071.375,44**	**544.252.765,44**	**40.822.083,33**	**27.503.119,56**	**22.743.980,11**
Percentage Demand	**13,1%**	**74,4%**	**5,6%**	**3,8%**	**3,1%**
Mean Supply (ton)	**13.040.497,00**	**36.675.187,78**	**4.139.886,78**	**4.711.490,56**	**1.319.201,11**
Percentage Supply	**21,8%**	**61,2%**	**6,9%**	**7,9%**	**2,2%**
R Square	**0,395**	**0,714**	**0,381**	**0,623**	**0,698**
P-value	**0,070**	**0,004**	**0,076**	**0,011**	**0,005**
Correlation	**Not ok**	**Ok**	**Not ok**	**Ok**	**Ok**

Based on regional data, the region with the most significant demand with many construction activities in a row is Java-Bali, Sumatra, Kalimantan, Sulawesi, and Maluku - Papua. Based on the analysis of supply-demand correlation in the previous section, it was obtained that there is a reasonably strong relationship between the increase in construction value and cement consumption. However, based on regional analysis, it turns out that some areas do not have a related relationship seen from the results of correlation analysis, namely the region of Sumatra and Kalimantan. This is possible because of the supply-demand activities between regions. The regional percentages of the total indicate an imbalance between demand and supply. However, it cannot be ascertained due to the lack of more specific data. However, in supporting construction activities in a region, the region can get supplies from other regions. When viewing the relationship between supply-demand regionally, there is no strong linkage.

5.3 Ratio of use of concrete material technology in the construction industry

Cement production absorbs large amounts of energy during its process, contributing to large CO2 emissions, especially from fuel and electricity use. This cement industry is estimated to have contributed 5-7% of the world's CO2 emissions.

The value of this ratio is sought to illustrate the sustainability aspect of the construction industry in whether there is an effort to reduce the use of cement or not, as cement is a material with a high carbon footprint. The ratio value will decrease if the increase in cement consumption is not greater than the increase in the value of construction works.

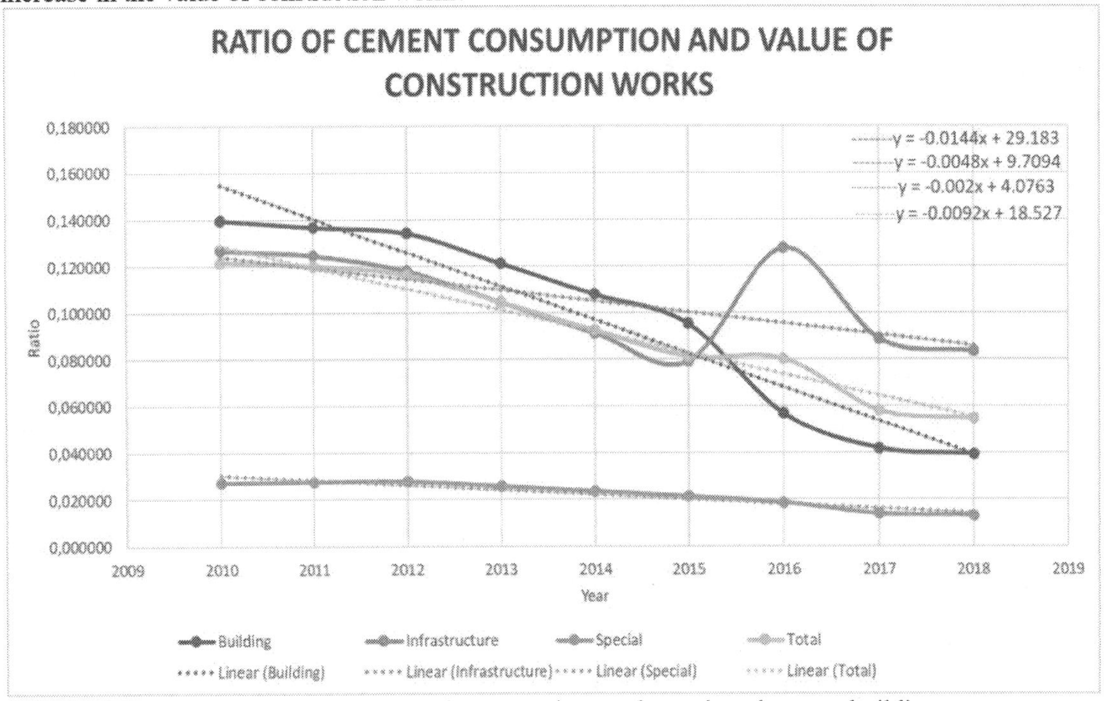

Figure 7. Graph of cement and construction market ratio values per building type

Based on the analysis conducted, the overall ratio of cement consumption to construction value decreased, indicating an effort to reduce the use of cement in the construction industry. Although the value of cement consumption and construction value is increasing every year, there is a decrease in growth from both parameters. Referring to figure 4, the decline in the growth of construction value tends to be stable compared to a more drastic decline in cement consumption growth. The ratio will illustrate the reduction in the growth rate of cement consumption, which illustrates the decrease in the use of cement in concrete material technology. In other words, the current use of concrete technology has led to sustainable development by using alternatives other than cement as the main component in applying concrete technology used in both ready mix and precast. Some efforts were made to develop technology in order to substitute cement with some materials that are more environmentally friendly, named green concrete.

This fact is in line with the commitments made by Indonesia at the G-20 Summit in Pittsburgh (2009) to reduce carbon emissions. A significant change in gradient occurred in 2012 - 2013 is a manifestation of the efforts made to support the implementation of sustainable construction by the establishment of the Indonesian Green Building Council in 2009, and support from the provincial government and municipality of Jakarta, with Governor decree No. 38/2012 [10]. A change in 2015 may also be affected by issuing a ministry-level regulation as guidance to implement sustainable construction principles, but only limited to building government-funded infrastructure [11]. The fact that building construction has a more significant decrease in cement consumption ratio than other types of construction is supported by the accreditation system for green building, which does not exist in other types of construction.

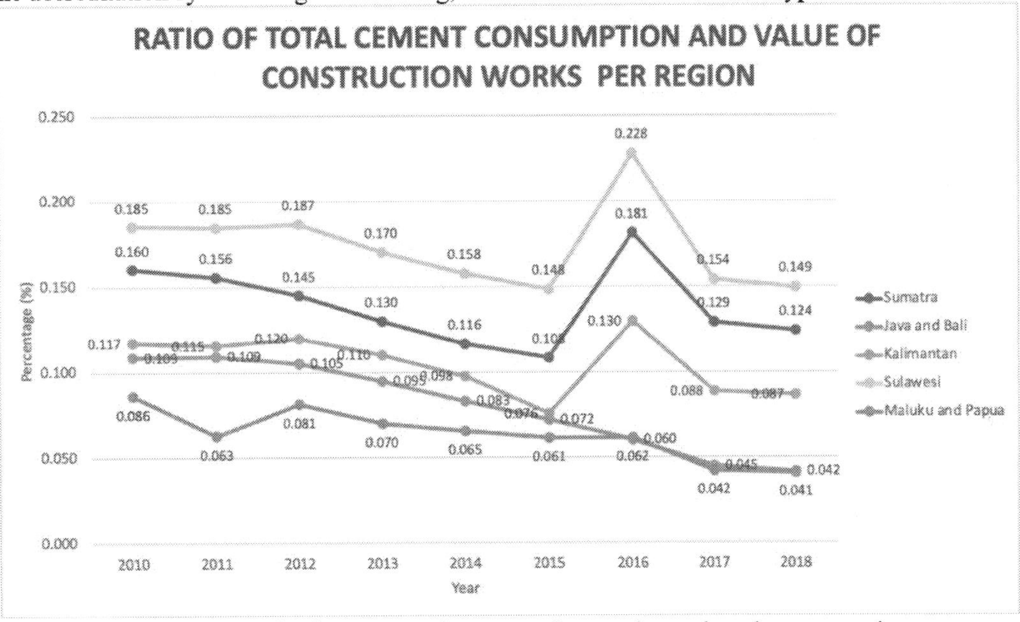

Figure 8. Graph of cement and construction market ratio values per region

The graph above illustrates the ratio of cement consumption as a supply for the use of technology in construction projects. The overall trend shows a number that decreases over time. It is seen that a significant decrease in the ratio of cement consumption and the value of completed construction projects occurred in the Java-Bali and Maluku-Papua regions. In the region of Kalimantan, Sumatra, and Sulawesi are heavily influenced by a relatively high ratio in 2016 where at that time the demand for concrete technology supply has surged because of the target of completion of projects that are quite a lot in that year, especially in infrastructure type construction work.

In order to realize the green concrete concept, cement substitution materials such as fly ash (as a recycle of coal's waste) and slag (as a recycle of steel's waste) have been developed. Thus, with the development of cement substitution materials, the use/demand of cement has decreased, which can be seen in the following graph.

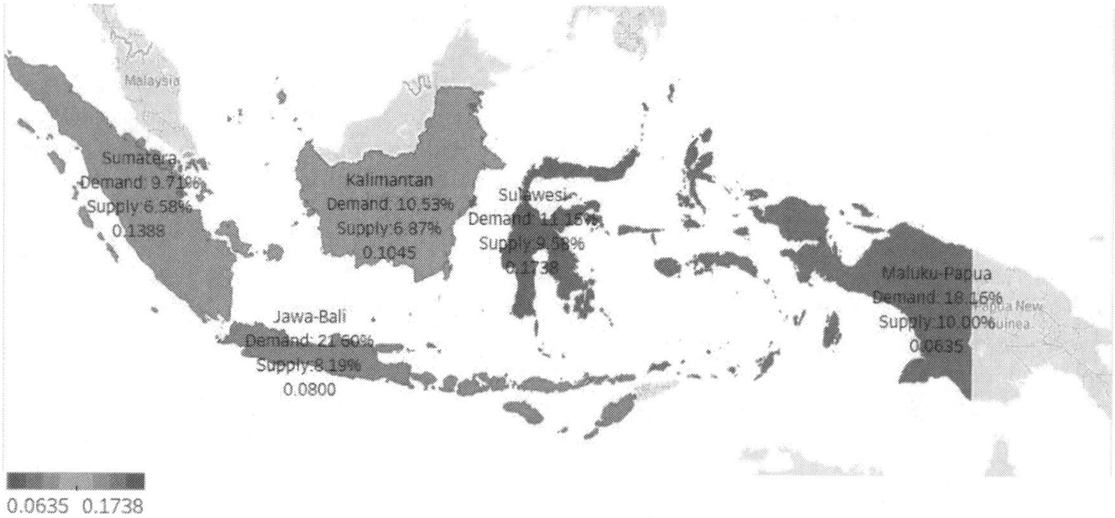

Figure 9. Average Supply-Demand & Cement Used Ratio

Currently, material technology development in the construction industry focuses on the sustainability aspect. Seen from Portland cement is rarely found in materials stores; now, more PCC (Portland composite cement) are sold in materials stores. Composite cements are cements in which a proportion of the Portland cement clinker is replaced by industrial by-products, such as granulated blastfurnace slag (GBS) and power station fly ash (also known as pulverized-fuel ash or PFA), certain types of volcanic material (natural pozzolans) or limestone. In addition, there are also efforts to recycle CO2 by injecting CO2 into the fresh concrete mixture. Apart from fulfilling the sustainability aspect, CO2 injection is proven to make concrete stronger and reduce the use of cement in concrete.

6. Conclusion

The implementation of concrete technology on construction projects is indicated through cement consumption at the value of completed construction. Cement consumption is defined as the supply activities carried out by batching plants to support the sustainability of construction projects. In contrast, the value of completed construction is defined as the demand of the construction market. Based on the data, cement consumption and construction value completed continued to increase from 2010 to 2018. Correlation analysis showed that these two variables have a significant association with R2 = 0.7. The supply-demand growth graph shows the relationship between the two variables. It can be concluded that the development of concrete technology in Indonesia is influenced by demand-pull, not supply-push. This is seen from the inconsistency of cement consumption growth, which influences the existing demand for construction projects. However, the concrete technology landscape reviewed against Indonesia's geographical position showed that some areas do not have a related relationship seen from the results of correlation analysis, namely the region of Sumatra and Kalimantan. However, in line with the development of concrete production from the distribution of batching plants, as the growth of regional development in Java-Bali and eastern Indonesia.

The sustainability issue of concrete technology can also be measured from existing economic indicators related to the ratio of cement consumption to completed construction values, which shows the cement used in construction projects. The results of this study show that since 2012 there has been a significant decrease in cement consumption for the supply of concrete material technology. The CO2 emissions that usually occur due to the use of ordinary Portland cement can eventually be reduced by developing the material composition that becomes the main component in concrete. There have been many developments related to the use of slag (waste from iron production) and fly ash (waste from coal-fired power plants) as an alternative to reduce the consumption of cement used in concrete technology. Even some construction projects have also begun to use geopolymer concrete material technology which is 80-90% more environmentally friendly than the use of normal Portland cement wherein one cubic of 40 MPa of concrete this alternative type can save CO2 emissions by 220 kg.

References

[1] BPS 2010-2020 *Konstruksi Dalam Angka* (Jakarta: Badan Pusat Statistik)
[2] Nugroho et al. 2020 Analysis if Input-Output Table : Integrated of Economic Development by Leading Sectors In Indonesia *2020 Asia-Pacific Statistics Week*
[3] Chrina Du Plessis 2006 A strategic framework for sustainable construction in developing countries *Construction Management and Economics* **25** 67-76
[4] Rika P, Ignatius M, Biemo W S 2021 Kajian Penerapan Teknologi Oleh Kontraktor dalam Menghadapi Kondisi Pandemi Covid-19 *Konteks* **15** 453
[5] B W Soemardi, K S Pribadi and E Warsidi *Construction Services Development Board 2020: 20 Years of LPJK : Indonesia Construction 2001-2020* (Bandung: ITBpress)
[6] Rani G K P, et al. 2021 The Technology Landscape of Construction Material in The Indonesian Construction Industry *Iconcess* **6**
[7] T Setiadi, M Abduh 2019 The study of sustainable procurement in the procurement of ready mixed concrete supplier ICRESBE **520**
[8] Ai-Hsuan Chiang, Ming-Yuan Huang 2021 Demand-pull vs supply-push strategy: the effects of organizational structure on supply chain integration and response capabilities *JMTM*
[9] Won J C, Ilyong J 2019 The performance of supply-push versus demand-pull technology transfer and the role of technology marketing strategies: The case of a Korean public research institute *Sustainability* **11**
[10] B Wiryomartono 2015 'Green building'and sustainable development policy in Indonesia since 2004 *International Journal of Sustainable Building Technology and Urban Development* **6(2)** 82-89
[11] Muhammad A, Reini D W, Yunita A M 2018 Framework Development Methodology for Sustainable Procurement of Construction Works in Indonesia *Matec Web of Conferences* **203** 02014
[12] Kerstin H 2021 Demand-pull and technology-push: What drives the direction of technological change? -- An empirical network-based approach *Arxiv* **2104.04813**

Acknowledgments

Authors wishing to acknowledge the support of Institut Teknologi Bandung for its P2MI funding that makes this research possible.

Study of Temporal and Spatial changes in the Estuarine Area of the Mayangan River, West Java – Indonesia

Hendra Achiari[1*], Rajin[1], Gurusu[1]

[1]Ocean Engineering Program, Faculty of Civil and Environmental Engineering, Institut Teknologi Bandung, Jl. Ganesha 10 Bandung 40132 Indonesia

*corresponding author's e-mail : h_achiari@itb.ac.id

Abstract. The estuarine of the Mayangan River is an area confluence the Mayangan River with the Java Sea. Pondok Bali beach which is about a half kilometer from River mouth of the Mayangan River has undergone a severe abrasion process since 2006. Pondok Bali's coastline changes are expected to affect the estuary area. It caused flooding and changes in water quality on the Mayangan water river area. This research was studied the factors that cause of river level rise and changes in water quality of the Mayangan River whether it has an impact of coastline geometry change. The results of hydrodynamic and water quality model simulations showed river water discharge and increase Sea level is a more influential factor compared to the change in coastline geometry or bathymetry of the river mouth. The increase in salinity at some observation points of the Mayangan River from 2006 to 2014 was only 0.2 ppt, from 2011 to 2014 the salinity increase was only 0.1 ppt, and from 2014 to 2020 only 0.03 ppt.

Keywords: abrasion, water quality, coastal process and modelling, coastal protection and environmental control, sensitivity analysis.

1. Introduction

Pondok Bali beach - Subang District, West Java is an estuarine where there is the Mayangan River (Part of Cipunegara main River) connected to the Java Sea. Pondok Bali has experienced quite severe abrasion since 2006 up to now (the timeline of shorelines can be seen in the **Figure *1***), characterized by changes in the use of rice fields into farmland and shrimp-fish pond. In addition, Pondok Bali also lost mangrove forests that are quite massive due to this abrasion process [1]. Changes in the coastline of Pondok Bali are expected to affect the water quality of the estuary area on the Mayangan River. Therefore, it needs in-depth analysis to get the best solution for citizens and government decision, especially in term of the ecosystems of the waters directly affected.

2. Problem description and Motivation

The change in coastline of Pondok Bali beach is predicted to affect the water quality on the Mayangan River. This study aims to find out whether there are changes in salinity in the Mayangan River estuary area affected by changes in coastline, sea level rise and changes in river discharge by using numerical model.

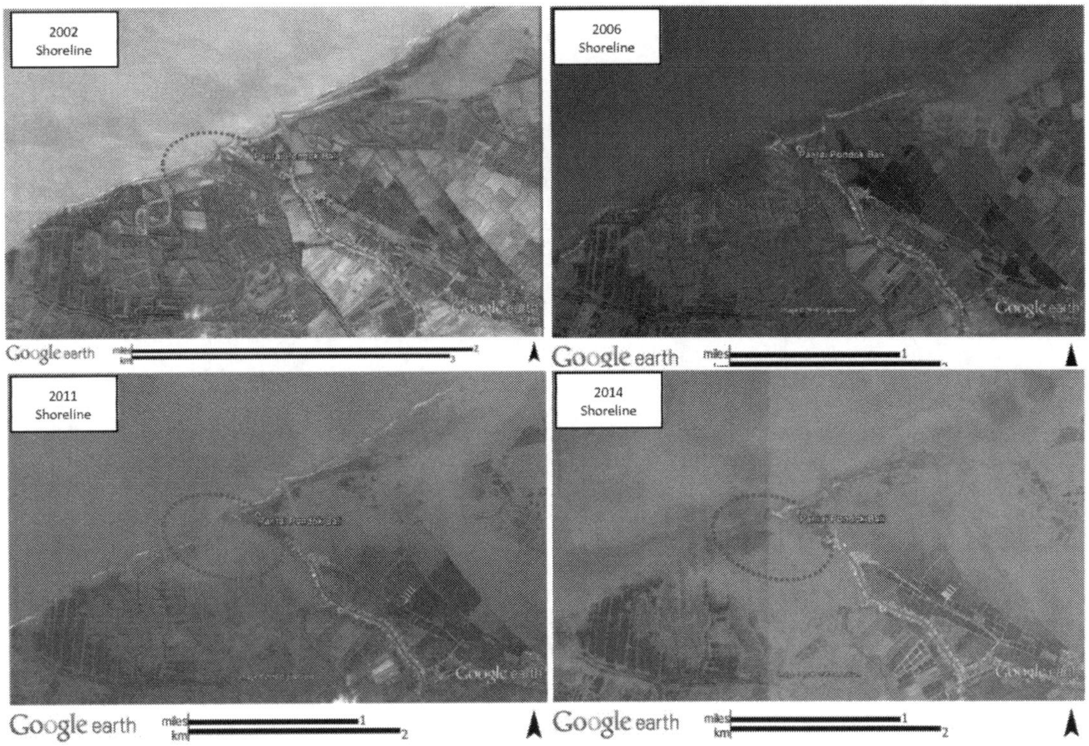

Figure 1 Timeline of Pondok Bali Beach coastline change (2002, 2006, 20011 and 2013) (source Google Earth).

3. Description of Model

RMA2 module (part of SMS 8.1 software) is a numerical model of two-dimensional average depth hydrodynamics module that uses the discrete method of finite-element method. RMA2 was developed by Norton, King and Orlob in 1973 [2]. This model can calculate the elevation of the water face and the horizontal components of current speed. The turbulence characteristics of the waters are modelled by the eddy viscosity coefficient while the frictional characteristics of the seafloor use the Manning numbers. RMA2 can calculate water in steady and unsteady state conditions.

In calculating the speed of horizontal directional current and water level elevation, the RMA2 model can include:

- Identify errors in mesh specifications.
- Consider the effects of the Earth's rotation (Coriolis Effect).
- Can consider the effects of wind stress.
- Can automatically determine Manning's number (based on water depth) and turbulent coefficient.

3.1. Hydrodynamics Model

RMA2 operates on the assumption that vertical current velocity is negligible. RMA2 is not recommended for studies that require precision at vertical current speeds. The RMA2 governing equation is composed of the flow continuity equation as shown at Equation (1) and the momentum conservation equation as given at Equation (2) and (3) as follows.

$$\frac{\partial h}{\partial t} + h\left(\frac{\partial u}{\partial x} + \frac{\partial v}{\partial y}\right) + u\frac{\partial h}{\partial x} + v\frac{\partial h}{\partial y} = 0 \quad (1)$$

$$h\frac{\partial u}{\partial t} + hu\frac{\partial u}{\partial x} + hv\frac{\partial u}{\partial y} - \frac{h}{\rho}\left[E_{xx}\frac{\partial^2 u}{\partial x^2} + E_{xy}\frac{\partial^2 u}{\partial y^2}\right] + gh\left[\frac{\partial a}{\partial x} + \frac{\partial h}{\partial x}\right] + \frac{gun^2}{\left(1.486h^{1/6}\right)^2}(u^2+v^2)^{1/2} - \zeta V_a^2 \cos\psi - 2hvw\sin\phi = 0 \quad (2)$$

$$h\frac{\partial v}{\partial t} + hu\frac{\partial v}{\partial x} + hv\frac{\partial v}{\partial y} - \frac{h}{\rho}\left[E_{yx}\frac{\partial^2 v}{\partial x^2} + E_{yy}\frac{\partial^2 v}{\partial y^2}\right] + gh\left[\frac{\partial a}{\partial y} + \frac{\partial h}{\partial y}\right] + \frac{gvn^2}{\left(1.486h^{1/6}\right)^2}(u^2+v^2)^{1/2} - \zeta V_a^2 \sin\psi + 2huw\sin\phi = 0 \quad (3)$$

where:

h	= total depth of water (η+d)
u,v	= current velocity in x,y Cartesian direction
x,y,t	= x, y Cartesian Coordinate and time
ρ	= mass density of the fluid
E	= Eddy Viscosity coefficient
g	= gravitation acceleration
a	= water elevation on the bottom
n	= Manning number
ζ	= wind stress coefficient
Va	= wind velocity
ψ	= wind direction
ω	= angular speed of Earth rotation
ϕ	= local latitude

3.2. Water Quality Model

The RMA4 water quality model was developed by Norton, King, and Orlob in 1973 [2]. RMA4 is designed to simulate the process of affection and diffusion at an average depth in the aquatic environment. This module can be used to evaluate concentrated substances dissolved in waters. In addition, RMA4 can also model the physical process of migration and mixing of substances in reservoirs, rivers, bays, and estuaries and coastal zones. The results of this model are useful as preliminary data on the process of mitigating oil spills or temperature increases due to steam power plants. The RMA4 water quality model has been applied in several cases including:

1. Modelling the distribution of the average salinity of each depth.
2. Observation of the pattern of heat distribution due to power plants.
3. Identify critical points when an oil spill occurs.
4. Determine the mixing zone area.
5. Identify the limits of salinity intrusion area due to tides.
6. Monitor water quality for fish habitats and recreational areas.

The RMA4 model is suitable for use in well-mixed waters or estuaries in vertical directions due to this model capable to produces a concentration distribution at an average depth.

The Governing equations in the RMA4 Module are given at equation (4) as follows:

$$h\left(\frac{\partial c}{\partial t} + u\frac{\partial c}{\partial x} + v\frac{\partial c}{\partial y} - \frac{\partial}{\partial x}D_x\frac{\partial c}{\partial x} - \frac{\partial}{\partial y}D_y\frac{\partial c}{\partial y} - \sigma + kc + \frac{R(c)}{h}\right) = 0 \quad (4)$$

where:

h	= total depth waters (η+d)
c	= pollutant concentration
t	= time
u,v	= fluid velocity in -x and -y direction
Dx and Dy	= dispersion coefficients
k	= 1st order pollutant decay
σ	= constituent for sink dan source
$R(c)$	= evaporation or rainfall.

4. Numerical Simulation

4.1. Configuration domain Model

The authors used SMS 8.1 software for simulating and visualization outputs. The mesh of waters domain was set by importing bathymetrical data file from RMA2 Hydrodynamics module, which covers the shoreline boundary and circulation offshore boundary and depth input. The detail mesh of domains model near the estuarine are shown at **Figure 2**.

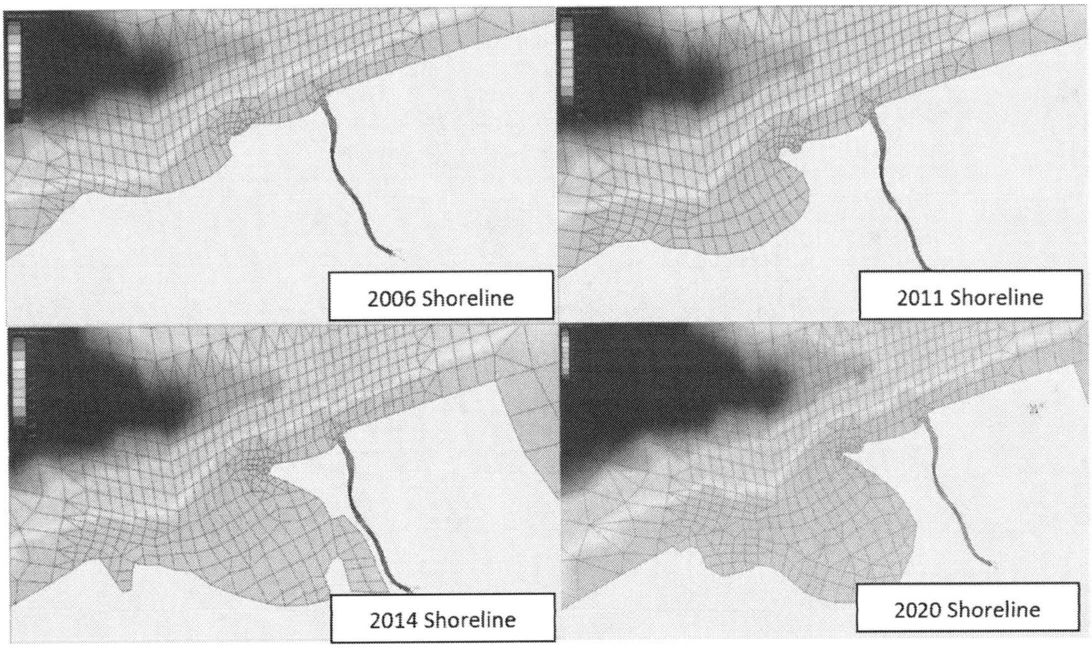

Figure 2. Mesh model for each scenarios simulation (year of 2006, 2011, 2014 and 2020)

4.2. Boundary conditions

The boundary conditions for both hydrodynamics and water quality model used two types, i.e. river cell boundary condition and Open sea offshore boundary condition. On the river cell boundary condition, it is entered the Mayangan River as the discharge data, and at the offshore open boundary condition, we set the sea level elevation data obtained from NAO-Tide software. The labels of open boundary condition are indicated by numbers 1 through 3, and the river boundary condition is indicated by the number 4 as shown in **Figure 3**.

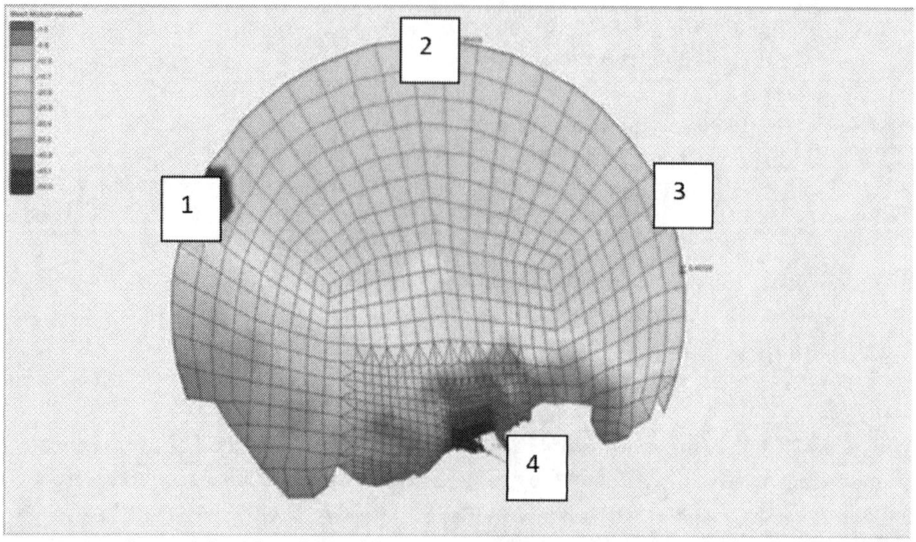

Figure 3. Mesh model with coloured bathymetric data and 4 labels of boundary conditions

The water level elevation data on three open seas boundary conditions (BC1, BC2 and BC3) were used in the simulation can be seen in **Figure 4**. This elevation data referred to the time from August 1, 2014 to August 1, 2015.

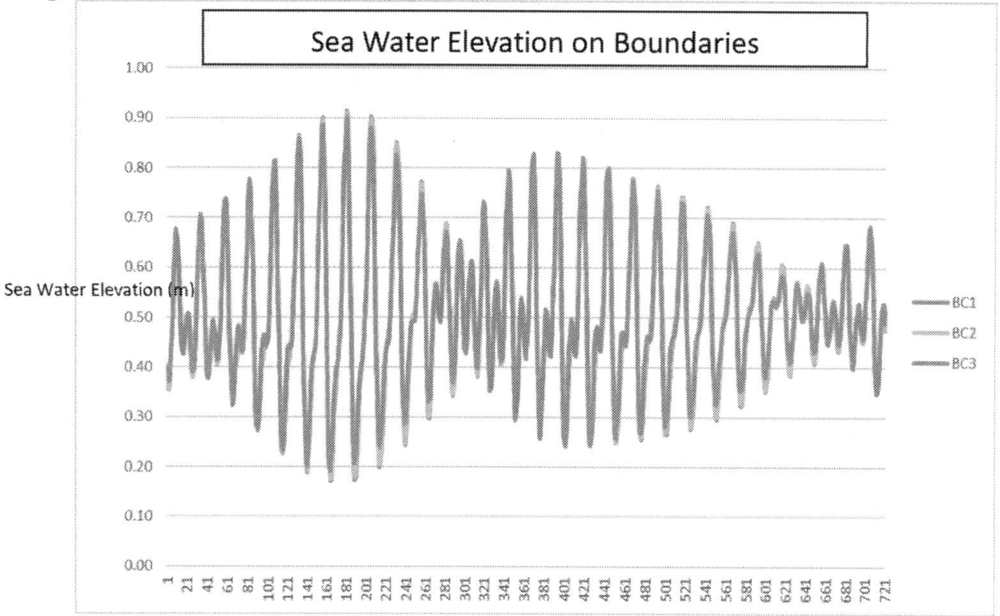

Figure 4. Elevation of water level on 3 open boundary conditions of model

4.3. Calibration

The Calibration of model is carried out to determine whether the model results can match to actual circumstances in the field data. This process is done by trial-and-error method by changing the parameters on the model control (bottom roughness) until it fitted to the desired result [3]. In this calibration was carried out for the output data of water level elevation, current speed, and salinity which is compared to the field data that located at Mayangan River mouth.

In the Preliminary model results before calibration, it showed the water level elevation condition were approached the actual condition with an N-RMSE value of 13%. During calibration, a comparison of the output data and actual of water level elevation can be seen in **Figure 5Figure 5**. And the N-RMSE value for the model's water level elevation can be seen in **Table 1.** As for the current speed condition, the initial model results have not fit to the actual results, therefore the calibration process will be carried out by changing the parameters of bed roughness. Bottom roughness values that used in the calibration process include 0.025, 0.001, and 0.0005, the results are shown at values and the N-RMSE values are given at **Table 2**.

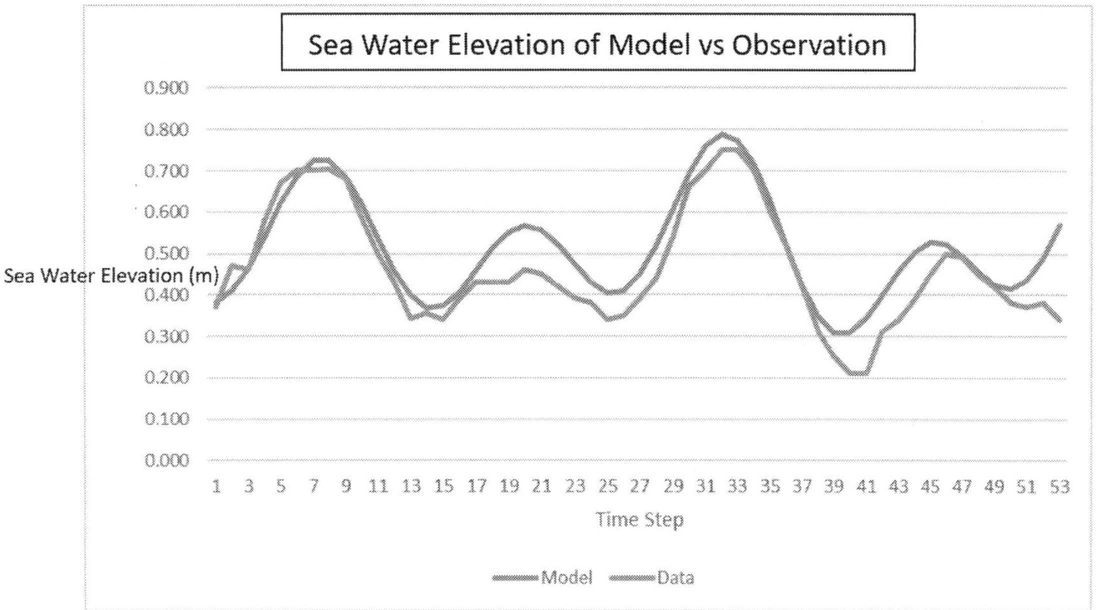

Figure 5. Comparison between model elevation and actual data elevation

Table 1 Comparison N-RMSE values of water elevation between Model and Field Data

Time Step	Model (m)	Observation (m)	Residuals	Squared Residuals	RMSE (m)	N-RMSE %
0	0.383	0.372	0.012	0.000	0.07	12.8
1	0.409	0.472	-0.063	0.004		
2	0.465	0.462	0.003	0.000		
3	0.541	0.582	-0.041	0.002		
4	0.621	0.672	-0.051	0.003		
5	0.687	0.702	-0.015	0.000		
6	0.724	0.703	0.021	0.000		
7	0.724	0.704	0.021	0.000		

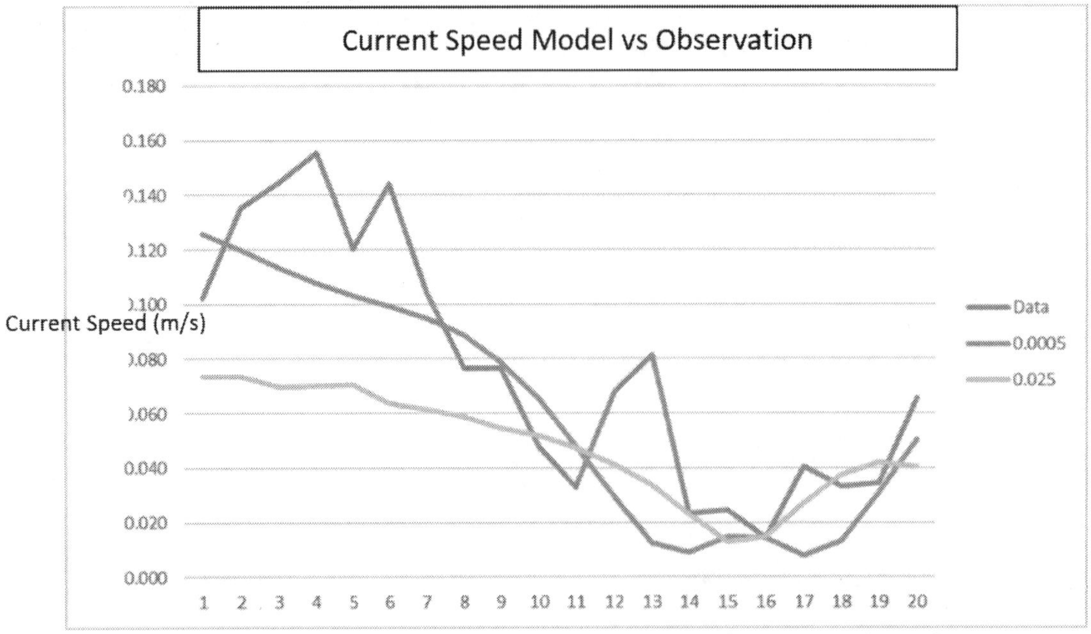

Figure 6. Comparison between model current speed and actual data values

Table 2 Comparison N-RMSE values of water elevation current speed between Model and Field Data

Bed Roughness	RMSE (m/s)	N-RMSE (%)
0.025	0.04	28.85
0.001	0.02	13.99
0.0005	0.02	12.54

Preliminary modelling results with the river boundary condition of the water quality at upstream of domain as the values of 0 ppt showed the distribution of water quality which similar enough to field data with an N-RMSE value of 10.2%. Then the water quality calibration was again carried out using the changing of river boundary condition at the upstream river as the values of 0.5 ppt and 1.5 ppt which resulted in a smaller N-RMSE value than the primary modelling before, which are in total values of 10.1% and 8.57% accordingly. The comparison of the results of water quality calibration with various river boundary values can be seen in **Figure 7** and **Table 3**.

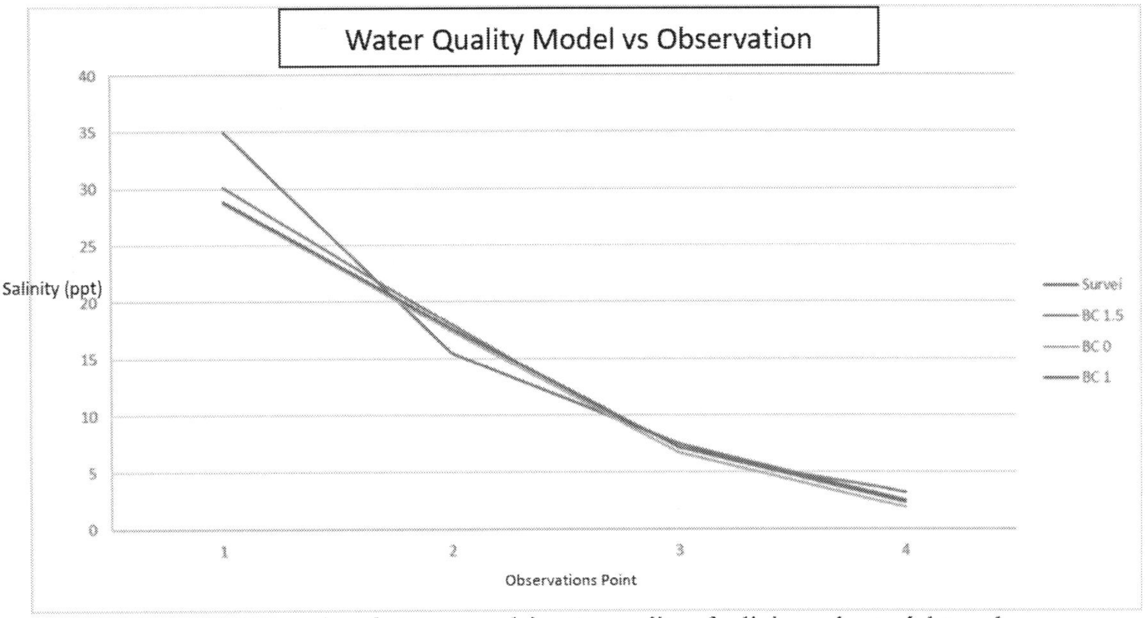

Figure 7. Comparison between model water quality of salinity and actual data values

Table 3 Comparison N-RMSE values of water quality-salinity between Model and Field Data

Observation Point	BC 0	BC 0.5	BC 1.5	Observation
Point 1 (ppm)	28.7	28.8	30.1	35
Point 2 (ppm)	17.4	17.7	18.0	15.5
Point 3 (ppm)	6.7	7.1	7.5	7.5
Point 4 (ppm)	1.9	2.4	3.2	2.5
RMS Error	3.3	3.3	2.8	-
N-RMSE (%)	10.20	10.10	8.57	

Based on the calibration process of the 3 parameters above, we finally selected to use the bed roughness as values of 0.0005 and the salinity of river boundary condition as a value of 1.5 ppm at the upstream of domain of model.

5. Model Results and Discussion

Based on the model simulation results, as shown in **Figure 8** and **Figure 9**, the salinities of the Mayangan River in various scenarios of changes in coastline are very identical and it does not change significantly in terms of the changing of domain model's shape respect to time. In the other word, although the Pondok Bali coastline has retreat up to 570 meters to the direction of land, the distribution of salinity on the Mayangan River has not undergone significant changes. The increase in salinity at some observation points of the Mayangan River from 2006 to 2014 was only 0.2 ppt, from 2011 to 2014 the salinity increase was only 0.1 ppt, and from 2014 to 2020 only 0.03 ppt according to the model results **Table 4**.

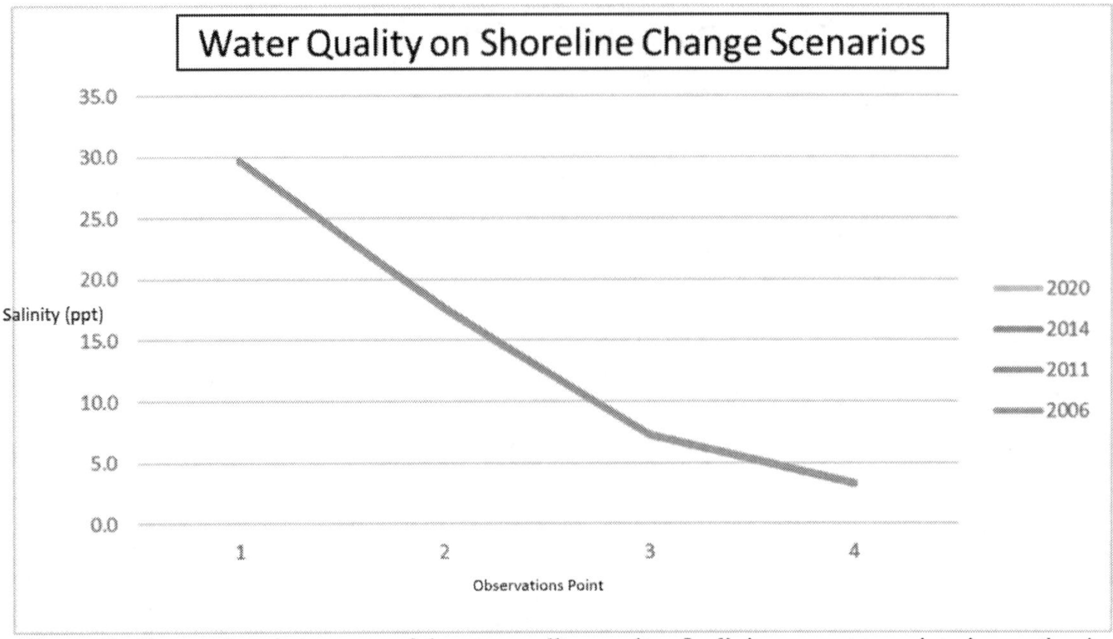

Figure 8. Comparison between model water quality results of salinity respect to point observation in Mayangan River with various coastline change scenarios

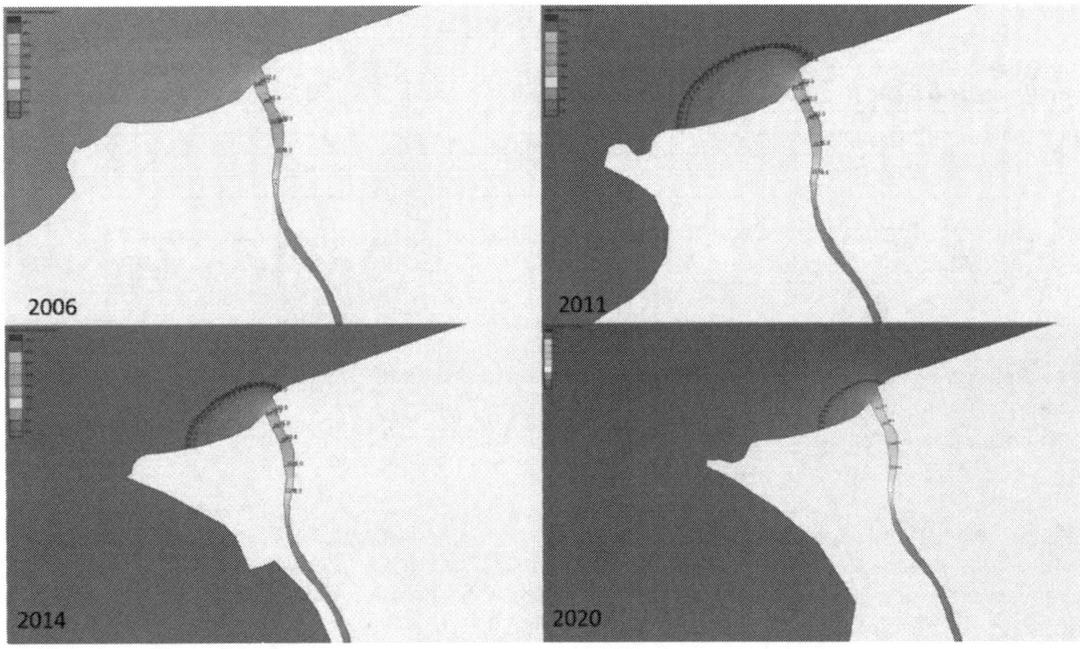

Figure 9. Comparison between contour of model salinity distribution with various coastline change scenarios at Mayangan River.

Table 4 Comparison averaged salinity changes values for various coastline geometry scenarios

Summary	2020	2014	2011	2006
Point 1 (ppt)	29.8	29.7	29.7	29.7
Point 2 (ppt)	17.8	17.7	17.6	17.6
Point 3 (ppt)	7.3	7.3	7.3	7.2
Point 4 (ppt)	3.3	3.3	3.3	3.2
Salinity Average Change (ppt)	**0.03**	**-**	**-0.1**	**-0.1**

6. Conclusion

The results of water quality simulations showed that bathymetry changes did not significantly affect the salinity distribution in the Mayangan River. The increase in salinity at some observation points of the Mayangan River from 2006 to 2014 was only 0.1 ppt as well as from 2011 to 2014 the increase in salinity was only 0.1 ppt and from 2014 to 2020 only 0.03 ppt. In general, it can be concluded that the changes in water quality properties, especially the salinity of the Mayangan River, are not affected by changes in the coastline geometry of Pondok Bali Subang, West Java.

References

[1] Achiari H, Wulandari N, Yustiani Y M, and Harlan D. 2015. *Study Erosion and Coastal Destruction at Pondok-Bali North Coast-West Java of Indonesia.* Proceeding of 34th IIER International Conference, Singapore, ISBN: 978-93-95465-79-6.

[2] Norton W R, King I P and Orlob G. T. 1973. *A Finite Element Model for Lower Granite Reservoir.* prepared for U.S. Army Corps of Engineers, Walla Walla, Washington.

[3] Lopez C L, Silva P A, Rocha A, and Dias J M. 2011. *Sensitivity Analysis of Ria de Aveiro Hydro-Morphodynamics to the Sea Level Rise Integration Period.* Journal of Coastal Research, SI 64 pp 230-234.

Acknowledgments

Authors wishing to acknowledge assistance from colleagues, and financial support from P2MI research funding Program of Coastal Engineering Research Group, FTSL ITB.

Evaluating the use of Google Maps as navigation application by identifying hazards and assessing risks using HIRA matrix

Wildi Kusumasari[1,2], Friska M Ilmi[1,3], Estiara Ellizar[1,4] and Yos Y Rabung[1,5]

[1]Ministry of Transportation, Republic of Indonesia, Jl. Medan Merdeka Timur No. 5 Jakarta Pusat, Indonesia
[2]email: wildi.kusumasari@kemenhub.go.id
[3]email: friska.ilmi@kemenhub.go.id
[4]email: estiara@kemenhub.go.id
[5]email: youssef.rabung@kemenhub.go.id

Abstract. Google Maps as a navigation application provides route information with the shortest route or fastest travel time, but this technology has a weakness as it does not consider the suitability of road condition with the type of vehicles, therefore the choice of the routes does not accommodate traffic safety aspects. This research aims to evaluate the use of Google Maps as a navigation application by identifying hazards and assessing risks. The method used is based on an analysis using the HIRA (Hazard Identification and Risk Assessment) matrix to determine the level of risk thus the step of mitigation can be executed. The data source obtained was from the results of questionnaires distributed to 1,114 respondents. As the results of analysis, it was found that 11 hazards from the usage of Google Maps came from the driver's aspect, Google Maps feature's aspect, and road conditions aspect. Evaluation of risks came from the three sources of hazard classified into 2 risks in the intolerable risk category and 9 risks in the tolerable risk category. After the mitigation efforts were arranged to control the risk, all risks are expected to meet the acceptable risk level. To anticipate the road traffic accident due to the usage of Google Maps, the recommendation for this navigation application is to add more features to warn the users about the road characteristics in the form of voice notification or pop up if not disturb the driver while driving and providing a specific route for truck/heavy vehicle as now the Google Maps is only available for light vehicle.

Keywords: Google Maps, hazard, risk, HIRA, safety

1. Introduction

The rapid development of technology is currently having an impact on various sectors, including the transportation sector where people is convenient to get access to information when traveling. One of the technologies which widely used by people is a navigation application that makes people easier to find information about routes and the fastest route to travel to their destination. The current position locating in a map, road navigation, vehicle tracking, etc. are the most used application using GPS systems [1]. Based on the survey conducted by GfK Indonesia for Baidu in 2015 entitled Mobile Apps Market Study Indonesia, the most popular navigation application on smartphones in Indonesia is Google Maps. It was stated that 3 (three) maps and navigation applications on smartphones were most widely by the Indonesian people were Google Maps (65.03%), then Waze (5.59%), and Google Earth (1.38%) [2].

However, every technology may have weaknesses that need continuous development and improvement to overcome the problem. Like other technologies, the navigation application is a technology that is supported by spatial data. Most of navigation application provides route information based on the shortest time and distance without considering the safety aspect for users. The service from navigation application does not provide warnings about the unsafe roads or information on road class that is adjusted to the type of vehicle as regulated in Law No. 22 of 2009 about Road Traffic and Transportation in Article 19 and Article 301 [3].

Every day, the news of accidents related to navigation applications as the driver's used route suggestions from it was increasingly reported by the media. This requires further analysis of the accident phenomenon, whether the convenience of technology provided by the navigation application is the cause of an accident or the user does not understand how to use it properly, so they get inappropriate advice that caused an accident. Based on this background, this research aims to improve safety features on the existing navigation applications to prevent traffic accidents caused by the drivers who follow route suggestions from the navigation application.

2. Literature Review

The information generated by Google Maps is traffic density and travel time. But in the information, there is a problem of discrepancies between the actual traffic conditions with the traffic density conditions generated by Google Maps. In addition, there is also a difference in actual travel time with travel time generated by Google Maps [4]. This shows that Google Maps has not covered all actual information, which includes one key parameter in road journey which is safety. The analysis of safety is closely related to potential hazards and risks. To prevent accidents, road users must recognize the hazards and be able to control them to avoid the risk of accidents. The occurrence of accidents is caused by unsafe conditions and unsafe act factors. An unsafe condition is a condition that has the potential to cause an accident, for example, the unsafe road conditions and hazards found in navigation applications. Meanwhile, an unsafe act is behavior from the road users that has the potential of road accidents [5].

The potential hazard is different from the danger. Definition of hazard is a condition or an object with the potential to cause death, injuries to road users, damage to equipment or structures, loss of material, or reduction of the ability to perform a prescribed function [6]. Moreover, the danger is a hazardous condition that is exposed to the surrounding environment and there is a high chance of an accident/incident [7]. Risk is a consequence of the existence of a hazard that can be estimated and quantified through the level of probability and severity of occurrence. The hazard identification process is one part of risk management where the risk assessment is a process to determine the priority of control over the level of risk of accidents or occupational diseases [8].

3. Methodology and Data Collection

In this research, we used a descriptive approach with qualitative design to identify the hazard and evaluate the risk of using Google Maps. Hazards are identified and grouped based on their sources which consist of sources of unsafe acts and unsafe conditions. The unsafe act is in the form of unsafe driver behavior, while the unsafe condition is in the form of unsafe road conditions and hazards in navigation applications. Each hazard found from the research variables that have been identified is then coded X1, X2, X3, and so on for hazards from driver behavior, Y1, Y2, Y3, and so on for hazards from navigation applications, and Z1, Z2, Z3 and so on for hazards from road conditions.

The analytical method that will be used is the Hazard Identification and Risk Assessment (HIRA) method which is presented in the form of a HIRA matrix by analyzing each hazard variable found using Google Maps. Broadly speaking, the Hazards that have been identified are then determined by the level of risk, as well as making control measures against the risks that will occur to avoid accidents and minimize the fatality rate due to accidents [7].

The next step of analysis is assessing each risk from the encountered hazard based on the probability of occurrence with the following criteria [6]:

- Risk value 5: always happened on a road trip (experienced by all the respondents)
- Risk value 4: often happened on a road trip (experienced by 70%-90% of respondents)
- Risk value 3: sometimes happened on a road trip (experienced by 40%-60% of respondents)

- Risk value 2: seldom happened on a road trip (experienced by 10%-30% of respondents)
- Risk value 1: never happened on a road trip (no one respondent has experienced it)

The severity of occurrence with the following criteria [6]:

- Risk value A: catastrophic events and tremendous damage and losses (high fatality)
- Risk value B: hazardous events and major damage and losses (1-2 deaths)
- Risk value C: events with great loss and damage
- Risk value D: events with minor loss and damage
- Risk value E: incident with no significant impact, damage, or loss

After an assessment has been carried out, the next step is to combine the probability of occurrence with the severity of occurrence to determine the risk level using the risk index shown in table 1 [6].

Table 1. Safety Risk Index.

Risk Probability		Risk Severity				
		Catastrophic A	Hazardous B	Major C	Minor D	Negligible E
Frequent	5	5A	5B	5C	5D	5E
Occasional	4	4A	4B	4C	4D	4E
Remote	3	3A	3B	3C	3D	3E
Improbable	2	2A	2B	2C	2D	2E
Extremely improbable	1	1A	1B	1C	1D	1E

Furthermore, the risk levels are grouped into 3 levels of safety risk tolerability shown in table 2. The last step of the HIRA process is to formulate risk mitigation efforts to reduce the level of risk to an acceptable level or eliminate the possibility of risk arising altogether [6].

Table 2. Safety Risk Tolerability.

Safety Risk Description	Safety Risk Index Range	Recommended Action
Intolerable	5A, 5B, 5C, 4A, 4B, 3A	Take immediate action to mitigate the risk or stop the activity. Perform priority safety risk mitigation to ensure additional or enhanced preventive controls are in place to bring down the safety risk to tolerable.
Tolerable	5D, 5E, 4C, 4D, 4E, 3B, 3C, 3D, 2A, 2B, 2C	Can be tolerated based on the safety risk mitigation. It may require management's decision to accept the risk.
Acceptable	3E, 2D, 2E, 1A, 1B, 1C, 1D, 1E	Acceptable as is. No further safety risk mitigation is required.

Primary data collections were done by recognizing the public's perception of using Google Maps (online questionnaires) and analyzing substandard road data sourced from The Ministry of Public Works and Housing. Secondary data collection was derived from online media regarding accident data due to the drivers' following instructions from Google Maps.

4. Result

4.1. Google Maps' Users Perception

The data source obtained was from the results of questionnaires distributed to 1,114 respondents on the Balitbanghub online survey. Surveys about the activities most frequently performed by respondents when using Google Maps are intended to find out the centers of activity that will be addressed by Google Maps users that have consequences for the characteristics of the route to be traversed. From the result, it can be concluded that Google Maps is usually used for traveling instead of daily activities (identified as hazard X1), with 31% of Google Maps used by drivers when traveling, and 31% used for visiting relatives/friends as shown on the pie chart in Figure 1.

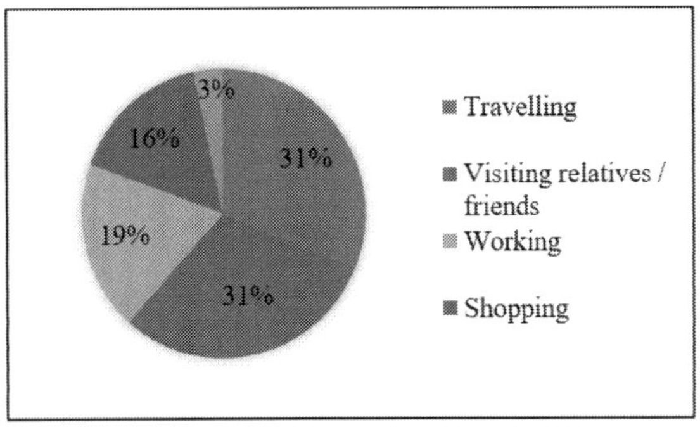

Figure 1. Activities using Google Maps.

Surveys about the frequency and reasons for using Google Maps are meant to find out if users rely on Google Maps to search for routes. Both data are already processed into cross-tabulation. The level of public trust in Google Maps is high. Based on the data shown in Table 3 as many as 48% of all respondents use Google Maps frequently (1-3 times a week) and they trust it because of its ease of use in finding a location. However, the application can potentially mislead the users because they depend heavily on Google Maps to guide them (identified as hazard X2).

Table 3. Cross-tabulation between reasons and frequency in using Google Maps.

Frequency	Reasons			Total
	Easier to find a place/ location	Trust Google Maps more than any other application	Trust Google Maps more than asking others	
Always (Every day)	18%	3%	2%	22%
Frequently (1-3 times a week)	48%	4%	3%	55%
Rarely (<3 times a week)	21%	1%	1%	23%
Total	87%	8%	5%	100%

The survey of vehicle type used by Google Maps users is intended to find out the suitability of vehicle type used with the characteristics of the road that users will cross. Similarly, locations, where drivers were lost when using Google Maps, are meant to find out if Google Maps provides inappropriate route suggestions between road characteristics and vehicle types. Based on the result, 63.38% of respondents got lost when they followed instructions from Google Maps. Both results of the survey conducted are processed into cross-tabulation as illustrated in Table 4. The table shows that 0.7% buses and 0.1% trucks were lost on a local road. The heavy vehicle passed the local road based on the instruction from Google Maps, the drivers choose the car option when clicking the mode option of Google Maps. Meanwhile, there is no distinction between heavy vehicles and light vehicles on Google Maps (identified as hazard Y1). Even though on Google Maps website has written disclaimer about heavy vehicles is not recommended to use Google Maps. But users of Google Maps are huge therefore every person can access Google Maps, whether they know or not about the disclaimer.

Table 4. Cross-tabulation between lost location and type of vehicle.

Type of Vehicle	Lost Location					Total
	Toll Road	Arterial Road	Local Road	Dead End	Others	
Motorbikes	2%	5%	33%	6%	4%	50%
Cars	2%	5%	34%	4%	4%	48%
Buses	0.1%	0.1%	**0.7%**	0.1%	0%	1%
Trucks	0%	0.0%	**0.1%**	0.0%	0%	0%
Others	0%	0.3%	0.4%	0.0%	0.1%	1%
Total	4%	10%	68%	10%	8%	100%

User's consistency in using Google Maps although users have experience got lost shown in cross-tabulation as illustrated in Table 5. Based on the table, Google Maps user is huge as many as 97% of respondents have consistency in using Google Maps, even though sometimes users got lost or misled when following instructions from Google Maps. Google Maps provides route suggestions based on the shortest route or fastest travel time but does not give warnings regarding the dangerous roads (identified as hazard Y2).

Table 5. Cross-tabulation between users' consistency in using Google Maps and impact when users get lost.

Impact when user get lost	User consistency in using Google Maps		Total
	Yes	No	
Travel distance becomes longer	75%	3%	77%
No meaningful impact	11%	0%	11%
Accident	0.3%	0%	0.3%
Met an unsafe road condition	11%	0.3%	11%
Total	97%	3%	100%

4.2. Accident Data Sourced from Online Media

The data about accidents due to the drivers following instructions from Google Maps is not available, but the news about traffic accidents because of it was reported by online media in Indonesia. In this research, data sourced from online media enhanced hazard identification and risk level determination. A motorcycle accident that happened in Bekasi on December 13, 2020, because of unsuitability between the type of road and type of vehicle (motorcycles pass through the toll road), it is possible that the user does not set the route choice to avoid the toll road (identified as hazard X3) and this case cause motorcyclist death [9]. A truck accident happened in Magetan on December 21, 2020, because the driver was speeding, might not comply with the speed limit (identified as hazard X4) so the turning route notification from Google Maps was late (identified as hazard Y4) and the shock driver could not maneuver properly and got an accident [10]. A car accident happened in Mojokerto on June 29, 2019, because the driver was not familiar with the road condition which the environment around the road is a ravine with 25 meters in depth (identified as hazard Z1) [11]. A car was drowned in the river in Semarang on February 26, 2021, because the driver was difficult to distinguish between road shoulder and river during the flood (identified as hazard Z2) [12].

4.3. Substandard Roads in Indonesia

Based on the data shown in Table 6, although the percentage of substandard national roads is not too large compared to the total existing national roads, dangerous (extreme) and high-risk roads might cause accidents. The hazard from substandard roads is the first, if the driver does not recognize the road conditions or it is the first time crossing the road, he cannot anticipate if he encounters the hazard. The second is if the road being traversed does not match the size of the vehicle, for example, buses that pass-through roads of substandard width, making it difficult to maneuver because the width of the road does not match the size of the bus. Another example is a truck loaded with overloading crossing the road with extreme vertical alignment will be difficult to climb with heavy loads or potentially experience braking failure when crossing long descents. Sometimes bus and truck drivers also get suggestions of alternative routes from navigation applications with shorter distances and travel times compared to passing through the main route which is a detour for reasons of time and cost-efficiency. This substandard road's condition in Indonesia can be identified as hazard Y3.

Table 6. Substandard national roads in Indonesia

Road Condition	Length	Percentage	Parameter
Substandard roads width	9.308,62 km	17,82%	Lane width < 2,75 m
Roads with dangerous conditions (extreme) are in hilly and mountainous areas and winding road terrain	422,23 km	0,81%	▪ Vertical alignment: 1. Hill grade: 5-45 m/km 2. Mountain grade: > 45 m/km ▪ Horizontal alignment: 1. Few turns (0,25 – 3,5 rad/km) 2. Many turns (>3,5 rad/km) ▪ Terrain: 1. 1,0 m < hill < 3,0 m 2. Mountain > 3,0 m
Roads with combined conditions between the width of substandard roads and are in hills and mountains and winding road terrain	250,13 km	0,48%	▪ Lane width < 2,75 m ▪ Vertical alignment: 1. Hill grade: 5-45 m/km 2. Mountain grade: > 45 m/km ▪ Horizontal alignment: 1. Few turns (0,25 – 3,5 rad/km) 2. Many turns (>3,5 rad/km) ▪ Terrain: 1. 1,0 m < hill < 3,0 m 2. Mountain > 3,0 m

The total length of national roads: 52.231,80 km
Source: The Ministry of Public Works and Housing, Indonesia

4.4. Hazard Identification and Risk Assessment

HIRA process consists of steps that will be described on the HIRA matrix. The steps are Hazard Identification was drawn by online survey results, online media, and substandard road data. Risk Assessment to the risk index based on probability and severity of occurrence to determine risk level. Mitigation to meet an acceptable level of risk.

Table 7. Hazard Identification and Risk Assessment (HIRA) Matrix.

Hazard Code	Hazard	Risk	Justification of Risk Value		Risk Index
			Probability	Severity	
X1	Google Maps is usually used for traveling instead of daily activities. Drivers are mostly not familiar with the road conditions	Drivers encounter a dangerous road that might lead to an accident	79% of respondents used Google Maps for non-daily activities (4)	Based on the accident from media indicating there were hazardous events and major damage and losses (1 death on motorcycle accident) (B)	**4B**

Hazard Code	Hazard	Risk	Justification of Risk Value		Risk Index
			Probability	Severity	
X2	The level of public trust in Google Maps is high. Therefore, users will be misled greatly if the app failed to deliver.	The driver got lost and lead to an accident	48% of respondents use Google Maps frequently (1-3 times a week) and they trust it because of its ease of use in finding a location (3)	63.38% of respondents got lost when they followed instructions from Google Maps (D)	**3D**
X3	Users do not understand/do not set route options to avoid certain routes	The suggested route does not correspond to the type of mode that might lead to an accident	0.3% of respondents had an accident due to following Google Maps instructions (2)	Based on the accident from media indicating there were hazardous events and major damage and losses (1 death on motorcycle accident) (B)	**2B**
X4	The driver is speeding.	Drivers did not receive notifications from Google Maps on time	A truck accident happened in Magetan on December 21, 2020, because the driver was speeding (2)	Based on the accident news, there were events with great loss and damage on heavy vehicle accidents (C)	**2C**
Y1	There is no distinction between heavy vehicles (HV) and light vehicles (LV) on Google Maps. Therefore, heavy vehicle often got disoriented on a small road	Heavy vehicles (HV) pass routes/classes of roads that did not match the size of their vehicles.	0.7% buses and 0.1% trucks were lost on a local road (2)	Based on the accident from media indicating there were events with great loss and damage on heavy vehicle accidents (C)	**2C**
Y2	Google Maps provides route suggestions based on the shortest time or distance	Misroute or accident on a dangerous road and the driver cannot anticipate it	77% of respondents' travel distance becomes longer when they got lost (4)	Based on the accident from media indicating there were hazardous events and major damage and losses (1 death on motorcycle accident) (B)	**4B**
Y3	There has been no warning on Google Maps about the width of substandard roads, steep inclines, and sharp bends, especially those leading to tourist attractions.	Drivers encounter a dangerous road that might lead to an accident	0.3% of respondents had an accident due to following Google Maps instructions (2)	Based on the accident from media indicating there were hazardous events and major damage and losses (1 death on motorcycle accident) (B)	**2B**
Y4	Notifications from Google Maps are late	Drivers did not receive notifications from Google Maps on time	A truck accident happened in Magetan on December 21, 2020, because the driver was speeding (2)	Based on the accident from media, the turning route notification from Google Maps was late and lead an accident with great loss and damage (C)	**2C**
Z1	The environment around the road is a deep cliff	Drivers encounter a dangerous road	11% of respondents met an unsafe road	Based on the accident from media indicating there were events with	**2C**

Hazard Code	Hazard	Risk	Justification of Risk Value		Risk Index
			Probability	Severity	
		that might lead to an accident	condition when getting lost (2)	great loss and damage (C)	
Z2	Road shoulder is closed to the river and difficult to distinguish it from the road shoulder, especially during the flood.	Drivers encounter a dangerous road that might lead to an accident	11% of respondents met an unsafe road condition when getting lost (2)	Based on the accident from media indicating there were events with great loss and damage (C)	**2C**
Z3	The condition of the road are substandard roads, steep inclines, and sharp bends, especially those leading to tourist attractions	Drivers encounter a dangerous road that might lead to an accident	11% of respondents met an unsafe road condition when getting lost (2)	Based on the accident from media indicating there were events with great loss and damage (C)	**2C**

Based on the HIRA matrix, it can be found 11 hazards and their risks consist of 2 risks in the intolerable category and 9 risks in the tolerable category. As the intolerable risks need to take immediate action to mitigate the risk, it can be recommended to add safety features (audible notification) about the hazardous road, without disturbing the driver while driving. The recommendation is expected lower risk levels from intolerable become tolerable or acceptable. Meanwhile, although 9 risks in the tolerable category can be tolerated based on the safety risk mitigation, it may require improvements to accept the risk. The recommendations to improve the safety aspect on Google Maps are providing voice notifications to remind users to choose the mode before clicking 'start' when accessing Google Maps, enabling high accuracy mode by calibrating a mobile phone, providing a specific route for truck/heavy vehicle as the Google Maps is only specified for the light vehicle currently, establishing a journey risk assessment as input for Google Maps safety feature/warning.

From this study, we could see that most people rely heavily on Google Maps even though it is lacking safety information such as road hazards or road class. These findings could explain various cases of a road accident, where the drivers were using Google Maps when venturing into an area that was not familiar to them. Accidents involving heavy vehicles are also recorded, meaning even professional drivers are relying on Google Maps with the high-risk vehicle.

To improve safety when using Google Maps, we must first understand two contributing factors on why accidents happen. First, unsafe act. One of the main reasons for accidents is due to human error, no matter how safe the instruments or the vehicle. For some drivers, using Google Maps is very convenient with the little restriction that they might get carried away. This proved to be a problem because a driver needs to keep focus along the road. Therefore, Google Maps need to design a better User Interface (UI) so that the driver will attend to its primary duties and disciplines such as mind which types of roads to use with their respective vehicle and obeying the speed limit. Second, unsafe conditions. In this case, Google Maps works by calculating the shortest route or fastest travel time including live traffics. Even though nowadays Google Maps already added some tweaks like a highway, toll, ferry, and some others, key information regarding warnings and extreme road conditions are not included.

As Google Maps become more crucial in vehicle transportation with its addition of safety feature, it might not only be thought that Google Maps is fully integrated into the vehicle itself. We have seen nowadays that smartphone is connected to vehicle via Android Auto or Apple Car Play, where the primary usage is listening to music and using Google Maps/Apple maps. The current implementation consists of using a cable to connect the phone to the car entertainment system, where some more advanced mode already uses wireless. In the future, we could find either a seamless connection between phone and vehicle, or the car will have its own independent unit of the digital map. The importance of Google Maps should not be underestimated in the modern world because no other means can provide a more efficient and reliable way to get to point of origin to the destination than Google Maps. That is why it is important to bring forward the issues of road safety to Google Maps.

5. Conclusion and Recommendation

Hazards of Google Maps usage that cause users to get lost and even have accidents come from a combination of hazards from the driver, the Google Maps navigation application itself, and road conditions. Based on the HIRA matrix, it can be found 11 hazards consist of 2 risks in the intolerable category and 9 risks in the tolerable category. As a risk control effort to meet acceptable risk based on the HIRA matrix, the recommendations of this study are as follows:

- Add warning features and information about road characteristics such as road width and geometric road conditions, either in the form of sound notifications, symbols, or pop up without disturbing the driver while driving.
- Provide voice notifications to remind users to choose the type of mode on Google Maps used before clicking 'start'.
- Enable high accuracy mode on Google Maps by calibrating mobile phones, but this needs to be more socialized to users because not all users understand.
- Provide special route options for heavy vehicle types such as buses and trucks that are distinguished by the type of route on light vehicles.
- Establish journey risk assessment to identify the risk along the road as the input to improve Google Maps safety features.

Acknowledgment
The authors thank the Head of Center for Research & Development on Road Transportation and Railway Ministry of Transportation, Google Indonesia, Google US, and all respondents. The authors also thank the reviewers for their constructive comments. The results and interpretation of the analysis are, however, completely the views of the authors.

References
[1] Islam R. *et al* (2010) *Mobile Application and Its Global Impact* (International Journal of Engineering & Technology IJET-IJENS Vol:10 No:06)
[2] GfK (2015) *Mobile Apps Market Study Indonesia* (accessed on: April 5, 2021) https://www.slideshare.net/BaiduIndonesia/mobile-apps-market-study-indonesia
[3] Law of Republic of Indonesia Number 22 Year 2009 Concerning Road Traffic and Transportation
[4] Paramesti L A, Atunggal D. (2019) *Suitability Level Analysis of Google Map's Travel Time and Traffic Density Classification*
[5] Satmiko H. (2017) *Investigasi Kecelakaan Transportasi* (Depok: Rajawali Press) p 57-58
[6] ICAO Doc. (2018) 9859 *Safety Management Manual (SMM) Fourth Edition* (Montreal: International Civil Aviation Organization)
[7] Panjaitan N. (2017) *Bahaya Kerja Pengolahan RSS (Ribbed Smoke Sheet) Menggunakan Metode Hazard Identification and Risk Assessment di PT. PQR* (Jurnal Sistem Teknik Industri, Vol. 19 No. 2, Juli 2017)
[8] Supriyadi *et al.* (2017) *Identifikasi Bahaya dan Penilaian Risiko pada Divisi Boiler Menggunakan Metode Hazard Identification Risk Assessment and Risk Control (HIRARC)* (Journal of Industrial Hygiene and Occupational Health. Vol. 1, No. 2, April 2017)
[9] Inews. *Kesasar Gara-gara Pakai Google Maps Pemotor Tewas Tenggelam di Kali Bekasi* (accessed on : April 5, 2021) (https://jabar.inews.id/berita/kesasar-gara-gara-pakai-google-maps-pemotor-tewas-tenggelam-di-kali-bekasi
[10] Nusa Daily. *Modal Google Map Truk Semen Terguling di Persawahan* (accessed on : April 5, 2021) https://nusadaily.com/regional/modal-google-map-truk-semen-terguling-di-persawahan.html

[11] Inews. *Andalkan Google Maps Bus Terperosok ke Sawah Saat Menuju Pantai Indrayanti* (accessed on : April 5, 2021) https://yogya.inews.id/amp/berita/andalkan-google-maps-bus-terperosok-ke-sawah-saat-menuju-pantai-indrayanti
[12] Tribun News. *Sopir Tuban Pakai Google Maps Gran Max Kecelakaan Tercebur Sungai Gebanganom Semarang* (accessed on : April 5, 2021) https://jateng.tribunnews.com/2021/02/26/sopir-tuban-pakai-google-maps-gran-max-kecelakaan-tercebur-sungai-gebanganom-semarang

Disaster and Resilient Infrastructures at Musi Riverside Settlement in Palembang

Bambang Wicaksono[1], Ari Siswanto[2], Widya Fransiska[2], Susilo Kusdiwanggo[3]

[1]Department Architecture, Faculty of Engineering Universitas Indo Global Mandiri Palembang, Jendral Sudirman Street, Palembang, South Sumatera
[2]Department Architecture Engineering, Universitas Sriwijaya, Srijaya Negara street, Bukit Lama, Ilir Barat I, Jendral Sudirman Street, Palembang, South Sumatera 30128
[3]Department Architecture, Faculty of Engineering, Universitas Brawijaya, Mayjen Haryono Street 167, Malang, East Java 65145

[1]bambang.wicaksono@uigm.ac.id, [2]ari_sisw58@yahoo.co.id, [2]widyafransiska@ft.unsri.ac.id, [3]kusdiwanggo@ub.ac.id

Abstract. The Settlement of the Musi Riverside in Palembang City is the strongest attraction for the community to build settlements. These settlement of the Musi Riverside, which have been inhabited for generations, has architectural and cultural aspects that reflect traditions that still survive. However, the community's need for housing and riverside infrastructure that extends to cover river catchment areas actually causes changes in land topography, changes in water management, and even river widths that cause urban flooding. The purpose of this study is to reconstruct the resilience of residential infrastructure on the banks of the Musi River, Palembang in the face of urban flooding. This study uses a quantitative method with descriptive analysis, which is made in cross tabulation and based on a geographic information system (GIS). The results showed that the resilience of riverbank infrastructure in the settlements of 5 Ulu and 7 Ulu in dry land was in the form of road expansion, while in wetlands it was in the form of expansion of concrete roads. The direction in reconstructing the disaster resilience of riverside settlements based on exogenous and endogenous aspects is to plan infrastructure criteria for good riverside land.

Keyword: Disaster, Resilience, Infrastructure, Musi Riverside, Settlements

1. Introduction (tambah referensi)

Settlements in Palembang actually develop based on the river environment and its people, this strengthens its character and potential as a city with riverside architecture. The natural environment of the Musi River could be felt by the community until the 1970s, because to fulfill their daily life they came from the river. Naturally, riverside life is a river that has floodplains on the left and right of the river, then linearly occupied by stilt dwellings of riverside communities. Current conditions indicate that the development of regional infrastructure and the growth of riverside settlement facilities lead to changes that are not in accordance with the natural conditions of the river and a decline in the quality of life on the banks of the river. The settlements were expanded without regard to the function of the river as a river border area, flood plain, and transportation route. This is what makes the city of

Palembang lose its local identity as a river city. In addition to these problems, another problem is the destruction of the river environment. Environmental damage is mainly caused by people throwing garbage in the river, so that it accumulates under the stage of the house. This causes river water to become polluted and obstructed river flow, and over time it causes flood disasters at several residential points (Rahman, 2019). The community is vulnerable to disasters because the community deliberately builds houses in flood-prone areas such as on the shores of rivers/rivers and in areas where the tides are frequent. The resilience of people living in flood-prone environments can be caused because they have no choice but to survive (Ariviyanti, 2014).

As a country that is very vulnerable to natural disasters, Indonesia still has major problems, namely the low performance of disaster management, low attention to disaster mitigation, and the weak role of schools/academics in the introduction of disaster mitigation education (Astuti and Sudaryono 2010). Flooding is indeed part of a natural process, but if the flood has crossed the land boundary, it is certain that the flood will bring losses and can threaten danger (Indasari, 2020). Due to the damage to public and household infrastructure, especially in low-lying areas in several areas along the Musi River in Palembang City (Faturrakhman, 2020).

The problems that occur in the city of Palembang, especially in the 5 Ulu and 7 Ulu settlements related to flooding and slums are caused by uncontrolled growth. This condition occurs continuously and tends to be more widespread, and its negative impact is increasingly felt. Changes in settlements occur in river land, housing and networks due to the actions of the community who continuously carry out riverbank development. Specifically, the Musi River tributary is a back area and a place for garbage disposal, the construction of land-patterned dwellings by stockpiling swamp areas that block the flow of water (Wicaksono, 2020). The decline in the functions and benefits of rivers due to pollution, narrowing and silting makes the rivers unable to hold water, then spills it around. Floods hit by carrying piles of garbage into the river. In addition to the expansion of residential areas on the surface, the expansion of riverbank land by piling up swamps is also a problem causing riverbank disasters. This is indicated by the expansion of infrastructure networks as access to the riverbanks by stockpiling to build roads. Then in overcoming this, how to reconstruct disaster resilience and riverside infrastructure?. To answer this problem, this study aims to reconstruct disaster resilience and settlement infrastructure on the banks of the Musi River, Palembang in the face of urban flooding.

One of the growth factors for riverside settlements is through the improvement and development of riverside land infrastructure. In terms of benefits, the improvement and development of this infrastructure is very helpful for the people on the banks of the Musi River in their activities, especially for people whose livelihoods are non-river. However, naturally for riverside environments, repairs or construction of new infrastructure such as concrete roads, inspection roads, riverside concrete sheeting, this actually has a negative impact on the river and tends to block the flow of water. Technically, in relation to the resilience of riverbank infrastructure, it is necessary to undertake infrastructure reconstruction efforts.

The condition of people living in riverbank areas is more vulnerable to all kinds of disaster risks from nature and as a result of the actions of the community itself. (Warsilah, 2018). Residents on the banks of the Musi River have a high vulnerability to fire disasters. The vulnerability of the community to fire disaster on the banks of the Musi river does not reduce the community's desire to stay. The emergence of a community effort in increasing resilience to the threat of fire disasters in the area for people who live on the banks of the Musi river (Meirizki, 2015). In determining the direction of community-based disaster adaptation and mitigation in accordance with the characteristics of the area and in accordance with the social conditions of the community in the area (Handayani, Surya, Anggraeni, & Setyono, 2019)

The concept of resilience has become an integral part of most development and humanitarian programs. The idea that communities are able to "anticipate, mitigate, mitigate, and recover from the effects of adversity without compromising long-term prospects" (Botle 2017). Resilience city refers to the ability of the community to absorb, adapt, and respond to changes that occur in the city system (ICLEI — Local Government for Sustainability, 2002). Vulnerability is related to several basic characteristics of society and its physical, economic, social and environmental characteristics.

2. Stuy Location

2.1. Riverside Flood

There are many factors causing the flood on the Musi River bank, including high rainfall, very low land, tidal waves, upstream deforestation, narrowing of the watershed and inappropriate land conversion. Flooding is not only a disaster caused by hydrometeorological conditions, but also human or community activities in the 5 Ulu and 7 Ulu settlements that cause flooding to occur, such as undirected land use, riverbank stockpiling and not based on land suitability. One of them is the problem of flooding and inundation in residential areas, which occur almost every rainy season, which is a problem that disturbs the community and the city government. The high intensity rain that occurred in the city of Palembang with a duration of more than 6 hours made most of the city of Palembang paralyzed by flooding, especially in settlements located close to the Musi River. Several points in the Palembang area experienced inundation with several variations in height from 0.5-1.5m with a fairly large area. Inundation due to rainwater runoff that is not accommodated by drainage or existing varies with inundation duration from 2 hours to 3 days.

The problem of flooding, especially in riverside residential areas, is indeed a serious challenge for stakeholders, including the Palembang City Government. This is because, the disruption of activities in urban areas, floods also have an impact on people's lives in flood-prone residential areas and their activities. Losses due to environmental imbalances such as the occurrence of puddles at several residential points. So far, disaster prevention as well as disaster management has been attached to the government's obligation so that people are not ready to face disasters and their prevention. Socialization and counseling related to disaster anticipation must be carried out continuously so that the community is able to prepare for disasters so that they can reduce the risk of the impact of the disaster. Flood disaster preparedness education, especially in the city of Palembang, needs to be done repeatedly and regularly so that it becomes a culture in the community.

Figure 1. Musi River tide After a rain with a duration of more than 6 Hours

2.2. Riverside Slums

The level of slums in the Seberang Ulu and Seberang Ilir settlements is classified into 1) not slum, 2) moderate slum, 3) heavy slum 4) very slum. In its development, tributary pollution is caused by the density of buildings. River pollution causes the slums of river settlements that can be seen in the current residential environment. In other words, the density of buildings in the watershed has a negative impact on the river ecosystem. The slums of residential areas, especially river banks, have environmental physical characteristics.

Figure 2. Musi River tributary waste heap

Slums in Palembang settlements are one of the sources of riverside disasters. The slums of settlements tend to be most severe towards settlements that lead to river areas. Slums generally occur under stilts and river areas as waste disposal areas. The reason for the existence of slum areas or the increase in the number of slum areas is that Seberang Ulu is due to economic factors such as poverty and the economic crisis which is higher than Seberang Ilir. In the Seberang Ilir settlement, the typology of the area which is more organized and the narrowing of wetlands due to the construction of non-stilt housing makes the level of slums in the river and under the stilts also lower. Basically, slums occur apart from the community factor throwing garbage into the river, slums also occur due to natural disasters such as floods that carry piles of garbage.

2.3. Riverside Infrastructure

With good infrastructure into the settlements (into the alleys) it is the main access for houses. This is what makes people change the direction towards the house. Network expansion is a form of availability of riverside settlement infrastructure in the form of roads and walkways (*Jerambah*). The expansion of the network is further expanding with trade routes on land which are marked by the construction of land roads and the permanent establishment of wooden rafters into concrete. The expansion of the riverside settlement network can have a positive impact, namely providing better accessibility of settlements and settlement infrastructure. Meanwhile, the negative impact caused the loss of floodplains and blocked river flows due to building roads over rivers. In addition, the imbalance in the pattern of infrastructure and network development for houses on stilts makes people want to make total or partial changes to the physical form of their houses. The natural environment on which humans build settlements is changing in less predictable ways. The condition of the Musi River and several of its tributaries in the 5 Ulu and 7 Ulu settlements are threatened with damage or loss due to extractive economic activities, illegal logging, such as large-scale plantations, mining, infrastructure development, both in the highlands and lowlands.

Figure 3. Infrastructure for the Musi Riverside tributary

3. Methodology

This research on the resilience disaster and infrastructure of settlement uses quantitative methods with a descriptive analysis approach, namely by scoring or assessing disaster resilience and settlement infrastructure in segment 1 or wet land areas and segment 2 or areas on dry land. Then the data was collected through a google earth view search in the settlements of 5 Ulu and 7 Ulu, then field observations and interviews were conducted. Interviews were conducted with residential communities to obtain in-depth information related to disasters faced by settlements. Observations were made more focused on the role of disaster resilience and riverbank infrastructure with variables Infrastructure, environment and riverside disasters and based on Geographic Information Systems (GIS).

The main data collection of this research is a questionnaire using a Likert scale. Questionnaire informants were obtained through purposive sampling with slovin sampling technique with a total of 50 informants in the 5 Ulu and 7 Ulu settlements. This division is based on residential segmentation, namely in communities close to the Musi River and those close to land roads on the banks of the river. Furthermore, the data analysis in this study is descriptive statistics and then cross tabulation is carried out. The output of this crosstab in the form of interpretation of data from the table is a score or assessment of disaster resilience and riverbank infrastructure in segment 1 and segment 2 in the 5 Ulu and 7 Ulu settlements which shows that on dry land it is in the form of road expansion, while in wet land it is an expansion of concrete roads

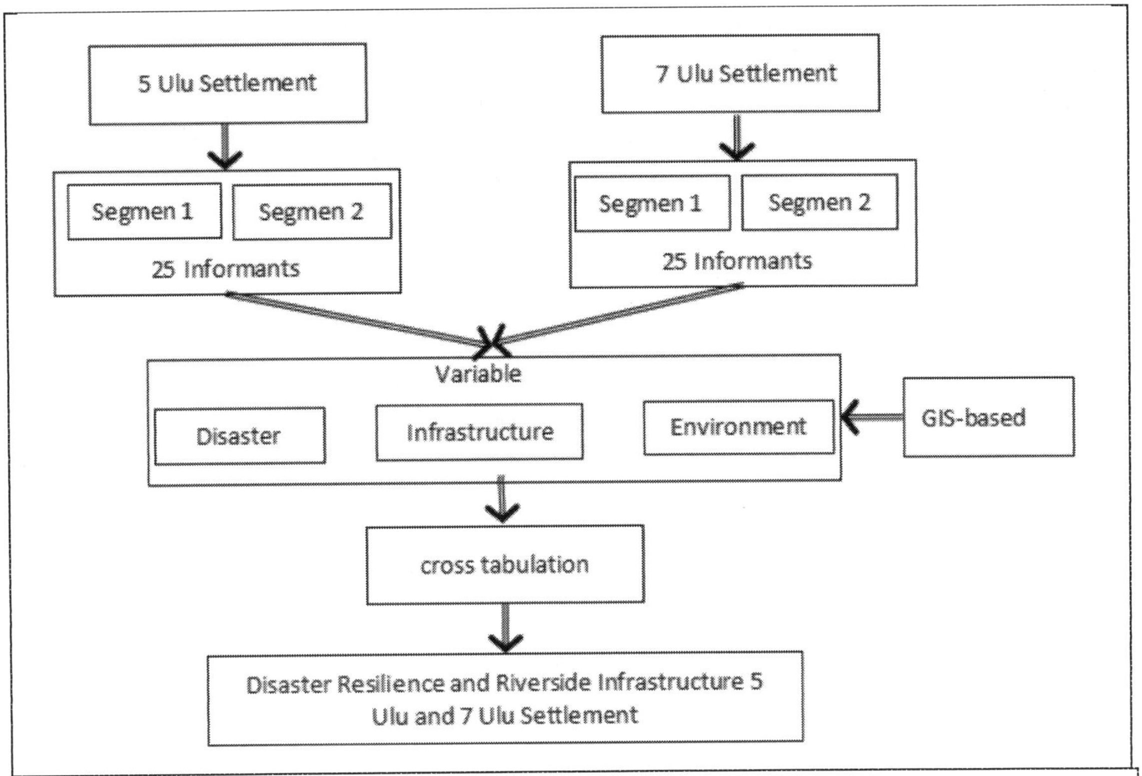

Figure 4. Flow chart Methodology Research

4. Result and Discusion

4.1. Research study area

The study area research is in a settlement of the Musi Riverside in Palembang City with case studies in two locations in the 5 Ulu and 7 Ulu settlements. This research focuses on the causes of disasters and the resilience capabilities of riverside infrastructure. The location of this research is divided into two segments. The allocation of first segment is considered based on the residential segments that are on the water's edge and the second segment that are not connected to the water's edge or lead to land roads.

In segment one, the location of the riverside that borders the water and faces the river that has river activities and culture, namely in settlements that still have a stilt pattern and have infrastructure in the form of wooden walkway (*Jerambah*). The two-character segment of this location is not bordered by the water's edge but on land. The area of this segment already looks more commercial than segment one, in terms of houses where the majority of the settlements are permanent, under the stilts being shops or residences and access roads that have been cast with concrete.

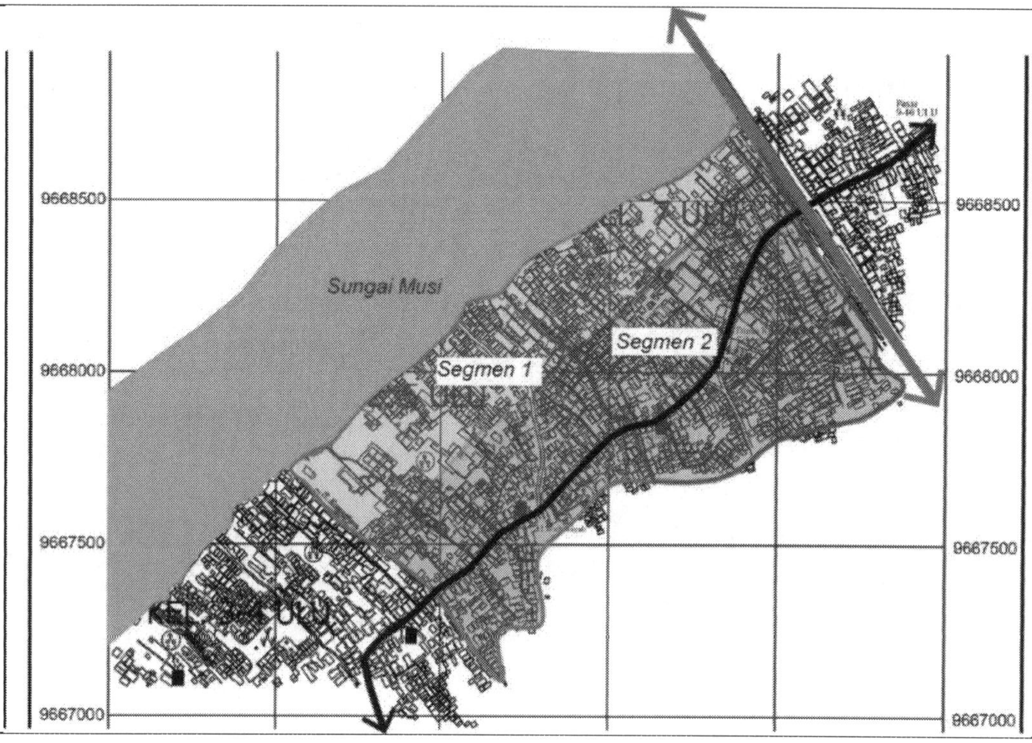

Figure 5. Research Map in 5 Ulu and 7 Ulu Settlement

4.2. Cross Tabulation Analysis

Cross tabulation analysis or cross tabulation is used to explain the level of resilience in the Infrastructure, environment and disaster variables which consist of high, medium and low categories.

Infrastructure: Riverside infrastructure is one of the variables in the identification of community resilience on the riverbanks in 5 Ulu and 7 Ulu Villages. The expansion of the residential surface is accompanied by the expansion of infrastructure by constructing inspection roads and concrete rafters on the river body. The most visible form of infrastructure resilience on the Musi River is the wooden walkway (jerambah). This jerambah is an infrastructure connection of riverside area that still exists and has survived from the beginning of the formation of settlements until now. But now the condition continues to be updated by changing the material to concrete walkway.

(a) (b) (c)

Figure 5. (a) creek bank crossing (b) Riverside inspection road (c) Residential land road

The results of the crosstab analysis on the categorization of infrastructure variables show that the majority of informants in segment 1 and segment 2 use dry land infrastructure which is in the high category, as many as 17 informants or 34%, namely land roads. Meanwhile, 16 people or 32% of informants used wetland infrastructure which was in the medium category, namely wooden walkway (jerambah) and 13 people or 26% of the informants who used river infrastructure which were in the low category, namely inspection roads.

Table 1. Infrastructure Variable Categorization Segment (Research Results, 2021)

	River	Wetland	Dryland	Total
Segmen 1	10	10	5	25
Segmen 2	5	3	12	25
Amount	15	13	17	50

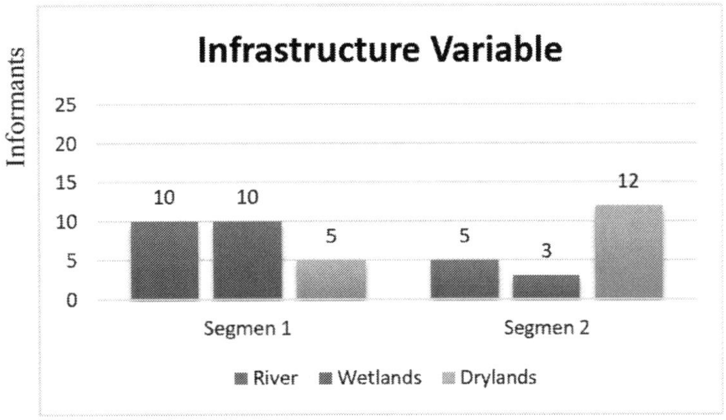

Figure 6. Riverside Infrastructure Variable Graph

The level of infrastructure resilience in each segment is included in the high category with a total score of 17 or 34%. In this category, the majority of people in segments 1 and 2 are more dominant in their activities on land, although there are still a small number of people who still make the river a priority that supports life. Therefore, the level of resilience of riverbank infrastructure is still categorized as high. People who live in riverside wetlands use concrete rafters as access between houses and to land areas.

Environment: The environment is categorized as important in a riverside residential community because environmental management will have an impact on the environment and humans. The riverside area is an area that is vulnerable to slums and disasters. The current condition of the Musi River and its tributaries is a back area and a garbage dump. The watershed is built by illegal houses. Until the 1970s, boats carrying merchandise could still sail on the tributaries that entered the 5 Ulu and 7 Ulu settlements. Entering the 1990s, there have been many community residential developments that tend to damage riverside ecosystems. The lack of public awareness of the riverside environment for waste management can affect the carrying capacity of the environment, especially at the research site there is no garbage dump/garbage container as a result the majority of people, especially those in the river, throw garbage into the river, although the visible garbage also comes from other river flows, but if people continue to throw garbage into the river it will further worsen the environmental quality on the banks of the river.

(a) (b)

Figure 7. (a) River slums (b) Garbage pile under the house

The results of the analysis show that the environmental area causing the disaster using crosstab shows that the environmental variable is high, namely 28 or 56%, namely the river. While the medium category is 16 or 32%, namely wetlands and the low category is 6 or 12%.

Table 2. Environmental Variable Categorization Segment (Research Results, 2021)

	River	Wetland	Dryland	Total
Segmen 1	15	8	2	25
Segmen 2	13	8	4	25
Amount	28	16	6	50

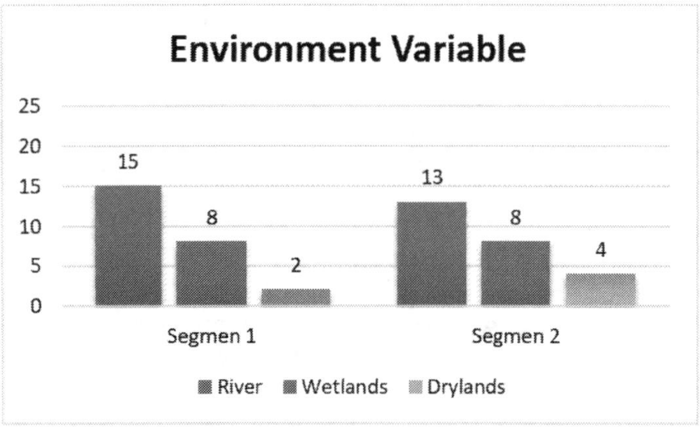

Figure 8. Riverside Environment Variable Graph

The results of the crosstab analysis calculation from segments one and two included in the environmental area that causes high disasters are rivers seen from the total value, namely 28 or 56%. The condition of the river environment is very polluted because of the behavior of people who throw garbage into the river. This shows that their awareness is still low to participate in improving environmental sustainability. For water sources, the majority of riverside communities depend on river water and mainland communities do not depend on river water.

Disasters: Many disasters that occur are caused by human activities, economics and environmental damage that have a negative impact. The research location which is close to the Musi River causes the water to overflow during high tide. the river water recedes, so that if the tides coincide with high rainfall intensity, it often causes flooding. During the dry season, there is a water quality crisis which makes it difficult for the surrounding community to get access to clean water. The occurrence of the dry season also causes the phenomenon of algal blooms or changes in the color of the water to black during the dry season.

Table 3. Segment Categorization of Disaster Variables (Research Results, 2021)

	River	Wetland	Dryland	Total
Segmen 1	10	12	3	25
Segmen 2	8	15	2	25
Amount	27	18	5	50

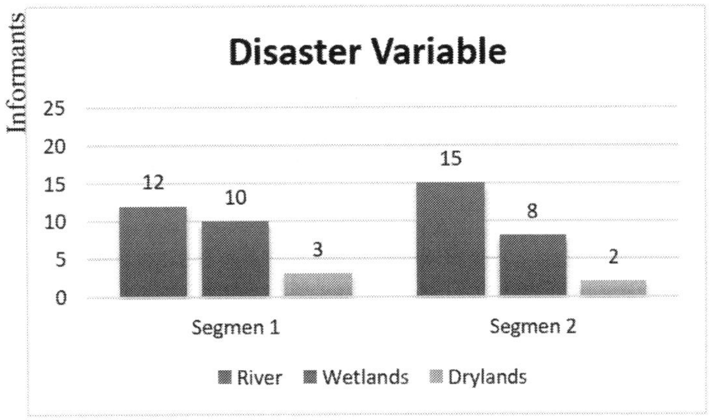

Figure 8. Riverside Disasters Variable Graph

The category of disaster resilience is also in the high category, namely in the river area with a total value of 27 or 54%. Flood disasters that occur around residential areas are very vulnerable as well as tidal flooding and strong winds that occur. Riverside disasters tend to originate from rivers, this is due to the condition of the river being shallow and clogged with piles of garbage, even tributaries have been cut off at several points due to landfills. So that when it rains heavily, the river cannot accommodate the water discharge and spills into the left and right areas of the river and is flooded. The majority of riverside communities still live on riverbanks, and even tend to expand on the surface. Factors that make people want to stay are close water access, activities on the banks of the river in the form of boat rentals, occupying vacant or non-owned land, and kinship or family factors.

Overall, the classification of the level of disaster and resilience of the 5 Ulu and 7 Ulu riverside infrastructure that is assessed based on the variables and conditions that occur in the field is categorized into high, medium and low, which can be seen in table 4.

Table 4. Classification of Disaster Level and Resilience of Riverside Infrastructure for 5 Ulu and 7 Ulu Settlements

Segmen	Infrastructure	Environtment	Disasters	Resistance Classification
Segmen 1	5	15	12	High

Segmen 2	12	13	15	
Amount	17	28	27	72
Segmen 1	10	8	10	Medium
Segmen 2	5	8	8	
Amount	15	16	18	49
Segmen 1	10	2	3	Low
Segmen 2	3	4	2	
Amount	13	6	5	24

The results of the classification in each segment are in the high category. In segment 1 and segment 2, the classification of the high level of resilience with a total score of 72, the medium category with a score of 49 and the low category with a value of 24. Judging from the statistical figures, the area that causes disasters from the aspect of other segments is the environmental and disaster aspects.

5. Conclusion

The results showed that the riverbanks in the 5 Ulu and 7 Ulu settlements consisted of infrastructure, environmental and disaster aspects. Segment 1 and segment 2 include a high level of resilience with a total score of 72. In principle, the nature of the river is to drain water and carry the remnants of river erosion. The river will not cause big problems, if humans or the community consciously disrupt the hydrological cycle and its ecosystem, where settlements form rivers that are not in accordance with the character of the river (exogenous). One of the problems is hoarding or building infrastructure that limits rivers and their settlements. This study has the limitation of not discussing in depth the handling of riverbank disasters with disaster mitigation because each layer of settlements has a different typology. In addition, this occurs due to limited data in the field (primary) and literature data (secondary). The existence of widespread riverside settlements has led to the growth of several new functions in the form of riverside infrastructure, land availability, and residential utilities. So this study shows that the resilience of riverbank infrastructure in the settlements of 5 Ulu and 7 Ulu in dry land is in the form of road expansion, while in wetlands it is in the form of expansion of concrete walkways. The direction in reconstructing the disaster resilience of riverside settlements based on exogenous and endogenous aspects is to plan criteria for good riverside land infrastructure. For the development of further research is to deepen the role of the community in the arrangement of adaptive settlements to overcome settlement disasters.

6. Acknowledgments
The author would like to acknowledge assistance or encouragement from the Chancellor of the Indo Global Mandiri University, fellow academics at the Faculty of Engineering, Department of Architecture, and colleagues in the field of riverside settlement studies who provided input and ideas in the implementation of this research.

7. References

[1] Ariviyanti, N. (2014). Faktor-faktor yang meningkatkan Resiliensi Masyarakat Dalam Menghadapi Bencana Rob di Kelurahan Tanjung Emas Semarang. Jurnal Teknik PWK UNDIP, 3 (4), 993-1002

[2] Astuti, S. I. dan Sudaryono, (2010). "Peran Sekolah dalam Pembelajaran Mitigasi Bencana" dalam Jurnal Dialog Penanggulangan Bencana, vol. 1, no. 1, tahun 2010, hlm. 30-42

[3] Bolte, P., Orlowsky, D.B., Marr, S., Moore, S., Rahmadana, M.F., & Sitompul, D. (2017). Resilience radar user manual.
[4] Faturrakhman A.S, Ari Siswanto, L. Teddy, (2020). Mitigasi Bencana Banjir Di Daerah Padat Penduduk Tepian Sungai Musi Palembang. Seminar Nasional AVoER XII 2020
[5] Handayani, W., Surya, P. I. H., Anggraeni, M., & Setyono, J. S. (2019). Community-based disaster management : Assessing Local Preparedness Groups (LPGs) to build a resilient community in Semarang City , Indonesia. Disaster Advances, 12(5), 1–14.
[6] Meirizki, Hanifa Carissa, (2015). Sebaran Tingkat Ketahanan Masyarakat Permukiman Padat Tepi Sungai Musi Terhadap Bencana Kebakaran Kelurahan 5 Ulu, Kecamatan Seberang Ulu 1, Kota Palembang. Skripsi. Universitas Gadjah Mada, Yogyakarta
[7] Rahman S, dkk, (2019) Identifikasi Karakteristik Permukiman Kumuh Tepian Sungai Di Kelurahan Sungai Bilu Kota Banjarmasin. Jurnal Arsitektur, Manusia, dan Lingkungan. Vol. 1 2. ISSN : 2656-7180
[8] Indasari, Mandalia, Iwan Rudiarto, (2020). Kemampuan Kebertahanan Masyarakat pada Permukiman Rawan Banjir di Kecamatan Barabai, Kabupaten Hulu Sungai Tengah. Jurnal Wilayah Dan Lingkungan. Volume 8 2, 116-129. P-ISSN: 2338-1604, E-ISSN: 2407-8751
[9] Warsilah, H. (2018). Ketahanan Sosial dalam Kota Tangguh Bencana.Jakarta: Yayasan Pustaka Obor Indonesia.
[10] Wicaksono, B. (2020). Changes in the elements of the Palembang Musi riverside settlement. Disertasi, Universitas Sriwijaya.

Impact analysis from visual improvement program of slums to sanitation management: case study Kampung Cibunut, Bandung City, Indonesia

I Kusumayanti[1], L P Defi[2], W M Fatimah[3], P W Sarli[4], P Soewondo[5]

[1] Water Supply and Sanitation Infrastructure Management, Institut Teknologi Bandung, Indonesia, kusumayantiintan@gmail.com;
[1] Water Supply and Sanitation Infrastructure Management, Institut Teknologi Bandung, Indonesia, larasatiputridefi3105@gmail.com;
[3] Water and Wastewater Engineering Research Group, Institut Teknologi Bandung, Indonesia, wikamf@gmail.com;
[4] Civil Engineering, Institut Teknologi Bandung, Indonesia, prasanti.wsarli@yahoo.com
[3] Water and Wastewater Engineering Research Group, Institut Teknologi Bandung, Indonesia, prayatnisoe@yahoo.com;

Abstract. Visual improvement of slum settlements in urban areas is a new breakthrough to increase public awareness of sanitation management; a capital to foster community participation and ensure the sustainability of implemented sanitation program. This study identified the impact from visual improvement program of slums to the sanitation management intention and behavior of the Kampung Cibunut community using the Theory of Planned Behavior. The sampling technique used in this research is cluster random sampling. A sample size of 86 households was obtained using the Yamane formula with a confidence interval of 10%, a population of 369 households, and a cluster technique proportional to size. The research used correlational design with Likert scale questionnaire. Based on the correlation test of the Spearman method, it was found that the visual improvement has a significant correlation with the intention to treat domestic wastewater. However, the visual improvement has no significant correlation with the behavior of treating domestic wastewater, intention and behavior of paying the cost of domestic wastewater management.

Keywords: domestic wastewater management, sanitation, slum settlement, theory of planned behavior, visual improvement program

1. Introduction

In 2035, the percentage of urban population in West Java Province is projected to reach more than 80% of the total population of West Java Province [1]. The increase of population and poverty in urban areas, if not followed by housing provision and job opportunities, will encourage the formation of slum settlements. Visual improvement of slum settlements in urban areas, such as in Kampung Jodipan-Ksatrian, Malang [2], Kampung Code, Yogyakarta [3], Kampung Wonosari, Semarang, Kampung Sindulang, Manado [4], and Kampung Pelangi 200, Bandung [5] became a new breakthrough in increasing public awareness of sanitation management. The painting of village which was originally intended only to improve the visuals of slum settlements and turn them into tourism destinations, turned out to have a positive impact on the behavior of community sanitation management including the management of domestic wastewater. Public awareness is the capital to grow community participation in sanitation programs that are implemented because when a person has a high level of awareness about sanitation management, one will have a sense of urgency and attach importance to a sanitation facility. In addition, someone who has a high level of awareness about sanitation management will also be more willing to contribute to the form of thoughts, energy, expertise, goods, and money for the operation and maintenance of the sanitation facilities owned or used [6].

The success of a visual improvement program for slums towards changes in community sanitation behavior is characterized by: 1) strategic location choices so that it is easier to attract public attention; 2) media attention that creates hype that results in an influx of tourists that triggers changes in people's behavior; 3) direct economic impact of village tourism objects; 4) support local government attention by allocating budget for sanitation purposes. Kampung Cibunut is said to have fulfilled the four indicators of success [2]. Therefore, it is necessary to conduct further research related to the impact of visual improvement program of slums to sanitation management, specifically domestic wastewater management.

2. Methodology

2.1. Research Approach

This study used a non-experimental quantitative approach with a correlational research design. The research will mainly focus on investigating the behavior of treating domestic wastewater and the behavior in paying the cost of wastewater management using the Theory of Planned Behavior. Based on the construct of Theory of Planned Behavior proposed by Ajzen [7], behavioral beliefs lead to positive or negative attitudes toward certain behaviors (Attitudes Toward Behavior), normative beliefs lead to the formation of perceptions of social pressure to take action (Subjective Norms), and control belief gives rise to one's perception in controlling/maintaining a behavior (Perceived Behavioral Control). The combination of attitudes towards behavior (ATB), subjective norms (SN), and perceptions of behavioral control (PBC) results in the formation of intention to conduct certain behavior. Intention is the closest determinant of behavior, or it can be said that the behavior displayed by a person will be consistent with his intention towards the behavior. Therefore, community's intention to treat domestic wastewater and pay the cost of domestic wastewater management is an independent variable and community's behavior to treat domestic wastewater and pay the cost of domestic wastewater management is the dependent variable. Details of the items in the dimensions of attitude towards the behavior, subjective norm and perceived behavioral control for the behavior of treating domestic wastewater and paying the cost of domestic wastewater management are shown in Table 1.

Table 1. Details of the items in the dimensions of attitude towards the behavior, subjective norm and perceived behavioral control for the behavior of treating domestic wastewater and paying the cost of domestic wastewater management

No.	Item	Dimension
1.	Public knowledge related to domestic wastewater treatment and the cost of domestic wastewater management	Attitude towards the behavior (ATB)
1.	Neighbors influence the behavior of treating domestic wastewater and paying the cost of domestic wastewater management	Subjective norm (SN)
2.	Family influence the behavior of treating domestic wastewater and paying the cost of domestic wastewater management	

Table 1 (continued)

No.	Item	Dimension
3.	Youth organization influence the behavior of treating domestic wastewater and paying the cost of domestic wastewater management	
4.	Family welfare empowerment organization influence the behavior of treating domestic wastewater and paying the cost of domestic wastewater management	
5.	Head of neighborhood influence the behavior of treating domestic wastewater and paying the cost of domestic wastewater management	
6.	Head of hamlet influence the behavior of treating domestic wastewater and paying the cost of domestic wastewater management	
7.	Head of region influence the behavior of treating domestic wastewater and paying the cost of domestic wastewater management	
1.	Avalailability of sanitation facillities and funds owmned by community influence the behavior of treating domestic wastewater and paying the cost of domestic wastewater management	*Perceived behavioral control (PBC)*
2.	The visual improvement program of Kampung Cibunut has an impact on additional income of village tourism objects affects behavior of treating domestic wastewater and paying the cost of domestic wastewater management	

2.2. Population

The number of households in Kampung Cibunut, Hamlet 07, Kebon Pisang Urban Village, Sumur Bandung Sub-district, Bandung City is 391 households. The visual improvement program was only carried out in 9 out of 10 neighborhoods, namely from neighborhood 1 to neighborhood 9 so that the population in this study amounted to 369 households.

2.3. Sampling Method

The sampling technique used is cluster random sampling. In this study, clusters were determined based on the number of neighborhoods who carried out the visual improvement program, so that there were 9 categories in this study. Because each of these categories has a different number of populations, the sampling method used is the cluster method with a proportional size/probability proportional to size, where the category with a large population has a greater chance of being selected as a sample [8].

2.4. Sample Size

The sample size in this study was determined using the Yamane formula [9].

$$n = \frac{N}{1 + Ne^2}$$

With a population size (N) = 369 households and a confidence level or e = 10%, the total number of research samples was 79 households, but to keep the number of samples in accordance with the calculation if an error occurred, the number of samples was added up to 86 households. Determination of the number of samples per category is shown in Table 1.

Table 2. Sample size of population

	RT 1	RT 2	RT 3	RT 4	RT 5	RT 6	RT 7	RT 8	RT 9	Total
Number of Households	30	33	44	39	51	50	48	38	36	**369**
Sample Proportion of Each Neighborhoods (RT)	9	9	10	10	10	10	10	9	9	**86**

3. Description of Study Area

3.1. General Description of Study Area
Kampung Cibunut is in Kebon Pisang Urban Village, Sumur Bandung Sub-district, Bandung City, West Java. The village which has an area of 34,857 m^2 is bordered by Baranangsiang Street and Cibunut River in the north, Cibunut River and Veteran Street in the south, Sunda Street in the west and Jenderal Ahmad Yani Street in the east. The main entrance to Kampung Cibunut is on Sunda Street. Kampung Cibunut is a unitary area of Hamlet 07 Kebon Pisang Urban Village which consists of 10 neighborhoods. Kampung Cibunut is in a strategic location, on the main road, only 1.6 km from the city center (Bandung City Square), only 1.8 km from the Bandung City government center (Bandung City Mayor's Office), only 2.9 km from the government center of West Java Province (Governor's Office) and is surrounded by various commercial facilities.

3.2. Description of Visual Improvement Program
The dream of realizing Kampung Cibunut become colorful village has been iniciated since 2010 from the youth of Kampung Cibunut who have a high interest in mural art. With their mural art skills, the youth organizations want to create an aesthetically pleasing visual appearance of Kampung Cibunut. In 2016, the Bandung Forum for Waste-Free Champion held a competition called "Adu Geulis" which is a competition to make proposals containing ideas in terms of structuring residential areas. Seeing a meeting point between the aspirations of the youth organizations and the Adu Geulis competition, namely the arrangement of the Kampung Cibunut area for the better living residential, these youth organization consolidated with stakeholders from neighborhood (rukun tetangga/RT), hamlet (rukun warga/RW), non-governmental groups (kelompok swadaya masyarakat/KSM) and family welfare empowerment organization (pemberdayaan kesejahteraan keluarga/PKK). This consolidation resulted in an agreement that the concept of zoning that would be proposed for the Adu Geulis competition was to create a mural that has educational value and paint all houses in Kampung Cibunut where each neighborhood has a different color from the others. The creative ideas of the people of Kampung Cibunut as outlined in the "Adu Geulis" proposal were appreciated by the Bandung Forum for the Free Waste of Waste and as a form of appreciation, Kampung Cibunut received assistance in the form of 4500 liters of paint of various colors from PT. ICI Paints Indonesia (Azkonobel Decorative Paints Indonesia) for coloring about 230 houses in Kampung Cibunut. The village painting activity started from 30 July to 17 November 2017. On 27 November 2017, the Kampung Cibunut Berwarna was inaugurated by the Mayor of Bandung as a creative village with an environmental perspective.

4. Results and Discussion

4.1. Mass Media Attention and Number of Visitors
Since the area was inaugurated as a creative village with an environmental perspective, there have been a total of 939 visits from 2017 to 2021. In period of 5 years, the most visits occurred in 2018 with 418 visits. The number of tourists that visit Kampung Cibunut can't be separated from the role of media attention that creates hype around people. Based on a search via Google, in a period of 5 years from 2017 to 2021, there are a total of 54 articles that has been published about Kampung Cibunut tourism with the highest number of articles being published was 37 articles in 2017.

4.2. Economic Impact of Visual Improvement Program
Based on the questionnaire results, there are 3 respondents opened a business after visual improvement program, along with those 12 respondents had opened a business before visual improvement program and the remaining 71 respondents did not have business at all. The average income increased by 44% after visual improvement program or around Rp687,000.

4.3. Government's Support Related to Sanitation

Access to domestic wastewater in the form of municipal domestic wastewater sewerage first built in Kampung Cibunut in 1987 through "Bandung Urban Development Program" or BUDP. Based on the questionnaire results, there are 51% of respondents who are connected to the sewerage system and the remaining 49% of respondents discharge their domestic wastewater directly into the river without prior treatment. In 2018, Kampung Cibunut became a gathering point for a series of "World Habitat Day and World Cities Day" events which were held from September 28 to October 3, 2018. The event was held by the Ministry of Public Works and Public Housing (PUPR). Knowing that the domestic wastewater in Kampung Cibunut has not been treated safely, PUPR stakeholders submitted an offer to build a communal domestic wastewater treatment plant. However, the stakeholders of Kampung Cibunut are still considering this offer due to unavailability of land and the condition of Kampung Cibunut which prone to flooding.

4.4. Community's behavior of treating domestic wastewater

All survey items were measured on a 5 – point Likert Scale. Number 1 represents "strongly disagree" and number 5 represents "strongly agree". In this study, the instrument validity test was carried out with expert's assessment and Cohen's Kappa Index type Intraclass Correlation Coefficients (ICC). By using the RStudio software, the ICC value is 0.691 with *p-value* of <0.05, based on the approach proposed by Koo and Li [9], the instrument is valid with the "acceptable" category. The reliability test was carried out using Cronbach's alpha. By using the RStudio software, the Cronbach's alpha value is 0.68. Based on the interpretation of Cronbach's alpha, according to De Vellis [11], the instrument is reliable with the "minimally acceptable" category. The impact of the visual improvement program in Kampung Cibunut on behavior of treating domestic wastewater in terms of the Theory of Planned Behavior components correlation value is shown in Table 3.

Table 3. The impact of the visual improvement program in Kampung Cibunut on behavior of treating domestic wastewater in terms of the Theory of Planned Behavior components correlation value

Correlation	p-value	Interpretation of p-value	Spearman ρ	Interpretation of Spearman ρ
SN – I	0.234	Not significant	-0.130	*Negligible relationship*
ATB – I	0.520	Not significant	-0.070	*Negligible relationship*
PBC – I	0.002	Significant	-0.329	*Moderate negative relationship*
SN – ATB	0.406	Not significant	-0.091	*Negligible relationship*
SN – PBC	1.28E-05	Significant	0.452	*Strong positive relationship*
PBC – ATB	0.255	Not significant	0.124	*Negligible relationship*
I – B	0.136	Not significant	0.162	*Negligible relationship*
PBC – B	0.276	Not significant	-0.119	*Negligible relationship*

SN: subjective norm, ATB: attitude towards the behavior, PBC: perceived behavioral control, I: intention, B: behavior, significant (p-value < 0,05), not significant (p-value > 0,05)

Based on Table 3, because the visual improvement program is one indicator of perceived behavioral control in treating domestic wastewater in Kampung Cibunut, the visual improvement program has a significant correlation with the intention to treat domestic wastewater in Kampung Cibunut. The perceived behavioral control and the intention to treat domestic wastewater has correlation value of -0.329 (*p-value* <0.05) and categorized as a moderate negative relationship. Based on the cross tabulation between perceived behavioral control and intention to treat domestic wastewater, there are 13% of respondents with a moderate perceived behavioral control value showed very low (2%), low (7%), high (1%) and very high (3%) intention to treat domestic wastewater value. There are 55% of respondents with high perceived behavioral value showed very low (12%), low (28%), high (10%) and very high (5%) intention to treat domestic wastewater value. There are 32% of respondents with very high perceived behavioral value showed very low (21%), low (7%), high (3%) and very high (1%) intention to treat domestic wastewater value.

The perceived behavioral control and the behavior of treating domestic wastewater has no significant correlation. Based on the cross tabulation between perceived behavioral control and behavior of treating domestic wastewater, there are 13% of respondents with a moderate perceived behavioral control value showed very low (12%) and low (1%) behavior of treating domestic wastewater value. There are 55% of respondents with high perceived behavioral value showed very low (29%) and low (26%) behavior of treating domestic wastewater value. There are 32% of respondents with very high perceived behavioral value showed very low (27%) and low (6%) behavior of treating domestic wastewater value. The plot of the correlation value on the Theory of Planned Behavior diagram for behavior of treating domestic wastewater is shown in Figure 1.

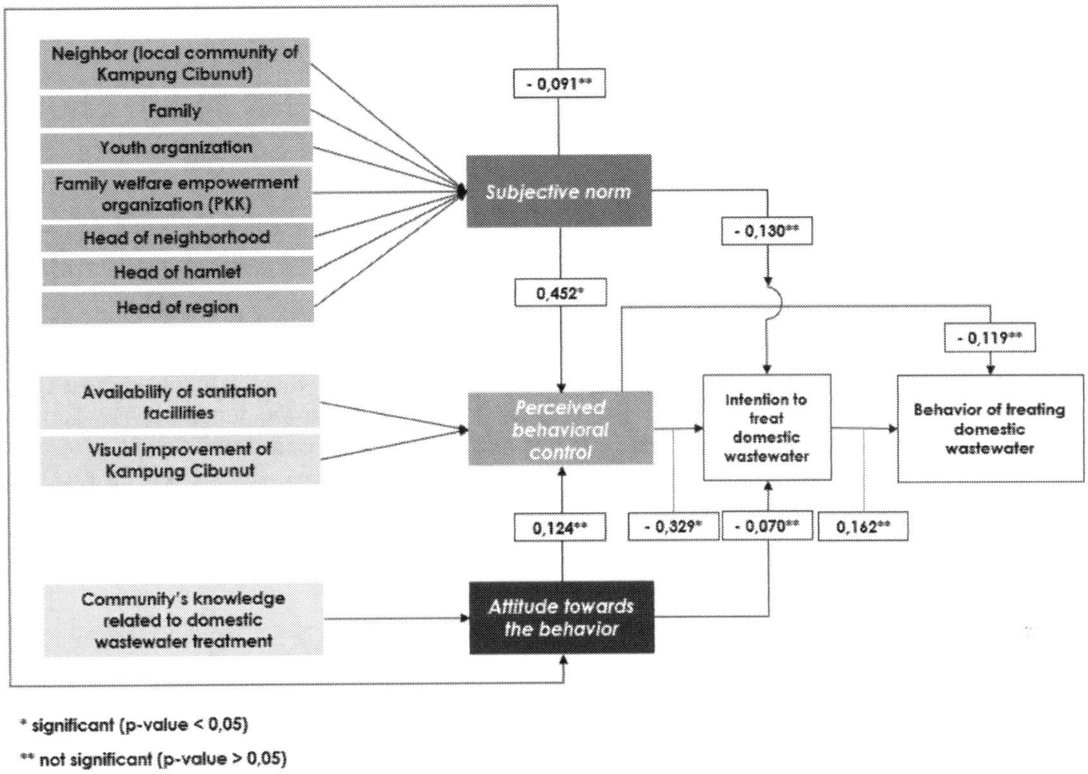

Figure 1. The plot of the correlation value on the Theory of Planned Behavior diagram for behavior of treating domestic wastewater

The low intention and behavior of treating domestic wastewater is because as many as 51% of respondents have been connected to the sewerage system so that the intention to treat domestic wastewater has emerged long before the visual improvement program is carried out, 24.5% of respondents who have not been connected to the sewerage system stated that they did not have the intention to treat domestic wastewater after the visual improvement program and only 24.5% of respondents had the intention to treat domestic wastewater.

4.5. Community's behavior of paying the cost of domestic wastewater management

The impact of the visual improvement program in Kampung Cibunut on behavior of paying the cost of domestic wastewater in terms of the Theory of Planned Behavior components correlation value is shown in Table 4.

Table 4. The impact of the visual improvement program in Kampung Cibunut on behavior of paying the cost of domestic wastewater in terms of the Theory of Planned Behavior components correlation value

Correlation	p-value	Interpretation of p-value	Spearman ρ	Interpretation of Spearman ρ
SN – I	0,017	Significant	0,256	*Weak positive relationship*
ATB – I	0,270	Not significant	-0,120	*Negligible relationship*
PBC – I	0,776	Not significant	0,031	*Negligible relationship*
SN – ATB	0,760	Not significant	0,033	*Negligible relationship*
SN – PBC	1,43E-04	Significant	0,399	*Strong positive relationship*
PBC – ATB	0,153	Not significant	0,156	*Negligible relationship*
I – B	0,934	Not significant	-0,009	*Negligible relationship*
PBC – B	0,841	Not significant	-0,022	*Negligible relationship*

SN: subjective norm, ATB: attitude towards the behavior, PBC: perceived behavioral control, I: intention, B: behavior, significant (p-value < 0,05), not significant (p-value > 0,05)

Based on Table 4, because the visual improvement program is one indicator of perceived behavioral control in paying the cost of domestic wastewater management in Kampung Cibunut, the visual improvement program has no significant correlation with the intention to pay the cost of domestic wastewater management in Kampung Cibunut. The perceived behavioral control and the intention to pay the cost of domestic wastewater management. Based on the cross tabulation between perceived behavioral control and intention to pay the cost of domestic wastewater management, there are 5% of respondents with a very low perceived behavioral control value showed very low (5%) intention to pay the cost of domestic wastewater management value. There is 1% of respondents with a moderate perceived behavioral control value showed very low (1%) intention to pay the cost of domestic wastewater management value. There are 22% of respondents with a moderate perceived behavioral control value showed very low (9%), low (2%), high (6%) and very high (5%) intention to pay the cost of domestic wastewater management value. There are 44% of respondents with high perceived behavioral value showed very low (19%), low (13%), high (7%) and very high (5%) intention to pay the cost of domestic wastewater management value. There are 29% of respondents with very high perceived behavioral value showed very low (12%), low (9%), high (5%) and very high (3%) intention to pay the cost of domestic wastewater management value.

The perceived behavioral control and the behavior of paying the cost of domestic wastewater management has no significant correlation. Based on the cross tabulation between perceived behavioral control and behavior of paying the cost of domestic wastewater, there are 5% of respondents with a very low perceived behavioral control value showed very low (5%) behavior of paying the cost of domestic wastewater. There is 1% of respondents with a moderate perceived behavioral control value showed very low (1%) behavior of paying the cost of domestic wastewater value. There are 22% of respondents with a moderate perceived behavioral control value showed very low (16%) and low (6%) behavior of paying the cost of domestic wastewater value. There are 44% of respondents with high perceived behavioral value showed very low (32%) and low (12%) behavior of paying the cost of domestic wastewater value. There are 29% of respondents with very high perceived behavioral value showed very low (24%) and low (5%) behavior of paying the cost of domestic wastewater value. The plot of the correlation value on the Theory of Planned Behavior diagram for behavior of paying the cost of domestic wastewater is shown in Figure 2.

The low intention of the community to pay the cost of domestic wastewater management is because 41% of respondents have been connected to the sewerage system and have already paid the cost of sewerage system management long before the visual improvement program was carried out, 10% of respondents who have been connected to the sewerage system but have not paid the cost of sewerage system management stated that the intention to pay the cost of sewerage system management had appeared before the visual improvement program was carried out, 7% of respondents who still discharged their wastewater directly into the river stated that they had no intention of paying the cost of domestic wastewater management, 12% of the respondents who still discharged their wastewater directly into the river stated that the intention to pay the cost of wastewater management had appeared before the visual improvement program was carried out and 30% of respondents who still discharged their wastewater directly into the river stated that the intention to pay the cost of domestic wastewater management appears after the visual improvement program was carried out.

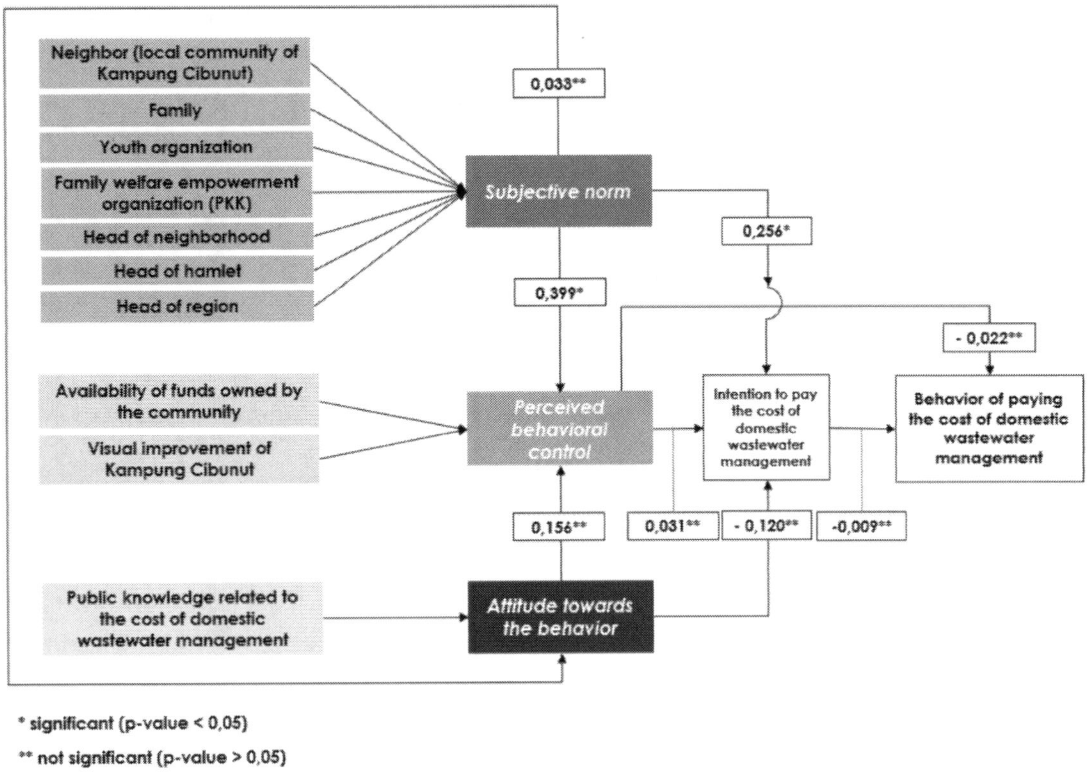

Figure 2. The plot of the correlation value on the Theory of Planned Behavior diagram for behavior of paying the cost of domestic wastewater management

5. Conclusion

Kampung Cibunut has met the indicators of success in visual improvement program of slum settlements to sanitation behavior including: 1) located in a strategic location; 2) media attention that creates hype which results in an influx of tourists; 3) direct economic impact of village tourism objects; 4) local government attention support. Based on the correlation test of the Spearman method, Based on the correlation test of the Spearman method, it was found that the visual improvement has a significant correlation with the intention to treat domestic wastewater. However, the visual improvement has no significant correlation with the behavior of treating domestic wastewater, intention and behavior of paying the cost of domestic wastewater management.

Acknowledgements

The authors would like to extend their gratitude toward the Research, Community Development and Study Group Innovation Program (*Program Penelitian, Pengabdian Masyarakat dan Inovasi Kelompok Keahlian* - P2MI) year 2021 of Institut Teknologi Bandung (ITB) for funding this research. This study was also carried out with the help of Water and Wastewater Engineering Research Group of the Faculty of Civil and Environmental Engineering, ITB and the Global Sanitation Graduate School Program in the Water Supply and Sanitation Infrastructure Management Study Programme, Faculty of Civil and Environmental Engineering, ITB.

Reference

[1] Badan Pusat Statistik 2018 *Indonesia Population Projection 2010-2035*. Volume number. (Jakarta: Badan Pusat Statistik) p 35

[2] Sarli P and Soewondo P 2017 *Strategic Thinking to Change Behavior and Improve Sanitation in Jodipan and Kesatrian, Malang, East Java, Indonesia* (World Academy of Science, Engineering and Technology, Open Science Index 130, International Journal of Urban and Civil Engineering) 11(10) p 1443

[3] Sarli P, Widiharso W, Zakiyya N M 2018 *Visual Improvement of Urban Slums as a Strategy to Improve Residents' Behavior towards Sanitation in Kampung Code, Yogyakarta, Indonesia* (Cambodia: The 11th Regional Conference on Environmental Engineering 2018 (RCEnvE-2018)) p 120

[4] P Soewondo, Windini, D Sulasiah, A K Putra, N M Zakiyya, P W Sarli, and M Handajani 2020 *Visual Improvement of Slum Areas to Accelerate Universal Access to Domestic Wastewater Treatment (Case study of Yogyakarta, Semarang and Manado) (IOP Conf. Ser.: Earth Environ. Sci.)* 409 012006 p 9

[5] W M Fatimah, P W Sarli, P Soewondo, and N M Zakiyya 2020 Strategic thinking to improve sanitation in Kampung Pelangi 200, Bandung City: in comparison with Kampung Jodipan-Ksatrian, Malang City (IOP Conf. Ser.: Earth Environ. Sci.) 592 012018 p 15

[6] Y Du, X Wang, D Brombal, A Moriggi, A Sharpley, and S Pang 2018 *Changes in Environmental Awareness and Its Connection to Local Environmental Management in Water Conservation Zones: The Case of Beijing, China* (Sustainability 10(6):2087) p 2

[7] I Ajzen 2005 *Attitudes, Personality and Behavior Second Edition* (New York) p 135

[8] P J Lavrakas 2008 *Encyclopedia of Survey Research Methods* (California: SAGE Publication) Volume 1 & 2 p 35

[9] Taro Y 1967 *Statistics, An Introductory Analysis, 2nd Ed.* (New York: Harper and Row) p 886

[10] Koo T K and Li M Y 2016 *A Guideline of Selecting and Reporting Intraclass Correlation Coefficient for Reliability Research* (Journal of Chiropractic Medicine) Volume 15 Issue 2 p 4

[11] DeVellis R F 2003 Scale Development: Theory and Applications Second Edition (California: Sage Publications) p 95

Redefining Role of Stakeholders in Restoring Hydrological Function and Conservation of Biodiversity of Nipa-Nipa Grand Forest Park, Southeast Sulawesi

Ira Ryski Wahyuni[1*], *Alfian Alfian*[2], *Mutmainnah Mutmainnah*[2], *La Ode Muhammad Erif*[3] and *Fitria Nur Indah Djafar*[2]

[1] Faculty of Science and Technology, UIN Sunan Gunung Djati Bandung, Jl. A.H. Nasution No 105 Bandung, West Java, 40614, Indonesia
[2] Komunitas Teras, Jl. Konasara No. 4, Kendari, Southeast Sulawesi, 93118, Indonesia
[3] Faculty of Forestry and Environmental Science, Halu Oleo University, Jl. Mayjen S. Parman, Lahundape, Kendari, Indonesia

Abstract. The Nipa-Nipa Grand Forest Park is a conservation forest area in Southeast Sulawesi. This park has potency of hydrological function and serves as the lungs of Kendari City. Its existence is currently used by residents for agriculture, clean water sources, settlements, and tourism without proper regulation and supervision. Therefore, it has potentially caused monopoly, unilateral control, and conflicts. In line with that, the lack of awareness and initiative from the community and stakeholders in maintaining the sustainability of the Nipa-Nipa Forest triggers uncontrollable damage. This study aims to restore hydrological function and conservation of the Nipa-Nipa by redefining role of stakeholders involved in its governance by using the Participatory Action Research (PAR). This method's actions consist of participatory mapping, planning, and agreements. The outputs of this approach were an increase of community participation in Nipa-Nipa management, the improved capacity building of Forest Conservation Farmers Groups (KTPH), the enactment of new regulations of the implementation of the Nipa-Nipa Forest Park, and the ratification of the MoU between the Farmers Group (KTPH) and the Nipa-Nipa Forest Management Unit (UPTD). The results of this study were developing a strategy for a better governance system, involving stakeholders in decision making, developing equal and harmonious government-community relations, and developing land-use planning that meets environmental conservation, and small sustainable business development.

Keywords: conservation forest, stakeholder, sustainable, Nipa-nipa grand forest park, water source

* Corresponding author: ira_ryski@uinsgd.ac.id

1 Introduction

The Nipa-Nipa grand forest park is one of the nature conservation areas in Southeast Sulawesi Province. Located at the edge of the island of Sulawesi as an upland area, it is surrounded by residential of Kendari City and Kendari Bay in the southern part, residential areas of Konawe Regency in the north part and the Banda Sea in the eastern part [1]. Administratively, this area crosses two districts that are Konawe Regency area of 5,391.6 Ha (±68.4%) and Kendari City area of 2,485.9 Ha (±31.6%.) with 7,877.5 Ha total covering area [2]. The history of the existence of the Nipa Nipa grand forest park as a conservation area began with being designated as a preserved forest through the Laiwoi Swapraja Order No. 125/58 of October 1, 1958 [3]. This area has various potential natural resources, one of which is hydrological potential, besides that, it also functions as the lungs of the city. Currently, it is used by residents for agricultural activities, meeting the needs of clean water and tourism but has not been supported by appropriate management. The water sources are used by individuals and institutions without proper regulation and supervision that triggered a monopoly and unilateral control of water sources and potentially create a conflict [4]. The reduced water discharge and even drought during the dry season are felt by the community as one of the impacts of unorganized management of water sources. In addition, the area used by residents to live and carry out agricultural activities is in an area directly adjacent to the forest area. The low concern and initiative from the community as direct users and related stakeholders in maintaining the potential of natural resources in the Nipa-nipa grand forest park area is a major obstacle to be faced. According to the great potential of this area as a support for the lives of the surrounding community, as well as its various potential conflicts, it is important to conduct research that aims to develop a strategy for managing forest and natural resources in an equitable, sustainable, and accountable manner. It is involving the Governance component (GOV) and developing a land-use planning strategy that meets the needs of development, environmental conservation and sustainable small business development.

2 Methodology

2.1 Research design

This study used the PAR (Participatory Action Research) method approach. PAR is a method of systematic investigation and social change that focuses on action research [5]. The core principles of PAR guide researchers to explicitly pay attention to power dynamics, interests served, and social networks throughout the research process by observing issues of voice, inclusion, cross-cutting and interdisciplinary collaboration [6]. The PAR process is designed to identify sources of conflict, gather sufficient information for negotiation, plan, organize and empower the parties. Then, invite the conflicting parties together to find a solution that is appropriate, legal, and sustainable [7]. Broadly speaking, research using the PAR method is divided into three stages of the process: data collection and validation, dialogue, and agreement. These stages are iterative to ensure flexibility to dynamics and changes in the field [6] [8] In this study, the PAR method used can be seen in Figure 1.

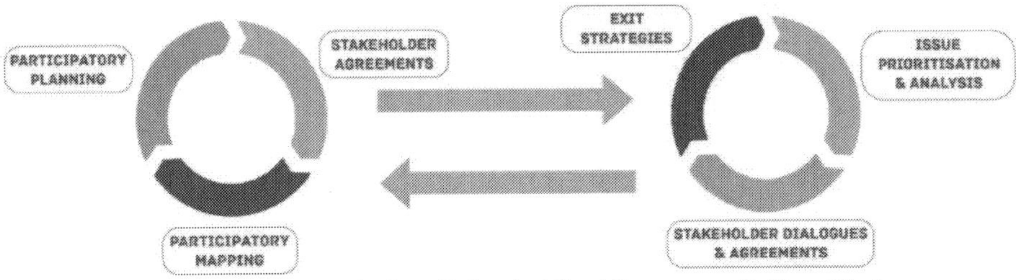

Fig 1 The flow of the PAR method applied to the Nipa-Nipa community Forest Park area

The research process stages in the first circle were participatory mapping (data collection, and validation), participatory planning (mapping of problems, potencies, and solutions), and agreements (Stakeholder's coordination meeting). In the second circle, the phases were identifying exit strategies, determining issue prioritization and analysis, facilitating stakeholder dialogs, and making agreements.

2.2 Study Area

This research was conducted in the Nipa-Nipa grand forest park, Kendari City, Southeast Sulawesi, Indonesia. Grand forest park Nipa-Nipa has several blocks, each with its functions and rules as shown in Figure 2.

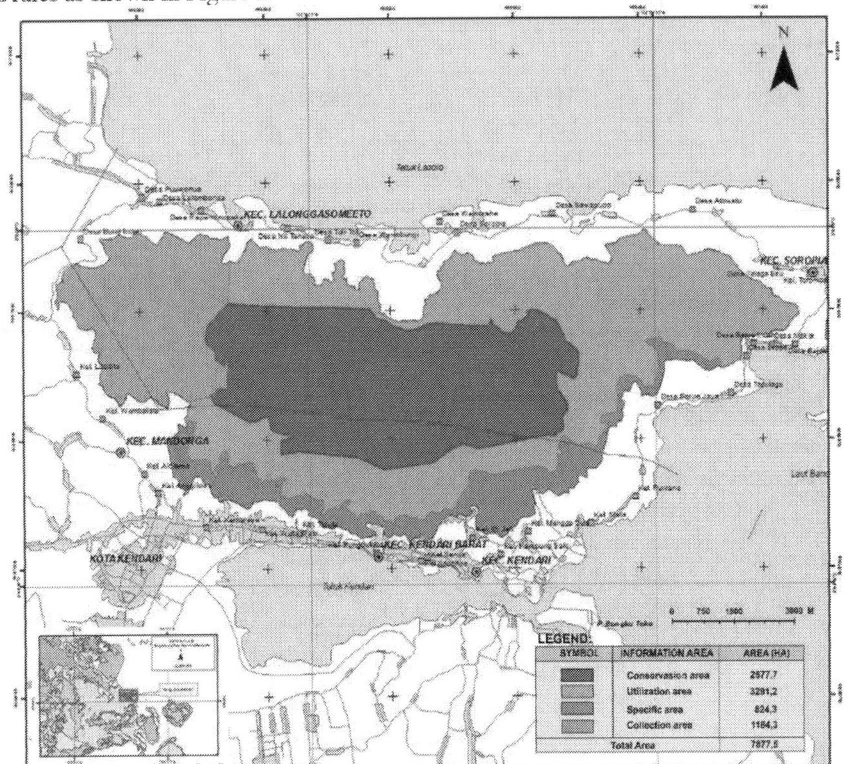

Fig 2 Map of Nipa-nipa grand forest park

IOP Conf. Series: Earth and Environmental Science 1065 (2022) 012049 doi:10.1088/1755-1315/1065/1/012049

Figure 2 shows the division of 4 blocks, consisting of protection block 2,577.7 ha, utilization block 3,291.2 ha, specific area block 824.3 ha, 1,184.3 ha plant collection block, with a total area of 7,877.5 ha.

3 Results and Discussions

3.1 Analysis assessment of environmental component

The initial stage in this research is to analyze the study of environmental components which consist of mapping of land cover and land-use changes in the Nipa-nipa grand forest park area. There were several steps at this stage such as survey methods and field observations, secondary data collection, discussions with local groups and communities as well as key informants' interviews.

3.1.1 Land cover and land-use changes in Nipa-nipa grand forest park

The history of land tenure of the Nipa-nipa grand forest park area began in 1918 with the entry of the Torete and Muna people who carried out land management for agriculture activities. More intensive land exploration began to occur in 1982. The processing of land into plantations was triggered by increasing economic needs then a source of income was needed from other sources, this is conducted by utilizing existing resources and land changes to the detriment of biodiversity [9] [10]. The changes in land cover that occurred in the last 20 years (1990-2010) include changes in primary forest to secondary forest by 48% of the total area of the cluster, an increase in approximately 5% of the area, the increase in residential area from 360 ha to 810 ha, then forest degradation by 2-12%. The land cover changes can be seen in Figure 3 and Table 1.

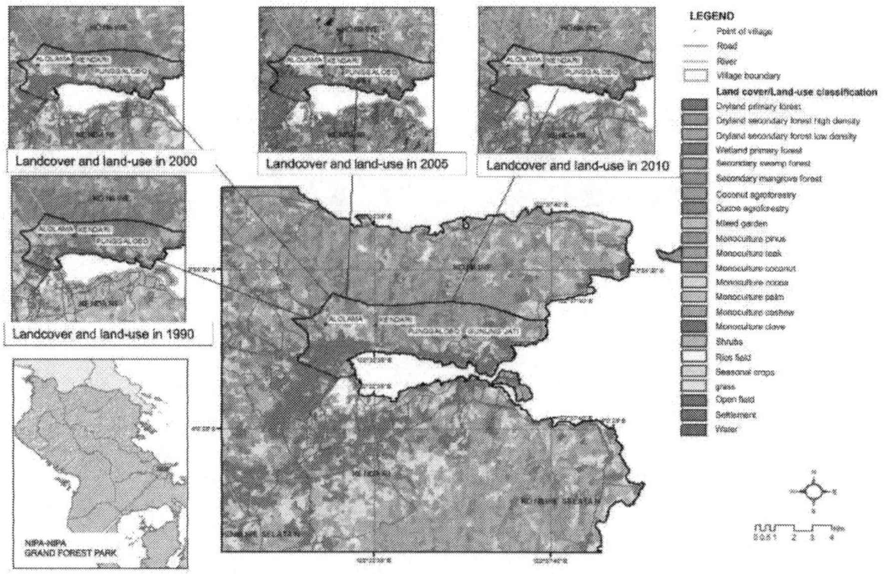

Fig 3 Map of changes in land cover and land-use in the Nipa-nipa area for a period of 20 years (1990-2010) [11]

Table 1 Area by year of land cover/land-use classification [11]

Land cover/Land-Use Classification	Area by year (Ha)			
	1990	2000	2005	2010
Dryland primary forest	2024	1331	1134	542
Dryland secondary forest high density	136	287	272	654
Dryland secondary forest low density	258	98	74	106
Primary mangrove forest	31	23	4	
Secondary mangrove forest	25	33	33	25
Coconut Agroforestry	7	1	34	103
Mixed chocolate garden	42	309	335	699
Mixed garden	3	57	53	129
Monoculture teak		65	53	157
Monoculture coconut	3	107	83	54
Monoculture chocolate		506	401	231
Monoculture cashew		19	54	14
Shrubs	661	109	151	96
Rice fields and seasonal crops	171	122	340	116

3.1.2 Existing condition of water resources management in the Nipa-Nipa grand forest park area

Increasing demand for water will turn water access and consumption into a competition that causes conflicts among beneficiaries, stakeholders, and groups at the national level and reduces the balance between supply and demand [12] [4]. In general, the pattern of water use in the Nipa-nipa grand forest park area has not a legal license yet, whether it is managed for business purposes or for individual needs.

Table 2 Utilization of water services by residents of Nipa-nipa grand forest park area

No	Services Area	Number of water consumer (Household)	Rate of payment (Rupiah/month)	licensing legality
1	Alolama district	2.332	7.000-50.000	none
2	Lahundape district	1.342	5.000-10.000	none
3	Gunung Djati district	2.285	5.000-10.000	none
4	Mangga dua district	1.918	9.000-50.000	none
5	Bumi Indah Resident	1.255	3.000-10.000	none
	Total consumer	9.132		

Table 2 shows the utilization of water services by residents living around the Nipa-nipa grand forest park area. A total of 9,132 households were successfully recorded by the mapping team spread over 5 district areas of water service utilization, consisting of of Alolama, Lahundape, Gunung Djati, Mangga Dua, and Bumi Indah Housing areas. Based on these data, there are variations in customer service rates paid by customers to access water services, even though there are no legal permits yet. The difference in services rates is distinguished based on personal use as well as for the use of water supply businesses such as refill drinking water depots.

3.1.3 Potential conflict in the Nipa-Nipa Forest Park Area

Conflict is a social situation, in which two or more actors seek to have more access to a resource or several sources at one time [13] [14]. In this study, conflict does not mean violence or armed conflict but rather an indicator of the status of disagreements over water access and distribution rights. Some of the conflicts that occurred in the Nipa-nipa grand forest park area:

Land-use conflicts. This area is used by residents to live and carry out agricultural activities in areas directly adjacent to the Nipa-Nipa Grand Forest Park area. Concern and initiative from the community to maintain the potential of natural resources in the Grand Forest Park Nipa Nipa area is very low and does not care about aspects of environmental sustainability

Conflict types of allowed plants. There is a conflict between UPTD Grand Forest Park and local farmer groups. Farmer groups and communities need access to fulfill their daily needs by making farming as a source of livelihood. However, without regard to the types of plants that are allowed to be planted in the Nipa-nipa Grand Forest Park area.

Conflicts over the use of water resources. Water sources in the Nipa-nipa area are used by individuals and institutions without any regulation and supervision, it is giving a monopoly opportunity and unilateral control of water sources. It also causes the reduce of water discharge and even some rivers experience drought in the dry season. Water-related conflicts are conflicts that occur between two or more areas or conflicts between groups and communities that compete for access and use of water resources [15] [16].

3.2 Redefining Role of Stakeholders in Restoring Hydrological Function and Conservation of Biodiversity of Grand Forest park Nipa-Nipa, Southeast Sulawesi

3.2.1 Institutions taking on roles in the Nipa-nipa grand forest park area

The stakeholder components that have a role in the management of the Nipa-nipa Grand Forest Park can be seen in Table 3:

Table 3 Institution roles in the Nipa-nipa grand forest park area

Stakeholders	The role in grand forest park management
Forest Conservation Farmer Group (KTPH)	Improve knowledge and skills in managing grand forest park land collaboratively and in collaboration with BP UPTD to manage land using livelihood and conservation principles.
Community (user)	Improve awareness of the importance of preserving the Nipa-nipa grand forest park as a support for the lives of the surrounding community so that they can cooperate with the authorities in preserving the area
Nipa-Nipa Grand Forest Management Unit (UPTD)	Socializing the function and position of the Grand Forest Park including the management plan of the grand forest park, especially to KTPH farmers. UPTD develops cooperation with communities represented by KTPH for collaborative land management in certain blocks to accommodate livelihood needs UPTD is actively strengthening the capacity of farmers.
Department of Agriculture and Forestry (Distanhut) Kendari City	Assisting and cooperating with the community for the provision of wood and multi-purpose tree seedlings, as well as efforts to conserve water and soil. All these tasks are carried out in coordination with the authorities in grand forest park.

3.2.2 Redefining the role of stakeholders in restoring the hydrological function and conservation of the biodiversity of grand forest park Nipa-nipa, Southeast Sulawesi

Stakeholder dialogue and agreements in the circle of PAR is an effort to redefine the role of stakeholders in restoring the hydrological function and conservation of the biodiversity of Nipa-Nipa grand forest park. Some of the activities carried out in this stage include:

Improving Community Participation in the Management of the Nipa-Nipa. This stage is the activities to build awareness of community to take part in the protection of the Nipa-nipa grand forest park and to improve the awareness of the real needs and the direct benefits that have been felt by the community towards the existence of the Grand Forest Park. The emerging challenges in this phase e.g., action plan for the park conservation, key relevant stakeholders to work with, and the roles of these actors particularly the head of neighborhoods.

Strengthening Forest Conservation Farmer Group (KTPH). The target of institutional strengthening of Forest Farmer Groups is the achievement of planned, measurable, conservation-minded, and sustainable activities if collaboration and accountability. Aspects that are part of institutional strengthening include land management, institutional management, and business management.

Encouraging the enactment of a new regulation which regulates the implementation of the Nipa-Nipa grand forest park collaboration. The output of this activity is the existence of a new regulation (the Governor's Regulation no. 18/2016) concerning guidelines for the procedure for collective agreements for the implementation of collaboration in the management of the Nipa-Nipa grand forest park. This regulation was created to regulate collaborative management of forest area utilization with the community and to manage the use of water environmental services and ecotourism as well.

Facilitating the Process of Memorandum of Understanding (Mou) Between Farmer Groups (KTPH) and Nipa-Nipa Grand Forest Management Unit (UPTD). One of the steps that can be taken by all stakeholders in resolving conflicts in the Nipa-nipa grand forest park is to build communication and planning that can accommodate all the interests in it without losing the function of the grand forest itself. It provides certainty for the community to use forest land by considering aspects of sustainability and welfare. In order to Implement the regional regulation on collaboration, a Memorandum of Understanding (MoU) process was designed between the KTPH and UPTD for land management of 4 groups of KTPHs, as well as encouraging the fulfillment of criteria mutually agreed upon between KTPH and UPTD including legal institutional aspects, organizational arrangements, plantation management, and business management plans.

4 Conclusions

Redefining the role of stakeholders in the restoration of hydrological functions and biodiversity conservation of the Southeast Sulawesi Nipa-Nipa Forest Park using the PAR method has created a better governance system by involving stakeholders in decision making, developing a more equal and harmonious relationship between the government and communities, as well as planning for environmentally friendly land uses, and supporting sustainable small businesses.

References

[1] B. I. d. P. Hutan, "Data dan Informasi," Balai Inventarisasi Dan Penataan Hutan Sulawesi Tenggara, Kendari, 2003.

[2] Balai UPTD TAHURA Nipa-Nipa, "Data dan Informasi," Balai UPTD TAHURA Nipa-Nipa, Kendari, 2003.

[3] D. K. K. W. S. Tenggara, "Informasi Kawasan Konsevasi Provinsi Sulawesi Tenggara, Proyek Pengembangan Kawasan Provinsi Sulawesi Tenggara," Departemen Kehutanan Kantor Wilayah Sulawesi Tenggara, Kendari, 1997.

[4] K. Veisi, M. Bijani and E. Abbasi, "A human ecological analysis of water conflict in rural areas: Evidence from Iran," Global Ecology and Conservation, vol. 23, p. e01050, 2020.

[5] O. Fals-Borda, Participatory action research in social theory: Origins and challenges. In P. Reason & H. Bradbury (Eds). Handbook of action research: Participatory inquiry and practice (pp.27-37), Thousand Oaks, CA: Sage Publication, 2001.

[6] L. Young, "Participatory action research (PAR): A research strategy for nursing?," Western Journal of Nursing Research, vol. 28, no. 5, pp. 499-504, 2006.

[7] C. MacDonald, "Understanding Participatory Action Research: A Qualitative Research Methodology Option," Canadian Journal of Action Research, vol. 13, no. 2, pp. 34-50, 2012.

[8] C. Elder, *Participatory Action Research (PAR) as a tool for transforming conflict, A case study from south central Somalia, Life & Peace Institute, 2016.*

[9] R. Muradian, *"Payments for Ecosystem Services as Incentives for Collective Action," Society & Natural Resources: An International Journal, vol. 26, pp. 1155-1169, 2013.*

[10] J. Terborgh, *"The Faith of Tropical Forest: A Matter of Stewardship," Conservation Biology, vol. 14, no. 5, pp. 1358-1361, 2000.*

[11] A. Widayati, J. Sirait, N. Khasanah and S. Dewi, *"Pengelolaan Lanskap Secara Kolaboratif di Sekitar Tahura Nipa-Nipa, Sulawesi Tenggara. Strategi Penghidupan dan Konservasi AgFor - 02," World Agroforestry Centre (ICRAF) Southeast Asia R, Bogor, Indonesia, 2014.*

[12] M. Bijani and D. Hayati, *"Farmers'perceptions toward agricultural water conflict: the case of Doroodzan dam irrigation network, Iran. J. Agric. Sci. Technol.," Agriculture Science Technology, vol. 17, no. 3, pp. 561-575., 2015.*

[13] J. L. LENNART, *"Integrating Swedish Water Resource Management: a multi-level governance trilemma," Local Environment, vol. 9, no. 5, pp. 413-424, 2004.*

[14] S. Mohammadinezhad and M. Ahmadvand, *"Modeling the internal processes of farmers' water conflicts in arid and semiarid regions: Extending the theory of planned behavior," Journal of Hydrology, vol. 580, pp. 124-241, 2020.*

[15] A. T. Wolf, *"Water Wars" and Water Reality: Conflict and Cooperation Along International Waterways, Wolf A.T. (1999) "Water Wars" and Water Reality: Conflict and Cooperation Along International Waterways. In: Lonergan S.C.NATO ASI Series (2. Environment), vol 65. Springer, Dordrecht, 1999.*

[16] B. Gholizadeh and M. Niknami, *"The causes and effects of water conflict: evidence from Damavand," Bulgarian Journal of Agricultural Science, vol. 26, no. 3, pp. 598-604, 2020.*

Acknowledgment

We would like to show our gratitude to the Center for International Forestry Research (CIFOR), Nipa-Nipa Grand Forest Management Unit, and Forest Conservation Farmer Group of Nipa-nipa grand forest park.

Optimizing rainwater harvesting systems for the dual purposes of water supply and runoff capture with study case in Bandung Area, West Java.

Muhammad Cahyono

Faculty of Civil and Environmental Engineering, Institute of Technology Bandung

cahyono@itb.ac.id

Abstract

Rainwater harvesting (RWH) systems use runoff to meet water supply demand and may also benefit runoff reduction. We develop the Rainwater Analysis and Simulation Program (RASP) model to simulate an RWH system using storage volume, roof area, irrigated area, and water demand as inputs. A cost-benefit model of RWH was developed with subject to reliability target of water supply and runoff capture in each simulation. Near-optimal solutions were identified for each case using a constrained nonlinear programming algorithm. We apply the model to evaluate the RWH system in Bandung Area, West Java. Various types of buildings, ranging from residential houses, schools, dormitories, offices, apartments, condominiums, mosques, and hospitals, are studied to provide several optimal parameters for tank volume and number of users for various land-use scenarios. The resulting parameters are obtained through optimization to get minimal RWH's capital costs and water purchase costs by applying several criteria related to the financial feasibility of the RWH system and by keeping the runoff coefficient not exceeding the pre-development value. The simulation results can be used as a guide for planners to consider the application of RWH in several types of buildings.

Keywords: rainwater harvesting, runoff capture, optimization.

1. Introduction

As the population grows, the demand for clean water increases. In several big cities in Indonesia, such as Bandung, Jakarta, and other big cities, many people still have difficulty getting access to clean water [11-13]. One alternative to help meet this need for clean water is through the Rainwater harvesting (RWH) system [11-13]. RWH can also provide additional benefits to reduce runoff [1]. In other words, RWH is a part of water conservation practice that can also provide environmental benefits through reducing runoff. Rainwater storage is usually used for outdoor irrigation or indoor use for flushing toilets, laundry, cleaning rooms, etc. However, if the RWH system is equipped with a simple water treatment widely available in the market, rainwater - after going through the processing process - can be used for other domestic needs except drinking. A conceptual RWH system is shown in Fig. 1. Components include a roof to capture rainwater, a filter to remove the initial fraction of runoff from the roof, a storage tank, a water treatment, a water tower, a water distribution tank, and a pump to supply water to the water treatment and distribution tanks.

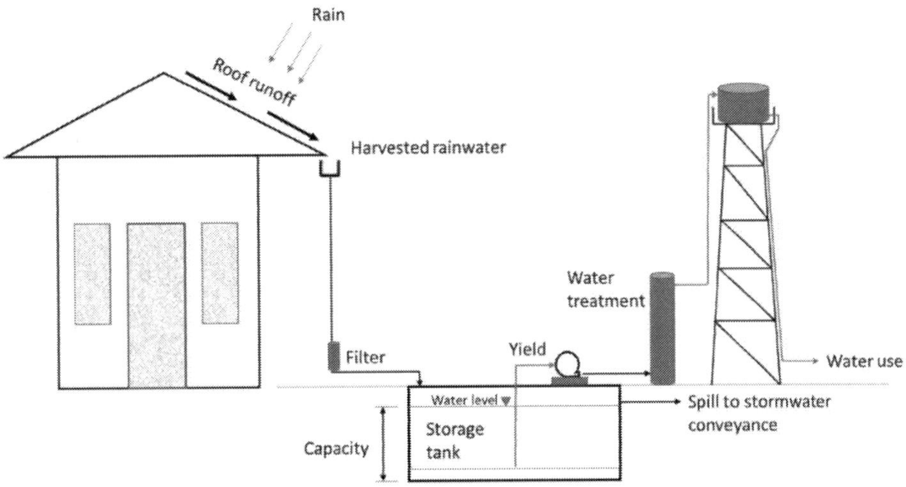

Figure1 Schematic of a typical rainwater harvesting system

In this paper, the study of the RWH system is limited to the case of one building unit (house, apartment, office, dormitory, schools, hospital, and mosque) in the Bandung area. The aim of this paper is to assess the cost-effectiveness of RWH for water supply and runoff capture. In this study, RWH criteria will be proposed in which the greatest benefit can be achieved with a minimum total cost consisting of capital costs for installing the RWH system and the cost of purchasing clean water and being able to capture rainwater runoff so that it does not disturb the balance of the hydrological environment. The parameters reviewed are the number of users, land area, roof area, green open land area, water tank volume.

2. Methodology

In rainwater harvesting system rainwater from the roof is collected and put into a storage tank. The water in the storage tanks can then be distributed using pumps upon request. However, to avoid excessive use of pumps, the water to be distributed is generally pumped into the distribution tank, which is generally placed in a relatively high position such as above a water tower. Furthermore, the water is distributed by gravity from this tank as needed. If the RWH system uses water treatment, the water in the storage tank is treated first before being supplied to the distribution tank. Excess water beyond the storage capacity will spill and enter the stormwater conveyance system.

Simulation of water mass balance can be carried out on various scales, from hours, days or months. However, simulations were carried out using a daily scale as recommended by Imteaz et al. [5] for more realistic results. The simulation is carried out using a combination of each input variable based on the Yield Before Spillage (YBS) algorithm (Fewkes and Butler [4]; Liaw and Tsai [7]; Chiu et al. [3]; Khastagir and Jayasuriya [6]. The daily water mass balance must meet the following equations (1) to (3).

$$V_t = Q_t + V_{t-1} - Y_t - O_t \quad (1)$$

$$Y_t = \begin{cases} D_t \\ V_{t-1} + \theta Q_t \end{cases} \quad (2)$$

$$V_t = \begin{cases} (V_{t-1} + Q_t - \theta Y_t) - (1-\theta)Y_t \\ S - (1-\theta)Y_t \end{cases} \quad (3)$$

where V_t and V_{t-1} = volume of water in the storage tank on day t and the day before (m^3), Q_t = amount of rainwater harvested on day t (m^3), and O_t = water spilled due to excess tank capacity on day t (m^3), Y_t= yield on day t, D_t = daily demand, and θ=1 for YBS. The amount of rainwater that can be accommodated and stored Q_t on a certain roof area on a certain day can be obtained from equation (4) proposed by Aladenola dan Adeboye [10].

$$Q_t = C_{roof} R_t A_{roof} \Delta t/1000 \quad (4)$$

where C_{roof}= roof runoff coefficient, R_t = rainfall depth (mm/day), A_{roof} = roof area (m^2), and Δt = time interval (day). Water requirements depend on the number of occupants,

$$D_t = a_P N_{pop} \quad (5)$$

where a_P = water need per capita per day and N_{pop} = number of residents or users. Simulations using mass balance Eqs. (1) – (5) were carried out for T years using daily rainfall data, where T = lifetime of the RWH system. From the simulation results, it can then be determined the average volume for Yield Y_w and water purchases V_w from the utility network for each particular month. The land runoff coefficient can be determined from the following relationship. i.e.

$$C_{ro} = \frac{\sum_{n=1}^{12}(Re_n - Y_{a,n} - I_{a,n})}{\sum_{n=1}^{12} Re_n} \quad (6)$$

where C_{ro} = land runoff coefficient, $R_{e,n}$, $Y_{a,n}$ and $I_{a,n}$ are the monthly average of rainfall, yield, and infiltration in open green spaces in units of m^3. Infiltration in open green open space is determined by the following equation

$$I_{a,n} = \sum_{n}^{12}(1 - C_{ra})Re_n A_{green} \quad (7)$$

where C_{ra} = land runoff coefficient in natural conditions before construction and A_{green} = green open area or irrigation area. The reliability of RWH to meet water needs can be calculated from the following.

$$Y_{eff} = \frac{\sum_{n=1}^{12} Y_{a,n}}{\sum_{n=1}^{12} D_t} \quad (8)$$

3. Optimization Model

The main design variables chosen were V_{tank} = the tank volume for rainwater storage, A_{land} = land area, A_{roof}, A_{green}, N_{pop}, C_{ro}, and Y_{eff}= yield reliability.The cost of purchasing water through utility networks in Bandung follows the Increasing Block Rate (IBR) scheme set by the Bandung city drinking water company (PDAM Tirtawening). According to the IBR tariff scheme, the unit price for purchased water increases due to increased volume. The IBR tariff scheme is an effective tool to prevent wasting scarce water resources. In Bandung, water usage payments are made monthly based on the month's water use. An illustration of block-level tariffs for monthly water use can be seen in Figure 2. In this Figure 2, C_{P1}, C_{P2}, C_{P3} and C_{P4} are unit prices of water per m^3 for use with a volume equal to V_{w1}, V_{w2}, V_{w3} and V_{w4} respectively. As an illustration, it can be explained as follows. If the volume of water consumption V_w is less than V_{w1}, the cost of using water $C_w = V_{w1}$ x C_{P1}. If V_w is between V_{w1} and V_{w2}, then $C_w = C_{P1}$ x $V_{w1} + C_{P2}$ (V_w-V_{w1}), and so on. The cost of water C_w if V_w is above V_{w3} is shown in Figure 2. Based on this IBR rate, in general, the average cost of water usage for a year is.

$$C_w = \sum_{n=1}^{12} min((V_{w,n}, V_{w1}) * C_{P1} + max(min((V_{w,n} - V_{w2}), V_{w2} - V_{w1}),0) * C_{P2} + max(min((V_{w,n} - V_{w3}), V_{w3} - V_{w2}),0) * C_{P3} + max(V_{w,n,} - V_{w3}, 0) * C_{P4} \quad (9)$$

where C_w = average annual cost of water purchases from utility networks (PDAM Tirtawening), $V_{w,n}$ = volume of water purchases in the nth month. By using the same procedure as for computation of C_w the cost of saving water per year due to the implementation of RWH can be stated as follows

$$B_w = \sum_{n=1}^{12} min((Y_{w,n}, V_{w1}) * C_{P1} + max(min((Y_{w,n} - V_{w2}), V_{w2} - V_{w1}),0) * C_{P2} + max(min((Y_{w,n} - V_{w3}), V_{w3} - V_{w2}),0) * C_{P3} + max(Y_{w,n,} - V_{w3}, 0) * C_{P4} \quad (10)$$

where B_w = annual cost savings, $Y_{w,n}$ = average yield volume in n^{th} month.

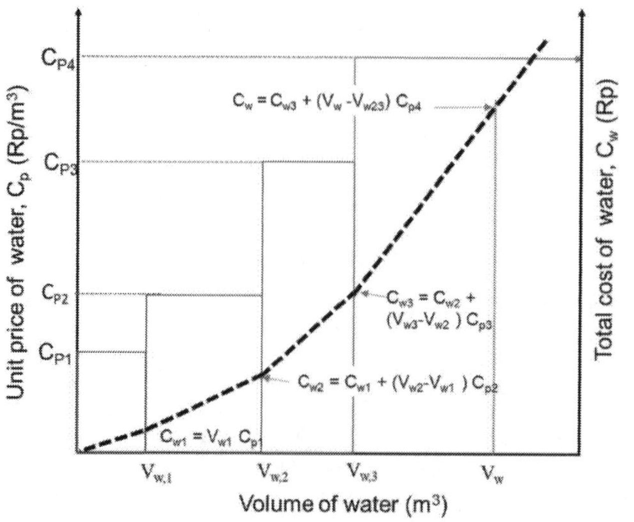

Fugure 2. Scheme of increasing block rate tariff of purchasing water

Capital costs F_{inv} or investment costs consists of the cost of tanks, pumps, pipelines, pressure control, simple water treatment (if necessary), filters, and installation. Other costs are maintenance costs C_M and operational costs C_{op} such as costs for replacing spare parts and water treatment materials, operator wages. Electricity costs for pumping are negligible because these costs are also included when purchasing water from the utility network (Okoye et al. [9], Sample and Liu [1]). Optimization is carried out to find the minimum total cost which consists of the capital cost to install the RWH system and the cost of purchasing clean water. Thus, the optimization model can be formulated as follows.

Objective function (Okoyea et al. [9]).

$$min\ F_o = F_{inv} + \sum_{\tau=1}^{T} \frac{c_\tau\, C_w}{(1+i)^\tau} \tag{11}$$

subject to

$$BCR = \frac{F_B}{F_C} = \frac{\sum_{\tau=1}^{T} \frac{(1+m)^\tau B_w}{(1+i)^\tau}}{F_{inv} + \sum_{\tau=1}^{T} \frac{(1+k)^\tau (C_M + C_{op})}{(1+i)^\tau}} > 1.0 \tag{12}$$

$$0.15 < C_{ro} < 0.3 \tag{13}$$

$$Y_{eff} > 0.4 \tag{14}$$

$$S_{min} < V_{tank} < S_{max} \tag{15}$$

$$N_{pop,min} < N_{pop} < N_{pop,max} \tag{16}$$

where m = increase in water price per year (%), i = discount rate per year (%), k = increase in maintenance and operational costs per year (%), S_{min} and S_{max} = minimum and maximum tank capacity, $N_{pop,min}$ and $N_{pop,max}$ = minimum and maximum number of occupants. Constraint (12) is the value of the benefit-cost ratio, BCR. The project lifetime is calculated for T = 10 years so that the BCR value should be greater than 1 at the 10th year. The limit (13) is related to the coefficient of rain runoff to comply with the environmental health condition. It should be emphasized that the runoff coefficient is not related to the concept of delta zero runoff. This idea applies that the peak discharge after the land development does not exceed the release in the condition predevelopment. The RWH system needs to be combined with infiltration wells or rainwater gardens to include the delta zero runoff. Limitation (14) is the reliability of the use of RWH, Y_{eff}. The Y_{eff} value is set above 40% so that the application of RWH will significantly help the scarcity of raw water. Limits (15) and (16) are to limit the volume of the tank and occupants within the minimum and maximum allowable values. This value is related to the area of land available for the installation of the tank and the provisions on the number of occupants that are justified for the building area.

The optimizations use a constrained non-linear programming algorithm to identify optimal or near-optimal solutions for each land use case. The parameters sought are the tank's volume and the number of occupants. The optimizations processes use a program code written in an excel spreadsheet and a solver facility with the non-linear Generalized Reduced Gradient (GRN) method.

4. Case Study

The case discussed in this paper is for RWH in Bandung. Rain data with a long record of more than 30 years was only available at the Bandung Meteorological Station, at Cemara Street, Number 66, Bandung. Therefore, the simulation results cannot accurately represent the entire Bandung area. The simulations use daily rainfall data from 1980 to 2019. The investment cost calculation uses the following unit prices: storage tank = Rp. 1,500,000,-/m^3, simple water treatment = Rp. 800,000,-/unit, with one water treatment unit, is required for every 5 m^3 tank volume, pipe Rp. 20,000,-/m with an estimated need of 20 m/5m^3 tanks, filters, etc., of Rp. 50,0000/unit RWH. The maintenance cost is approximately Rp.250,000/year for cleaning every 5m^3 tank, maintaining the pipes and pumps, and water treatment about Rp.200,000,-/unit per year for replacing spare parts and materials. Operational costs are Rp. 500,000,-/month or Rp. 6,000,000,-/per year for a part-time operator. The operator is required especially for the RWH system in large buildings and with many users such as offices, schools, apartments, mosques, and hospitals.

For RWH on a residential scale, operational costs are neglected because these costs are also included when buying water from utility networks. The increase in water prices is 5% per year. This increasing cost is estimated based on water price for last 10 years. The increase in maintenance and operating costs is 2% per year. The interest rate used for calculating financial feasibility is 7%. The runoff coefficient of the land before construction is 30%, which assumes that 60% of the volume of rainwater per year enters the soil through infiltration [14, 15]. The irrigation requirement is considered equal to the average potential evaporation in Bandung, 4 mm/day. Irrigation is only needed if the rainfall intensity is less than 4 mm/day. Simulations were carried out for a roof area of 70% of the land area. Thus, 30% of the open green space was available for gardens. The water requirement uses values given by the Indonesian National Standard (SNI 03-7065-2005) regarding planning the plumbing system.

4.1 Residential house

The water requirement per capita for residential houses is 120 l/capita/day. The cost of purchasing water for residential houses is C_{P1} = Rp. 3300,-/m^3, C_{P2} = Rp. 6000,-/m^3, C_{P3} = Rp. 9400,-/m^3 and C_{P4} = RP. 12,600,-/m^3. The constraint for volume is V_{w1} = 10 m^3, V_{w2} = 20 m^3, V_{w3} = 30 m^3. Simulations are carried out for land areas ranging from 200 m^2 to 1000 m^2. Several design parameters of the simulation results can be seen in Table 1. The simulation results show that a land area of over 300 m^3 with a minimum building area of 210 m^2 can produce optimal solutions.

Table 1. Optimization results of RWH system for residential home

		1	2	3	4	5
Land Area	m^2	300	400	500	700	1000
Roof Area	m^2	210	280	350	490	700
Number of residence	person	12	14	17	23	33
Water demand	m^3/day	1.81	2.22	2.64	3.65	5.21
Tank volume	m^3	7.75	12.48	17.35	25.26	36.28
Runoff Coefficient, Cr_o	-	0.30	0.30	0.30	0.30	0.30
Befenit-Cost Ratio, B/C	-	1.00	1.00	1.00	1.08	1.17
Yield Realibiliy, η	%	0.47	0.52	0.54	0.55	0.55

4.2 School and Office

According to SNI 03-7065-2005, water demand per student or per employee varies from 40 liters/student/day: (elementary school), 50 liters/day (lower midle school), 80 liters/day (upper middle school) and 50 liters/day/employee. In this simulation, the average value of 60 liters/student/day is used. The tariff for purchasing water for schools and offices is C_{P1} = Rp. 2,100,-/m^3, C_{P2} = Rp. 3,800,-/m^3, C_{P3} = Rp. 6,000,-/m^3 and C_{P4} = RP. 8,500,-/m^3. School and office buildings are large buildings so simulations are carried out for land areas ranging from 1000 m^2 to 5000 m^2. The simulation results are given in Table 2. Applications of RWH for school and office buildings are feasible for land areas above 2000 m^2 with many students ranging from 150 to 375 persons.

Table 2. Optimization results of RWH system for School and Office

	unit	1	2	3	4	5
Percil Area	m^2	1000	2000	3000	4000	5000
Roof Area	m^2	700	1400	2100	2800	3500
Number of pilg	person	156	178	243	308	375
Water demand	m^3/day	10.53	13.09	18.15	23.25	28.52
Tank volume	m^3	6.60	41.92	77.36	112.56	147.85
Runoff Coefficient, Cro	-	0.30	0.30	0.30	0.30	0.30
Befenit-Cost Ratio, B/C	-	1.00	1.00	1.00	1.00	1.00
Yield Realibiliy, η	%	0.26	0.43	0.47	0.49	0.50

4.3 Apartment, Dormitory, and Condominium

Simulations were also carried out for apartment buildings, dormitories, and condominiums. The need for clean water per capita is the same as the need for housing, 120 liters/capita/day. The water tariff used is C_{P1} = Rp. 4,600,-/m^3, C_{P2} = Rp. 7,200,-/m^3, C_{P3} = Rp. 10,700,-/m^3 and C_{P4} = RP. 14,400,-/m^3. This rate is relatively expensive compared to residential houses. Apartments, dormitories, and condominiums are large buildings thus the simulations were carried out for land from 1000 m^2 to 5000 m^2. Table 3 shows the simulation results. The results indicate that the application of RWH for apartments, dormitories, and condominiums is very feasible, meeting all specified criteria.

Table 3. Optimization results of RWH system for apartment & condominium

		1	2	3	4	5
Land Area	m^2	1000	2000	3000	4000	5000
Roof Area	m^2	700	1400	2100	2800	3500
Number of residence	person	38	60	90	120	150
Water demand	m^3/day	5.76	9.65	14.40	19.20	23.99

Tank volume	m^3	28.69	87.80	133.65	178.20	222.74
Runoff Coefficient, C_{ro}	-	0.30	0.30	0.30	0.30	0.30
Befenit-Cost Ratio, B/C	-	1.00	1.00	1.09	1.14	1.18
Yield Realibiliy, η	%	0.50	0.59	0.60	0.60	0.60

4.4 Mosque

Mosques in busy activity areas such as campus mosques, mosques in office areas, and business centers are generally filled with worshippers. The size of the mosque is also generally wide. Water needs are estimated for 4 liters of ablution per person and 5 liters of toilet per congregation. There are five prayers per day, so the water requirement is 25 liters/congregation/day. The number of congregations counted is the same as the number of congregations at the time of one obligatory prayer. The purchase price of water for mosques follows the social tariff, namely C_{P1} = Rp. 900,-/m^3, C_{P2} = Rp. 900,-/m^3, C_{P3} = Rp. 1,400,-/m^3 and C_{P4} = RP. 2,900,-/m^3. The mosque building under review is a large mosque building with a land area of over 3,000 m^2. The simulation results can be seen in Table 4. Financially, the application of RWH for mosques on a land area of more than 5,000 m^2 is feasible, but the use of rainwater can only meet water needs of approximately 25%, below the specified criteria of 40%.

Table 4. Optimization results of RWH system for Mosque

Land Area	m^2	3000	4000	5000	7500	10000
Roof Area	m^2	2100	2800	3500	5250	7000
Number of pilgrims	person	3674	4358	5119	7061	9052
Water demand	m^3/day	36.67	44.02	52.07	72.55	93.47
Tank volume	m^3	10.89	22.61	34.34	63.70	93.13
Runoff Coefficient, Cr_o	-	0.30	0.30	0.30	0.30	0.30
Befenit-Cost Ratio, B/C	-	1.00	1.00	1.00	1.00	1.00
Yield Realibiliy, η	%	0.23	0.25	0.27	0.29	0.30

4.5 Hospital

The need for water is very large for hospitals. According to the Regulation of the Minister of Health of the Republic of Indonesia Number: 986/Menkes/Per/XI/1992, it is required 500 liters/bed/patient. The purchase price of water for hospitals is C_{P1} = Rp. 4,600,-/m^3, C_{P2} = Rp. 7,200,-/m^3, C_{P3} = Rp. 10,700,-/m^3 and C_{P4} = RP. 1,400,-/m^3. Hospital buildings are also generally large at above 1000 m^2. The simulation results for hospital buildings with a land area of more than 1000 m^2 can be seen in Table 5. The RWH system is very feasible to be applied in hospitals to meet water needs with the number of patients between 9 and 90 per day because the simulation results provide an optimal solution for all land areas considered.

Table 5. Optimization results of RWH system for Hospital

Land Area	m^2	1000	2000	3000	5000	10000
Roof Area	m^2	700	1400	2100	3500	7000
Number of bed patient	person	9	14	22	36	90
Water demand	m^3/day	5.76	9.65	14.40	23.99	56.78
Tank volume	m^3	28.69	87.80	133.65	222.74	300.00
Runoff Coefficient, Cr_o	-	0.30	0.30	0.30	0.30	0.30

Befenit-Cost Ratio, B/C	-	1.00	1.00	1.09	1.18	1.82
Yield Realibiliy, η	%	0.50	0.59	0.60	0.60	0.50

5. Conclusion

RWH studies for various types of buildings, ranging from residential houses, schools, dormitories, offices, apartments, condominiums, mosques, and hospitals, provide several optimal parameters for tank volume and number of users for various land-use scenarios. The resulting parameters are obtained through optimization to get minimal RWH's capital costs and water purchase costs and by applying several criteria related to the financial feasibility of the RWH system and keeping the runoff coefficient not exceeding the pre-development value. The simulation results can be used as a guide for planners to consider the application of RWH in several types of buildings.

6. References

[1] Sample, D. J. and Liu, J., 2014. Optimizing rainwater harvesting systems for the dual purposes of water supply and runoff capture. Journal of Cleaner Production 75 174-194.

[2] Campisano, A. and Modica, C., 2012. Optimal sizing of storage tanks for domestic rainwater harvesting in Sicily. Resour. Conserv. Recycl. 63, 9e16 http:// dx.doi.org/10.1016/j.resconrec.2012.03.007.

[3] Chiu, Y. R., Liaw, C.-H. and Chen, L. -C., 2009. Optimizing Rainwater Harvesting Systems as an Innovative Approach to Saving Energy in Hilly Communities," Renewable Energy, Vol. 34, No. 3, 2009, pp. 492-498. http://dx.doi.org/10.1016/j.renene.2008.06.016

[4] Fewkes, A. and Butler, D., 2000. Simulating the performance of rainwater collection and reuse systems using behavioural models. Build. Serv. Eng. Res. Technol. 21 (2), 99e106 http://dx.doi.org/10.1177/014362440002100204.

[5] Imteaz, M.A., Shanableh, A., Rahman,A., and Ahsan, A., 2011. Optimisation of rainwater tank design from large roofs: a case study in Melbourne, Australia. Resour. Conserv. Recycl. 55 (11), 2011, 1022 -1029. https://doi.org/10.1016/j.resconrec.2011.05.013

[6] Khastagir, A.and Jayasuriya, N., 2010. Optimal sizing of rain water tanks for domestic water conservation. J. Hydrol. 381 (3e4), 181e188 http://dx.doi.org/10.1016/ j.jhydrol.2009.11.040

[7] Liaw, C.-H. and Tsai, Y.-L., 2004. Optimum storage volume of rooftop rain water harvesting systems for domestic use. J. Am. Water Resour. Assoc. 40 (4), 901e912 http://dx.doi.org/10.1111/j.1752-1688.2004.tb01054.x.

[8] Sample, D. J, Liu, J. and Wang, S., 2013. Evaluating the dual benefits of rainwater harvesting systems using reliability analysis. J. Hydrol. Eng. 18 (10), 1310e1321 http://dx.doi.org/ 10.1061/(ASCE)HE.1943-5584.0000748.

[9] Okoyea,C. O., Solyalı, O. and Akıntug, B.˜2015. Optimal sizing of storage tanks in domestic rainwater harvesting systems: A linear programming approach. Resources, Conservation and Recycling 104 (2015) 131–140. https://doi.org/10.1016/j.resconrec.2015.08.015.

[10] Aladenola, O. O, and Adeboye, O., 2010. Assessing the Potential for Rainwater Harvesting. Water Resour Manage (2010) 24:2129–2137 DOI 10.1007/s11269-009-9542-y

[11] Juliana, I.C., Kusuma, MSB, Cahyono, M., Kardhana, Martokusumo. 2017. *Performance of rainwater harvesting system based on roof catchment area and storage tank capacity,* MATEC Web Conf., 101, https://doi.org/10.1051/matecconf/201710105014

[12] Juliana, I.C, Kusuma, MSB, Cahyono, M., Kardhana, Martokusumo. 2017. *The Effect Of Differences Rainfall Data Duration and Time Period In The Assessment of The Rainwater Harvesting System Kinerjanc, AIP Conference Proceeding,* **1903**, 100006, DOI: https://doi.org/10.1063/1.5011616.

[13] Juliana, I.C., T.A., Gunawan, Fadel. 2019. Feasibility Analysis of Rainwater Harvesting System Implementation For Public Facilities in Palembang, IOP Conf. Series: Earth and Environmental Science, **248**, https://doi:10.1088/1755-1315/248/1/012005

[14] Lacombe, G. et al. .2016. Contradictory hydrological impacts of afforestation in the humid tropics evidenced by long-term field monitoring and simulation modellin. Hydrol. Earth Syst. Sci., 20, 2691–2704, 2016 www.hydrol-earth-syst-sci.net/20/2691/2016/ doi:10.5194/hess-20-2691-2016.

[15] Allocca, V'; Manna, F.; De Vita, P. 2013. Estimating annual effective infiltration coefficient and groundwater recharge for karst aquifers of the southern Apennines. Hydrol. Earth Syst. Sci. Discuss., 10, 10127–10159, 2013 www.hydrol-earth-syst-sci-discuss.net/10/10127/2013/ doi:10.5194/hessd-10-10127.

Analysis of Biodegradation Characteristics Based on Visual Observation and Mass Loss Percentage in Soil Burial Test of Sustainable Geo-bag Materials as Temporary Structures in Natural Coastal Protection System

W Gunawan[4], Rildova[1], N Yuanita[2], A Kurniawan[2], K Kahdar[3], and V Herbenita[4]

[1] Offshore Engineering Research Group, Faculty of Civil and Environmental Engineering, Institut Teknologi Bandung, Indonesia
[2] Coastal Engineering Research Group, Faculty of Civil and Environmental Engineering, Institut Teknologi Bandung, Indonesia
[3] Craft and Tradition Research Group, Faculty of Art and Design, Institut Teknologi Bandung, Indonesia
[4] Ocean Engineering Master Program, Faculty of Civil and Environmental Engineering, Institut Teknologi Bandung, Indonesia

Corresponding author: william.gunawan09@gmail.com

Abstract. Coastal erosion is a significant hazard that causes heavy loss of land and damage to structures in coastal areas globally. To mitigate this problem, coastal engineers worldwide are focusing on sustainable coastal development practices. To this end, many countries are using natural protective methods, including vegetation such as mangroves. However, this practice is not without challenges; to ensure that mangroves could grow strongly required at least two years. A natural coastal protection system is the proposed solution, which combines mangrove vegetation as the main structure and geo-bag-dykes as temporary structures to solve this problem. The role of geo-bag-dykes is needed until mangrove plants mature. So, the length of time for geo-bags to degrade naturally is vital to get considered because otherwise, it will become hazardous coastal debris. Today, commonly used geo-bags are made from synthetic fibers that take a long time to degrade. Thus, this study tested alternative materials made from natural fibers, such as lyocell, kenaf, cotton, and rayon. The objective is to determine whether these alternative materials can be used as sustainable geo-bag materials or not in terms of biodegradation. Based on the soil burial test results contained in ISO 11721-1, these alternative materials degrade much faster when compared to commonly used geo-bag materials.

Keyword. Natural coastal protection system, synthetic fibers, alternative materials, biodegradation, soil burial test.

1. Introduction

Coastal zone is an interface between the land and sea, which is comprised of a continuum of coastal land, intertidal area, aquatic systems including the network of rivers and estuaries, islands, transitional and intertidal areas, salt marshes, wetlands, and beaches [1]. The coastline is a line that separates the land and the sea.

Coastal erosion is a result of human activities and natural environment changes making the coastal dynamic action (wave, current, wind) lose balance in the coastal process, and the long-term loss of sediments of coastal zones results in the destruction process of coastline retreat and beach erosion. At present, coastal erosion is very common, it is worldwide, and it has about 70% of the sandy coast marked erosion in the world. Coastal land is being swallowed by seawater and coastal villages and houses are forced to move inland, which squeezes the living space of humans, destroys the beach biodiversity and ecological balance, and has direct or indirect harm on human life and the natural environment. So, coastal erosion has changed from a natural environment change to a severe hazard [2]. As seen in Figure 1, a natural coastal protection system is a proposed solution to coastal erosion, which combines the main natural protection (vegetation such as mangroves) and temporary artificial structures as sustainable coastal development practices. Geo-bag-dikes were chosen as the temporary structures to protect the growth of mangrove seedlings from sea waves, at least two years are required to ensure that the mangroves grow strongly [3].

Fig. 1. Natural coastal protection system (presentation material of Yuanita et al, 2019).

In today's practice, the geo-bag-dikes are made of polymeric synthetic fibers (nonwoven geo-bags), such as polypropylene, polyester, polyethylene, or polyamide which take a relatively long time to degrade naturally. Thus, this study focused on finding alternative materials from natural fibers, such as lyocell (Tencel), kenaf, cotton, and rayon. The objective is to determine whether these alternative materials can be used as sustainable geo-bag materials or not in terms of biodegradation based on soil burial tests (ISO 11721-1).

2. Methodology

To achieve the research objective, the scope of works as seen in Figure 2 consists of:

- Literature study of natural fibers, mechanical properties, and testing methods.
- Tools and materials preparation. Four natural fiber materials were selected based on the market's availability and the potential to be applied as a sandbag, e.g. Tencel (PT. Rehal Traco Pekalongan Production), kenaf (Innatex Pak Tomo Bandung Production), cotton (local market production), and rayon (local market production). Two samples of commercial geo-bags from international (PT. Tirta Citra Bara Persada Geo-bag) and local (Buana Paksa Indonesia Geo-bag) companies.
- Soil burial test based on ISO 11721-1. All of the fabric samples were buried in the soil at different depths. The soil used in this test is Lembang soil with the Green Flourist brand.
- Observation of soil burial treatments. Each degradation rate was observed within specific periods.
- Comparison and analysis.

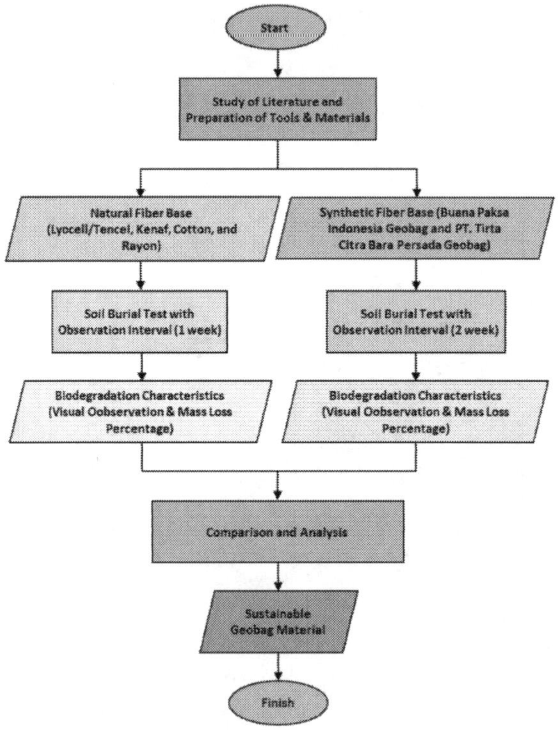

Fig. 2. Methodology.

For the soil burial test based on ISO 11721-1 "Textiles – Determination of Resistance of Cellulose – Containing Textiles to Microorganisms – Soil Burial Test," the first step is cutting the samples to a size of 30 cm × 20 cm. Then, the mass samples are measured using an analytical balance sheet. The next step is the burial treatments. There are three depths of the burial treatments, at the bottom of a pot (the pot base) with a depth of 10 cm, at the middle with the depth of 5 cm, and on the soil surface (control samples). During the burial treatments, the temperature of the soil must be 25±5°C and the humidity must be 95±5%. Then on each observation, the buried specimens are cleaned with water, then soaked with 70% alcohol for 30 minutes, while the control samples were directly soaked with water for 30 minutes. The next step is drying by putting them in an oven at 45±5°C for 24 hours. After drying, conditioning them at 20±2°C for the temperature and 65±4% for the humidity for 24 hours. Finally, the mass is measured again to get the mass loss percentage.

The test method has ever been done with cotton (CO), viscose (CV), modal (CMD), Tencel (CLY), polylactic acid (PLA), polyethylene terephthalate (PET), and polyacrylonitrile (PAN) as the specimens. The weight loss of fabrics after 1 and 4 months is given in Figure 3. The weight change is a direct way to measure biodegradability [4]. The weight losses for cellulosic fibers are higher in comparison with synthetic fibers, as expected.

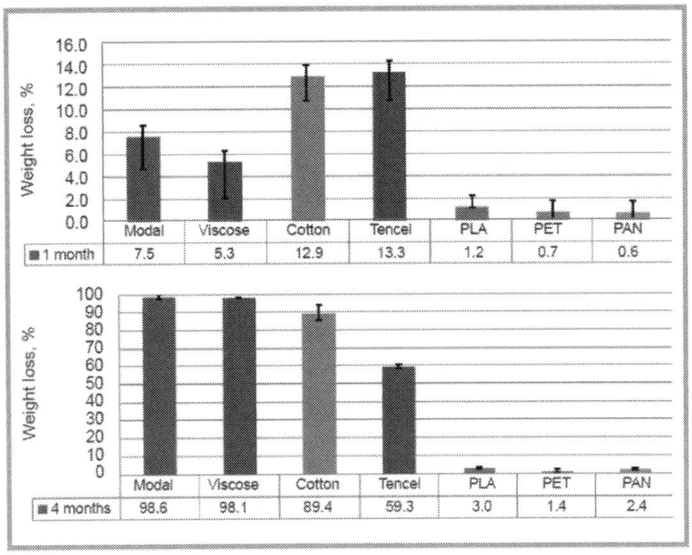

Fig. 3. Weight losses of fabrics for 1 month and 4 months burial intervals [5].

Besides that, biodegradation can be assessed firstly by their physical appearance. Changes in physical appearance can be observed by naked eyes. The visual changes are listed in Table 1.

Table 1. Visual observations were determined for the fabrics after different soil periods [5].

Fabric code	Observations of fabrics after soil burial period (in comparison with control samples)	
	After 1 month interval	After 4 month interval
CMD	thin	small fabric pieces in soil
CV	thin and brittle	small fabric pieces in soil
CO	brittle and colour change	small fabric pieces in soil
CLY	brittle, colour change, small hole	colour change, brittle
PLA	no change	slight colour change by taking moisture
PET	no change	slight colour change with soil particles
PAN	no change	slight lightening of colour

3. Result and Discussion

The result of the soil burial test is divided into two parts (natural fiber base and synthetic fiber base). Figure 4 shows the results of the changes that occur during the soil burial testing between the natural fabrics. It also compares visual observation and mass loss percentage before and after the soil burial test. Cotton fabrics experienced the most significant changes. They are noticeably degraded after 15 weeks of soil burial testing. It also means that cotton fabrics have the fastest rate of biodegradation.

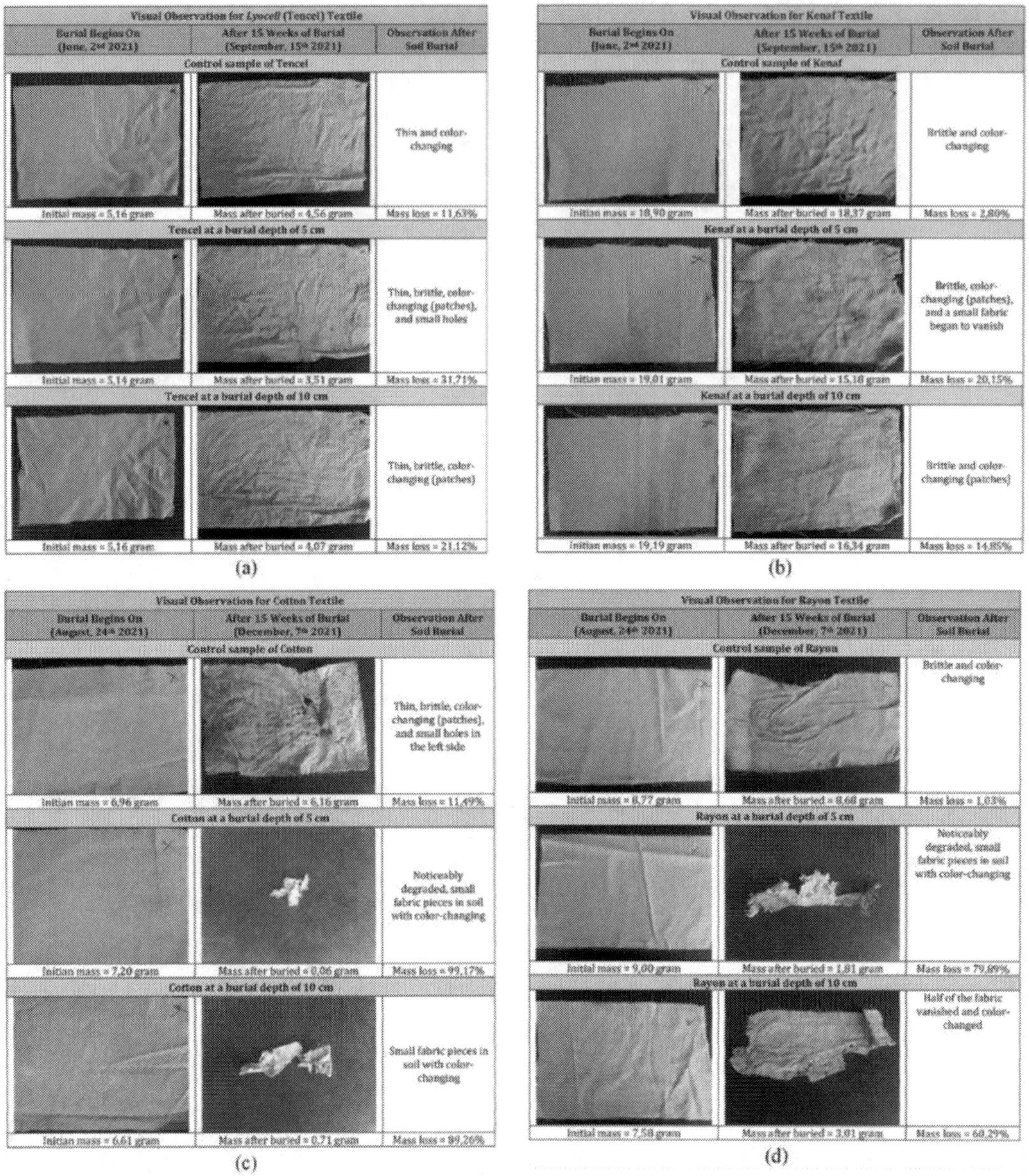

Fig. 4. The results after the soil burial test between the natural fabrics: Tencel (a), kenaf (b), cotton (c), and rayon (d).

Different things happened in the soil burial test of the synthetic geo-bags. Figure 5 shows that both of the synthetic geo-bags did not change significantly. After 22 weeks of soil burial testing, the synthetic geo-bags just reached approximately 10% for the mass loss percentage.

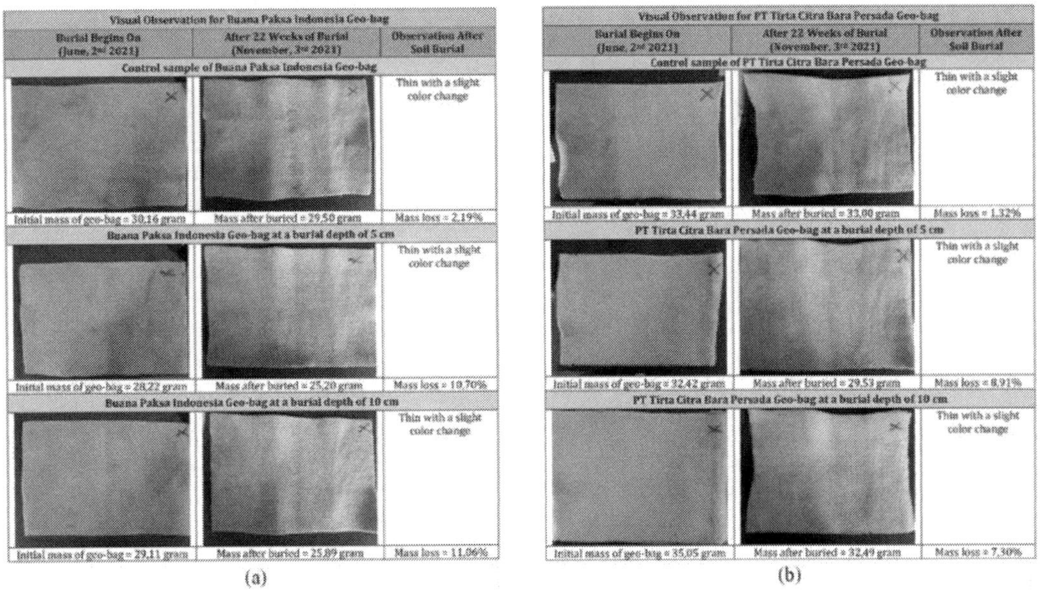

Fig. 5. The results after the soil burial test between the synthetic geo-bags: Buana Paksa Indonesia (a) and PT Tirta Citra Bara Persada (b).

The mass loss percentage is summarized in Figure 6. Natural fabrics have a much higher mass loss percentage than synthetic geo-bags. It implies that the natural fabrics are degraded faster than the synthetic geo-bags. The fastest degraded natural fabric is cotton textile that is buried in the middle of the pot, while the slowest degraded natural fabric is kenaf. Because the mangroves need approximately 2 years to can act as a beach protector, the sustainable geo-bag material is kenaf. Therefore, kenaf textile is concluded as a good alternative to replace synthetic fibers and offer sustainable advantages.

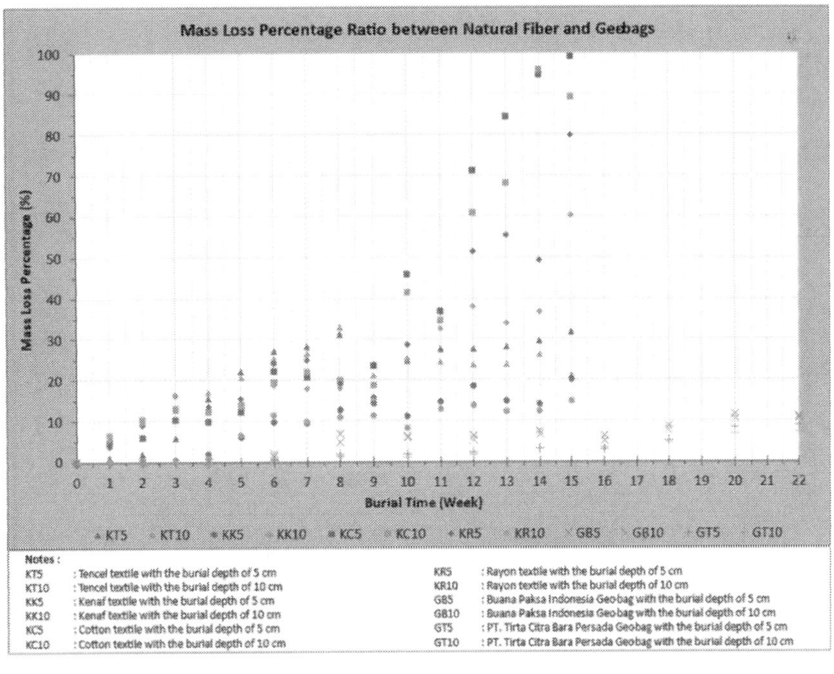

Fig. 6. Mass loss percentage ratio.

4. Conclusion

The natural fabrics are degraded faster than the synthetic geo-bags. Nonetheless, the mass-loss rate of the natural fabrics is too fast due to the mangrove growth, which takes at least two years until maturity. Special treatment is required to reduce the biodegradation rate. Future research objectives will focus on how to get the effect of the special treatment on natural fabrics due to the biodegradation rate. In addition, it is also necessary to test the content of the soil used.

Acknowledgment

This research is funded by the ITB Research and Community Service (P2MI) grant through the Offshore Engineering Research Group and the Coastal Engineering Research Group, Faculty of Civil and Environmental Engineering.

References

[1] B. Cincin-Sain, R. W. Knecht, Integrated Coastal and Ocean Management: Concepts and Practices. (1998)

[2] Y. Yincan, Marine Geo-Hazards in China. (United Kingdom, Elsevier, 2017)

[3] M. R. R. Nugraha, R. I. A. Gemilang, Rehabilitas Pantai Berlumpur Dengan PEGAR Geobag Rangka Bambu di Desa Sriwulan Kabupaten Demak. (Jurnal Teknik Hidraulik, 2017)

[4] ISO 11721-1, Textile – Determination of Resistance of Cellulose – Containing Textiles to Microorganisms – Soil Burial Test (2001)

[5] V. Sular, G. Devrim, Biodegradation Behaviour of Different Textile Fibres: Visual, Morphological, Structural Properties and Soil Analyses. (Dokuz Eylul University, Izmir, Turkey, 2019)

[6] N. D. Yaacob, H. Ismail, S. S. Ting, Soil Burial of Polylactic acid/Paddy straw powder biocomposite, **11(1)**, pp. 1255-1269. (Bioresources, 2016)

[7] N. Yuanita, A. Kurniawan, H. Setiawan, F. Hasan, M. Khasanah, Physical Model of Natural Coastal Protection System: Wave Transmission Over Mangrove Seedling Trees. The 3rd International Water Safety Symposium. (J. Coastal Research Special Issue Number 91, Coconut Creek, Florida, 2019, pp. 176-180)

Assessment of water quality and challenges for vegetable irrigation in Greater Bandung Area, Indonesia

M Firdayati[1*], R Anindita[1], P A Krisnamurti[1], M Handajani[1]

[1]Department of Environmental Engineering, Faculty of Civil and Environmental Engineering, Institut Teknologi Bandung, West Java, Bandung, Indonesia

*mayrina@tl.itb.ac.id

Abstract. The high demand of water mainly used for industry and domestic purposes puts pressure on water for agriculture. In addition, the change of massive land use and poor sanitation condition affect the availability of water sources and quality for irrigation. These are very important due to the increase of consumer needs for good, healthy, and safe food, especially for raw vegetables. This study identified water sources and then examined and assessed water quality for raw vegetables irrigation in Greater Bandung Area (GBA) as one of the main vegetable producers in West Java Province. This produced around 2.26-million tons of vegetables in 2019 whose harvest was distributed for local and regional consumption. The research was conducted by field surveys in four vegetable plantations having about 350 farmers, spreading over 3 regencies, Bandung City (city boundary), West Bandung Regency, and Bandung Regency. Water quality was analysed based on National Water Quality Standard Class II and other parameters such as SAR (Sodium Adsorption Ratio), RSC (Residual Sodium Carbonate) and %Na. Field survey showed that farmers used different water sources for agriculture, namely groundwater, springs, lake, and greywater. Assessment of water quality using Pollution Index method showed that the dominant water source was slightly polluted with pollutant parameters of conductivity, BOD, %Na, ammonia, chloride, and sulphate exceeding the standards. It showed possibility of domestic wastewater contamination to irrigation water source. Based on the measured parameters, water used for irrigation, even groundwater, was polluted and might cause potential health risk for consumer.

Keywords: irrigation water, water quality, water source, vegetable plantation, pollution index

1. Introduction

Around 70 percent of global water consumption is in the agricultural sector, including for horticulture, forestry, plantation, livestock and fisheries, and food supply chain activities [1]. The same profile is also established in Indonesia, especially in Java Island [2]. Rapid rate of urbanization, industrialization, and population growth in Indonesia causes competition for water use to become more stringent. Lack of freshwater resources and degradation of water quality become big challenge for irrigation water [3], [4]. High water quality is required for industry and drinking, and it also supposed to be for water irrigation.

The low chain of safe sanitation services in Indonesia has an impact on the low quality of water, especially surface water which is become one place for domestic wastewater disposal. The threat also comes from wastewater of livestock activities. Fecal contaminants were found to be present in mixed drainage and surface water such as rivers and lakes. This condition can be exacerbated if the industry is around the water body.

The existence of high-water quality always influences the health people surrounding since most of them eat raw vegetables and fruits, which is quite popular in Indonesia and several other Asian countries [5]. People often consume raw vegetables and fruits as side dishes, including in Great Bandung Area (GBA). Raw vegetables and fruits have been known as carriers of human disease for more than a century. Foodborne illness due to food contamination can occur at any point from agriculture to food processing and end-user [6]. However, most foodborne outbreaks have been traced to contamination in agriculture and use of contaminated water [7].

In line with this, water quality is an important factor that is often neglected compared to water quantity. Hence, this study aimed to examine and assess water quality for vegetable irrigation in GBA as one of the main vegetable producers in West java Province. This area produced around 2.26 million tons of vegetables in 2019. Thus, this study expected to get an overview related to sources, conditions, and challenges of water quality used to produce the raw vegetables and fruits.

2. Methodology

2.1. Study Area

Bandung is a city that has a strategic position and role in the national context. In Government Regulation No. 47 of 1997 concerning the National Spatial Planning (RTRWN), Bandung City was designated as one of the National Activity Centers (PKN). In addition, in the RTRWN, Bandung City and parts of Bandung Regency are designated as the Mainstay Area of Bandung Basin and Surroundings with leading sectors of industry, food crop agriculture, tourism, and plantations.

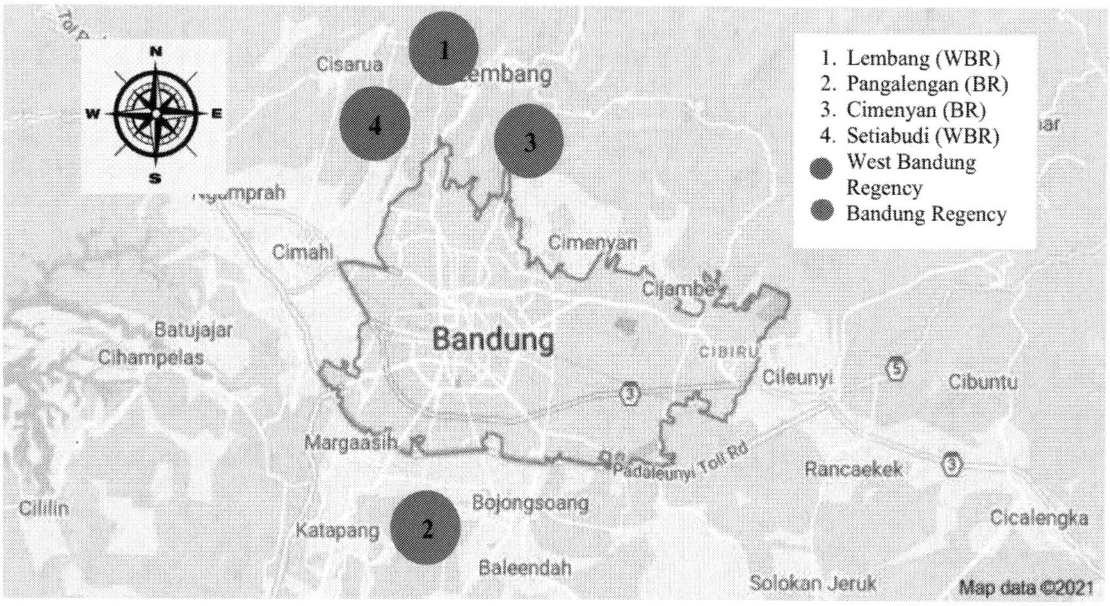

Figure 1. Plantation location (Google, 2021)

Geographically, West Bandung Regency is located between 6°,373' to 7°1,031' South Latitude and 107°.1 10' to 107°4.40' East Longitude. The administrative area of West Bandung Regency is divided into 16 sub-districts with an area of 1,305.77 Km^2. This area has fertile land and many rivers flow causing most of its land area to be used for the agricultural sector. This has more potential because West Bandung Regency has a tropical climate, with 119 rainy days and an average monthly rainfall of 70.41 mm/month in 2015. The relatively sufficient rainfall and fertile soil conditions make this area a very potential prospect in the development of the agricultural sector. Agricultural cultivation takes the largest area in West Bandung Regency land, covering an area of 66,500,294 hectares, non-agricultural cultivation covering an area of 12,159,151 hectares, protected areas covering an area of 50,150,928 hectares, and others covering an area of 1,768,654 hectares. The use of agricultural land in West Bandung Regency is very wide because its original and economic potential is agriculture whose products are marketed to big cities such as Bandung and Jakarta. In this study, there are two locations in West Bandung Regency being studied, namely plantations in Lembang (Fig.1, no.1) and Setiabudi (Fig.1, no.4) areas. The water sources for Lembang plantation come from artesian wells, forest springs, and storage ponds, while for Setiabudi's, only come from artificial wells.

Bandung Regency is located at 6° 49 '- 7° 18' South Latitude and 107° 14'-107° 56' East Longitude. The geographical location of Bandung Regency area is in the eastern and southern parts of Bandung Basin which is the edge of an ancient lake with a geological type in the form of volcanic rock content. Meanwhile, based on the topography, most of the areas in Bandung Regency are mountainous or hilly areas with altitudes above sea level varying from 500 m to 1,812 m. Bandung Regency is also drained by the Citarum River, the existence of this river is profitable from the agricultural sector, industry, and as a raw material for water, but if the rainfall is high enough in certain areas there will be puddles of water. In general, the location of Bandung Regency in the highlands or mountains makes the air temperature in Bandung Regency quite cool, ranging from 12°C-21°C. The area of Bandung Regency is divided into several categories including the area of rice field agricultural land (technical irrigation, non-technical irrigation, and non-irrigated), non-rice field agricultural land, and non-agricultural land area. The sub-district with the widest area of agricultural land is Rancaekek District around 3,051.6 Ha, the village with the largest non-rice field agricultural land is Pangalengan District around 17,898.81 Ha, and the village with the largest non-agricultural land area is Pangalengan District around 8,435. 90 Ha. In this study, there are two locations in Bandung Regency being studied, namely plantations in Pangalengan (Fig.1, no.2) and Cimenyan (Fig.1, no.3) areas. Water sources for Pangalengan plantations come from Situ Cileunca, mountain springs, and storage ponds, while for Cimenyan's comes from greywater and artesian wells.

Table 1. Summary of locations, coordinates, water sources, and types of vegetables grown on the plantation

Plantation	Location	Coordinates	Water Sources	Commodities
1	Lembang	6°48'42.6"S 107°39'26.6"E	Artesian Well, Forest Spring, Storage Pond	Lettuce, broccoli, Kenyan beans, chickpeas, tomatoes, and peppers.
2	Pangalengan	7°12'08.6"S 107°32'38.7"E	Situ Cileunca, Mountain Spring, Storage Pond	Chilies, tomatoes, beans, broccoli, scallions, lettuce, potatoes, mustard greens, carrots, and radishes.
3	Cimenyan	6°50'30.0"S 107°40'35.8"E	Artesian Well, Greywater	Red lettuce, romaine, eggplant, broccoli, bok choy, choy sum, red spinach, mustard greens, radishes, and sweet potatoes.
4	Setiabudi	6°50'27.4"S 107°34'51.9"E	Artesian Well	Green lettuce, red lettuce, beets, and chilies.

2.2. Samplings

A total of nine samples of plantation water sources were taken aseptically using glass sample bottles and plastic sample bottles. Aseptic treatment of samples aimed to prevent samples from being contaminated by microbes and interfered with laboratory test results. All samples were preserved by placing them in a cool box for later laboratory analysis in less than 24 hours. A summary of the location, coordinates, water sources, and commodity of each plantation are shown in Table 1.

2.3. *Analysis*

This study was conducted to identify and analyse the possible presence of microorganisms in each plantation water sources. The quality of water sources was based on parameters of temperature, electrical conductivity, total dissolved solids (TDS), total suspended solids (TSS), pH, BOD, COD, DO, SAR, %Na, RSC, ammonia, chloride, sulphate, boron, oils, and fats, MBAS, total coliforms, fecal coliforms, and *Escherichia coli*. Water quality data from plantations was compared with class 2 quality standards based on Government Regulation of the Republic of Indonesia Number 22 of 2021 concerning Implementation of Environmental Protection and Management and the excellent quality standards class based on the Study of Irrigation Water Quality Criteria by scientific examination based on other country regulations [8]. The two quality standards were then used as a standard for assessing the status of water quality using Pollution Index (PI) method based on the Decree of the Indonesian Minister of the Environment Number 115 of 2003 concerning Guidelines for Determining the Status of Water Quality.

The pollution index of irrigation water can be calculated by the following equation.

$$PI_j = \sqrt{\frac{\left(\frac{C_i}{L_{ij}}\right)_M^2 + \left(\frac{C_i}{L_{ij}}\right)_R^2}{2}} \qquad (1)$$

Information:
PIj = Pollution index
ci = Measured parameter concentration
Lij = Parameter concentration according to quality standard
M = Maximum
R = Average

After calculating the pollution index value, the results would be evaluated with the provisions in Table 1.

Table 2. Category according to pollutant index value

PI value	Information
0 < PIj < 1.0	Meet Quality Standard
1.0 < PIj < 5.0	Light Polluted
5.0 < PIj < 10	Medium Polluted
PIj > 10	Heavy Polluted

3. Result and Discussion

Interviews with farmers of four plantations area in Cimenyan and Lembang District, West Bandung Regency and Pangalengan District, and Bandung Regency show that most farmers have been growing vegetables and fruit for average 10 (ten) years. Farmers are still applying the type of conventional farming with an intercropping pattern. This is done due to limited land, and farmers need to optimize the available land. The vegetable products are marketed locally to traditional markets, modern markets, restaurants, and online sales. Around 20-30% of vegetables are also sent to Jakarta, outside Java and exported to other ASEAN countries.

3.1. *Irrigation water source*

As shown in the Table 1, water sources for plantation irrigation come from artesian wells, forest springs, mountain springs, and greywater. In general, water sources are distributed through the main pipe which is then divided into several branch pipes. Those branches then accommodated into dividing tanks or holding/storage ponds in plantation. The type of watering applied in plantations is by manual watering with a water hose from storage pond. In rainy seasons, irrigation can be affected by rainfall, especially related to the quality and quantity of water in the pond. Precipitation in both districts is quite high, so it is quite helpful for farmers for saving water. However, in the dry season, the situation can change.

Cimenyan plantation using artesian wells (10 meters deep) and greywater, would suffer from dry season. In rainy season, plants would irrigate once daily and sometimes no need for watering at all. In contrast, in dry season, water source with only greywater and they must irrigate twice daily. Then, their strategy for saving water was to choose plant commodity with drought-tolerant plant such as sweet potato. Cimenyan plantation is located in the middle of Bandung basin and not far from Bandung City. The sub-urban area suffers from a short shortage of water reserve and drying up of artesian wells in the dry season [9].

(a) (b) (c) (d) (e) (f)

(g) (h) (i)

Figure 2. Plantation water sources; (a-c) Lembang Plantation, a. Artesian Well, b. Forest Spring, c. Storage Pond; (d-f) Pangalengan Plantation, d. Situ Cileunca, e. Mountain Spring, f. Storage Pond; (g-h) Cimenyan, g. Greywater storage pond, h. Artesian Well; (i) Setiabudi, i. Artesian Well

Lembang plantation had different problem. Water sources for the plantation come from artesian wells with an average depth of 45 m, and construction started in 2017 with government assistance for an artesian well. Previously, their water sources came from shallow well, forest springs, and rainwater. Along with the deteriorating quality of surface water and shallow well commonly used by residents around the plantation, there had been competition for the use of artesian wells and forest springs. Besides that, during dry season, quantity of water also decreases. Then sometimes they can only rely on water from the storage pond.

Pangalengan plantation, the oldest plantation in this study, use water from Situ Cileunca (Cileunca Lake) for irrigation. In the past few years, farmers had to combine water source with mountain water and artesian well to ensure the safety of water used for raw vegetable irrigation. The change of land use in Situ Cileunca with tourism developments [10] and their domestic wastewater affect quality of lake water and irrigation water. Similar situation with Lembang plantation, there is competition with local community for water resource as well.

Climate change affects drought condition and fluctuation of precipitation in West Java [9]. Bandung Area experienced climate change phenomenon and climate-related hazards such as floods and drought. Annual average rainfall was fluctuating in significant value in 2013 and 2016 by 223.45 mm and 295.80 mm respectively. Lack of clean water and sanitation infrastructure development in Greater Bandung Area increased competition between domestic and other sectors. Necessity in protected cultivation due to degradation of water quality irrigation source increased the vulnerability of agricultural water use. It will be difficult to provide a stable supply of agricultural water [11]. Even though it is realized, no research and GOI (Government of Indonesia) policy specifically addressed the situation and comprehensive solutions related to limitation choice of water sources for agriculture, especially for raw vegetables plantation requiring better water quality than other kind of agriculture.

3.2 Irrigation Water Quality

Irrigation has been considered as one of the important pathways for contamination of human enteric pathogens to crops [12]. Hence, irrigation water quality become important. Table 3 shows the laboratory results for samples of irrigation water sources in the plantations studied. It can be seen that parameter exceeding standard were conductivity, BOD, DO, %Na, ammonia, chloride, sulphate, oil and grease.

High conductivity, chloride, and sodium on water could cause high amount of salt in water. The main salt in greywater from Cimenyan plantation could come from detergent used for laundry. Powdered detergents that are common in Indonesia, generally have more salt compare with liquid detergent. This may be the cause of sodium presence that accumulated in storing pond at Pangalengan that influence by Situ Cileunca. In addition, the sodium content can come from the characteristics of igneous rocks in the volcanic deposition area in the Bandung Basin which affect the hydrochemistry of the aquifer [13]. While sodium is needed by soil organism and plants, high sodium particularly in clay soil can reduce soil drainage with degradation of soil composition and permeability. It will inhibit infiltration by plants [14]. High conductivity water could disturb water uptaken by plant because the plant unable to compete with ions in the soil solution for water. Whereas high concentration of chloride could cause plant toxication indicated by burning leaf and necrosis.

Low concentration of DO and high value of BOD on water can be caused by animal manure and domestic sewage. In some countries, organic matter concentration not really important even can affect the odor and become nutrients for microbe [11]. High BOD indicated low concentration on oxygen and low oxygen concentration caused disruption of the photosynthesis process by plants and absorption of nutrients. Standards of DO and BOD in Government Regulation (PP 22-year 2021) because of the emphasis on the regulation in terms of stream water quality management.

Sources of ammonia, sulphate are domestic, industrial wastewater or run off from fertilized agricultural land. High ammonia concentrations can be toxic to plants indicated by burn leaves, blackened roots, and death. Sulphate can be naturally occurring as the result of the breakdown of leaves that fell into a stream, of water passing through rock or soil containing gypsum and other common minerals, or of atmospheric deposition. It caused a decrease in phosphorus in the soil causing plants to lack phosphorus, thereby inhibiting growth. Oil and grease concentration parameters in almost all source of water except mountain spring water of Pangalengan exceed standard parameter. However, high values of oil and grease in the greywater and Situ Cileunca shows that the presence was more due to contamination from domestic wastewater coming from laundry and kitchen. Oils and fats that accumulate in the soil cause the soil pores to close and interfere with the absorption of water by plants.

Table 3. Water quality status assessment results using the Pollution Index (PI)

Plantation	Water Source	*Pollution Index* (PI)		Information
		PP 22/2021	Irrigation R&D	
Lembang	Artesian Well	1.2	4.2	Lightly polluted
	Forest Spring	1.1	2	Lightly polluted
	Storage Pond	3.1	3.7	Lightly polluted
Pangalengan	Situ Cileunca	3.7	2.3	Lightly polluted
	Mountain Spring	2.1	1.8	Lightly polluted
	Storage Pond	2.6	1.2	Lightly polluted
Cimenyan	Greywater	1.67	2.61	Lightly polluted
	Artesian Well	1.14	2.6	Lightly polluted
Setiabudi	Artesian Well	1.22	2.33	Lightly polluted

Table 4. Laboratory results of water quality in plantation

Parameter	Unit	Standards		Lembang			Pangalengan			Cimenyan		Setiabudi
		PP 22/2021	Irrigation R&D	Artesian Well	Forest Spring	Storage Pond	Situ Cileunca	Mountain Spring	Storage Pond	Greywater	Artesian Well	Artesian Well
PHYSICAL												
Temperature	°C	±3	Normal	21.11	22.78	24.44	22.22	22.20	22.78	24.17	22.12	23.15
Conductivity	μmhos/cm	-	275	268.00	58.56	256.30	96.00	25.50	107.86	316.00	180.00	243.00
TDS	mg/L	1000	200	133.80	29.30	128.30	47.90	12.76	53.86	173.80	99.00	133.65
TSS	mg/L	50	100	6.30	7.60	12.30	10.60	0.00	11.30	35.00	2.00	2.00
CHEMICAL												
pH		6-9	6-8.5	7.13	7.62	9.45	8.63	7.54	7.08	7.46	6.62	6.88
BOD	mg/L	3	10	4.10	3.90	3.05	9.15	3.05	2.80	6.86	1.72	1.70
COD	mg/L	25	50	5.00	5.00	5.00	15.00	5.00	5.00	9.80	4.90	4.90
DO	mg/L	4	-	6.03	6.96	10.96	6.30	6.60	3.30	3.60	5.00	6.80
SAR		-	1	0.10	0.50	0.11	0.17	0.17	0.31	0.70	0.12	0.28
%Na	%	-	20	5.64	55.91	4.91	14.95	15.00	25.26	27.92	8.78	16.88
RSC		-	1	-1.53	-0.04	-0.89	-0.55	-0.44	-0.38	0.18	-0.31	0.03
Ammonia	mg/L	0.2	0.5	0.10	0.14	0.94	1.36	0.48	0.66	0.32	<0.038	0.08
Chloride	mg/L	300	4	38.40	1.46	27.70	4.38	1.46	1.46	23.30	14.60	8.75
Sulphate	mg/L	300	4	<4	<4	<4	<4	8.30	4.33	13.40	6.96	13.00
Boron	mg/L	1	0.2	0.05	<0.01	<0.01	<0.01	<0.01	<0.01	<0.01	<0.01	<0.01
Oil and Grease	mg/L	1	-	1.4	1.2	1.6	2.4	1	2.6	2.2	1.6	1.6
MBAS	μg/L	200	-	0.1	0.123	0.036	0.047	0.036	0.064	<0.01	<0.01	<0.01
MICROBIOLOGICAL												
Total Coliform	MPN Index/100 ml	5000	10000	540	70	46	350	23	110	≥2400	240	93
Fecal Coliform	MPN Index/100 ml	1000	2000	350	70	46	240	4.5	110	460	4	0
Escherichia coli	MPN Index/100 ml	1000	2000	17	70	46	240	2	110	170	0	0

Table 5. Comparison of greywater in Cimenyan Plantation with the characteristics of domestic wastewater

Parameter	Unit	Cimenyan	Firdayati, et al. (2009)	Qasim (2018)	Randra (2021)
pH	-	7.46	5.5-8.8	6.7 – 7.5	6.12 – 6.86
Temperature	^{0}C	24.17	-	10 – 21	-
Conductivity	µmhos/cm	316	243-1860		
TDS	mg/L	173.8	-	250 – 700	-
TSS	mg/L	35	27-194	150 – 400	92 – 171
Chloride	mg/L	23.2		30 – 100	-
Oil and Grease	mg/L	2.2	28-146	50 – 150	2.6 – 158.2
BOD	mg/L	6.86	111-690	110 – 400	222.53 – 412.25
COD	mg/L	9.8	189-1171	200 – 780	313.6 – 940.8
SAR	-	0.70	0.8-4.5	-	-
MBAS	mg/L	<0.01	0.9-24	-	-
Boron	mg/L	<0.01	0.02-0.47	-	-
Fecal Coli	MPN Index/100 ml	460	(1.6-2.9) x10^{13}	-	-
Total Coliform	MPN Index/100 ml	>=2,400	-	$10^6 – 10^9$	>12,000

Based on Tables 3-4, it is found that the dominant water source was lightly polluted with the most significant parameters polluting the source, namely BOD, %Na, ammonia, chloride, and sulfate.

The storage ponds in each plantation came from the main water source of each plantation and showed an increase in the number of *Escherichia coli* from the main water source to the storage pond. This indicated that microbial contamination occurred during the journey from the main water source to the reservoir. The water source reservoirs in both plantations did not use covers and were left open allowing wild animals to pass and becoming vectors of various pathogens through feces [15]. Water storage facilities should be closed to prevent contamination or irrigation water should be treated prior to use. The secondary water source of plantation 1 originating from forest springs shows values *Escherichia coli* as much as 70/100 mL and the secondary water source of plantation 2 originating from Situ Cileunca shows values *Escherichia coli* as much as 240/100 mL indicating fecal contamination [16]. Therefore, better water sanitation processes should be developed to treat water before being used for irrigation to reduce the risk of contamination [17].

The results of laboratory analysis showed that the aforementioned nine plantation water sources were lightly polluted, but contained a number of microbes that still met the quality standards. In this case, positive results of microbes in irrigation water sources can still potentially contaminate planted vegetables and fruit [18], [19].

Based on Table 5, the difference between the results of greywater test in Cimenyan and the quality of domestic wastewater could be influenced by environmental conditions. When sampling was carried out, it rained at the sampling location so that it affected the samples taken. Rainwater that entered the grey water caused it to dilute so that the concentration of wastewater was lower. The storage pond in Cimenyan was a simple pond. The soil in the Cimenyan district was andosol soil containing allophane, which is a mineral that can function as an adsorbent so that it can eliminate *Escherichia coli* and phosphate. In addition, there are water hyacinths which can reduce BOD, COD, and surfactants.

4. Conclusion

Study from four areas of vegetable plantation in Greater Bandung Area shows that water sources for irrigation came mainly from artesian well followed by spring water, surface water, and reuse of greywater. Precipitation helps maintain water availability for irrigation in the rainy season. However, in the dry season, all plantations suffered drying up of artesian well and had to depend on other water sources such as limited storage pond that keeps all sources before being used as irrigation water. Some sources had experienced degradation of water quality caused by the change of land use and then became entry of domestic wastewater discharge from the local community. Analysis of water source irrigation quality show that some parameters exceeded water quality standards especially irrigation quality criteria that was recommended based on scientific examination on standards of water irrigation in many other countries. The parameters were conductivity, BOD, DO, %Na, Ammonia, Chloride, Sulphate and Oil and Grease. This showed possibility contamination of wastewater from domestic activities to irrigation water source. Based on the measured parameters, waster used for irrigation, even using groundwater, was polluted and could become worst in the future. It may cause potential health risk for consumer.

References

[1] FAO, *Water for Sustainable Food and Agriculture A report produced for the G20 Presidency of Germany*. 2017. [Online]. Available: www.fao.org/publications
[2] B. S. Wignyosukarto, "Pengelolaan Sumberdaya Air Dan Kesejahteraan Rakyat (Water Resources Management and People's Welfare)," *Pusat Studi Ekonomi Kerakayatan Universitas Gadjah Mada*, Feb. 09, 2021.
[3] J. Winpenny, I. Heinz, and S. Koo-Oshima, "The economics of wastewater use in agriculture The Wealth of Waste The wealth of waste-The economics of wastewater use in agriculture 2," Rome, Italy, 2010.
[4] A. Y. Hoekstra and M. M. Mekonnen, "The water footprint of humanity," *Proc. Natl. Acad. Sci. USA*, vol. 109, pp. 3232–3237, 2012, doi: 10.1073/pnas.1109936109/-/DCSupplemental.
[5] N. T. van Ha et al., "Bacterial contamination of raw vegetables, vegetable-related water and river water in Ho Chi Minh City, Vietnam," *Water Science and Technology*, vol. 58, no. 12, pp. 2403–2411, 2008, doi: 10.2166/wst.2008.435.
[6] WHO, *WHO estimates of the global burden of foodborne diseases*. 2015.
[7] L. (Lise) Korsten, E. M. (Erika M.) du Plessis, South Africa. Water Research Commission, and F. South Africa. Department of Agriculture, *Investigation into the link between water quality and microbiological safety of selected fruit and vegetables from farming to processing stages. (Volume 1) : report to the Water Research Commission and Department of Agriculture, Forestry and Fisheries*. 2015.
[8] I. A. Yusuf, "Kajian Kriteria Mutu Air Irigasi (Review of Water Quality Criteria for Irrigation)," 2014.
[9] S. Sagala, N. Pranindita, D. Azhari, N. K. Ishadi, J. M. Kanai, and A. Lechner, "Hydro-meteorological Hazards in Bandung City, Indonesia N O. 1 (N B S) 2 0 2 1 0 3 0 4 RDI Team," 2021.
[10] A. N. K. Latif, W. D. Pratiwi, and S. Samsirina, "Analisis Perubahan Permukiman Akibat Pariwisata di Kawasan Wisata Situ Cileunca Kabupaten Bandung (Analysis of Settlement Changes Due to Tourism in the Situ Cileunca Tourism Area, Bandung Regency)," *Jurnal Lingkungan Binaan Indonesia*, vol. 8, no. 2, pp. 70–78, Jun. 2019, doi: 10.32315/jlbi.8.2.70.
[11] H. Jeong, H. Kim, and T. Jang, "Irrigation water quality standards for indirect wastewater reuse in agriculture: A contribution toward sustainablewastewater reuse in South korea," *Water (Switzerland)*, vol. 8, no. 4, 2016, doi: 10.3390/w8040169.
[12] S. Park *et al.*, "Multifactorial effects of ambient temperature, precipitation, farm management,

and environmental factors determine the level of generic Escherichia coli contamination on preharvested spinach," *Applied and Environmental Microbiology*, vol. 81, no. 7, pp. 2635–2650, 2015, doi: 10.1128/AEM.03793-14.

[13] O. B. Sunarwan, N. Kamal, and M. Luthfi, "Identification of Physical and Chemical Parameters of Groundwater in Volcanic Sedimentary Aquifers (Case study: Bandung Groundwater Basin)," *Jurnal Teknik Universitas Pakuan*, vol. 16, no. 1, 2015, [Online]. Available: http://portal.vsi.esdm.go.id

[14] T. Bauder, R. Waskom, P. Sutherland, and J. G. Davis, "Irrigation Water Quality Criteria," 2011. [Online]. Available: www.ext.colostate.edu

[15] L. Beuchat and J.-H. Ryu, "Produce Handling and Processing Practices," 1997.

[16] S. G. Jackson *et al.*, "Escherichia coli O157 : H7 diarrhoea associated with well water and infected cattle on an Ontario farm," Cambridge University Press, 1998.

[17] A. Allende and J. Monaghan, "Irrigation water quality for leafy crops: A perspective of risks and potential solutions," *International Journal of Environmental Research and Public Health*, vol. 12, no. 7, pp. 7457–7477, 2015, doi: 10.3390/ijerph120707457.

[18] P. Amoah, P. Drechsel, and R. C. Abaidoo, "Irrigated urban vegetable production in Ghana: Sources of pathogen contamination and health risk elimination," *Irrigation and Drainage*, vol. 54, no. SUPPL. 1, Jul. 2005, doi: 10.1002/ird.185.

[19] J. A. Thurston-Enriquez, P. Watt, S. E. Dowd, R. Enriquez, I. L. Pepper, and C. P. Gerba, "Detection of Protozoan Parasites and Microsporidia in Irrigation Waters Used for Crop Production," 2002. [Online]. Available: http://www.ncbi.nlm.nih.gov

Tsunami Risk Assessment in Business Continuity Planning for Palu Special Economic Zone

Muhammad Rizki Purnama[1†], Mohammad Bagus Adityawan[2], Krishna Suryanto Pribadi[3], Mohammad Farid[4], Widyaningtias[2], and Arno Adi Kuntoro[5]

[1]Graduate School of Water Resources Management, Institut Teknologi Bandung
Jalan Ganesha Nomor 10, Bandung, Jawa Barat, Indonesia
†Corresponding author, E-mail: mrizkipurnama@outlook.com
[2] Department of Water Resources Engineering and Management, Institut Teknologi Bandung
Jalan Let. Jen. Purn. Dr. (HC). Mashudi Nomor 1, Jawa Barat, Indonesia
[3] Research Center on Disaster Mitigation
Institute of Research and Community Services, Insitut Teknologi Bandung
Jalan Ganesha Nomor 10, Bandung, Jawa Barat, Indonesia
[4] Center for Coastal and Marine Development
Faculty of Civil and Environmental Engineering, Insitut Teknologi Bandung
Jalan Ganesha Nomor 10, Bandung, Jawa Barat, Indonesia
[5] Center for Water Resources Development
Faculty of Civil and Environmental Engineering, Insitut Teknologi Bandung
Jalan Ganesha Nomor 10, Bandung, Jawa Barat, Indonesia

Abstract. On September 28, 2018, an earthquake with a magnitude of 7.4 Mw struck Central Sulawesi, Indonesia and triggered a major tsunami. At least 3475 people lost their lives as reported by the Indonesian National Board for Disaster Management (BNPB), while a large unknown number is still buried under the rubble and presumed dead. This event also damaged the facilities and infrastructure of the Palu City coastal area. Currently, The Government of Palu is developing the Palu Special Economic Zone (PSEZ) in Tawaeli District, Palu, Central Sulawesi to attract investors and spur economic growth in neighbouring areas. Thus, mitigation plans are urgently needed to reduce damage and losses due to the impact of future tsunamis on PSEZ. This study assesses the impact of tsunamis around the Palu Special Economic Zone and its supporting infrastructure. Hypothetical earthquake-induced tsunamis were simulated using Delft3D and DelftDashboard. The fault scenario is developed based on data from PUSGEN. Further parameters were developed with the empirical earthquake source-scaling laws. The model simulates the tsunami generation, propagation, and inundation from the source to the coastal area. A tsunami impact scenario is then developed as a basis for analysing the business impact to the transportation infrastructure and critical utilities serving the PSEZ, as well as to the operation of the industries and businesses within the PSEZ. The lessons learned from this study can assist stakeholders in developing an areawide business continuity plan as a commitment of the stakeholders for implementing areawide business continuity management for the PSEZ, which will improve the economic resilience of the Palu City as well as South Sulawesi region.

Keywords: tsunami, mitigation, business continuity plan, Palu.

1. Introduction

On September 28, 2018, an earthquake with a magnitude of 7.4 Mw struck Central Sulawesi, Indonesia and triggered a major tsunami with a height of 10 meter [1]. At least 3475 people lost their lives as reported by the Indonesian National Board for Disaster Management (BNPB) [2]. The tsunami event of 2018 in Palu made a massive destruction along the coastal area of Palu City

Currently, The Government of Palu is developing the Palu Special Economic Zone (PSEZ) in Tawaeli District, Palu, Central Sulawesi to attract investors and spur economic growth in neighbouring areas. Thus, mitigation plans are urgently needed to reduce damage and losses due to the impact of future tsunamis on PSEZ. There have been many studies focusing on reducing the impact of future tsunami such as tsunami hazard assessment, tsunami countermeasures, and development of tsunami early warning system [3, 4, 5]. However, this study assesses the impact of tsunamis around the PSEZ through the development of ABCP Strategy that can assist stakeholders in implementing areawide business continuity management (ABCM).

2. Study Area

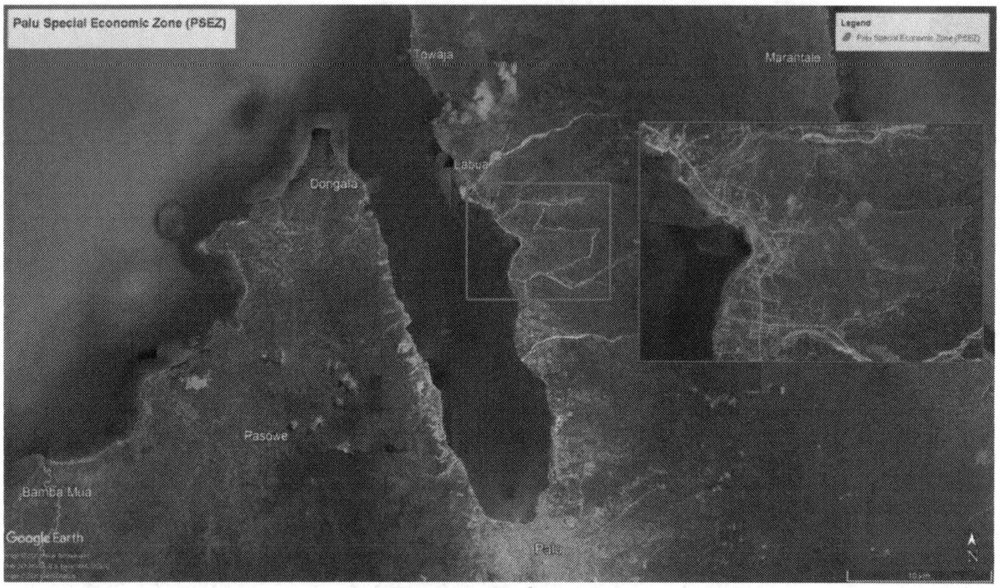

Figure 1. Study Area

This study was carried out in the developed area of Palu Special Economic Zone (PSEZ). This area was also affected by the 2018 Sulawesi Earthquake and Tsunami. The event caused high financial and physical losses for the community of Palu City. PSEZ was planned in 2007 as a 3000 hectares industrial park, and proposed to become SEZ in 2011 and began its operation in 2017. It was designed as an integrated logistics center and mining product processing (nickel, iron ore, and gold) in the Sulawesi Economic Region. The PSEZ is integrated with the Port of Pantoloan which is a strategic route for maritime transportation. Port of Pantoloan is a feeder port that connecting one to another main port between the eastern and western regions of Indonesia. The PSEZ is projected to spur an invesment of IDR 92.4 trillion and created at least 97,500 employments by 2025 [6]

Palu City and its nearby area has been hit by a number of tsunamis in the past. The highest frequency of tsunami events for the Indonesian archipelago occurs in the Makassar Strait, a zone of major geological importance. Most of its faults are located under the sea and produce large shallow earthquakes that have a high possibility of generating tsunami [7].

3. Methodology

3.1. Tsunami Numerical Simulation

Table 1. Earthquake-induced tsunami Parameter

Scenario	Fault	Mag. M_w (max)	Depth (km)	Focal Mechanism Strike (deg)	Dip (deg)	Rake (deg)	Dimension Length (km)*	Width (km)	Dislocation (m)
1	Makassar Strait North and Makassar Strait Central	8.1	14	199	45	90	270	52	10.2

Hypothetical earthquake-induced tsunamis were simulated based on previous study which is considered as the worst-case tsunami scenario for Palu Bay [8]. The parameter is shown in Table 1. Thus, its scenario is developed based on data from PUSGEN. Further parameters were developed with the empirical earthquake source-scaling laws [9]. Tsunami generally consist of three phases, namely generation, propagation, and inundation. All of the tsunami process were simulated using Delft3D and DelftDashboard. Delft3D has been widely used for tsunami numerical simulation [10,11,12,13]. The model capable to simulate the tsunami generation, propagation, and inundation from the earthquake source to the coastal area. Using the fault segment data, DelftDashboard computes a spatially varying vertical water level displacement along the fault segment and creates an initial tsunami wave to be propagated along the sea [14]. Once the initial generation wave is generated, the front wave would start moving towards the deep ocean and towards the nearby shoreline. The propagation and inundation process are simulated using the hydrodynamic module in Delft3D-FLOW [15]. Furthermore, the flowchart of this study is given in Figure 2.

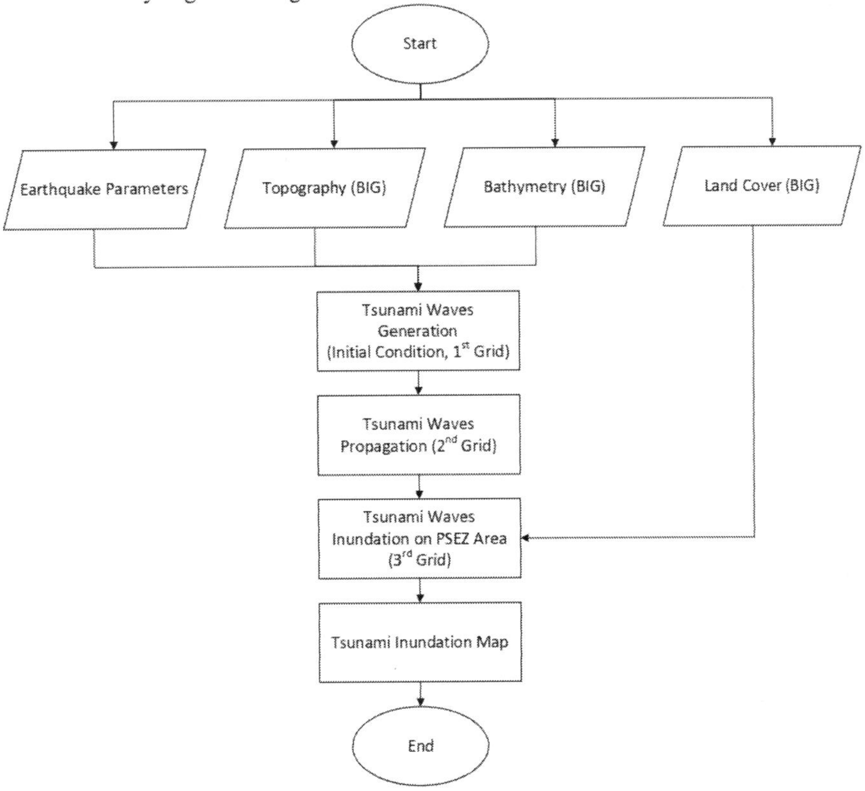

Figure 2. Methodology

The model run in three different grids and connected by process named nesting to minimize error and computational time [16]. The first-level grid, which has resolution of 0.01 arc degrees or 1 km, covers the area from the earthquake sources through the PSEZ. The second-level grid, which has resolution of 0.001 arc degrees or 100 m, covers the area of Palu Bay. The finest one, which has resolution of 0.0001 arc degrees or 10 m, covers the area of PSEZ. The grids are shown in Figure 3.

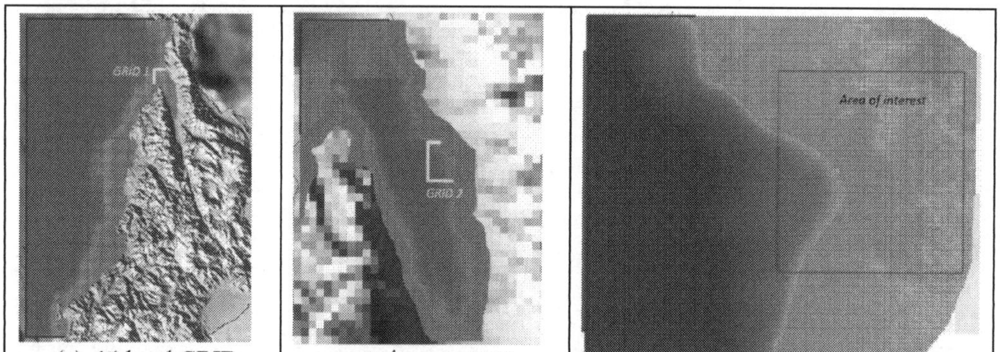

(a) 1st-level GRID (b) 2nd-level GRID (c) 3rd-level GRID

Figure 3. Certain type of GRID that is used in tsunami numerical simulation

3.2. Bathymetry, Topography, and Land Cover Data

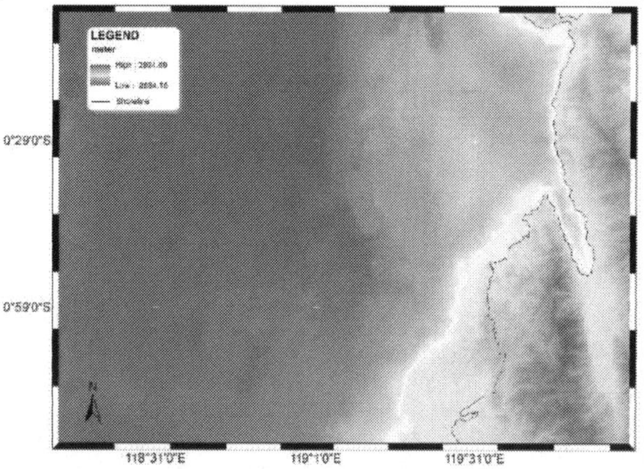

Figure 4. Bathymetry

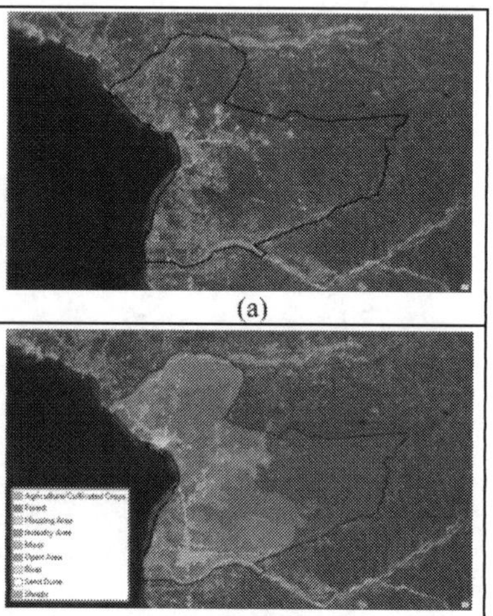

(a)

Figure 5. (a) PSEZ Boundary (b) Land Cover of PSEZ

Bathymetry and topography data around Palu Bay were obtained from Geospatial Information Agency of Indonesia. Bathymetry and topography data are available at resolution of 20 m and 8 m, respectively. However, the bathymetry data from BIG did not cover large area. Thus, we combine the bathymetry of GEBCO08 (30 m resolution) to overcome this gap. The bathymetry of the study area shown in Figure 4.

The wave propagation overland is highly influenced by its land cover [17]. The surface roughness value of Manning (n) is determined based on the land cover of the study area. Manning's value is referenced on various studies, as shown in Table [18,19].

Table 2. Manning's n value

Land Cover	Manning's n
Agricultural/Cultivated Crops	0.035
Forest	0.10
Housing Area	0.10
Industry Area	0.10
Moor	0.030
Open Area	0.010
River	0.025
Sand Dune	0.020
Shrubs	0.08

3.3. Business Continuity Planning

The tsunami impact scenario, which is based on the numerical modelling, is then developed as a basis for analysing the business impact to the PSEZ. The tsunami impact scenario is used as a basis for analyzing the business impact of the tsunami to the transportation infrastructure and critical facilities serving the PSEZ, which will affect the operation of the industries and businesses within the PSEZ. Based on the business impact analysis, improvement to the infrastructure and critical facilities will be proposed in order that disruptions to the operation of the PSEZ can be minimized bellow maximum tolerable period of disruption (MTPD) for the operation of the PSEZ. The approach for the development of business continuity plan for the PSEZ is using the methodology of Areawide BCP (ABCP) initially proposed by Hitoshi Baba et al. in [20] and further developed by AHA Center and JICA in 2015 [21] and [22], focusing on the cooperative effort between government agencies, infrastructure operators and the industry, to ensure the resilience of the critical infrastructure supporting the PSEZ.

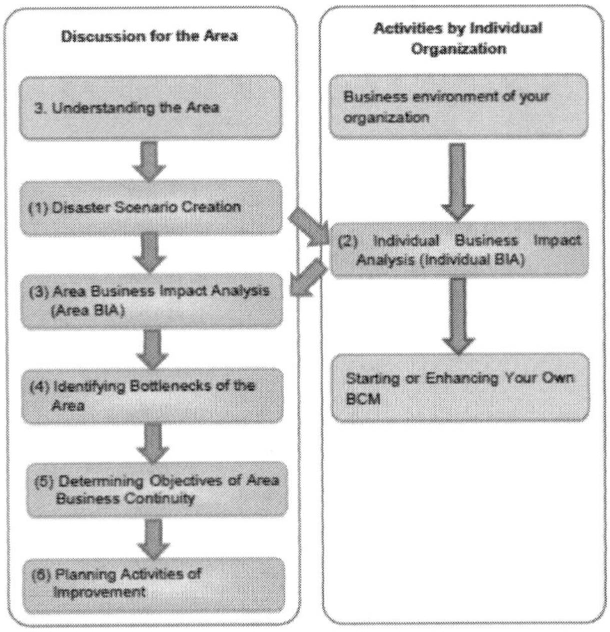

Figure 6. Steps to determine Area BCM Strategy (from [16])

The analysis of business impact to the PSEZ is conducted using the approach proposed by Benavente et al. [23] for port logistic business continuity management.

4. Result and Discussion

4.1. Tsunami Numerical Simulation

Initial sea surface levels as a result of earthquake generation and the simulated tsunami wave propagation are presented in Figure 7 and Figure 8, respectively. The generated initial wave height of the tsunami is about 7 meters. Tsunami propagation and inundation simulation were carried out from the earthquake source, which was Makassar Strait Central and Makassar Strait North faults, to Palu City Coast.

This scenario gives the most significant impact to the PSEZ. The tsunami maximum inundation map is shown in Figure 9. Area closer to the coastline may experience inundation with a depth of more than 6 meters.

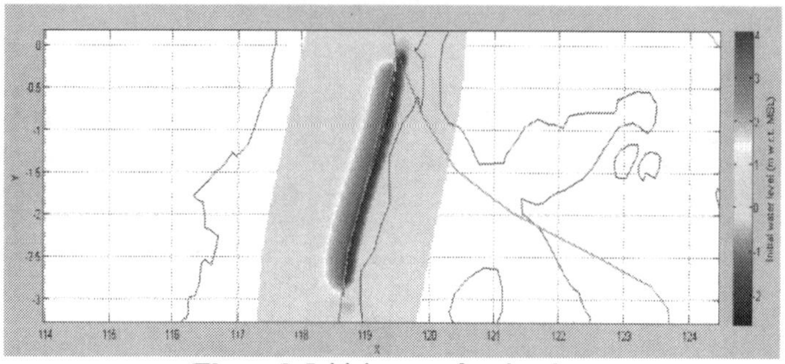

Figure 8. Initial sea surface level

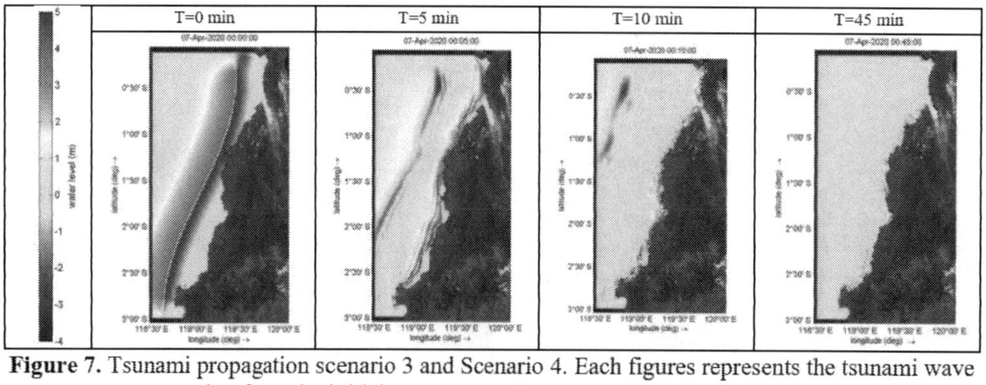

Figure 7. Tsunami propagation scenario 3 and Scenario 4. Each figures represents the tsunami wave propagation from the initial wave generation to 120 mins after the generation

Six point of observation was defined as shown in Figure 9 (b) until (g). This point was taken as a consideration based on critical facilities around the PSEZ area, such as Port of Pantoloan, Container Yard, PSEZ main road access, developed area, and customs and excise facility. Each points have time arrival and its range of inundation depth as shown in Table 3.

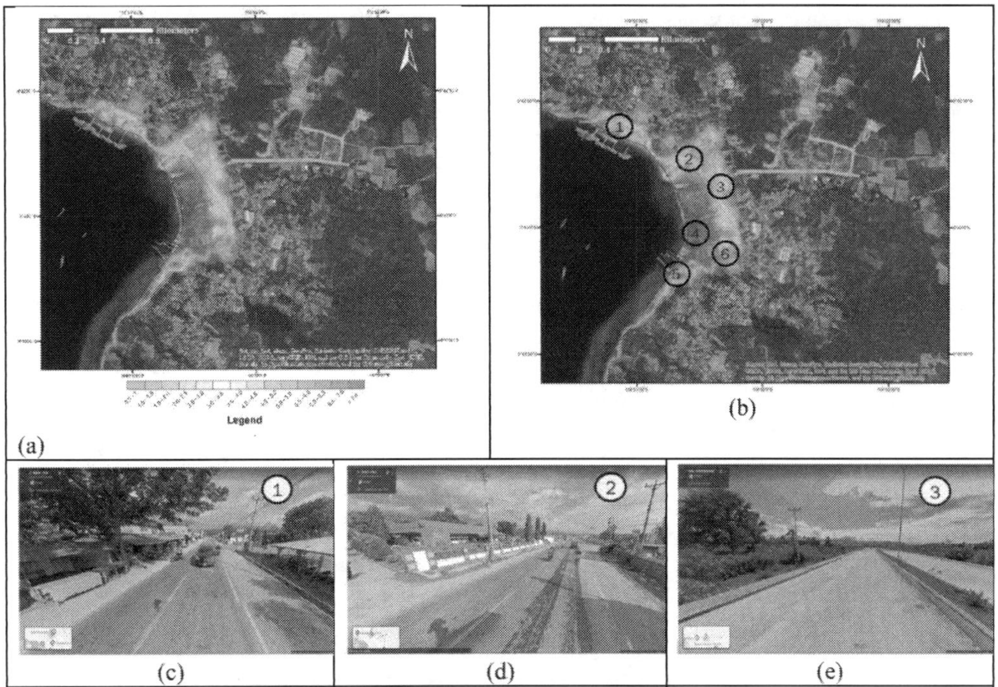

Figure 9. (a) Tsunami inundation map (b) Point of observation at certain critical infrastructures (c) Port of Pantoloan Area (d) Container Yard (e) Mangu Street, as one of main access to PSEZ (f) Toli-toli Street, developed area (g) Customs and Excise Facility

Table 3. Tsunami arrival time and inundation depth at point of observation

Point of Observation	Inundation Depth (m)	Arrival Time (mins)
1	4.0 – 5.0	38.0
2	3.0 – 4.0	43.0
3	3.0 – 4.0	44.0
4	7.0 – 8.0	39.0
5	6.0 – 7.0	39.0
6	4.0 – 5.0	44.0

4.2. Business Continuity Plan

The business continuity planning process for the implementation of an Areawide Business Continuity Management (ABCM) system was initiated and facilitated by the Coordinating Ministry of Economic Affairs under the Deputy for Coordination of Regional Development and Spatial Planning. The process commenced by establishing the Palu City Working Group for ABCP development of PSEZ (PCWG-ABCP), led by the Assistant of Economic Affairs and Development of the Municipality of Palu City. Its members consist of representatives from various local government agencies of Palu City, infrastructure operators, academics and industry tenants. The industrial park of the PSEZ registered already 32 tenants, but up to 2021 there are only 5 tenants already operating.

As discussed before, the assumed disaster is an earthquake generated tsunami, caused by the displacement due to the movement of fault at North and Central Makassar Strait. The earthquake is assumed to cause an intensity of VI-VIII MMI around Palu City, slight to moderate damages incurred by the earthquake to some of the infrastructures and buildings in the industrial parks. In this analysis, the impact of the assumed tsunami run up and inundation as well as the assumed earthquake intensity to the critical infrastructures is analyzed. Table 4 shows the impact of the assumed tsunami and earthquake, based on the experience from September 28, 2018 earthquake and tsunami disaster in Palu area. The analysis of the impact is conducted through a series of workshops organized by the PCWG-ABCP. Based on the result of the analysis, a recovery performance curves for the PSEZ are established, with 60% function recovery at the 30 days after the disaster, shown in Figure 10.

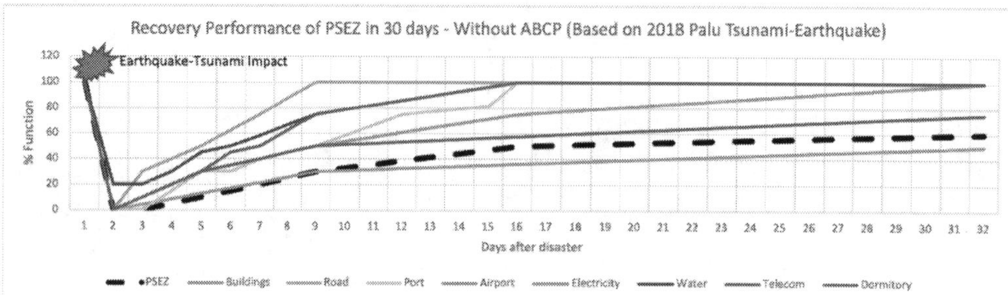

Figure 10. Step Recovery Performance of PSEZ in 30 days- Without ABCP (Based on 2018 Palu Tsunami-Earthquake Disaster) to determine Area BCM Strategy (from [16])

An analysis of the strategy to improve the recovery performance is developed by the PCWG-ABCP, resulting in the performance improvement strategy for the infrastructure and facilities supporting the operation of the PSEZ, as shown in the Table 5 and 6. Table 5(a) presented the definition of the core business of the PSEZ to be protected by the strategy and Table 5(b) presented the determination of the risk level of the impacts that may be caused by the PSEZ disruption due to the disaster. Tabel 5(c) shows the process to define the MTPD (Maximum Tolerable Period of Disruption), RTO (Recovery Time Objective) and RLO (Recovery Level Objective) of the ABCP for PSEZ The MTPD for the overall PSEZ is determined at 30 days, so the RTO is taken at 29 days, assuming that a one-day mobilization process of the ABCP is required before recovery can start. RLO is defined at 75% in order that tenants will not move out from PSEZ due to the disruption.

Based on the analysis, a strategy for ABCP is developed to ensure the recovery of PSEZ to 75% function within the defined RTO, as shown in Table 6. The recovery performance of the PSEZ based on the strategy is shown in Figure 11.

Table 4. Impact of tsunami to PSEZ operation

No.	Affected infrastructure & facilities	Damage *	Impact to PSEZ	Response	Time of Recovery*(in days)			
					30%	50%	75%	100%
1	PSEZ administrative office building	Heavy damage due to EQ	Service to tenant disrupted	Move to temporary office	7d	30d	60d	120d
2	Factory buildings	Heavy damage in 1 factory due to EQ	Production stopped	Repair of the factory	7d	14d	30d	90d
3	Access road to PSEZ	Medium/heavy damage (1.5km)	Access to PSEZ lightly affected	Clearing and repair by PSEZ	1d	3d	5d	7d
4	Access road to Pantoloan Port	Not affected	Pantoloan Port still accessible	Debris Clearing	1d	2d	3d	5d
5	Pantoloan Port	Moderate damage to port facilities (including crane) due to tsunami,	Export/import through port is stopped, logistic disrupted	Coordination with PELINDO (Port Company) for recovery of import/export services by provision of cranes from other ports	3d	7d	10d	14d
6	Access Road to Airport	Roads blocked by tsunami debris	Mobility of PSEZ business actors and workers reduced	Alternative roads	2d	3d	7d	14d
7	Airport	Moderate damage due to EQ, closed to public use	Mobility of PSEZ business actors and workers reduced	Waiting for recovery and opening again	3d	7d	14d	30d
8	Electric Power	Moderate damage to 20 kv transmission, sub-stations, coal power plant	Power outage in the PSEZ and surrounding areas, telecommuni-cation and water supply stopped, tenant production disrupted	Communicate with PLN (Power Company) to accelerate power recovery. Limited power from tenants' backup generators	3d	7d	14d	30d
9	Water Supply	Heavy damage to intake and distribution system in Palu City	PSEZ is not affected by city water supply disruption as tenants have their own deep well sources However quality of water is affected by the earthquake	Water filter installation and improvement of deep well	2d	4d	7d	14d
10	Telecom	Damage to BTS	Quality of telecommunica-tion service reduced, PSEZ administration and services degraded	Waiting for recovery	3d	5d	7d	14d
11	Workers' dormitory	Heavy damage due to earthquake	Production process disrupted	Provision of temporary housing using product warehouses	3d	7d	30d	60d

*Note: Based on the experience from the 28 September 2018 Earthquake and Tsunami in Palu

Table 5. Development of ABCP Strategy for PSEZ

(a) Definition of Core Business of PSEZ

Core Business	Stakeholders
Provision of industrial lands	Palu city government

(b) Determination of risk level as the impact from PSEZ disruption

Impact \ Risk	Low (L)	Medium (M)	High (H)
Tenants leave location	No impact/ very low	Some tenants/ recoverable	All tenants/ unrecoverable
Loss of employment		Temporary/ recoverable	Long term/ unrecoverable
Investment in PSEZ decreasing		Temporary/ recoverable	Long term/ unrecoverable
Regional economic decreasing		Temporary/ recoverable	Long term/ unrecoverable

(c) Determination of MTPDs and RTOs

Impact of PSEZ Disruption \ Duration of PSEZ disruption	3d	1w	2w	1m	3m	6m	1y	MTPD days	RTO days	RLO % of PSEZ function recovery
Tenants leave location	L	L	M	M	H	H	H	30	29	75%
Loss of employment	L	L	M	M	H	H	H	30	29	50%
Investment in PSEZ decreasing	L	L	L	L	M	H	H	90	89	50%
Regional economic decreasing	L	L	L	L	M	M	H	180	179	30%

Note:
MTPD = Maximum Tolerable Period of Disruption (day/week/month/year)
RTO = Recovery Time Objective (in days)
RLO = Recovery Level Objective (in % of recovered function)

Table 6. Proposed ABCP Strategy for PSEZ Palu

No	Facilities/ Infrastructure	MTPD (days)	Target		Proposed improvement	Period of Impleme-ntation	Responsible organization
			RTO (days)	RLO (%)			
Buildings and Housing							
1	PSEZ administrative building	30	29	**100**	New building in accordance with SNI 1726-2019	2022-2023	PSEZ
2	Factory buildings	30	29	**100**	• Retrofit existing buildings to suit SNI 1726-2019 • New buildings must be designed in accordance to earthquake resistance standards	2022-2025	Tenants
3	Workers' dormitory	30	29	**100**	New building/retrofit in accordance with SNI 1726-2019	2022-2023	• PSEZ • Tenants
Infrastructure/utilities							
4	Access road to PSEZ	6	5	**75**	• Strengthen existing road • Build alternative road • Landslide protection	2022-2023	Public Work (Bina Marga)
5	Roads inside PSEZ	4	3	**75**	Strengthen existing and build new roads to standards	2022-2025	PSEZ
6	Drainage inside PSEZ	6	5	**75**	Design and build drainage system according to standard	2022-2023	PSEZ
7	Waste water treatment plant	6	5	**75**	Design and build sewerage and earthquake resistant water treatment plant according to standard,	2023-2025	• PSEZ • Tenants
8	Access road to Pantoloan Port	4	3	**75**	• Strengthen and improve existing road	2022-2023	Min of Public Works and Housing

					• Dikes for coastal protection		(Bina Marga and SDA)
9	Pantoloan Port	7	6	**75**	• Rebuild damaged facilities emphasizing on tsunami mitigation approach • Establish Disaster Management Department	2022-2025	• PELINDO IV/ • Min. of Transport (DJPL)
10	Access Road to Palu Airport	7	6	**75**	• Repair and strengthen existing • Landslide protection • Alternative road	2022-2023	Public Work & Housing (Bina Marga)
11	Palu Airport	7	6	**75**	• Repair/rebuild for better emergency preparedness • Improve coordination for emergency support	2022-2023	Ministry of Transport (DJPU)
12	Electric Power	7	6	**75**	• Strengthen substations • Alternative loop for 20 kV network • Prepare back-up generators i	2022-2023	PLN PSEZ/Tenants
13	Water Supply	4	3	**75**	• Build transmission line from Wombo dam to PSEZ taking into consideration earthquake hazard • Build distribution network inside PSEZ	2023-2025	• PDAM Palu • Min. of Public Works and Housing (Cipta karya) • PSEZ
14	Telecommunica-tion network	5	3	**75**	• Strengthen BTS for earthquake resistance, • improve fiber optic	2022-2023	• Operators: PT Telkom, Telkomsel, Indosat, Tri, XL, Smartfren • Min.of Info. Com.
Resources and Services							
15	PSEZ Fire Department	1	0	**75**	Establish Fire Department, Train personnel	2023	• PSEZ • BPBD • Fire Department of Palu City
16	PSEZ Human Resources	1	0	**75**	• Establish Emergency Response and Recovery Division (DRU) within PSEZ and develop SOPs • Train PSEZ personnel for emergency recovery and management	2022-2025	• PSEZ • Tenants • BPBD
17	Human resources in local government agencies	1	0	**75**	Train related agencies' human resources for disaster mitigation and emergency response	2022-2025	Local government of Palu, including BPBD
18	Heavy equipment	1	0	**75**	• Identify source of heavy equipments to be mobilized in emergency response • Procure minimum amount of heavy equipment for Disaster Recovery Unit and train personnel.	2022-2025	PSEZ
19	Evacuation shelters and equipment	1	0	**75**	• Design and build save evacuation shelters, • procure evacuation equipment • train personnel for evacuation	2022-2025	• PSEZ • BPBD

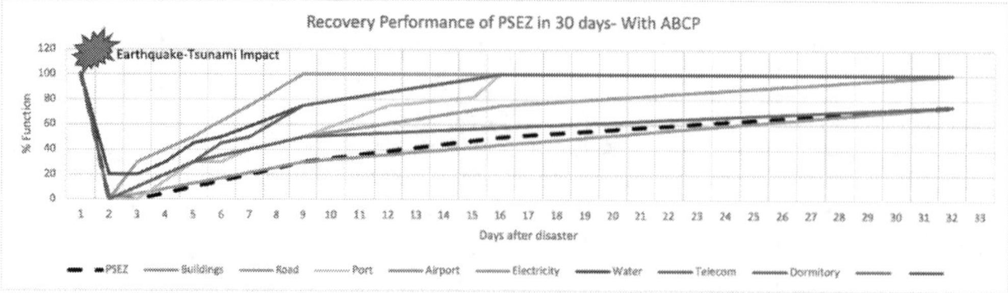

Figure 11. Recovery Performance of PSEZ in 30 days- With ABCP

5. Conclusion

The study shows that tsunami as well as the earthquake that generated the tsunami are the hazards that must be considered in the effort to plan and implement an areawide business continuity management (ABCM) the Palu Special Economic Zone (PSEZ), which is needed to ensure the resilient of the industry in the area, which in turn will contribute to the economic resilience of the region. Strategies to improve the resilience of the infrastructure and critical facilities in supporting the continuity of operation of the PSEZ need to be developed and implemented consistently, in order to prevent the dire experience of the September 28, 2018 Palu earthquake and tsunami. The ABCP development is currently still in progress and the PCWG-ABCP continues to work to produce an implementable plan. However, the plan will not be useful unless it is implemented within the ABCM cycle which will requires the leadership of the city as well as the PSEZ management, supported by strong coordination and commitments from all the concerned stakeholders.

Acknowledgments
The authors would like to acknowledge the support given by the Coordinating Ministry for Economic Affairs and Working Group (POKJA) for the Preparation of the Palu Special Economic Zone ABCP, who provided the opportunity to conduct the exercise of area business continuity planning with Palu City. Appreciation is also due to P2MI ITB 2021 and Institute for Research and Community Services of Institut Teknologi Bandung, who provided research funding through Riset ITB 2021 with the title of "Development of Beach Protection Concept against Tsunami Hazard based on Characteristic of Coastal Area".

6. References

[1] Sihombing Y I, Adityawan M B, Widyaningtias, Chrysanti A, Farid M, Nugroho J, Kuntoro A A, Kusuma M A, "Tsunami Overland Flow Charachteristic and Its Effect on Palu Bay Due to the Palu Tsunami 2018" Journal of Earthquake and Tsunami pp.1-22, 2020.

[2] Gusman, et al., "Source Model for the Tsunami Inside Palu Bay Following the 2018 Palu Earthquake, Indonesia" Geophysical Research Letters, pp46, 2019.

[3] Hosrpool, et al., "A probabilistic tsunami hazard assessment for Indonesia" Natural Hazards and Earth System Sciences pp.3105-3122, 2014.

[4] Syamsidik, Tursina, Suppasri A, Al'ala M, Luthfi M, Comfort L K, "Assessing the tsunami mitigation effectiveness of the planned Banda Aceh Outer Ring Road (BORR), Indonesia", Natural Hazards and Earth System Sciences Vol.19(1), pp. 299–312, 2019

[5] Nurendyastuti, A K, Marta Dinata M M, Mitayani A, Purnama M R, Adityawan M B, Farid M, Kuntoro A A, & Widyaningtias. Tsunami Early Warning System Based on Maritime Wireless Communication. Journal of Civil Engineering Forum, 2001, *in press*.

[6] Annon., Regional Development in Numbers of Sulawesi Region 2020., BAPPENAS (Ministry of National Development Planning of the Republic of Indonesia)., 2020. https://simreg.bappenas.go.id/assets/temaalus/document/Publikasi/DokPub/Gabung_PDDA_2020_Sulawesi_FINAL_100121.pdf
[7] Prasetya G and De Lange W P, "The Makassar Strait Tsunamigenic Region, Indonesia", Natural Hazards, Vol.24, pp.295-307, 2001.
[8] Purnama, M R, Adityawan M B, Farid M., Chrysanti A., 2021 "Effectiveness of Elevated Road in Reducing Inundation Area in Palu City Coast, Sulawesi, Indonesia". Journal of Disaster Research, in review
[9] Papazachos B, Scordilis E, Panagiotopoulos D, Papazachos C, Karakaisis G, "Global relations between seismic fault parameters and moment magnitude of earthquakes", Bulletin of the Geological Society of Greece 36, 1482–1489, 2004.
[10] Fachrurrazi, Syamsidik, Al'ala M, and Mahardi W, "Numerical simulations of tsunami waves impacts on Ulee Lheue Harbour in Banda Aceh-Indonesia", IOP Conf. Series: Earth and Environmental Science Vol.56, 2017.
[11] Apotsos A, Buckley M, Gelfenbaum G, Jaffe B, and Vatvani D, "Nearshore Tsunami Inundation Model Validation: Toward Sediment Transport Applications", Pure Applied Geophysics, Vol.168, pp.2097–2119, 2011.
[12] Purnama M R, Adityawan M B, Farid M, Chrysanti A, Rayadi B T, and Suryadi Y, "Development of Tsunami Inundation Map for the Coast of Palu City", IOP Conference Series: Earth and Environmental Science, Vol.737(1), 2021
[13] Van Veen B A D, Vatvani D, and Zijl F, "Tsunami Flood Modelling for Aceh & West Sumatra and Its Application for an Early Warning System", Continental Shelf Research, Vol.79, pp.46-53, 2014.
[14] Ormondt M V, Nederhoff K, Dongeren A V, "Delft Dashboard: a quick set-up tool for hydrodynamic models", Journal of Hydroinformatics, Vol.22 (3): 510–527. 2020
[15] Deltares, "Delft3D-FLOW, User Manual". Boussinesqweg: Netherlands, 2019.
[16] Tang L, Titov ssVV, and Chamberlin CD, "Development, testing, and applications of site-specific tsunami inundation models for real-time forecasting", Journal of Geophysical Research, Vol:114, pp. 1-22, 2009.
[17] Adityawan M B, Rahmasari P, Chrysanti A, Farid M, Yakti B P, and Purnama M R,"Effect of Land Cover on Tsunami Overland Flow Propagation: A Case Study of Painan, West Sumatra, Indonesia," International Journal on Advanced Science, Engineering and Information Technology, vol. 11, no. 5, pp. 1704-1712, 2021.
[18] Prawirabhakti A, & Andiese V W, "Numerical Model of Manning Roughness Coefficient Effect on Water Level Rise in Palu Bay" Jurnal Infrastruktur pp.1-8, 2015.
[19] Kalyanapu AJ, Burian SJ, McPherson TN, "Effect of land use-based surface roughness on hydrologic model output", Journal of Spatial Hydrology, Vol.9(2), 2009
[20] Baba H, Watanabe T, Nagaishi M, Matsumoto H, "Area Business Continuity Management, a new opportunity for building economic resilience", Procedia Economics and Finance 18, pp.296–303, 2014
[21] Annon., Planning Guide for Area Business Continuity ~ Area BCM Toolkits ~Version 2, Main Volume, AHA Center, JICA, OYO International Corporation, Mitsubishi Research Institute, Inc., CTI Engineering International Co., Ltd., March 2015, https://openjicareport.jica.go.jp/pdf/1000023389.pdf
[22] Annon., Planning Guide for Area Business Continuity ~ Area BCM Toolkits ~Version 2, Supplemental Volume - Tools for Area BCM, AHA Center, JICA, OYO International Corporation, Mitsubishi Research Institute, Inc., CTI Engineering International Co., Ltd., March 2015, https://openjicareport.jica.go.jp/pdf/1000023390.pdf
[23] Benavente F C, Gallardo M R, Esquivel M B, Akakura Y and Ono K, "Methodology and procedure of business impact analysis for improving port logistics business continuity management', IDRiM (2016) 6 (1), ISSN: 2185-8322, DOI10.5595/idrim.2016.0114, 2016

Groundwater salinity and its effect on elevated concentrations of other contaminants in a coastal plain: a perspective for infrastructure protection

A F Rusydi[1*], S-I Onodera[2], S Ioka[3], R Maria[4], F Firmansyah[5], M Saito[2], W Purwoko[6], and D Sukmayadi[6]

[1]Research Center for Limnology and Water Resources, National Research and Innovation Agency, Cibinong 16911, Indonesia
[2]Graduate School of Advanced Science and Engineering, Hiroshima University, Higashi Hiroshima 7398521, Japan
[3]Institute of Regional Innovation, Hirosaki University, Hirosaki 0368560, Japan
[4]Bureau for Organization and Human Resources, National Research and Innovation Agency, Jakarta 10340, Indonesia
[5]Research Center for Geological Disaster, National Research and Innovation Agency, Bandung 40135, Indonesia
[6]Research Center for Geological Disaster, National Research and Innovation Agency, Bandung 40135, Indonesia

*anna.fadliah.rusydi@brin.go.id

Abstract. Salinization is a serious threat to coastal groundwater because it degrades water quality and results in trace metal dissolution. Meanwhile, groundwater quality is critical for human and development, as contaminated groundwater cause a negative effect on human`s health and infrastructure. Our investigation of a specific area, Indramayu, on West Java's north coast revealed that most groundwater samples are brackish to saline in existence. TDS concentrations in groundwater samples were predominantly in the >1,000 to nearly 30,000 mg/L range. Concentration contours were disproportionately high in the north tip area. Numerous factors contribute to the high salinity of this area, including the interaction of saline water and sediment with low-salinity groundwater and evaporation. Additionally, high TDS levels were significantly correlated with Fe^{2+} and Mn^{2+} concentrations ($p < 0.001$). The area's high salinity was thought to facilitate the dissolution of trace metals from minerals such as siderite and rhodochrosite (SI of the minerals > 1). Concentrations of Fe^{2+} and Mn^{2+} were very likely to damage the water piping system. In terms of water quality, less than 30% of samples were fit for human consumption. Finally, the government should take into account the groundwater environment when developing the area in order to protect human health and infrastructure.
Keywords: Coastal groundwater, Indramayu, Iron, Manganese, Salinity

1. Introduction

Indonesia's coastal areas offer numerous advantages for economic development. There are several reasons for this, including its convenient transportation location, fertile soil, and abundant groundwater potential. As a result, the area has become a target area for urbanization. Numerous urbanized areas in Indonesia and other countries, for example, Jakarta in Indonesia, Bangkok in Thailand, Manila in the Philippines, and Bangkok in Vietnam, and Recife in Brazil, are located on the coast [1]–[3].

Former research has shown that urbanization has contributed to environmental pressure, particularly water resources [1]–[4]. Water resources themselves are critical to the development of a region. Until now, Indonesia's best source of clean water has been groundwater. On the other hand, it was revealed that coastal groundwater is vulnerable to development. It has been discovered that development exacerbates groundwater stress, both in terms of quantity and quality [3], [5].

Groundwater salinization is a common issue in coastal area, posing a serious threat to groundwater quality. Salinization may occur because of "modern" seawater intrusion caused by excessive groundwater pumping [3], [4]. Apart from human activities, coastal groundwater salinization may result from the interaction of marine-derived sediment and saline paleo water with fresh groundwater [3], [6]. Exposure to high salinity water may have a detrimental effect on the cardiovascular system and other health problems [7]. Additionally, increased salinity in groundwater promotes the dissolution of other contaminants, such as metals, from the rock/sediment into the groundwater [2], [8]–[10]. Certain metals, such as dissolved iron and manganese, have impacted negatively on both human health and infrastructure [11].

The purpose of this article is to evaluate the level and source of groundwater salinization in Indonesia's coastal region, as well as the resulting environmental impact. This groundwater quality data is essential for area development in terms of clean water resources and infrastructure protection.

2. Methods

The interpretation of hydro-chemical data was used to determine the quality of groundwater. The salinity parameter that was examined is total dissolved solids (TDS). Then, trace metals, specifically dissolved iron (Fe^{2+}) and manganese (Mn^{2+}), were used to determine the effect of high salinity groundwater. The concentrations of major cations and anions have been reported previously [9], [12]. Then, trace metals, specifically dissolved iron (Fe^{2+}) and manganese (Mn^{2+}), are used to determine the effect of high salinity groundwater.

Additionally, correlation analysis was used to determine the influence of salinity on trace metals variables. strength of the association between the salinity and trace metals variables. Further, the significance of the correlation between those two variables is indicated by the p-value [13], which simply states that if the p-value for a certain independent variable is less than (0.05), the variable is strongly correlated with the dependent variable. In contrast, if the *p*-value is greater than 0.05, the variable is weakly or not correlated with the dependent variable. In other words, a larger *p*-value indicates that the relationship is less significant.

2.1. Study Site

The study area is Indramayu, which is located on West Java's north coast (see figure 1's index map). The study site's predominant land uses are agriculture and brackish-water aquaculture, with residences concentrated along the Cimanuk River (figure 1). Indramayu's geological composition is dominated by Flood Plain Deposition (Qaf). The coastal zone is composed of two types of deposits: Delta (Qad) and Coastal (Qac) [9], [12].

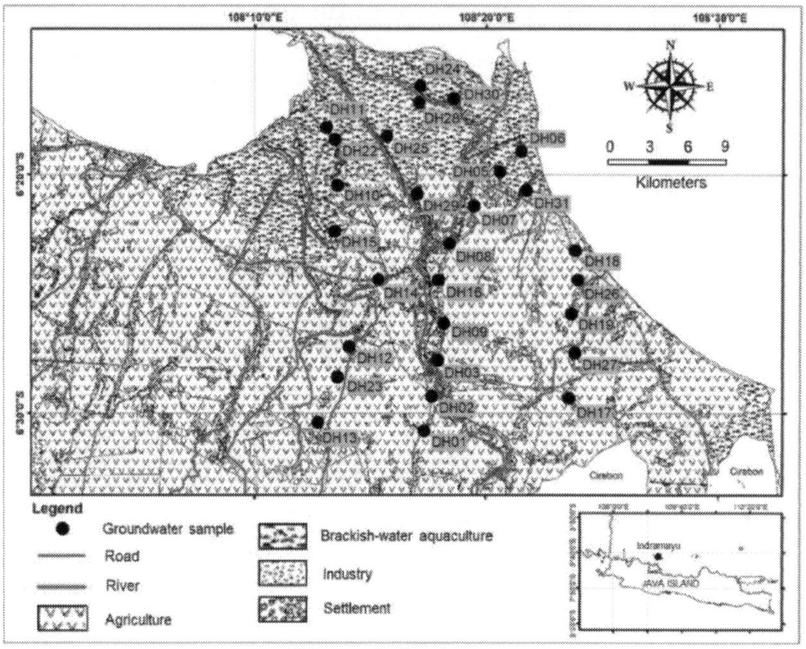

Figure 1. Study area

The study site has a relatively shallow groundwater table. Groundwater levels average 3.5 meters with a flow direction toward the sea [9]. The simplified geological cross-section of the Indramayu groundwater basin reveals the presence of multiple layers, including unconfined and confined aquifers. Rusydi et al. (2021) and Maria et al. (2018) discussed the stratigraphic description of the groundwater basin's geological sediments in detail [9], [14]. As a developing region, an environmental assessment that supports regional development is vital.

2.2. Samples collection and analysis

We collected a total of 28 groundwater samples from a variety of land uses (figure 1). The samples from the agriculture and settlements areas were taken mostly from unconfined aquifer at a depth of ≤20 meters [12]. Meanwhile, samples from brackish-water aquaculture were collected primarily from confined aquifers with a depth of ≥20 m [12]. Previous studies were used to determine the concentrations of major ions, iron, and manganese in groundwater [9], [12]. Meanwhile, TDS concentrations were deduced using Geochemist's Workbench software by estimating from the contents of major cations (sodium/Na^+, potassium/K^+, calcium/Ca^{2+}, and magnesium/Mg^{2+}) and anions (chloride/Cl^-, sulphate/SO_4^{2-}, bicarbonate/HCO_3^-).

The concentrations of TDS and trace metals were plotted on a graph to demonstrate their vertical and spatial distribution. ArcGIS and Surfer 16 with kriging gridding method were used to create the concentration contours. Further, Fe^{2+} and Mn^{2+} sources were predicted using the saturation indices of minerals in sediments, which were calculated using the PHREEQC software. The groundwater quality for drinking water purposes was then determined in terms of TDS, Fe^{2+}, and Mn^{2+} contents.

3. Result and Discussion

3.1. Salinity distribution and potential sources

The salinity of groundwater can be expressed in terms of TDS concentrations, as TDS is a parameter for the amount of inorganic salt in water [15]. The groundwater samples have TDS contents range from 490 to 29,126 mg/L, with an average of 6,746 mg/L and 8,810 mg/L standard deviation. The highest TDS concentrations were found in the northern area, which is dominated by brackish-water aquaculture (figure 2a). Simultaneously, TDS concentrations in groundwater samples tended to increase with depth of well (figure 2b). The TDS concentration ranged from 3,262 to 29,129 mg/L at brackish-water aquaculture sites, with an average of 15,156 mg/L and a standard deviation of 10,243 mg/L. Meanwhile, TDS concentrations were relatively low in agriculture and settlement land uses, at 490, 5806, 2073, and 1731 mg/L for the minimum, maximum, average, and standard deviation, respectively.

The majority of samples contained significant amounts of TDS, ranging from >1000 to 30,000 mg/L, indicating the groundwater samples were mostly classified as brackish water to saline water [16]. According to Rusydi et al. (2021) the salinization problem in the area was caused by geological mixing and saline water intrusion along the coastal line [9].

Table 1. Correlation between salinity parameters and trace metals in groundwater samples. The SO_4^{2-}, Cl^-, Fe^{2+}, and Mn^{2+} are originated from Rusydi et al., 2021 [9]

	TDS	SO_4^{2-}	Cl^-	Fe^{2+}	Mn^{2+}
TDS	1.00				
SO_4^{2-}	0.64	1.00			
Cl^-	0.51*	0.9	1.00		
Fe^{2+}	0.76	0.94	0.86*	1.00	
Mn^{2+}	0.46	0.88	1.00	0.81	1.00
Confidence level (α)	= 95%				
Significance level (*p* value)	=< 0.001, *<0.02				

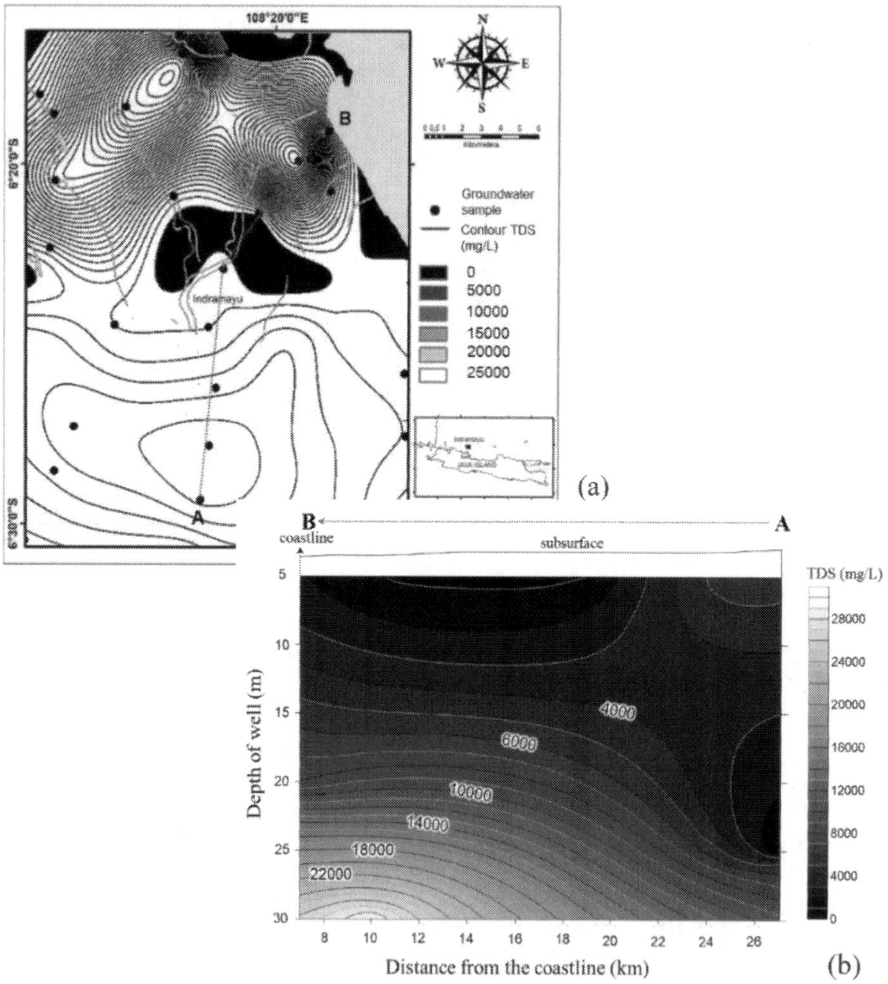

Figure 2. (a) Spatial and (b) vertical (A-B cross section) distribution of TDS in the groundwater samples

Previous research has established a strong correlation between salinity and elevated trace metal contents [8], [10]. Similarly, Table 1 reveals a positive and significant correlation ($p < 0.001$) between high contents of Cl^- and SO_4^{2-}, which are characteristic of salinity, with high quantities of Fe^{2+} and Mn^{2+}. Salt concentrations can enhance ionic strength and then reduce the ion activity coefficient, allowing for the easy dissolution of Fe and Mn in inorganic complexes [10]. According to a recent study, in addition to salinity, the presence of Fe^{2+} and Mn^{2+} was caused by the decomposition of organic materials, which resulted in a reducing groundwater environment [9].

3.2. Trace metals distribution and potential origins

As with TDS, Fe^{2+} and Mn^{2+} concentrations were highest along the coast, where brackish-water aquaculture is the predominant land use type [9]. Additionally, as the depth of the well increases, the Fe^{2+} and Mn^{2+} levels tended to increase. (figure 3a and b). The distribution of trace metals was comparable with TDS. Fe^{2+} and Mn^{2+} concentrations in brackish-water aquaculture site range between 0.1 and 252 mg/L and 0.1 and 22 mg/L, respectively. Furthermore, Fe^{2+} and Mn^{2+} concentrations in other land uses range between 0.1 and 11 mg/L and 0.1 to 8.1 mg/L, respectively. In comparison to TDS concentrations, trace metals concentrations vary more widely, indicating a local influence, such as heterogeneous geological material in the aquifer, that promotes Fe^{2+} and Mn^{2+} dissolution into groundwater [9].

The interaction of water with the surrounding mineral/soil, which includes iron and manganese and is characterized by a high salinity, was considered to be a critical factor in the release of Fe^{2+} and Mn^{2+} to the groundwater in Indramayu. The saturation index (SI) was used to identify possible dissolved mineral/soil. The SI was calculated using PHREEQC Interactive Programming [17] in conjunction with a principal computation as given in equation 1:

$$SI = \log IAP / k_t \quad \ldots\ldots (1)$$

where IAP expresses the ion activity product of the dissociated chemical species in solution, and k_t represents the chemical's equilibrium solubility product.
SI can be interpreted in the following states: SI = 0 indicated that the mineral is in equilibrium with the solution (saturated), SI 0 indicated that the mineral is undersaturated, and SI > 0 indicated that the mineral is supersaturated. If a solution's saturation index indicates supersaturation (SI >> 0), a mineral can precipitate.

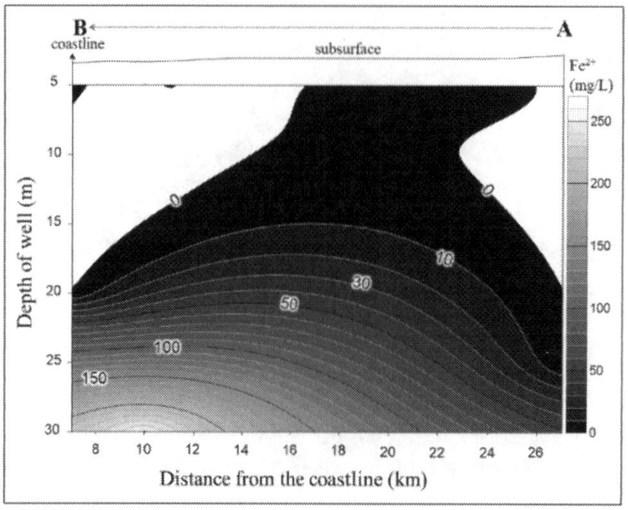

(a)

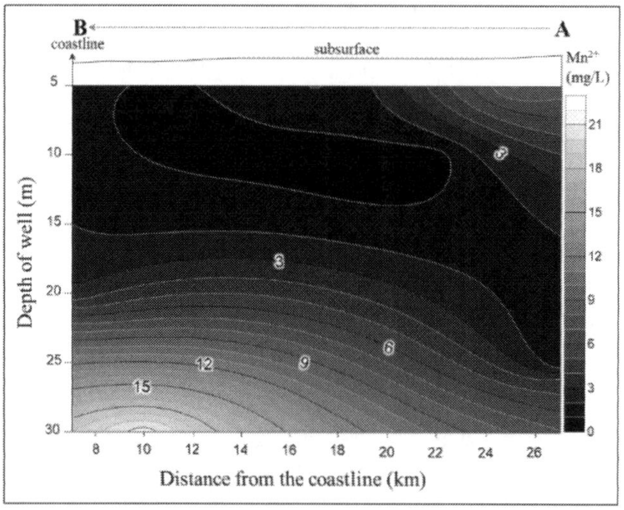

(b)

Figure 3. Distribution of iron (a) and manganese (b) in relation to the depth of the well

Besides concentrations of dissolved Fe and Mn, the PHREEQC inputs were groundwater temperature (°C); pH; and alkalinity as bicarbonate (HCO_3^-). All the inputs were sourced from Rusydi, et al., 2021 [9], [12]. The SI values were determined for a variety of minerals, including melanterite ($FeSO_4.H_2O$); siderite ($FeCO_3$); manganite ($Mn_2O_3.H_2O$); pyrochroite ($MnOH_2$); pyrolusite (MnO_2); and rhodochrosite ($MnCO_3$). These minerals are found in naturally occurring Fe^{2+} and Mn^{2+} sources.

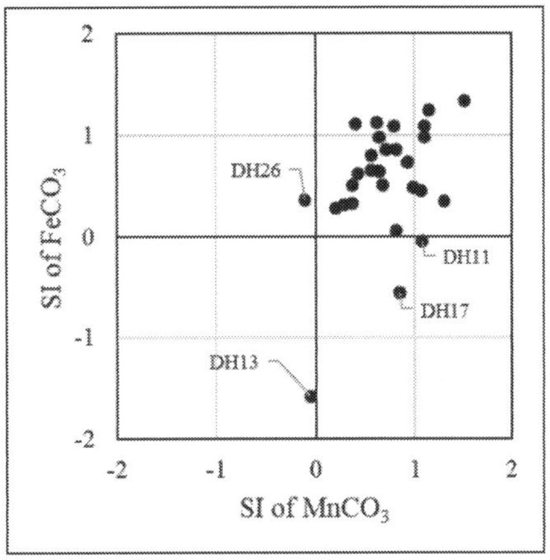

Figure 5. Saturation Indices of FeCO3 and MnCO3. PHREEQC Interactive 3.4.0 is used to perform the calculation.

Based on SI values, only $FeCO_3$ and $MnCO_3$ were typically found in an oversaturated state (SI > 0) (figure 5) or were capable of releasing Fe^{2+} and Mn^{2+} into groundwater, respectively. Meanwhile, the remaining minerals were all in an unsaturated state (SI < 0), which indicated that they are more likely to exist in precipitation form.

3.3. Groundwater quality for drinking water and challenge for infrastructure

The Indonesian government regulates TDS, Fe, and Mn levels in water sources to ensure that they are fit for human consumption [18]. Based on this regulation, concentrations of TDS, Fe^{2+}, and Mn^{2+} in a supply of drinking water should not exceed 1000, 0.3, and 0.1 mg/L, respectively [18]. Therefore, only about 30% of groundwater samples were suitable for drinking water when compared to TDS threshold limit. Additionally, nearly 50% and greater than 60% of samples were include inappropriate Fe^{2+} and Mn^{2+} concentrations [9].

The Water Quality Index (WQI) can be used to explain the quality of groundwater based on the three parameters listed above [19]. WQI is the most effective method for determining the acceptability of water for drinking purposes. It was estimated by utilizing the water quality rating scale, relative weight, standards of quality, and concentration for each parameter by following Eq. 2 to 4 [19], [20].

The first step is calculation of quality rating of each parameter (q_i):

$$q_i = {C_i}/{S_i} \; x \; 100 \qquad (2)$$

where C_i and S_i are concentration and standard value of each parameter, respectively.

The second calculation is determination of sub index (SI) of each parameter

$$S_i = \frac{w_i}{\sum w_i} \; x \; q_i \qquad (3)$$

where Wi is the weight of each parameter. TDS, Fe^{2+}, and Mn^{2+} are all weighted at 4 for w_i [20].

Finally, the overall WQI for TDS, Fe, and Mn are determined by using the expression:

$$WQI = \sum SI_{i-n} \quad (4)$$

n is the number of chemical parameters.

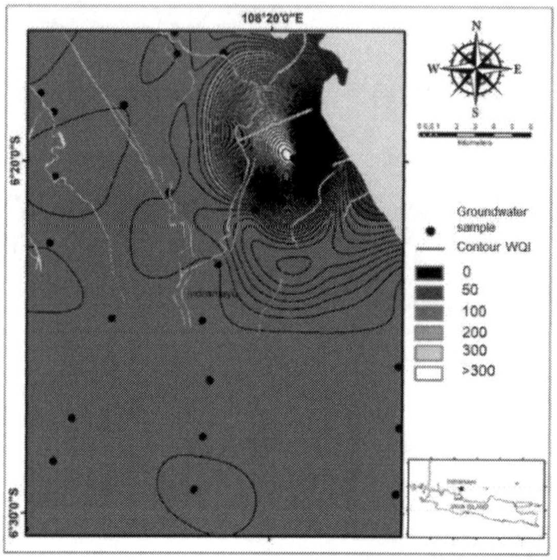

Figure 5. Spatial distribution of WQI in the groundwater samples

Water is categorized into five categories depending on WQI results : (i) WQI <50 as excellent water, (ii) WQI = 50 to 100 as good water, (iii), WQI = 100 to 200 as poor water, (iv) WQI = 200 – 300 as very poor water, and (v) WQI > 300 is unsuitable for drinking. The groundwater samples have WQI range from 38 to 37,371 with the spatial distribution as presented in figure 5. Based on WQI classification, there are respectively 5, 3, 5, 4, and 11 samples categorized as excellent, good, poor, very poor, an unsuitable water for drinking water. The vital area with very high WQI (>>300) was in the brackish-water aquaculture site.

The salinity and trace metals concentrations suggested that the groundwater cannot sustain human life. Field observations showed that the groundwater, particularly in the north tip area, is reddish in color due to the presence of iron and manganese. Direct use of groundwater without treatment will result in significant damage to the infrastructure in the area. Iron and manganese deposits will form in the piping system, causing scalding and tubing damage. As a result, the condition of the groundwater should be considered before the area is developed [11].

4. Conclusion

The purpose of this study is to assess the groundwater quality in a coastal area in order to facilitate development. The study was undertaken in Indramayu, Indonesia, on the north coast of West Java. The area required a groundwater environment investigation due to its development status. Groundwater samples were collected from the confined and unconfined aquifer. The findings indicate that the groundwater is primarily brackish to saline in character, with a high TDS concentration. Correspondingly, the Fe^{2+} and Mn^{2+} concentrations were high and largely exceed the standard for clean water. As a result of these conditions, the WQI calculation indicated that only a few samples are suitable for use as drinking water. Meanwhile, the majority of samples were classified as poor, very poor, or unsuitable for drinking water purpose. Additionally, the salt content of the groundwater aids in the dissolution of iron and manganese from the sediment. The SI calculations of suspected minerals in sediments indicated that rhodochrosite and siderite could be trace metal sources. Furthermore, the analysis of Fe and Mn in groundwater implied that these metals are naturally abundant in the area. This situation was disadvantageous to development, as trace metals are corrosive to infrastructure.

5. Acknowledgements

Authors would like to thank to thank Indonesian Institute of Sciences (LIPI), National Research and Innovation Agency (BRIN) Republic of Indonesia; Japanese Society for the Promotion of Sciences Grant-in-Aid for Scientific Research B (Grant No 17H04494); and the Asia-Pacific Network for Global Change Research (Grant No. CRRP2019-09MY-Onodera) for facilitating and funding this study.

6. References

[1] Y. Umezawa *et al.*, "Sources of nitrate and ammonium contamination in groundwater under developing Asian megacities," *Science of the Total Environment*, vol. 404, pp. 361–376, 2008, doi: 10.1016/j.scitotenv.2008.04.021.

[2] M. Berg *et al.*, "Hydrological and sedimentary controls leading to arsenic contamination of groundwater in the Hanoi area, Vietnam: The impact of iron-arsenic ratios, peat, river bank deposits, and excessive groundwater abstraction," *Chemical Geology*, vol. 249, pp. 91–112, 2008, doi: 10.1016/j.chemgeo.2007.12.007.

[3] L. Cary *et al.*, "Origins and processes of groundwater salinization in the urban coastal aquifers of Recife (Pernambuco, Brazil): A multi-isotope approach," *Science of the Total Environment*, vol. 530–531, pp. 411–429, Oct. 2015, doi: 10.1016/j.scitotenv.2015.05.015.

[4] S.-I. Onodera *et al.*, "Effects of intensive urbanization on the intrusion of shallow groundwater into deep groundwater: Examples from Bangkok and Jakarta," *Science of the Total Environment*, vol. 404, pp. 401–410, 2008, doi: 10.1016/j.scitotenv.2008.08.003.

[5] M. Kagabu, J. Shimada, R. Delinom, M. Tsujimura, and M. Taniguchi, "Groundwater flow system under a rapidly urbanizing coastal city as determined by hydrogeochemistry," *Journal of Asian Earth Sciences*, vol. 40, no. 1, pp. 226–239, Jan. 2011, doi: 10.1016/j.jseaes.2010.07.012.
[6] A. Cardona, J. J. Carrillo-Rivera, R. Huizar-Álvarez, and E. Graniel-Castro, "Salinization in coastal aquifers of arid zones: An example from Santo Domingo, Baja California Sur, Mexico," *Environmental Geology*, vol. 45, no. 3, pp. 350–366, Jan. 2004, doi: 10.1007/s00254-003-0874-2.
[7] R. Chakraborty, K. M. Khan, D. T. Dibaba, A. Khan, A. Ahmed, and M. Z. Islam, "Health Implications of Drinking Water Salinity in Coastal Areas of Bangladesh," *International Journal of Environmental Research and Public Health*, vol. 16, no. 3746, 2019, doi: 10.3390/ijerph16193746.
[8] H. X. Weng, Y. C. Qin, and X. H. Chen, "Elevated iron and manganese concentrations in groundwater derived from the Holocene transgression in the Hang-Jia-Hu Plain, China," *Hydrogeology Journal*, vol. 15, no. 4, pp. 715–726, Jun. 2007, doi: 10.1007/S10040-006-0119-Z.
[9] A. F. Rusydi *et al.*, "Vulnerability of groundwater to iron and manganese contamination in the coastal alluvial plain of a developing Indonesian city," *SN Applied Sciences*, vol. 3, no. 4, Apr. 2021, doi: 10.1007/s42452-021-04385-y.
[10] Z. Zhang, C. Xiao, O. Adeyeye, W. Yang, and X. Liang, "Source and mobilization mechanism of iron, manganese and arsenic in groundwater of Shuangliao City, Northeast China," *Water (Switzerland)*, vol. 12, no. 2, Feb. 2020, doi: 10.3390/w12020534.
[11] W. H. Organization, *WHO guidelines for drinking-water quality*, 4th ed. Geneva: World Health Organization, 2011. doi: 10.1016/S1462-0758(00)00006-6.
[12] A. F. Rusydi *et al.*, "Potential Sources of Ammonium-Nitrogen in the Coastal Groundwater Determined from a Combined Analysis of Nitrogen Isotope, Biological and Geological Parameters, and Land Use," *Water*, 2021, doi: 10.3390/w13010025.
[13] R. E. Walpole, R. H. Myers, S. L. Myers, and K. Ye, *Probability & statistics for engineers & scientists : MyStatLab update*.
[14] R. Maria, A. F. Rusydi, H. Lestiana, and S. Wibawa, "*Hidrogeologi Dan Potensi Cadangan Airtanah Di Dataran Rendah Indramayu* (Hydrogeology and Potential of Groundwater Reserves in the Indramayu Lowlands)," *RISET Geologi dan Pertambangan*, vol. 28, no. 2, p. 181, 2018, doi: 10.14203/risetgeotam2018.v28.803.
[15] WHO, "Total dissolved solids in Drinking-water," 2003.
[16] P. Stuyfzand, "A new hydrochemical classification of water types," *IAHS Publication*, no. 182, pp. 89–95, 1989.
[17] D. Parkhurst and C. Appelo, "Description of Input and Examples for PHREEQC Version 3-A Computer Program for Speciation, Batch-Reaction, One-Dimensional Transport, and Inverse Geochemical Calculations," *U.S. Geological Survey Techniques and Methods, Book 6, Chap. A43*, 2013, Accessed: Jan. 31, 2022. [Online]. Available: http://www.hydrochemistry.eu

[18] Indonesian Government Regulation No 82/2001, "Guidelines for water quality management and pollution control."

[19] T. Akter *et al.*, "Water Quality Index for measuring drinking water quality in rural Bangladesh: A crosssectional study," *Journal of Health, Population and Nutrition*, vol. 35, no. 1, Feb. 2016, doi: 10.1186/s41043-016-0041-5.

[20] A. K. Batabyal and S. Chakraborty, "Hydrogeochemistry and Water Quality Index in the Assessment of Groundwater Quality for Drinking Uses," *Water Environment Research*, vol. 87, no. 7, pp. 607–617, Mar. 2015, doi: 10.2175/106143015x14212658613956.

Comparative of material properties between natural fibers and geo-bag synthetic fibers as sustainable material of temporary structure in natural coastal protection systems

V Herbenita[4*], Rildova[1], N Yuanita[2], A Kurniawan[2], K Kahdar[3], and W Gunawan[4]

[1] Offshore Engineering Research Group, Faculty of Civil and Environmental Engineering, Institut Teknologi Bandung, Indonesia
[2] Coastal Engineering Research Group, Faculty of Civil and Environmental Engineering, Institut Teknologi Bandung, Indonesia
[3] Craft and Tradition Research Group, Faculty of Art and Design, Institut Teknologi Bandung, Indonesia
[4] Ocean Engineering Master Program, Faculty of Civil and Environmental Engineering, Institut Teknologi Bandung, Indonesia

*Corresponding author: inivivieherbenita@gmail.com

Abstract. The natural coastal protection system is a proposed solution to the coastal erosion problem, including in Indonesia's coastal areas, which combines primary natural protection (vegetation such as mangroves) and temporary artificial structures as sustainable coastal development practices. The geo-bag-dike was chosen as a possible temporary structure to protect the growth of mangrove seedlings from sea waves for about two to five years before they are strong enough to prevent erosion. In today's practice, the geo-bag-dike is made of polymeric synthetic fibers (nonwoven geo-bag), which take a relatively long time to degrade naturally. Thus, this study focused on finding sustainable alternative materials from natural fibers, such as lyocell (Tencel), kenaf, cotton, and rayon, that can be applied as the primary material of geo-bag-dikes by comparing their material properties to nonwoven geo-bag. The results of the laboratory testing method show that kenaf textile has the highest values of tensile and tearing strength, but not higher than nonwoven geo-bags (768 N, 1080 N, and 2140 N for the tensile strength of kenaf, local nonwoven geo-bag, and imported nonwoven geo-bag, respectively). Therefore, kenaf textiles have the potential as a sustainable alternative material. By extrapolating, it is necessary to increase the layer number of kenaf textiles to at least three layers of fabric to resemble the tensile strength of nonwoven geo-bags.

Keywords: Natural coastal protection system, Material properties, Sustainable alternative materials, Natural fibers, Nonwoven geo-bag.

1. Introduction

Coastal environments are transitional zones between terrestrial and marine ecosystems that are constantly influenced by changes on land and sea. The coastline is the line that separates the land from the ocean. In 2005, an estimated 1.7 billion people, or 38% of the world's population (up to six billion people), resided within 50 km of the shore, with up to 45% living within 150 km of the shore. This is similar to the population in the mid-1950s [1]. Coastal erosion is a major issue that affects all of the world's coastal areas, including Indonesia. Coastal erosion, also known as abrasion, is the process of material or sediment eroding from the coastal profile (along the coastline) as a result of an imbalance in material or sediment supply. This is generally caused by natural coastal dynamics (such as waves, currents, tides, and sea level rise), storms, climate change, and tectonic forces, including land subsidence. Erosion can cause coastline erosion, threatening natural ecosystems and surrounding human communities, infrastructure, and other public facilities.

Hard structures, such as groins, breakwaters, revetments, seawalls, jetties, or a combination of these types of coastal protection, have been widely used in erosion control. However, the construction of hard structures sometimes only secures one area of the coastal section while causing erosion in other sections. Therefore, to mitigate coastal erosion problems, coastal regions worldwide are focusing on sustainable coastal development practices by planting natural vegetation, such as mangroves. However, the planting of young mangrove seedlings is frequently damaged by sea currents or waves before the mangroves can grow strong enough to protect coastlines [2]. At least three years are required to ensure that mangroves can grow strongly [3]. The natural coastal protection system is a proposed solution that combines mangrove vegetation as the main structure and geo-bag-dike as a temporary artificial structure to solve this problem. The geo-bag-dike was chosen as a possible temporary structure required until mangrove plants mature.

A geo-bag is a bag construction system that wraps a fill material such as sand, gravel, asphalt, or mortar and can be used to replace conventional structures in hydraulic and coastal engineering. In today's practice, geo-bags, also known as geotextile containment, are made of polymeric synthetic fibers (nonwoven materials) such as polypropylene (PP), polyester (PET), polyethylene (PE), and polyamide (PA). Geo-bags have been widely used as coastal protection in recent years because they are a relatively simple and cost-effective type of construction that can be used in distant regions without using heavy equipment, stone, or concrete. The cost of concrete armour per meter can reach 40–50 million IDR per meter, while PEGAR bamboo frame geo-bags cost between 1-2 million IDR per meter [3]. However, geo-bag materials take a long time to degrade naturally when used as a temporary artificial structure and are considered to become harmful coastal debris beyond their service life. Thus, this study focused on identifying alternative materials from natural fibers, such as lyocell (Tencel), kenaf, cotton, and rayon, that can be used as the primary material for a sustainable geo-bag by comparing their material properties to those of nonwoven geo-bags. Two different geo-bag manufacturers provide samples of nonwoven geo-bags.

2. Materials and methods

2.1. *Materials*

Four natural fiber textiles were used in this study in sheets or rolls. To better understand the material properties, no finishing or dyeing treatment was applied to the material before testing and analysis. Meanwhile, for synthetic geotextiles, two nonwoven geo-bag test materials from different manufacturers were used in the form of bags or sacks. Table 1 lists the sizes required for all test materials to meet the needs of specimens for all laboratory material properties testing, and an example of materials used is shown in Figure 1.

Table 1. Test material specifications required for testing the properties of materials in the laboratory.

Test materials	Length (MD)	Width (CD)	Number of needed	Production
Lyocell (Tencel)	6 m	1.10 m	1 sheet	PT Rehal Traco, Pekalongan, Central Java
Kenaf	6 m	1.20 m	1 sheet	Innatex Pak Tomo, Bandung, West Java
Cotton	6 m	1.20 m	1 sheet	Local Market
Rayon	6 m	1.30 m	1 sheet	Local Market
Local nonwoven geo-bag (400 gsm)	0.60 m	0.40 m	19 sacks	Buana Paksa Indonesia (BPI), Semarang, Central Java
Imported nonwoven geo-bag (600 gsm)	2.40 m	1.45 m	3 sacks	PT Tirta Citra Bara Persada, Jakarta

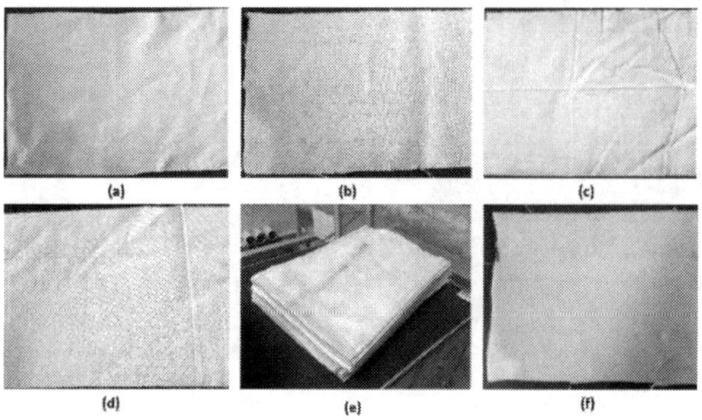

Figure 1. Examples of materials used: (a) Lyocell (Tencel) (b) Kenaf (c) Cotton (d) Rayon (e) Local nonwoven geo-bag (f) Imported nonwoven geo-bag.

2.2. Methods

Laboratory testing was used in this study to determine and compare the strength of materials, which includes the physical, mechanical, and hydraulic properties of natural and synthetic fiber geo-bags as the main material for sustainable geo-bags in natural coastal protection systems. The physical, mechanical, and hydraulic properties were tested in the Textile Testing Laboratory of Balai Besar Tekstil (BBT) Bandung using the test criteria and standards shown in Table 2. Testing the physical properties of a material consists of measurements of thickness and weight per area (unit mass). The mechanical properties of materials testing consist of tensile strength, tearing strength, puncture resistance, and dynamic puncture behavior (cone drop). Testing the hydraulic properties of the material consists of permittivity and pore size. When considering the type of fiber and its manufacturing techniques, the natural fiber material used as a test sample is classified as a fabric that only meets the physical and mechanical properties, such as the thickness and unit mass of the fabric for physical properties, and tensile strength and tearing strength for mechanical properties. Thickness and unit mass were tested for lyocell (Tencel), cotton, and rayon using SNI ISO 5084:2010 [4] and SNI ISO 3801:2010 [5], respectively. The tensile strength of lyocell (Tencel) was tested using SNI 0276:2009 [6] and kenaf, cotton, and rayon were carried out according to ASTM D 4632/D4632M–15a [7]. Five different tests in the direction of warp (MD) and weft (CD) were carried out for each fabric. The tearing strength of kenaf was tested using the trapezoidal method according to ASTM D4533/D4533M–07 [8] and the Elmendorf method for lyocell (Tencel), cotton, and rayon as specified by SNI ISO 13937–1:2010 [9].

Meanwhile, hydraulic property testing is applicable for nonwoven geo-bags, which is compared to the material specification data provided by the manufacturer. In total, 6 measurements were taken for natural fiber textiles and 12 measurements were taken for synthetic fiber geo-bags. Figure 2 presents flowcharts of the methodological strategies proposed to achieve this research purpose.

Table 2. Testing method for the physical, mechanical, and hydraulic properties of the material.

Test criteria	Standards and method of analysis					
	Tencel	Kenaf	Cotton	Rayon	Local nonwoven geo-bag (400 gsm)	Imported nonwoven geo-bag (600 gsm)
Physical properties of materials						
Thickness	SNI ISO 5084 : 2010 [4]	ASTM D5199-12 [10]	SNI ISO 5084 : 2010 [4]	SNI ISO 5084 : 2010 [4]	ASTM D5199 – 12 [10]	
Unit mass	SNI ISO 3801 : 2010 [5]	ASTM D5261-10 [11]	SNI ISO 3801 : 2010 [5]	SNI ISO 3801 : 2010 [5]	ASTM D5261 – 10 [11]	
Mechanical properties of materials						
Tensile strength (Grab test)	SNI 0276 : 2009 [6]	ASTM D4632/D4632 M-15a [7]	ASTM D4632/D4632 M-15a [7]	ASTM D4632/D4632 M-15a [7]	ASTM D4632 / D4632M – 15a [7]	
Tearing strength	SNI ISO 13937 – 1 : 2010 [9]	ASTM D4533/D4533 M-07 [8]	SNI ISO 13937 – 1 : 2010 [9]	SNI ISO 13937 – 1 : 2010 [9]	ASTM D4533 / D4533M – 07 [8]	
Static puncture resistance	Not tested				ASTM D6241 – 14 [12]	
Dynamic puncture behaviour (Cone drop method)	Not tested				ISO 13433:2006 [13] or SNI 08 – 4650 – 1998 [14]	
Hydraulic properties of materials						
Permittivity	Not tested				ASTM D4491 / D4491M – 15 [15]	
Permeability (m/s)	Not tested					
Pore size, O_{95} (mm)	Not tested				ASTM D4751 – 16 [16]	

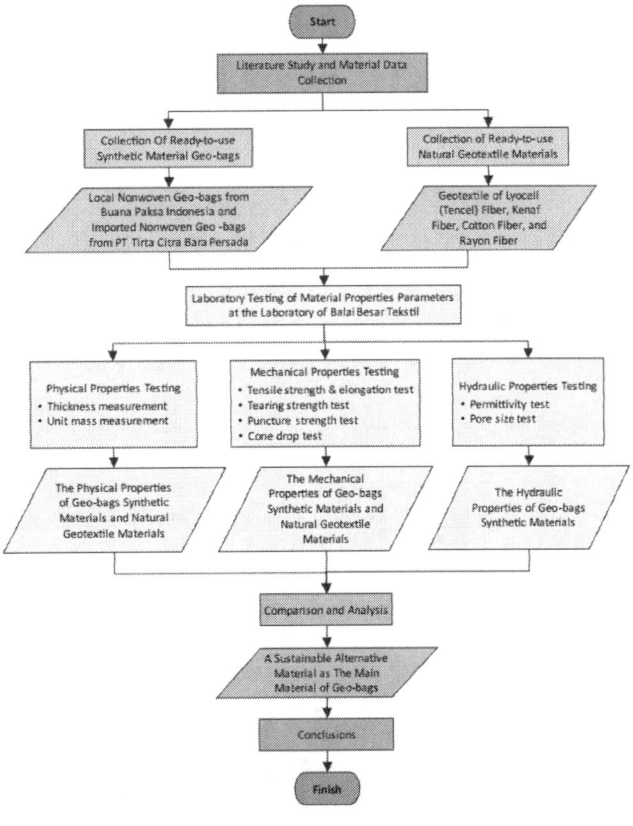

Figure 2. A flow diagram of the research methodology.

3. Result and discussion

3.1. Result

The testing of material properties for natural fiber textiles took approximately one month, from May 07, 2021 to June 3, 2021 for lyocell (Tencel) and kenaf, and from June 18, 2021 to July 21, 2021 for cotton and rayon. Meanwhile, material properties testing for synthetic geotextiles (nonwoven geo-bags) took about a month and a half, from May 6, 2021, to June 18, 2021. Table 3 presents the test results of material properties of natural fiber textiles and synthetic geotextiles (nonwoven geo-bag). It shows that kenaf has the highest thickness of all natural fibers tested. It is followed by cotton, rayon, and lyocell (Tencel). Thickness is used to control the quality of textiles and can affect other properties such as mass per unit area (unit mass), puncture resistance, tensile strength, and tearing strength [17]. It also shows that kenaf has the highest unit mass, followed by rayon, cotton, and lyocell (Tencel). The mass per unit area (unit mass) is also a good indicator that can affect tensile strength and tearing strength. Tensile and tearing strength typically increase with increasing unit mass [17]. Kenaf exhibits the highest tensile strength and tearing strength in the warp (MD) and weft (CD) directions. Kenaf has the highest tensile strength at 768 N in the warp direction (MD) and 454 N in the weft direction (CD). It is followed by rayon, lyocell (Tencel), and cotton with 346 N, 310 N, and 281 N, respectively, in the warp direction (MD), and by lyocell (Tencel), rayon, and cotton with 254 N, 221 N, and 204 N, respectively, in the weft direction (CD). In the warp direction (MD), the tearing strength of kenaf, rayon, cotton, and lyocell (Tencel) is 185 N, 25 N, 9.20 N, and 4.40 N, respectively, while in the weft direction (CD), it is 133 N, 15 N, 7.10 N, and 3.50 N, respectively.

Meanwhile, the imported nonwoven geo-bag has the highest material properties in all parameters compared to the local nonwoven geo-bag. The results of dynamic puncture behavior for imported nonwoven geo-bags "exceed the capacity of the test equipment," according to Table 3. It means that the absence of holes occurs when the steel cone is dropped on the geotextile test sample and does not penetrate the geotextile. In other words, imported nonwoven geo-bags have better dynamic puncture than local nonwoven geo-bags when it comes to withstanding pressure induced by the impact and penetration of objects that fall unexpectedly. Geotextiles with a higher unit mass have the maximum impact resistance to dropping load pressure and are not easily damaged by the increased amount of fiber in the geotextile cross-section [18].

3.2. Discussion

The discussion part contains three main sections: the comparison of typical values and laboratory test results for natural fiber and synthetic fiber material properties (nonwoven geo-bags); comparison of specification data and laboratory test results for material properties of each nonwoven geo-bag; and comparison of material properties for natural fiber fabrics to local nonwoven geo-bags and imported nonwoven geo-bags.

3.2.1. Comparison of typical values and laboratory test results for natural fiber and synthetic fiber material properties (nonwoven geo-bags). According to An Introduction to Geosynthetic Engineering [19] and the GRI-GT12(a) – ASTM Version Standard Specification [20], the test results of laboratory material properties will be evaluated to determine the fit between the range of values of laboratory test results and the typical values of geotextile material properties in Table 3. Before comparing, the units for permittivity to permeability of local and imported nonwoven geo-bags have been changed in order to display better and more understandable comparative results by multiplying the value of nonwoven geo-bag permittivity from the test results with the nominal thickness of geotextile based on ASTM D 4632/D4632M–15a [7]. The permeability of local and imported nonwoven geo-bags is 0.002 m/s and 0.001 m/s, respectively. Table 3 shows that the overall value of material test results from natural fiber textiles is still below the typical geotextile value range, especially in the MD and CD directions for lyocell, cotton, and rayon. The thickness, unit mass, tensile strength, and tearing strength of kenaf in the MD and CD directions are also found to be within the typical value range of geotextile material properties. In this study, kenaf has the highest material property value among other natural fibers, but it is not higher than the material property of nonwoven geo-bags, and its material property value has fulfilled the typical values of geotextile material properties [19]. In other words, kenaf has the most satisfactory mechanical properties and can potentially be used as a sustainable alternative primary material for temporary artificial structures when compared to other natural fiber textiles based on the value of material properties. Meanwhile, only the puncture strength of local nonwoven geo-bags is lower than the typical value of geotextiles [20]. The typical puncture strength of a geotextile with a unit mass of 406 g/m^2 is 3560 N, but the puncture strength of local nonwoven geo-bags with a higher unit mass of 422 g/m^2 is 2580 N. However, the overall value of the material properties for local nonwoven geo-bags fulfilled the typical geotextile material properties. None of the material properties testing results for imported nonwoven geo-bags were below the typical value range for geotextile material properties, indicating that imported nonwoven geo-bags with 600 gsm achieved the typical values for geotextile material properties.

Table 3. Results of laboratory testing material properties of natural fiber textiles and synthetic geotextiles (nonwoven geo-bag).

Material properties testing		Typical geotextile values	Test results					
			Tencel	Kenaf	Cotton	Rayon	Local nonwoven geo-bag (400 gsm)	Imported gonwoven geo-bag (600 gsm)
Thickness (mm)		0.254 – 7.62 [19]	0.17	0.77	0.34	0.32	3.26	4.36
Unit mass (g/m^2)		100 – 1000 [19]	78	309	111	127	422	527
Tensile strength	Warp strength (MD)[a]	300N – 3000N [19]	310 N (31.60 kg)	768 N (78.30 kg)	281 N (28.60 kg)	346 N (35.30 kg)	1080 N (110 kg)	2140 N (218 kg)
	Weft strength (CD)[b]		254 N (25.90 kg)	454 N (46.30 kg)	204 N (20.80 kg)	221 N (22.50 kg)	856 N (87.30 kg)	2140 N (218 kg)
Tearing strength	Warp strength (MD)[a]	90 N – 1300 N [19]	4.4 N (0.45 kg)	185 N (18.90 kg)	9.2 N (0.94 kg)	25 N (2.50 kg)	439 N (44.80 kg)	814 N (83 kg)
	Weft strength (CD)[b]		3.5 N (0.36 kg)	133 N (13.60 kg)	7.1 N (0.72 kg)	15 N (1.50 kg)	373 N (38 kg)	895 N (91.20 kg)
Static puncture strength		For 406 g/m^2: 3560 N For 542 g/m^2: 4000 N [20]	Not tested				2580 N (263 kg)	7070 N (721 kg)
Dynamic puncture behaviour (Cone drop) (mm)		–	Not tested				16	(Exceed the capacity of the test equipment)[c]
Permittivity (s^{-1})		–	Not tested				0.4767	0.3005
Permeability (m/s)		10^{-5} – 1 [19]	Not tested				0.002	0.001
Pore size, O_{95} (mm)		Woven: 0.15 – 0.60 Nonwoven: 0.05 – 0.35 [19]	Not tested				0.134	0.096

[a] MD (machince direction) = the direction of the machine or the direction of the length of the roll or sack.
[b] CD (cross machine direction) = the direction across the machine or the direction of the width of the roll or sack.
[c] There is no holes created or the steel cone does not penetrate the geotextile when dropped onto the geotextile test sample.

3.2.2. Comparison of specification data and laboratory test results for material properties of each nonwoven geo-bag. In this study, the ratio value of each material property is used to compare specification data and laboratory test results for the material properties of local and imported nonwoven geo-bags. The ratio value is used to evaluate and determine the difference between a material's actual strength and its allowable strength. The actual strength of a material is expected to exceed the necessary or allowed strength. Calculate the value of the ratio using equation (1).

$$Ratio = \frac{Actual\ strength}{Allowable\ strength} = \frac{Test\ result\ value}{Specification\ data\ value} \tag{1}$$

Table 4. Comparison of specification data and laboratory test results for the material properties of local nonwoven geo-bag Buana Paksa Indonesia (400 gsm).

Material properties	Specification data	Laboratory test results	Ratio
Thickness	3 mm	3.26 mm	1.09
Tensile strength (Warp direction, MD)	18 kN/m	1080 N (10.80 kN/m) [a]	0.60
Elongation (Warp direction, MD)	≥ 50 %	63.50%	1.27
Tearing strength (Warp direction, MD)	680 N	439 N (44.80 kg)	0.65
Static puncture strength	580 N	2580 N (263 kg)	4.45

[a] The tensile strength unit kN/m was calculated by dividing the tensile strength value by the specimen's nominal width, which is 100 mm [7].

Table 4 and Table 5 compare the results of laboratory testing of material properties for local nonwoven geo-bags with a 400 gsm to the material data specifications from Buana Paksa Indonesia and imported nonwoven geo-bags with a 600 gsm to the material data specifications manufactured by PT Tirta Citra Bara Persada, respectively. According to Table 4, the range of ratio values for comparing specification data and laboratory test results for local nonwoven geo-bag material properties is 0.60 to 4.45, with thickness, elongation, and static puncture strength values having a ratio value of more than 1.

Table 5. Comparison of specification data and laboratory test results for the material properties of imported nonwoven geo-bag PT Tirta Citra Bara Persada (600 gsm).

Material properties		Specification data	Laboratory test results	Ratio
Thickness (pressure 2 kPa)		5 mm	4.36 mm	0.87
Unit mass		600 g/m^2	527 g/m^2	0.88
Tensile strength	Warp strength (MD)	30 kN/m	2140 N (21.40 kN/m) [a]	0.71
	Weft strength (CD)	45 kN/m	2140 N (21.40 kN/m) [a]	0.48
Elongation	Warp strength (MD)	50%	68.30%	1.37
	Weft strength (CD)	40%	82.50%	2.06
Static puncture strength		7000 N	7070 N (721 kg)	1.01
Permeability		0.03 m/s	0.001 m/s	0.04
Pore size, O_{95}		0.08 mm	0.096 mm	0.83 [b]

[a] The tensile strength unit kN/m was calculated by dividing the tensile strength value by the specimen's nominal width, which is 100 mm [7].
[b] The ratio value for pore size is determined by dividing the value of the specification data by the value of the laboratory test results, because the value of the pore size in the test results exceeds the value of the specification.

This implies that the actual thickness, elongation, and puncture strength of local nonwoven geo-bags are higher than the allowable strength in the specification data, with thickness, elongation, and puncture strength ratios of 1.09, 1.27 (MD direction), and 4.45, respectively. Meanwhile, the range of ratio values for comparing specification data and laboratory test results for the material properties of imported nonwoven geo-bags based on Table 5 is between 0.04 and 2.06. It also shows that elongation and static puncture strength have a ratio value greater than 1, which implies both properties are higher than the allowable strength in the specification data.

3.2.3. Comparison of material properties for natural fiber fabrics to local nonwoven geo-bags and imported nonwoven geo-bag. Tensile strength is one of the material properties that must be evaluated and is the most essential property in determining geotextile materials [21]. The mechanical properties of tensile strength are a significant parameter in designing geotubes and geo-bags as wave retainers. Tensile strength is utilized as a geotextile material attribute to withstand excessive pressure or traction during the handling process, the filling process, and the pressure of the filled material (such as sand) after filling. According to Table 3, natural fiber material property values, including tensile strength values in the MD and CD directions, are still lower than those of local and imported synthetic nonwoven geo-bags. This can be caused by a variety of factors that influence the tensile strength of the material, including the thickness and layers of the material [22], the material's unit mass [18], the number of threads, yarn orientation, and the material layer stacking orientation [23]. The influence of material thickness on the tensile strength value, as shown graphically in Figure 3, reveals that imported synthetic nonwoven geo-bags have the highest tensile strength at 2140 N in both the MD and CD directions. It is followed by local synthetic nonwoven geo-bags, kenaf, rayon, lyocell (Tencel), and cotton with 1080 N, 768 N, 346 N, 310 N, and 281 N, respectively in the warp direction (MD), and by local synthetic nonwoven geo-bags, kenaf, lyocell (Tencel), rayon, and cotton with 856 N, 454 N, 254 N, 221 N, and 204 N, respectively in the weft direction (CD). The thickness of the material was found to affect the cross-sectional area exposed to tensile loads in this study. When compared to other natural fiber textiles, kenaf has a thicker thickness of 0.77 mm, resulting in a bigger total cross-sectional area that can bear the tensile load applied during the test. This also occurred with imported nonwoven geo-bags, which were used as the test material with the greatest thickness of 4.36 mm. The tensile cross-sectional area is determined by multiplying the fabric thickness by the clamp width of 25 mm [7]. Overall, Figure 3 illustrates that the tensile strength load increases as the thickness of the material utilized increases. As a result, the relationship between material thickness (mm) and tensile strength load (N) is positive linear.

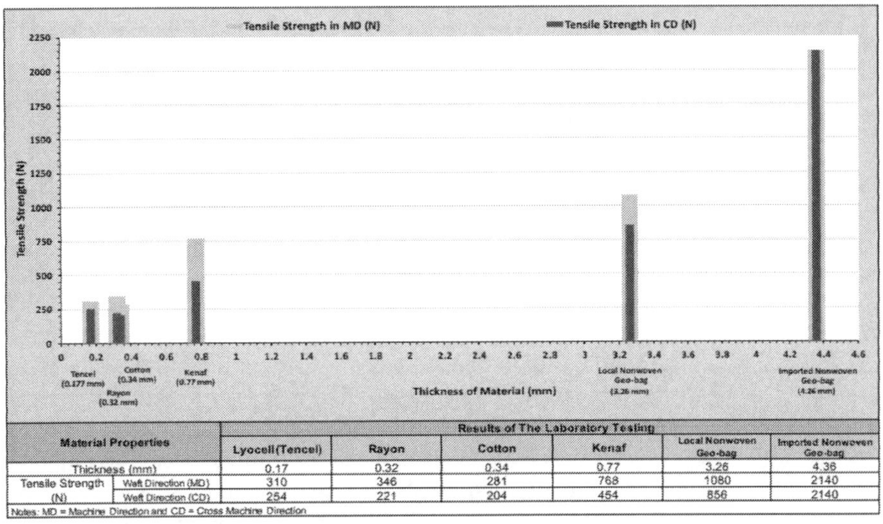

Material Properties		Results of The Laboratory Testing					
		Lyocell(Tencel)	Rayon	Cotton	Kenaf	Local Nonwoven Geo-bag	Imported Nonwoven Geo-bag
Thickness (mm)		0.17	0.32	0.34	0.77	3.26	4.36
Tensile Strength (N)	Waft Direction (MD)	310	346	281	768	1080	2140
	Weft Direction (CD)	254	221	204	454	856	2140

Notes: MD = Machine Direction and CD = Cross Machine Direction

Figure 3. Comparison of material thickness (mm) and tensile strength (N) between natural fiber and geo-bags to determine the optimal thickness or additional layer.

The influence of material thickness on tensile strength can alternatively be interpreted as the effect of adding a layer to tensile strength. This was confirmed in the research of Mohammed et al. (2018), which found that adding a kenaf layer from a composite mixture of kenaf and unsaturated polyester resin (UPR) had a positive effect on increasing the tensile load along with the increase in thickness [22]. If a linear regression is performed on the increase in tensile strength due to layer addition [22], the linear regression equation for tensile strength is obtained as $y = 1090\,x + 3970$, with the variable x representing the number of layers and the variable y representing the tensile strength in N, so that the average percent increase in tensile strength for each layer was 17.53%. By applying the percent increase in tensile strength to the value of the tensile strength of natural fiber in the MD direction shown in Figure 3, the number of layers of natural fiber textile should be increased, which is 15 layers (34 layers) for lyocell, 3 layers (11 layers) for kenaf, 17 layers (38 layers) for cotton, and 13 layers (30 layers) for rayon to approach the value of the tensile strength of local nonwoven geo-bags (imported nonwoven geo-bags).

The usage of the knitting method in stacking or increasing the number of layers of fabric needs to be explored further to identify its effect on the value of material properties, specifically the value of the material's tensile strength. One method of layering the material is by knitting numerous layers of carbon fiber fabrics with polyester or PET DTY yarn (white yarn) with 1+1 rib loops [24]. Another factor that affects the tensile strength is the unit mass of the fabric.

In this study, the unit mass of the material is measured in mass per unit area of the test sample, with a minimum total area of 100000 mm^2 and is expressed in units of g/m^2. In this study, the effect of material unit mass on the tensile strength values for the MD and CD directions in units of kN/m is presented in Figure 4. Tensile strength is expressed in kN/m by dividing it by the nominal width of the specimen, which is 100 mm [7].

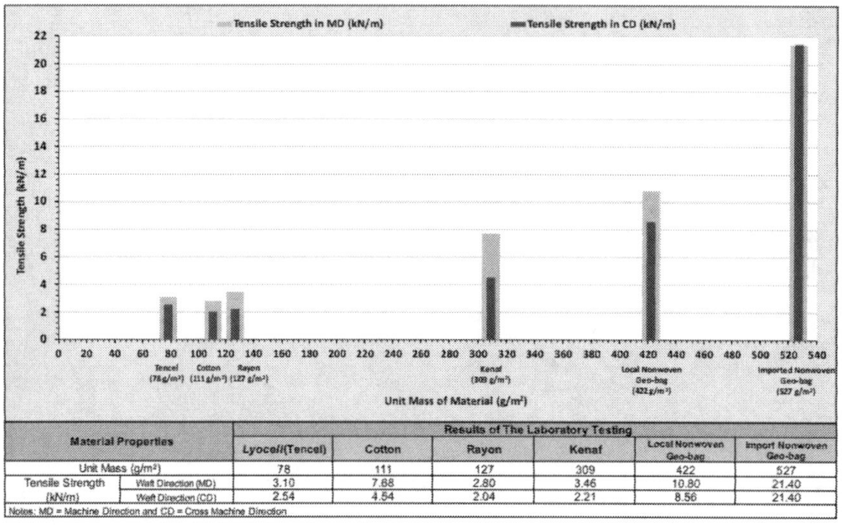

Material Properties		Results of The Laboratory Testing					
		Lyocell(Tencel)	Cotton	Rayon	Kenaf	Local Nonwoven Geo-bag	Import Nonwoven Geo-bag
Unit Mass (g/m^2)		78	111	127	309	422	527
Tensile Strength (kN/m)	Watt Direction (MD)	3.10	7.68	2.80	3.46	10.80	21.40
	Weft Direction (CD)	2.54	4.54	2.04	2.21	8.56	21.40

Notes: MD = Machine Direction and CD = Cross Machine Direction

Figure 4. Comparison of unit mass (g/m^2) and tensile strength (kN/m) between natural fiber and geo-bags.

Imported nonwoven geo-bag material with a unit mass of 527 g/m^2 has the highest tensile strength, which is 21.40 kN/m for both MD and CD directions. Meanwhile, for natural fiber materials, kenaf with a unit mass of 309 g/m^2 produces a higher tensile strength than other natural fiber materials, which is 7.68 kN/m and 4.54 kN/m for the MD and CD direction, respectively. Thus, materials with a higher unit mass will produce a higher maximum tensile strength as well, because the variation in thickness of the material can affect the unit mass to tolerated tensile strength. Furthermore, the number of fibers and fiber density at a higher unit mass will increase, as will their ability to resist higher tensile loads [18]. Therefore, the relationship between the material unit mass (g/m^2) and the tensile strength (kN/m) has a positive linear correlation. The number of threads, yarn orientation, and material layer stacking orientation must also be considered when increasing the number of natural fiber layers. This was confirmed in the study of A. Hamdan et al. (2016), who observed that sample 1 (S1) of an epoxy resin composite mixture with three layers of A/A/A orientation woven kenaf could withstand the highest tensile load compared to other test samples, proving that the orientation of the stacking of layers of fabric and the highest number of threads in each area of the fabric has a large influence on the tensile strength [23]. Based on Table 3, the tensile strength value in the MD direction for kenaf and other natural fiber fabrics is higher than the tensile strength in the CD direction. This indicates that the number of threads in the MD direction exceeds the number of threads in the CD direction. So that MD orientation can be used for each layer of fabric when stacking or increasing the number of layers for sacks of kenaf fabric. Furthermore, the tensile load to be received by the sack, such as the load in the sack handling process, must be in the same direction as the stacking orientation and retained on the side of the sack with the highest thread number or yarn density.

4. Conclusion

Based on laboratory testing results of material properties in the Textile Testing Laboratory of Balai Besar Tekstil (BBT) Bandung using different test criteria and standards, kenaf textile has the potential to be a sustainable alternative material for temporary artificial structures in natural coastal protection systems, such as geo-bags. At least three layers of kenaf textile are required to match the tensile strength of nonwoven geo-bags. The addition of kenaf textile layers increases the thickness and unit mass, which affects the tensile strength produced. When increasing the number of layers, it is important to consider the number of threads, the orientation of the yarn fibers, and the orientation of the stacking layers of fabric that will receive the tensile load. However, additional research is required to investigate the effects of increasing the number of layers of natural fiber material, knitting method, or layer stacking of natural fiber material on material properties, particularly the tensile strength, and biodegradation characteristics in order to determine alternative high-quality sustainable materials made from natural fibers. Future research will also explore the economic aspects of the proposed material and how to obtain it at the lowest cost. Aside from the material properties, the biodegradation characteristics of the materials tested in this study are presented and evaluated by the authors in a separate paper.

Acknowledgement

This research is funded by the ITB Research and Community Service (P2MI) grant through the Offshore Engineering Research Group and the Coastal Engineering Research Group, Faculty of Civil and Environmental Engineering.

References

[1] Kay R and Alder J 2005 *Coastal Planning and Management* vol 2 (London and New York: Taylor & Francis)
[2] Yuanita N, Kurniawan A, Hakim M L A, Irawan K R and Saputra N W 2020 Physical model of natural coastal protection system: geobag-dyke performance to effectiveness of natural coastal protection system *J. of Sust. Sci. and Manag.* **15** 85–99
[3] Nugraha M R R and Gemilang R I A 2017 Rehabilitas pantai berlumpur dengan PEGAR geobag rangka bambu di Desa Sriwulan Kabupaten Demak *J. Tek. Hidraulik* **8** 71-86
[4] SNI ISO 5084 2010 *Tekstil - Cara Uji Tebal Tekstil dan Produk Tekstil* (Indonesia: BSN)
[5] SNI ISO 3801 2010 *Tekstil - Kain Tenun - Cara Uji Berat Kain per Satuan Panjang dan Berat Kain per Satuan Luas* (Indonesia: BSN)
[6] SNI 0276 2009 *Cara Uji Kekuatan Tarik dan Mulur Kain Tenun* (Indonesia: BSN)
[7] ASTM D4632/D4632M–15a 2015 *Standard Test Method for Grab Breaking Load and Elongation of Geotextiles* (USA: ASTM International)
[8] ASTM D4533/D4533M–07 2013 *Standard Test Method for Trapezoid Tearing Strength of Geotextiles* (USA: ASTM International)

[9] SNI ISO 13937–1 2010 *Tekstil - Kekuatan Sobek Kain - Bagian 1: Cara Uji Kekuatan Sobek Menggunakan Metoda Pendulum (Elmendorf)* (Indonesia: BSN)
[10] ASTM D5199–12 2019 *Standard Test Method for Measuring the Nominal Thickness of Geosynthetics* (USA: ASTM International)
[11] ASTM D5261–10 2018 *Standard Test Method for Measuring Mass per Unit Area of Geotextiles* (USA: ASTM International)
[12] ASTM D6241–14 2014 *Standard Test Method for Static Puncture Strength of Geotextiles and Geotextile-Related Products Using a 50-mm Probe* (USA: ASTM International)
[13] ISO 13433 2006 *Geosynthetics–Dynamic Perforation Test (Cone Drop Test)* (Switzerland: Technical Committee ISO)
[14] SNI 08–4560 1998 *Cara Uji Daya Tahan Geotekstil Terhadap Pelubangan Cara Kerucut Jatuh* (Indonesia: BSN)
[15] ASTM D4491/D4491M–15 2020 *Standard Test Methods for Water Permeability of Geotextiles by Permittivity* (USA: ASTM International)
[16] ASTM D4751–16 2016 *Standard Test Methods for Determining Apparent Opening Size of a Geotextile* (USA: ASTM International)
[17] Torosian G T and Millan P A M 2016 Physical properties, behavior, and testing of geotextiles *Geotextiles: From Design to Applications* number 175, ed R M Koerner (Amsterdam: Woodhead Publishing) pp 105-113
[18] Carvalho R, Fangueiro R and Soutinho F 2011 Mechanical properties of needle-punched nonwovens for geotechnical applications *Int. Conf. on Engineering UBI (Covilhã)*
[19] Shukla S K 2016 *An Introduction to Geosynthetic Engineering* (London: CRC Press/Balkema)
[20] Geosynthetic Research Institute (GRI)–GT12(a) *Standard Specification for Test Methods and Properties for Nonwoven Geotextiles Used as Protection (or Cushioning) Materials* (USA: Geosynthetic Institute)
[21] Leão A L, Cherian B M and De Souza S F 2012 Natural fibres for geotextiles *Handbook of Natural Fibres* vol 2, ed R M Kozłowski (New Delhi: Woodhead Publishing Limited) pp 280-311
[22] Mohammed M, Rahman R, Mohammed A M, Osman A F, Adam T, Dahham O S, Hashim U, Noriman N Z and Betar B O 2018 Effect of kenaf fibre layers on mechanical and thermal properties of kenaf/unsaturated polyester composites *Int. Conf. on Materials Engineering and Science* (Malaysia: IOP Publishing) pp 1–6
[23] Hamdan A A, Mustapha F, Ahmad K A, Rafie A S M, Ishak M R and Ismail A E 2016 The effect of customized woven and stacked layer orientation on tensile and flexural properties of woven kenaf fibre reinforced epoxy composites *Int. J. of Polymer Science* **2016** 1-11
[24] Pei X, Shang B, Chen L, Li J and Tang Y 2016 Compression properties of multilayer-connected biaxial weft knitted carbon fiber fabric reinforced composites *Composite Part B: Engineering* **91** 296-305

Integration Of Mass Public Transport Fare In The Jakarta Area

S Triana[1], A Sjafruddin[2], R H Karsaman[2], S Kaderi[3]

[1]Doctoral Program in Transportation Engineering, Faculty of Civil and Environmental Engineering, Institut Teknologi Bandung
[2]Transportation Engineering Expertise Group, Faculty of Civil and Environmental Engineering, Institut Teknologi Bandung
[3]School of Business and Management, Institut Teknologi Bandung

E-mail: sofyantriana@gmail.com, adesjaf@gmail.com, ruherkar@yahoo.com, sudarso_kw@sbm-itb.ac.id

Abstract. This study specifically discusses the concept of integrating mass public transport fares in Jakarta. Based on the identification of problems and also based on several studies of mass transportation in various cities in the world, the integration of fares can increase the number of mass public transportation ridership. This causes a change in the movement pattern of mass public transport users. An interim analysis has been carried out on a simple public transport network with the number of movements in the origin and destination matrix between stations/stops was 19,611 passengers/hour and variations in the amount of fare between IDR 2,500,-/passenger up to IDR 7,500,-/passenger (Existing average fare is IDR 4,188,-/passenger). In the optimization process, an existing operator's revenue is an objective function. The optimization process for determining mass public transport fares is carried out recursively because it is influenced by changes in the integrated fare amount to the number of mass public transport users. The results of the optimization process show that the integrated fare of IDR 3,850/passenger (9% lowest than the average existing fare) that increasing public transport users by 3% compared to the number of passengers in the existing condition for 1 operating hour.

Keywords: Fare integration, Mass Transport, Optimization, Dynamic Programming, Backward Recursion

1. Introduction

Mass public transportation in urban areas is one of the answers to the problem of the need for passenger movement in carrying out daily routine activities. The operation of mass transportation in urban areas is one solution to serve the high movement, reduce traffic congestion levels, reduce pollution levels, and fuel saved.

Based on data from Badan Pengelola Transportasi Jabodetabek (BPTJ) that in 2019 the average daily movement in the Jabodetabek area was ± 100 million trips per day. In 2019 the number of Transjakarta Bus passengers was 264.6 million passengers (http://data.jakarta.go.id) with an average daily movement of 998,658 people trips/weekday and the number of Indonesian Commuter Train passengers was 340.65 million passengers (http://www.krl.co.id) with an average daily movement of 979,853 trips per person/working day. The percentage of daily mass transit users is only ± 2% of the total movement in

IOP Conf. Series: Earth and Environmental Science 1065 (2022) 012056 doi:10.1088/1755-1315/1065/1/012056

the Greater Jakarta area. This shows the low interest in using mass public transportation in the Greater Jakarta area. The causes of these problems include the ease of owning a private vehicle, private-public transportation services such as taxis or online motorcycle taxis, government policies related to transportation management, as well as the connectivity problem of mass public transport as the backbone/main network of Jabodetabek transportation which is not connected to one another, resulting in this happening. difficulty to access between mass public transport.

One way to overcome the operational problems of mass public transportation is to implement the operation integration and integration of mass public transportation fares in the Jabodetabek area. Technically operational with the process of merging or integrating terminal points and or stations between mass transit in order to create the concept of the connectedness of movement on the main public transport network in the Greater Jakarta area. The connectedness concept is part of the integration concept in the form of physical integration, institutional integration, service & operational integration as well as fare integration. Based on the experience of other countries regarding the integrated operation of several types of mass transportation, it shows an increase in transportation performance (an increase in the number of transport users or ridership). The integration process does not run directly but is carried out in stages. The earliest stage is regarding the physical integration of the station points or terminals of several modes of mass transportation, after that institutional integration, systems (services and operations), and fare rates can be carried out.

One of the points of the integration concept being analyzed at this time is the planning and development process for setting integrated fares between mass transit. The fare is highly dependent on the amount of investment, operational costs, and profits by taking into account the ability and desire of the community to use mass transportation as well as the ability of the Government to provide subsidies. Based on previous studies regarding the integration process of mass transit fares, the advantages of various aspects of intermodal integration, one of which is service fares that can be reduced through the fare integration mechanism (May and Roberts, 1995). Another research on the concept of fare integration according to Takahashi, T., (2014) from the University of Tokyo is analyzed based on the shape of the route serving urban and suburban areas, the mass transit operators involved, the travel time with case studies of the operation of several MRT operators in Tokyo City.

2. Objective

The objective of this research is to determine the concept and scheme of integration of mass transportation fares with a case study of mass transportation in the Greater Jakarta area. Either a fare integration concept and scheme can be applied in other cities in Indonesia.

The concept of fare integration is to determine the optimum integrated fare in the mass transportation service zone which is formed based on daily passenger movements. The optimization carried out is based on the operator's income, the profit of mass transit users (the deviation between the new fare and the old fare), and the amount of Public Service Obligation/PSO.

3. Framework Analysis

The optimization analysis is carried out using the Dynamic Program Method. The choice of this method is because it can perform step-by-step optimization in order to obtain the minimum integrated fare value and the minimum subsidy amount (close to the existing subsidy amount, due to avoid increasing expenditure burden from the Government). The Dynamic Programming Method is Backward Recursion, by setting the Objective Function in the form of Existing Mass Public Transportation Total Revenue (Fare and Subsidy Income) as a reference for new income that will be obtained when integrated fares are implemented. Changes in fares will certainly affect the amount of ridership of users of mass public transportation and non-mass public transportation. The framework analysis flow chart can be shown in Figure 1.

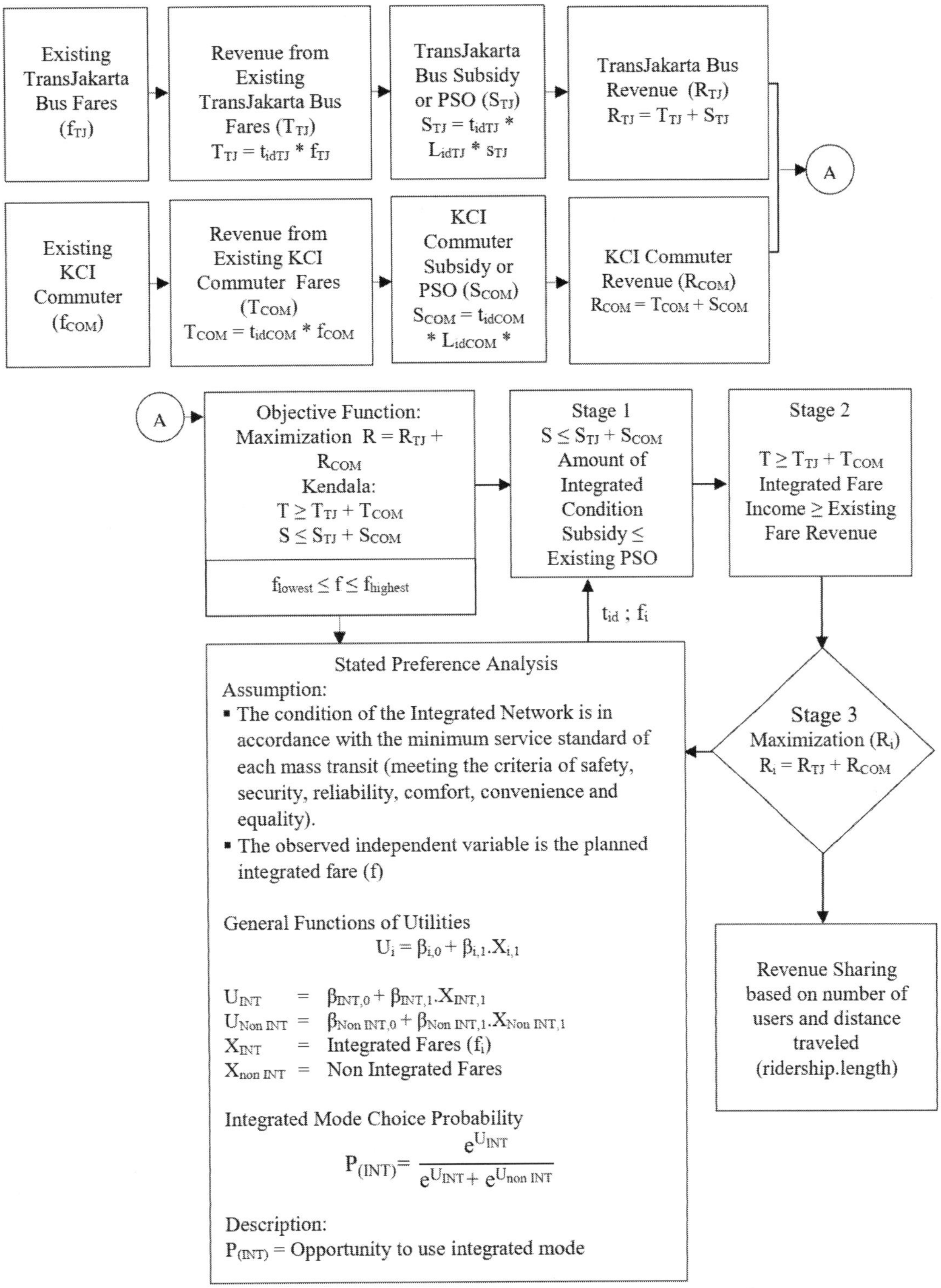

Figure 1. Framework Analysis

4. Data and Analysis

This Stated Preference analysis is applied to the mass transportation network in the case of 2 Mass Transportation Modes with a service network as shown in Figure 2. Based on the percentage of mass public transport users in Jakarta which is 19% of the total movement (Data Source: statistik.jakarta.go.id) then the total movement of all movements in the study area is 181,986 people/hour (Cumulative of Origin Destination Trip of existing mass transit users is 42,688 passengers/hour).
Assumptions of the analysis of the implementation of the Integrated Fares for a Single Fare:
a. Changes in fares do not take into account the choice of transportation mode
b. The lowest fare limit and the highest fare are determined based on the applicable fares.

Basic Commuter Line and TransJakarta Bus Fares
a. Basic Commuter Line Fare
 1. PM 17 of 2018 (Ministry of Transportation, Guidelines for Calculation and Determination of Fares for Transportation of People by Train)
 2. Basic Commuter Line fare = IDR 453.-/passenger/km.
b. Basic TransJakarta Bus Fares
 1. PM no. 89 of 2002 (Calculation of TransJakarta Bus Operating Costs)
 2. Basic TransJakarta bus fare = IDR 965.-/passenger/km.

Amount of Subsidy (IDR/passenger.km) based on existing condition
a. Amount of Commuter Line Subsidy = IDR 314.-/passenger/km
b. TransJakarta Bus Subsidy Amount = IDR 577.-/passenger/km.

Existing TransJakarta KRL and Bus Fares (subsidized fares)
a. Existing Commuter Line Fare = IDR 2,500.-/15 km (IDR 500.-/5 km further)
b. TransJakarta Bus Basic Fare = IDR 3,500.- (Flat Fare).

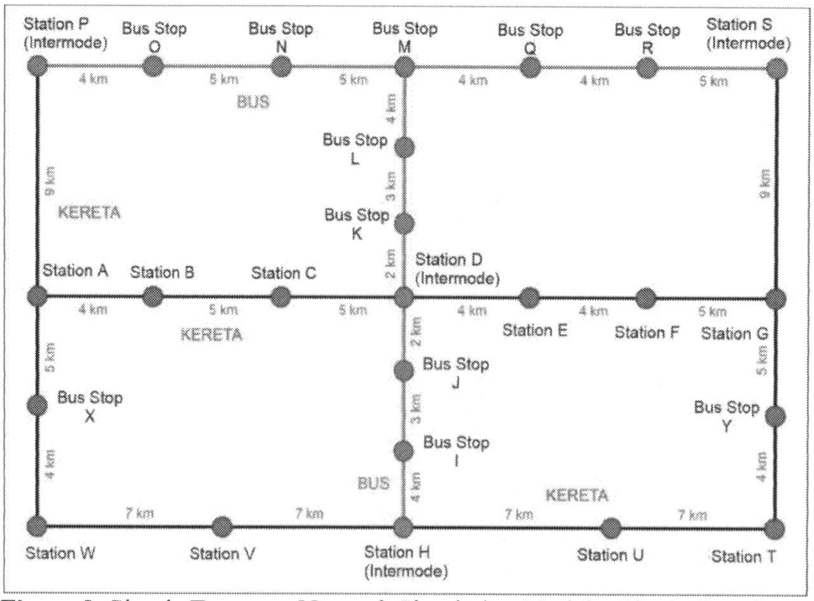

Figure 2. Simple Transport Network Simulation Model – Single Fare for 2 Mass Transport Modes

a. Income Calculation from Existing Fare
 1) Revenue from Existing Commuter Line Rates (T_{COM})

$$T_{COM}=\sum_{i=1}^{d} t_{id\ COM} * f_{COM} \tag{1}$$

T_{COM} = IDR 41,624,000.-/hour

2) TransJakarta Bus (TJ)

$$T_{TJ}=\sum_{i=1}^{d} t_{id\ TJ} * f_{TJ} \tag{2}$$

T_{TJ} = IDR 33,971,000.-/hour

b. Income from Existing Subsidies
   1) Revenue from Existing Commuter Line Subsidies (S_{COM})

$$S_{COM}=\sum_{i=1}^{d} t_{id\ COM} * l_{id\ COM} * s_{COM} \tag{3}$$

S_{COM} = IDR 59,844,318.-/hour

2) Revenue from Existing TransJakarta Bus Subsidies (S_{TJ})

$$S_{TJ}=\sum_{i=1}^{d} t_{id\ TJ} * l_{id\ TJ} * s_{TJ} \tag{4}$$

S_{TJ} = IDR 59,870,674.-/ hour

c. Total Revenue from Existing Fares and Existing Subsidies
   1) Total Commuter Line Revenue (R_{COM})
   $R_{COM} = T_{COM} + S_{COM}$ (5)
   R_{COM} = IDR 41,624,000.-/hour + IDR 59,844,318.-/hour
   R_{COM} = IDR 101,468,318.-/hour

   2) TransJakarta Bus Total Revenue (R_{TJ})
   $R_{TJ} = T_{TJ} + S_{TJ}$ (6)
   R_{TJ} = IDR 33,971,000.-/hour + IDR 59,870,674.-/hour
   R_{TJ} = IDR 93,841,674.-/hour

   3) Total Revenue (R) Commuter Line (R_{COM}) and TransJakarta Bus (R_{TJ})
   $R = R_{COM} + R_{TJ}$ (7)
   R = IDR 101,468,318.-/hour + IDR 93,841,674.-/hour
   R = IDR 195,309,992.-/hour

Scenario analysis of the Integration of Single Fare Commuter Line and TransJakarta Bus was carried out using the Backward Recursion Dynamic Program Method. The steps are taken in determining the Single Fare:

a. Determining the Objective Function as a condition so that the applicable fare does not burden passengers and does not reduce operator income
   - The objective function model is to maximize total revenue (*Revenue*)
     Maximization $R = R_{COM} + R_{TJ}$ (8)

Maximization $R = [T_{COM} + S_{COM}] + [T_{TJ} + S_{TJ}]$ (9)
Maximization $R = T_{AM} + [S_{COM} + S_{TJ}]$ (4.10)

$$T_{AM} = \sum_{i=1}^{d} t_{id\ AM} * f_{single} \quad (10)$$

- Fare Constraint Model to Single Fare:
  $f_{COM} = f_{TJ} = f_{single}$
  $f_{terendah} \leq f_{single} \leq f_{tertinggi}$
  IDR 2,500.- $\leq f_{single} \leq$ IDR 7,500.-

$$\sum_{i=1}^{d} t_{id\ AM} * f_{lowest} \leq f_{single} \leq \sum_{i=1}^{d} t_{id\ AM} * f_{highest}$$

$$\sum_{i=1}^{d} t_{id\ AM} * \text{IDR } 2{,}500.- \leq f_{single} \leq \sum_{i=1}^{d} t_{id\ AM} * \text{IDR } 7{,}500.- \quad (11)$$

- Subsidy Constraint Model:
  $S_{COM} \leq R_{COM}$

$$\sum_{i=1}^{d} t_{id\ COM} * l_{id\ COM} * s_{COM} \leq R_{COM} \quad (12)$$

$S_{TJ} \leq R_{TJ}$

$$\sum_{i=1}^{d} t_{id\ TJ} * l_{id\ TJ} * s_{TJ} \leq R_{TJ} \quad (13)$$

$S_{COM} + S_{TJ} \leq R$

b. Dynamic Program Model becomes:
- Maximization $R = T + S$

$$R = \left[\sum_{i=1}^{d} t_{id\ AM} * f_{single}\right] + \left[\sum_{i=1}^{d} t_{id\ COM} * l_{id\ COM} * s_{COM}\right] + \left[\sum_{i=1}^{d} t_{id\ TJ} * l_{id\ TJ} * s_{TJ}\right]$$

$$\text{IDR } 195{,}309{,}992.\text{-/hour} = \left[\sum_{i=1}^{d} t_{id\ AM} * f_{single}\right] + \left[\sum_{i=1}^{d} t_{id\ COM} * l_{id\ COM} * s_{COM}\right] + \left[\sum_{i=1}^{d} t_{id\ TJ} * l_{id\ TJ} * s_{TJ}\right]$$

Description: Known Variables = $t_{id\ AM}$, $l_{id\ AM}$, $t_{id\ COM}$, $l_{id\ COM}$, $t_{id\ TJ}$, $l_{id\ TJ}$
Unknown Variables = f_{single}, s_{COM}, s_{TJ}?

- Constraint
  Subsidy: $S_{COM} + S_{TJ} \leq R$
  IDR 314.- /passenger/km $\leq s_{COM} \leq$ IDR 532.-/passenger/km
  IDR 577.- /passenger/km $\leq s_{TJ} \leq$ IDR 904.-/passenger/km

  Fare: $f_{COM} = f_{TJ} = f$
  $f_{lowest} \leq f \leq f_{highest}$
  IDR 2,500.- $\leq f \leq$ IDR 7,500.-

c. Completing the Dynamic Program Model with the Backward Recursion Method
Step 1 : Determine the flow of the stages of finding the optimum value. Stages are carried out based on the objective function and boundary conditions at each stage
Step 2 : Visualization of Backward Recursion Dynamic Program
Step 3 : Mathematical Model of Objective Functions (States) and Boundary Conditions
Step 4 : The Value Search Process for each stage.

The steps are carried out as follows:
Step 1 : The objective function is Total Revenue (Revenue) from Commuter Line and TransJakarta Bus as Maximum Revenue Rmax. The maximum revenue value is obtained from the existing conditions where the Commuter Line and TransJakarta Bus operate respectively, which is IDR. 195,309,992,-/hour.
Step 2 : Visual Program Dynamic Model – Backward Recursion.

Mathematical Equation of Backward Recursion Dynamic Model in Simple Transportation Network Simulation:

$$f_i(x_i) = \text{Max}\ \{d(x_i, x_{i+1}) + f_{i+1}(x_{i+1})\},\ i = 1, 2, 3 \ \rightarrow \text{as Objective Function}$$

Stage 3 of Backward Recursion Dynamic Model
$f_4(x_4) = 0,\ x_4 =$ IDR 195,309,992.-/hour

Boundary Conditions:
- $R \rightarrow R \leq$ IDR 195,309,992.-/hour
- KRL Commuter Subsidies
  IDR 314.- /passenger/km $\leq s_{COM} \leq$ IDR 532.-/passenger/km
- Bus TransJakarta Subsidies
  IDR 577.- /passenger/km $\leq s_{TJ} \leq$ IDR 904.-/passenger/km
- $S_{COM} + S_{TJ} \leq R$
  $S_{COMi} + S_{TJi} = Ri$

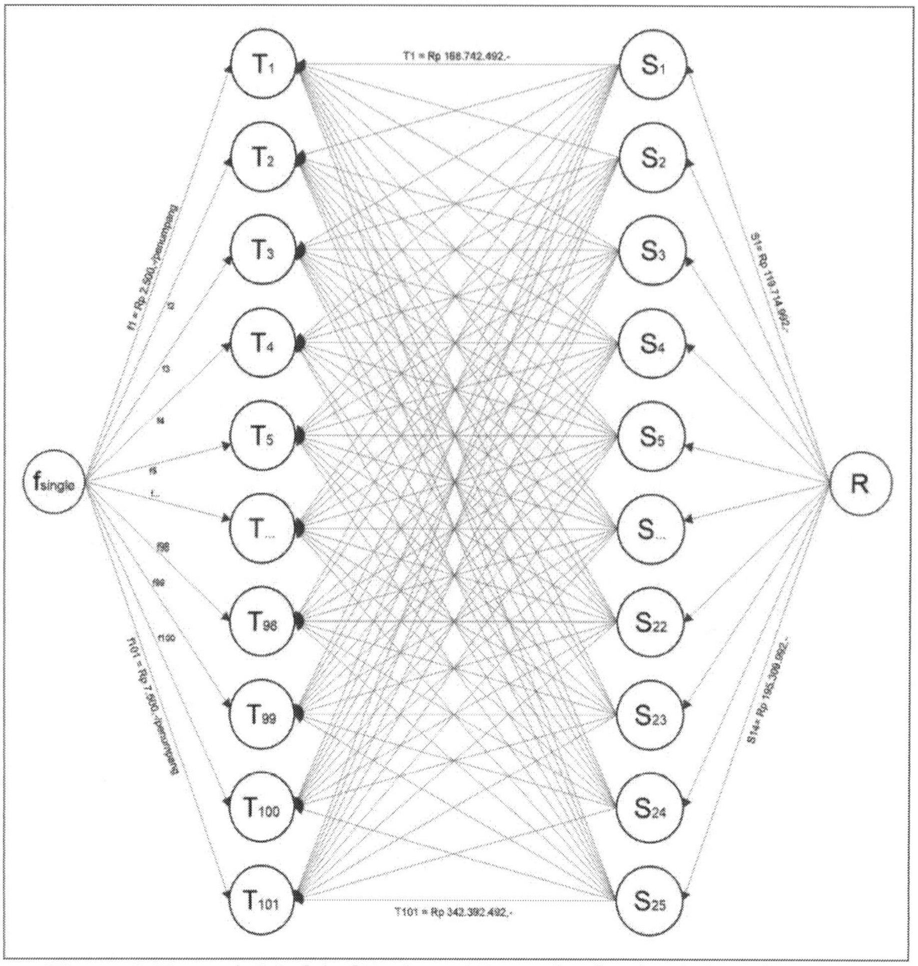

Figure 3. Dynamic Model of Backward Recursion in Simulation of Simple Transportation Network (Single Fare) for 2 Mass Transportation Modes

Table 1. Stage 3 (Single Fare) 2 Modes of Mass Transport

x_3	S_i		$d(x_3, x_4) + f_4(x_4)$	Optimum Solution	x_4
	S_{COM}	S_{BUS}		$f_3(x_3)$	
S_1	314	577	119,714,992	119,714,992	R
S_2	314	600	122,101,518	122,101,518	R
S_3	314	700	132,477,718	132,477,718	R
S_4	314	800	142,853,918	142,853,918	R
S_5	314	904	153,685,992	153,685,992	R
S_6	400	577	136,105,474	136,105,474	R
S_7	400	600	138,492,000	138,492,000	R
S_8	400	700	148,868,200	148,868,200	R
S_9	400	800	159,244,400	159,244,400	R
S_{10}	400	904	170,076,474	170,076,474	R
S_{11}	450	577	145,634,824	145,634,824	R
S_{12}	450	600	148,021,350	148,021,350	R
S_{13}	450	700	158,397,550	158,397,550	R

Table 1. Stage 3 (Single Fare) 2 Modes of Mass Transport

x_3	S_i		$d(x_3, x_4) + f_4(x_4)$	Optimum Solution	x_4
	S_{COM}	S_{BUS}		$f_3(x_3)$	
S_{14}	450	800	168,773,750	168,773,750	R
S_{15}	450	904	179,605,824	179,605,824	R
S_{16}	500	577	155,164,174	155,164,174	R
S_{17}	500	600	157,550,700	157,550,700	R
S_{18}	500	700	167,926,900	167,926,900	R
S_{19}	500	800	178,303,100	178,303,100	R
S_{20}	500	904	189,135,174	189,135,174	R
S_{21}	532	577	161,338,992	161,338,992	R
S_{22}	532	600	163,725,518	163,725,518	R
S_{23}	532	700	174,101,718	174,101,718	R
S_{24}	532	800	184,477,918	184,477,918	R
S_{25}	532	904	195,309,992	195,309,992	R

Stage 2 of Backward Recursion Dynamic Model
Boundary Conditions:

- R → R ≤ IDR 195,309,992.-/hour
- Boundary Conditions
  IDR 2,500.- /passenger ≤ f_{single} ≤ IDR 7,500.-/passenger

Table 2. Stage 2 (Single Fare) 2 Modes of Mass Transport (IDR)

x_2		$d(x_2, x_3) + f_3(x_3)$					Optimum Solution	x_3
		$x_3 = S_1$	$x_3 = S_2$	$x_3 = S_3$	$x_3 = S_{...}$	$x_3 = S_{25}$	$f_2(x_2)$	
T1	2500	168,742,492	171,129,018	181,505,218	...	244,337,492	194,662,324	S11
T2	2550	169,723,042	172,109,568	182,485,768	...	245,318,042	192,861,968	S4
T3	2600	170,703,592	173,090,118	183,466,318	...	246,298,592	193,842,518	S4
T4	2650	171,684,142	174,070,668	184,446,868	...	247,279,142	194,823,068	S4
T5	2700	172,664,692	175,051,218	185,427,418	...	248,259,692	191,441,700	S7
T6	2750	173,645,242	176,031,768	186,407,968	...	249,240,242	192,422,250	S7
T7	2800	174,625,792	177,012,318	187,388,518	...	250,220,792	193,402,800	S7
T8	2850	175,606,342	177,992,868	188,369,068	...	251,201,342	194,383,350	S7
T9	2900	176,586,892	178,973,418	189,349,618	...	252,181,892	192,977,374	S6
T10	2950	177,567,442	179,953,968	190,330,168	...	253,162,442	193,957,924	S6
T11	3000	178,547,992	180,934,518	191,310,718	...	254,142,992	194,938,474	S6
T12	3050	179,528,542	181,915,068	192,291,268	...	255,123,542	192,291,268	S3
T13	3100	180,509,092	182,895,618	193,271,818	...	256,104,092	193,271,818	S3
T14	3150	181,489,642	183,876,168	194,252,368	...	257,084,642	194,252,368	S3
T15	3200	182,470,192	184,856,718	195,232,918	...	258,065,192	195,232,918	S3
T16	3250	183,450,742	185,837,268	196,213,468	...	259,045,742	185,837,268	S2
T17	3300	184,431,292	186,817,818	197,194,018	...	260,026,292	186,817,818	S2
T18	3350	185,411,842	187,798,368	198,174,568	...	261,006,842	187,798,368	S2
T19	3400	186,392,392	188,778,918	199,155,118	...	261,987,392	188,778,918	S2
T20	3450	187,372,942	189,759,468	200,135,668	...	262,967,942	189,759,468	S2
T21	3500	188,353,492	190,740,018	201,116,218	...	263,948,492	190,740,018	S2
T22	3550	189,334,042	191,720,568	202,096,768	...	264,929,042	191,720,568	S2

Table 2. Stage 2 (Single Fare) 2 Modes of Mass Transport (IDR)

x_2		$d(x_2, x_3) + f_3(x_3)$					Optimum Solution	x_3
		$x_3 = S_1$	$x_3 = S_2$	$x_3 = S_3$	$x_3 = S_{...}$	$x_3 = S_{25}$	$f_2(x_2)$	
T23	3600	190,314,592	192,701,118	203,077,318	...	265,909,592	192,701,118	S2
T24	3650	191,295,142	193,681,668	204,057,868	...	266,890,142	193,681,668	S2
T...	...	...	...	...	...	...	...	
T100	7450	265,816,942	268,203,468	278,579,668	...	341,411,942	...	
T101	7500	266,797,492	269,184,018	279,560,218	...	342,392,492	...	

Stage 1 of Back Recursion Dynamic Model
Boundary Conditions:
$f_{lowest} \leq f \leq f_{highest}$
IDR 2,500.- $\leq f \leq$ IDR 7,500.-

Table 3. Stage 1 (Single Fare) 2 Modes of Mass Transport

x_1	$d(x_1, x_2) + f_2(x_2)$								Optimum Solution	
	$x_2 =$ T1	$x_2 =$ T2	$x_2 =$ T...	$x_2 =$ T14	$x_2 =$ T15	$x_2 =$ T16	$x_2 =$ T...	$x_2 =$ T28	$f_1(x_1)$	x_2*
f_{single}	2,500	2,550	...	3,150	3,200	3,250	...	3,850	3,200	T11

OPTIMUM SOLUTION
From Stage 1 going to T_5 and S_5 become R
$F_{single} \rightarrow T_{15} \rightarrow S_3 \rightarrow R$
IDR 3.200 $\rightarrow T_{15}$ = IDR 59,127,444.-/hour $\rightarrow S_5$ = IDR 136,105,474.-/hour
$\rightarrow$ R = IDR 195,232,918.-/hour

d. Distribution of Revenue and Subsidies on the Integrated Mass Transportation Network 2 Mass Transportation Modes with Single Fare is carried out based on:
- For passengers who use 1 mode of mass transportation, the amount of income from fares and subsidies is given to the mass transportation operator concerned
- For passengers who use more than 1 mode (in this case 2 modes of mass transportation), the distribution of Revenue from Fares and Subsidies is carried out based on the Proportion of Passenger Travel Distance
- The proportion of fare revenue is obtained based on the weight (percentage) of distance fare revenue. Likewise with the amount of subsidies, obtained based on the weight (percentage) of the amount of subsidies/distance from each mode.

Table 4. Recapitulation of Revenue Sharing for each Operator

No.	Recapitulation	Ammount	
1	Tid (Passenger/hour)		
	TransJakarta Bus		4,759
	KRL Commuter Line		9,905
	TransJakarta Bus + KRL Commuter Line		4,947
2	Integrated Fares (f_{single}) (passenger/km)	IDR	3,200
	TransJakarta Bus Subsidies (passenger/km)	IDR	577

Table 4. Recapitulation of Revenue Sharing for each Operator

No.	Recapitulation	Ammount	
	KRL Commuter Line Subsidies (passenger/km)	IDR	400
3	Travel Distance (take Bus and KRL Commuter when traveling):		
	- Bus Travel Distance, km		3,003
	- KRL Commuter Line Travel Distance, km		2,712
4	Total Revenue (per hour):	IDR	195,232,918
	- TransJakarta Bus Revenue (1)	IDR	15,228,800
	- KRL Commuter Line Revenue (1)	IDR	31,696,000
	- TransJakarta Bus Subsidy	IDR	59,870,674
	- KRL Commuter Subsidy	IDR	76,234,800
	- TransJakarta Bus Revenue (2)	IDR	7,504,407
	- KRL Commuter Line Revenue (2)	IDR	4,698,237
5	Recapitulation Revenue		
	- TransJakarta Bus (per hour)	IDR	82,603,881
	- KRL Commuter Line (per hour)	IDR	112,629,037

Description: (1) user with single mode of transportation, (2) user with 2 modes of transportation

Assumptions set in the preparation of stated preference analysis:

- Operations of mass public transport have met the Transport SPM
- Vehicle Operation Costs outside the mass public transport service network are represented by motorcycle users (percentage of motorcycle users in Jabodetabek is ± 80%) with the assumption that motorcycle operating costs from previous studies are IDR 750.-/km s.d. IDR 1,000.-/km, set at IDR 875.-/km
- Integrated Mass Public Transport Travel Costs are carried out in the range of IDR. 3,500.-/passenger to d. IDR. 15,000.-/passenger (with variations in rates starting from IDR. 3,500.-/passenger; IDR. 5,000.-/passenger; IDR. 7,500.-/passenger; IDR. 10,000.-/passenger; IDR. 15,000.-/passenger)
- Travel time for motorcycle users is calculated based on the assumed driving speed of 50 km/hour
- The travel time of mass public transport users is in accordance with the operational speed of transportation (according to the SPM of each mass public transport, the maximum operational speed of the TransJakarta Bus is 50 km/hour, the maximum operational speed of the KCI Commuter Line is 70 km/hour, the maximum operational speed of the MRT is 80 km/hour).

The utility functions of each modeled mode are grouped into 2 categories of transportation, namely mass public transportation with integrated fares (AM) and non-mass public transportation (N):

$$V_{AM} = \beta_{AM,0} + \beta_{AM,1} * TC + \beta_{AM,2} * TT \quad (14)$$

with $\beta_{AM,0} = 0$

$$V_{NA} = \beta_{NA0} + \beta_{NA,1} * TC + \beta_{NA,2} * TT \quad (15)$$

Mode Probabilitas:

$$P(AM) = \frac{e^{V_{AM}}}{e^{V_{AM}} + e^{V_{NA}}} \quad (16)$$

$$P(NA) = \frac{e^{V_{NA}}}{e^{V_{AM}} + e^{V_{NA}}} \quad (17)$$

Description:

AM	=	Mass Public Transport (Commuter Line, TransJakarta Bus)
NA	=	Non-Mass Public Transport (Ojeg Online, Conventional Taxi, Paratransit, and Private Vehicle)
TC	=	Operating Cost or Fare (as a Travel Cost)
TT	=	Time Travel
β_0	=	Constant
β_i	=	Coefficient of i

Binomial Logit Modeling is assisted by Biogeme software.
Biogeme output recapitulation:

AM utility: ASC_AM + B_TT * AM_TT + B_TC * AM_TC

NA utility: ASC_NA + B_TT * NA_TT + B_TC * NA_TC

Estimated parameters

ASC_NA	=	-0,585
B_TC	=	-0,000129
B_TT	=	0,00443

$$P\ (AM) = \frac{e^{(-0{,}000129.TC\ +\ 0{,}00443.TT)}}{e^{(-0{,}000129.TC\ +\ 0{,}00443.TT)} + e^{(-0{,}585\ -0{,}000129.TC\ +\ 0{,}00443.TT)}}$$

$$P\ (NA) = \frac{e^{(-0{,}585\ -\ 0{,}000129.TC\ +\ 0{,}00443.TT)}}{e^{(-0{,}000129.TC\ +\ 0{,}00443.TT)} + e^{(-0{,}585\ -0{,}000129.TC\ +\ 0{,}00443.TT)}}$$

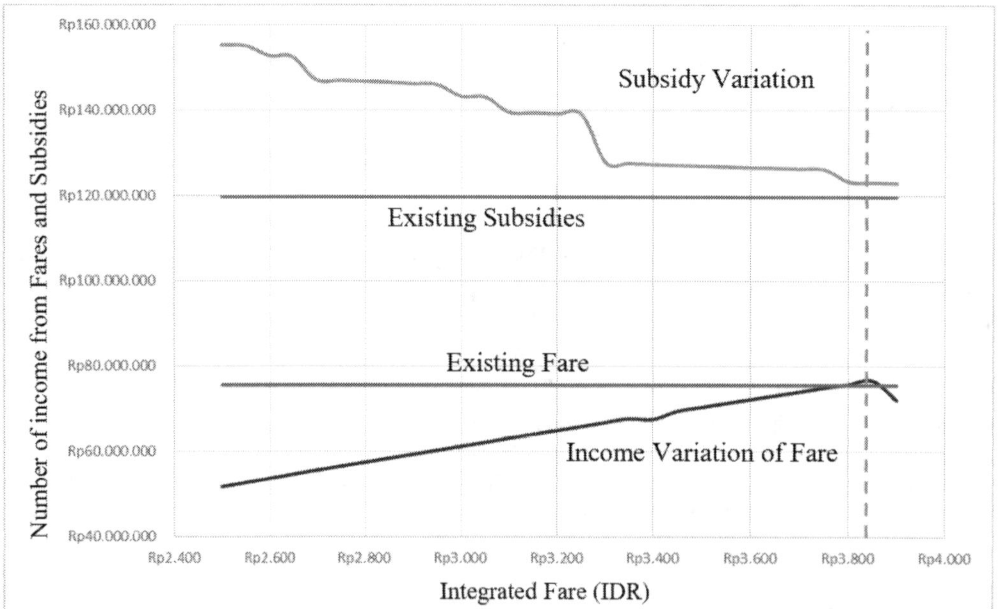

Figure 4. Comparison of Revenues between TransJakarta Bus and Commuter KRL in Existing Conditions and Conditions of Integrated Fares Variations

Figure 4 shows the increase in integration tariffs along with a decrease in the amount of subsidies. Each variation in the amount of integrated fares and subsidies set refers to the existing total revenue of KCI

Commuter and TransJakarta Bus. The size of the integrated tariff affects the choice of mass and non-mass transportation modes. Tariff-based revenue will be close to existing tariff revenue. The amount of the subsidy given to each integrated tariff shows the amount of the subsidy that is close to the amount of the existing subsidy income. The optimum value of the integrated fare shows the value of IDR 3,850 per passenger.

5. Conclusion

The integrated fare which can be concluded as an integrated fare that can be applied is of course in accordance with the initial plan that:

a. The integrated fare does not burden mass public transport users, which can be seen from the amount of ridership that is still above the existing condition, from the fare range between ± IDR 2,500 to IDR 3.850,-
b. The subsidy provided by the government is as small as possible so as not to burden the amount of the subsidy that has been given
c. The amount of riderships with the application of integrated fares can be greater than the existing conditions.

This research will still be continued on the mass public transport service network model with the basic model of movement from origin to destination in the form of a zone system. This process is carried out to determine the process of determining the operational costs of movement with movement from one zone to another by comparing the use of mass public transportation and non-mass public transportation based on travel costs and travel time.

6. References

[1] G Abrate, M Piacenza, D Vannoni 2005: The Impact of Integrated Tariff Systems on Public Transport Demand: Evidence from Italy (The University of Eastern Piedmont and HERMES)
[2] Central Bureau of Statistics for the Capital of Jakarta 2015 Book of the Capital of Jakarta in Figures
[3] R Bellman 1972 Dynamic Programming, Princeton University Press
[4] Governor of the Special Capital Region of Jakarta 2014 Regulation of the Governor of DKI Jakarta number 35 of 2014 concerning Minimum Service Standards for TransJakarta – Busway Management Units
[4] Government of the Republic of Indonesia 2014 Government Regulation of the Republic of Indonesia, number 74 of 2014 concerning Road Transportation
[5] Government of the Republic of Indonesia 1999 Law number 22 of 1999 concerning Regional Autonomy
[6] A Schöbel 2004 Design of Zone Tariff Systems in Public Transportation (Technische Universität Kaiserslautern)
[7] T Takahashi 2014 Economic Analysis of Tariff Integration in Public Transport (University of Tokyo)
[8] V R Vuchic 2005 Urban Transit: Operations, Planning, and Economics, John Wiley & Sons., Inc.
[9] TransJakarta Bus Route Data, www.transjakarta.co.id/peta-rute/, download time: February 2nd, 2019
[10] Commuter Line Jabodetabek Route, diperoleh dari http://www.krl.co.id/#m_perjalanankrl, download time: Desember 21th, 2018

Palu Housing Reconstruction Process: Reviewing and Learning after the 2018 Earthquake

R G K Pradoto[1], A Oktavianus[1,2], K S Pribadi[1], I M A B Rasmawan[1], and L D Wulandari[1]

[1] Construction Engineering and Management Research Group, Faculty of Civil and Environmental Engineering, Institut Teknologi Bandung, Indonesia
[2] Construction Engineering and Management Division, Civil Engineering Department, National Taiwan University, Taiwan

ranipradoto@itb.ac.id

Abstract. The strong earthquake that hit the Province of Central Sulawesi on 28 September 2018 (M7.4), followed by tsunami and liquefaction has caused many casualties, injuries as well as damages in the housing sector. In the rehabilitation and reconstruction (R&R) program, the Government policy was to provide the affected community with aid to rebuild their houses (for houses outside the red zone), the Government also will build new settlements for relocation (for houses inside the red zone), which both are expected to be more earthquake-resistant. In order to understand what happens in the field related to various issues, an investigation to study the whole processes and to document the lessons learnt from the reconstruction process is conducted in Central Sulawesi Province. One of the highlights of this study is the impact of the three types of disasters that occurred in Palu Municipality (earthquake, tsunami, and liquefaction) on the housing reconstruction process after the disaster occurred. The primary field survey and Focus Group Discussion were conducted through interviews with various parties in the earthquake-affected areas to reveal the findings of problems and obstacles in the field during its two years of implementation, from 2018 to 2020. This study focuses on reviewing and learning from the post-disaster housing sector reconstruction process. The result of this study is expected to provide future lessons for the post-disaster housing reconstruction process in Indonesia and other countries.
Keywords: post disaster housing reconstruction, housing reconstruction project, housing recovery process.

1. Introduction

Striking a magnitude 7.4 earthquake on 28 September 2018 which was followed by liquefaction and a tsunami with a height between 2.2 to 11.3 meters, the Province of Central Sulawesi in Indonesia was severely damaged. Most of the area of Palu Municipality, the capital of Central Sulawesi Province, along with the surrounding regencies, namely Sigi, Donggala, and Parigi Moutong Regencies around Palu Bay, were devastated by the earthquake, tsunami and liquefaction. Data released on October 20, 2018, of the total population of 1.2 million people before the disaster, the death toll reached 2,256, while 164,626 people had to flee to informal refugees and to tents outside their homes, and 20,257 people desperately needed temporary shelter (1, 2).

The government immediately declared an emergency situation until the end of October 2019 for an emergency response period that focused on evacuating victims and meeting the basic needs of refugees. The next six months the emergency situation was gradually handled to enter the recovery phase. Housing was the most affected due to the liquefaction event which destroyed many houses and the tsunami that swept housing along the coast in Palu Bay. As of April 2019, the total number of damaged houses recorded was 110,214 houses (3). Then, the Government developed a rehabilitation and reconstruction (R&R) program to deal with affected housing and infrastructure. In the rehabilitation and reconstruction program, the Government policy was to provide the affected community with aid to rebuild their houses (for houses outside the red zone), the Government also will build new settlements for relocation (for houses inside the red zone), which both are expected to be more earthquake-resistant.

In order to understand what is happening on the ground in relation to various issues, this study aims to observe the entire process and document lessons learned from the housing reconstruction process in Central Sulawesi Province. The study covers various aspects such as legal framework, program management, institutional coordination, program management, reconstruction strategy, technical design, quality assurance and community participation. Research into post-disaster housing reconstruction believe that effective participation of all stakeholders, particularly the inclusion of affected communities in reconstruction, is an important principle for successful projects (4).

This study is also set to provide an understanding of the complex conditions and challenges faced in the planning, management and implementation of housing reconstruction in Palu. By studying the housing reconstruction process and the complex conditions resulting from the impact of multiple disaster events, this study is expected to provide future lessons for the post-disaster housing reconstruction process in Indonesia and other countries.

2. Research Methodology and Description of Case Study

In order to get a comprehensive understanding, this study is divided into three stages. The first part is a phase to review the literature related to post-disaster housing reconstruction and documents related to the implementation of the rehabilitation and reconstruction (R&R) program for housing in Palu. An understanding of housing reconstruction activities in Palu leads to three parts, namely program management, reconstruction strategies, and technical aspects.

The second part uses the field survey method to find out the actual conditions and obtain primary data in the field related to the implementation of the housing reconstruction program. The field survey was carried out through interviews with communities affected by the earthquake about how to propose a house construction model to be chosen. The field survey also made observations on several types of houses that have been built by the Government, non-government organizations (NGOs), and communities independently. On the other hand, the field survey also observed the quality of the house and the precast concrete components used in making the house earthquake resistant.

The field survey conducted interviews with various parties in the earthquake-affected area, which included homeowners, builders and craftsmen, builders/applicators (a special term used for companies or individuals proposing to build houses following a certain approved prototype model), and task forces. from governments involved in the recovery program, non-governmental organizations, material suppliers, and other respondents.

Then, discussions through Focus Group Discussions with various stakeholders were also carried out afterwards to reveal the findings of problems and obstacles in the field. The results of the field visit were used as material for the discussion. Several issues related to the implementation of the housing reconstruction program are noted and found in this study.

IOP Conf. Series: Earth and Environmental Science 1065 (2022) 012057 doi:10.1088/1755-1315/1065/1/012057

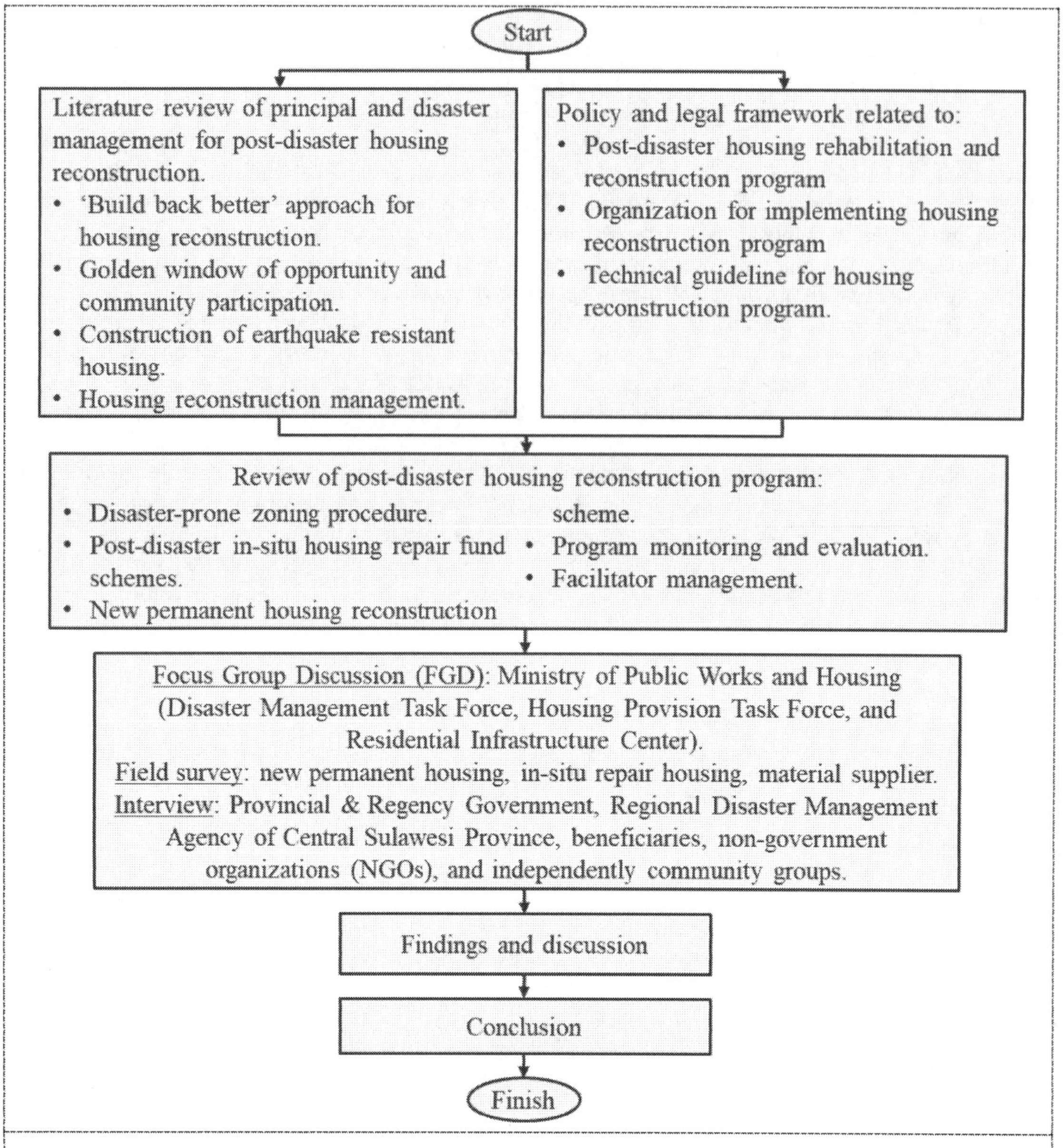

Figure 1. Research Methodology.

This study focuses on housing reconstruction programs in Palu which connected to some surrounding regencies, namely Sigi, Donggala, and Parigi Moutong Regencies. One of the main characteristics that stands out in this study is the occurrence of multiple disasters (earthquake, tsunami, liquefaction). The location of this study is along the bay of Palu with administrative areas from different local governments. Geographically, the location of this study based on the areas of the regency and municipality affected by the disaster on 28 September 2018 which be shown in the following figure.

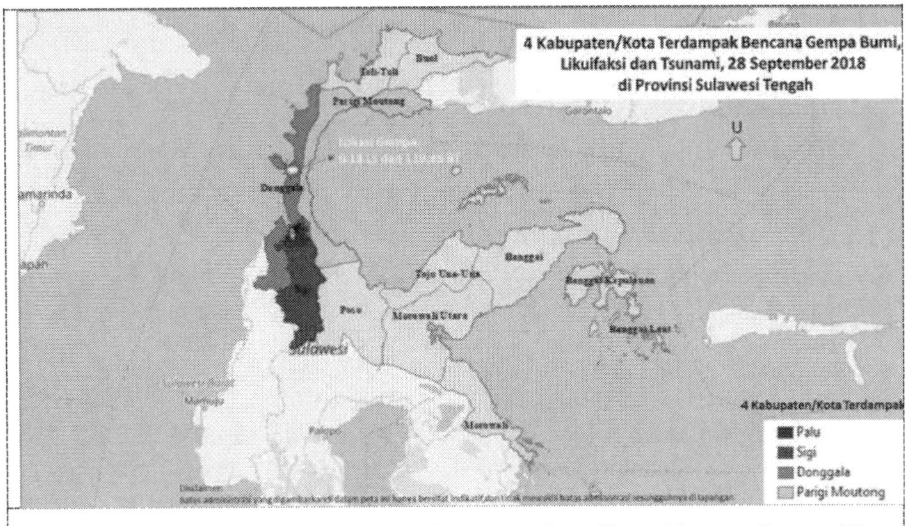

Figure 2. Location of the regency and municipality affected by the disaster (5)

3. Post-Disaster Housing Reconstruction

3.1. "Build Back Better" Approach for Post-Disaster Reconstruction

Research for post-disaster housing reconstruction states that the principle of project success involves community participation from all stakeholders, especially affected communities in reconstruction (6). It is generally believed that participatory among communities is a very efficient method for housing reconstruction and disaster awareness of participants about 10 years after "building back better" (7). The "Build Back Better" approach also plays a role in mitigation because it is not only rebuild destroyed houses or buildings, but also to build disaster-resilient communities. Affected communities who are involved in the reconstruction process with proper guidance and supervision will gain better knowledge and skills regarding the construction of earthquake-resistant houses.

3.2. Golden Window of Opportunity and Community Participation

The most important thing for disaster-prone areas like Indonesia is socialization to increase public understanding and awareness in dealing with disasters (8). post-disaster utilization is the right time to design a housing reconstruction program with earthquake-resistant buildings in a way that is affordable, easy to implement and takes advantage of the window of opportunity that exists right after the earthquake disaster. Various efforts need to be made to train affected communities to engage in construction methods to make permanent houses resistant to earthquakes in the future. For ease of implementation, the post-disaster housing reconstruction management framework should be a practical guide for policy makers and practitioners in the field. Obstacles to the reconstruction process towards the construction of permanent housing must be identified, so that it becomes one of the important strategies for community readiness in rebuilding a safer environment in the future (9, 10). A good initiation of the reconstruction program is needed by considering the needs of the affected communities, especially increasing the capacity of earthquake-resistant buildings. The reconstruction program was made as a whole, involving the participation of all actors from various levels from the district to small community groups.

The participation of disaster-affected communities is very important in post-disaster reconstruction because they are considered the parties who best understand disaster-prone conditions. Davidson (11) stated that, when beneficiaries are integrated into the early stages, they can have a significant impact on the project with long-term benefits to them and to other stakeholders. Therefore, participation is very necessary in order to receive all useful information related to reconstruction as well as community involvement in decision making. The conditions of collective agreement to produce decisions are a coordinating step during the reconstruction process and also to formulate the best solutions for problems that arise during the reconstruction process.

3.3. *Construction of Earthquake Resistant House*

As in all disaster-prone areas in Indonesia, the concept of building earthquake-resistant houses on a massive scale continues to be introduced. Several concepts of earthquake-resistant houses were informed as an alternative choice for the community. the concept of building an earthquake-resistant house using instant, healthy and simple home technology. Education programs must also be disseminated to affected communities to believe that the construction of the house will provide a sense of security and comfort for the community (8). In addition, training was provided on basic knowledge of earthquakes, building earthquake-resistant houses and selecting building materials according to the standards set by the Ministry of Public Works and Housing in Indonesia.

There are several aspects that must be considered in the construction of residential houses in order to meet the provisions for earthquake resistance or at least minimize the impact of damage caused by the earthquake. The next explanation will be devoted to the structure of the Reinforced Concrete Frame House, as a residential structure that is commonly built in Indonesia. The aspects that will be explained include building safety, building location, building plans, material quality requirements, and technical requirements for building structures.

3.4. *Management of Housing Reconstruction*

Housing reconstruction management. Housing reconstruction management plays an important role in rebuilding communities. The framework for post-disaster project management should provide practical guidance for policymakers and practitioners. Several management issues for permanent housing reconstruction were identified that are of strategic importance to community preparedness; The need for the initiation of a good reconstruction program based on a comprehensive damage and loss assessment, and taking into account the needs of the affected communities; establishment of a legislative, regulatory and policy framework that provides direction for stakeholders; The need for critical assessment of community vulnerability and identification of existing and underlying disaster hazards and risks; The need for the involvement of beneficiary communities in the reconstruction process is critical to giving them ownership of the program; Provision of education and capacity building for stakeholders during the permanent housing reconstruction management process (10, 12). In addition, it has been shown that the development of a common framework enhances data collection for further research in order to improve management practices for the post-disaster housing sector (9).

In terms of post-disaster programs, good governance plays an important role in influencing the housing sector recovery process. Disaster rehabilitation has the potential to be a long and complex process in the disposition and responsibilities between the national and local government. To overcome the lengthy process, good governance must reduce the distance between them (13).

4. Post-Disaster Housing Reconstruction Program in Palu

4.1. *Policy and Legal Framework*

Indonesia has a law that regulates disaster management, namely Law 24 of 2007. This regulation becomes the basis for implementing disaster management activities, both in the emergency and recovery and reconstruction stages so that they are coordinated by the Government and Regional Governments. In the disaster situation in Central Sulawesi Province, the regulations used as guidelines in the rehabilitation and reconstruction efforts are as follows:

1. Presidential Order of the Republic of Indonesia No. 5 of 2018 for acceleration of post-earthquake and tsunami rehabilitation and reconstruction in Central Sulawesi Province.
2. National Disaster Management Agency Regulation No. 7 of 2018 for arrange direct foreign grants fund for natural disaster management in Central Sulawesi Province.

3. Governor Regulation of Central Sulawesi Province No. 10 of 2019 for post-disaster rehabilitation and reconstruction plans.
4. Governor Decree of Central Sulawesi Province No. 369/105.1/BPBD-G.ST/2019.
5. Ministerial Decree of Public Works and Housing No. 1176/KPTS/M/2019 for the task force for the implementation of disaster management of the Ministry of Public Works and Housing.
6. Mayor Regulation of Palu Municipality No. 24 of 2020 and 35 of 2019 concerning technical guidelines for the rehabilitation and reconstruction.

4.2. *Housing Reconstruction Program*

4.2.1. Procedure for Determining Disaster-Prone Zoning. The occurrence of the tsunami and liquefaction phenomena which were quite large caused reconstruction not to be started right away, there was a factor in mapping the area to be considered so that rebuilding/new development would not be carried out in areas with the potential for similar disasters to occur in the future. The determination of disaster-prone zoning in Palu Bay involves many ministries and institutions at the national level, this is because the determination of disaster-prone zones is a cross-sectoral effort. The process of determining the zoning map takes a long time due to a comprehensive study and agreement from various stakeholders.

The disaster-prone zoning map of Palu Bay was listed in Governor Regulation of Central Sulawesi Province No. 10 of 2019. The map contains four zone categories, namely:

1. Forbidden zone (ZRB 4): forbidden zone for rebuilding and new construction. This zone prioritizes the use of space for protected areas, green open space, and monuments.
2. Restricted zone (ZRB 3): forbidden zone for new construction. Empty land is prioritized for protected areas and agricultural/plantation/forestry cultivation. Redevelopment in this zone must comply with the latest design standards.
3. Conditional zone (ZRB 2): New developments must comply with the latest building design standards. This zone must pay attention to the level of disaster vulnerability.
4. Development zone (ZRB 1): This zone is allowed for new construction by following the latest building design standards.

The disaster-prone zoning map can be accessed by the public through a layer on Google Maps in the following figure and link.

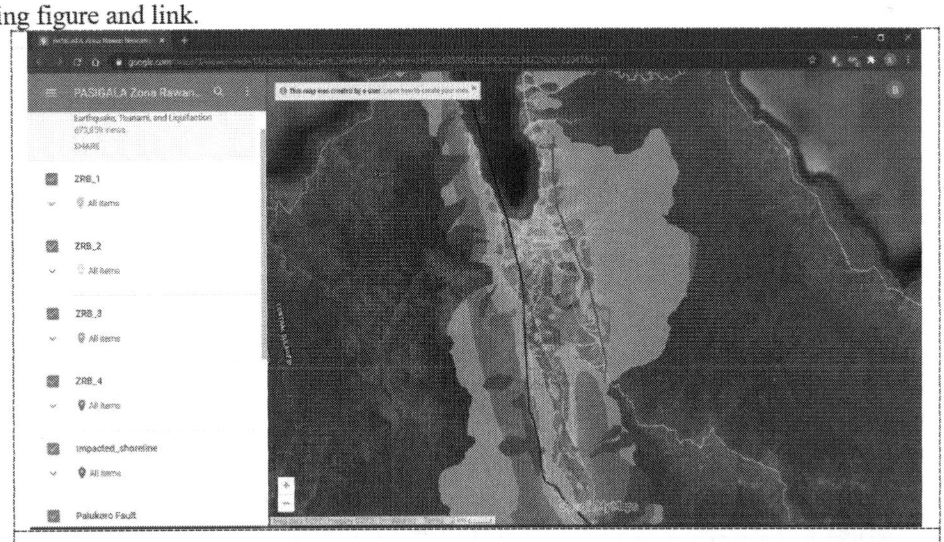

Figure 3. Screenshot of Palu Bay's area disaster prone zoning on Google Map. (https://www.google.com/maps/d/viewer?mid=1IUL2o02rOia3q9LwbLZlAvW4f50FgK1s&ll=-0.9765263385261225%2C119.94727626123047&z=11)

4.2.2. Fund Schemes for Post-Disaster In-Situ Housing Repair. Based on the Presidential Order of the Republic of Indonesia No. 10 of 2018 concerning the Acceleration of the Rehabilitation and Reconstruction of Central Sulawesi, the Head of National Disaster Management Agency of Indonesia proposed a budget allocation to the Minister of Finance for the funding of post-earthquake rehabilitation and reconstruction, including assistance for the rehabilitation and reconstruction of housing carried out by the community.

Table 1. Category of post-disaster housing rehabilitation and reconstruction.

House damage level	Location	Rehabilitation and reconstruction	Stimulus assistance for house repair funds
Heavy damage	ZRB3 and ZRB 4	It is forbidden to rebuild the house. Relocated to a permanent residential location.	IDR 50 million for heavy damage.
	ZRB 1 and ZRB 2	House repair/rebuild with assistance of the house repair funds	
Medium and light damage	ZRB3 and ZRB 4	House repairs but the owner is required to make a statement of responsibility	(1) IDR 25 million for moderate damage. (2) IDR 10 million for minor damage.
	ZRB 1 and ZRB 2	House repair with assistance of the house repair funds	

In the implementation of post-disaster rehabilitation and reconstruction phase I in Palu Municipality, the national government through an international grant has allocated funds for the rehabilitation and reconstruction of housing for the category of heavily damaged houses and in phase II through a national government grant for the regional government, funds for rehabilitation and reconstruction assistance have been allocated. for the category of heavily damaged houses, moderately damaged houses and lightly damaged houses.

In an effort to accelerate the recovery of post-disaster conditions, the Government seeks to immediately channel funds to repair/rebuild affected houses. There are four patterns, namely:

1. Contractual manufacturing model.
2. Contractual conventional model (built by contractor).
3. Independent manufacturing and/or conventional models (built independently by owner).
4. Heavy damaged house has been built in advance (reimbursement).

The whole process is carried out with the assistance of Assistance Team for Acceleration of Housing Development.

4.2.3. Scheme for New Permanent Housing Reconstruction. The housing reconstruction program in Central Sulawesi province does not only mean repairing damaged houses but reconstruction of affected housing by moving them to safe locations. The tsunami and liquefaction caused the need to relocate housing reconstruction outside the red zone. The new permanents housing reconstruction is located on the new location in the ZRB 1 and ZRB 2, so this scheme is for relocating residents from the red zone. There are three new permanent housing reconstruction schemes, i.e.:

1. Relocation to a large residential area from the Government. Several locations have been determined by the Governor to become new housing locations, there are district of Duyu, Tondo-Talise, and Pombewe Olobuju. Communities are given the option to relocate to these housing locations with priority for affected communities living close to each location. This is important to minimize the negative impact on their livelihoods.
2. Relocation to satellite land. Relocation on a smaller scale than the first option was proposed by the prospective beneficiaries in groups. This satellite land option makes it easier for the government because residents have identified their population, usually a group of one region, so that the government can focus on finding land according to that amount.

3. Independent relocation. This relocation scheme is considered as an alternative option if individual residents or very small groups have land that they can use for their new permanent homes. Financing for this type of relocation follows the self-help housing scheme of the 'One Million Houses Program' (national housing provision program), it is also called by 'Rekompak' program.

The relocation mapping of the communities affected by the disaster is presented in the **Fig.4**.

Regarding financing for the reconstruction of new permanent housing, there are two schemes that come from the government and the private sector or NGO. The first scheme is the housing provision program from national government. The Directorate General of Human Settlements of the Ministry of Public Works and Housing was in charge for the rehabilitation and reconstruction program. The program is a community-based post-disaster settlement redevelopment that is oriented towards disaster risk reduction. Housing construction in this program uses RISHA earthquake-resistant houses.

The second scheme is a donor program by the private sector and non-governmental organizations. This pattern is the result of the involvement of the private sector in supporting the government to rebuild housing using earthquake-resistant houses that are technically approved by the Government. One of the donors for this housing development program in Central Sulawesi Province is the Buddhist Tzu Chi Foundation.

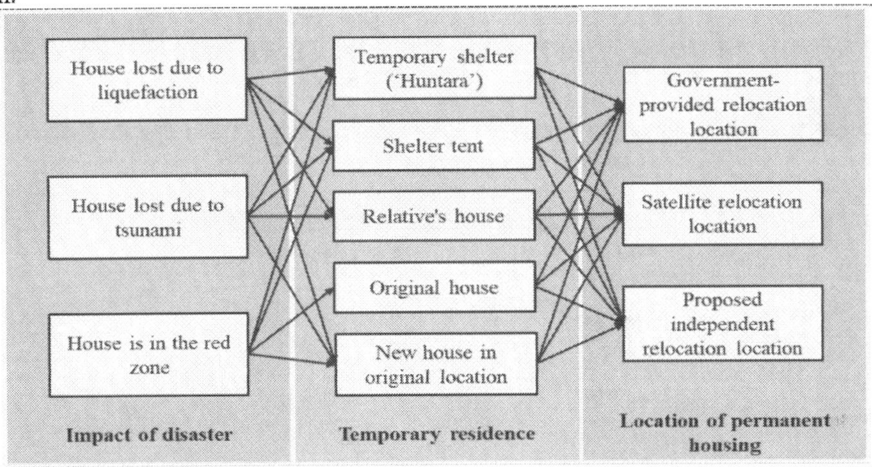

Figure 4. Location Mapping of temporary and permanent housing for disaster-affected communities (14).

4.2.4. Program Monitoring and Evaluation. Monitoring the progress and evaluation of the housing rehabilitation and reconstruction program in the province of Sulawesi is being carried out in stages, from evaluation of implementation at the project site to evaluation of the program in coordination with all stakeholders. Periodically coordination and evaluation are carried out at the regency or municipality level. This evaluation serves as input for coordination at the provincial level to make decisions on major and inter-regional issues. The monitoring and evaluation aim to discuss progress and problems faced in the implementation of rehabilitation and reconstruction, as well as to formulate recommendations needed to improve the performance and quality of the implementation of the rehabilitation and reconstruction program.

The person in charge of monitoring at the provincial level is the Regional Disaster Management Agency of Central Sulawesi Province with joint evaluations with stakeholders from the ministry task force and stakeholders at the provincial and regency/municipality levels, as well as NGOs. Meanwhile, the Regional Disaster Management Agency of the municipalities (Palu, Sigi, Donggala, and Parigi Moutong) are responsible for municipality -level monitoring based on the progress of the project in the site. Program monitoring in the site is always carried out with stakeholders, affected communities and representatives of community leaders.

4.2.5. Facilitator Management (HDAA Team). The government formed small teams of 3-5 people who functioned as facilitators, called the Housing Development Acceleration Assistance Team (HDAA Team). HDAA Team is a regional assistance team to assist the implementation of post-earthquake, tsunami and liquefaction rehabilitation and reconstruction in Palu Municipality, Central Sulawesi Province. This team was divided into two groups, namely HDAA Team for Construction of Heavy Damaged Houses and HDAA Team for Rehabilitation of Moderately Damaged and Lightly Damaged Houses. The team is responsible to RDMA at the regency or municipality level.

HDAA Team aims to guide the technical repair and construction of houses carried out by the community independently. This is also to ensure the quality of housing repair and reconstruction work so that it is earthquake resistant. Moreover, this team assists the community in reporting the progress of work and administrative work to funders or the Government.

5. Findings

5.1. Implementation

The housing reconstruction program in Central Sulawesi Province was planned for April 2019 to December 2020 (15). The progress of permanent housing reconstruction as per September 2020 shown in Table 2. The data in those table show reconstruction progress for heavily damaged houses based on the type of houses built. The total of 15,633 unit of houses were finished and still under construction at the time.

Table 2. Number of houses built in the permanent housing reconstruction program based on the type of houses per December 2020 (14)

No	Type of house	In situ (stimulating fund to beneficiary)			Relocation	Total
		Phase I	Phase II			
		Group-managed	Contractual	Individual		
1	RISHA	665	313		1,762	2,740
2	Confined masonry (RIKO)	3618	587	5,923	-	10,128
3	RISBA	3	380	-	-	383
4	Ruspin	-	-	-	52	52
5	RISMA	43	138	-	-	181
6	RICAG	-	22	-	-	22
7	RIDHA/Conwood	-	-	-	2,000	2,000
8	RIKA	127	-	-	-	127
	Total	4,456	1,440	5,923	3,814	15,633

As mentioned earlier, there are two schemes in permanent housing reconstruction, relocation and in situ. The government of national, provincial, and municipality assumed the responsibility for relocation, with permanent housing built in relocation areas by contractor through open tender. Land provision were done through several scheme, mainly ex concession land owned by the government, other schemes are grant of land (agriculture and farming land, the government will later compensate land owner with land in different location) from community to government, and land purchasing by the government. Relocation facing slow progress as land preparation, main infrastructure, and construction tender for housing take some time to complete.

In situ reconstruction so far proceed in two phases, phase I began in 28th of February 2019, with the source of fund from foreign grant. Phase II began in 16th of December 2019, with the source of fund from national government grant to provincial government (14). The implementation of in situ reconstruction, house rebuilding, was handed over to the beneficiaries of stimulating fund. The beneficiaries then proceed to rebuild their houses through community self-managed within group-managed, individual self-managed, or contractual with builder/contractor/housing applicator, while The Government provided supervision and assistance through facilitators.

Based on the delivery mechanism, most of the houses rebuilt and still under construction, as shown in Table 2 were done by group and individual-managed scheme. Figure 5 show the proportion of houses built and still under construction based on the delivery mechanisms.

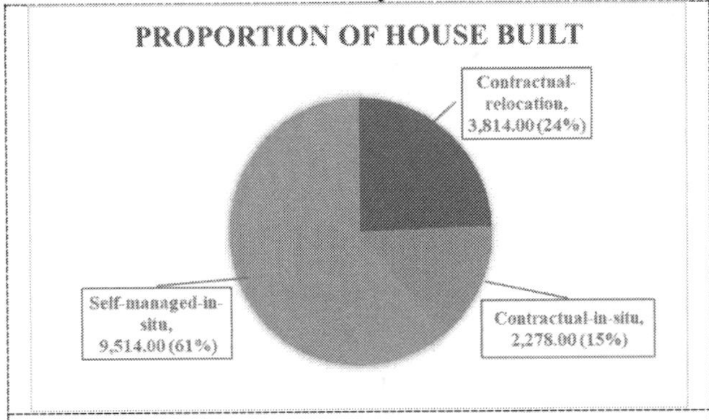

Figure 5. Proportion of houses built based on delivery mechanism

Beneficiaries' interest to conventional confined masonry house increased in phase II, particularly in Palu Municipality, where they do not necessarily need to form a group to obtain the stimulating fund, thus create individual self-managed scheme. The characteristics of conventional confined masonry houses that resembled public houses in general, mainly because there is no new material and technology needed, makes it not depend on contractor/builder/house applicator. Some beneficiaries saw this as an opportunity to save contractor fees.

5.2. *Housing Design and Specification*

Government policy in housing sector reconstruction is to build earthquake resistant house. In the implementation, type of houses available and approved to build vary. There are 8 types of earthquake resistant house finished and still under construction at the time, as shown in Table 2. The interesting finding is that, almost all types of house available in Palu were similar to that in Lombok in term of technical design, it makes sense since the reconstruction program of Lombok started earlier than in Palu despite both disasters occur in the same year (2018).

The standard design for earthquake resistant houses available for the reconstruction program is type-36 house (36 m^2 floor size), with estimated budget of IDR 50 million per house. This 50 million in budget are actually the same amount as stimulating fund given to heavily damaged house (in situ). Since beneficiaries in relocation category were given house instead of fund, in order to maintain the principal of equality, the Government impose the IDR 50 million budget for each house, both relocation and in situ. The budget may be higher than IDR 50 million provided the beneficiaries add their own money to that budget.

In relocation case, determination for the type of house to build are up to the project owner (government or donor), as long as it conforms with the standard for earthquake resistant house. In our field finding, relocations fund by the Government built RISHA, while relocation fund by donor vary, some built RISHA, other built RIDHA. Meanwhile for in situ case, determination for the type of house to build are up to each beneficiary, beneficiaries then may proceed to arrange agreement with contractor/builder/house applicators (if using contractual delivery mechanism).

Different from houses built by donors and the government in the relocation program, beneficiaries who build houses in-situ can choose their own type of house. There are a few types of houses in which come to the beneficiaries' interest. Those houses are RISHA, RIKO, and RISBA house type, as shown in Table 2. This finding is similar to the condition in Lombok, despite the many new type of houses introduced, RISHA, conventional confined masonry house (RIKO), timber structured house (RIKA), and steel structured house (RISBA) were the most popular prototypes chosen by the community (16).

Some types of earthquake resistant houses built in Central Sulawesi Province, in brief explanation are as follow:

- Instant dan simple precast concrete house (RISHA), one of the most popular earthquake resistant house in Indonesia, developed by the Ministry of Public Works and Housing, using knock down system with precast concrete as structural element and bolt as joint element.
- Conventional confined masonry house (RIKO), typical house often found in Indonesia with brick as main structural component with simple reinforced concrete structure as the confinement.
- Safe and instant house (RIDHA), type of house that become well-known since Lombok Earthquake, RIDHA used conwood as main material, made from mixed of cement and cellulose fibre. Conwood material manufactured in board components, this board component then attached to each other with bracing components.
- Stilt house (combination of RISHA as lower structure in first storey and timber structure in second storey), so far can only be found in relocation of new permanent housing in Mamboro District. The idea was to accommodate the fisherman community who want to live close to coastline, but since the coastline of Palu Bay considered as red zone, stilt house was built to mitigate tsunami.

Figure 6. RISHA in relocation of new permanent housing (14)

Figure 7. Conventional confined masonry houses (RIKO) (14)

Figure 8. Conwood houses (RIDHA) built by donor.

Figure 9. stilt house built by fishermen community in Mamboro District.

The other types of earthquake resistant house built in Central Sulawesi Province are steel structured house (RISBA), instant house with panel system (Ruspin) which similar to RISHA, and timber structured house, which resembled many traditional houses in Indonesia. Although there is some type of earthquake resistant house not mentioned, the number of those houses are considered insignificant.

5.3. Community Perception

Until the third trimester of 2020, the community has occupied new housing locations in district of Tondo-1, Balaroa, and Mamboro. Based on team interviews in the field, the community admitted that they really appreciated the relocation program by the government, so that they could immediately move from temporary shelters. Some of the community stated that relocation to new housing brought many benefits to them, although others still did not want to move because the location was far from the original location of their previous house. They are willing to move to a satellite housing location.

In addition to appreciating the new housing relocation, the community also appreciates the financial assistance to repair their houses in-situ. Communities can use the funds to repair or rebuild their homes. People prefer to build conventional type houses (RIKO) compared to other types of houses because the implementation of RIKO type construction is easier for ordinary people to understand. From the government's perspective, they actually recommend the RISHA type earthquake-resistant house because the quality can be guaranteed. However, there was an incident of fraud by the applicator to the community at the beginning of the program so that public trust fell on RISHA.

Figure 10. Interview to permanent housing's beneficiaries

Having several options for communities can help them find earthquake-resistant housing which is very important for disaster-prone areas such as in the Central Sulawesi Province. However, the government must supervise the implementation of house construction carried out by applicators so that public trust is maintained properly.

5.4. Community Contribution

The role of the community in the housing reconstruction program is very important to be able to provide an understanding of relocation to a safer place and to get earthquake-resistant houses for them. Communities have different roles for the type of reconstruction or relocation. For relocation to a large residential area, the Government always involves the community to provide input from planning and implementation. For relocation to satellite land and independent relocation, the community can actively propose a location, choose the type of house, and even participate in building their house.

In in-situ development with funding from the government, the community's role is bigger by building their own houses. The community can create groups as a concept of community collective participation for the management of housing development assistance funds. The implementation of community groups is an alternative that is applied to post-disaster in Indonesia, as applied to post-disaster housing reconstruction in West Nusa Tenggara Province (16). The groups can function very effectively in in-situ reconstruction program, especially for people who are used to doing communal work.

In addition, the team noted that the role of women is very important as social capital in a faster disaster recovery. Women, especially housewives, will take on the role of planning family life when the recovery period begins, including ensuring that the family can live in a strong and suitable home for activities. Another very influential social capital is the role of youth groups and the smallest administrative unit of community. The powerful groups make community contributions to housing reconstruction after a disaster more rapid.

5.5. Problem and Challenges

Development for relocation in a large residential area from the Government faces complex land supply problems. The government provides land through three schemes, namely using land that was ex-Cultivation Rights (concession), land grants, and land purchases. Although it has been handled directly by the Ministry of Agrarian Affairs and Spatial Planning/National Land Agency of Indonesia, it turns out that the use of the ex-concession land is still having problems with claims from the concession holder and several people. These obstacles resulted in the Ministry of Public Works and Housing being unable to immediately build housing. Meanwhile, development for relocation to satellite land has problems in financing. The local government (regency or municipality) is responsible for providing land. Some of them are hampered by bureaucratic administrative processes for legal land provision and budgeting.

The local government is responsible for collecting data on the people affected by the disaster. However, the large number of people affected by the disaster, including the staff from the Regional Government, made the data continue to change. The resources of the local government are not sufficient to collect data quickly. On the other hand, there is no initial data that shows the exact number of people who live. For example, the people who live in Palu Municipality come from Palu Municipality and other areas where they are not registered by the municipality government. This makes it difficult to register the affected people.

Availability of skilled labor. Prior to the disaster, Palu Municipality and the surrounding area did not have a sufficient number of skilled workers because most of them worked as fishermen and traders. In the post-disaster period, the number of skilled workers becomes a problem because the need is very large. Many skilled workers are imported from Java, but the number is still insufficient. The involvement of local communities to become skilled workers is also very minimal because they are not used to construction work that is not friendly to the very hot weather of Palu. Although construction materials are abundant in Palu, the limited number of skilled workers is a major obstacle to the progress of the housing reconstruction program, both in terms of quantity and quality.

The pandemic condition due to Covid-19 has worsened the housing reconstruction situation. Although Palu Municipality and some regencies in Central Sulawesi Province are categorized as a green zone of COVID-19 spreading, the pandemic situation has had a major impact on the implementation of reconstruction due to the presence of workers from outside the area of Palu and reconstruction management involving many stakeholders from various backgrounds. other regions in Indonesia.

The housing reconstruction program in Central Sulawesi Province also faces challenges with the involvement of many stakeholders. These stakeholders come from the community group level, local regency or municipality, provincial, national and international levels. These parties are part of the reconstruction program management and donors from the private sector and the World Bank. Coordination between stakeholders is a formidable challenge because of the procedures that each organization has. For example, each party has different procedures to ensure the legality of providing land in preparation before construction begins. This is a big challenge for implementing housing reconstruction programs in Central Sulawesi Province to meet the needs for the number of houses while ensuring the quality of the houses.

6. Discussion

Context of local and national conditions that affect the decision-making process at the national and local levels. The phenomenon of multi-event disasters in Central Sulawesi Province has a big impact that is very different from disasters in other regions in Indonesia. The local government lost a lot of resources and the phenomenon of liquefaction in a large area forced the national government to take disaster management steps. The government is taking the first step to formulate a disaster-prone zoning map to identify safe locations for housing residents and this requires experts from various expertise at the national level. Regarding housing relocation, decision-making requires coordination between local and national level because of the responsibility for recovery from the government at the national, local, and private levels or non-governmental organizations. Having many parties involved certainly brings great benefits to the restoration and reconstruction of housing, but on the other hand, decision-making procedures also require careful coordination.

Regarding funding, housing reconstruction in Central Sulawesi Province received financial assistance from international donors outside of the state budget allocation. The government's budget allocation is still divided with housing recovery due to the earthquake in West Nusa Tenggara so that the concentration of funding has not been fully allocated to Central Sulawesi Province. However, the tsunami and liquefaction in a large area attracted international attention, so technical assistance and funds from international donors greatly helped the reconstruction of housing in Central Sulawesi Province.

Context of multi-event disaster requires handling with consideration of good technical aspects. The earthquake, liquefaction, and tsunami resulted in the existence of a red zone that was forbidden for housing. Public awareness is needed of the prohibition in the red zone, while the government is expected to immediately develop the zone into a protected forest area or public area so that it cannot be re-inhabited by the community. Housing reconstruction also requires technical support in the form of earthquake-resistant houses. Both housing being relocated by the Government and repairing houses in-situ independently by the community require technical assistance to ensure that houses meet the technical standards of earthquake-resistant houses. Public awareness and technical assistance are the keys to building earthquake-resistant communities.

Context of socio-cultural is one of the opportunities that must be used in building an earthquake-resistant community. By taking advantage of the golden window of opportunity; government, local non-governmental organizations, and local community groups must play an active role in building public awareness in obtaining earthquake-resistant housing and be involved in recovering the situation through construction. Culture in the form of communal mutual support becomes social capital to be able to speed up recovery through better recording of the number of affected people, determining locations in groups, and building housing reconstruction with group involvement.

On the other hand, the situation of Covid-19 pandemic has opened up opportunities for the community to contribute more in housing reconstruction. When the situation is difficult and it is more expensive to bring in skilled construction workers from area of Java island, the government and community have the opportunity to build local skilled workers by increasing their capacity to build earthquake-resistant houses.

7. Conclusion

Housing rehabilitation and reconstruction program in the area of Palu faced some problems and challenges in decision-making process, land provision, technical assistance in community, construction workers, and pandemic situation. The right community-based strategy and good collaboration between stakeholders are very important to lead the housing reconstruction program. Strong coordination between national and local governments is a force in restoring the situation after a multi-event disaster that has paralyzed the region and its government apparatus. On the other hand, efforts to build awareness of the importance of earthquake-resistant houses and live safely from disaster challenges require contributions from communities, activists, and the government by utilizing social capital in the community. The reconstruction program for safe and disaster-resilient housing should be an opportunity to build disaster-resilient communities with awareness and participation.

The authors wish to extend their gratitude and appreciation to the Institute of Research and Community Outreach of ITB who provided the grant for this research under the 2020 of P3MI Program. The author would also like to thank the Ministry of Public Works and Housing's Disaster Management Task Force: Mr. Arie Setiadi Moerwanto, Mr. Deddy Permadi, and in particular to Mr. Rezki Agung Hakiki for their assistances in collecting research data.

References

[1] Indonesia NDMAo. Losses and Damage from Disaster Impacts in Central Sulawesi Reached IDR 13.82 Trillion (in Indonesian): National Disaster Management Agency of Indonesia; 2018 [Available from: https://bnpb.go.id/berita/kerugian-dan-kerusakan-dampak-bencana-di-sulawesi-tengah-mencapai-1382-trilyun-rupiah
[2] UNOCHA. Central Sulawesi Earthquake Response Plan. Indonesia Humanitarian Country Team, United Nations Office for the Coordination of Humanitarian Affairs; 2018.
[3] Province RDMAoCS. Progress of the Post-Earthquake, Tsunami and Liquefaction Rehabilitation and Reconstruction in Central Sulawesi (in Indonesian). September 2020.
[4] Barenstein JED. Housing Reconstruction in Post-earthquake Gujarat: A Comparative Analysis. January 2006.
[5] Governor Regulation of Central Sulawesi Province No. 10 of 2019 for Post-Disaster Rehabilitation and Reconstruction Plans, (2019).
[6] Sadiqi Z, Trigunarsyah B, Coffey V. A framework for community participation in post-disaster housing reconstruction projects: A case of Afghanistan. International Journal of Project Management. 2017;35:900-12.
[7] Abe M, Ochiai C, Okazaki K. Is post-disaster housing reconstruction with participatory method effective to increasing people's awareness for disaster prevention? Procedia Engineering. 2018;212:411-8.
[8] Pribadi K, Kusumastuti D, Sagala S, Wimbardana R. Post-Disaster Housing Reconstruction in Indonesia: Review and Lessons from Aceh, Yogyakarta, West Java and West Sumatera Earthquakes. 2014. p. 197-223.
[9] Bilau AA, Witt E, Lill I. A Framework for Managing Post-disaster Housing Reconstruction. Procedia Economics and Finance. 2015;21:313-20.
[10] Bilau AA, Witt E, Lill I. Practice Framework for the Management of Post-Disaster Housing Reconstruction Programmes. Sustainability. 2018;10(11):3929.
[11] Davidson CH, Johnson C, Lizarralde G, Dikmen N, Sliwinski A. Truths and myths about community participation in post-disaster housing projects. Habitat International. 2007;31(1):100-15.
[12] Pribadi KS, Kusumastuti DI, Rildova, editors. Leaning from Recent Indonesian Earthquakes: An Overview to Improve Structural Performance2008.
[13] Gjerde M, de Sylva S. Governance and recovery: comparing recent disaster recoveries in Sri Lanka and New Zealand. Procedia Engineering. 2018;212:527-34.
[14] Ministry of Public Works and Housing GoI. Stakeholder Engagement Plan (in Indonesian). 2019.
[15] Presidential Order of The Republic of Indonesia No. 5 of 2018 for acceleration of post-earthquake and tsunami rehabilitation and reconstruction in Central Sulawesi province (in Indonesian), (2018).
[16] Pribadi K, Pradoto R, Hanafi E, Rasmawan IM. Lombok earthquake, one year later: housing sector recovery. E3S Web of Conferences. 2020;156:01012.

Comparative Analysis of Municipal Solid Waste to RDF Pretreatment Methods in Indonesia

I M W Widyarsana[1*] and D Saraswati[1]

[1]Department of Environmental Engineering, Faculty of Civil and Environmental Engineering, Bandung Institute of Technology, Jl. Ganesha 10, Bandung

*Corresponding author
Email: imwwresearch@gmail.com

Abstract. Municipal solid waste in Indonesia has a high potential to be used as an alternative energy source. One of the methods is by producing refuse-derived fuel (RDF) as coal substitution. Pre-treatment is needed to reduce high water content in domestic waste, especially in Indonesia, to produce high-quality RDF. This study compares and analyzes the optimal pre-treatment method to produce RDF from domestic waste based on each RDF product characteristic. The research reviewed and analyzed the data of RDF characteristics of various pre-treatment methods applied in Indonesia from pilot-scale experiments as primary data and previous research as secondary data. The methods compared are fermentation method, sun-drying, and bio drying. RDF from each method was tested to collect the proximate and ultimate characteristics data. Statistical analysis of RDF characteristics is carried out to determine whether there is a significant difference from different pre-treatment methods. This analysis shows no significant effect on the difference in the characteristics of the RDF. RDF from bio drying provides the highest average number of calorific values and lowest moisture content. The study concluded that the bio drying method is recommended in more waste treatment facilities in Indonesia to increase waste as an alternative energy resource.

Keywords: refuse-derived fuel, briquetting, pre-treatment, characteristics, waste to energy.

1. Introduction

Based on data from the Indonesia Central Statistics Agency, there is an increment in population growth of 1.2% per year [1]. This condition resulted in the rising waste generation in Indonesia, which is 67.8 million tons per year [2]. Most developing countries, including Indonesia, face this problem. Furthermore, some regions in Indonesia still apply the conventional method of waste management. Waste is considered dirty. It emits an unpleasant odor and causes disease vectors. It also reduces any comfort and gives a bad impression. Waste management in Indonesia also still refers to the 'collect-haul-dump' method and ends up in a landfill. The landfill has a limited-service period and cannot always accommodate waste generated from all people. Most landfills are operated by an open-dumping method, resulting in high polluting potential toward the environment. Several landfills are also not equipped with an appropriate facility such as drainage system, methane gas collection, and leachate management, which indicate a highly probable risk of polluting the environment [3]. Waste management improvement is needed to avoid environmental and social problems in later years [4].

On the other hand, the limited availability of fossil fuels as the main energy source has resulted in developing alternative energy sources to meet energy consumption. The implementation of waste-to-energy is expected to overcome this problem. Waste-to-energy (WtE) technology is a waste treatment that converts waste into alternative fuels to produce energy. Waste-to-energy transforms waste into energy or raw materials that are more beneficial, environmentally friendly [5], and significantly reduce waste volume. It is considered to solve waste problems quickly while at the same time producing energy resources to achieve a reduction in greenhouse gas emissions target by 29% in 2030 [6].

Most waste-to-energy technologies require some method of waste processing to maximize the performance and potential of energy generation. Therefore, the majority of waste-to-energy facilities are located near a material recovery facility or landfill. Waste is separated from unprocessed or inert materials and processed by specific technology or method based on its materials [7].

The biggest obstacles of municipal solid waste-to-energy treatment are the high-water content, heterogeneous shapes and sizes, and the mixed conditions of organic waste, plastic waste, and other materials in waste that made them difficult to be sorted. Direct utilization of waste in thermal installations such as incineration is hard to apply because of this high-water content condition [7]. Municipal solid waste in Indonesia has high-water content characteristics – above 50%, requiring a pre-treatment method to reduce its water content before being processed to produce energy [8]. This pre-treatment method aims to improve the quality of raw material by reducing the water content to produce high-quality briquette or refuse-derived fuel (RDF) as an alternative fuel and can qualify the requirements standards.

Various methods of solid waste drying have been applied in various countries to improve the quality of waste [9]. Drying methods that have been applied in waste treatment facilities in Indonesia, such as fermentation of waste at the local waste processing site (TOSS) Klungkung, Bali, and RDF (refuse-derived fuel) bio drying in Cilacap, Central Java.

2. Objective

Several previous studies have investigated the process, performance efficiency, and economic feasibility of various pre-treatment drying methods for improving the quality of solid waste from domestic activities, industry, and agriculture. However, studies on municipal solid waste (MSW) quality optimizing with different pre-treatment drying methods for municipal waste, particularly in Indonesia, have not been carried out lately. Therefore, this study highlights the comparison of drying methods for optimizing the quality of municipal solid waste (MSW) to be applied effectively, based on local conditions and available technology, and recommends the optimal pre-treatment method to apply in more waste facilities in Indonesia.

3. Materials and Method

This study reviews various methods of drying domestic waste, drying parameters, and RDF characteristics as the end product. Data required in this study are gathered using primary data and secondary data. Statistical analysis was carried out to compare different drying methods regarding the moisture reduction, drying process, and briquette as RDF output characteristics. Figure 1 below illustrates the method done in this study.

3.1. Drying Pre-treatment Method

Based on the literature reviews, there are several options for waste drying methods in Indonesia. The selected methods to analyze in this study were fermentation, bio drying, and sun drying. Each method will be explained as follows.

3.1.1. Waste Fermentation (Study Case: Klungkung, Bali). This drying method highlights the process of organic compounds degradation by microorganisms that produce heat so that the temperature increases and causes the water content in the waste to evaporate. Waste fermentation is a pre-treatment process of waste materials similar to microbiological food preservation that can produce energy [9]. The waste fermentation process has been applied at *Gema Santi* Local Waste Processing Site (TOSS) in Klungkung Regency, Bali Province. Local people also call this process *peyeumisasi* because it adapts an aerobic fermentation process in fermented cassava production, or people in Indonesia usually called *peuyeum*. The primary data was collected by fermentation pilot experiment.

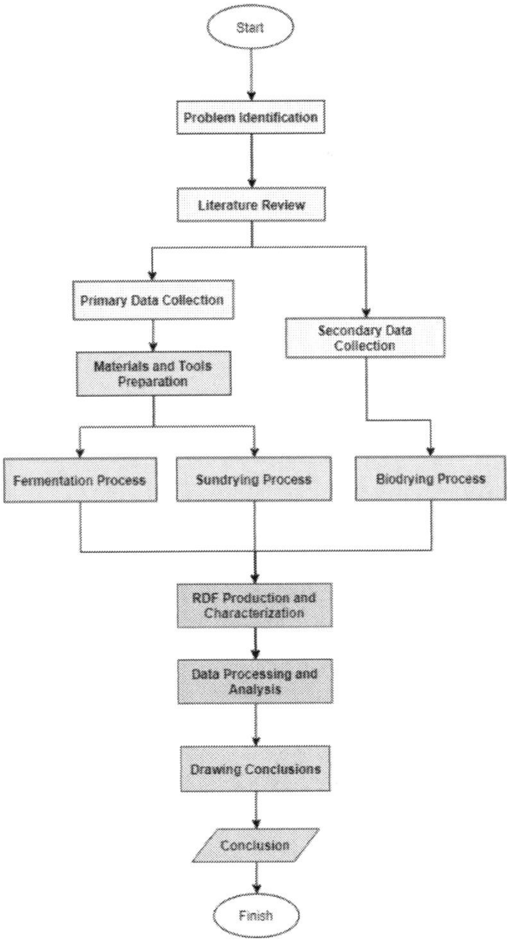

Figure 1. Research methodology flow diagram

Before being pre-treated, the waste must be sorted from any inert material such as gravel, ceramic, metal, glass, etc. The waste fermentation process used bamboo baskets with dimensions (80×80×80) cm^3. Subsequently, 25 kg of mixed fresh waste was placed in the baskets. The compaction process is carried out by trampling every layer of waste added to the basket while pouring it with bio activator liquid. Referring to the technical fermentation manual [10], the dose of bio activator used is 1 L of bio activator dissolved in 40 L of water for 1 ton of waste. Thus, 25 mL of bio activator was obtained dissolved in 1.25 L of water for 25 kg of waste. The experiment measured temperature and pH every day for at least ten days [10]. On the 10th day, the dry waste can be harvested and processed as briquettes or RDF. If the waste has reduced its moisture content, it will enter the next process. This waste fermentation mechanism method was applied and referred to the waste processing procedure at the waste treatment facility in Klungkung Regency, Bali Province [10].

3.1.2. Sun Drying. Sun-drying can be done by placing the waste in direct sunlight and using its heat to evaporate and reduce water content in waste [11]. From the literature review, no waste treatment facilities in Indonesia used sun-drying as a waste drying method. However, this method has been applied in other countries, such as Jordan and Egypt, to reduce moisture content in food waste. Both countries have similar conditions with Indonesia's weather conditions [12]. In Indonesia, the high intensity of sunlight is a suitable condition to apply waste drying using sunlight. Therefore, this study proposed sun drying as one of the feasible options for waste pre-treatment. This study applied a pilot experiment of 25 kg waste sun drying. Mixed fresh waste is placed inside the heating unit under direct sunlight. The sun heating process was done for 7-9 hours in two days. The experiment measured the fluctuation of heating temperature over two days of the drying process. This method referred to implementing the food waste drying process in Egypt and Jordan from previous research [12].

3.1.3. Waste Biodrying (Study Case: Cilacap, Central Java). The bio drying process aims to remove water content in organic waste with a high-water content by utilizing the heat generated from the aerobic degradation process by organic components in aeration [13]. Bio drying method for waste pre-treatment in Indonesia has been implemented at Jeruk Legi Landfill in Cilacap. Although this study did not perform any pilot experiment of this method, there was sufficient data to analyze using secondary data from the previous research [14]. The composition of the mixed fresh waste used in this process is similar to the waste sample used in the pilot-scale of fermentation method and sun-drying experiment. Fresh mixed waste is placed in a big pile and covered with a special fabric layer. An exothermic aerator will inject this pile to escalate biochemical reactions that can increase the temperature of the material in the range of 50 –70°C [14].

3.2. RDF Analysis
Dried waste from the fermentation and sun drying process is shredded and mixed with adhesive to be molded into briquettes as one of the RDF forms. This briquetting process refers to the technical guidance literature of biomass briquetting according to Njenga [15] and Temmerman [16]. RDF characteristics (proximate analysis) data from fermentation and sun drying methods are tested in the laboratory, while characteristics of bio drying RDF are collected from secondary data.

3.3. Data Analysis
Statistical analysis was performed to compare the RDF characteristics from analyzed methods. Each method is analyzed to determine the most feasible and effective to reduce the moisture content of waste. Independent T-Test analysis is carried out on the waste characteristics (proximate) between the various methods to determine whether the difference of method gives the significant difference in their RDF characteristics.

4. Result and Discussion
4.1. Waste Fermentation Method
The fermentation basket conditions did not emit any odor in observations that lasted ten days from the fermentation process. The final condition of waste produced is black and undergoes weight reduction. The average results of temperature and pH measurements in the fermentation process for ten days and weight reduction measurements are shown in the following table.

Table 1. Fermentation Methods Process Result

(Waste/Sample)	Initial Weight (kg)	Final Weight (kg)	Temp. Average (°C)	pH Average	Weight Reduction (%)
1	25.8875	15.31	51.16	5.99	40.87
2	25.39	15.18	45.97	5.88	40.23

(Waste/Sample)	Initial Weight (kg)	Final Weight (kg)	Temp. Average (°C)	pH Average	Weight Reduction (%)
3	24.63	15.18	44.86	6.02	38.36

The advantage of using this fermentation process is this process does not require any external energy. It only utilized microbiological energy to increase heat catalyzed by bio-activator liquid. This process also requires a more efficient land area because, in a 1 m^3 basket, it can process up to 100-200 kg of waste [10]. This process does not require any expert or skilled workers to operate the unit. However, the disadvantages of this method are they require a longer frequency of process, up to 10-12 days, to reach 40-50% of moisture content reduction.

After going through the pre-treatment process and briquettes production, the characteristics of the briquettes in the form of proximate analysis of moisture content are tested in the laboratory, for instance, ash content, volatile content, and fixed carbon in the laboratory. Table 2 shows characteristics of briquette with fermentation drying waste pre-treatment method.

Table 2. Laboratory Analysis Result of Briquette Characteristics from Fermentation Pretreatment Method

Sample	Moisture Content (%)	Volatile Content (%)	Ash Content (%)	*Fixed Carbon* (%)	Calorific Value (Kcal/kg)
Fermentation 1	33.47	33.47	66.53	21.85	2674.76
Fermentation 2	36.24	36.24	63.76	12.05	2633.62
Fermentation 3	34.92	34.92	65.08	11.24	2634.21
Average	**34.9**	**34.9**	**65.1**	**15.0**	**2647.5**

From comparing before and after the RDF production, there is a significant increase in the value of calorific waste value before and after briquetting. Based on raw waste characteristics, the calorific value of the waste sample is only about 1600 – 1700 kcal/kg. The increment of calorific value from RDF production can reach up to 27- 35%.

4.2. Sun Drying Method

From the sun-drying pilot experiment, the average process temperature ranged at 37° - 59°C. The highest temperature happened at approximately 12 am – 1 pm. Table 3 shows characteristics of briquette with sun drying waste pre-treatment method.

Table 3. Laboratory Analysis Result of Briquette Characteristics from Sun-drying Pretreatment Method

Sample	Moisture Content (%)	Volatile Content (%)	Ash Content (%)	*Fixed Carbon* (%)	Calorific Value (Kcal/kg)
Drying 1	28.13	36.85	63.15	11.82	2884.29
Drying 2	26.85	20.63	79.37	24.41	3009.66
Drying 3	37.42	11.01	88.99	19.23	2992.41
Average	**30.8**	**22.8**	**77.2**	**18.5**	**2962.1**

Compared to the fermentation method, sun-dried waste had lower moisture content. There is also a significant increase in calorific waste value before and after briquetting. Based on raw waste characteristics, the calorific value of the waste sample is only about 1600 – 1700 kcal/kg. The increment of calorific value from RDF production can reach approximately 30- 43%. This explanation can be concluded for only two days, and the sun drying method can give a similar final condition of waste as the fermentation method.

The disadvantage of the sun drying method is this method needs a more spacious area than fermentation. Because to optimize the sunlight exposure to the waste, the waste pile has to be placed in a thin layer — this condition cause it requires more area to treat more waste simultaneously.

4.3. Biodrying Methods

The experiment of sun bio drying was reviewed from research done by Paramita et al. [15]. This experiment investigates the effectiveness of bio-drying to increase the calorific value of waste. The result of the experiments is shown in Table 4 [15].

Table 4. Biodrying Methods Process Result

Sample	Initial Moisture (%)	Final Moisture (%)	Moisture Reduction (%)	Duration (days)	Calorific Value (Kcal/kg)
Biodrying 1	58	26	32	46	3686
Biodrying 2	58.2	20.3	37.9	41	3602
Biodrying 3	60	25.1	34.9	27	3026
Biodrying 4	58.2	21.5	36.7	27	4064
Biodrying 5	59.6	28	31.6	40	3435
Average	**58.8**	**24.2**	**34.6**	**36.2**	**3536**

4.4. Statistical Analysis

Research data of moisture content, volatile content, ash, and *fixed carbon* were analyzed using independency T-test analysis, as shown in Table 5. This analysis finds the effect of briquette characteristics from different pre-treatment methods. Table 5 describes a statistical analysis of briquette characteristics from fermented and sun-dried waste.

Table 5. The results of the independency T-test analysis of the effect of *pre-treatment* (fermentation or sun-drying) on the characteristics of briquettes

		Independence t-test for Equality of Means						
		t	df	Sig. (2-tailed)	Mean Difference	Std. Error Difference	95% Confidence Interval of the Difference	
							Lower	Upper
Moisture Content	Equal variances assumed	-1,343	6	**0,228**	-2,903	2,162	-8,193	2,387
	Equal variances not assumed	-1,343	5,960	**0,228**	-2,903	2,162	-8,201	2,396
Volatile Content	Equal variances assumed	1,673	6	**0,145**	10,634	6,355	-4,916	26,184
	Equal variances not assumed	1,673	3,936	**0,171**	10,634	6,355	-7,124	28,392
Ash Content	Equal variances assumed	-1,664	6	**0,147**	-3,995	2,401	-9,870	1,881
	Equal variances not assumed	-1,664	5,577	**0,151**	-3,995	2,401	-9,980	1,991

		Independence t-test for Equality of Means						
		t	df	Sig. (2-tailed)	Mean Difference	Std. Error Difference	95% Confidence Interval of the Difference	
							Lower	Upper
Fixed Carbon	Equal variances assumed	-0,563	6	**0,594**	-3,737	6,634	-19,970	12,496
	Equal variances not assumed	-0,563	4,631	**0,599**	-3,737	6,634	-21,207	13,733
Calorific Value	Equal variances assumed	-0,619	6	**0,559**	-151,872	245,487	-752,558	448,813
	Equal variances not assumed	-0,619	5,185	**0,562**	-151,872	245,487	-776,216	472,471

The statistical analysis results using independency T-test with a confidence level of 95%, the significance value of each characteristic parameter is shown in Table 5. The significance value is used to test the hypothesis. In this case, the hypothesis is determined as follows.

- H0: there is no significant difference in the characteristics of RDF with the variation in *pre-treatment*
- H1: there is a significant difference in RDF characteristics with the variation in *pre-treatment*. H0 is accepted if the significance value is > 0.05.

Table 5 shows the overall significance value of each characteristic parameter greater than 0.05. There is no significant difference in the characteristics from different waste pre-treatment. The difference in the waste pre-treatment method does not significantly affect the differences in the characteristics of the RDF.

4.5. *Optimal Methods Analysis*

By comparing the duration of heating and the calorific value of RDF produced, bio drying is the best method because it only requires a short drying time and results in the best characteristics of RDF. Based on the National Standard of Indonesia (SNI) 8966:2021 of Solid Recovery Fuel [17], there are several requirements of briquette quality below.

a. Briquette size
b. Non-organic percentage in raw material
c. Net calorific value
d. Chemical compounds: moisture, ash, volatile matter, sulfur, chlorine, mercury, potassium, sodium.
e. Hardgrove grindability index (HGI)
f. Ash fusion temperature (AFT)

Due to limited available data, this study only highlights the most important briquette characteristics: moisture content and calorific value. The average characteristic value of each pre-treatment method discussed is shown in Table 6 below.

Table 6. Average Characteristics Value of Compared Pretreatment

Briquette (Sample)	Final Moisture Content (%)	Calorific Value (Kcal/kg)
Fermentation	34,9	2647,5
Sundrying	30,8	2962,1
Biodrying	24,2	3536

Based on analysis results in Table 6 above, the best RDF characteristic is indicated by the average value from the bio drying process. We can conclude that bio drying gives the highest average calorific value and the lowest moisture content.

5. Conclusion

From the research results, it is concluded that there are various methods of pre-treatment to reduce the moisture content of waste with the difference of each performance and RDF characteristics result. RDF production may increase the calorific value up to 40% compared to fresh waste. The difference in pre-treatment treatment has no significant effect on the difference in the characteristics of the waste. The recommendation from this research is to determine the best methods of pre-treatment based on specific moisture content and characteristics of waste. The average value indicates the best RDF characteristic from the bio drying process because it gives the highest average number of calorific values and lowest moisture content. From this study, it can be proposed that bio drying is one of the recommended methods to apply in more waste treatment facilities in Indonesia to increase waste as an alternative energy resource.

6. Reference

[1] Indonesia Central Statistics Agency 2020 *Population in Indonesia* (Jakarta: Indonesia Central Statistics Agency

[2] Ministry of Environment and Forestry of Indonesia 2020 *Sistem Informasi Pengelolaan Sampah Nasional* (Jakarta: Ministry of Environment and Forestry of Indonesia)

[3] Widyarsana I M W, Damanhuri E, Agustina E and Aulia R N 2019 *Int. J. GEOMATE* vol 17 (Tsu: GEOMATE International Society) p 164

[4] Widyarsana I M W and Rahayu M A I 2021 *J. Mater. Cycles Waste Manag.* vol 23 (Berlin: Springer Verlag) p 2

[5] Confederation of European Waste-to-Energy Plants 2020 *What is waste-to-energy* (Bilbao: Confederation of European Waste-to-Energy Plants)

[6] Ministry of Energy and Mineral Resources of Indonesia 2020 *Laporan Capaian Kinerja Kelistrikan* (Jakarta: Ministry of Energy and Mineral Resources of Indonesia)

[7] Triyono B, Prawisudha P, Pasek A D and Mardiyati 2018 *Int. Conf. on Thermal Science and Technology* (Maryland: AIP Conf. Proc. American Institute of Physics) p 3

[8] Pasek A D, Gultom K W and Suwono A 2013 *J. Eng. Technol. Sci.* vol 45 (Bandung: Bandung Institute of Technology) 243

[9] Legino S, Hidayawanti R, Wirantika I 2019 *Sriwijaya Int. Conf. on Basic and Applied Science (Palembang)* vol 1282 (Bristol: IOP Publishing) p 1

[10] Djaja S D S 2019 *Enhanced Waste Management and Sustainable Material Management*

[11] Tun M M and Juchelkova D 2019 *Environ. Eng. Res.* vol 24 (Seoul: Korean Society of Environmental Engineers) p 529

[12] Hii C L and Law C L 2012 *Sun Drying: Fundamentals Applications and Innovations* (Singapore: The Centre of Food and Bio-product Processing Group)

[13] Tom A P Pawels R and Haridas A 2016 *Waste Manag.* vol 49 (New York: Elsevier Ltd.) p 64

[14] Paramita W, Hartono D M and Soesilo T E B 2018 *Environ. Earth Sci. vol* 159 (Bristol: IOP Publishing) p 1-6

[15] Njenga M 2014 *Charcoal briquettes production. A practical training manual* (Somalia: UN-Habitat Somalia)

[16] Temmerman M 2019 *Briquette Production Manual* (Gambia: The Climate Technology Centre & Network)

[17] National Indonesian Standart 2021 *SNI 8966:Bahan bakar jumputan padat untuk pembangkit listrik* (Jakarta: National Indonesian Standart)

7. Acknowledgment
This study is funded by Research, Community Service, and Innovation Program (P3MI) Air and Waste Management Research Group, Civil and Environmental Engineering, Bandung Institute of Technology.

Removal of Sulphate and Heavy Metals from Acid Mine Drainage using Permeable Reactive Barrier Technique

Rissa Anungstri[1,] , Himawan Tri Bayu Murti Petrus[1,2] , Agus Prasetya[1,2,*]**

1. Department of Chemical Engineering (Sustainable Mineral Processing Research Group), Faculty of Engineering, Universitas Gadjah Mada, Jalan Grafika No. 2 Kampus UGM Bulaksumur, D.I. Yogyakarta, 55281, Indonesia

2. Unconventional Geo-resources Research Group, Faculty of Engineering, Universitas Gadjah Mada, Jalan Grafika No. 2 Kampus UGM Bulaksumur, D.I. Yogyakarta, 55281, Indonesia

Corresponding author: *aguspras@ugm.ac.id **rissa.anungstri@mail.ugm.ac.id or rissa.anungstri@amman.co.id

Abstract. Containing sulphates and heavy metals, acid mine drainage (AMD) should be managed strategically to mitigate and control the migration of the contaminants to the downstream area. Conventional treatment techniques such as using lime to increase pH levels and metal precipitation or using imported material are usually inefficient and unsustainable. The AMD treatment investigated in this study uses the permeable reactive barrier (PRB) technique to enhance bacterial sulphate reduction and metal sulphide precipitation. The AMD treated is seepage water from industrial mining waste rock dump. This study aims to calculate the removal efficiency percentage of reactive materials to reduce contaminants in a batch test. Reactive materials used were organic waste generated locally i.e. domestic sewage sludge (SE), municipal compost (CO), cocopeat (CP), and the inorganic waste material is fly ash (FA) from a coal-firing power plant. A batch test was conducted in 56 days in an anaerobic chamber using nitrogen gas to support an anaerobic environment during subsampling. Mine water used in this test has a low pH level of 3.2, alkalinity (as CaCO3) < 1 mg/L, high sulphate 3280 mg/L, and contains Fe 46 mg/l, Al 54 mg/L, Cu 2.3 mg/L and Zn 3.4 mg/L. The test result at day 56 from using individual reactive material shows increased pH levels to 6.9; 5.6; 3.7; and 11.6 for sewage (SE), compost (CO), cocopeat (CP), and fly ash (FA), respectively. Alkalinity was increased to 1450 mg/L (SE), 323 mg/L (FA), 15 mg/L (CO), 1 mg/L for CP. The highest sulphate removal was measured in 85% from addition of FA. Sulphate removed from organic material reactor were 52% by (SE), 17% by (CO), 20% by (CP). %RE of dissolved metals (Al, Cd, Co, Mn, Ni, Fe, Cu, Zn), from SE reactor was 80%, CO reactor 80%, CP reactor 52%, FA reactor 94%. Oxidation-Reduction Potential (ORP) was measured to determine reducing conditions. ORP were measured at -551 mv, 255 mv, 156 mv, and -113 mv for SE, CO, CP and FA respectively. SE has the potential to remove metals and favour reducing conditions for sulphide precipitation at medium pH levels. Meanwhile, metal precipitation from addition of FA is mainly due to hydroxide precipitation at high pH levels. FA was able to decrease the most sulphate due to ion adsorption.

Keyword: Acid Mine Drainage, Fly ash, Organic Waste, Permeable Reactive Barrier

1. Introduction

The oxidation of sulphide minerals from waste rock dump releasing acid mine drainage (AMD) has low pH levels, high sulphate and heavy metal concentrations. Controlling AMD by source control combined with migration control will be more effective in managing the environmental impacts to ambient surface water or groundwater [1]. Permeable reactive barrier (PRB), consisting of reactive materials, is installed in the path of groundwater migration to treat contaminated groundwater. PRB is an alternative option in replacing the pump and treatment method. Lime treatment techniques are usually inefficient due to armoring effect [2]. Biological based treatments will result in high sulphate recovery and heavy metal reduction to treat AMD [3][4]. Using the PRB principle, anaerobic cells has been applied underground to treat AMD [5]. It can be combined with other treatments, such as wetland and reactive drainage [5] [4] [6]. PRB is should be specifically designed based on its contaminant quality, hydrological, geological speciation, construction, and reactive material selection [6]. Experiments using various reactive materials will be conducted using batch and column studies [7]. The selection of materials was carried out to consider their availability, cost and environmental sustainability [9]. This research is a batch study that focuses on the different reactivities of selected materials as a single source. The single batch study is carried out prior to conducting mixture batch and the column test. Organic substrates were locally obtained from a sewage treatment plant (SE), coco-net industry waste (CP), and mess hall compost (CO). While fly ash (FA), selected as an inorganic material, was collected from a coal-firing power plant. Target of removal efficiency (%RE) for sulphate in groundwater is 70% to achieve the standard of sulphate concentration based on class I water stipulated in Indonesian Government regulation number 22/2021 [10] while for metals reduction, targeted %RE for Mn, Fe and Cu are 92%, 50%, 76% respectively. Organic material acts as the electron donor and source for sulphate reducing bacteria (SRB) to reduce sulphates and favor metal sulphide precipitation. The sulphates reduction mechanism remediated by SRB activities according to the generalized reactions (1) and (2)[11] for inorganic and organic substrate. Reduction of SO_4 releases H_2S, HCO_3^-(or fully become CO_2), NH_3^+, and PO_4^- [12]. Cationic metal (Me) will be precipitated as sulphide (equation 3). Where, Inorganic substrate such as H_2 and CO_2,. Organic matter represents for $(CH_2O)_x.(NH_3)_y.(H_3PO_4)_z$ undergoing oxidation.

$$4\,H_2 + SO_4^{2-} \xrightarrow{SRB} H_2S + 2\,H_2O + 2\,OH^- \qquad \text{Inorganic substrate} \tag{1}$$

$$Organic\ matter + SO_4^{2-} \xrightarrow{SRB} HS^- + HCO_3^- + H_2O \qquad \text{Organic Substrate} \tag{2}$$

$$Me^{n+} + HS^- \rightarrow Me(S)_n + H^+ \tag{3}$$

In equations (4) and (5), under acidic pH of AMD, CaO in fly ash will undergo dissolution and release OH^- and Ca^{2+} will react with SO_4^- to form gypsum [13]. Cationic metal (Me), like Fe^{2+}, Al^{3+} and Mn^{2+}, will be precipitated as hydroxide (equation 6) [2].

$$CaO + H_2O \rightarrow Ca^{2+} + 2\,OH^- \tag{4}$$

$$Ca^{2+} + SO_4^{2-} \rightarrow 2H_2O + Ca.SO_4.2H_2O \tag{5}$$

$$Me^{n+} + OH^- \rightarrow Me(OH)_n \tag{6}$$

2. Materials and Methods

2.1 Properties of Acid Mine Drainage

AMD was collected from a seepage waste rock dump and placed in a refrigerator in 4° C until needed. AMD from the waste rock is chosen for its high sulphate and metal concentration that suits a short period batch reaction test to simulate the reaction in an aquifer. The characteristics of AMD used in this study as described in **Table 1**, AMD has low pH levels at 3.2; no alkalinity and high sulphate concentrations at 3280 mg/L. Fe dominates metal content in AMD with 46 mg/L, Mn with 26 mg/L, Zn with 3.4 mg/L, Cu with 2.52 mg/L and Al with 54 mg/L.

Table 1. Characteristic of Acid Mine Drainage from Waste Rock Dump

Parameter	Unit	Value
pH	Unit pH	3.2
Electrical Conductivity (EC)	us/cm	3637
Dissolved Oxygen (DO)	mg/L	1.4
Oxidation Reduction Potential (ORP)	mV	528
Alkalinity (as CaCO3)	mg/L	<1
Sulphate (SO_4^-)	mg/L	3280
Aluminium (Al)	mg/L	54
Arsenic (As)	mg/L	0.002
Barium (Ba)	mg/L	0.053
Boron (B)	mg/L	0.011
Cadmium (Cd)	mg/L	0.0138
Cobalt (Co)	mg/L	0.649
Copper (Cu)	mg/L	2.32
Chromium (Cr)	mg/L	0.001
Iron (Fe)	mg/L	46
Lead (Pb)	mg/L	0.003
Lithium (Li)	mg/L	0.01
Manganese (Mn)	mg/L	26
Molybdenum (Mo)	mg/L	0.0005
Nickel (Ni)	mg/L	0.102
Selenium (Se)	mg/L	0.01
Strontium (Sr)	mg/L	0.467
Zinc (Zn)	mg/L	3.4

2.2 *Properties of Organic Substrate*

Municipal sewage sludge (SE), mess hall compost (CO) and cocopeat (CP) were analyzed based on the dry weight. They were measured for pH, moisture content, elemental metal, total organic carbon (TOC) C, C/N, S, TOC, semi quantification of sulphate reducing bacteria (SRB) present in the substrate, and leaching characteristic [7]. Solid materials were digested using EPA 6020 method followed by metal analysis using *inductively coupled plasma* (ICP-MS) with APHA 3125 (2017) method, TOC analysis using a TOC analyzer with APHA-5310-B (2017) method, and SRB was estimated using SRB-BART Kits. Leaching tests of each organic substrate were carried out to determine the leachable component in Mili-Q water for dissolved metal, alkalinity, sulphate and dissolved organic carbon (DOC) [7].

2.3 *Chemical Properties of Fly Ash*

The chemical composition of coal fly ash is highly heterogeneous and dependent on the origin of the source coal and combustion system in a power plant. Fly ash was digested using the EPA method and analyzed for elemental metal concentration using APHA 3125 (2017) method, while its oxide content Al_2O3, Fe_2O3, Ca_2O_3, K_2O, MgO were analyzed using ASTM D4326 method.

3. Batch Test

3.1 Experimental Set Up

The batch test was conducted in a capped 1000 ml glass bottle (Figure 1). Each material was wet weighed 200 g for SE, 100 g for CO, 50 g for CP, 100 g for FA and then poured and mixed well with AMD in the bottle. If corrected to dry mass, the ratio between SE, CO, CP to AMD was 0.03. The mixture purged with nitrogen gas for 40 minutes to remove dissolved oxygen from the solution. All bottles were kept in the dark place until sampling and measurement. No inoculum bacteria were used in this experiment. In addition, one bottle full of AMD was prepared for control.

Figure 1. Batch Test Reactor

3.2 Sampling and Analysis

Sampling and sample preparations were conducted in an anaerobic chamber using Aldrich Atmost bag. Prior to working in the chamber, N_2 gas was purged until dissolved oxygen (DO) concentration measured inside the chamber was less than < 0.5 mg/L (Figure 2). Water samples were measured to determine the temperature, pH level, electrical conductivity (EC), oxidation-reduction potential (ORP), and dissolved oxygen (DO) using YSI multimeter professional plus on days 0, 1, 2, 7, 14, 21, 28 and 56 which was calibrated each day of sampling.

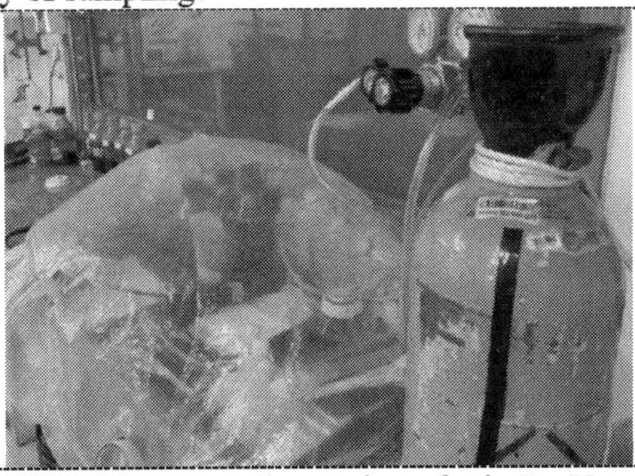

Figure 2. Sampling and Analysis Set Up

Sampling was taken on days 0, 7, 14, 28 and 56 for alkalinity, SO_4^+, and dissolved metals (Al, As, B, Ba, Ca, Cd, Co, Cr, Cu, Fe, K, Pb, Li, Mg, Mn, Hg, Mo, Ni, Na, Se, Si, Ag, Sr, and Zn). For H_2S, TOC and NH_3 analysis, samples were taken on days 0 and 56. At the end of treatment, the residual sediment was taken from each reactor and froze until analysis. The pH was determined based on APHA 4500-H+ (2017) method, APHA 2500 (21017) for EC, APHA 4500-O (2017) for DO, and APHA 2580 for ORP. Unfiltered samples were analyzed for Alkalinity, SO_4^-, NH_3, TOC and H_2S. Alkalinity was measured using APHA 2320-B (2017). H_2S sample is preserved with NaOH and Zn Acetate and analyzed using ion selectivity method, while NH_3N and TOC samples are preserved with H_2SO_4 and analyzed with APHA 4500 NH_3 (2017) for NH_3N. TOC, Cl^- and SO_4^- were measured using spectrophotometry method. Dissolved metal samples were filtered using a Nalgene filter set and Millipore filter of 0.45 μm and acidified with HNO_3 1:1. The metal analysis was conducted using ICP-MS APHA-3125-B (2017) method, The sediment underwent an acid digestion using EPA Method 6020 and followed by elemental analysis same with analytical methods for water analysis.

3.3 Removal Efficiency

Removal efficiency of batch treatment is calculated using equation (7).

$$\% RE = \frac{\text{Cout}-\text{Cin}}{\text{Cin}} * 100\% \tag{7}$$

Where, % RE is the percentage of removal, C_{out} = final concentration of reaction (mg/L) and C_{in}= initial AMD concentration (mg/L).

4. Result and Discussion

4.1 Organic Substrate

From **Table 2**, SE has the lowest C/N ratio of 5 among other organic materials, which has 18 and 97 for CO and CP, respectively indicating excess nitrogen over carbon. Although, the ideal C/N ratio for organic degradation by bacteria is 10 [4], Nevertheless, only observing the C/N ratio does not correctly show the potential carbon degradation capacity [7]. Among other organics, SE has a relatively higher total metal concentration of As, B, Co, Cr, Cu, Pb, Mn, Ni and Zn and will possibly be leached out during treatment.

Table 2 Main Composition of Reactive Materials

Parameter	Unit	SE	CO	CP	FA
pH	-	7,2	7,6	5,5	12,3
Moisture	%	83	74	45	<10
Carbon (C)	%	22	31	44,49	<0,01
Nitrogen total (N)	%	4,4	1,7	0,46	<0,01
C/N	-	5	18	97	
Total Organic Matter (TOM)	%	16	57	29	<0.01
Phosphor (P)	mg/kg	14200	506	2710	197
Sulfur (S)	mg/kg	13600	380	4060	2200
Cation Exchange Capacity	0.01 meq/100g	6.03	10.2	18.2	0.68
Calcium (Ca)	mg/kg	32900	5650	1510	43500
Magnesium (Mg)	mg/kg	4090	2840	943	17000
Sodium (Na)	mg/kg	1050a	603	772	1140[a]
Potassium (K)	mg/kg	3060	15700[a]	18600	1040
Aluminum (Al)	mg/kg	3920	2790	240	10200[a]
Arsenic (As)	mg/kg	4	2	0,1	11[a]
Boron (B)	mg/kg	80*	19	16	188[a]
Barium (Ba)	mg/kg	57	22	2,6	150
Cadmium (Cd)	mg/kg	0.75	0.14	0.07	0.11
Cobalt (Co)	mg/kg	2.2	4.7	<0.5	16.8[a]
Chromium (Cr)	mg/kg	12.1	4.2	0.5	18.8[a]
Copper (Cu)	mg/kg	283	103	45.1	14.6
Iron (Fe)	mg/kg	4940	9500	689	26100
Lead (Pb)	mg/kg	13.0[a]	3.4	2.1	4.7
Manganese (Mn)	mg/kg	290[a]	160	120	320a
Nickel (Ni)	mg/kg	14.1	2.9	1.4	26.2[a]
Selenium (Se)	mg/kg	6.4	<0.5	<0.5	0.8
Strontium (Sr)	mg/kg	108[a]	36	10	170[a]
Zinc (Zn)	mg/kg	592[a]	64	8	34

[a]significant amount or highest value among other material

According to **Table 3** leaching test characteristics to organic material using Mili-Q water indicates that some metals, such as As, Cu, Ba, B, Fe, Pb, Mn, Sr, and Zn have the potential to leach out; Alkalinity was generated from SE and CO but not from CP, and SE also generates sulphate. From the indicative test for SRB presence in organic material, SE contained 325 cfu/ml SRB with moderate aggressivity while low SRB presence in CO and CP material.

Table 3. Leachate Characteristic of Organic Material

Parameter	Unit	SE	CO	CP
pH	pH unit	7.43	6.0	6.0
Alkalinity	mg/L	48	51	4
Sulphate (SO_4^-)	mg/L	14	<5	<5
Dissolved Organic Carbon (DOC)	mg/L	4	24	33
As	mg/L	0.0030	0.0030	< 0.0005
Ba	mg/L	0.004	0.012	0.003
Cu	mg/L	0.003	0.054	0.002
Sr	mg/L	0.049	0.025	0.004
Zn	mg/L	0.013	0.042	0.009
SRB population	cfu/ml	325	20	-
	Aggressivity	moderate	low	-

4.2 Fly Ash Properties

Fly ash used in this test contains 22% CaO and a sum of SiO_2, Al_2O_3 and Fe_2O_3 is 61.4%, which classifies fly ash as a class C ASTM, that if exposed to water, will be hardened within 45 minutes [16]. With a pH of 12.3, fly ash also has corrosive properties. Metallic contents such as Al, B, Ba, Ca, Cr, Fe, Mn, Mg, Ni and Na are the highest compared to SE, CO and CP.

4.3 pH, DO, Conductivity and ORP

From Figure 3,Figure 4 all reactors except CP showed a sharp increase in pH levels, which occurred from the first two weeks and then flattened or slightly increase from day 28th to 56th. The SE reactor provides a pH increase of up to 6.9, while the CO reactor can increase the pH of AMD to 5.6. The increase in pH in SE and CO reactors corresponds to equations (1) and (2) influenced by SRB activities as electron acceptors. The pH value of treated water from the FA reactor rose from 3.2 to 11.6. The increase in pH from the FA reactor is caused by CaO on the surface of FA, which is highly soluble in releasing OH^-, as stated in equation (4)

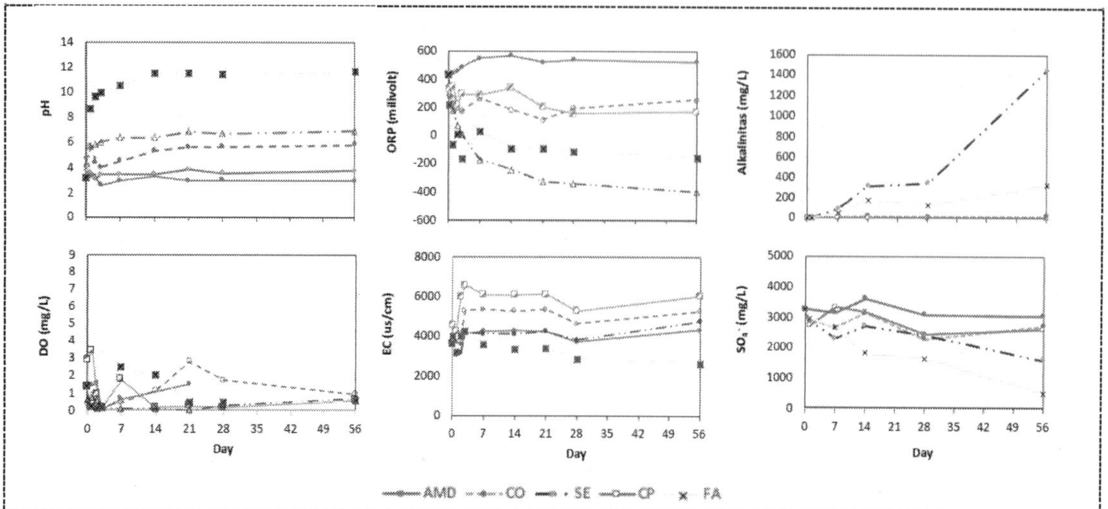

Figure 3. Value of pH, DO, ORP, EC, SO_4^-, and Alkalinity of Each Reactor during Batch Test.

The ORP on the SE reactor reached -200 mV on day 7th and reached -396 mV on day 56th. Eh value between -150 to -200 mV is the primary metabolic condition for SRB in reducing sulphate [17] [8]. CO and CP reactors can only lower the ORP but not until the reducing condition (ORP negative), although CP did supports anaerobic condition [18]. This different reducing process and anaerobic condition remediated by bacteria on SE reactor consistent with higher availability of SRB on SE material which also more aggressive than in CO and CP material (Table 3).

Table 4. Physical Characteristic of Treated Water after 56 Days

Parameter	Unit	AMD	Reactor SE_D56	Reactor CO_D56	Reactor CP_D56	Reactor FA_56D
pH	Unit pH	3.2	6.9	5.6	3,7	11.6
EC	us/cm	3637	4763	5236	6029	2640
DO	mg/L	1.4	0.7	0.6	0.6	0.5
ORP	mV	528	-396	255	168	-156
Eh [a]	mV	596	-328	329	242	-87

a. Eh = Eh (mv) = ORP measured + (-0.716*T + 220) (Groundwater Geochemistry and Isotopes, Ian Clark, 2015)

The lowest dissolved oxygen concentration (DO) is found in the SE reactor. DO in other bottles are slightly above 0.5 mg/L which indicates a decreasing anaerobic condition [19], but DO in AMD is still in the range of the actual field condition. Major anion and cation concentrations (Ca+, Na+, Mg^{+}, K^+, $SO4^{2-}$, HCO^{3-}, OH^-, Cl^-) reflects to the total dissolves value. The Increase of EC in organic reactors are due to some cation release from material. However, in the FA reactor, decreasing EC indicates ion adsorption in precipitates [13].

4.4 Alkalinity, Sulphate, H_2S, Ammonia and TOC

The SRB activities on reducing conditions strengthen by the data from SO_4^-, alkalinity and H_2S concentration. On the SE reactor, SO_4^- could be reduced from 3280 mg/L to 1580 mg/L, with %RE reached 52%, the alkalinity increasing to 1450 mg/L, with the presence of 0.653 mg/L H_2S on Day 56th, high NH_3N (73 mg/L) was also detected. Compared with %RE SO_4^- from CO and CP reactor that only

17% and 20% respectively, and producing low alkalinity (15 and 1mg/L) with H_2S detected only <0.001 mg/L and ammonia <1 mg/L (**Table 5**).

Table 5. Final Concentration of Alkalinity, Sulphate, H_2S, and NH_3^+ In Batch Test

Parameter	Unit	AMD	Reactor SE_D56	Reactor CO_D56	Reactor CP_D56	Reactor FA_56D
Alkalinity	mg/L	1	1450	15	1	322
Sulphate (SO_4^-)	mg/L	3280	1580 (52%)[a]	2730 (17%)	2610 (20%)[a]	489 (85%)[a]
H_2S	mg/L	<0.001	0.653	<0.001	<0.001	<0.001
Ammonia (NH_3^+)	mg/L	0.35	73	0.63	0.30	0.53
TOC	mg/L	<1	162	25	43	2

a. In bracket is %RE value.

Contrary to the decrease of SO_4^- in the FA reactor to 489 mg/L or %RE was 85%, medium alkalinity was generated up to 322 mg/L, indicates chemical precipitations at high pH and the deposition of sulphate caused by ionic adsorption $FeOH_3$ and $AlOH_3$ [13]. Increasing alkalinity in the FA reactor was mainly in the form of OH^- according to equation (5) and SiO_2 decay. TOC concentration reflects total organic carbon content in water. SE produces highest TOC, 162 mg/L TOC at the end of treatment. The CP reactor produced a higher TOC concentration than the CO reactor (43 over 25 mg/L), but the % RE of SO_4^- for both CO and CP reactors were similar (20% and 17%). This fact was due to the high lignin content in CP [20], which Carbon will be degraded more slowly than Carbon from CO [20]. The profile of changes in significant ion concentrations within 56 days showed in Figure 4. There were indications of cation exchange processes in which K, Na & Cl arose compared to the initial concentration of AMD water. Elevated values from them were obtained from leached-off reactive material.

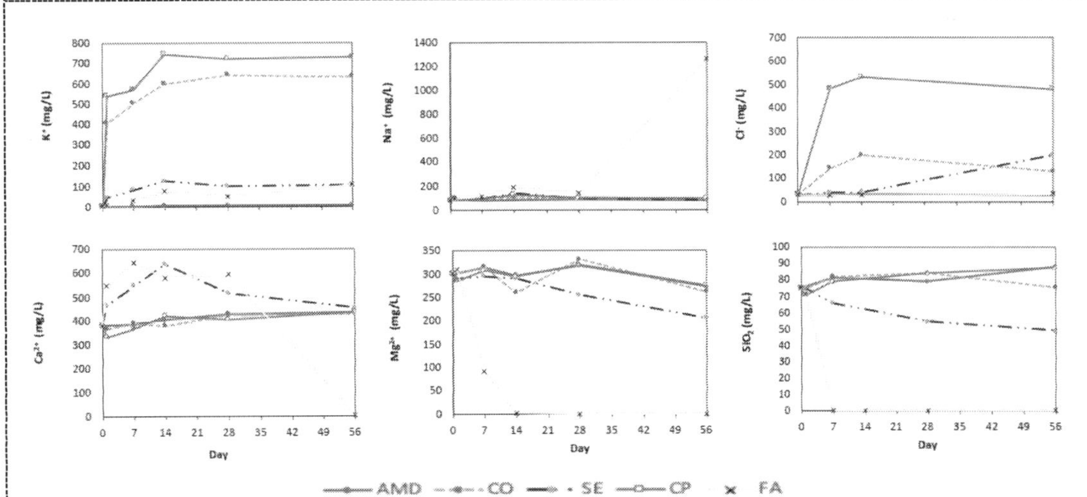

Figure 4. Concentraion changes of K^+, Na^+, Cl^-, Ca^{2+}, Mg^{2+}, SiO_2 for Each Reactor during the Batch Test

4.5 Dissolved Metal

4.5.1 Removal Efficiency

From Table 6, The main mechanism of dissolved metals removal on organic reactor occurred within 3 hours of AMD treatment, which reached an average %RE of 41%, 45%, 34% respectively. Over a period of 56 days the efficiency reached 80%, 80%, 52% and 94%.

Table 6. Removal Efficiency Percentage in of Batch Test

Parameter	AMD (mg/L)	Reactor SE %RE		Reactor CO %RE		Reactor CP %RE		Reactor FA %RE	
		H-3 [a]	D-56 [b]	H-3	D-56	H-3	D-56	H-3	D-56
As	0.002	35%	-650%	45%	-55%	60%	-85%	-50%	88%
Al	54	62%	100%	70%	100%	52%	73%	78%	99%
Cd	0.0138	49%	100%	42%	86%	26%	11%	6%	98%
Co	0.649	29%	100%	36%	81%	22%	16%	9%	100%
Cu	2.32	64%	100%	83%	100%	79%	90%	48%	100%
Ni	0.102	33%	81%	47%	91%	26%	14%	6%	100%
Zn	3,4	39%	100%	34%	84%	22%	44%	17%	100%
Cr	0.001	50%	0%	50%	50%	0%	50%	-800%	-30000%
Fe	46.3	58%	100%	70%	100%	59%	-3%	43%	100%
Pb	0.003	67%	83%	-33%	83%	-100%	83%	-100%	83%
Li	0.01	10%	30%	10%	-30%	0%	-80%	-80%	-970%
Mn	26	28%	84%	20%	15%	16%	-6%	6%	100%
Mo	0.0005	0%	-100%	0%	-500%	-300%	-300%	-900%	-27900%
Se	0.01	57%	-9%	79%	89%	77%	85%	-18%	-1%
Sr	0.467	-91%	-315%	-59%	-258%	3%	-104%	-93%	-168%
B	0.053	-96%	-96%	-238%	-521%	-109%	-334%	-2334%	64%
Ba	0.011	-82%	-91%	-709%	-573%	-55%	-618%	-400%	-91%
Elements Removed		14	11	14	11	13	9	8	11
Elements Increasing		3	6	3	6	4	8	9	6
% RE average		41%	80%	45%	80%	34%	52%	27%	94%

a. H-3 : 3 hours after mixing
b. D-56 : day 56 after mixing
c. %RE targeted for Mn, Fe and Cu are 92%, 50%, 76%.

From Figure 5, dissolved metals such as Al, Cu, Cd, Co, Fe, Mn, Ni, Zn show a decreased concentration While the decrease of metals concentration in FA is related to the high pH level and hydroxide deposits [2], the decline of metals in SE occurs at a moderate pH, with the primary control mechanism being the deposition sulfide at ORP < -150 mv. Metal removal at moderate pH level depends on anaerobic condition (ORP negative), organic carbon dan SRB [22]. Otherwise, the metal reduction mechanism in the CO and CP reactor is indicated more by the ion adsorption process[19].

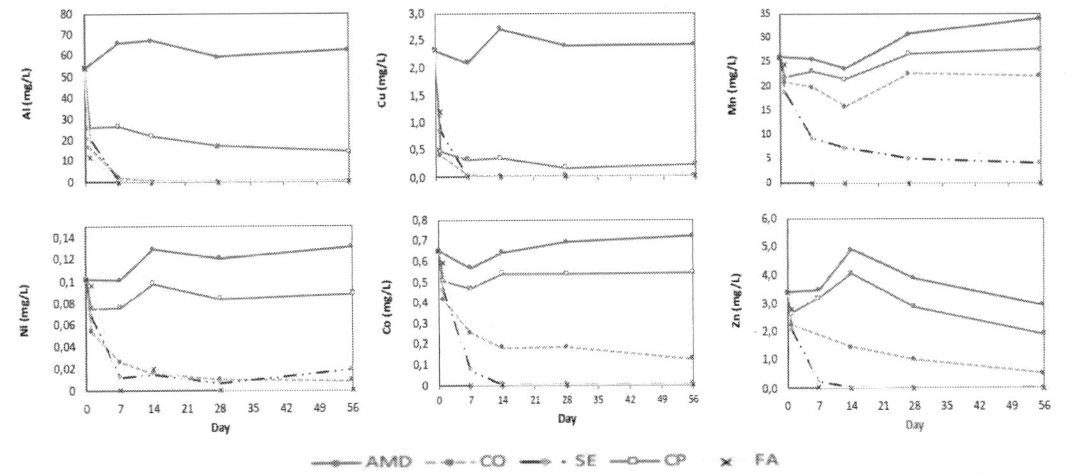

Figure 5. Concentration changes of of Al, Cu, Mn, Ni, Co, and Zn in Treated Water Concentration For Each Reactor in the Batch Test

4.5.2 *Leached Metals*

Some elements were leached out from the reactive materials, which were indicated by higher concentrations in AMD treated water and showed negative results on the %RE calculation (Table 6). From Figure 6, At FA reactor, Ba, Se and Sr were leached out from fly ash at pH <11or less than 14 day but then precipitated in the end, but Cr, Li & Mo continued to leach during treatment. At CO Reactor, As, Li, Mo, Se, B, and Ba were elevated. At the SE reactor, As Mo, Se, Sr, B and Ba. At CP reactor, As, Fe, Li, Mn, Mo, Se, Sr, B, Ba cannot be reduced during treatment.

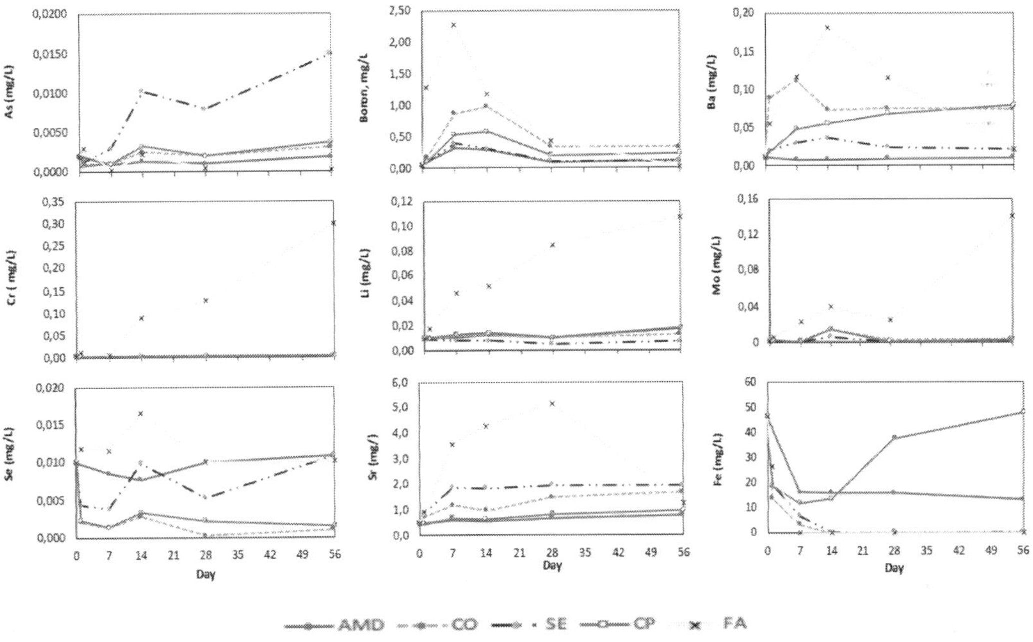

Figure 6. Concentration changes of As, B, Ba, Cr, Li, Mo, Se, Sr and Fe in in the Bacth Test

From 17 dissolved metals measured (Al, As, B, Ba, Ca, Cd, Co, Cr, Cu, Fe, Pb, Li, Mn, Ni, Se, Sr, and Zn), the SE reactor managed to reduce the concentration of 11 elements. While the CO decreased 11, the CP reactor lowered 9 elements, and the FA reactor reduced 11 elements.

4.6 Sediment Quality

On the 56th day, the batch tests were ended. The floating and sedimental in CP reactor deposits stayed light brown as the original material. The SE reactor has greyish water and black and foul-smelling sediment, indicating intense bacteria and hydrogen sulphide gas activities. Sediments in FA have a thin white gypsum layer above a grey hardened cement-like material due to the pozzolanic nature of FA. CO deposits were unchanged black, and odorless. In a glass bottle containing only AMD, there is a thin orange layer inside the glass bottle that indicates the deposition of Fe.

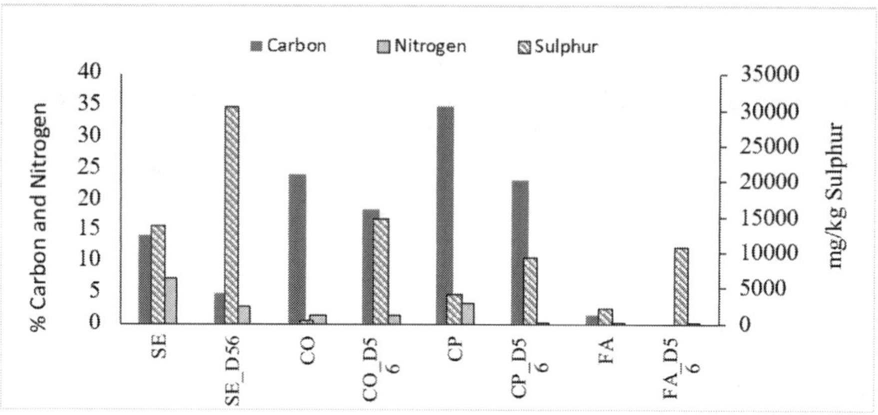

Figure 7. Concentration of Carbon, Nitrogen and Sulphur after Treatment AMD with Sewage Sludge (SE), Compost (CO), Cocopeat (CP) and Flyash (FA).

From Figure 7 the sewage sludge lost 65% of C during the reaction and 62% of N and resulted in 124% S from SE material precipitated in sediment during treatment. At CO reactor, 23% C was used, 11% of N, and added 3768% of S in sediment. For CP reactor, 34% C was used, 93% of N, and -130% S. While in FA, S added to sediment was calculated as 391%. SE could be categorized as fast reactive material, CO is a medium reactive material, and CP is low reactive material.

5. Conclusion

The sulphate and metal removal capability at moderate pH levels remediated by SRB is only favored in SE reactor while CO and CP did not support anaerobic reducing condition due to the low availability of SRB that exist in the original material. The metal removal mechanisms in the FA reactor occurred at high pH while sulphate removal is due to ion adsorption on hydroxide precipitates. From the four studied materials, every one of them can lower AMD's primary contaminants, namely Sulphate, Al, Cd, Co, Cu, Pb, Ni, and Zn. Although, some elements will be leached out from each of them. Sulphate reduction is expected to be 98% by FA, 52% by SE, 17% by CO, and 20% by CP. If compared to targeted 70% removal sulphate in groundwater, only FA and SE reactor met the target. The average %RE from dissolved metals (Al, Cd, Co, Mn, Ni, Fe, Cu, Zn), from SE reactor was 80%, 80% from CO reactor, 52% from CP reactor, and 94% from FA reactor. If compared to %RE target for main contaminant Fe, Mn and Cu, only FA reactor has successfully achieved the target of metals removal. Some adjustments need to be considered to achieve the target %RE in the following study. They are including additional of SRB source, the composition of the mixtures based on organic carbon release characteristics (fast, medium and slow) to sustain longer carbon sources to SRB in the barrier, amount of fly ash added to system, and also combined treatment possibility. The right amount of fly ash needs to be determined to be just enough to increase pH for metal sulphide precipitation and favor the optimum life of SRB, without producing more leached metals into the environment.

.

Acknowledgments
The Authors acknowledges the full support and funding from PT Amman Mineral Nusa Tenggara in treatability test before implementation of passive treatment using PRB principle on-site and special assistances and support from colleagues and Laboratory technical staffs.

References

[1] D. B. Johnson and K. B. Hallberg, "Acid mine drainage remediation options: A review," *Science of the Total Environment*, vol. 338, no. 1-2 SPEC. ISS., pp. 3–14, Feb. 2005, doi: 10.1016/j.scitotenv.2004.09.002.
[2] V. Vadapalli, L. Felicia Petrik, and W. Gitari, "Neutralization of Acid Mine Drainage using Fly Ash and Strength Development of the Resulting Solid Residues." [Online]. Available: www.eskom.ac.za/annreport
[3] K. Rambabu, F. Banat, Q. M. Pham, S.-H. Ho, N.-Q. Ren, and P. L. Show, "Biological remediation of acid mine drainage: Review of past trends and current outlook," *Environmental Science and Ecotechnology*, vol. 2, p. 100024, Apr. 2020, doi: 10.1016/j.ese.2020.100024.
[4] K. K. Kefeni, T. A. M. Msagati, and B. B. Mamba, "Acid mine drainage: Prevention, treatment options, and resource recovery: A review," *Journal of Cleaner Production*, vol. 151. Elsevier Ltd, pp. 475–493, May 10, 2017. doi: 10.1016/j.jclepro.2017.03.082.
[5] S. M. Doshi, "Bioremediation of Acid Mine Drainage Using Sulfate-Reducing Bacteria," 2006. [Online]. Available: www.clu-in.org
[6] E. J. Clyde, P. Champagne, H. E. Jamieson, C. Gorman, and J. Sourial, "The use of a passive treatment system for the mitigation of acid mine drainage at the Williams Brothers Mine (California): Pilot-scale study," *Journal of Cleaner Production*, vol. 130, pp. 116–125, Sep. 2016, doi: 10.1016/j.jclepro.2016.03.145.
[7] A. K. Thakur, M. Vithanage, D. B. Das, and M. Kumar, "A review on design, material selection, mechanism, and modelling of permeable reactive barrier for community-scale groundwater treatment," *Environmental Technology and Innovation*, vol. 19. Elsevier B.V., Aug. 01, 2020. doi: 10.1016/j.eti.2020.100917.
[8] Itrc, "Prepared by The Interstate Technology & Regulatory Council PRB: Technology Update Team," 2011. [Online]. Available: www.itrcweb.org.
[9] I. Moodley, C. M. Sheridan, U. Kappelmeyer, and A. Akcil, "Environmentally sustainable acid mine drainage remediation: Research developments with a focus on waste/by-products," *Minerals Engineering*, vol. 126. Elsevier Ltd, pp. 207–220, Sep. 01, 2018. doi: 10.1016/j.mineng.2017.08.008.
[10] "REPUBLIK INDONESIA."
[11] A. Luptakova and E. Macingova, "Alternative substrates of bacterial sulphate reduction suitable for the biological-chemical treatment of acid mine drainage," 2012.
[12] D. W. Blowes, C. J. Ptacek, S. G. Benner, C. W. T. McRae, T. A. Bennett, and R. W. Puls, "Treatment of inorganic contaminants using permeable reactive barriers," *Journal of Contaminant Hydrology*, vol. 45, no. 1–2, pp. 123–137, 2000, doi: 10.1016/S0169-7722(00)00122-4.
[13] S. Saha and A. Sinha, "A review on treatment of acid mine drainage with waste materials: A novel approach," *Global Nest Journal*, vol. 20, no. 3. Global NEST, pp. 512–528, Dec. 01, 2018. doi: 10.30955/gnj.002610.

[14] G. J. Zagury, V. I. Kulnieks, and C. M. Neculita, "Characterization and reactivity assessment of organic substrates for sulphate-reducing bacteria in acid mine drainage treatment," *Chemosphere*, vol. 64, no. 6, pp. 944–954, 2006, doi: 10.1016/j.chemosphere.2006.01.001.
[15] Y. Vasquez, M. C. Escobar, C. M. Neculita, Z. Arbeli, and F. Roldan, "Selection of reactive mixture for biochemical passive treatment of acid mine drainage," *Environmental Earth Sciences*, vol. 75, no. 7, Apr. 2016, doi: 10.1007/s12665-016-5374-2.
[16] · Book, "Pemanfaatan Fly Ash dan Bottom Ash untuk Pengelolaan Batuan dan Air Asam di Tambang Batubara Environment Indicators on Sustainability Report of Mining Industries View project Candra Nugraha," 2021. [Online]. Available: https://www.researchgate.net/publication/355586409
[17] O. Gibert, J. de Pablo, J. L. Cortina, and C. Ayora, "Treatment of acid mine drainage by sulphate-reducing bacteria using permeable reactive barriers: A review from laboratory to full-scale experiments," *Reviews in Environmental Science and Biotechnology*, vol. 1, no. 4. pp. 327–333, 2002. doi: 10.1023/A:1023227616422.
[18] P. K. Mondal *et al.*, "Evaluation of peat and sawdust as permeable reactive barrier materials for stimulating in situ biodegradation of trichloroethene," *Journal of Hazardous Materials*, vol. 313, pp. 37–48, Aug. 2016, doi: 10.1016/j.jhazmat.2016.03.049.
[19] G. Yim, S. Ji, Y. Cheong, C. M. Neculita, and H. Song, "The influences of the amount of organic substrate on the performance of pilot-scale passive bioreactors for acid mine drainage treatment," *Environmental Earth Sciences*, vol. 73, no. 8, pp. 4717–4727, Apr. 2015, doi: 10.1007/s12665-014-3757-9.
[20] M. A. Tanchuling, A. C. Resurreccion, and D. Ong, "A RESEARCH PROJECT REMOVING HEAVY METALS FROM WASTEWATER OF SMALL-SCALE GOLD MINERS OF CAMARINES NORTE (PHILIPPINES) USING COCO-PEAT AS SORBENT MATERIAL."
[21] J. Skousen, "Review of Passive Systems for Acid Mine Drainage Treatment", doi: 10.1007/s10230-016-0417-1/Published.
[22] C. D. Church, R. T. Wilkin, C. N. Alpers, R. O. Rye, and R. B. Mccleskey, "Microbial sulfate reduction and metal attenuation in pH 4 acid mine water," 2007, doi: 10.1186/1467-4866-8-10.
[23] H. Prokkola, E. T. Nurmesniemi, and U. Lassi, "Removal of metals by sulphide precipitation using na2s and hs−-solution," *ChemEngineering*, vol. 4, no. 3, pp. 1–10, Sep. 2020, doi: 10.3390/chemengineering4030051.

The Impact of Pavement Distress to Vehicle Maneuver in Bandung City

Jongga Jihanny[1], Kardina Nawassa Setyo Ayuningtyas[1], Anju David[1], Firnandia Zahwa Salsabila[1], Murti Kamajaya[1], Difa Salmadhia[1]

[1]Civil Engineering Department, Faculty of Civil and Environmental Engineering, Institut Teknologi Bandung, Indonesia

*kardina@ftsl.itb.ac.id

Abstract. Pavement distress can interfere with the vehicle's maneuver, causing the vehicle to reduce its speed and change lane or do an evasive maneuver to avoid it. These behaviors could impact traffic conflict significantly. This study aims to identify the impact of pavement distress on vehicle maneuver, specifically the speed and evasive maneuver of the vehicle when approaching poor pavement conditions. The results show differences in responding to pavement distress for each type of vehicle. Motorcycles tend to maneuver or change lanes to avoid the poor pavement condition and it may be because of its dimension, while heavy vehicles tend to reduce the speed while confronting the pavement distress. Besides, light vehicles tend to maneuver, reduce their speed, and do a combination of those responses. From this study, we can investigate that it could be dangerous if the vehicle takes an opposite lane while there is a vehicle on the opposite lane. Some vehicles reduce their speed while confronting poor pavement thus they could be more careful on roads, nevertheless, in the rainy season, it could be dangerous, especially when the pothole is covered with rainwater. The improvement of road conditions is vital to enhance travel time and reduce obstacles on road.

Keywords: Pavement distress, Potholes, Vehicle Maneuver, Speed

1. Introduction

Road crashes are considered a global problem and efforts are increasing to enhance road safety to avoid their occurrences. One of the attributes influencing road safety is surface condition on roads [1]–[6] Pavement condition parameters including skid resistance, roughness, and rutting are the most common indicators of road safety problems [7]–[9]. In general, a number of studies have shown that an improvement of pavement condition can have a considerably impact on safety benefits on roads and related to ride quality [3], [10], [11]. Furthermore, other studies show that poor pavement condition was associated with proportionally more severe crashes, nonetheless, the very poor pavement condition was associated with less severe crashes. Very good pavement condition might induce speeding behavior, therefore could have caused more severe crashes [3]. Pavement deterioration could causes several losses including increased travel time, congestion, pollution, and road crash [12].

Several earlier studies investigated that the impact of pavement conditions on vehicle maneuver mostly due to skidding [3], [12]. Besides, it could lead to road crash if a vehicle tries to avoid the poor pavement condition on a lane by taking an opposite lane or do a sudden break that could cause to rear-end crashes. Around 1.39% road crash occurred on poor pavement conditions in Indonesia between 2012-2019. The most common pavement condition or pavement distress on road infrastructure is pothole. Potholes is one of pavement deterioration that has bowl-shaped depression on pavement surface that has negative impact on roadway safety (damaged wheels, flat tires, and road crashes) [13], [14] Pothole on the road surface is also one of the main factors causing accidents, especially for 2-wheeled vehicles/motorcycles [15]. Road damage can cause a road crash and increase the casualty severity level, especially for vulnerable road users and for high-speed vehicles. On one perspective, drivers could lower their speed and increase their focus because of the existence of potholes. Nevertheless, drivers could do dangerous behavior e.g sudden braking, avoiding pothole by taking the opposite lane and could lead to road crash. Road surface improvement including maintenance for pothole corresponding to lower road fatalities [5]. Based on these facts, this study focuses only on pothole as a main pavement distress that can interfere the vehicle's maneuver.

Regarding the issue that pavement condition is one of the factors that could lead to road accident, this research aims to observe the existing condition of how the vehicle may respond on poor pavement condition. This research aims to indicate the impact of poor pavement conditions on vehicles movement. Attributes related to pavement conditions and road safety considered are vehicle speed, changing lane and maneuver, and others dangerous movements such as sudden brake, move to the roadside, and others that could lead to road crash.

2. Methodology

Based on the issue that pavement distress could have a negative impact on vehicle respond on roads, it is necessary to identify how the vehicle may respond while passing through poor pavement area and identify the possibility of dangerous driving behavior because of it. This research used primary data, collected on local roads in Bandung City. Several attributes are considered to be observed: speed reduction, vehicle maneuver and changing lane to opposite direction, and other dangerous movement including sudden braking, or changing lane to the roadside.

The primary survey was conducted in 8 locations in Bandung City which have poor pavement conditions. The duration of primary survey for each location is about 1.5-2.5 hours based on traffic condition. Data were collected and recorded while there is no traffic congestion on inspected road and vehicles are quite freely passing the road without any significant obstacle except the pavement condition of road itself. The locations were chosen based on the road condition, which may have an impact to vehicle movement especially with poor pavement condition. In order to observe whether the vehicle do a lane change especially to the opposite lane, then the chosen locations do not have a road damage on opposite lane.

This research used descriptive statistic to describe the respond of vehicle on poor pavement conditions. The considered attributes will be combined and analyzed to see whether it may cause a dangerous behavior and could lead to traffic accident.

3. Data And Analysis

Data were collected using a primary survey in Bandung city, especially on 8 local roads with poor pavement conditions. The data collection comprises the type of vehicle (motorcycle, light vehicle, heavy vehicle), speed reduction while approaching the poor pavement location, changing lane, presence of a vehicle in the opposite direction, and other dangerous behavior that could lead to a road crash. The survey was conducted on 1-15 December 2021 while there is no traffic congestion. The location of survey can be seen in **Figure 1** and the number of vehicles surveyed on each location can be seen in **Table 1**.

Figure 1 Survey location

Table 1 Number of vehicles surveyed on each location

Location	MC	% MC	LV	% LV	HV	% HV	Total	Total (veh/hour)
Terusan Suryani Street (1.5 hour)	445	96%	15	3%	4	1%	465	310
Cikutra Street (1.5 hour)	1132	91%	99	8%	9	1%	1241	827
Hasanuddin Street (2.5 hour)	564	75%	188	25%	5	1%	758	303
Ciumbuleuit Street (1.5 hour)	1519	77%	433	22%	22	1%	1975	1317
Cijerah Street (1.5 hour)	1002	84%	117	10%	77	6%	1197	798
Neglasari Street (2.15 hour)	310	79%	77	20%	5	1%	393	183
Purwakarta Street (2 hour)	1355	74%	417	23%	47	3%	1820	910
Rereng Adumanis Street (1.5 hour)	508	83%	97	16%	8	1%	614	409

Light vehicles in this study are minibuses and cars which have four wheels and length low than 5 meter. Heavy vehicles are buses and trucks which have four wheels or more with the length more than 5 meter.

The majority of vehicles on all surveyed roads are motorcycles, with more than 75 percentagewise, followed by light vehicles and heavy vehicles, respectively. Ciumbuleuit street can be categorized into high traffic condition because it has more than 1000 veh/hr, while Purwakarta, Cijerah, and Cikutra street have medium traffic (500 – 1000 veh/hr) and others are low traffic. Locations of the survey and their existing condition regarding pavement conditions are shown in Table 2.

Table 2 shows the location of the survey and pavement conditions in more detail.

Table 2 Location of Survey and Pavement Condition

No	Location	Existing Condition (Documentation)	Pavement Condition
1	Terusan Suryani Street 		Pavement condition: - Road deterioration are spread on one lane with shallow hole - there are some loose material (ravelling) on the road Traffic condition: - low traffic - the majority of traffic is motorcycle
2	Cikutra Street 		Pavement condition - Road deterioration with quite deep hole and spread on one way with large area - There are some loose material (ravelling) on the road Traffic condition - medium traffic - the majority of vehicle is motorcycle
3	Hasanuddin Street 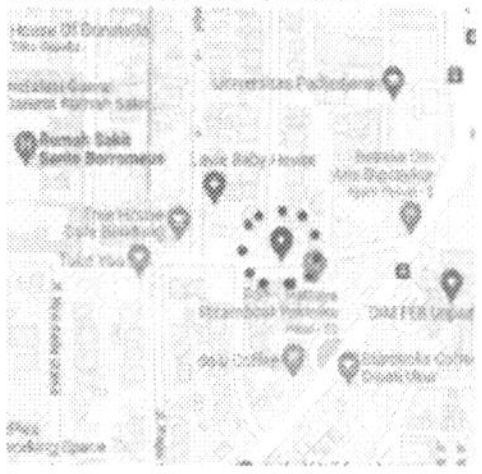		Pavement condition - Road deterioration with small pothole on half of a lane - There are no loose material on road Traffic condition - low traffic - the majority of vehicle are motorcycle and car
4	Ciumbuleuit Street 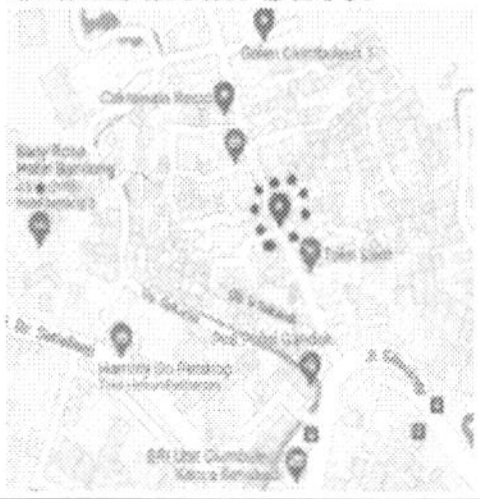		Pavement condition - Road deterioration with small pothole near roadside - There are no loose material on road Traffic condition - high traffic - the majority of vehicle are motorcycle and car

IOP Conf. Series: Earth and Environmental Science 1065 (2022) 012060 doi:10.1088/1755-1315/1065/1/012060

No	Location	Existing Condition (Documentation)	Pavement Condition
5	Cijerah Street 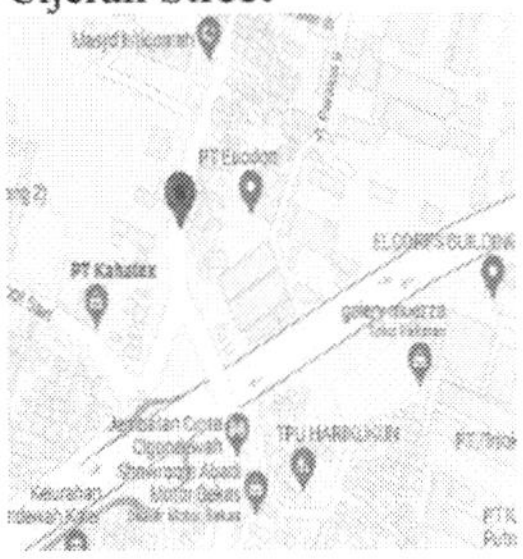		Pavement condition - Road deterioration with big pothole in the middle of a lane - There are many loose material (ravelling) on the road Traffic condition - medium traffic - the majority of vehicle are motorcycle and car
6	Neglasari Street 		Pavement condition - Road deterioration with quite big pothole in the middle of a lane - There are many loose material (ravelling) on the road Traffic condition - low traffic - the majority of vehicle is motorcycle
7	Jalan Purwakarta 		Pavement condition - Road deterioration with shallow hole which spread in one lane - There are a few loose material (ravelling) on the road Traffic condition - medium traffic - the majority of vehicle is motorcycle
8	Jalan Rereng Adumanis 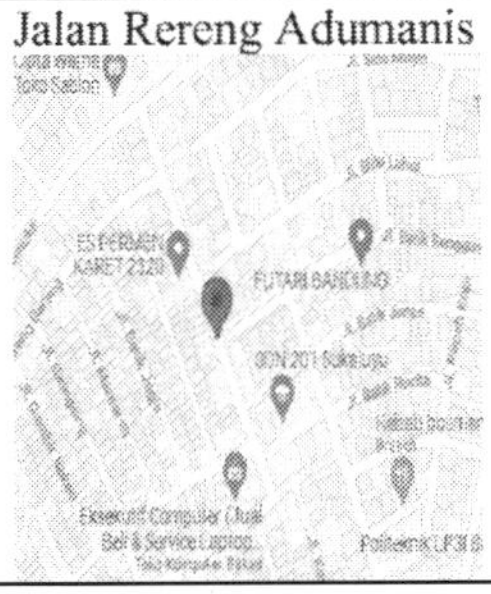		Pavement condition - Road deterioration with small pothole in the middle of one lane - There are no loose material (ravelling) on the road Traffic condition - low traffic - the majority of vehicle is motorcycle

The majority of location has significant road deterioration that could lead to vehicle disruption. The result of survey can be seen in **Figure 1** and Appendix A. Driver behaviors that are considered in this study are:

- speed reduction
- maneuvering to avoid the poor pavement conditions
- the existence of vehicle on the opposite way

- do nothing
- combination of the behavior
- other dangerous behavior: sudden braking, changing lane to the opposite way

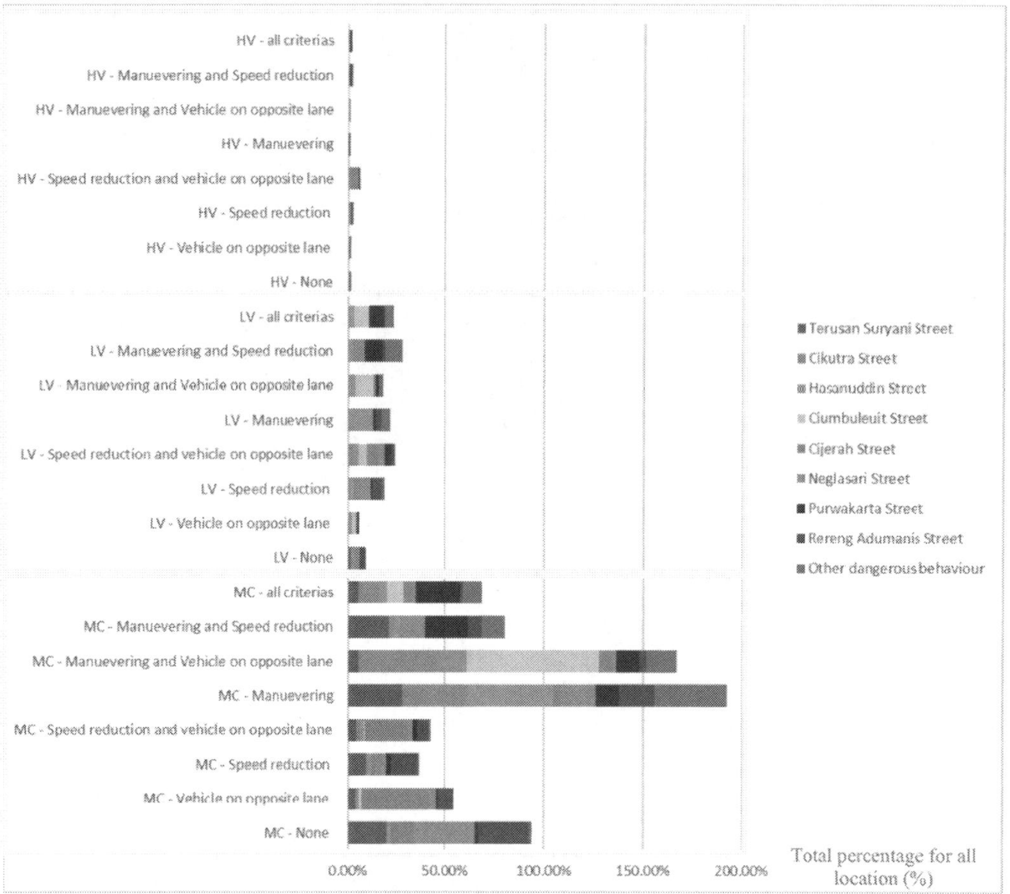

Figure 2 Survey results

Based on survey result in Figure 2, the majority of vehicle on all roads is motorcycle, followed by light vehicles and heavy vehicle, respectively. The Ciumbuleuit street has the biggest number of motorcycles for maneuvering whether there is a vehicle on opposite direction. This may because the high number of traffic on Ciumbuleuit street, and small dimension of motorcycle, thus motorcycles could still maneuver with or without vehicle on the opposite direction and did not influence significantly.

In general, the most common behavior for motorcycle on a poor road pavement is only maneuvering to avoid the pothole/any pavement distress, followed by doing maneuver or changing lane while there is a vehicle on opposite lane and do nothing while confront the poor pavement condition. Additionally, many of them do both responds while passing the poor pavement condition: movement to avoid poor pavement condition and reducing speed near the poor pavement area. Furthermore, many of motorcycles do other dangerous behavior including sudden braking and moving to the roadside/opposite lane while doing the maneuver whether there is or there is not an existence of a vehicle on opposite lane.

Based on these results, maneuvering is the most preferable option for motorcycles compared to other responds to avoid poor pavement condition. The reason for this behavior is probably because the dimension of motorcycle is small, therefore, it could easily do the maneuver to avoid poor pavement conditions. Should the poor pavement area is small and not in their track, the motorcycle could avoid it without any difficulties and do nothing because of this.

For light vehicles, there is an insignificant number of differences for vehicles in responding the poor pavement conditions. Light vehicles do the maneuvering and speed reduction, and several number of them still do the maneuver while there is a vehicle on opposite lane. Many light vehicles do other dangerous respond confronting poor pavement condition while maneuvering and reducing speed at the same time. This may because the poor pavement condition could have an impact on light vehicle while passing it and for the pothole that may not be seen from long distance, the light vehicle may do the dangerous behavior including sudden break and taking a roadside.

Heavy vehicles are the lowest number of vehicles among the type of vehicles. Heavy vehicles consist of buses, and trucks. The highest number of responding the poor pavement condition for heavy vehicles is doing speed reduction, especially when there is a vehicle on opposite lane. This may because of the dimension of heavy vehicles, they could not maneuver freely to avoid poor pavement area.

4. Conclusion

In this study, primary survey was conducted to identify vehicle respond regarding poor pavement conditions. The biggest number of vehicles on all of the survey location in Bandung City local road is motorcycle (more than 75%), followed by light vehicles and heavy vehicle, respectively.

There are some differences for responding the poor pavement conditions for each type of vehicle. Motorcycles tend to do maneuver or changing lane to avoid poor pavement condition area. It may because of its dimension is small thus it can move freely on road compared to others. On the other hand, heavy vehicles tend to reduce the speed while confronting the poor pavement area, and it may because of large dimension of trucks, therefore it cannot move freely to avoid the poor pavement area. Beside of that, light vehicles tend to do maneuver, reduce their speed, and do the combination of those responses. The number of differences for each respond and combination is not significant for light vehicle.

Motorcycles and light vehicles tend to do dangerous behavior while maneuvering and reducing their speed. It could be dangerous especially if the vehicle does the maneuver to the opposite lane while there is a vehicle on opposite lane.

Based on this study, we can indicate that it could be dangerous if the vehicle is taking an opposite lane while there is a vehicle on that opposite lane. Some vehicles reduce their speed while confronting a poor pavement are and they could be more careful on roads, nevertheless, in the rainy season, it could be dangerous, especially when the pothole of poor pavement conditions are covered with rainwater. The improvement of road condition is vital to enhance travel time and reduce obstacles on road.

Acknowledgments
This research is supported by P2MI Research Grant Institut Teknologi Bandung, Indonesia in 2021.

References

[1] A. Montella, F. Mauriello, M. Pernetti, and M. Rella Riccardi, "Rule discovery to identify patterns contributing to overrepresentation and severity of run-off-the-road crashes," *Accid. Anal. Prev.*, vol. 155, no. March, p. 106119, 2021, doi: 10.1016/j.aap.2021.106119.
[2] Q. Yuan and J. Wang, "Goods movement, road safety, and spatial inequity: Evaluating freight-related crashes in low-income or minority neighborhoods," *J. Transp. Geogr.*, vol. 96, no. February, p. 103186, 2021, doi: 10.1016/j.jtrangeo.2021.103186.

[3] Y. Li, C. Liu, and L. Ding, "Impact of pavement conditions on crash severity," *Accid. Anal. Prev.*, vol. 59, pp. 399–406, 2013, doi: 10.1016/j.aap.2013.06.028.
[4] M. S. B. Shaheed, K. Gkritza, W. Zhang, and Z. Hans, "A mixed logit analysis of two-vehicle crash severities involving a motorcycle," *Accid. Anal. Prev.*, vol. 61, pp. 119–128, 2013, doi: 10.1016/j.aap.2013.05.028.
[5] Y. Sari and M. H. Yudhistira, "Bad light, bad road, or bad luck? The associations of road lighting and road surface quality on road crash severities in Indonesia," *Case Stud. Transp. Policy*, vol. 9, no. 3, pp. 1407–1417, 2021, doi: 10.1016/j.cstp.2021.07.014.
[6] A. D. Saputra, "Study of Traffic Accident Rate in Indonesia Base on KNKT (Komite Nasional Keselamatan Transportasi) Database from 2007-2016," *War. Penelit. Perhub.*, vol. 29, no. 2, pp. 179–190, 2017, doi: 10.1016/j.injury.2011.11.002.
[7] N. Hussein, R. Hassan, and M. T. Fahey, "Effect of pavement condition and geometrics at signalised intersections on casualty crashes," *J. Safety Res.*, vol. 76, pp. 276–288, 2021, doi: 10.1016/j.jsr.2020.12.021.
[8] F. S. Carvalho, C. A. S. Machado, P. H. C. Araújo, L. L. B. Bernucci, and J. A. Quintanilha, "Safety effects of pavement resurfacing: A Brazilian experience," *Case Stud. Transp. Policy*, vol. 9, no. 4, pp. 1860–1867, 2021, doi: 10.1016/j.cstp.2021.10.006.
[9] T. Tsubota, C. Fernando, T. Yoshii, and H. Shirayanagi, "Effect of Road Pavement Types and Ages on Traffic Accident Risks," *Transp. Res. Procedia*, vol. 34, pp. 211–218, 2018, doi: 10.1016/j.trpro.2018.11.034.
[10] S. Cafiso, A. Montella, C. D'Agostino, F. Mauriello, and F. Galante, "Crash modification functions for pavement surface condition and geometric design indicators," *Accid. Anal. Prev.*, vol. 149, no. May 2020, p. 105887, 2021, doi: 10.1016/j.aap.2020.105887.
[11] T. U. Saeed, T. Hall, H. Baroud, and M. J. Volovski, "Analyzing road crash frequencies with uncorrelated and correlated random-parameters count models: An empirical assessment of multilane highways," *Anal. Methods Accid. Res.*, vol. 23, p. 100101, 2019, doi: 10.1016/j.amar.2019.100101.
[12] A. Setyawan, I. Kusdiantoro, and Syafi'i, "The effect of pavement condition on vehicle speeds and motor vehicles emissions," *Procedia Eng.*, vol. 125, pp. 424–430, 2015, doi: 10.1016/j.proeng.2015.11.111.
[13] J. S. Miller and W. Y. Bellinger, "FHWA, Distress Identification manual for the Long-Term Pavement Performance Program. Report FHWA-HRT-13-092," *Fed. Highw. Adm.*, no. May, p. 142, 2014.
[14] Y.-C. (James); Tsai and Anirban Chatterjee, "Pothole Detection and Classification Using 3D Technology and Watershed Method," *J. Comput. Civ. Eng.*, vol. 32, no. 2, 2018, [Online]. Available: https://ascelibrary.org/doi/epdf/10.1061/%28ASCE%29CP.1943-5487.0000726#.
[15] Marsaid, M. Hidayat, and Ahsan, "Identifikasi kecelakaan lalu lintas pada pengendara sepeda motor di wilayah polres kabupaten malang," *J. Ilmu Keperawatan*, vol. 1, no. 2, pp. 98–112, 2013.

APPENDIX A Survey Results in Detail

Type of Vehicle	Manuevering	Speed reduction	Vehicle on opposite lane	Terusan Suryani Street	Cikutra Street	Hasanuddin Street	Ciumbuleuit Street	Cijerah Street	Neglasari Street	Purwakarta Street	Rereng Adumanis Street	Other dangerous behaviour	Average
MC	No	No	No	19.83%	0%	2%	0%	12%	32%	2%	25%	2%	11.55%
	No	No	Yes	3.45%	1%	2%	1%	35%	4%	1%	8%	0%	6.75%
	No	Yes	No	9.05%	1%	2%	0%	2%	6%	3%	14%	1%	4.46%
	No	Yes	Yes	4.31%	2%	2%	1%	22%	3%	2%	6%	1%	5.22%
	Yes	No	No	28.02%	34%	43%	0%	3%	18%	12%	18%	36%	19.49%
	Yes	No	Yes	5.17%	41%	15%	66%	6%	3%	11%	4%	15%	18.90%
	Yes	Yes	No	20.69%	3%	4%	0%	0%	13%	22%	7%	12%	8.61%
	Yes	Yes	Yes	5.39%	10%	5%	9%	4%	2%	22%	1%	10%	7.37%
LV	No	No	No	1.51%	1%	2%	0%	1%	2%	0%	2%	0%	1.05%
	No	No	Yes	0.22%	1%	1%	2%	1%	0%	0%	1%	0%	0.72%
	No	Yes	No	0.22%	1%	2%	0%	1%	8%	1%	5%	1%	2.20%
	No	Yes	Yes	0.00%	3%	3%	4%	6%	3%	3%	2%	0%	2.99%
	Yes	No	No	0.65%	0%	11%	0%	0%	1%	1%	3%	5%	2.08%
	Yes	No	Yes	0.00%	1%	3%	8%	2%	0%	1%	0%	2%	1.99%
	Yes	Yes	No	0.43%	0%	2%	0%	0%	6%	9%	2%	9%	2.38%
	Yes	Yes	Yes	0.22%	1%	2%	7%	0%	1%	7%	1%	4%	2.36%
HV	No	No	No	0.22%	0%	0%	0%	0%	0%	0%	0%	0%	0.14%
	No	No	Yes	0.22%	0%	0%	0%	0%	0%	0%	0%	0%	0.14%
	No	Yes	No	0.00%	0%	0%	0%	1%	0%	0%	0%	0%	0.31%
	No	Yes	Yes	0.22%	0%	0%	0%	5%	0%	0%	0%	0%	0.71%
	Yes	No	No	0.22%	0%	0%	0%	0%	0%	0%	0%	0%	0.06%
	Yes	No	Yes	0.00%	0%	0%	0%	0%	0%	0%	0%	0%	0.14%
	Yes	Yes	No	0.00%	0%	0%	0%	0%	1%	1%	0%	1%	0.19%
	Yes	Yes	Yes	0.00%	0%	0%	0%	0%	0%	1%	0%	0%	0.18%

Evaluation of Pedestrian Facilities Performance

N Hidayat*, I Haryanto, and R P Isheka

Civil Engineering Department, Vocational School, Gadjah Mada University, Yogyakarta, Indonesia

*e-mail: nursyamsu.h@ugm.ac.id

Abstract. Most of sidewalks, especially in developing countries, are roled as public area with many social activities besides as the main function for serving pedestrian movement. Therefore, sidewalks may need some facilities to support user activities, namely benches, planting, including stalls for street vendors. The aims of this research is to evaluate pedestrian facilities in two different characteristic zone y.i. residential and business area in Yogyakarta city using Pedestrian Index (P-Index) method. The data are collected in KHA Dahlan-H Agus Salim-Kauman coridors, and Sudirman-Suroto coridors, that both coridors are considered residential and business area in downtown of Yogyakarta. Some variables are taken in to account, namely mobility, safety, facility, and accessibility. The results reveal that the pedestrian facilities are supportive towards pedestrians. The sidewalk in Sudirman-Suroto corridors are better in safety aspect, but must be improved.

Keywords: sidewalk, public area, mobility, safety, accessibility

1. Introduction

Yogyakarta city is well known as a tourist destination and has enormous potential to develop cultural and traditional tourism. Even though there have been so many tourists both domestic and foreign who have visited, the number that can be withdrawn is still very large. Tourist objects have sprung up in various locations, therefore supporting facilities needs to be considered. Complaints that are often heard during the peak holiday season are difficulties in accessing tourist attractions because of the difficulty of getting vehicle parking. So it is necessary to find a solution so that tourists can get a decent parking point, and then have no trouble getting to tourist sites. So the government is trying to provide access from parking lots to tourist destination areas.

In addition to tourism potential, there are some business center points and commercial areas that need attention. The affordability of these areas continues to be improved with the development of transportation modes, especially public transportation. The improvement of public transportation services must be simultaneously with the improvement of facilities for pedestrians as passengers.

The aims of this study is to evaluate the condition of pedestrian facilities/sidewalks in two areas in Yogyakarta City which have different activity characteristics, namely the business area represented by Sudirman-Suroto Streets, and tourist-residential areas, namely on KHA Dahlan-H Agus Salim-Kauman Streets.

2. Methodology

2.1 Pedestrian Index (P-Index) Method

This method was developed in Malaysia with the aim of getting the rate value of a sidewalk (Ghani et.al, [1]). To determine the rating, a codification of "star" is used, provided that the more stars you get, the better the pavement quality. A good sidewalk is defined as a sidewalk that is equipped with pavement, without obstacles, providing access to the surrounding environment according to its designation.
The P-Index method has four variables, namely mobility (M), facility (F), safety (S), and accessibility (A), which have a relationship as described below:

$$P = f(M, S, F, A) \tag{1}$$

If it is assumed that the relationship between these functions is a linear equation, it can be written as follows:

$$P = b_1M + b_2S + b_3F + b_4A \tag{2}$$

where $b_{1\text{-}4}$ is the weighting coefficient of each criterion
The weights take into account the conditions found at the site. It could be that the weight given to each criterion is the same, namely 0.25. Thus the above equation becomes

$$P = \frac{M+S+F+A}{4} \tag{3}$$

Table 1 shows the stars value in the P-Index method

Table 1 Interpretation of the P-Index Value

P-Index	Star Rating	Description
0 - 20	*	Hostile towards pedestrians
21 - 40	**	Unfavorable to pedestrians
41 – 60	***	Walkable
61 – 80	****	Supportive towards pedestrians
81 - 100	*****	Very pedestrian friendly

Source: Ghani et al [1]

2.2 Mobility (M)

Mobility (M) describes the surface condition/availability of sidewalks compared to the total length of the road. Mobility is calculated by the formula:

$$M = \left(\frac{0.5D_C}{D}\right)x100 \tag{4}$$

were,
D_C : total length of pavement on both sides of the road (m)
D : road length (m)

2.3 Safety (S)

Safety (S) is a condition where there is a barrier on the sidewalk with the road body both physically and spatially. Safety is calculated by the following equation:

$$S = \left(\frac{0.5D_{SP}}{D}\right)x100, \quad 0 \le D_{SP} \le D_C \tag{5}$$

were,
D_{SP} : the total length of the sidewalk equipped with barriers, calculated on both sides of the road

D_C : total length of pavement on both sides of the road (m)
D : road length (m)

2.4 Facility (F)
This variable indicates the availability of facilities for pedestrians crossing the road, at least the availability of facilities in the form of bollards, ramps, and zebra crosses.

$$F = \frac{\sum_{i=1}^{n}(0.5B_i+0.5R_i+Z_i)}{3n} x100 \quad (6)$$

were,
B_i : availability of bollards
R_i : availability of ramps
Z_i : availability of zebra cross

$$B_i, R_i = \begin{cases} 2, \text{ if available on both sides} \\ 1, \text{ if available on one side} \\ 0, \text{ not available} \end{cases}$$

$$Z_i = \begin{cases} 1, \text{ if available} \\ 0, \text{ not available} \end{cases}$$

2.5 Accessibility (A)
Accessibility describes the extent to which environmental access/land use is close to pedestrian facilities within walking distance. The ability and willingness to walk is very dependent on the situation and conditions encountered, but under normal conditions it is set as far as 500 m.

$$A = \frac{\sum_{j=1}^{k} L_j}{k} x100 \quad (7)$$

were,
L_j : percentage of total land use within walking distance of land use j
j : 1, 2, 3 ..... k

Tabel 2 shows interpretation of Mobility, Safety, Facility, and Accessibility Values
Tabel 2 Interpretation of Indicator Values

Score	Star	M	S	F	A
0 – 20	*	Little/without mobility	Very unsafe	Not-exixtence or negligible ped. facility	Land uses inaccessible by walking
21 – 40	**	Inadequate mobility	Unsafe	Inadequate ped. facility	Poor accessibility to land uses by walking
41 – 60	***	Fair mobility	Walk with caution	Moderate availability of ped. facility	Moderate accessibility to land uses by walking
61 – 80	****	Good mobility	Safe	Good pedestrian facility	Good accessibility to land uses by walking
81 - 100	*****	Excellent mobility	Excellent safety	Excellent pedestrian facility	Excellent accessibility to land uses by walking

Source: Ghani et al [1]

IOP Conf. Series: Earth and Environmental Science 1065 (2022) 012061 doi:10.1088/1755-1315/1065/1/012061

3. Study Area

This study takes sidewalk on the main road at two locations in Yogyakarta City which have different land use characteristics. One location is a business area, while the other is a tourist and residential area. Table 3 presents characteristic of study area, and Figures 1-2 show location of the study.

Table 3 Study Area

Code	Streets	Length (m)	Description
1	KH Ahmad Dahlan	910	tourist and residential
2	H Agus Salim	461	tourist and residential
3	Kauman	303	tourist and residential
4	Jendral Sudirman	285	business
5	Suroto	422	business
	Total length	2381	

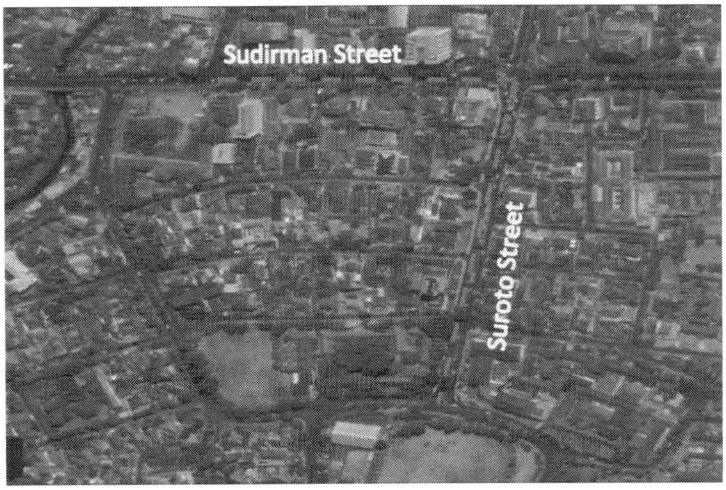

Figure 1 Sudirman-Suroto Streets

Figure 2 KHA Dahlan-H Agus Salim-Kauman Streets

4. Results and Discussion

Firstly, the data for variable of Mobility (M) are presented as be seen in Table 4. The sidewalk data collection was carried out on both sides of the road.

Table 4 Data and Analyses of Mobility (M) Variable

Land Use	Code	Streets	Road Length (m) (D)	Sidewalk Length (m) (D_C)	Mobility (M)
tourist and residential	1	KHA Dahlan	910	1800	98.90
	2	H Agus Salim	461	922	100.00
	3	Kauman	303	606	100.00
		Sub Total	**1674**	**3328**	**99.40**
business	4	Sudirman	285	570	100.00
	5	Suroto	422	755	89.45
		Sub Total	707	1325	93.71

Above data were calculated with, then the overall value of Mobility is obtained. Mobility value for KHA Dahlan street, H Agus Salim street, and Kauman street (tourist and residential area) is 99.40. While the Mobility value for Sudirman street and Suroto street (business district) is 93.71. These represent a 5-stars rating which means that both road corridors have very good mobility.

The Safety value is calculated using the survey data as shown in Table 5. The barrier between the sidewalk and the road in this context is the availability of space between the sidewalk and the road/buffer (spatial), as well as physical barriers/fence (physical).

Tabel 5 Data and Analyses of Safety (S) Variabel

Land Use	Code	Streets	Road Length (m) (D)	Length of sidewalk with buffer (m) (D_{SP})	*Safety* (S)
tourist and residential	1	KHA Dahlan	910	141.66	7.78
	2	H Agus Salim	461	91.17	9.89
	3	Kauman	303	0	0.00
		Sub Total	**1674**	232.83	6.95
business	4	Sudirman	285	435	76.32
	5	Suroto	422	422	50.00
		Sub Total	**707**	857	60.61

Equation 5 is used to calculate the value of S with the data above, then the overall value of Safety is obtained. S value for KHA Dahlan, H Agus Salim, and Kauman Streets (tourist and residential area) is 6.95, or a 1-star rating in other words is in the very unsafe category. While the value of S for Sudirman and Suroto Streets (business district) is 60.61. This represents a 4-stars rating which means the condition of the safety aspect is safe.

The calculation of the Facility (F) variable uses survey data which is then analyzed as shown in Table 6.

Tabel 6 Data and Analyses of Facility (F) Variabel

Land Use	Code	Streets	Σ Junction	Σ Bollard (B)	Σ Ramps (R)	Σ Zebra Cross (Z)	0.5B+0.5R+Z	Facility
tourist and residential	1	KHA Dahlan	11	2	2	1	3	100
	2	H Agus Salim	3	2	2	1	3	100
	3	Kauman	3	2	2	1	3	100
		Sub Total	17	2	2	1	3	100
business	4	Sudirman	0	2	2	1	3	100
	5	Suroto	6	2	2	1	3	100
			6	2	2	1	3	100

Using equation (6), the Facility (F) value for all the roads reviewed is 100, meaning that the facilities are very good or excellent, and get 5-stars.

Accessibility (A) analysis uses the data collected including the characteristics of land use in the area under review. There are 4 types of land use that can be identified, namely tourist areas, shopping/commercial, educational, and residential areas. Figure 3 and Table 7 present land use data on Sudirman-Suroto Corridors, whereas Figure 4 and Table 8 show the similar things on KHA Dahlan-H Agus Salim- Kauman Streets.

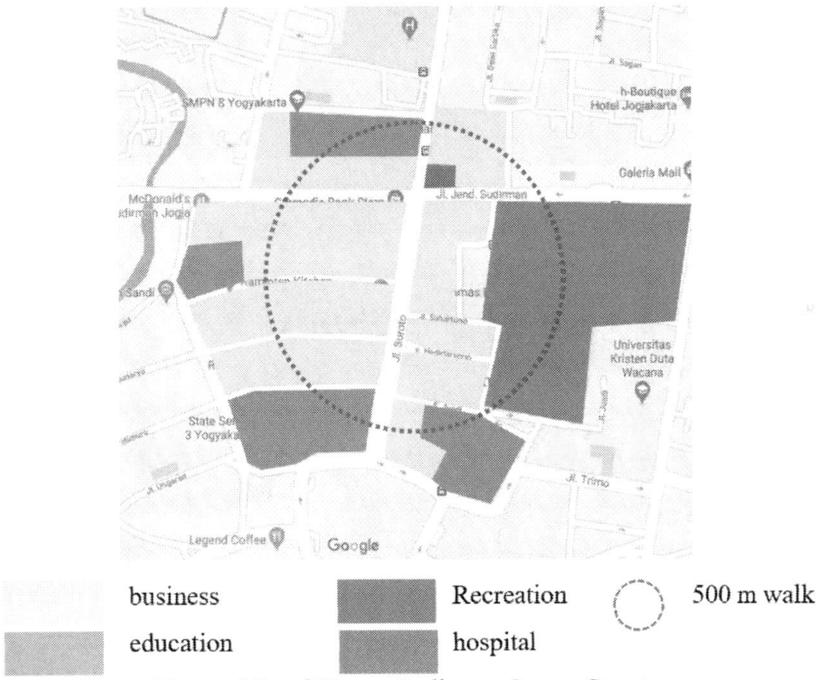

Figure 3 Land Use on Sudirman-Suroto Streets

Tabel 7 Variable of Accessibility (A) in Sudirman-Suroto Streets Area

Land Use	Percentage within 500 m walking distance
Recreational	10 %
Education	50 %
Business/Commercial	90 %
Hospital	25 %

Using equation (7) the Accessibility value is obtained as below

$$A = \frac{\sum_{j=1}^{k} L_j}{k} x100$$
$$= 43.75$$

The Accessibility value of 43.75 represents 3 stars, which means that the access to usability is sufficient.

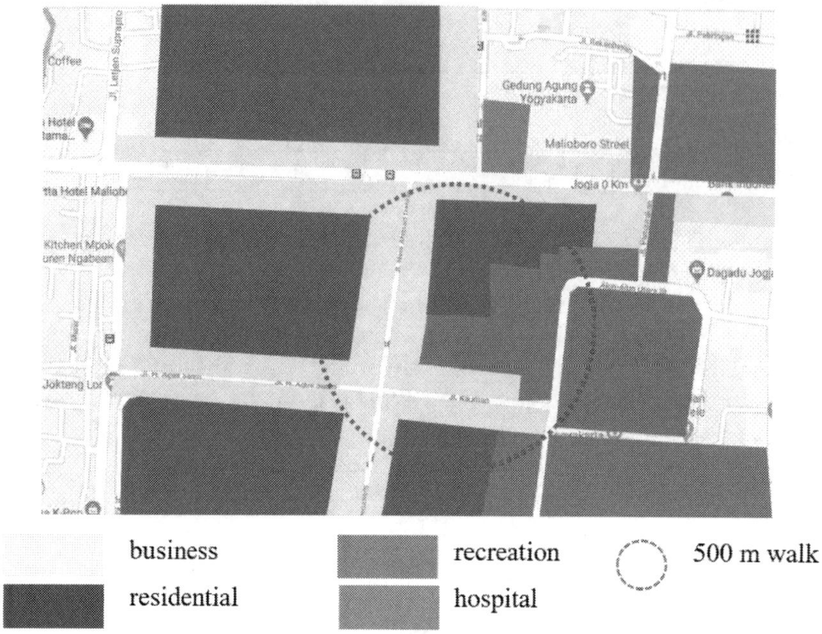

Figure 4 Land Use on KHA Dahlan- H Agus Salim- Kauman Streets

Tabel 8 Variable of Accessibility (A) in KHA Dahlan-H Agus Salim-Kauman Streets

Tata Guna Lahan	Percentage within 500 m walking distance
Recreational	90 %
Hospital	5 %
Business/Commercial	50 %
Residential	80 %

Using equation (7) the Accessibility value is obtained as below

$$A = \frac{\sum_{j=1}^{k} L_j}{k} x100$$
$$= 56.25$$

The Accessibility value of 56.25 represents 3-stars, which means that access to usability is sufficient. Analysis of the entire road segment can be calculated with the data values of M, S, F, and A on each road segment. The four variables are weighted according to local/location considerations. In this study, it is assumed that the weights of the four variables are the same, namely 0.25 each. Thus, the overall quality of pedestrian facilities with these four indicators can be summarized in Table 9.

Table 9 Result of Analyses using P-Index Method

Code	Streets	M	S	F	A	P-Index	Star
1	KH Ahmad Dahlan	99	8	100	56.25	66	****
2	H Agus Salim	100	10	100	56.25	67	****
3	Kauman	100	0	100	56.25	64	****
		99	**7**	**100**	**56.25**	**66**	****
4	Sudirman	100	76	100	43.75	80	****
5	Suroto	89	50	100	43.75	71	****
		94	**61**	**100**	**43.75**	**75**	****

Table 9 shows that overall pedestrian facilities in the research area get a score of 66 and 75 or a 4-stars rating, which means that pedestrian facilities are comfortable and can support travel activities, so it is appropriate to serve activities in the area which are access to tourist areas, as well as surrounding business activities. The P-Index value of Sudirman-Suroto Sreets is higher than KHA Dahlan-H Agus Salim-Kauman Streets, in this case the area of Sudirman-Suroto Streets is superior in terms of safety. However, KHA Dahlan-H Agus Salim-Kauman Streets is superior in terms of accessibility.

Ghani et al [1] implemented P-Index Method in heritage area in Malacca Historical City, Malaysia. The research revealed that most of the roads are not pedestrian friendly because of getting low score of indicator value, namely, inadequate mobility, very unsafe for walking, and non-existence or negligible pedestrian facility.

The results obtained in this study are different from the research conducted by Ayu [2], who conducted research on sidewalk facilities in the South Square area of Yogyakarta and its surroundings using the visual observation method. The results obtained are that almost all roads in the research object get LOS C and LOS D values, which means that although pedestrians are still acceptable, the conditions are not comfortable. Likewise, the results of Hidayat and Haryanto [3] study in the Yogyakarta Kraton area stated that the facilities provided were still not feasible. This means that access to the Kraton area from the results of this study is good and proper, but the facilities in the Jeron Beteng area still need to be improved. Another research by Hidayat and Suwardo [3] in Malioboro area, Yogyakarta also stated that the priority of handling sidewalk facilities according to pedestrian perceptions stated that one of the priorities for handling requested by sidewalk users is increasing comfort.

P-Index Method take into account physical characeristics of the sidewalk/roadway especially on roadway and sidewalk length. Therefore, some additional indicators relating safety, pedesetrian behavior, perception should be considered in the next study. On the other hand, P-Index provide a simple method for measuring conformance level and objective feedback to planning criteria. It can be used as an instrument to determine feasibility of pedestrian infrastructure.

5. Conclusion

This study is able to evaluate the condition of pedestrian facilities in two areas in Yogyakarta City which have different activity characteristics, namely commercial/business areas and tourist access areas. The results showed that the condition of the sidewalks in the two areas was comfortable and able to serve pedestrians well.

Based on achievement of star rating, both street corridors have no problem with Mobility and Facility variables. In terms of Safety, Sudirman-Suroto streets have better achievement than KHA Dahlan-Agus Salim-Kauman streets, 4-stars compared to 1-star. Buffer zone in the form of planting tree and gardens exist almost along the Sudirman-Suroto streets, that separate sidewalk and roadway spatially. Something that will not meet in another corridors. Regarding Accessibility, both street corridors attain 3-stars or moderate closeness to land use by walking. For ideal accessibility can be reached when 80% of the land uses are located within walking distance (500 m).

Sudirman-Suroto street corridor with a dominant business activity is better from the safety aspect. Thus, the condition here needs to be maintained and even improved, especially in the KHA Dahlan-H Agus

Salim-Kauman street corridor, which is access to the famous tourist areas of Malioboro and the Sultan Palace/Kraton.

6. Refferences

[1] Ghani, Nabila Abdul, Shimizu, Tetsuo, Mokhtar, Safizahanin, 2015. Assessment of Pedestrian Facilities in Malacca World Heritage Site, Malaysia using P-Index Method, Jounal of The Eastern Asia Society for Transportation Studies, Vol. 11.

[2] Ayu, Anita Indah P, 2019. Analisis Fasilitas Jalur Pejalan Kaki Berdasarkan Pengamatan Visual Wilayah Selatan Kawasan Beteng Keraton Yogyakarta, Thesis Civil Engineering Department, Vocational School, Gadjah Mada University (not published)

[3] Hidayat, Nursyamsu, and Haryanto, Iman, 2019. Kelayakan Fasilitas Pejalan Kaki di Kawasan Wisata Kraton Kasultanan Yogyakarta, National Conferences of Applied Technology, Yogyakarta

[4] Hidayat, Nursyamsu, and Suwardo, 2015. Prioritas Penanganan Trotoar di Area Perkotaan Berdasarkan Persepsi Pengguna dengan Metode Faktor Analisis, National Conferences of Applied Technology, Yogyakarta

Coastal protection system design at the Indonesian Mangrove Center in Pekalongan, Central Java – Indonesia

Yola[1], N Yuanita[2], A Kurniawan[2], E O Nugroho[3], R J Martin[1], I F Muhammad[1], F S Wijaya[1], and N Fauzi[1]

[1]Ocean Engineering Master Program, Faculty of Civil and Environmental Engineering, Bandung Institute of Technology, Bandung 40132, Republic of Indonesia
[2]Coastal Engineering Research Group, Faculty of Civil and Environmental Engineering, Bandung Institute of Technology, Bandung 40132, Republic of Indonesia
[3]Water Resources Engineering Research Group, Faculty of Civil and Environmental Engineering, Bandung Institute of Technology, Bandung 40132, Republic of Indonesia

*yolarohimah@gmail.com

Abstract. The Pekalongan area suffered from coastal flooding and heavy coastal erosion due to land subsidence and increasing wave condition in the last decade. Pekalongan government built a land dike in 2018 to protect Pekalongan city from coastal flooding. This study aims to determine a suitable coastal protection system, consisting of hard-structures and mangroves, to solve the erosion problem at the Pekalongan coast, specifically at the future Pekalongan Mangrove Information Centre (PIM). Environmental data, i.e., wind, wave, current, tidal, sediment, and soil data, were obtained by field survey and secondary data collection from open-source data. The selection of the most suitable coastal protection design concepts was performed by comparing several alternatives of structure materials. Coastal protection layout was obtained based on a numerical sediment transport model result using Delft3D software. From environmental data processing and analysis with various design aspects, the top elevation of the coastal protection structure is +4.5 m MSL (Mean Sea Level). The configuration of the mangrove planting media consists of 'guludan' structure technique with dimension 10 m x 5 m x 4 m.
Keywords: *PIM Pekalongan, Coastal Protection, Numerical Modeling, Groin, Guludan.*

1. Introduction

Pekalongan is one of the districts in Indonesia that experienced severe beach erosion and land subsidence problems within the last decade [1]. Pekalongan Mangrove Information Center (PIM) is in the Pekalongan district which is directly adjacent to the Pekalongan coast [2]. As a result of this erosion and land subsidence, the coastal area of North Pekalongan has a low land elevation value of < 1 meter. This causes the site to be inundated by tidal flooding. The tidal flood that occurred in the area was frequent and resulted in damage to infrastructure and housing, causing considerable losses. Therefore, it is necessary to carry out mitigation efforts to reduce the amount of damage [3]. This city has high development potential with the characteristics of a sloping and shallow topography area, and there is growing mangrove vegetation [4]. Currently, the City of Pekalongan is planning the Indonesian Mangrove Center (Pekalongan Mangrove Park) with areas such as those that continue to be developed as one of the environmental destinations of the coastal area. Pekalongan is expected to become a central area for developing mangrove forest ecosystems with various functions such as ecological, educational, disaster mitigation, economic, social, and cultural. It will increase the empowerment of local communities. It is hoped that mangroves can help overcome the erosion problem in Pekalongan [5]. The purpose of this study is to determine the most suitable design of coastal protection system, which consists of hard structures and mangroves, to solve erosion problems at Pekalongan coast, specifically at the location of the Pekalongan Mangrove Information Center (PIM) [6].

2. Research Method and Problem Scope

The methodologies used to finish this study are literature study, environmental data collection. selection coastal protection design and concepts consider with environmental data, erosion and land subsidence problem, Selection mangrove planting media design and concepts consider with environmental data, erosion and land subsidence problem, and interpretation of sediment transport model result

For the limitations used in this study are The area under review is only in the Pekalongan North Coast area around the Pekalongan Indonesian Mangrove Center area, it is assumed that the structural material is land in the modeling in Delft3d, the primary consideration was the hydraulic aspect, not the ecological perspective, in determining the type and layout of the mangrove planting media structure.

3. Data Collection

Primary data collection was carried out by conducting a field survey directly in the coastal area of Pekalongan by the FTSL ITB team in 2021 as well as data from observation stations around the coastal area of Pekalongan.

Primary data collected includes bathymetry, tidal elevation, wind speed, wave, current velocity, and sediment data (TSS). Location of the data collection point is shown at figure 1.

Figure 1. Data collection points at the Pekalongan coastal area.

3.1. Bathymetric Data

Bathymetric data from primary data that has been obtained includes data on river cross-sectional depths and bathymetric surveys for the Pekalongan waters area with the lowest low tide datum (LWS).

3.2. *Current Velocity Data*

In the field data collection process, the current velocity measurement process using ADCP was carried out in the range of April 6, 2021, at 15:00 WIB to April 9, 2021, at 17:10 WIB with a measurement range every ten minutes at coordinates 109.689317 East Longitude and 6.828245 South Latitude at five depths. From the data obtained, the majority of the u-direction currents have speeds of 0 – 0.4 m/s while the v majority have 0 – 0.1 m/s speeds. It can also be concluded that the current at the measurement point leads to the southeast and northwest.

3.3. *Sediment Data*

After performing a sieve test for the bottom sediment, the results of the test are processed and analyzed to determine the grain size of the bottom sediment. In addition to testing the bottom sediment, testing is also carried out in the laboratory for testing floating sediments. From the field data, it was obtained that the type of sediment found in the waters of the prospective location of the Pekalongan mangrove park was of non-cohesive type with the type of grain in the form of sand. In addition to testing the bottom sediment, testing is also carried out in the laboratory for testing floating sediments. The results showed that the largest average for Total Suspended Solid (TSS) was at the observation point MA-08 at 407.5 ppm and the lowest TSS content was at the observation point MA-07 at 35.9 ppm.

3.4. *Tidal Data*

Tidal data from measurements in the field was taken from April 06 to 10, 2021 with one hour intervals at five points on the river, namely MA02, MA07, MA08, MA10, and MA12. The tidal range is 0.75 m.

3.5. *Wave Data*

Field data collection also includes measurement of wave parameters consisting of measurements of wave height and period. The wave parameter data was taken on April 6, 2021, at 15:00 WIB until April 9, 2021, at 17:10 WIB at coordinates 109.689317 East Longitude and 6.828245 South Latitude. Measurements of wave height and period are carried out at 10 minute intervals. From the field data, it can be concluded that the wave height at the measurement location is in the range of 0.4 – 1.3 m with a period of 3.5 – 7 seconds. From the results of the measurement of the wave parameters, it can be obtained a significant wave height value of 0.89 m with a period of 5.57 seconds.

4. The Election of Mangrove Planting Structure

The type of mangrove planting structures which will be used as an options are bronjong and guludan as shown in figure 2, figure 3, and figure 4.

Figure 2. Bamboo gabion.

Figure 3. Wire gabion.

Figure 4. Guludan.

4.1. Selection of Structure Type

Between the two structures that will be used as a mangrove planting medium, both are good choices for environments such as the Pekalongan coast where the area can be submerged in water for more than 1 meter at high tide, each with its own advantages.

Disadvantages of bronjong include:

- If using bamboo bronjong, the construction of bronjong will easily change shape because the material of manufacture is woven bamboo, and the use of wire in environments with high salt water causes a high corrosion rate will affect the bronjong resistance.
- The embankment material from the bottom to the surface is rock which will provide very high settlement to the soil beneath it, while there has been quite a high land subsidence in that area.

Disadvantages of guludan include:

- The guludan uses bamboo as the basic material for making the frame. Bamboo wood is easily weathered, especially if it continues to be submerged in water.
- Embankment material in the form of clay and sand which will provide settlement to the embankment itself and the soil beneath it.

From this analysis, the structure of the mangrove planting media used was the guludan technique. Some of the reasons are:

- One of Pekalongan coastal problems is land subsidence, which if you use bronjong with the basic material in the form of stone, it will increase the rate of Pekalongan land subsidence when compared to mounds whose backfill material is soil.
- The use of bamboo is also more economical than the use of gabion wire.
- Bamboo woven gabions are more flexible than mounds, so it is feared that they will be easily damaged.

4.2. Dimension of Mangrove Planting Structure

There are 3 things that affect the numbers in determining the elevation of the mounds, namely:

- Ups and downs

Based on the environmental data that has been obtained, the tides at the research site are around 1m.

- Land subsidies

So, it is assumed that the Pekalongan area experiences a land subsidence of 10 cm/year.

- Settlement

Soil data was obtained from the Geological and Soil Mechanics Report for the Detailed Design of Flood Control and Pekalongan Rob which can be seen in figure 5.

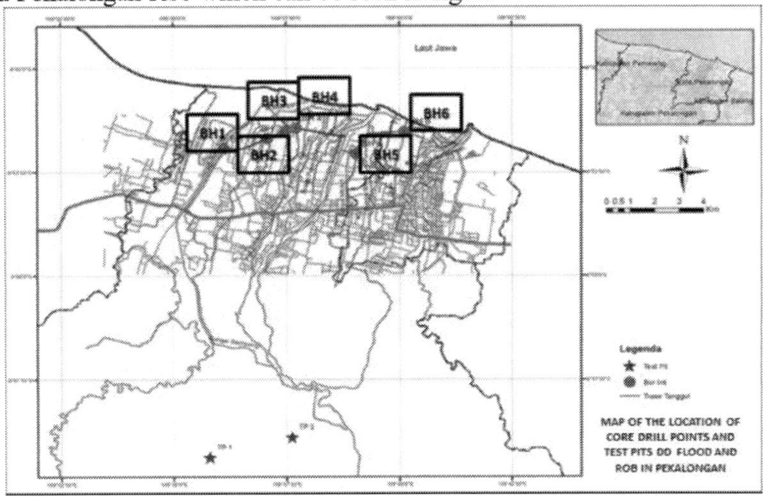

Figure 5. Drill point location map and test pit.

Table 1. Soil mechanics laboratory analysis results (BH-06).

Depth (m)	N-SPT	Classification	Clay or Sand	γ_{total} kN/m^2
0-3	0	Tufaceous silt	clay	16
3-6	4	Sand	sand	18
6-9	4	Silty sand	sand	18
9-12	5	Silty sand	sand	18
12-15	7	Silty sand	sand	18
15-18	6	Silty sand	sand	18
18-21	8	Silty sand	sand	18
...		...	...	...
72-75	45	sand	sand	18

The step of calculation for settlement that occurred due to piles of mounds in the Pekalongan PIM area.

1) Calculate σ_0, the stress of the layer above it with equation 1.

$$\sigma_0 = \gamma_{eff} \times H \quad (1)$$

$\sigma_0 = (16 - 10) \times (0.5 \times 3) = 9\ kN/m^2$

2) Calculate $\Delta\sigma$, embankment stress

Assuming 2 meters with a value of γ = 17 kN/m^3, 1.5 meters submerged in water and 0.5 meters above the water surface.

$$\Delta\sigma = \gamma_{timbunan} \times H_{timbunan} \quad (2)$$

$\Delta\sigma = (7 \times 1.5) + (17 \times 0.5) = 19 kN/m^2$

3) Calculate Settlement with equation 3.

$$S = C_c \frac{H}{1+e_0} log \frac{\sigma_0+\Delta\sigma}{\sigma_0} \quad (3)$$

$$S = 1.5 \frac{3}{1+2} log \frac{9+19}{9} = 0.74\ m$$

Settlement due to embankment of guludan is 0.74 m.

4) Calculate consolidation time with equation 4.

$$t = \frac{T_v \times {H_d}^2}{C_v} \quad (4)$$

Table 2. Calculation of consolidation time in seconds.

U	Tv	t (s)
U=25%	0.04	18000000
U=50%	0.197	88650000
U=75%	0.432	194400000
U=90%	0.848	381600000

Table 3. Calculation of consolidation time in a year.

U	Tv	t (year)
U=25%	0.04	0.58
U=50%	0.197	2.85
U=75%	0.432	6.25
U=90%	0.848	12.27

Where U is percent of settlement. The following in figure 6 is a graph of the relationship between time vs U (25%, 50%, 75%, 90%)

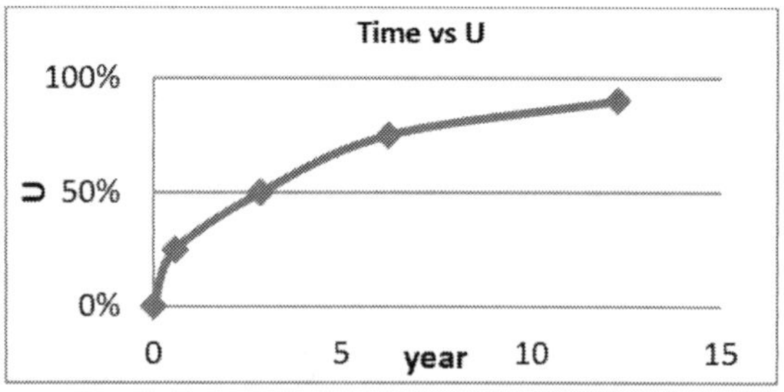

Figure 6. Graph of relationship between time vs U.

The following in figure 7 is a graph of relationship between time vs Settlement (25%,50%,75%,90%).

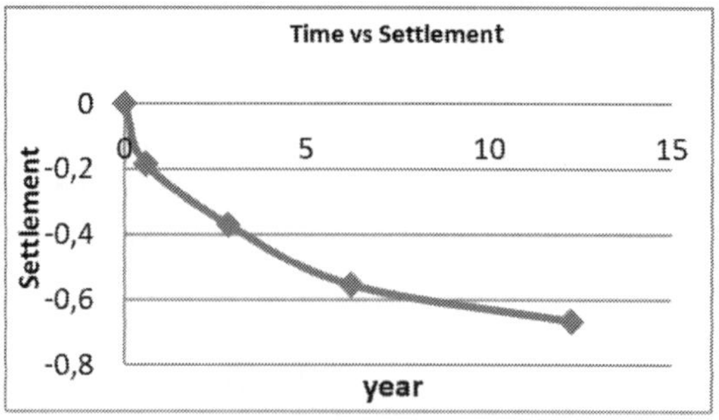

Figure 7. Graph of relationship between time vs settlement.

- Pessimistic elevation = d+ Average tidal height + Land subsidance 13 thn + Sett 90% = 1.5 m + 0.4 m +1.3 m + 0.67 m = 3.87 m ≈ 4 m
- Optimistic elevation = d+ Average tidal height + Land subsidance 3 thn + Sett 50% = 1.5 m + 0.4 m + 0.3 + 0.37 m= 2.57 m ≈ 2.6 m

So that the design of guludan and its dimensions can be illustrated as shown in figure 8.

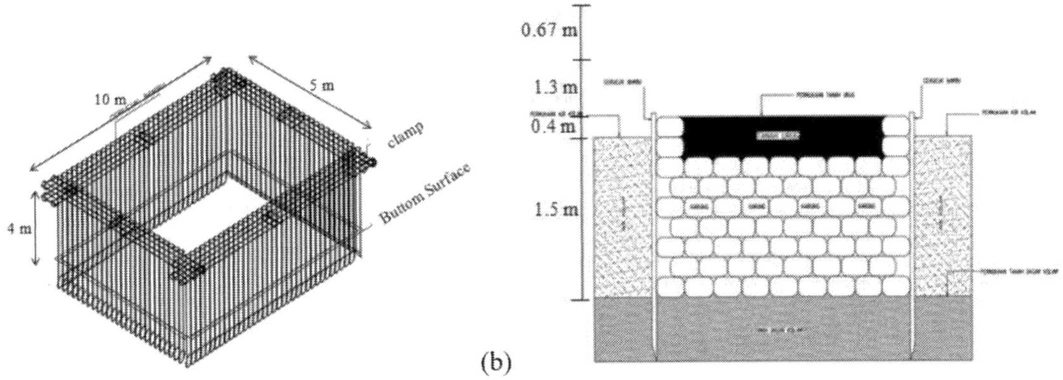

(a) (b)

Figure 8. (a) dimention of guludan (b) Cross section of the guludan that has been filled with soil.

5. The Election of Coastal Protection Layout and Design

The following are factors that must be considered in planning the coastal protection structure in the gaps found on the Pekalongan beach:

- Aims to prevent the erosion of Pekalongan from increasing.
- To prevent erosion from increasing, a structure that functions to catch sediment is needed.
- Because a mangrove park will be built at the research site, water must remain in that area,
- So that water can still enter but the amount can be controlled, a structure that still has access to the meeting between the high seas and the mangrove area is needed.
- The problem faced besides erosion is land subsidence, so what can be done is to add the value of land subsidence in the calculation of the elevation of the coastal protection structure or create a structure that does not depend on the impact of land subsidence, namely a floating structure. However, floating structures cannot capture sediment to overcome erosion problems.
- Environmental data needed to determine the layout of the structure include the direction of the waves and currents.
- The dominant direction of the waves comes from the North and East.
- The highest current velocity is 0.29 m/s 0.3 m/s and 1.12 m/s with a direction towards the Northwest or West along the coast.

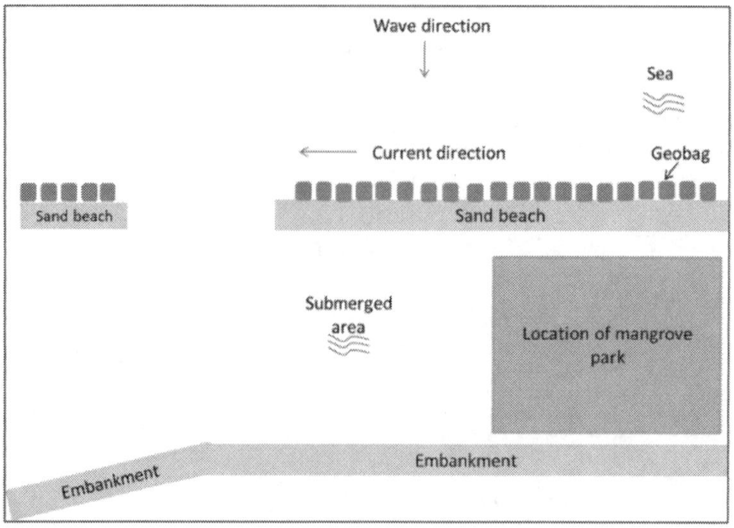

Figure 9 . Top view of the existing research site.

5.1. Selection of Structure Type

Several marine building options was determined as solutions to solve the problems in Pekalongan.

Option 1. Groyne

The groyne structure is the most common type of coastal building in the Java Coast, as indicated in [4].

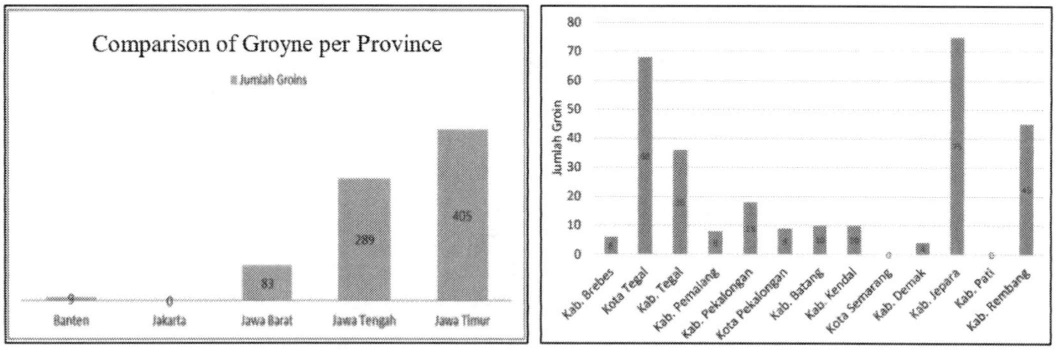

Figure 10. Comparison of the number of groins in Pantura Java.

A rubble mound groyne structure consists of granular stone. The structure layout required special consideration to ensure that it can overcome erosion problems by slowly trapping sediment into the submerged area and exploiting diffraction events to prevent erosion on the outer side of the groyne.

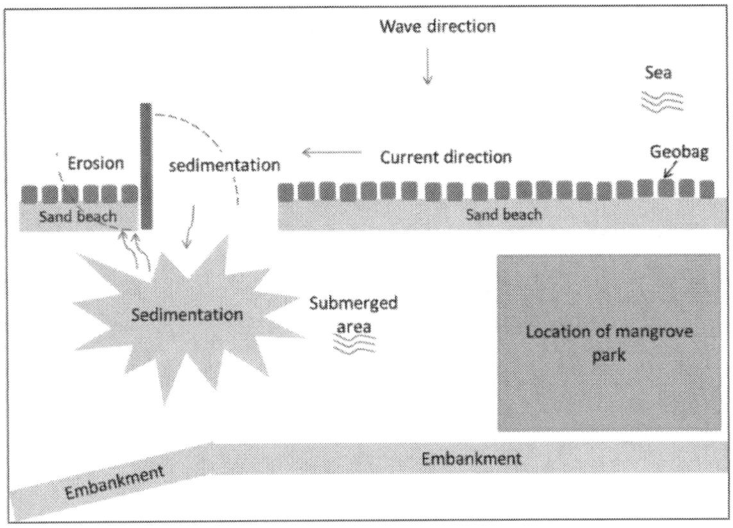

Figure 11. Groin structure sketch of the research site.

Option 2. Gap Breakwater
Gap breakwater is one type of coastal protection building that is quite often found in the Java Coast.

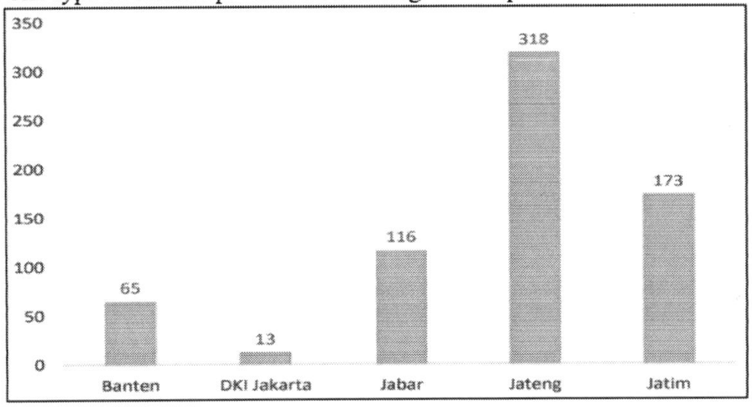

Figure 12. Distribution of gap breakwater per province.

The rubble mound gap breakwater using stone material can overcome erosion problems by utilizing diffraction events (waves through the gap) to slowly capture sediment into the submerged area so that over time it will reduce the impact of erosion experienced by Pekalongan. Breakwater gap also take advantage of gap so as not to block access to seawater as a whole.

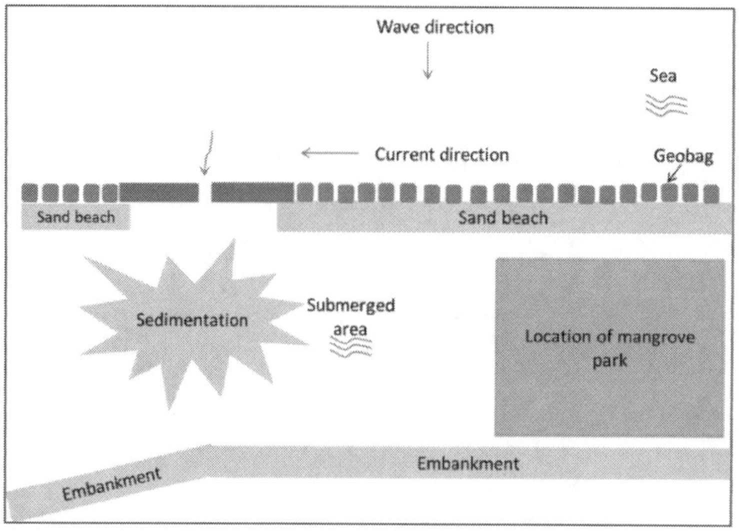

Figure 13. Structural sketch of the gap breakwater at the research site.

Option 3. Offshore Breakwater
In theory, the change in the coastal environment from the use of an offshore breakwater is the formation of a knob on the back side of the breakwater. But the research area is a water-buried area that cannot hold sediment into a sedimented area. Then the sediment will spread into the pond buried area over time so that erosion will be controlled periodically.

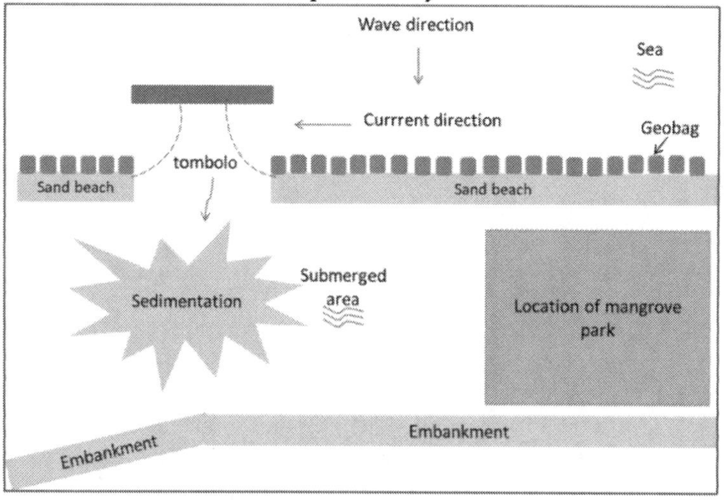

Figure 14. Sketch of the offshore breakwater structure at the research site.

The coastal protection building chosen to overcome the problems that occur in the coastal area of Pekalongan is Groin. The groins used in this thesis have been selected for the rubble mound type using stone materials with the location of the material procurement in Rembang, Central Java, and the distribution using the sea route.

5.2. Dimension of Coastal Protection Layout and Design
Before calculating the dimensions of the groyne structure, the following is the calculation of the breaking waves and the width of the surfzone if Ho = 3 m and T=6 s.

- Calculate wavelength with equation 5.

$$Lo = 1.56\ T^2 \tag{5}$$

$Lo = 1.56 \times 6^2 = 56.16$ m

- Calculate *breaker heigh* indeks (Ω_b) with equation 6.

$$\Omega b = 0.3\left(\frac{H_0}{L_0}\right)^{-\frac{1}{3}} \tag{6}$$

$0.3\left(\frac{3}{56.16}\right)^{-1/3} = 0.797$

- Calculate the value of Kr with equation **7**

$$K_r\sqrt{\frac{\cos\alpha_0}{\cos\alpha}} \tag{7}$$

$\sqrt{\frac{\cos 45}{\cos 15.36695}} = 0.856$

- Calculate height of breaking wave (H_b)

$$\Omega_b = Hb/H' \tag{8}$$

Hb= 0.797 (3 x 0.856) = 2.047 m

- Calculate depth of breaking wave (d_b)

$$\gamma_b = \frac{H_b}{d_b} \tag{9}$$

$0.78 = \frac{2.047}{d_b}$, d_b = 2.624 m

- Calculate wide of surfzone

Surfzone wide = d_b / coastal slope= 2.624 / 0.02 = 131.2 m

1. Groyne length

   Groyne length = 60% x surfzone wide = 60% x 131.2 = 78.72 m

2. Groyne elevation with equation 12.

   Looking for wave set up value with equation 10.

$$S_w = 0.19\left(1 - 2.82\sqrt{\frac{H_b}{gT^2}}\right)H_b \tag{10}$$

$S_w = 0.19\left(1 - 2.82\sqrt{\frac{1.12}{9.81\,(6)^2}}\right)1.12 = 0.179\ m$

Calculate wave run up (Ru) value with Irribaren number as equation 11.

$$\text{Ir} = \frac{tg\,\theta}{\left(\frac{H}{Lo}\right)^{0.5}} \tag{11}$$

$Ir = \frac{1/2}{\left(\frac{1.216}{56.16}\right)^{0.5}} = 3.4$

After knowing the Irribaren (Ir) value, plot it on the graph wave run up to get the $\frac{Ru}{H}$value in order to find the wave run-up value.

$\frac{Ru}{H} = 1.2$, Ru= 1.2 x 1.216 = 1.4592 m

Sea level rise that will occur in 2031 with the best estimate being 20 cm = 0.2 meters (planned building age = 10 years).

The height of land subsidance is assumed to be 10 cm/year with a design age of 10 years, so the land subsidence value for groyne elevation is 1 meter.

Groyne elevation= HHWL+*Sw* +Ru +SLR +*Freeboard+Land subsidance* (12)
Groyne elevation = 0.7 m + 0.179 m + 1.459 m + 0.2 m + 1 m + 1 m = 4.538 m ≈ 4.5 m

Where: HHWL = Higher High Water Lavel, Sw = Wave Set up, SLR = Sea Level Rise, Fr = Freeboard, Ru = Run up, LS = Land subsidence

3. Armor weight

In this plan the groynes will use a protective layer of natural crushed stone. For the protection layer from crushed stone with rough angles n=2 as shown in stability coeffisient table by Triatmodjo.

The Hudson formula is on equation 13.

$$W = \frac{\gamma_r H d^3}{K_D (S_r - 1)^3 \cot\theta} \text{ for } S_r = \frac{\gamma_r}{\gamma_\alpha} \tag{13}$$

With groyn slope 1:1.5 the primary layer value is 360 kg, secondary layer value is 36 kg, and core layer is 3.6 kg.

4. Upper Width and Layer Thickness

The minimum upper width is the same as the thickness of the primary layer, or adjusted according to the needs of the groyne surface function. The thickness of the armor layer based on Hudson (1953) (in Triatmojo, 1990) can be shown in equation 14.

$$t = n\, k_\Delta \left(\frac{W}{\gamma_r}\right)^{1/3} \tag{14}$$

From the calculation the thickness of primary layer is 2 m, secondary layer is 0.98 m, and core layer is 0.56 m.

5. Toe protection

Toe protection hight (t_{toe}) plan 0.5 m.
Toe protection wide (B) = 2H = 2 x 0.5 = 1 m

Based on the cross section of the groynes above, the layout of groyne can be shown at figure 15 and its cross section is found at figure 16.

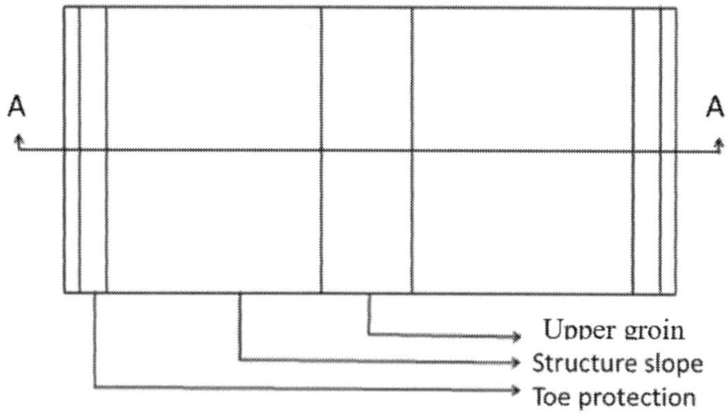

Figure 15. The top view of breakwater.

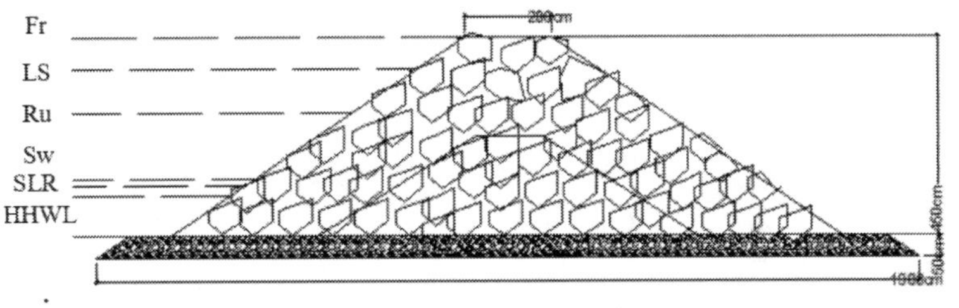

Figure 16. Cross section A-A.

6. Result of Sediment Transport Simulation

The following is the condition of sedimentation and erosion that occurred after a simulation of a numerical model of sediment transport was carried out for 15 days and a year in the scenario conditions before and after the existence of a coastal protection structure.

The scenario after the land subsidence occurs is the current existing condition, while the scenario before the land subsidence is the current condition where the bathymetry is increased 50 cm higher in the area of interest. The modelling domain is based on a grid as shown in figure 17.

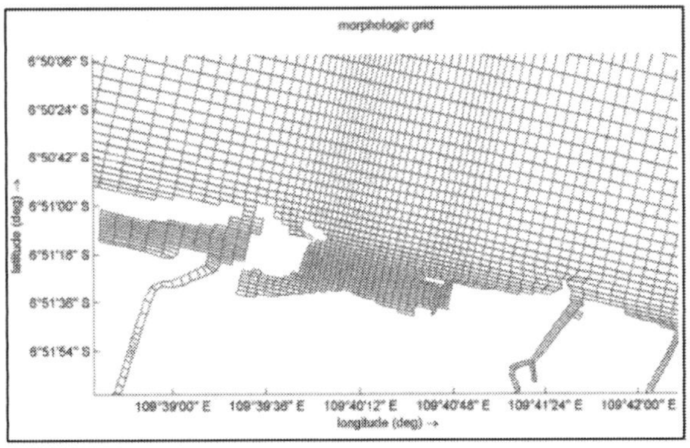

Figure 17. Domain modelling and grid design.

6.1. Sediment Transport Simulation

Single groyne which is located to the west of the gap can be seen at figure 18 and figure 19.

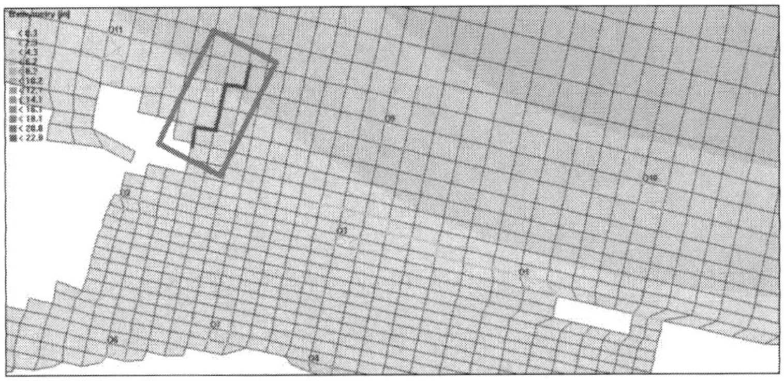

Figure 18. Groyne 1 on the grid.

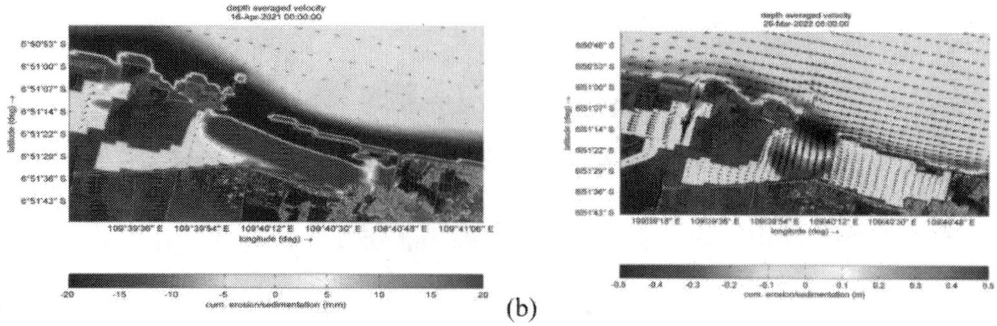

(a) (b)

Figure 19. **(a)** results of the 15 day simulation of sedimentation and erosion and **(b)** results of one year simulation.

Offshore breakwater which is located about 400 m from the coastal line can be seen in figure 20 and figure 21.

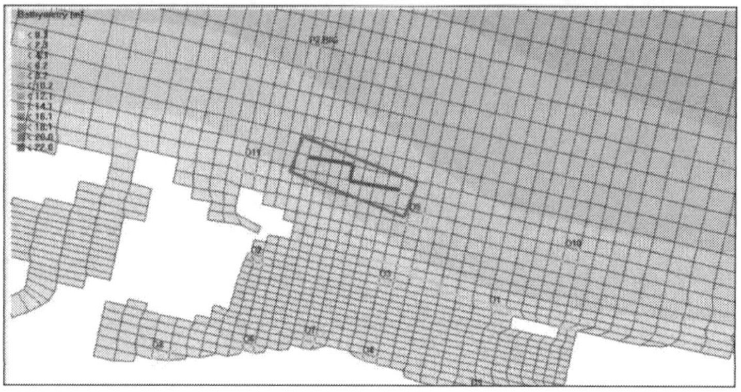

Figure 20. Detached breakwater on the grid.

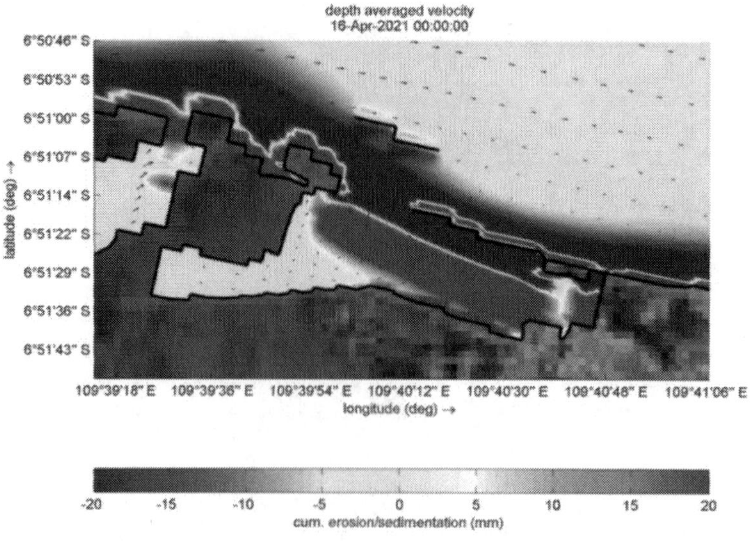

Figure 21. Sedimentation/erosion conditions at the end of day 15 simulation of detached breakwater.

Groin series with a distance between groin is about 300 m can be seen in figure 22 and figure 23.

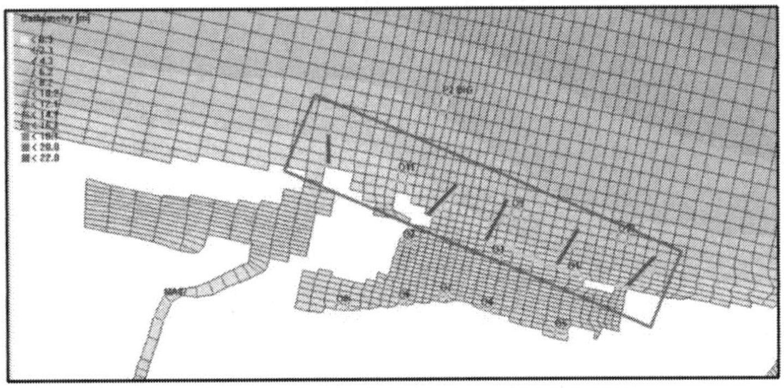

Figure 22. Groin series on the grid.

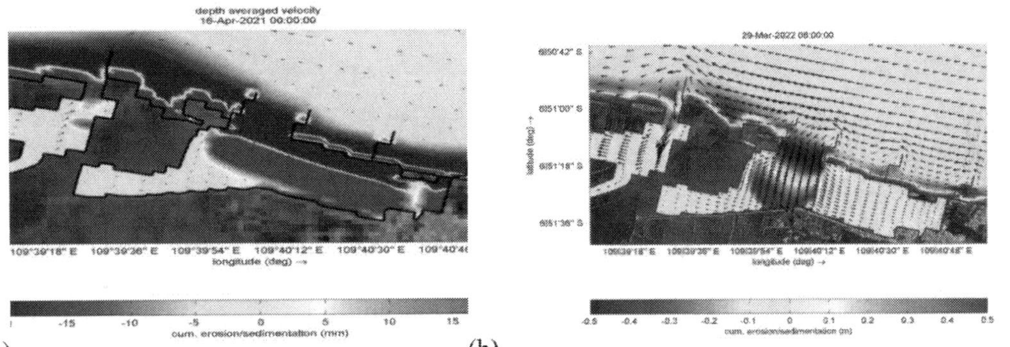

(a) (b)

Figure 23. (a) sedimentation/erosion conditions at the end of day 15 and (b) results of one year simulation.

Green to brown is an indicator of sedimentation and green to blue is an indicator of erosion. By the end of one year simulation, the erosion and sedimentation conditions are getting bigger in certain areas, it shows that the sediment concentration is still quite a lot when compared to conditions after the structure exsist.

6.2. Sediment Transport Analysis

Model results were observed at each cross-section to see the sedimentation and erosion profiles along the cross-section of the review. The location of the created cross-section is shown in figure 24 . The results of the sedimentation and erosion profiles formed along the previously created reference lines are shown in figure 25 until figure 28 . There is a significant influence on erosion and sedimentation along the cross section CD effect of changes in depth causes the sediment transport because section CD is the section that located near the structure.

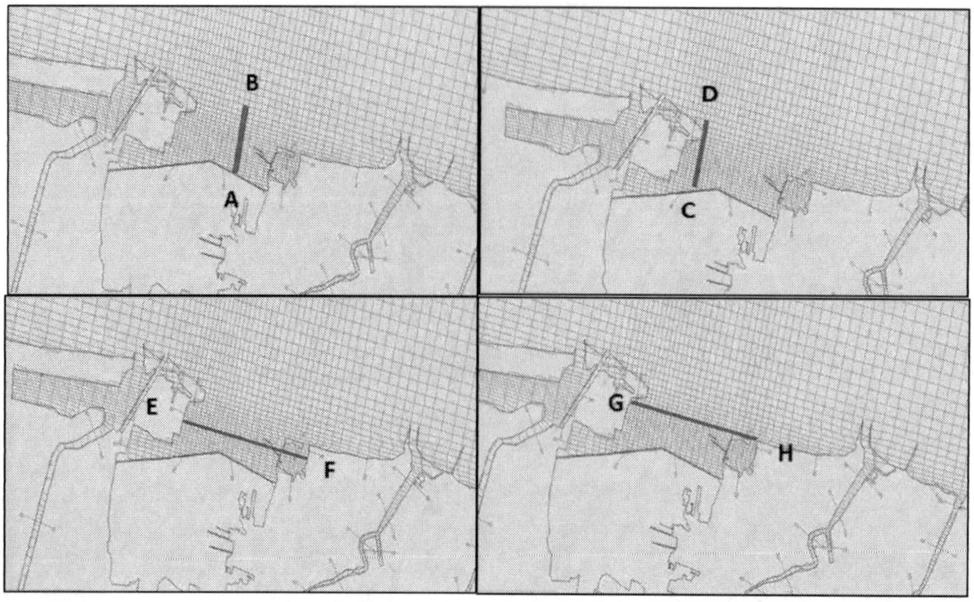

Figure 24. Cross sectional locations in the modelling domain.

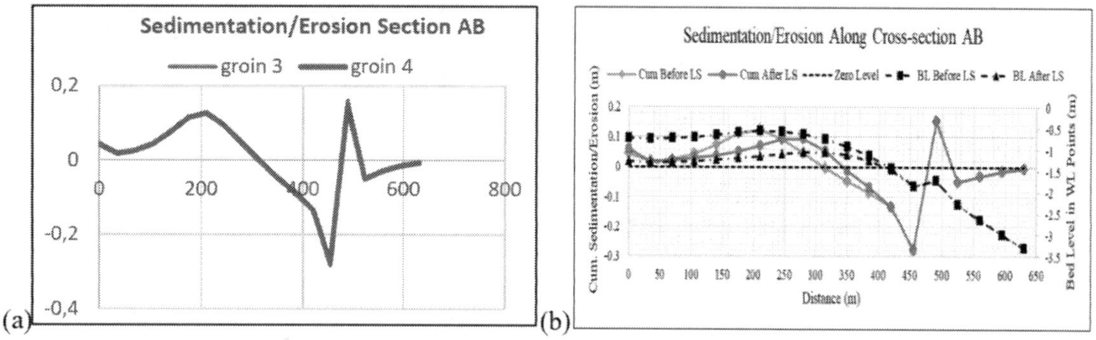

Figure 25. (a) sedimentation/erosion conditions section AB at the end of day 15 after structure existing and (b) before structure existing.

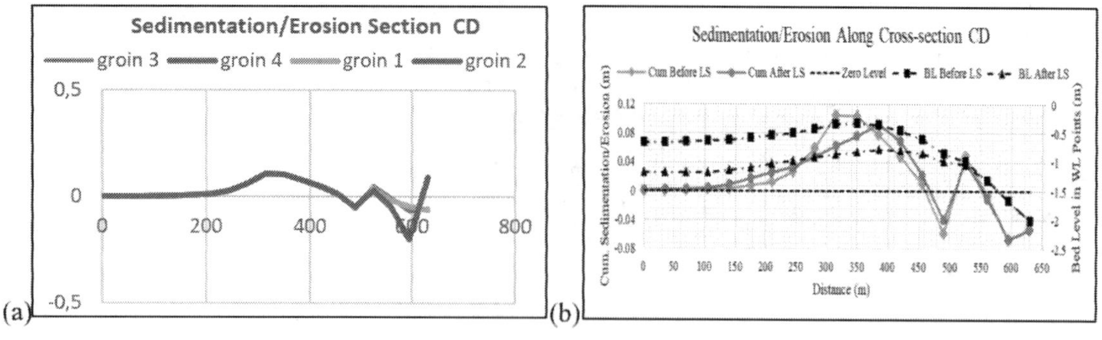

Figure 26.(a) sedimentation/erosion conditions section CD at the end of day 15 after structure existing and (b) before structure existing.

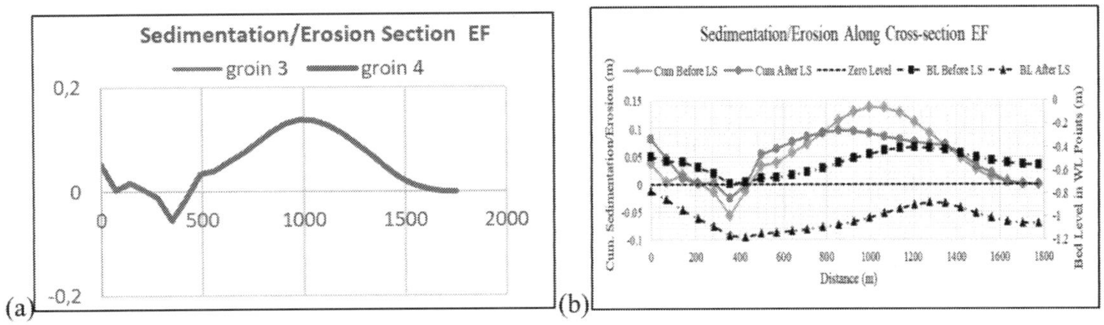

Figure 27. (a) sedimentation/erosion conditions section EF at the end of day 15 after structure existing and (b) before structure existing.

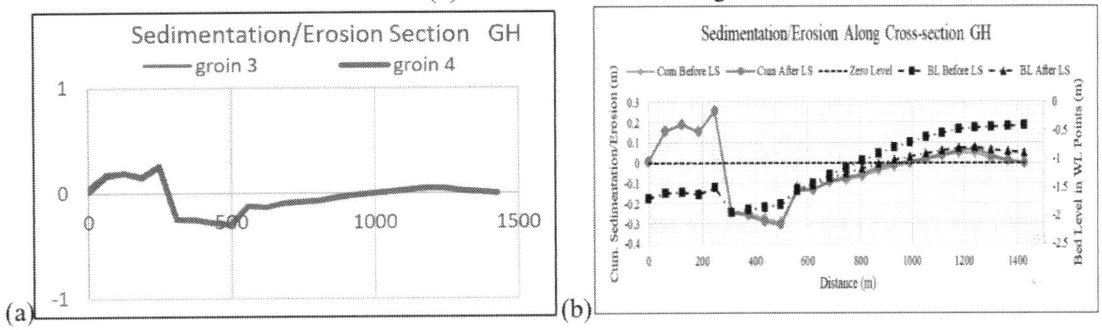

Figure 28. (a) sedimentation/erosion conditions section EF at the end of day 15 after structure existing and (b) before structure existing.

In addition to use the cross-sectional line method, the sampling point method is also used in the research area that represents the embankment area, gap area, and the area around the coast as shown in Figure 29. The results of the analysis at the sample points are shown in Figure 30 for result without structure and Figure 31 for result after structure exist. There is a significant influence on erosion and sedimentation in point O3 effect of the structure when compared with the results of the time series before there is a structure. As it is shown from the time series graph one year after there is structure, the groin series has the maximum effect in increasing sedimentation.

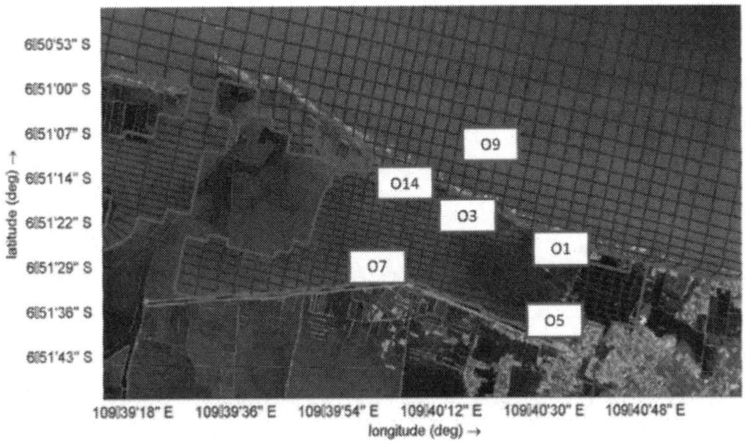

Figure 29. Location of sedimentation and erosion sample points in the area of interest.

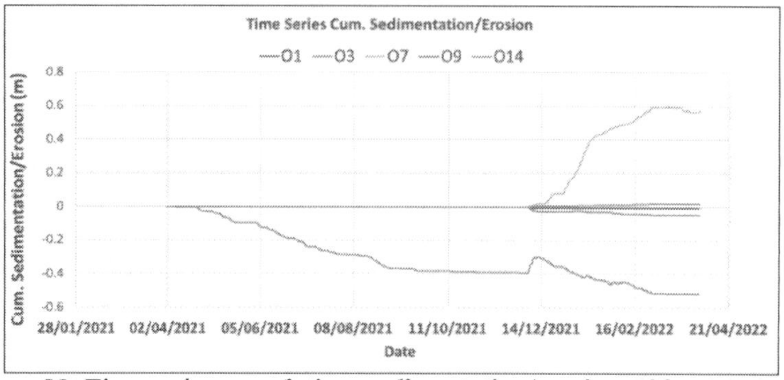

Figure 30. Time series cumulation. sedimentation/erosion without structure.

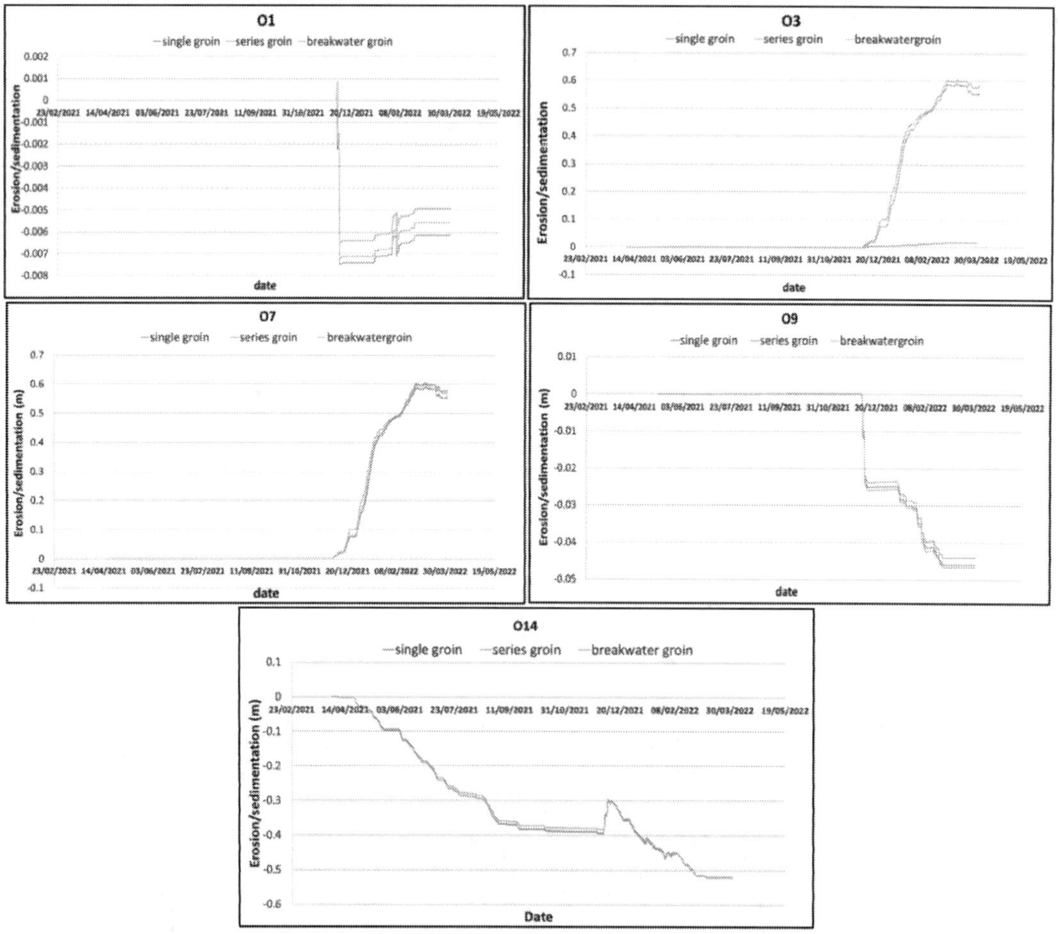

Figure 31. Time series cumulation. sedimentation/erosion one year simulation with structure.

7. Conclusion

From the whole series of observation and analysis processes, it can be concluded several important things related to the planning of coastal protection buildings in Pekalongan, including:

1. Based on the analysis results using existing data, the concept of coastal protection structure is a rubble mound groin with stone material. Groynes were chosen for coastal protection in

Pekalongan because it captures sediment fix the eroded area, and stone material was chosen because it resistance is more than ten years. It is suitable with the long required time to overcome coastal erosion.

2. Based on the results of the analysis using environmental data that has been processed, the structural planning of the building dimensions is as follows: Building Type : Groin; Building Height: 4.5 meters; Upper Width: 2 meters; Building Length: 45 meters; Scope : 1: 1.5; Building Function: Captures sediment to solve erosion problem.
3. Groyne material consisted of rough angled stone with uniform primary, and secondary armor layers weighing 360 kg to minimise construction cost budget. The groyne is designed to have enough pores so that excessive erosion does not occur on the other side of the groyne. The core layer and toe protection are used stones weighing 3.6 kg. The nearest stone material can be taken from the quarry located in Rembang regency and transport by barge.
4. Based on the results of the analysis using existing environmental data, it is obtained the planning of the structure of the planting media for mangrove mounds. And obtained the planning dimensions of the building as follows: Structure Type : Guludan; Height: 4 meters; Width: 5 meters; Length: 10 meters; Settlement value: 0.7 meters.
5. Based on the 1-year modelling simulation from April 1 2021, the groyne series has the maximum effect in increasing sedimentation.

Acknowledgments
This research is partially funded by the ITB Research and Community Service (P2MI) grant through the Coastal Engineering Research Group, Faculty of Civil and Environmental Engineering, and the Indonesia Collaboration Research Program grant from Bandung Institute of Technology initiated by the Ministry of Education and Culture Republic of Indonesia.

References

[1] M.A. Marfai, D. Mardiatno, A. Cahyadi, F. Nucifera, H. Prihatno 2013 *Pemodelan spasial bahaya banjir rob berdasarkan skenario perubahaan iklim dan dampaknya di pesisir Pekalongan*, Jurnal Bumi Lestari, **13(2)** 244 – 256

[2] M.A. Marfai 2014 *Impact of sea level rise to coastal ecology: A case study on the northern part of Java island, Indonesia*, Quaestiones Geographicae, **33(1)** 107 – 114

[3] Elias R, Roelvink D J A, Walstra D J, Stive M and Klein M D 2000 Hydrodynamic Validation of Delft3D with Field Measurements at Egmond 27th International Conference on Coastal Engineering 'ICCE 2000' in Sydney. 1 – 14

[4] Pamfilia C Sangasari, Jansen T, Tawas H 2019 *Perencanaan Bangunan Pengaman Pantai di Pantai Kalinaung Kabupaten Minahasa Utara* Jurnal Sipil Statik 7(8) 975-984

[5] Li L and Huang Z 2013 Modelling the Change of Beach Profile Under Tsunami Wave: A Comparison of Selected Sediment Transport Models Journal of Earthquake and Tsunami. 7(1) 1350001-1 – 1350001-29

[6] Basyir M A S 2015 *Masterplan forMangrove in Pekalongan City*

[7] Nashrullah S, Aprijanto, Pasaribu J M, Hazarika M K and Samarakoon L 2013 Study on Flood Inundation in Pekalongan, Central Java International Journal of Remote Sensing and Earth Sciences. 10(2) 76 – 83

[8] Novico F, Astjario P and Bachtiar H 2013 Erosion-Sedimentation and Tidal Current Conditions Around Coastline in the front of Steam Power Plant Tarahan Lampung Using Delft3D Version 3.28 Jurnal Geologi Kelautan. 11(1) 39 – 54

[9] Pratama M B 2019 Tidal Flood in Pekalongan: Utilizing and Operating Open Resources for Modelling The 2nd EPI International Conference on Science and Engineering. 1 – 10

[10] Prihantono J, Fajrianto I A and Kurniadi Y N 2018 Hydrodynamics and Sediment Transport Modelling Around Tanjung Pontang Coastal Waters, District of Serang-Banten Jurnal

Kelautan Nasional. 13(2) 75 – 88
[11] A.R. Maharlika, S.P. Hadi, Kismartini, A.L. Hoya 2020 *Tidal flooding and coastal adaptation responses in Pekalongan city*, in Proceedings of the 6th International Conference on Energy, Environment, Epidemiology, and Information System
[12] Syahputra I and Yuliana H 2019 *Pemodelan Kecepatan Arus dan Tinggi Gelombang Pada Rencana Breakwater Kolam Pelabuhan Meulaboh Dengan Menggunakan Program Delft3D Jurnal Teknik Sipil Unaya.* 5(2) 59 – 67
[13] Semeidi H, Agung D P, Armyanda T, Aristiya P W dan Afif F N 2020 *Bangunan Pantai di Pantura Jawa, PT Penerbit IPB Press*
[14] Yudowaty S O, Atmodjo W and Wulandari S Y 2012 *Studi Transpor Sedimen di Pantai Slamaran Pekalongan* Journal of Oceanography. 1(2) 197 – 204
[15] Triatmodjo Bambang 1999 *Teknik Pantai* 5(7) 979-8541

Bioprocess of Oil Contaminated Water At Gathering Station Through A Consortium Of Endogenous and Exogenous Bacteria

Sandra Madonna [1,2], Dzalika Nurperbangsari[2], Astrid Meilasari Sugiana[3], and Zulkifliani[4]

[1] Department of Environmental Engineering, Faculty of Civil and Environmental Engineering (FTSL), Bandung Institute of Technology, Bandung, Indonesia
[2] Department of Environmental Engineering, Faculty of Engineering and Computer Science, Universitas Bakrie, Jakarta, Indonesia
[3] Department of Political Science, Faculty of Economics and Social Sciences, Universitas Bakrie
[4] Research and Development Center of Oil and Gas Technology "LEMIGAS", Jakarta,

E-mail: s.ramadonna@gmail.com

Abstract. Production activities of petroleum have resulted in extensive environmental pollution, due to accidents, leaks, and oil spills during these activities. Gathering Station is one production facility at the oil field that serves as a gathering place for several liquids produced from production wells to measure the production flow rate. Wastewater at the Gathering Station that still contains oil, so before being discharged into water bodies should be treated to match the existing quality standards. One of the promising methods to remove the oil-polluted at gathering stations is the use of bioprocess technology, which is an eco-friendly, cost-effective, and sustainable approach. A laboratory study was carried out on the Bioprocess of wastewater contaminated oil at the gathering through a consortium of endogenous bacteria and *Bacillus cereus* and *Pseudomonas aeruginosa* as consortium exogenous bacteria. The major objective of this research is to determine the effectiveness of consortium endogenous bacteria and consortium exogenous bacteria (*Bacillus cereus* and *Pseudomonas aeruginosa*) in degrading oil at the gathering station. Bioprocess was done by using 10% (v/v) inoculum consortium of endogenous bacteria and 10% (v/v) inoculum *Bacillus cereus* and *Pseudomonas aeruginosa* as exogenous bacteria and using oil spill dispersant (OSD) 0.2% (v/v) in media. Oil concentrations were determined by the gravimetric method (SNI 6989.10-2011). The result showed that the mixture of *Bacillus cereus* and *Pseudomonas aeruginosa* as exogenous bacteria and consortium endogenous bacteria in media produced the highest value of efficiency in degradation oil that up to 64.29% during 21 days (from 1062 mg/l to 379.25 mg/l), with a total population 65 x 10^6 cfu/ml, this indicated a synergistic relationship between *Bacillus cereus* and *Pseudomonas aeruginosa* as exogenous bacteria and consortium endogenous bacteria at room temperature (21±1°C), pH range value 8.34-9.68 and Total Dissolved Solids (TDS) range value from 553-167 mg/l.

Keywords: Bioprocess, Endogenous bacteria, Exogenous bacteria, Gathering station, Oil and grease.

1. Introduction

Increasing production activities of petroleum at oil fields contribute to increasing wastewater at gathering stations. Gathering station is one production facility at the oil field that serves as a gathering place for several liquids produced from production wells to measure the production flow rate. It should be treated to match the existing quality standards before ditching argued into water bodies. Petroleum contains a complex combination of aliphatic compounds (e.g. alkanes, alkanes, and cycloalkanes), monoaromatic hydrocarbons (e.g. xylenes, toluene, phenol, benzene, ethylbenzene), and polycyclic aromatic hydrocarbons (e.g. anthracene, and naphthalene). In the environment, these compounds could cause damage to living organisms and human health. Contamination of water and soil from various types of petroleum products results in the migration of hydrocarbons to the unsaturated zone, therefore, because of the toxicity of these pollutants, they have become a primary environmental concern over the last decades. (Cabral et al., 2021)[1]. In soil, contamination of hydrocarbon often leads to possibilities of uptake of the contaminants by the plants that are grown in the contaminated sites. Moreover, from those plants, the hydrocarbon contaminants can further be introduced to animal and human populations through the food chain. (Patowary et al., 2018[2]).

The produced water that containing crude oil flow to gathering station caused by the fluid from the drilling process. Oil and water separation process at gathering station currently uses physical chemical processing using a coagulant in large quantities to help the process of settling small particles that cannot settle by themselves. The use of coagulants in large quantities however requires a lot of costs and associated with a range of problems such as environmental toxicity and resistance to biodegradation (Mulligan et al., 2001[9]; Yeung and Gu, 2011[4]). Bioprocess on waste pit at gathering station using microorganisms can treated residue of petroleum in waste water before being discharged into water bodies. One of the promising methods to treat the oil polluted water is the use of bioprocesses, which is an eco-friendly, cost-effective and sustainable approach. Many researchers have conducted studies involving microorganisms to treat oil contaminated environments. A suitable method that can be adopted to fasten the biodegradation process of such organic compounds (Patowary et al., 2018[2]).

Endogenous microorganisms that live in locations contaminated with petroleum can be used as degrading microbes because they have the ability to degrade contaminants and reduce the degradation power of petroleum (Patowary et al., 2018[2]; Madonna, S et al., 2007[6]). Beside that exogenous bacteria can also used to degrade petroleum contaminants. There are two approaches in the application of the bioprocess: bio stimulation and bioaugmentation. Bio stimulation uses the metabolic capacity of indigenous microbial populations, in this way bioprocess can be carried out by modifying the supporting environment by carrying out bio stimulation to overcome limiting factors that can limit the rate of hydrocarbon biodegradation by indigenous microbial populations. or electron donors, to enhance the biodegradation of contaminants by native microbes. The second approach is in the form of bioaugmentation, namely by adding exogenous microbes that have special abilities to degrade contaminants. A number of factors can inhibit the bioprocess, including limited physical interactions between microorganisms and substances (bioavailability of contaminants and/or bio accessibility) (Lima et al., 2017[3]).

A laboratory study was carried out on the Bioprocess of oil contaminated water at the gathering station through a consortium of endogenous bacteria and *Bacillus cereus* and *Pseudomonas aeruginosa* as consortium exogenous bacteria. The major objective of this research is to determine the effectiveness of consortium endogenous bacteria and consortium exogenous bacteria (*Bacillus cereus* and *Pseudomonas aeruginosa*) in degrading oil at the gathering station.

2. Materials and Methods

2.1. Preparation Media

Nutrient Agar (NA) and Nutrient Broth (NB) media using for culture media. Nutrient Agar (NA) contained the following: 1 g of beef extract, 1 g of peptone, 15 g of agar in 1 L distilled water. Nutrient Broth (NB) contained the following: 1 gram of meat extract, 1 gram of peptone, oil spill Dispersant (OSD) Aquaquick 2000 with a concentration of 0.2% and crude oil of 0.1% (v/v) in 1 L distilled water.

2.2. Preparation Endogenous Bacteria and Exogenous bacteria

Endogenous bacteria were isolated from water at gathering station. The consortium of exogenous bacteria in this study was *Bacillus cereus* and *Pseudomonas aeruginosa* from the Microbiology and Bioprocess Laboratory of Research and Development Center of Oil and Gas Technology "LEMIGAS", Jakarta.

2.3. Activation and Preparation of inoculum

Activation and Preparation of inoculum were carried on erlenmeyer flasks contained 250 ml Nutrient Broth (NB) media on shaker incubator at 100 rpm for 24 hours.

2.4. Bioprocess

The bioprocess was carried out using 32 erlenmeyer flasks containing 200 ml produce water and 10 % (v/v) bacterial suspension was inoculated at 27±1°C and 150 rpm at shaker incubator for 21 days. Periodically (0 day, 7 day, 14 day, and 21 day) measurements of oil & grease concentration, Total population of Bacteria, pH, and Total Dissolved Solid (TDS) was carried out.

2.5. Oil & Grease Test

Oil & Grease contain was carried out based on SNI 6989.10 in 2004[7] with observations on days 0, 7, 14, and 21. Calculation of oil & grease residue (mg/l) using the formula: $((W1-W0) \times 1000)/V$
Where:
W1 = weight of flask filled with oil (mg)
Wo = weight of empty pumpkin (mg)
V = sample volume (200 ml test medium)

The efficiency value of the bioprocess in reducing oil & grease is calculated by the following formula: $((A-B))/B \times 100$
Where:
A = early bioprocess (mg/l)
B = end of bioprocess (mg/l)

2.6. Total Population

Total population was carried out using the Total Plate Count (TPC) method based on SNI 2897, 2008[5]. Every 1 ml of the bioprocesses product in an Erlenmeyer flask was diluted up to 10^{-7} by inserting the 1 ml into 9 ml sterile distilled water. The solution was homogenized using a vortex, then from a dilution of 10^{-3} to 10^{-7}, each was taken as much as 0.1 ml to be inoculated on the solid NA agar medium. This inoculation was carried out twice (Duplo), incubated at 36°C for 48 hours.

2.7. pH

Measurement of pH is based on SNI 06-6989.11 of 2004[8] with observations on days 0, 7, 14, and 21. Performed using a pH meter that has been calibrated with a buffer solution of pH 4, 7, and 10.

2.8. Total Dissolved Solid (TDS)

Total Dissolved Solid (TDS) testing was carried out at the beginning and the end of the observation time, using a calibrated OEM digital TDS meter. Recording the results of reading the scale or numbers on the TDS meter display.

3. Result and Discussion

3.1. Preliminary Treatment

Before the bioprocess, the initial condition of the water is measured, the results of the measurement of these parameters are listed in Table 3.1.

Table 3.1. Parameters Value Before Bioprocess

Treatment	Oil & Grease (mg/l)	Total Population of Bacteria (cfu/ml)	pH	TDS (mg/l)
A	1067.75	43×10^7	9.36	553
B	1053.5	40×10^7	9.25	261
C	1062	39×10^6	8.86	190
Control	1000	30×10^4	8.76	167

Note:
A = 10% (v/v) consortium endogenous bacteria;
B = 10% (v/v) consortium exogenous bacteria;
C = 10% (v/v) consortium exogenous and endogenous bacteria;
Control = without addition bacteria

Based on (Table 3.1), it is known that the oil & grease content ranges from 1000 mg/l to 1067.75 mg/l. The initial content of oil & grease still exceeds The Quality Standards for Wastewater Produced for Oil and Gas Exploration and Production Activities from New On-shore Facilities in 2010[4] which is 25 mg/l, so a bioprocess is needed to reduce oil & grease levels by using consortium endogenous and exogenous microorganisms. Meanwhile, the total population ranged from 30×10^4 cfu/ml – 43×10^7 cfu/ml. The pH value of the media ranges from 8.76 to 9.36, the pH value which tends to be high is influenced by the addition of alkaline (alkaline) Oil Spill Dispersant (OSD). Total Dissolved Solid (TDS) levels in each treatment were influenced by the oil degradation process that occurred, where TDS levels ranged from 553 – 167 mg/l.

3.2. Bioprocess Oil & Grease

Observations on the decrease in oil & fat during bioprocessing were carried out every 7 days for 21 days, observations were made on days 0, 7, 14, and 21 days. The decrease in oil & fat (mg/l) in each treatment is shown in Table 3.2.

Table 3.2. Oil & Grease Concentration (mg/l) and Efficiency Value

Treatment	Observation Time (Days)				Efficiency Value (%)
	0	7	14	21	
A	1067.75	888	685.5	509	52.33
B	1053.5	915.5	670	495.75	52.94
C	1062	873	547.75	379.25	64.29
Control	1000	916	797.25	678.75	32.13

Note:
A = 10% (v/v) consortium endogenous bacteria;
B = 10% (v/v) consortium exogenous bacteria;
C = 10% (v/v) consortium exogenous and endogenous bacteria;
Control = without addition bacteria

Based on Table 3.2 show that decreasing concentration of oil & grease at each treatment. The highest efficiency value of oil & grease biodegradation was 64.29% which degraded from 1062 mg/l to 379.2 mg/l, occurred in treatment C with the addition of an exogenous consortium (10%) to the wastewater containing endogenous bacteria, it is known that there is a synergistic mutualism relationship between the consortium of exogenous bacteria and the consortium of endogenous bacteria in the media. In treatment A using a consortium of endogenous bacteria and B using a consortium of exogenous bacteria, it can be seen that the value of efficiency on reducing oil & grease is not much different in the two treatments, which are 52.33% and 52.94%. This shows that consortiums of exogenous bacteria and endogenous bacteria can degrade oil & grease. In the control that was not given the addition of a bacterial consortium, the efficiency value of reducing the concentration of oil & grease was the lowest of all treatments at 32.13%. This decrease occurred due to the influence of evaporation that occurred during incubation and due to the presence of endogenous bacteria present in the media, as well as the presence of OSD in the treatment which could cause oil & grease to be dispersed into smaller lumps into the water.

The Table 3.2 also shows that the consortium of endogenous bacteria (Treatment A) on the 7th day of observation degraded faster than the consortium of exogenous bacteria (Treatment B) with the concentration of decreasing oil & grease that occurred was 888 mg/l at treatment A and 915.5 mg/l at treatment B. This shows that endogenous bacteria was adapted more quickly because these bacteria come from the wastewater environment at gathering station, while exogenous bacteria are still adapting on the 7th day. On the 14th day of observation, the consortium of exogenous bacteria had adapted to the new environment and was able to degrade oil & grease faster than endogenous bacteria with the decreased concentration of oil & grease degradation that occurred was 685.5 mg/l (Treatment A) and 670 mg /l (Treatment B).

3.3. Total Population of Bacteria

Total Population of Bacteria was observed every 7 days for 21 days to find out how many populations of bacteria degraded oil & grease in this study. Total Population Bacteria of each observation are shown in Table 3.3.

Table 3.3. Total Population of Bacteria (cfu/ml) During Bioprocess

Treatment	Observation Time			
	0 Day	7th Day	14th Day	21st Day
A	43×10^7	57×10^7	30×10^7	85×10^6
B	40×10^7	39×10^7	44×10^7	30×10^6
C	39×10^6	61×10^7	43×10^7	65×10^6
Control	30×10^4	34×10^4	41×10^4	31.5×10^6

Note:
A = 10% (v/v) consortium endogenous bacteria;
B = 10% (v/v) consortium exogenous bacteria;
C = 10% (v/v) consortium exogenous and endogenous bacteria;
Control = without addition bacteria

Based on Table 3.3, it is known that in each treatment there was an increase in the number of populations. In treatments A and C the population tends to increase in the same way on the 7th day and decrease on the 14th day. In treatment A with the addition of the endogenous consortium the initial population was 43 x 107 cfu/ml, on the 7th day the bacterial population increased by 57 x 10^7 cfu/ml, and on the 14th day, it decreased with the bacterial population as much as 30 x 10^7 cfu/ml, then on day 21, it decreased again to 85 x 10^6 cfu/ml. In the initial population in treatment C with the addition of a consortium of exogenous bacteria containing endogenous bacteria, the initial bacterial population was 39 x 10^6 cfu/ml, the same as treatment A increased on the 7th day with a bacterial population of 61 x 10^7 cfu/ml and decreased on the 14th day of 43 x 10^7 and the 21st day of 65 x 10^6 cfu/ml. Treatment B showed different fluctuations. Treatment B with the addition of an exogenous consortium at the beginning of the observation (day 0) the initial population was 40 x 10^7 cfu/ml, on day 7 it decreased by 39 x 10^7 cfu/ml and increased with a population of 44 x 10^7 cfu/ml , and on day 21 it decreased again by 30 x 10^6 cfu/ml. This shows that exogenous bacteria are still in the adaptation phase to the environment and then begin to enter the log phase on the 14th day.

The control media that contained water from the gathering station and only contained endogenous bacteria in water with an initial population of 30 x 10^4 cfu/ml and there was a not increase too high on the 7th to 14th day of observation, but on the 21st day the population increased rapidly that is 31.5 x 10^6 cfu / ml. This indicates that the endogenous bacteria on the control medium undergoes adaptation phase until the 14th day and increased population growth rapidly at the end of the observation on 21st day. The population increases due to that bacteria using hydrocarbons as carbon source for bacteria nutrient.

The results showed that at the end of the observation (21 days) the bacteria was in the death phase due to unfavorable environmental conditions where the pH was too high which affected the growth of bacteria, in line with a decrease in the number of populations bacteria in all treatments. The increase in pH value occurs due to the decomposition process of nitrogen compounds carried out by a consortium of endogenous and exogenous bacteria during biodegradation.

3.4. pH

The pH value affects the ability of bacteria to maintain the continuity of cellular activity, cell membrane transport, and the balance of reactions that catalysed by enzymes. The pH value of each treatment during bioprocess can be seen at Table 3.4.

Table 3.4. The pH Value During Bioprocess

Treatment	Observation Time			
	0 Day	7th Day	14th Day	21st Day
A	9.36	9.44	9.24	9.41
B	9.25	9.42	9.19	9.68
C	8.86	8.45	8.83	9.03
Control	8.76	8.34	8.73	8.72

Note:
A = 10% (v/v) consortium endogenous bacteria;
B = 10% (v/v) consortium exogenous bacteria;
C = 10% (v/v) consortium exogenous and endogenous bacteria;
Control = without the addition bacteria

The pH value of treatment A with the endogenous consortium at the beginning (day 0) had a pH value of 9.36 then, on the 7th day it increased to 9.44, and decreased on the 14th day to 9.24 and on the 21st day increased to 9.41. Treatment B with addition exogenous consortium at the beginning of the observation (day 0) the pH value was 9.25 and fluctuated like the endogenous consortium with the pH values being 9.42 (7th day), 9.19 (14th day), and 9.68 (21st day). Meanwhile, the pH value in treatment C with addition consortium exogenous and endogenous bacteria, had a smaller value than treatment A and B. the pH value at the beginning of the observation (day 0) was 8.86, and than it decreased to 8.45 on the 7th day, and reach to 8.83 on the 14th day and at the end of the observation (21st day) reach to 9.03. The high pH value due to addition of the alkaline OSD in each treatment.

3.5. *Total Dissolved Solid (TDS)*

The Total Dissolved Solid (TDS) test is carried out as supporting data to determine the solids dissolved in water during the degradation process of oil & grease by a consortium of bacteria, where these solids consist of organic substances, organic salts, and dissolved gases. Data on the reduction in TDS levels are shown in Table 3.5.

Table 3.5. The Total Dissolved Solid (TDS) (mg/l)

Treatment	Observation Time	
	0 Day	21st Day
A	553	263
B	261	259
C	190	170
Control	167	168

Note:
A = 10% (v/v) consortium endogenous bacteria;
B = 10% (v/v) consortium exogenous bacteria;
C = 10% (v/v) consortium exogenous and endogenous bacteria;
Control = without addition bacteria

At the beginning of the observation (day 0) the TDS level in treatment A with the addition a consortium of endogenous bacteria had TDS level of 553 mg/l and treatment B with the addition a consortium of exogenous bacteria, with a TDS of 261 mg/l, and in treatment C with the addition a consortium of exogenous bacteria and endogenous bacteria with a TDS of 190 mg/l, while the control without the addition of a bacterial consortium had a TDS level of 167 mg/l. The addition a consortium of bacteria to the media causes the dissolved organic matter levels increased.

During 21 days of observation, TDS levels in each treatment decreased, respectively 262 mg/l in the consortium of endogenous bacteria, 259 mg/l on the consortium of exogenous bacteria, 170 mg/l on the consortium of exogenous bacteria containing endogenous bacteria, while the control without the addition of a bacterial consortium increased by 168 mg/l. The decrease in TDS occurred due to the effect of biodegradation of oil & grease by a consortium of bacteria, biomass that occurs in endogenous bacteria which causes the amount of dissolved solids increased.

4. Conclusions

The result showed that the mixture of *Bacillus cereus* and *Pseudomonas aeruginosa* as exogenous bacteria and consortium endogenous bacteria in media produced the highest value of efficiency in degradation oil that up to 64.29% during 21 days (from 1062 mg/l to 379.25 mg/l), with a total population 65 x 10^6 cfu/ml, this indicated a synergistic relationship between *Bacillus cereus* and *Pseudomonas aeruginosa* as exogenous bacteria and consortium endogenous bacteria at room temperature (21±1°C), pH range value 8.34-9.68 and Total Dissolved Solids (TDS) range value from 553-167 mg/l.

References

[1] Cabral, L., Giovanella, P., Pellizzer, E. P., Teramoto, E. H., Kiang, C. H., & Sette, L. D. (2022): Microbial communities in petroleum-contaminated sites: Structure and metabolisms. *Chemosphere*, ***286*** (March 2021). https://doi.org/10.1016/j.chemosphere.2021.131752

[2] Patowary, R., Patowary, K., Kalita, M. C., & Deka, S. (2018): Application of biosurfactant for enhancement of bioremediation process of crude oil contaminated soil. *International Biodeterioration and Biodegradation*, ***129*** (November 2017), 50–60. https://doi.org/10.1016/j.ibiod.2018.01.004

[3] Lima, A. T., Hofmann, A., Reynolds, D., Ptacek, C. J., Van Cappellen, P., Ottosen, L. M., Pamukcu, S., Alshawabekh, A., O'Carroll, D. M., Riis, C., Cox, E., Gent, D. B., Landis, R., Wang, J., Chowdhury, A. I. A., Secord, E. L., and Sanchez-Hachair, A. (2017): Environmental Electrokinetics for a sustainable subsurface, *Chemosphere,* **181**, 122–133. https://doi.org/10.1016/j.chemosphere.2017.03.143

[4] Yeung, A.T., Gu, Y.Y., 2011. A review on techniques to enhance electrochemical remediation of contaminated soils. J. Hazard. Mater. **195**, 11–29.

[5] SNI 2897:2008, 2008. *Metode Pengujian Cemaran Mikroba dalam Daging, Telur, dan Susu, serta Hasil Olahannya*. Jakarta: BSN.

[6] Madonna, S.Effendi, A. J., & Mulyono, M. (2007): Bioremediasi Tanah Terkontaminasi Hidrokarbon Minyak Bumi TotalFinalElf dengan Menggunakan Teknik Biopile pada Skala Laboratorium, *Jurnal Biologi Lingkungan,* **1(2)**, 78–88.

[7] SNI 06-6989.10-2004, 2004. *Cara Uji Minyak dan Lemak secara Gravimetri*. Jakarta: BSN.

[8] SNI 06-6989.11-2004, 2004. *Cara Uji Derajat Keasaman (pH) dengan Menggunakan Alat pH Meter*. Jakarta: BSN.

[9] Mulligan, C.N., Yong, R.N., Gibbs, B.F., 2001. Surfactant-enhanced remediation of contaminated soil: a review. Eng. Geol. **60**, 371–380.

Health risk-based prioritization approaches of pharmaceuticals in the Upper Citarum River Basin

R R Utami[1,2,*], I R S Salami[1], G W Geerling[3,4], S Notodarmojo[1] and A M J Ragas[4]

[1]Department of Environmental Engineering, Faculty of Civil and Environmental Engineering, Institut Teknologi Bandung, Jl. Ganesha no. 10, Bandung 40132, Indonesia
[2]Institute for Science in Society, Faculty of Science, Radboud University, Nijmegen, The Netherlands
[3]Department of Water Quality and Ecology, Deltares, Delft, the Netherlands
[4]Department of Environmental Science, Radboud Institute for Biological and Environmental Sciences, Faculty of Science, Radboud University, Nijmegen, The Netherlands
*Corresponding Author

rosetyatiretno@gmail.com

Abstract. Two methods were developed to prioritize human health risks of pharmaceuticals based on defined daily doses (DDDs) obtained from the WHO and acute toxicity data on pharmaceuticals (TDLo) obtained from the PubChem database. A major advantage of these methods is that it requires no acceptable daily intake (ADI), reference dose (RfD), or tolerable daily intake (TDI) values, which are often unavailable or difficult to obtain for pharmaceuticals group. Both methods were applied to prioritize 16 pharmaceuticals detected by passive samplers in the water source of Ciwalengke Village, a peri-urban area in the Upper Citarum River Basin. Ten out of 16 pharmaceuticals could be prioritized by using both methods. The risk indicator values (RI_{hum}) showed good agreement between the two approaches, especially for Caffeine and Trimethoprim. The priority ranking of Caffeine and Lidocaine were always in the top 3 highest human health risks. The ranking comparison showed considerable consistency, indicating that both prioritization methods are reasonably in line with each other. The pharmaceutical intake through the exposure of contaminated river water was expected to pose negligible human health risks (i.e. $RI_{hum} < 1$) but can be refined more by conducting supplementary effects studies for other individual pharmaceuticals or pharmaceuticals mixtures.
Keywords: DDD; HRV; Pharmaceuticals; Prioritization; Health Risk; TDLo.

1. Introduction

Pharmaceuticals are used to prevent and treat animal and human diseases. After use, pharmaceuticals and their metabolites can be released to the environment, e.g. through excretion and emission into sewerage systems, from aquaculture and meadow livestock, or application of manure to agriculture [1]. The pharmaceuticals that are discharged to water can cause adverse effects for aquatic ecosystems [2, 3, 4, 5, 6] and human health [7, 8, 9, 10, 11]. Not all impacts are directly visible and observable; some impacts are difficult to detect, such as delayed and subtle effects (e.g. behavioural changes, chronic effect, bioaccumulation, etc.) [12, 13, 4, 2, 14, 64, 65]. One of the major challenges for risk assessment of pharmaceuticals is that information on potential adverse impacts of pharmaceuticals in water is still limited [15]. This is partially due to the fact that toxicity testing of pharmaceuticals is time-consuming, complicated and expensive [16]. Moreover, there are more than 1500 active pharmaceutical ingredients (APIs) being used by society, some in substantial quantities [60]. It is therefore hardly surprising that APIs and their major metabolites frequently appear in the surface waters around the globe [17, 61]. To deal with this myriad of APIs and metabolites, prioritization approaches are needed to reduce the number of pharmaceuticals to be studied in order to identify and focus on pharmaceuticals that are likely to pose the greatest risk [18, 19, 16].

Several pharmaceutical prioritization methods have been applied around the world [20, 16, 21, 22, 23, 62]. A study of risk prioritization assessed 99 pharmaceuticals in cities of Iraq (i.e. Basrah, Mosul, Baghdad). The result revealed that only analgesics, antidepressants, and antibiotics groups which had high risk for aquatic and terrestrial ecosystem [20]. Another study showed 67 pharmaceuticals and personal care products (PPCPs) were detected in the effluent from municipal wastewater treatment plants of Beijing. A risk prioritization scenario was applied and 17 PPCPs were classified as high risk priority with the highest risk were ofloxacin, 17a-ethynylestradiol, dibutyl phthalate, dioctyl phthalate, and sulfamethoxazole [23]. Most prioritization studies indicated that less than 10 compounds had high risk [24]. It signified that only a restricted number of pharmaceuticals cause the major part of the risk. However, these prioritizations only assessed the risk for aquatic ecosystems. In some Indonesian areas, river water is still being used a water source for daily domestic activities and drinking water [25, 26, 27]. Therefore, prioritization based on human health risk is crucial for appropriate risk management of river basins where the communities are dependent on the use of river water, like in Indonesia.

Health risk-based prioritization approaches require health-based reference values (HRVs) such as the tolerable daily intake (TDI), acceptable daily intake (ADI), or reference dose (RfD) [28, 29, 19, 30]. Pharmaceuticals prioritization can be particularly challenging because HRVs are often lacking. In order to overcome limited availability of HRVs, this study proposes two methods to prioritize pharmaceuticals in the absence of HRVs, i.e. based on defined daily dose (DDD) and based on acute toxicity data. These methods may be used by the authorities in proposing suitable mitigation and/or monitoring measures for health protection in connection with pharmaceuticals contamination in surface water. The prioritization approaches were applied in a tributary of the Citarum River in Ciwalengke Village, Upper Citarum River Basin, where the residents still use river water as their main water source for domestic activities and drinking water [31, 32, 33, 34].

2. Materials and method

The procedure of human health risk-based prioritization for pharmaceuticals is illustrated in figure 1. The prioritization was based on a non-dimensional indicator of human health risk (RI_{hum}) that was estimated for each substance individually using the equation (1) below:

$$RI_{hum} = \frac{CDI}{HRV} \qquad (1)$$

where CDI is the estimated chronic daily intake (mg/kg/day), and HRV is the health-based reference value (mg/kg/day). The HRVs used in the present study were derived using two methods, i.e.:

a. HRVs were derived from the therapeutic concentration/oral dosing. It was assumed that for most pharmaceuticals no toxic effects will occur at the normal dosing regimen. The normal dosing has been defined by the World Health Organisation in the Defined Daily Dose (DDD). A safety factor of 1000 was applied to derive an HRV from the DDD.

b. HRVs were derived from acute toxicity data (TDLo) obtained from the PubChem database. Assuming all Toxic Dose Low (TDLo) values had the same exposure duration, the lowest TDLo value was used as HRV after division by an assessment factor of 1000.

The CDI value (mg/kg/day) value was calculated with the assumption that the river water was directly ingested as drinking water and by implementing the following equation (2):

$$CDI = \frac{C \times IR}{BW} \tag{2}$$

C is the pharmaceuticals concentration (mg/L) in the surface water (obtained by passive sampling); IR is the ingestion rate (with the assumption of 2 L/day); also BW as the body weight (assumed to equal 60 kg for adults; [38]).

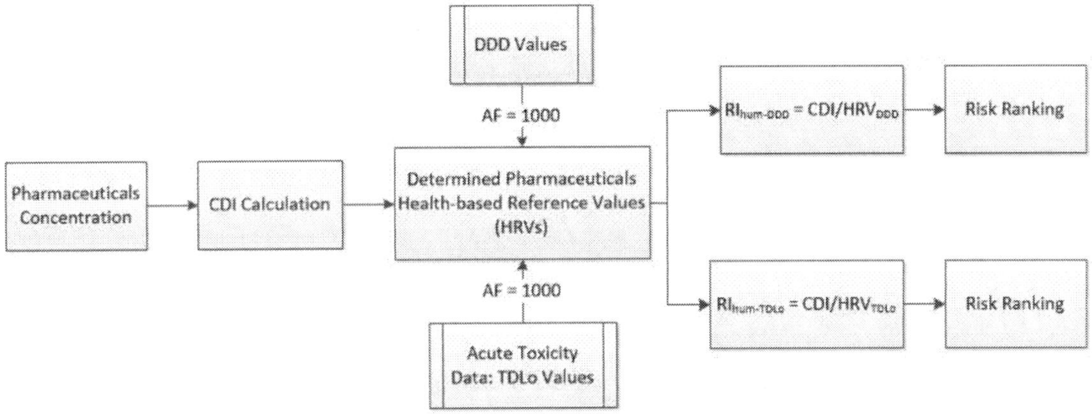

Figure 1. Diagram of pharmaceuticals health risk-based prioritization approaches.

Pharmaceutical concentrations were obtained by passive sampling. Passive sampling is a cost-efficient environmental monitoring technique for long-term on-site pollutant monitoring [39, 40, 41, 42, 43, 44]. The passive sampling technique and its methods had been subjected to extensive testing compared to traditional sampling [45, 46, 47, 48, 49].

We deployed two types of passive samplers, i.e. speedisk and silicon sheets, to cover the polar range of substances (hydrophobic-hydrophilic). Prior to field deployment, the silicon sheets were rinsed with performance-reference-compounds (PRCs) for estimating the sampling rate (L/d) and aqueous phase concentrations of pharmaceuticals in both speedisks and silicon sheets [50, 51]. After the period of sampling, the samplers were cleaned, cooled, and the active pharmaceutical ingredients were extracted and analyzed through gas chromatographic (GC) and high performance liquid chromatographic (HPLC).

In this study, we used similar sampling procedures and passive samplers analyses as described in Utami et al. (2020; [28]). The passive samplers were deployed for four weeks in the dry season (August 18th, 2016 – September 20th, 2016). For the deployment, three speedisks and six silicon sheets were linked on a iron bar and were kept immersed for the duration of sampling period. In spite of polar ranges differences, the concentration uptake of silicon and speedisk samplers might be partly overlaps. Each time the measurement produced two concentration values, we used the highest value as the input of risk prioritization procedure.

The passive samplers were deployed in two locations in Ciwalengke Village. The first one is in the water source inlet of the village and the second location is in the water channel in the middle of the village. Ciwalengke is a neighborhood located in Sukamaju Village of Majalaya District, Bandung Regency and included in the Upper Citarum River Basin (UCRB) area (7.0522° S, 107.7533° E; figure 2).

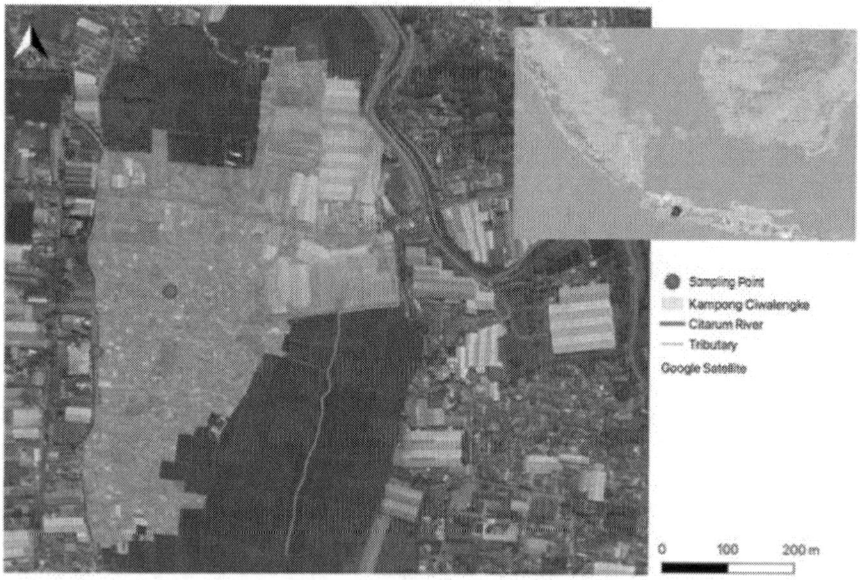

Figure 2. Location of the passive sampler in the Ciwalengke Village.

Ciwalengke Village covers an area of about 90 ha and is surrounded by factories and rice fields (figure 2). It is one of the slums in the UCRB with approximately 4000 inhabitants [33]. The lack of a clean water service system and the limited access to piped water causing the residents to use unprotected water sources from the irrigation channel, originating from a tributary of the Citarum River [31, 32]. This channel is contaminated with waste discharged by the surrounding industries [33, 34]. The water flows directly into the village through pipes and is stored in a tub and shallow well. The residents use the river water for their daily activities. The water does not fulfil the criteria of water source quality based on Indonesian Regulation [52, 53, 33, 54, 55].

3. Results and discussion

3.1 Pharmaceuticals HRV derivations

Sixteen pharmaceuticals were detected in the sampling area. The highest concentration was found for Lidocaine with 350 ng/L, followed by Caffeine and Ibprofen with 310 ng/L and 60.8 ng/L, respectively. The lowest concentration was found for Sulfadiazine with 0.3 ng/L. Table 1 lists the measured concentrations and the health-based reference values (HRVs) derived for the 16 pharmaceuticals using DDD values from WHO and acute toxicity data (TDLo values) from the PubChem Database.

Table 1. Measured water concentrations, chronic daily intake (CDI), and health-based reference values (HRVs) derived from DDD and TDLo values.

Pharmaceuticals	CAS	Concentration (ng/L)	CDI (mg/kg.day)	HRVs (mg/kg/day)[a]	
				DDD	TDLo
BPA	80-05-7	45.00	1.5.E-06	NA	NA
Caffeine	58-08-2	310.00	1.0.E-05	6.7.E-03	7.0.E-03
Chloramphenicol	56-75-7	22.69	7.6.E-07	5.0.E-02	2.5.E-02
Chloroquine	54-05-7	0.82	2.7.E-08	8.3.E-03	2.0.E-03
Cytarabine	147-94-4	2.88	9.6.E-08	NA	9.5.E-05
Estrone	53-16-7	18.00	6.0.E-07	1.7.E-05	NA
Furazolidone	67-45-8	0.62	2.1.E-08	NA	NA
Ibuprofen	51146-56-6	60.80	2.0.E-06	2.0.E-02	8.0.E-03
Lidocaine	137-58-6	350.00	1.2.E-05	5.0.E-02	1.6.E-02
Lincomycin	154-21-2	7.85	2.6.E-07	3.0.E-02	NA
Paracetamol	103-90-2	38.93	1.3.E-06	5.0.E-02	5.0.E-03
Progesterone	57-83-0	2.18	7.3.E-08	5.0.E-03	NA
Salicylic acid	69-72-7	2.61	8.7.E-08	4.7.E-02	1.1.E-02
Sulfadiazine	68-35-9	0.30	9.9.E-09	1.0.E-02	2.9.E-02
Sulfamethoxazole	723-46-6	15.43	5.1.E-07	3.3.E-02	1.6.E-02
Trimethoprim	738-70-5	21.00	7.0.E-07	6.7.E-03	6.0.E-03

[a] NA: Not Available

From the WHO DDD values, 13 out of 16 pharmaceuticals could be processed to obtain HRVs. BPA, Cytarabine and Furazolidone did not have DDD values, so the HRVs could not be derived. From acute toxicity data from the PubChem Database, 11 out of 16 HRVs could be derived. BPA, Estrone, Furazolidone, Lincomycin, and Progesteron did not have a TDLo value.

No toxic effects are expected at the DDD. However, a safety factor of 1000 was applied because the DDD applies for short term exposure (e.g. 3-10 days in a row) while the human reference values such as the ADI, TDI and RfD reflect a lifetime daily exposure. The height of the assessment factor does not affect the ranking of the drugs, but it does affect the absolute value of the risk indicator.

Whether an HRV should be derived from acute toxicity data obtained from literature (PubChem Database) is a matter of debate among risk assessors. In total, 250 acute toxicity data were extracted from the PubChem Database (i.e. LD50, LDLo, TDLo, EC50, LC50), but only TDLo values (56 data) were used since lethality was considered an inappropriate effect indicator for the present prioritization exercise. An assessment factor of 1000 was applied to reduce the uncertainty in the HRV.

In an exposure risk assessment, uncertainties can be present in the characterization of the exposure scenario, the parameter estimates, model predictions, etc [63]. To cover those uncertainities, we need to apply an assessment factor. If an assessment factor equals to or higher than 100 is used, it signifies a high level of uncertainty [37]. In this study we faced the uncertainity of pharmaceuticals toxicity data. Practically, the uncertainity of toxicity data in a substance could be reduced by performing additional toxicity tests for other species which can be quiet complicated and also time-consuming. Instead of doing that, this study assumed that the level of uncertainity of the detected pharmaceuticals were quiet high and use 1000 as single assessment factor value for all pharmaceuticals. This assumption was applied since only short term toxicity data were available for all measured pharmaceuticals in this study, irrespective of whether or not the species tested was a standard test organism. The use of a factor of

1000 on acute or short-term toxicity data is a protective and conservative factor and is intended to assure that pharmaceuticals with the potential to cause negative impacts are identified [36, 37]. Pharmaceuticals typically are biologically active compounds that interact with specific targets in the human body [64]. The effect triggered can be very specific and can be difficult to detect in a toxicity study, e.g. a behavioural change of an antidepressant [13, 14, 65]. As such, it is defendable to use a high assessment factor to exptrapolate TDLo values to HRVs.

The methods proposed in the present study resulted in more HRVs than the common method of using ADIs, TDIs and RfDs. In total, HRVs could be derived for 14 out of 16 pharmaceuticals while only one HRV value is available when using ADIs, TDIs and RfDs, i.e. for BPA (RfD of 0.05 mg/kg/day from IRIS or TDI of 0.004 mg/kg/day from EFSA). Prioritization would have been impossible if it would have been based on HRVs from ADIs, TDIs and RfDs only.

3.2 Health risk prioritizations

Table 2 lists the results of the risk prioritization based on DDD data from the WHO and acute toxicity data (TDLo) extracted from PubChem. The first prioritization using DDD values shows that the highest risk rank was obtained for Estrone with a $RI_{hum\text{-}DDD}$ of 3.6.E-02 (1st), followed by Caffeine with a $RI_{hum\text{-}DDD}$ of 1.6.E-03 (2nd) and Lidocaine with a $RI_{hum\text{-}DDD}$ of 2.3.E-04 (3rd). The last risk rank or lowest $RI_{hum\text{-}DDD}$ value was Sulfadiazine with 9.9.E-07 (13th). Three out of 16 pharmaceuticals could not be prioritized because their HRVs could not be derived. From a total of 13 $RI_{hum\text{-}DDD}$ calculations, there was no $RI_{hum\text{-}DDD}$ value was above 1, indicating negligible human health risks.

Table 2. Health risk-based prioritizations of pharmaceuticals based on DDD and TDLo values.

Pharmaceuticals	CAS	DDD[a]		TDLo[a]	
		RI_{hum}	Rank	RI_{hum}	Rank
Estrone	53-16-7	3.6.E-02	1	NA	NA
Caffeine	58-08-2	1.6.E-03	2	1.5.E-03	1
Lidocaine	137-58-6	2.3.E-04	3	7.3.E-04	3
Trimethoprim	738-70-5	1.1.E-04	4	1.2.E-04	6
Ibuprofen	51146-56-6	1.0.E-04	5	2.5.E-04	5
Paracetamol	103-90-2	2.6.E-05	6	2.6.E-04	4
Sulfamethoxazole	723-46-6	1.5.E-05	7	3.2.E-05	7
Chloramphenicol	56-75-7	1.5.E-05	8	3.0.E-05	8
Progesterone	57-83-0	1.5.E-05	9	NA	NA
Lincomycin	154-21-2	8.7.E-06	10	NA	NA
Chloroquine	54-05-7	3.3.E-06	11	1.4.E-05	9
Salicylic acid	69-72-7	1.8.E-06	12	7.8.E-06	10
Sulfadiazine	68-35-9	9.9.E-07	13	3.4.E-07	11
Cytarabine	147-94-4	NA	NA	1.0.E-03	2
BPA	80-05-7	NA	NA	NA	NA
Furazolidone	67-45-8	NA	NA	NA	NA

[a] NA: Not Available

The second prioritization approach was performed by using HRVs from TDLo data extracted from PubChem. The result showed that 11 out of the 16 pharmaceuticals could be prioritized. The first risk rank was obtained by Caffeine with a $RI_{hum-TDLo}$ of 1.5.E-03 (1st), followed by Cytarabine with a $RI_{hum-TDLo}$ of 1.0.E-03 (2nd) and Lidocaine with a $RI_{hum-TDLo}$ of 7.3.E-04 (3rd). The lowest $RI_{hum-TDLo}$ value was Sulfadiazine with 3.4.E-07 (11th). Five phramaceuticals could not be prioritized because of lacking HRVs. All risk indicators from 11 $RI_{hum-TDLo}$ estimation were under 1, implying negligible human health risks.

3.3 Comparison between prioritization methods

Based on the prioritization results (table 2), 10 out of 16 pharmaceuticals could be prioritized by using DDD and TDLo methods, i.e. Caffeine, Lidocaine, Trimethoprim, Ibuprofen, Paracetamol, Sulfamethoxazole, Chloramphenicol, Chloroquine, Salicylic acid, and Sulfadiazine. In assessing the performance of prioritization approaches, the result of $RI_{hum-DDD}$ were plotted against the $RI_{hum-TDLo}$ values (figure 3). The graph consist of 10 dots, representing pharmaceuticals that could be prioritized by using both methods.

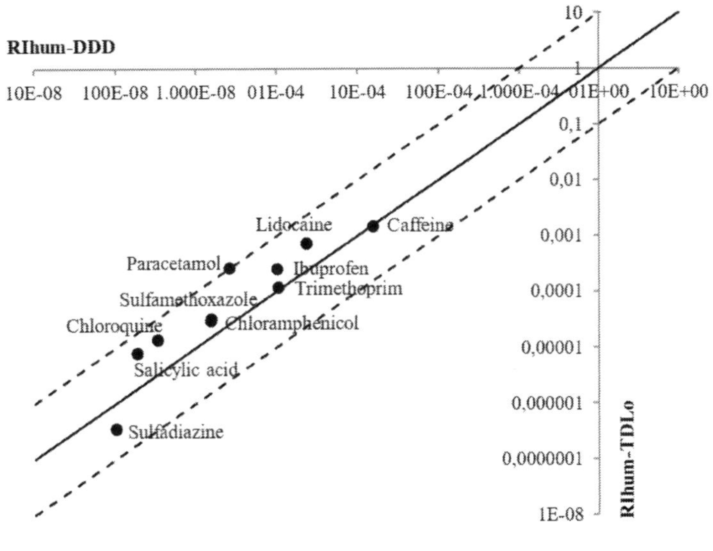

Figure 3. Comparison of $RI_{hum-DDD}$ (x-axis) and $RI_{hum-TDLo}$ (y-axis) of pharmaceuticals in the study location. The solid line indicates a 1:1 match between $RI_{hum-DDD}$ and $RI_{hum-TDLo}$, while the dashed lines indicate a ten-fold difference.

The solid line in figure 3 indicates a perfect match between two RI_{hum} values of a pharmaceutical. The dashed lines indicate a factor of 10 difference. The developed approaches performs relatively good from the comparison between $RI_{hum-DDD}$ and $RI_{hum-TDLo}$ values, especially for Caffeine and Trimethoprim (see figure 3). Both RI_{hum} values of Caffeine and Trimethoprim were very close to each other. Generally, the $RI_{hum-TDLo}$ values tend to be slightly higher than the $RI_{hum-DDD}$ values for almost all pharmaceuticals, except for Caffeine and Sulfadiazine. Overall, RI_{hum} comparison of 10 pharmaceuticals were within a factor of 10, indicating that both methods produced resemble risk indicators. The ranking comparisons of 10 pharmaceuticals between two methods were shown in table 3.

Table 3. Comparison of health risk ranking of 10 pharmaceuticals between two prioritization methods.

Pharmaceuticals	CAS	$RI_{hum-DDD}$	$Rank_{DDD}$	$RI_{hum-TDLo}$	$Rank_{TDLo}$
Caffeine	58-08-2	1.6.E-03	1	1.5.E-03	1
Lidocaine	137-58-6	2.3.E-04	2	7.3.E-04	2
Trimethoprim	738-70-5	1.1.E-04	3	1.2.E-04	5
Ibuprofen	51146-56-6	1.0.E-04	4	2.5.E-04	4
Paracetamol	103-90-2	2.6.E-05	5	2.6.E-04	3
Sulfamethoxazole	723-46-6	1.54.E-05	6	3.2.E-05	6
Chloramphenicol	56-75-7	1.51.E-05	7	3.0.E-05	7
Chloroquine	54-05-7	3.3.E-06	8	1.4.E-05	8
Salicylic acid	69-72-7	1.8.E-06	9	7.8.E-06	9
Sulfadiazine	68-35-9	9.9.E-07	10	3.4.E-07	10

Comparing the results of both prioritization approaches (table 3), Caffeine and Lidocaine were the two pharmaceuticals that ranked in the top two of both methods. The rank differences between two methods were only found in Trimethoprim and Paracetamol. Overall, the comparison of the DDD and TDLo-based rankings shows considerable consistency in results, indicating that both prioritization methods are reasonably in line with each other.

All pharmaceuticals detected in the study area had risk indicator values under unity. It shows that the intake of the investigated individual pharmaceuticals with the consumption of Citarum River water causes negligible health risk. Nonetheless, it should be remembered that this study did not consider other exposure routes such as dermal exposure, inhalation exposure, nor the consumption of contaminated fish. Furthermore, there might be potential interactions between pharmaceuticals that could lead to adverse effects [56, 57, 58]. For instance, a study by Denton et al (2003) showed that a combination of esfenvalerate and diazinon resulted in a toxic effect on larvae of *Pimephales promelas*, while no toxicity was observed when only esfenvalerate was given [59]. Therefore, caution is warranted when estimating toxicity of pharmaceutical mixtures in the water.

4. Conclusions

Two approaches of health risk-based prioritization were developed and applied to 16 pharmaceuticals detected by passive samplers in the water source of Ciwalengke Village in the UCRB. Each method produced 13 and 11 pharmaceutical HRVs respectively, while only one HRV could be derived when applying the common method by using ADI, TDI, or RfD value. This study has proven that the developed methods are able to overcome the limited availability of ADI, TDI, and RfD values in pharmecuticals prioritization.

In total, 10 out of 16 pharmaceuticals could be prioritized by using both methods. The comparison between methods showed that risk indicator values of 10 pharmaceuticals were within a factor of 10, showing that both methods resulted resemble risk indicators. The prioritization results of both methods also showed considerable consistency in ranking with Caffeine and Lidocaine that were always in the top 3 highest risks, indicating that the risk prioritization approaches in this study are reasonably robust.

The risk indicators resulting from the pharmaceuticals intake from the Citarum River water were far below unity, implying negligible human health risks ($RI_{hum} < 1$) for all individual pharmaceuticals toxicity. However, it should be remembered that we did not consider other exposure routes (e.g. dermal exposure, inhalation, dietary of contaminated fish, etc.) nor possible interactions between pharmaceuticals that might provoke adverse effects. The health risk-based prioritization methods in this study can be a beneficial tool for the authorities to determine which pharmaceuticals with the highest risk and increase the efficiency of risk assessment in the river basin.

References

[1] Boxall A B A, Kolpin D, Halling-Sørensen B and Tolls J 2003 *Environ. Sci. Technol.* **37(15)** 286A–294A.
[2] Yeh A, Marcinek D J, Meador J P and Gallagher E P 2017 *Aquat. Toxicol.* **190** 21–31.
[3] Marigoudar S R, Mohan D, Nagarjuna A and Karthikeyan P 2018 *Ecotoxicol. Environmen. Saf.* **148** 327-335.
[4] Meador J P, Yeh A, Young G and Gallagher E P 2016 *Environ. Pollut.* **213** 254-267.
[5] Brumovský M, Bečanová J, Kohoutek J, Borghini M and Nizzetto L 2017 *Environ. Pollut.* **229** 976-983.
[6] Jorgenson Z G, Thomas L M, Elliott S M, Cavallin J E, Randolph E C, Choy S J, Alvarez D A, Banda J A, Gefell D J, Lee K E, Furlong E T and Schoenfuss H L 2018 *Environ. Pollut.* **236** 718-733.
[7] Post G B, Cohn P D and Cooper K R 2012 *Environ. Res.* **116** 93–117.
[8] Flint S, Markle T, Thompson S and Wallace E 2012 *J. Environ. Manage.* **104** 19-34.
[9] Kim K H, Kabir E and Jahan S A 2017 *Sci. Total Environ.* **575** 525-535.
[10] Wang L, Liu Z, Zhang J, Wu Y and Sun H 2016 *Environ. Res.* **149** 164-170.
[11] Ma Y, He X, Qi K, Wang T, Qi Y, Cui L, Wang F and Song M 2019 *J. Environ. Sci.* **77** 210 – 217.
[12] Vörösmarty C J, McIntyre P B, Gessner M O, Dudgeon D, Prusevich A and Green P 2010 *Nature* **467** 555-561.
[13] Vasquez M I, Lambrianides A, Schneider M, Kümmerer K and FattaKassinos D 2014 *J. Hazard Mater* **279** 169–189.
[14] Mitchell P J 2005 *Eur. J. Pharmacol.* **526** 147-162.
[15] Roos V, Gunnarsson L, Fick J, Larsson D G J and Ruden C 2012 *Sci. Total Environ.* **421** 102–110.
[16] Aubakirova B, Beisenova R and Boxall A B A 2017 *Integr. Environ. Assess.Manag.* **13(5)** 832–839.
[17] Monteiro S C and Boxall A B A 2010 *Rev. Environ. Contam. Toxicol.* **202**.
[18] Besse J P, Kausch-Barreto C and Garric J 2008 *Hum. Ecol. Risk Assess.* **14(4)** 665–695.
[19] Guo J, Sinclair C J, Selby K and Boxall A B A 2016 *Environ. Toxicol. Chem.* **35(6)** 1550-1559.
[20] Al-Khazrajy O S A and Boxall A B A 2016 *Environ. Sci. Pollut. Res.* **23** 15712–15726.
[21] Burns E E, Thomas-Oates J, Kolpin D W, Furlong E T and Boxall A B A 2017 *Environ. Toxicol. Chem.* **36** 2823–2832.
[22] Burns E E, Carter L J, Snape J, Thomas-Oates J and Boxall A B A 2018 *J. Toxicol. Environ. Health* **21(3)** 115-141.
[23] Lyu S, Chen W, Qian J, Wen X and Xu J 2019 *Sci. Total Environ.* **697** 133850.
[24] Saunders L J, Mazumder A and Lowe C J 2016 *Environ. Toxicol.Chem.* **35** 919–929.
[25] Salim A G, Dharmawan I W S and Narendra B H 2019 *Jurnal Ilmu Lingkungan Program Studi Ilmu Lingkungan Sekolah Pascasarjana UNDIP* **17(2)** 333-340.
[26] BBWS Citarum (Balai Besar Wilayah Sungai Citarum) 2019 Profil BBWS Citarum http://sda.pu.go.id/balai/bbwscitarum/profil-bbws-citarum/.
[27] Bappenas (Badan Perencanaan Pembangunan Nasional) 2014 Kondisi Fisik dan Spasial http://citarum.bappenas.go.id/tentang-kami/sekilas-citarum/kondisi-fisik-dan-spasial.html.
[28] Utami R R, Geerling G W, Salami I R S, Notodarmojo S and Ragas A M J 2020 *Sci. Total Environ.* **738** 140130.
[29] Melnyk L J, Wang Z, Li Z and Xue J 2016 *Food Chem. Toxicol.* **96** 167-173
[30] Baken K A, Sjerps R M A, Schriks M and van Wezel A P 2018 *Environ. Int.* **118** 293–303.
[31] Mulyani N 2015 *Evaluation of water service system existing condition in Ciwalengke, Majalaya District, Bandung Regency* (Bandung: Bandung Institute of Technology).
[32] Maryam S 2015 *Evaluation of sanitation services planning system in Bandung Regency, (Case study: Ciwalengke Village)* (Bandung: Bandung Institute of Technology).

[33] van Dorssen A J 2017 *Localised sanitation with nutrient harvesting in semi-urban slums, The case of Ciwalengke, Majalaya District, Indonesia* (Wageningen: Wageningen University)
[34] Kleijn P and Tjiong W 2017 *Mapping the community water system in Kampong Ciwalengke, A water system survey towards a better understanding of the distribution of pollution in Kampong Ciwalengke* (Leeuwarden: University of Applied Sciences Van Hall Larenstein)
[35] RIVM (National Institute for Public Health and the Environment) 2015 *Guidance for the derivation of environmental risk limits, Part 3. Derivation of ERLs for freshwater and marine water* version 1.0 (Bilthoven: National Institute for Public Health and the Environment)
[36] EC (European Comission) 2003 *Technical Guidance Document on Risk Assessment, Part I & Part II* (Italy: European Communities)
[37] EC (European Comission) 2011 *Common Implementation Strategy for the Water Framework Directive (2000/60/EC), Guidance Document No. 27, Technical Guidance For Deriving Environmental Quality Standards* (Italy: European Communities)
[38] Gordon B, Callan P and Vickers C 2008 *WHO Chronicle* **38(3)** 564.
[39] Smedes F, Bakker D and de Weert J 2010 *The Use of Passive Sampling in WFD Monitoring* (Delft: Deltares)
[40] Kot-Wasik A, Zabiegała B, Urbanowicz M, Dominiak E, Wasik A and Namie´snik J 2007 *Anal. Chim. Acta* **602** 141–163.
[41] Amato E D, Covaci A, Town R M, Hereijgers J, Bellekens B, Giacometti V, Breugelmans T, Weyn M, Dardenne F, Bervoets L and Blust R 2018 *Chemosphere* **209** 363-372.
[42] Gao X, Xu Y, Ma M, Rao K and Wang, Z 2019 *Ecotoxicol.Environ. Saf.* **178** 25-32.
[43] Valenzuela E F, Menezes H C and Cardeal Z L 2019 *Anal. Chim. Acta* **1054** 26-37.
[44] Zillien C, van Loon C, Gülpen M, Tipatet K, Hanssen B, Beeltje H, Roex E, Oldenkamp R, Posthuma L and Ragas A M J 2019 *Sci. Total Environ.* **694** 133733.
[45] Vrana B, Mills G A, Allan I J, Dominiak E, Svensson K, Knutsson J, Morrison G and Greenwood R 2005 *Trends Anal. Chem.* **24** 10.
[46] Vrana B, Smedes F, Prokeš R, Loos R, Mazzella N, Miege C, Budzinski H, Vermeirssen E, Ocelka T, Gravell A and Kaserzon S 2016 *Trends Anal. Chem.* **76** 153-165.
[47] Allan I J, Booij K, Paschke A, Vrana B, Mills G A and Greenwood R 2009 *Environ. Sci. Technol.* **43(14)** 5383-5390.
[48] Rusina T P, Smedes F, Koblizkova M and Klanova J 2010 *Environ. Sci. Technol.* **44(1)** 362-367.
[49] Rusina T P, Smedes F, Brborić M and Vrana B 2019 *Sci. Total Environ.* **696** 133935.
[50] Smedes F and Booij K 2012 *Guidelines for passive sampling of hydrophobic contaminants in water using silicone rubber samplers* (Copenhagen: ICES Techniques in Marine Environmental Sciences).
[51] Hamers T, Legradi J, Zwart N, Smedes F, de Weert J, van den Brandhof E J, van de Meent D and de Zwart D 2018 *Environ. Toxicol. Pharmacol.* **64** 48-59.
[52] Regulation of the Government of the Republic of Indonesia number 82 of 2001 on Water Quality Management and Water Pollution Control.
[53] Regulation of the Government of the Republic of Indonesia number 22 of 2021 on Implementation of Environmental Protection and Management.
[54] Venema R 2015 *Modeling Land Use and Water Quality Management Scenarios for the Upper Citarum River, Java, Indonesia* (Enschede: University of Twente).
[55] Lufiandi, Geerling G, Effendi A J, Roosmini D, Sudradjat A and Smits A J M 2015 *Water quality management and water pollution control under decentralization in Indonesia: The case of Upper Citarum River* (Bandung: *The 5th Environmental Technology and Management Conference "Green Technology towards Sustainable Environment"*) ISBN: 978-979-98278-5-2.
[56] Thompson H M 1996 *Interactions between pesticides; a review of reported effects and their implications for wildlife risk assessment* vol. 5 (Slough: Central Science Laboratoo).
[57] Mukono H J 2005 *Toksikologi Lingkungan* (Surabaya: Airlangga University Press).

[58] Woods M, Kumar A and Correll R 2002 *Bull. Environ. Contam. Toxicol.* **68** 801–808.
[59] Denton D L, Wheelock C E, Murray S A, Deanovic L A, Hammock B D and Hinton D E 2003 *Environ. Toxicol. Chem.* **22(2)** 336–341.
[60] Guo J, Sinclair C J, Selby K and Boxall A B A 2016 *Environ. Toxicol. Chem.* **35** 1550-1559.
[61] Wilkinson J L, Boxall A B A, Kolpin D W, (+123) and Teta C 2022 *PNAS* **119(8)** e2113947119.
[62] Gunnarsson L, Snape J R, Verbruggen B, Owen S F, Kristiansson E, Margiotta-Casaluci L, Österlund T, Hutchinson K, Leverett D, Marks B and Tyler C R 2019 *Environ. Int.* **129** 320-332.
[63] USEPA 2011 *Exposure Factors Handbook: 2011 Edition* (Washington, DC: U.S. Environmental Protection Agency, Office of Research and Development, National Center for Environmental Assessment).
[64] Boxall A B A, Rudd M A, Brooks B W, (+38) and Kraak G V D 2012 *Environ. Health Perspect.* **120(9)** 1221-1229.
[65] Ramasubbu R 2004 *Prog. Neuropsychopharmacol. Biol. Psychiatry* **28(7)** 1201-1207.

Acknowledgements
The authors would like to thank Radboud University and Deltares for funding this study.

Preliminary studies of Bandung City Health System Resilience (case study : Covid-19 pandemic)

D Roosmini[1,2], T F Kanisha[3], A Nastiti[1,2], S W Kusumah[1], I R S Salami[1,2]

[1] Centre of Environmental Studies, Institute for Research and Community Services, Institut Teknologi Bandung

[2] Environmental Management Technology Research Group, Faculty of Civil and Environmental Engineering, Institut Teknologi Bandung

[3] Environmental Engineering Department, Faculty of Civil and Environmental Engineering, Institut Teknologi Bandung

E-mail: dwinaroosmini@yahoo.com

Abstract. The SARS-CoV-2 (Severe Acute Respiratory Syndrome Coronavirus-2) virus causes the COVID-19 pandemic globally. The city of Bandung is the city that became the cluster of Covid-19 spread after DKI Jakarta at the beginning of the pandemic. In 2019, the Global Health Security Index (GHSI) was developed to assess countries' readiness to face a pandemic. The index obtained shows that globally the value of this readiness is still minimal. In improving community health system services, health system resilience has been developed and studied in recent years. This study conducted a preliminary evaluation of the Bandung City health system in dealing with the Covid-19 pandemic using the concept of resilience. This research uses the West Java and Bandung City Health Office website. Methods used for evaluating the health system resiliency are comparing the cases of Covid-19 at the sub-district to the health facilities, the incidence of infectious diseases in 2019, and the socio-economic conditions of each sub-district. Research shows that the health system in Bandung City has not met the targets in the 2018-2023 Regional Medium-Term Development Plan. Overall, the initial research on the Bandung City health system still does not meet the criteria for resilience.

Keywords: Covid-19, Bandung, Resilience, Health-System

1. Introduction

World Health Organization (WHO) reported the health system's performance in 2000 and universal health coverage in 2013. It highlighted the health system in low and middle-income countries (LMICs-Low Middle-Income Countries). The WHO finds that it still faces various challenges, causing shortages in performance, responsiveness, and resilience [1]. The term health system resilience has become a topic that has been discussed quite a lot in recent years: the conceptual analyses, resilience index, the development of resilience health system and measuring the resilient [1,2,3,4,5,6]. Most research related to health system resilience evaluates the response to health emergencies due to pandemics and disasters in terms of service continuity, time to recovery, and drop-in service [4,7,8].

The global development of the Covid-19 pandemic has shown the unpreparedness of various countries worldwide, including Indonesia. It already predicted by the results of the 2019 Global Health Security Index study, GHSI measures readiness to face an epidemic or pandemic. The results of the GHSI study indicate that most countries do not show evidence of the health security capacities and capabilities needed to prevent, detect, and respond to significant outbreaks of infectious diseases. The average readiness score of countries worldwide (195 countries) is 40.2 out of 100; among 60 high-income countries, the average GHS Index score is 51.9. Furthermore, the average value of 116 high-and-middle income countries did not reach 50 [9]. The GHS Index document found various weaknesses, among others are: the weaknesses in preventing, detecting, and responding to health emergencies; the vast gaps in the health system; the vulnerable to political, socio-economic, and environmental risks that contribute to outbreak preparedness and response and non-compliance with international norms.

Indonesia's overall score in the 2019 GHSI is at number 30 out of 195 countries with a score of 56.6; among ASEAN countries. Thailand achieved the highest readiness score at number 6 with 73.2, then Malaysia in 18th with 62.2. Singapore is at 24th place with 58.7, Vietnam at 50th place with 49.1, and the Philippines at 53rd with 47.6 [9]. This GHSI study shows how important it is for a country to prepare a health system that is resistant to the threat of emergency conditions, both due to disasters and the threat of infectious diseases such as the Covid-19 pandemic. GHSI score of 56.6 for Indonesia distributed in several aspects, the ability of Prevention, Detection and Reporting, Rapid Respond, Health System, Compliance with International Norms, and Risk Environment. The high point for Indonesia is for the Compliance with International Norm's aspect, 72.5 compare to an average of 195 countries that is 48.5. The lowest score for Indonesia is in the aspect of Health System with 39.4, although still higher than the global average of 26.4. The weak part of the Health system in Indonesia is in the component Infection control practices and availability of equipment, with a score of 0, compared to a global score of 20.8. Health capacity in clinics, hospitals, and community care centres also gives a low score of 28.4, even though it is still higher than the global average score of 24.4.

2. Bandung City

Bandung city is the capital of West Java Province, adjacent to and bordering other cities and regencies in West Java Province, namely: Cimahi City, West Bandung Regency, and Bandung Regency, as shown in figure 1. Bandung has 30 sub-districts with 151 urban villages, 1,591 hamlet (Rukun Warga/RW), and 9,904 neighbourhood (Rukun Tetangga/RT), with a total population in Bandung City is 2,444,160. The highest population number is in Babakan Ciparay Sub-district, and the lowest population number is in Cinambo Sub-district [10]. The densest population area is in Bojongloa Kaler District for 41,492 residents/km2, and the least dense population is Gede Bage for 4,232 residents/km2 [10].

The ratio of health facilities per capita in Bandung City compared to other cities and regencies in West Java Province is described in Figure 1. Figure 1 describes the facilities of health facilities ratio per population; the Ministry of Health of Indonesia state ideal number of hospital bed for people is one bed for 1000 people and one Sub-district level Health center for 30000 people. Compared to sixteen (16) other cities/regencies in West Java Province, Bandung City has a hospital bed ratio of 2.2, the second-highest ratio after Cirebon City. Some other cities and regencies need to fulfill the ideal condition state by the Ministry of Health with a current ratio of 0.2 and 0.3. The ratio of Sub-district level Health centers to the population in Bandung City is 1.0 as required by the Ministry of Health Indonesia. The highest ratio in West Java Province is Cirebon City 2.1, followed by Tasikmalaya City 1.8 and Bogor City 1.3 [11].

GRDP (Gross Regional Domestic Product) of Bandung City is the second-largest in West Java Province after the Bekasi Regency. The GRDP of Bandung City is 289,312.25 billion rupiahs, with the Regional Revenue in 2019 of 6,552,675,659. The Human Development Index in Bandung City is 81.62.

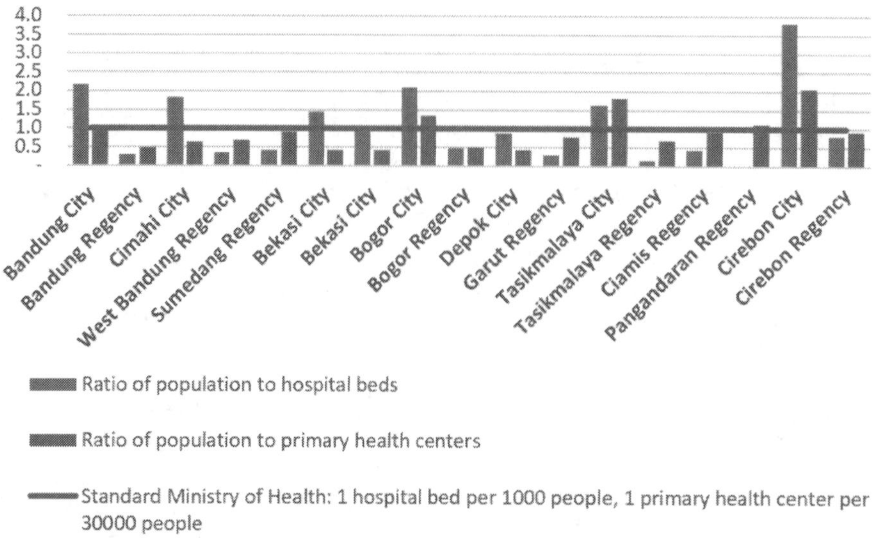

Figure 1. West Java Health Facilities Ratio to residents of several Cities and Regencies (analyzed from West Java Health Profile 2020 [12])

3. Methods

The study on health system resilience in Bandung City covered the community health condition as a measure of health system performance and the environmental health condition. These two evaluation goals are to evaluate the public health condition in Bandung City, using data available online as well as through correspondence with the Bandung City Health Office. The evaluation compares the community's health condition in different subdistricts in Bandung City and relates it to the environmental health condition in every subdistrict. The principle correlation analysis method looks at the correlation between community health and environmental health conditions variables for base data to see the health system resilience. The preliminary studies of Health System Resilience in Bandung City evaluation use Bandung City strategy document for Public Health, Bandung in Number, Health Profile of Bandung and West Java Province. Data and results from secondary data are compared to some community resilience indicators developed by Kruk [4], Bhandari [6], and Haldane [8]. Findings on data analyses are used to recommend Bandung City for their strategy in increasing their health system resilient.

4. Result and Discussion

4.1. Community health evaluation of Bandung City

The health profile of Bandung City [12] showed that the infant mortality rate (IMR) in Bandung is 2.73 per 1000 population, with the under-5 mortality rate (Angka Kematian Bawah Lima Tahun/AKABA) at 3.37 per 1000 population. Andir and Sukajadi sub-districts have the highest IMR and AKABA, while Cibeunying Kaler and Panyileukan sub-districts have no death cases. Figure 2., showed further that Bandung Kulon and Sukajadi sub-districts are the two areas with the highest number of under-five mortality cases compared to sixteen (16) sub-districts with no deaths. Bandung City Health Profile [12] reports three (3) highest infectious diseases cases in the year 2015-2019: diarrhea, pneumonia under five years, and DHF (Dengue Hemorrhagic Fever). The incidences for each disease are 25 per 1000 population, 43.35 per 1000 population, and 1.76 per 1000 population. The sub-district Bandung Kulon, Astanaanyar, and Kiaracondong are three areas with a high case of diarrhea. Meanwhile, Rancasari,

Regol, and Cinambo are the sub-districts with the least number of diarrhea cases. The sub-districts of Bandung Kulon, Cibeunying Kidul, and Cicendo are the three areas with the highest number of cases of pneumonia in children under-5 years, with minor cases in Gedebage, Bandung Wetan, and Cidadap sub-district. Kiaracondong, Coblong, and Arcamanik sub-districts have the highest dengue cases in Bandung City; meanwhile, Cinambo, Bandung Wetan, and Sumur Bandung are the three areas that have the least number of cases. Overall, the highest incidence of infectious diseases in the city of Bandung is shown in Figure 3.

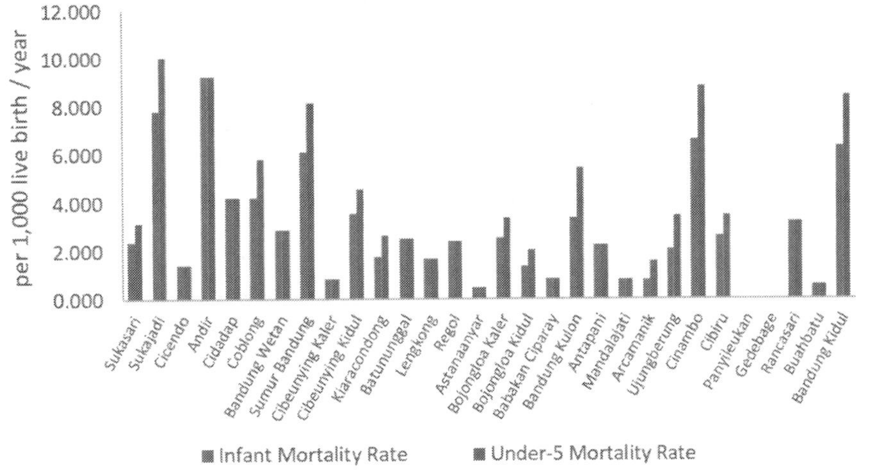

Figure 2. Infant Mortality Rate and Under-5 Mortality Rate in Sub-districts (analyzed from Bandung Health Profile [12])

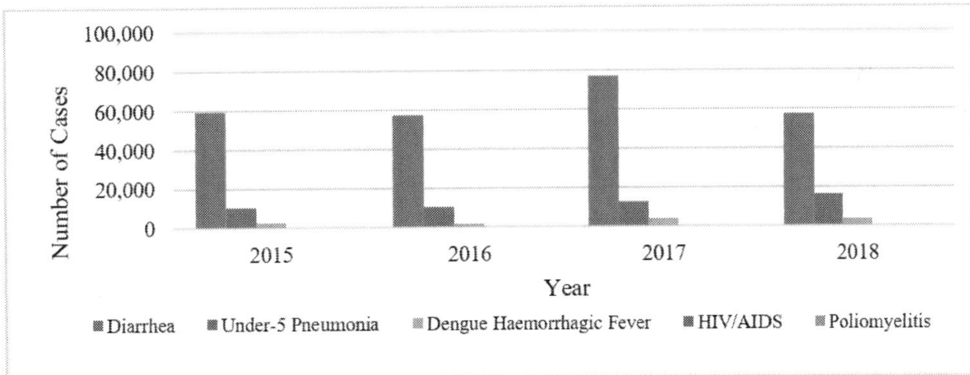

Figure 3. The five (5) highest infectious diseases in Bandung City in 2015-2018

The sub-district Andir has a higher under-5 mortality case than the average case in Bandung City. Most infant deaths cases are related to LBW (Low Birth Weight), asphyxia, and other diseases. For under-5 mortality cases, the most common causes are fever and pneumonia. Although pneumonia and diarrhea in children under-5 are the two types of diseases with the highest number of cases in Bandung during the last five years, it turns out that the number of deaths caused does not contribute significantly to the under-5 mortality number.

The life expectancy in 2019 is 74 years, increasing every year from 2015 to 2019. With a value of 81.62 in 2019, the Human Development Index (HDI) increased from 2015 to 2019. It shows improvement in health, education, and community welfare yearly. However, when compared to other big cities on the island of Java, Bandung's HDI is still below several cities, such as Surabaya (82.22), Semarang (83.19), and Yogyakarta (86.65).

4.2. Environmental health evaluation

The environmental health condition was evaluated based on data from the Health Profiles of Bandung City 2019. The highest population density was found at the subdistrict of Bojongloa Kaler (41.493 people/km2), Andir (27,233 people/km2), and Astanaanyar (24.079 people/km2). In comparison, Gedebage (4.232 people/km2), Cinambo (6.896 people/ km2), and Panyileukan (7.747 people/km2) are the areas with the lowest value.

Clean and Healthy Living Behavior (Perilaku Hidup Bersih Sehat/PHBS) are other indicators used in Environmental Health status. The PHBS measured as the percentage from monitored households. Data in Health Profiles showed that Panyileukan, Bandung Kidul, and Cicendo subdistricts have a high percentage of clean and healthy living behavior households. At the same time, Arcamanik, Cinambo, and Batununggal have the smallest percentage.

Another environmental health condition indicator is the water supply access. The access in each sub-district is described in Figure 3. There are four (4) types of water supply in the community: piping system from the municipal water company or nonmunicipal water company, borehole, and dug well. With the water supply system types, the dug well is the system with poor water quality compared to other systems. The subdistrict with a high percentage of dug well is in Andir and Cibiru, which means that the community in this subdistrict has potential exposure to poor water quality. On the other hand, the piping water and borehole give better water quality. The high percentage of municipal water company access is in subdistrict Mandalajati with 80%, followed by Antapani 76% and Sumur Bandung 75%. Panyilekan and Ujung Berung are two sub-districts with the community's highest percentage of boreholes. Some of the subdistricts that showed a low percentage of total access to water supply are Bojong Loa Kaler 76% and Arcamanik 78%.

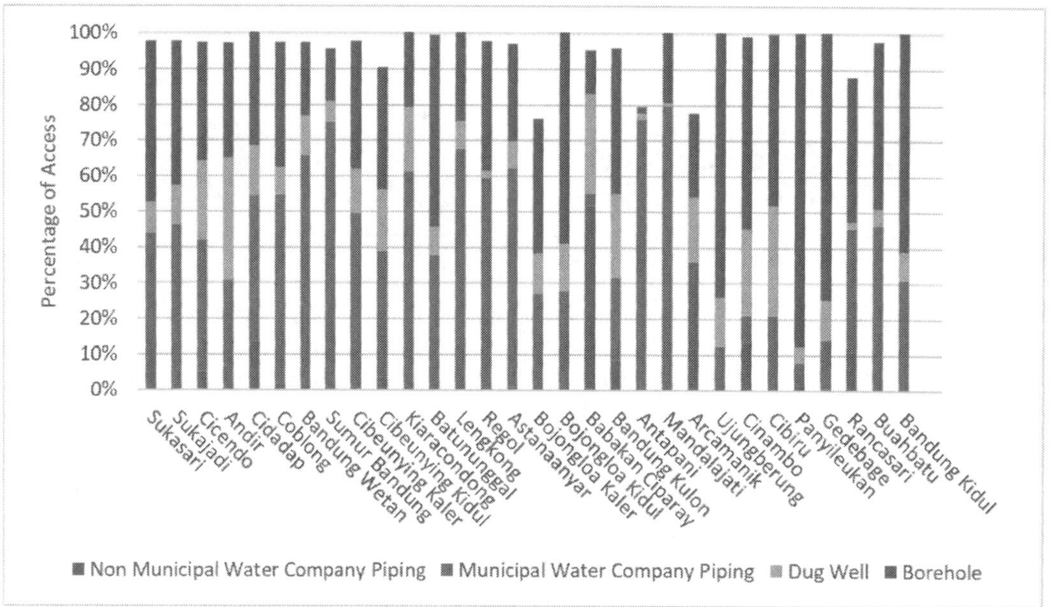

Figure 3. Bandung City Health Facilities Ratio to Residents in each Subdistricts

4.3. Bandung Health System

Evaluation of the health system, in general, is carried out by adjusting indicators of public health, environmental health, and health resources to the Strategic Planning (Rencana Strategis/Renstra) and Regional Medium Term Development Plan (Rencana Pembangunan Jangka Menengah Daerah/RPJMD) of Bandung City (2018-2023). Adjustment to existing conditions in 2019, it is known that four indicators do not meet the target of health system development. The target that is not fulfilled is under-5 mortality rates, the percentage of elderly health services, the ratio of sub-district public health to city population, and the percentage of the population with access to safe drinking water.

The strategic planning and RPJMD of the Bandung health system need to be compared with the Global Health Security (GHS) Index to prevent communicable diseases. Looking at the achievement of Bandung City in 2019, it only meets five (5) out of nineteen (19) points of international standards. Namely the doctor's ratio to the population (1:100.000), the ratio of nurses and midwives to the population (1:100.000), healthcare facilities' capacity for isolation of patients with highly infectious diseases risk, the enactment of regulations related to universal health service coverage, and monitoring of disease cases in healthcare facilities. In anticipation of a pandemic, Bandung's health system needs to be improved. In comparing the health system in Bandung City to GHS Index, this study analyses the system's response to the Covid-19 pandemic from March 2020 to June 2021. Some of the identified aspects are shown in Table 5.

Tabel 5. The comparison of some GHS Index element to Bandung Health System

No	Elements	Indicator	Notes	Fulfilled/Not
1	Public knowledge (Anticipative)	There is training for the community continuously related to disaster emergencies	The form of training can also be packaged in an educational structure	No
2		Percentage of households that have access to community-level disaster early warning	Simulations can be run periodically as a respond to emergencies	No
3	Community Relations (Transformative)	The proportion of individuals actively involved in community associations or community events	The actual condition of the community level, for example, NGO (Non-Governmental Organizations)	No
4		Percentage of volunteers involved in disaster management (more than 15 persons for each district)	Volunteers can help people who are in trouble or experiencing a disaster crisis	No
5	Communication (Adaptive & Transformative)	Community can easily access information needs related to disasters	The process of distributing information can be done through one door or easily accessible institution, for example, people who need oxygen cylinders immediately or hospitalization (checking BOR value)	Need to be optimized
6		There is a coordination between government and community regarding regional policies or infrastructure development	Public has the opportunity to express their opinions through open and periodic public discussions	No
7	Health (Absorptive & Adaptive)	Inventory of disease history neatly to anticipate disasters or biological threats in the future	100% of health facilities, especially public health centre have a well diagnosis equipment so that early detection can be done	Need to be optimized

No	Elements	Indicator	Notes	Fulfilled/Not
8		Every health facility, especially hospitals, already have a good reporting system regarding BOR	A good reporting system is defined when the BOR is approaching 60%, there are reports for additional bed units or plans for other isolation facilities	Need to be optimized
9		Residents who are monitored carry out clean and healthy lifestyles in anticipation of infectious disease spreads	100% of the population is monitored for each district so that representative data can be evaluated easily	Need to be optimized
10		Ratio of public health number to the total population	All sub-districts meet the ideal ratio of public health centre to population (1:30.000)	Need to be optimized
11		Percentage of meeting the minimum criteria for health workers	All health facilities meet the minimum standard of health workers that must be available	Need to be optimized
12		There is assistance in supplying basic needs (food, water, etc) by the community level	Increased awareness and social interaction at the sub-district level to facilitate distribution of aid	Need to be optimized
13	Resources (Absorptive & Adaptive)	Residents get access to proper drinking water and minimize the use of dug wells	100% of the population can consume water according to health requirements and optimize the use of adequate facilities (PDAM)	Need to be optimized
14		The existence of emergency service facilities that can be easily accessed at public facilities and at the sub-district level by the community	Provision of first aid facilities for accidents at public facilities, including cardiac shock devices, first aid box, etc	Need to be optimized

Source : Processed from GHS index document (2019)

Kruk, 2107 proposed the health system resilience index, comprised of five indexes: Aware (know health system capacity, knowing risk and population, communicate), Diverse (effectively respond to a range of health needs, adequately finance health system; prevent financial harm), Self Regulating (isolate the threat and maintain core function, leverage outside capacity), Integrated (coordinate with non-health actors, engage citizens and communities to built trust, link healthcare provision to public health, coordinate primary and referral care), and Adaptive (shift resources to meet the need, promote rapid local decision making, evaluate to improve).

The pandemic Covid-19 in Bandung City gives a significant experience for the future development of health system resilience. A considerable effort has been spent to tackle the pandemic during the Covid-19 pandemic, with vertical and horizontal coordination at the provincial and national levels. Good cooperation between various parties, the community, the private sector, business people, and the government is constructive in overcoming the Covid-19 pandemic, although it is still far from perfect. The tag line together to recover together is stronger, showing that several indexes proposed by Kruk can be identified in resolving the effect of the pandemic on the social economy. In developing the Health System Resilience in Bandung City, some good practices need to be elaborate in the frame of Health System Resilience more precisely and put on the Strategic Plan agenda of the City Health Office of Bandung City Government.

The recommendations for developing Bandung health system resilience, in general, are as follows:
1. Improving health services for the elderly by ensuring that health screening is carried out according to national standards. This action aims to determine the disease history of vulnerable ages and control infectious disease spreading. Besides, this recommendation can be used as a reference for determining health programs in anticipation of extreme conditions ahead, including a case study of COVID-19.
2. Addition of public health center (facility) units in areas that do not meet the ideal ratio (1:30.000) to facilitate public health services with affordable access.
3. Increased monitoring of population access to safe drinking water and the percentage of households that behave in a clean and healthy life consistently and making improvements based on monitoring and data collection. This action needs to be done to ensure community health status in Bandung, mainly controlling the spread of infectious diseases.
4. Equipping sub-district health center with qualified diagnostic equipment and experts is an important aspect of improving surveillance at the sub-district level.

5. Conclusions

1. Some of the public health, environmental health, and health resources targets have not met all the target indicators in Strategic Planning and Regional Medium Term Development Plan Kota Bandung for 2018-2023.

2. The important targets that need to be fulfilled are under-5 mortality rates, the percentage of elderly health services, the ratio of the public health centre to the population, and the percentage of the population with access to safe drinking water and sanitation.

3. Evaluation of healthcare facilities' conditions during the COVID-19 pandemic shows that Bandung's health system only met five (5) of the nineteen (19) indicators based on the GHS index. The indicators are
   - the doctor's ratio to population (1:100.000),
   - the ratio of nurses and midwives to population (1:100.000),
   - the capacity of healthcare facilities to the isolation of patients with highly infectious diseases risk,
   - the enactment of regulations related to universal health service coverage, and
   - Monitoring the disease cases in healthcare facilities.

4. To reach a resilience health system, Bandung City needs to improve their Strategic Planning and Regional Medium Term Development Plan to accommodate the index and indicator of health system resilience compatible with local specific character.

6. References

[1] Turenne C P, Gautier l, Degroote S, Guillard E, Chabrol F and Ridde V 2019 Conceptual analysis of health systems resilience: A scoping review *Social Science & Medicine* **232** 168-180
[2] Wulff K, Donato D and Lurie N 2015 What is Health Resilience and how can we built it ? *Annual rev Public Health* **36** 361-174
[3] Lawal S A 2016 Health sector Development and sustainability in Nigeria using the resilient health systems framework *Ibadan Journal of Sociology* **4** 81-115
[4] Kruk ME 2017 Building resilient health systems: a proposal for a resilience index *BMJ* 357 1-8
[5] Durski K N, Osterholm M, Majumdar S, Nilles E, Bausch D G and Atun R 2020 Shifting the paradigm: using disease outbreak to build resilient health systems *BMJ Global Health* **5** 1-3

[6] Bhandari, S. and Alonge, O., 2020 Measuring the resilience of health systems in low- and middle-income countries: a focus on community resilience *Health Research Policy and Systems* **18:81** 1-19

[7] Mills A 2017 Resilient and responsive health systems in a changing world *Health Policy and Planning* **32** iii1-iii2

[8] Haldane V et al 2021 Health systems resilience in managing the COVID-19 pandemic: lessons from 28 countries *Nature Medicine* **27** 964-980

[9] NTI, Johns Hopkins Bloomberg School of Public Health 2019 *Global Health Security Index* Nuclear Threat Initiative

[10] Bandung City Central Bureau of Statistics 2020 *Bandung City in Number 2020*

[11] West Java Health Office 2020 West Java Province Health Office

[12] Bandung City Health Office 2010 *Bandung City Health Profile 2019*

[13] silience : Key Concepts and Strategies. Retrieved from WHO International Website, euro.who.int.

Acknowledgments
This Applied Research is funded by the Directorate of Research and Community Service, Directorate General of Research and Development of the Ministry of Research, Technology, and Higher Education.

AUTHOR INDEX

Achiari, Hendra....458
Aditramulyadi, D. D. 340
Adityawan, M. B....367, 535
Alam, H. N....125, 137, 327
Alfian, Alfian....499
Alkas, M. Jazir....92
Amelia, M....149
Anggraina, P. L. 229
Anindita, R....525
Anungstri, Rissa....608
Apriadi, D. 1
Ariawan, I. M. A. 386
Arum, Sekar....198
Asrurifak, M. 102
Ayuningtyas, Kardina N. S. 621
Bratanata, Jonathan....35
Cahyono, Muhammad....508
Chaeriyah, S....149
Dahlan, M. Z....149
David, Anju....621
Defi, L. P....490
Djafar, Fitria N. I. 499
Ellizar, Estiara....468
Erif, L. O. Muhammad....499
Faisal, B....149
Fani, Ayudia M. 306
Farid, M. 367, 535
Fatimah, W. M. 375, 490
Fauzani, D....11
Fauzi, N. 81, 217, 240, 639
Firdayati, M. 525
Firmansyah, F. 548
Fransiska, Widya....478
Gazali, A. 447
Geerling, G. W. 667
Gunawan, W....518, 560
Gurusu....458
Habibbullah, Muhammad Y. 375
Handajani, M. 11, 525
Handoko, B. 350
Hanisah, M....318
Hapsari, R. K. 367
Hardaya, Adi P. 205
Haryanto, Budi....92
Haryanto, I. 630
Hasibuan, N. 318
Hasiholan, Budi....417
Helmy, Q....11
Hendriyawan....340
Herbenita, V....518, 560
Hermawan, M. W. 327
Hermawan....417
Hidayat, N. 630
Humam, Afif....167
Hutriani, I. W....149
Ilmi, Friska M. 468
Ilmi, Nisa F. F....167
Indrayana, D. V. 19
Indriastuti, A. K....178
Ioka, S. 548
Iqbal, R....350
Irawan, Diki S. 156
Iridiastadi, H....19
Irsyam, M. 102
Isheka, R. P....630
Jamal, Mardewi....92
Jihanny, Jongga....621
Juangsa, Firman B. 306
Kaderi, S. 571
Kahdar, K. 518, 560
Kamajaya, Murti....621
Kamila, Putri S. 397
Kanisha, T. F....678
Kardiansyah, T. 11
Karjadi, Entin A. 275
Karsaman, R. H. 358, 571
Kawahara, Y. 28
Krisnamurti, P. A. 525
Krisnanto, S....1, 35
Kuntoro, A. A. 367, 535
Kurniawan, A. 81, 217, 264, 518, 560, 639
Kusdiwanggo, Susilo....478
Kusumah, S. W. 678
Kusumasari, Wildi....468
Kusumastuti, D....407
Kusumayanti, I. 490
Lambang-Goro, G. 102
Lugina, F. P....28
Madonna, Sandra....659
Mahardika, I. 447
Maria, R. 548
Marselina, M. 434
Martdianto, Rommy....125
Martin, R. J. 81, 217, 240, 639
Marzuki, P. F. 19
Maulina, Sarah....156

Meilano, I. 102
Moestopo, M. 407
Muhammad, I. F. 81, 217, 240, 639
Muin, M. 42, 54, 66, 125, 137, 327
Mukhaiyar, Utriweni 167
Mursyida, Nurdianti 375
Muslim, A. B. 42, 54, 125, 137, 327
Mutmainnah, Mutmainnah 499
Nadapdap, L. 318
Nasrul, N. 188
Nastiti, A. 287, 678
Nathanael, Rinaldy J. 167
Nawir, H. 1
Ngii, E. 188
Notodarmojo, S. 11, 350, 667
Nova, Maria C. 156
Nugroho, E. O. 81, 217, 240, 639
Nur, A. 117
Nurperbangsari, Dzalika 659
Nursetyowati, Prismita 156
Oei, Natanael I. W. 417
Oginawati, K. 117, 167
Oktavianus, A. 584
One, L. 188
Onodera, S.-I. 548
Onof, C. 430
Paramashanti 264
Pasaribu, Udjianna S. 167
Petrus, Himawan T. B. M. 608
Philander, Evan 306
Pradoto, R. G. K. 447, 584
Prasetya, Agus 608
Prawisudha, Pandji 306
Pribadi, K. S. 19, 535, 584
Primaswari, G. 297
Purba, R. P. 447
Purnama, Muhammad R. 535
Purwanta, Wahyu 306
Purwoko, W. 548
Puspitasari, T. A. 42
Putra, Atanasius A. P. 417
Putri, A. T. 447
Putri, N. M. 434
Qomariyah, S. 253
Rabung, Yos Y. 468
Radhika 367
Ragas, A. M. J. 667
Rajin 458
Rakhmadi, Adhitya 287
Ramadhanti, D. T. 358
Ramdani, Yusep 375
Rasmawan, I. M. A. B. 584
Rasyid, Ahmad 92
Rildova 264, 340, 518, 560
Rismantojo, E. 1
Rivani, Y. 178
Roosmini, D. 678
Ruadelia, M. 318
Rusydi, A. F. 548
Sabrina, Marsha 275
Sahadewa, A. 1
Saito, M. 548
Salami, I. R. S. 397, 667, 678
Salmadhia, Difa 621
Salsabila, Firnandia Z. 621
Saraswati, D. 599
Sarli, P. W. 490
Sengara, I. Wayan 35
Setiadji, B. H. 178
Setiyawan, A. S. 117
Sigalingging, J. 318
Sihombing, R. 66, 327
Sioné, L. 430
Siswanto, Ari 478
Sjafruddin, A. 571
Soemardi, B. W. 447
Soewondo, P. 117, 490
Sucipto, V. N. B. 264
Sugiana, Astrid M. 659
Suharyanto 287
Sukmayadi, D. 548
Suryati, I. 318
Susila, E. 1
Tawekal, R. L. 66, 125, 137, 340
Templeton, M. R. 430
Thanaya, I. N. A. 386
Triana, S. 571
Uchida, T. 28
Udo, M. I. 188
Utami, R. R. 667
Utomo, A. B. 297
Utomo, B. 253
Wahyudi, A. H. 253
Wahyuni, Ira R. 499
Wibowo, S. 358, 407
Wicaksono, Bambang 478
Widiyarta, I. M. 386
Widyaningtias 367, 535
Widyarsana, I. M. W. 599
Wijaya, F. S. 81, 217, 240, 639
Wikaningrum, T. 198, 229
Wulandari, L. D. 584
Wurjanto, Andojo 205
Yatmadi, Denny 137

Yola ... 81, 217, 240, 639
Yuanita, N. ... 217, 240, 518, 560, 639
Yudistira, I. D. G. R. ... 386
Zevi, Yuniati ... 375
Zulkifliani ... 659